CW01111668

NUMERICAL MATHEMATICS AND SCIENTIFIC COMPUTATION

Series Editors

A. M. STUART   E. SÜLI

NUMERICAL MATHEMATICS AND SCIENTIFIC COMPUTATION

Books in the series

Monographs marked with an asterisk (*) appeared in the series 'Monographs in Numerical Analysis' which is continued by the current series.

For a full list of titles please visit
http://www.oup.co.uk/academic/science/maths/series/nmsc

* J. H. Wilkinson: *The Algebraic Eigenvalue Problem*
* I. Duff, A. Erisman, and J. Reid: *Direct Methods for Sparse Matrices*
* M. J. Baines: *Moving Finite Elements*
* J. D. Pryce: *Numerical Solution of Sturm–Liouville Problems*

C. Schwab: *p- and hp- Finite Element Methods: Theory and Applications in Solid and Fluid Mechanics*
J. W. Jerome: *Modelling and Computation for Applications in Mathematics, Science, and Engineering*
A. Quarteroni and A. Valli: *Domain Decomposition Methods for Partial Differential Equations*
G. Em Karniadakis and S. J. Sherwin: *Spectral/hp Element Methods for Computational Fluid Dynamics*
I. Babuška and T. Strouboulis: *The Finite Element Method and its Reliability*
B. Mohammadi and O. Pironneau: *Applied Shape Optimization for Fluids*
S. Succi: *The Lattice Boltzmann Equation for Fluid Dynamics and Beyond*
P. Monk: *Finite Element Methods for Maxwell's Equations*
A. Bellen and M. Zennaro: *Numerical Methods for Delay Differential Equations*
J. Modersitzki: *Numerical Methods for Image Registration*
M. Feistauer, J. Felcman, and I. Straškraba: *Mathematical and Computational Methods for Compressible Flow*
W. Gautschi: *Orthogonal Polynomials: Computation and Approximation*
M. K. Ng: *Iterative Methods for Toeplitz Systems*
M. Metcalf, J. Reid, and M. Cohen: *Fortran 95/2003 Explained*
G. Em Karniadakis and S. Sherwin: *Spectral/hp Element Methods for Computational Fluid Dynamics, Second Edition*
D. A. Bini, G. Latouche, and B. Meini: *Numerical Methods for Structured Markov Chains*
H. Elman, D. Silvester, and A. Wathen: *Finite Elements and Fast Iterative Solvers: with Applications in Incompressible Fluid Dynamics*
M. Chu and G. Golub: *Inverse Eigenvalue Problems: Theory, Algorithms, and Applications*
J.-F. Gerbeau, C. Le Bris, and T. Lelièvre: *Mathematical Methods for the Magnetohydrodynamics of Liquid Metals*
G. Allaire and A. Craig: *Numerical Analysis and Optimization: An Introduction to Mathematical Modelling and Numerical Simulation*
K. Urban: *Wavelet Methods for Elliptic Partial Differential Equations*
B. Mohammadi and O. Pironneau: *Applied Shape Optimization for Fluids, Second Edition*
K. Boehmer: *Numerical Methods for Nonlinear Elliptic Differential Equations: A Synopsis*
M. Metcalf, J. Reid, and M. Cohen: *Modern Fortran Explained*
J. Liesen and Z. Strakoš: *Krylov Subspace Methods: Principles and Analysis*
R. Verfürth: *A Posteriori Error Estimation Techniques for Finite Element Methods*

A Posteriori Error Estimation Techniques for Finite Element Methods

RÜDIGER VERFÜRTH

Fakultät für Mathematik, Ruhr-Universität Bochum

OXFORD
UNIVERSITY PRESS

OXFORD
UNIVERSITY PRESS

Great Clarendon Street, Oxford, OX2 6DP,
United Kingdom

Oxford University Press is a department of the University of Oxford.
It furthers the University's objective of excellence in research, scholarship,
and education by publishing worldwide. Oxford is a registered trade mark of
Oxford University Press in the UK and in certain other countries

© Rüdiger Verfürth 2013

The moral rights of the author have been asserted

First published in 2013
Impression: 2

All rights reserved. No part of this publication may be reproduced, stored in a retrieval system, or transmitted, in any form or by any means, without the prior permission in writing of Oxford University Press, or as expressly permitted by law, by licence or under terms agreed with the appropriate reprographics rights organization. Enquiries concerning reproduction outside the scope of the above should be sent to the Rights Department, Oxford University Press, at the address above

You must not circulate this work in any other form
and you must impose this same condition on any acquirer

British Library Cataloguing in Publication Data
Data available

ISBN 978–0–19–967942–3

Printed and bound in Great Britain by
Clays Ltd, St Ives plc

PREFACE

In the past three decades self-adaptive discretisation methods have gained enormous importance for the numerical solution of partial differential equations that arise from physical and technical applications. The aim is to obtain a numerical solution within a prescribed tolerance using a minimal amount of work. The main tools in achieving this goal are a posteriori error estimates which give global and local information on the error of the numerical solution and which can easily be computed from the given numerical solution and the data of the differential equation.

In this monograph we review the most frequently used a posteriori error estimation techniques. Although there are various approaches to adaptivity and a posteriori error estimation, they are all based on a few common principles. Our main goal is to clearly elaborate these basic principles and, thus, to give guidelines for developing adaptive schemes for new problems.

In Chapter 1 we present the most frequently used a posteriori error estimates within the framework of a simple model problem: the linear or bi-linear conforming finite element discretisation of the two-dimensional Poisson equation with mixed Dirichlet and Neumann boundary conditions. The intention is to present the basic principles of a posteriori error estimation using a minimal amount of notation and techniques. Chapter 1 in particular addresses the following topics.

Residual estimates (Section 1.4 (p. 10)): These consist of weighted L^2-norms of element and edge residuals. The element residuals are obtained by inserting element-wise the computed discrete solution into the strong form of the differential equation. The edge residuals consist of inter-element jumps of the trace operator which links the strong and weak form of the differential equation. These terms are obtained by an integration by parts element-wise. For the model problem the trace operator is the normal derivative. The weighting factors are suitable powers of the local mesh-size. The powers reflect the different regularity requirements for strong and weak solutions. Up to multiplicative constants, the residual estimates provide upper and lower bounds for the error of the computed discrete solution. The upper bounds are global with respect to the computational domain. The lower bounds are local in that the error indicator attributed to an element is bounded by the error on the given element and adjacent neighbours. The different nature of the upper and lower bounds has a structural reason. The upper bounds are based on the stability of the variational problem and use the inverse of the differential operator which is a global operator. The lower bounds on the other hand are based on the continuity of the variational problem and use the differential operator itself which is a local operator.

Vertex oriented residual estimates (Section 1.5 (p. 17)): This type of residual estimate is obtained by regrouping the element and edge residuals and by attributing the results to the vertices of the mesh. This results in sharper error bounds in that the multiplicative constants are smaller and can be determined explicitly. On the other hand, the results of this section require a slightly more involved analysis.

Edge residuals (Section 1.6 (p. 20)): These are obtained from the residual estimates mentioned first by dropping the element contributions. This is possible since a refined analysis reveals that, for the lowest order conforming finite element discretisation, the edge residuals are dominant. The main tool for proving this result is a suitable H^1-stability estimate for the L^2-projection onto the finite element space.

Auxiliary local problems (Section 1.7 (p. 25)): Here, the a posteriori error estimates are obtained by evaluating suitable norms of solutions of appropriate discrete local auxiliary problems. The auxiliary problems are usually posed on a single element or on a small patch of elements which either share a given vertex or which are adjacent to a given element. The local problems are higher order finite element discretisations of Poisson equations with Dirichlet or Neumann boundary conditions. The data of the auxiliary problems are given by the computed discrete solution and its element and edge residuals. This approach again yields global upper and local lower bounds for the error up to multiplicative constants. The resulting error indicators are equivalent to the residual ones in that, up to multiplicative constants, they can locally be bounded by the residual indicators and vice versa.

Hierarchical estimates (Section 1.8 (p. 31)): The key idea of this approach is to solve the differential equation approximately using a more accurate finite element space and to compare this solution with the original discrete solution. In order to reduce the computational cost of the new problem, the new finite element space is decomposed into the original one and a nearly orthogonal higher order complement. Then, only the contribution corresponding to the complement is computed. To further reduce the computational cost, the original bi-linear form is replaced by an equivalent one which leads to a diagonal stiffness matrix. In contrast to the previous techniques, this approach yields, up to multiplicative constants, upper and lower bounds for the error both of which are global. The main tools for analysing hierarchical estimates are a saturation assumption and a strengthened Cauchy–Schwarz inequality.

Gradient recovery (Section 1.9 (p. 36)): The basic idea of this approach is to compute a piecewise linear recovery of the piecewise constant gradient of the computed discrete solution and to use the difference between both as an error indicator. The recovery can be computed by taking the average of the constant gradients associated with the elements sharing a given vertex. This approach too only yields global upper and lower bounds on the error.

Equilibrated residuals (Section 1.10 (p. 41)): The basic steps of this approach are as follows. The error of the computed discrete solution is first characterised as the minimum of a suitable quadratic energy functional. Then, this functional is extended to a larger space with weaker regularity assumptions. Next, the minimum of the extended functional is characterised as the saddle-point of a suitable Lagrange functional. Here, the corresponding constraint specifies the difference between the original function space and the larger one. Then, the Lagrange functional is localised to the elements of the mesh or small patches of elements. This gives rise to local infinite-dimensional variational problems. To render these computable, they are finally replaced by suitable discretisations. The resulting auxiliary discrete problems can often be interpreted as discretisations of Poisson equations with modified edge and element residuals as input data. The modifications are such that the input data satisfy additional compatibility or equilibration conditions, whence the name of the approach.

Dual weighted residuals (Section 1.11 (p. 45)): Up to now all a posteriori error estimates were devised to provide upper and lower bounds for the error measured in a suitable norm, typically the energy norm. Sometimes the solution of the differential equation is not of primary interest but a functional of the solution. An example for this situation is the computation of the drag and lift of a body. Usually the functional is continuous on the space corresponding to the weak formulation of the differential equation. Then every a posteriori error estimate for the error measured in the norm of this space also provides an error estimate for the error of the functional. This bound, however, often overestimates the error of the functional. Dual weighted residuals are tailored to this situation. Usually they are based on the approximate discrete solution of a suitable dual variational problem having as input the error of the functional.

The hyper-circle method (Section 1.12 (p. 48)): This approach has become very popular recently since it yields upper bounds for the energy norm of the error with constant 1 and does not contain any hidden or unknown constants. It is based on a dual variational principle which was originally proposed in 1947 by W. Prager and J. L. Synge for problems of linear elasticity. When looking at the hyper-circle method from this perspective, it is a completely different approach from the methods considered before. We, however, present a different point of view for this method. This reveals that the hyper-circle method can be analysed within the same framework as the error estimation techniques considered before. The key observation is to interpret the hyper-circle method as a lifting of the element and edge residuals to the space of vector fields which together with their divergence are square integrable element-wise.

Efficiency and asymptotic exactness (Section 1.13 (p. 53)): The quality of an a posteriori error estimate is often measured by its efficiency index, i.e. the ratio of the estimated error and of the true error. The estimate is called efficient if the efficiency index and its inverse remain bounded for all mesh-sizes. It is called asymptotically exact if its efficiency index tends to one when the mesh-size converges to zero. Every a posteriori error estimate which provides upper and lower bounds for the error is efficient. Yet, in general it is not asymptotically exact. This property usually depends on additional properties of the mesh and often requires the super-convergence of the finite element discretisation.

Convergence of the adaptive process (Section 1.14 (p. 58)): The key idea of adaptive mesh-refinement based on a posteriori error estimation is to refine those elements which give a large contribution to the estimated error. Intuitively this should yield a better mesh and discrete solution. Numerical experiments support this idea. Yet, for a long time, it was an open question whether this could also be proved rigorously. In 1996 W. Dörfler was the first to solve this problem. We give a slightly simplified version of his proof. It is worth mentioning that it requires the error estimate to provide both global upper and local lower bounds for the error.

All a posteriori error estimates of Chapter 1 are based on the following principles:

- the equivalence of the error and a suitable dual norm of the residual,
- the Galerkin orthogonality of the error, i.e. the residual annihilates the lowest order conforming finite element space,
- an L^2-representation of the residual by square integrable functions defined on the elements, viz. the element residuals, and on the inter-element boundaries, viz. the edge residuals,
- error estimates for a suitable quasi-interpolation operator of Clément-type,
- inverse estimates for suitable local cut-off functions.

The first principle reveals a fundamental difference of a priori and a posteriori error estimates: a priori error estimates are based on the stability of the discrete problem; a posteriori error estimates require the stability of the infinite-dimensional variational problem. In particular, the first step only requires properties of the infinite-dimensional variational problem and is completely independent of the particular discretisation.

The analysis of Chapter 1 is performed for the simplest model problem. This reduces the technical apparatus and requires only minimal knowledge of variational and finite element methods. The model problem has enough structure to exhibit the aforementioned basic principles of a posteriori error analysis and the fundamental properties of reliability and efficiency of a posteriori error estimates. Yet, it is not complex enough to exhibit the third fundamental property: robustness with respect to singular perturbations by lower order terms of the differential operator. This will come into play when considering more complex equations in Chapters 4–6.

In Chapter 2 we show how to use a posteriori error estimates for adaptive mesh-refinement and how to implement a simple adaptive discretisation scheme. We in particular address the following topics.

Mesh-refinement (Section 2.1 (p. 64)): The refinement of meshes requires two ingredients: first, a marking strategy that decides which elements should be refined and, second, refinement rules which determine the subdivision of elements. We present two of the most popular marking strategies: the maximum strategy and the equilibration strategy. The latter one is also called bulk chasing or the Dörfler strategy. For the refinement of elements, we discuss the regular refinement, the irregular refinement, and the marked edge bisection. The irregular refinement is needed to avoid hanging nodes.

Mesh-coarsening (Section 2.2 (p. 69)): Sometimes elements must be coarsened in the course of an adaptive process. For time-dependent problems this is obvious: a critical region, e.g. an interior layer, may move through the spatial domain in the course of time. For stationary problems this is less obvious. Yet, for elliptic problems one can prove that coarsening must be allowed to ensure the optimal complexity of the adaptive process. The basic idea of the coarsening process is to go back in the hierarchy of partitions and to cluster elements with too small an error.

Mesh-smoothing (Section 2.3 (p. 70)): Mesh-smoothing algorithms try to improve the quality of a mesh while retaining its topological structure. The vertices of the mesh are moved but the number of elements and their adjacency remain unchanged. Most strategies use a process similar to the well-known Gauss–Seidel algorithm to optimise a suitable quality function over the class of all meshes having the same topological structure. They differ in the choice of the quality function. Mesh-smoothing strategies do not replace the mesh-refinement methods, but they do complement them. In particular an improved mesh may thus be obtained when a further refinement is impossible due to exhausted storage.

Data structures (Section 2.4 (p. 74)): To get an idea of the data structures required for an adaptive finite element code, we briefly describe the classes implemented in the `Java` demonstration applet `ALF` (`Adaptive Linear Finite` elements) and the `Scilab` function library `AFEM` (Adative `Finite` Element Methods). Both are available at the address `www.rub.de/num1/softwareE.html` together with short user guides in pdf format. We stress that both `ALF` and `AFEM` are devised for demonstration purposes only. Elaborate research and production codes for adaptive finite element methods are available in the literature and on the Internet, cf. the references given in Section 2.4.

Chapter 2 is completed by a few numerical examples in Section 2.5 (p. 76) that demonstrate the advantages of adaptive mesh-refinement.

Chapters 1 and 2 are essentially self-contained and can be read with a minimal knowledge of partial differential equations and finite element methods. They may serve as the basis for an introductory course or seminar on adaptive finite element methods. Chapters 3–6 extend the results of Chapter 1 to more complex linear and nonlinear elliptic and parabolic equations and require more elaborate mathematical tools. On the other hand, to a large extent, they can be read independently of Chapters 1 and 2.

Chapter 3 collects the technical prerequisites for the a posteriori error estimates of Chapters 4–6. After introducing the required notation for function spaces and finite element meshes and spaces in Sections 3.1 (p. 79) and 3.2 (p. 81), we in particular provide the following tools.

Trace inequalities (Section 3.3 (p. 87)): With the help of judiciously chosen affine vector fields we prove a trace equality linking function values on the boundary of a simplex or parallelepiped to function values and derivatives inside the element. Using this trace equality we then establish trace inequalities with respect to weighted and unweighted L^p-norms. The weight function is always the nodal shape function associated with a given vertex of the element. The trace inequalities are equalities in the space L^1. These results will enable us to efficiently bound the errors of quasi-interpolation operators on the skeleton of a finite element partition by corresponding quantities on the elements. They are also crucial for the results of Section 3.8 where we derive upper bounds for general residuals which in particular contain contributions from the skeleton of a finite element partition.

Poincaré and Friedrichs' inequalities (Section 3.4 (p. 91)): The upper bounds in all our a posteriori error estimates can be reduced to Poincaré and Friedrichs' inequalities more or less explicitly. Moreover, the constants in these inequalities essentially influence the multiplicative constants in the upper bounds for the error. Therefore, Poincaré and Friedrichs' inequalities and the explicit knowledge of the corresponding constants is crucial for all subsequent results. Having in mind this application, we strive for sharp and explicit bounds for Poincaré and Friedrichs' constants on finite element stars, i.e. on patches of elements sharing a given vertex. For convex stars we can resort to the literature and obtain generic bounds which only depend on the Lebesgue exponent. For the general case, however, we have to develop special tools: transformation, reduction, and decomposition. All these tools take advantage of the fact that finite element stars are always star-shaped with respect to the common vertex. They yield upper bounds for the Poincaré constants which can explicitly be computed in terms of elementary geometrical data of the finite element stars. In particular, for this general class of domains, there cannot exist generic Poincaré constants. All estimates must suitably measure the lack of convexity. It is noteworthy that our bounds for the Poincaré and Friedrichs' constants are robust with respect to a possible anisotropy of the elements.

Interpolation error estimates (Section 3.5 (p. 108)): The previous results for Poincaré and Friedrichs' inequalities are used to derive error estimates for a quasi-interpolation operator of Clément type. These results extend those of Section 1.3.3 to arbitrary space dimensions and general L^p-spaces and, most important, give explicit and sharp bounds for the constants in the error estimates.

Inverse estimates (Section 3.6 (p. 112)): The lower bounds in all our a posteriori error estimates rely on inverse estimates for weighted L^p-norms involving suitable local cut-off functions as weight functions. Accordingly, explicit and sharp bounds for the constants in these inverse estimates are crucial for evaluating the multiplicative constants in the lower a posteriori error bounds. The inverse estimates are usually proved in a non-constructive way by transforming to a suitable reference element and using the equivalence of norms on finite-dimensional spaces there. This proof is rather simple and gives insight into the dependence of the constants on the geometry of the elements. Yet, it provides no information on the dependence on the polynomial degree. We obtain this information by applying a dimension reduction argument and by resorting to properties of Legendre polynomials in one space dimension. Readers who are not interested in the dependence on the polynomial degree may skip the slightly technical Sections 3.6.3–3.6.6.

Decomposition of affine functions in $L^p(0,1;Y^*)$ (Section 3.7 (p. 130)): The residuals associated with space-time finite element discretisations of parabolic equations can be split into two contributions associated with a spatial and a temporal residual. All a posteriori error estimates of Chapter 6 rely on the fact that suitable dual norms of the sum of these contributions can be bounded from above and from below by the sum of the corresponding norms. This result is based on the abstract technical result of Section 3.7. Basically it is a strengthened Cauchy–Schwarz inequality. Yet, it is proved in a different way which is better adapted to general L^p- and dual spaces.

Estimation of residuals (Section 3.8 (p. 132)): This section is at the heart of all subsequent results. The guiding principle of all our a posteriori error estimates is to first prove the equivalence of the norm of the error to a suitable dual norm of the residual associated with the current discrete approximation.

This step only requires structural properties of the variational problem and is completely independent of the actual discretisation. The residuals can always be decomposed into a regular and a singular part. This decomposition is labelled the L^p-representation. The regular part can be represented by an L^p-function defined on the computational domain. The singular part, on the other hand, can be associated with an L^p-function defined on the skeleton of the finite element partition. For the model problem of Chapter 1, the regular and singular parts correspond to the element and edge residuals, respectively. The L^p-representation is a structural property of second-order systems in divergence form. It must be modified when passing to higher order equations such as, e.g. the bi-harmonic equation or when violating the divergence form. The L^p-representation is the fundamental structural property which, using the technical tools provided in Sections 3.3, 3.4, and 3.6, enables us to derive computable upper and lower bounds for dual norms of the residuals and to efficiently compute a lifting of the residual to the space of piecewise $H(\text{div})$-vector fields. In order to avoid unnecessary technical difficulties and to simplify the exposition, we first establish the upper bounds and construct the lifting under the additional assumption that the residuals annihilate the lowest order conforming finite element space. This assumption is labelled Galerkin orthogonality. Contrary to the L^p-representation, the Galerkin orthogonality is not a structural property but merely a technical one. We later prove that it can be dispensed with at the expense of additional terms in the upper bounds and the lifting. These terms measure the consistency error related to the missing Galerkin orthogonality. We briefly indicate how to bound the additional terms in an explicit and efficient way for two important sources of missing Galerkin orthogonality: stabilisation by SUPG-discretisations and inexact solution by nested iterative solvers.

Chapter 4 is devoted to a posteriori error estimates for linear elliptic equations. In Section 4.1 (p. 151) we put these equations and their discretisation into a general abstract framework. This in particular accentuates the generic equivalence of the error to the associated residual and highlights that this is a structural property of the variational problem which is completely independent of the discretisation. Thus, the derivation of a posteriori error estimates amounts to the task of establishing computable bounds for dual norms of residuals. This is achieved using the general tools provided in Chapter 3. The concrete realisation of this task depends on the differential equation and discretisation under consideration. To better appreciate this approach, we first revisit the model problem of Chapter 1 and formulate it in the present abstract framework. This is the topic of Section 4.2 (p. 157). We then address the following problems.

Reaction–diffusion equations (Section 4.3 (p. 159)): When applying the results of Chapter 3 in a naive way to reaction–diffusion equations, one obtains error estimates which are not robust: the ratio of estimated and true error is not uniformly bounded with respect to the size of the diffusion relative to the reaction. We show that this undesirable phenomenon can be avoided by a more judicious choice of the weighting factors of the element and face residuals. The modified factors incorporate the local mesh-Péclet number and thus take into account the relation of the local mesh-size to the relative size of the diffusion and reaction. With this modification the general results of Chapter 3 and Section 4.1 immediately yield robust a posteriori error estimates.

Convection–diffusion equations (Section 4.4 (p. 163)): The results of the previous section suggest that we measure the error in the energy norm associated with the diffusion and reaction part of the differential operator and determine the weighting factors of the element and face residuals accordingly. Unfortunately, the convective part of the differential operator is not uniformly bounded with respect to the energy norm. As a consequence, the a posteriori error estimates are not fully robust. Instead, the ratio of estimated and true error depends on the mesh-Péclet number. This drawback can be avoided by measuring the error in a modified norm which in addition contains a dual norm of the convective derivative. At first sight the new norm may look strange. Yet, it is the stationary limit of the natural parabolic energy norm associated with the corresponding time-dependent convection–diffusion equation. This reflects the fact that the material derivative is the sum of the temporal and the convective derivative. Thus, the new norm is quite natural. The fact that the same a posteriori error estimate is robust for the new norm and not robust for the old energy norm can be explained as follows: an overestimation of the energy norm is linked to a large convective derivative of the error and thus indicates an unsatisfactory resolution of interior or boundary layers. Similar results are established for the hyper-circle method and error estimates based on the solution of auxiliary local discrete Dirichlet or Neumann problems.

Anisotropic meshes (Section 4.5 (p. 177)): When considering singularly perturbed elliptic equations with dominant reaction or convection terms as in Sections 4.3 and 4.4, one often encounters solutions with boundary or interior layers. These layers can most efficiently be resolved by so-called anisotropic elements which are very small in the direction perpendicular to the layer and relatively large in the directions parallel to the layer. These elements no longer satisfy the condition of shape regularity which was always required up to now. Here, as usual, shape regularity means that the ratio of the diameter of any element to the diameter of the largest ball inscribed into the given element is uniformly bounded with respect to all elements. Nevertheless, to obtain reasonable a posteriori error bounds for meshes containing anisotropic elements, we must appropriately define the notion of anisotropy, identify new geometric quantities which suitably characterise the elements, and modify the results of Sections 3.5 and 3.6 concerning quasi-interpolation operators and local cut-off functions such that they take account of the anisotropy.

Non-smooth coefficients (Section 4.6 (p. 191)): In this section we consider diffusion equations with either discontinuous or anisotropic diffusion coefficients. In the first case, the diffusion matrix is only smooth on a finite number of subdomains and may have large jumps across the interface of the subdomains. This models for instance several layers of fluids with rather different viscosity which weakly depend on the depth. To obtain good a posteriori error estimates for this class of problem, the finite element mesh has to be consistent with the partition induced by the subdomains. Moreover, we must suitably modify the quasi-interpolation operator of Section 3.5 such that it takes account of the jumps of the diffusion. In the second case, the diffusion matrix varies only slightly over the computational domain but has eigenvalues of very different size resulting in an anisotropy of the differential equation. This models for instance elastic materials in thin layers. For this class of problem, the derivation of good a posteriori error estimates is inspired by the following observation. A suitable transformation of the computational domain leads to an isotropic differential equation. The discrete problem is thus transformed into a finite element discretisation of the isotropic problem on an anisotropic mesh. The results of the preceding section show how to obtain reasonable a posteriori error estimates for the transformed differential equation and its discretisation. Transforming back to the old computational domain yields a reasonable a posteriori error estimate for the original problem. For the practical computation of the error estimates one of course wants to bypass the transformation. In fact this can be achieved quite simply. A detailed analysis reveals that one only has to measure length in a metric depending on the diffusion matrix.

Eigenvalue problems (Section 4.7 (p. 205)): As a simple but representative example, we consider the eigenvalue problem for reaction–diffusion equations and its conforming finite element discretisation. Given the results of the previous sections, the estimation of the residual is straightforward. Its relation to the error, however, is not standard since it involves perturbation terms which depend on the error of the eigenvalues and on the L^2-norm of the error of the eigenfunctions. In order to control these terms and to prove that they are higher order perturbations, we have to resort to a priori error estimates. Thus we encounter the novel situation that the derivation of a posteriori error estimates requires a priori error estimates. In Section 5.3, we will consider the eigenvalue problem from a non-linear perspective. Then we can dispense of the perturbation terms. This, however, has to be paid for by the condition that the analytical and discrete solutions have to be sufficiently close. This essentially means that the approximation for the eigenvalue has to be closer to the eigenvalue than the latter's distance to its neighbouring eigenvalues.

Mixed formulation of the Poisson equation (Section 4.8 (p. 208)): The interest in mixed formulations of the Poisson equation is twofold. First, the mixed formulation allows the direct approximation of the solution's gradient without resorting to numerical differentiation. Second, the a posteriori error analysis of this problem lays the ground for the analysis of the equations of linear elasticity in the next section. The mixed formulation of the Poisson equation and its discretisation directly fit into the abstract framework of Section 4.1. The estimation of the residual, however, is not straightforward since we now have to work with $H(\text{div})$-spaces which have less regularity than the classical Sobolev spaces. Due to this lack of regularity the first a posteriori error estimates for this problem were not efficient. The upper and lower bounds on the error differed by a factor corresponding to the maximal mesh-size. This drawback was overcome by C. Carstensen and E. Dari, R. Duran, C. Padra, and V. Vampa by taking advantage of the Helmholtz decomposition of L^2-vector fields. We give a slightly modified presentation of this approach which fits into our general framework.

Linear elasticity (Section 4.9 (p. 223)): The displacement formulation of the equations of linear elasticity and the corresponding finite element discretisations directly fit into the abstract framework of Sections 3.8 and 4.1. Correspondingly, the derivation of a posteriori error estimates is as simple as for the model problem of Chapter 1. Yet, the displacement formulation of the equations of linear elasticity and the corresponding finite element discretisations suffer from serious drawbacks. The displacement formulation and its discretisations break down for nearly incompressible materials which is reflected by the so-called locking phenomenon. The resulting a posteriori error estimates are not robust in that the ratio of estimated and true error depends on the Lamé parameters and tends to infinity in the incompressible limit. Often, the displacement field is not of primary interest, but the stress tensor is of physical interest. This quantity, however, is not directly discretised in displacement methods and must a posteriori be extracted from the displacement field which often leads to unsatisfactory results. These drawbacks can be overcome by suitable mixed formulations of the equations of linear elasticity and appropriate mixed finite element discretisations. Correspondingly these are the focus of this section. We are primarily interested in discretisations and a posteriori error estimates which are robust in the sense that their quality does not deteriorate in the incompressible limit. This is reflected by the need for estimates which are uniform with respect to the Lamé parameters. The main tools for achieving this goal are the Hellinger–Reissner principle, the properties of the Brezzi–Douglas–Fortin–Marini and Raviart–Thomas spaces, and the Helmholtz decomposition. We also provide a posteriori error estimates based on the solution of auxiliary local discrete elasticity equations with Dirichlet or Neumann boundary conditions.

Stokes equations (Section 4.10 (p. 237)): The Stokes equations are another important example of saddle-point problems. In contrast to the problems of the preceding two sections, the main technical difficulty is now associated with the Lagrange multiplier, viz. the pressure, associated with the constraint on the primal variable, viz. the velocity. The Stokes equations and their mixed finite element discretisations fit into the abstract framework of Sections 3.8 and 4.1. This applies to a large class of discretisations including stable mixed finite approximations and stabilised schemes of SDFEM or SUPG type. All these discretisations are analysed in a single common framework. In contrast to scalar linear elliptic equations, the corresponding spaces are now Cartesian products of suitable Sobolev spaces associated with the primal variable and the Lagrange multiplier having different regularities.

This results in different scaling factors for the residuals associated with the momentum and the mass equation. Apart from this difference, the a posteriori error analysis of the Stokes equations is rather similar to that of the model problem of Chapter 1. This also applies to error estimates based on the solution of auxiliary local discrete Stokes problems.

Higher order differential equations (Section 4.11 (p. 248)): We consider in this section the bi-harmonic equation as a representative example for higher order differential equations. Its variational formulation based on an energy principle and corresponding conforming finite element discretisations fit perfectly into the abstract framework of Section 4.1. The only modification arises when establishing lower bounds for the error. Since the variational problem is posed in $H^2(\Omega)$, we must now use the smooth cut-off functions of Section 3.2.5. Conforming finite element discretisations of the bi-harmonic equation, however, need C^1-elements which are rather expensive due to the required high polynomial degree. To overcome this difficulty one often tries to relax the C^1-constraint by either considering a mixed variational formulation or non-conforming approximations each having its own benefits and drawbacks. The a posteriori error analysis of the saddle-point formulation of the bi-harmonic equation would be completely straightforward if the corresponding bi-linear form were inf-sup stable. Unfortunately this is not the case. This deficit is reflected in sub-optimal a priori error estimates and non-standard a posteriori error estimates. Contrary to the previous results the latter now need an a priori bound for the solution of the variational problem, contain a data error incorporating a global mesh-size and require some assumptions on the smoothness of the local mesh-size which slightly restrict the adaptive process. In the last part of this section we also consider a representative non-conforming discretisation of the bi-harmonic equation. Here, the main difficulty consists in establishing upper bounds for the error which are not spoilt by the consistency error of the discretisation. The key idea in achieving this goal is to introduce a suitable lifting or averaging of the non-conforming finite element approximation to a suitable conforming finite element space and to control the error of this lifting.

Non-conforming discretisations (Section 4.12 (p. 261)): To simplify the exposition and to better exploit the fundamental differences from conforming methods, we first consider the two-dimensional Poisson equation with homogeneous Dirichlet boundary conditions as a model problem. We afterwards discuss the necessary modifications for general diffusion problems, mixed boundary conditions, and three-dimensional problems. The discretisation is arbitrary up to the following two fundamental conditions. First, the discrete space should encompass the lowest order conforming finite element space. Second, the discrete functions should be weakly continuous and should satisfy the homogeneous Dirichlet boundary condition weakly in the sense that the averages of their inter-element jumps and their averages on boundary faces vanish. The key ingredient for the a posteriori error estimation is a Helmholtz-type decomposition of the element-wise gradient of the discrete solution. This splits the error into a conforming and a non-conforming part. The conforming part can be represented as the gradient of a suitable H^1-function. Its estimation is completely standard, since the non-conforming space encompasses the lowest order conforming finite element space. The non-conforming part, on the other hand, can be represented as the curl of a stream function or, in three space dimensions, vector potential. The estimation of this term relies on two crucial observations. First, the curl is orthogonal to the gradient of H^1-functions. Second, due to the weak continuity of the discrete function and the weak Dirichlet boundary condition, the element-wise gradient of the discrete solution is orthogonal to the curl of continuous piecewise linear functions. The non-conforming part of the error contributes to the error indicator inter-element jumps of the tangential derivative of the discrete solution and its tangential derivatives on boundary faces. These contributions do not change when passing to general diffusion problems, i.e. the particular form of the diffusion only enters into the conforming part of the error indicator.

Convergence of the adaptive process (Section 4.13 (p. 264)): We extend the results of Section 1.14 and give a general convergence proof for a generic adaptive algorithm. This establishes the convergence of the adaptive algorithm for all problems and all error indicators considered in this chapter using any marking and refinement strategy of Chapter 2. The basic ideas of the convergence proof can be described as follows. Using the properties of the variational problem and its discretisation, we first prove that, when dropping the stopping criterion, the adaptive algorithm produces sequences of mesh-functions and discrete solutions which converge in L^∞ and in the space of the variational problem, respectively to some limit functions. The limit of the mesh-functions may not be zero and the limit function of the discrete solutions is the solution of *some* infinite-dimensional variational problem. We must still identify this limit function with the solution of the variational problem under consideration. This is achieved indirectly by first proving the convergence to zero of the error indicators and then using the reliability of the indicators. The convergence proof of the error indicators is the crucial step. It relies on the fact that the limiting function solves some variational problem and that the error indicators yield local lower bounds for the dual norm of the corresponding residual. It proceeds by splitting the partitions and the error indicators into three contributions. The first one corresponds to elements that are subdivided sufficiently many times. The third one refers to elements that are no longer refined. The second one corresponds to the remaining elements. For the first group, the convergence of the error indicators is derived from the convergence of the mesh-functions and the discrete solutions using once more the efficiency of the error indicators. For the second group, the convergence is established by proving that the mesh-functions converge to zero on the corresponding subdomain. The third group finally is treated by using the results for the first two groups and properties of the marking strategies. In summary, the convergence result crucially hinges on the efficiency of the error indicators, i.e. their property to yield local lower bounds for the error. The size of the multiplicative constants in the upper and lower bounds for the error does not play any role as long as these constants are independent of the partitions. Thus the convergence proof does not need the robustness of the error indicators. This is due to the fact that the convergence result is a qualitative and not a quantitative one, i.e. it gives no information whatsoever on the convergence speed. Estimates of the latter, on the other hand, crucially hinge on the robustness of the error indicators. Such estimates of the convergence rate together with bounds for the growth of the partitions are mandatory to establish the optimal complexity of the adaptive process. Up to now corresponding results are known only for symmetric coercive problems and thus exclude the mixed discretisations of this chapter.

In Chapter 5 we extend the results of the previous chapter to nonlinear elliptic equations. Similarly to Section 4.1 we derive in Section 5.1 (p. 281) an abstract framework for nonlinear elliptic equations and their discretisations. The basic tool now is the implicit function theorem. It is used to prove once more the equivalence of error and residual. The implicit function theorem applies to solutions which are regular in the sense that a suitable linearisation of the nonlinear problem admits a unique weak solution. This excludes limit and bifurcation points. These, however, can be fitted into the general framework by suitable modifications of the spaces and functionals. Thus, for a large class of nonlinear problems, we are again back to the task of estimating suitable dual norms of residuals. We apply the general results to the following problems.

Quasilinear equations of second order (Section 5.2 (p. 290)): Examples of equations considered in this section are the equations of prescribed mean curvature, the α-Laplacian, the subsonic potential flow of an ideal compressible gas, convection–diffusion equations with a nonlinear diffusion, and Bratu's equation. We give residual a posteriori error estimates, estimates based on the solution of auxiliary local discrete linear problems, and estimates for L^p-norms.

Eigenvalue problems (Section 5.3 (p. 299)): Eigenvalue problems for linear differential operators can be reformulated within the general nonlinear framework of this chapter. The nonlinearity then stems from the normalisation condition for the eigenfunction. The detour to nonlinear problems is recompensed by readily available a posteriori error estimates for simple eigenvalues and associated eigenfunctions without any restriction on the position of the eigenvalue within the spectrum. Within this framework, multiple eigenvalues correspond to bifurcation points and can therefore be treated by modifying the spaces and functionals as described in Section 5.1. When compared with the linear theory of Section 4.7, we can now dispense with the higher order perturbation terms and the required a priori error estimates. This, however, has to be paid for by the condition that the analytical and discrete solutions have to be sufficiently close. This essentially means that the approximation for the eigenvalue has to be closer to the eigenvalue than the latter's distance to its neighbouring eigenvalues.

Stationary Navier–Stokes equations (Section 5.4 (p. 301)): Mixed finite element discretisations of these equations directly fit into the present general nonlinear framework. This applies both to standard stable discretisations and general stabilised schemes of SDFEM or SUPG type. The main technical difficulty is the divergence constraint for the velocity field. This can be handled with the methods developed for the linear Stokes equations. We consider residual estimates as well as estimates based on the solution of auxiliary local discrete Stokes problems and also provide estimates with respect to the L^2-norm for the velocity and the H^{-1}-norm for the pressure.

The final Chapter 6 is concerned with a posteriori error estimates for linear and nonlinear parabolic equations. In particular we consider the following problems.

The heat equation (Section 6.1 (p. 309)): For parabolic problems, the heat equation has the same model character as the Poisson equation for elliptic equations. Using a parabolic energy argument, we first prove that a suitable energy norm of the error of any space–time discretisation of the heat equation is equivalent to the corresponding dual norm of the associated residual. As for elliptic problems, this is a structural property of the variational problem and is completely independent of the discretisation. The latter comes into play when estimating the residual. Contrary to stationary problems, the residual is now decomposed into two contributions which are associated with a temporal and a spatial error. These contributions are correspondingly labelled temporal and spatial residual. The technical result of Section 3.7 enables us to bound the dual norms of these contributions separately. The estimation of the temporal residual is elementary. Taking advantage of the general results of Section 3.8, the spatial residual can also be estimated easily. Combining both estimates yields computable upper and lower bounds for the error. Moreover, the a posteriori error estimates split into two contributions associated with the two types of residuals. This splitting allows for a separate adaptation of the temporal and spatial mesh.

Time-dependent convection–diffusion equations (Section 6.2 (p. 317)): As for stationary convection–diffusion equations, the major challenge is to obtain estimates which are robust with respect to the relative size of the diffusion compared to the convection. The first step, the equivalence of error and residual, is achieved in the same way as for the heat equation. The same applies to the splitting of the residual and to the estimation of the spatial residual. The estimation of the temporal residual, however, requires more care. Doing this in a naive way results in estimates which are not robust. As for the stationary case, this is due to the fact that the convective term is not uniformly bounded with respect to the energy norm associated with the diffusion and reaction terms. This difficulty can again be solved by augmenting the energy norm by a suitable dual norm of the convective derivative. In contrast to stationary problems, this now gives a contribution to the error estimate involving the dual norm which cannot be evaluated explicitly. Nevertheless, to obtain an explicitly computable and robust a posteriori error estimate, we have to solve an additional auxiliary discrete reaction–diffusion problem associated with the lowest order conforming finite element space. This is motivated by the observation that the dual norm of a function is the same as the energy norm of the solution of a reaction–diffusion equation with the given function as input datum. The auxiliary problem is now global with respect to the spatial domain. Thus the evaluation of the error estimate is as costly as an additional time-step.

General linear parabolic equations of second order (Section 6.3 (p. 326)): The results of the preceding section immediately extend to this general problem. We thus obtain robust a posteriori error estimates for a broad regime of problems ranging from diffusion dominated over reaction dominated to convection dominated problems. If the convection is not dominant, the auxiliary problems mentioned above can be omitted.

Method of characteristics (Section 6.4 (p. 329)): This method is based on the observation that the sum of the temporal and of the convective derivative is the total derivative along the characteristics of the differential operator. This allows us to perform the discretisation in two stages. In the first stage the characteristics and the total derivative along these are approximated. The second stage consists in solving a reaction–diffusion problem with the result of the first stage as input. The method of characteristics has the appealing effect that it automatically introduces some up-winding and thus has a stabilising effect. Moreover, in each time-step, one only has to solve a symmetric positive definite discrete problem. For the method of characteristics, the equivalence of error and residual, the splitting of the residual, and the estimation of the temporal residual are achieved in the same way as for the heat equation. The estimation of the spatial residual, however, is more intricate. This is due to the fact that this term reflects the difference between the method of characteristics and the standard backward Euler scheme. To control the spatial residual, we resort to the same trick as for the convective derivative in Section 6.2 and solve an additional discrete problem associated with the lowest order conforming finite element space. The auxiliary problem is again global with respect to the spatial domain and amounts in an extra cost corresponding to an additional time-step.

Time-dependent Stokes equations (Section 6.5 (p. 335)): The time-dependent Stokes equations could be formulated as a heat equation in the space of solenoidal velocity fields. Since this space has the same analytical properties as the standard H^1-space, the results of Section 6.1 would carry over immediately. This approach, however, is not feasible since virtually all finite element discretisations used in practice are non-conforming in the sense that the discrete velocities are either discontinuous and thus not contained in H^1 or are not solenoidal. Therefore, we must develop particular techniques for the time-dependent Stokes equations. The crucial point here is the equivalence of error and residual. Once this property is established, the a posteriori error estimation follows standard lines and is quite simple. The splitting of the residual into temporal and spatial contributions is established as for the heat equation and can be controlled using the general result of Section 3.7. The estimation of the temporal residual is elementary. The spatial residual finally can be bounded using the techniques developed for the stationary Stokes equations.

Nonlinear parabolic equations of second order (Section 6.6 (p. 347)): The a posteriori error analysis of nonlinear parabolic equations is based on the abstract results of Section 5.1. These in particular yield the equivalence of error and residual. The nonlinearity of the problem is reflected by the need to further decompose the temporal residual into an affine part and a non-affine part. The first part can be interpreted as a linearisation. The second part is of higher order when compared to the first one. Once this splitting is achieved, the estimation of the different contributions to the residual is standard. Yet, the affine part of the temporal residual contributes a dual norm to the a posteriori error estimates. To render this term computable, we once more have to solve an additional auxiliary linear problem associated with the lowest order conforming finite element space. When working in a Hilbert space setting, the H^1-norm of the solution of the auxiliary problem is equivalent to the dual norm of the affine part of the temporal residual. Unfortunately, nonlinear problems often require us to leave the

Hilbert space setting and to work in general L^p-spaces. In order to extend the Hilbert space result to the general L^p-setting, we have to prove the stability of the Laplacian with respect to $W^{1,p}$-norms both in the analytical and the discrete case.

Finite volume methods (Section 6.7 (p. 360)): Finite volume methods are a different popular approach for solving parabolic problems, in particular those with large convection. For this type of discretisations, the theory of a posteriori error estimation and adaptivity is much less developed than for finite element methods. Yet, there is an important particular case where finite volume methods can easily profit from finite element techniques. This is the case of so-called dual finite volume meshes which is briefly described in this section. For these, there is a natural one-to-one correspondence between piecewise constant functions on the dual finite volume mesh and continuous piecewise (multi-) linear functions associated with the primal finite element mesh. Therefore, the errors of both approximations are equivalent. Consequently, the finite volume mesh can be adapted by estimating the finite element error and adapting the finite element mesh correspondingly as described in Chapter 2.

Convergence of the space–time adaptive process (Section 6.8 (p. 362)): While there are several quite satisfactory convergence and optimality results for the adaptive process applied to linear elliptic problems, the corresponding theory for parabolic equations is still in its infancy. We report on a recent result of C. Kreuzer, C. A. Möller, A. Schmidt, and K. G. Siebert which proves that, for every positive tolerance ε and every finite final time T, a suitable space–time adaptive process yields a discrete solution of the heat equation with an $L^2(0, T; H_0^1(\Omega))$-error less than ε. The main difficulty in establishing this result may be described as follows. Every space–time adaptive algorithm has to ensure that the Euclidean sum of the errors committed while advancing in time is below a given tolerance. This is an L^2-type condition which requires the knowledge of the committed errors at *all* intermediate times. At the n-th time-step, however, we only have access to current values of the errors and not to future ones. Correspondingly, the actual error control is of an L^∞-type. This difference leads to severe problems. To overcome these, we must explicitly control the effect of replacing the L^2-criterion by an L^∞-one. This must happen at two instances. The first one is related to the right-hand side f and can be handled quite easily assuming the extra regularity $f \in H^1(0, T; L^2(\Omega))$. The second one is related to the possible increase in energy due to a potential coarsening of the partitions. This task is more delicate and is achieved with the help of an explicit a priori estimate for the discrete energy which only depends on the right-hand side f and the final time T irrespective of the time-steps, spatial partitions, and finite element spaces. Notice that a coarsening of the spatial partitions while advancing in time is mandatory for an efficient space–time adaptive process. The result of this section can easily be extended to linear parabolic equations with arbitrary diffusion and reaction terms. Currently, however, it is not known how to treat parabolic equations with a non-symmetric or indefinite spatial term such as the time-dependent Stokes equations. Similarly we are currently not aware of any result establishing the optimal complexity of the space–time adaptive process.

Most sections close with a bibliographical remark which indicates the historical development and hints at further results.

My sincere thanks are due to Miss F. Tantardini for her careful reading of the manuscript and her very helpful suggestions and remarks, and to the staff at Oxford University Press.

Bochum, 17th November 2012

CONTENTS

1

A Simple Model Problem

In this chapter we present the most frequently used a posteriori error indicators for the lowest order conforming finite element discretisation of the two-dimensional Poisson equation with mixed Dirichlet and Neumann boundary conditions. We prove that all indicators are reliable, i.e. yield, up to multiplicative constants, global upper bounds for the error, and that most of them are efficient, i.e. yield, again up to multiplicative constants, local or global lower bounds for the error. We also show that a suitable adaptive algorithm based on any of these error indicators yields a sequence of discrete solutions which converges to the weak solution of the Poisson equation.

1.1 Motivation and Overview

In the numerical solution of practical problems in physics or engineering such as, e.g. computational fluid dynamics, elasticity, or semiconductor device simulation, one often encounters the difficulty that the overall accuracy of the numerical approximation is deteriorated by local singularities arising, e.g. from re-entrant corners, interior or boundary layers, or sharp shock-like fronts. An obvious remedy is to refine the discretisation near the critical regions, i.e. to place more grid-points where the solution is less regular. The question then is how to identify those regions and how to obtain a good balance between the refined and unrefined regions such that the overall accuracy is optimal.

Another closely related problem is to obtain reliable estimates of the accuracy of the computed numerical solution. A priori error estimates, as provided by the standard error analysis for finite element or finite difference methods, are often insufficient since they only yield information on the asymptotic behaviour of the error and require regularity properties of the solution which are not satisfied in the presence of singularities as described above.

These considerations show the need for error indicators which can a posteriori be extracted from the computed numerical solution and the given data of the problem. Of course, the calculation of the a posteriori error indicator should be far less expensive than the computation of the numerical solution. Moreover, the a posteriori error estimate should be local and should yield reliable upper and lower bounds for the true error in a user-specified norm. In this context one should note that global upper bounds are sufficient to obtain a numerical solution with an accuracy below a prescribed tolerance. Local lower bounds, however, are necessary to ensure that the grid is correctly refined so that one obtains a numerical solution with a prescribed tolerance using a (nearly) minimal number of grid-points.

Disposing of an a posteriori error indicator, an adaptive mesh-refinement process has the following general structure.

Algorithm 1.1 (General adaptive algorithm). *Given: the data of a partial differential equation and a tolerance* ε.

Sought: a numerical solution with an error less than ε.

(1) *Construct an initial coarse mesh* $\mathcal{T}_0$ *representing sufficiently well the geometry and data of the problem; set* $k = 0$.

(2) *Solve the discrete problem associated with* $\mathcal{T}_k$.

(3) *For every element* K *in* $\mathcal{T}_k$ *compute the a posteriori error indicator.*

(4) *If the estimated global error is less than* ε stop, *otherwise decide which elements have to be refined and construct the next mesh* $\mathcal{T}_{k+1}$. *Increase* k *by* 1 *and return to step* (2).

The above algorithm is best suited for stationary problems. For transient calculations, some changes have to be made:

- the accuracy of the computed numerical solution has to be estimated every few time-steps,
- the refinement process in space should be coupled with a time-step control,
- a partial coarsening of the mesh might be necessary,
- occasionally, a complete re-meshing could be desirable.

In both stationary and transient problems, the refinement and unrefinement process may also be coupled with or replaced by a moving-point technique which keeps the number of grid-points constant but changes their relative location.

In order to make Algorithm 1.1 operative, we must specify:

- a discretisation method,
- a solution method for the discrete problems,
- an error indicator which furnishes the a posteriori error estimate,
- a refinement strategy which determines which elements have to be refined or coarsened and how this has to be done.

The first two points are standard ones and are not the objective of this monograph. The last point will be addressed in Chapter 2. The third point is the objective of this chapter and of Chapters 4–6.

In this chapter we will describe various possibilities for a posteriori error estimation. In order to keep the presentation as simple as possible we will consider a simple model problem: the two-dimensional Poisson equation, problem (1.1) (p. 4), discretised by continuous linear or bi-linear finite elements, equation (1.3) (p. 5). We will review several a posteriori error estimates and show that, in a certain sense, they are all equivalent and yield lower and upper bounds for the error of the finite element discretisation. The estimates can roughly be classified as follows.

- *Residual estimates*: Estimate the error of the computed numerical solution by a suitable norm of its residual with respect to the strong form of the differential equation (Sections 1.4 (p. 10), 1.5 (p. 17), and 1.6 (p. 20)).

- *Solution of auxiliary local problems*: On small patches of elements, solve auxiliary discrete problems similar to but simpler than the original problem and use appropriate norms of the local solutions for the error estimation (Section 1.7 (p. 25)).
- *Hierarchical error estimates*: Evaluate the residual of the computed finite element solution with respect to another finite element space corresponding to higher order elements or to a refined grid (Section 1.8 (p. 31)).
- *Averaging methods*: Use some local extrapolate or average of the gradient of the computed numerical solution for error estimation (Section 1.9 (p. 36)).
- *Equilibrated residuals*: Evaluate approximately a dual variational problem posed on a function space with a weaker topology (Section 1.10 (p. 41)).
- *Dual weighted residuals*: Approximately solve a dual variational problem with the residual of the primal problem as right-hand side (Section 1.11 (p. 45)).
- *Hyper-circle method*: Use a functional analytic relationship of H^1- and $H(\mathrm{div})$-spaces in order to evaluate the residual of the numerical solution with respect to the original primal variational problem (Section 1.12 (p. 48)).

In Section 1.13 (p. 53), we briefly address the question of efficiency, i.e. whether the ratio of the estimated and the exact error remains bounded or even approaches 1 when the mesh-size converges to 0. In Section 1.14 (p. 58), we finally prove that an adaptive method based on a suitable error indicator and a suitable mesh-refinement strategy converges to the true solution of the differential equation.

In order to get a first impression of the capabilities of such an adaptive refinement strategy, we consider a simple, but typical example. We are looking for a function u which is harmonic in the interior Ω of a circular sector centred at the origin with radius 1 and angle $\frac{3}{2}\pi$, which vanishes on the straight parts of the boundary $\partial\Omega$, and which has normal derivative $\frac{2}{3}\sin(\frac{2}{3}\varphi)$ on the curved part of $\partial\Omega$, cf. problem (1.1) (p. 4). Using polar coordinates, one easily checks that $u = r^{2/3}\sin(\frac{2}{3}\varphi)$. We compute the Ritz projections $u_{\mathcal{T}}$ of u onto the spaces of continuous piecewise linear finite elements corresponding to the two triangulations shown in Plate 1 in the colour plate section, cf. problem (1.3) (p. 5). The triangulation in the left part of Plate 1 is obtained by 10 uniform refinements of an initial triangulation $\mathcal{T}_0$ which consists of six isosceles triangles with short sides of unit length. In each refinement step, every triangle is cut into two new ones by connecting the midpoint of its longest edge to the vertex opposite to this edge, cf. Section 2.1.4 (p. 68). Moreover, the midpoint of a longest edge having its two endpoints on $\partial\Omega$ is projected onto $\partial\Omega$. The triangulation in the right part of Plate 1 is obtained from $\mathcal{T}_0$ by applying 15 steps of the adaptive refinement strategy of Algorithm 1.1 using the error indicator $\eta_{R,K}$ of Theorem 1.5 (p. 15). A triangle $K \in \mathcal{T}_k$ is divided into two new ones as described above if $\eta_{R,K} \geq 0.5 \max_{K' \in \mathcal{T}_k} \eta_{R,K'}$, cf. Section 2.1 (p. 64). Midpoints of longest edges having their two endpoints on $\partial\Omega$ are again projected onto $\partial\Omega$. For both meshes we list in Table 1.1

Table 1.1 Number NT of triangles, number NV of vertices, number NN of unknowns, and relative error ε for the uniformly and adaptively refined meshes of Plate 1

refinement	NT	NV	NN	ε
uniform	3072	1601	1473	0.5%
adaptive	514	361	237	0.5%

the number NT of triangles, the number NV of vertices, the number NN of unknowns, and the relative error $\varepsilon = \frac{\| |\nabla(u-u_{\mathcal{T}})| \|}{\| |\nabla u| \|}$, where $\|\cdot\|$ and $|\cdot|$ denote the $L^2(\Omega)$- and Euclidean norm, respectively. It clearly shows the advantages of the adaptive refinement strategy.

1.2 The Model Problem and its Discretisation

As a model problem we consider the *Poisson equation* with mixed Dirichlet–Neumann boundary conditions

$$\begin{aligned} -\Delta u &= f && \text{in } \Omega \\ u &= 0 && \text{on } \Gamma_D \\ \frac{\partial u}{\partial n} &= g && \text{on } \Gamma_N \end{aligned} \tag{1.1}$$

in a connected, bounded, polygonal domain $\Omega \subset \mathbb{R}^2$ with boundary Γ consisting of two disjoint parts Γ_D and Γ_N. We assume that the *Dirichlet boundary* Γ_D is closed relative to Γ and has a positive length and that f and g are square integrable functions on Ω and Γ_N, respectively. The *Neumann boundary* Γ_N may be empty.

For any open subset ω of Ω with Lipschitz boundary γ, we denote by $L^2(\omega)$ and $L^2(\gamma)$ the standard Lebesgue spaces equipped with the usual norms [**3**], [**109**, Section I.2]

$$\|\varphi\|_\omega = \left\{ \int_\omega \varphi^2 \, \mathrm{d}x \right\}^{\frac{1}{2}}, \quad \|\varphi\|_\gamma = \left\{ \int_\gamma \varphi^2 \, \mathrm{d}s \right\}^{\frac{1}{2}}.$$

For vector fields $\mathbf{v} : \omega \to \mathbb{R}^2$, we use the convention $\|\mathbf{v}\|_\omega = \| |\mathbf{v}| \|_\omega$, where $|\cdot|$ is the Euclidean norm in $\mathbb{R}^2$. If $\omega = \Omega$, we omit the subscript Ω.

As usual, $H^1(\Omega)$ denotes the Sobolev space of all functions in $L^2(\Omega)$ which have all their first-order derivatives in $L^2(\Omega)$, too. Set

$$H^1_D(\Omega) = \left\{ \varphi \in H^1(\Omega) : \varphi = 0 \text{ on } \Gamma_D \right\}.$$

The standard weak formulation of problem (1.1) is then:

find $u \in H^1_D(\Omega)$ such that

$$\int_\Omega \nabla u \cdot \nabla v = \int_\Omega f v + \int_{\Gamma_N} g v \tag{1.2}$$

holds for all $v \in H^1_D(\Omega)$.

The Lax–Milgram lemma [**109**, Theorem 1.1] implies that problem (1.2) has a unique solution.

Throughout this chapter we denote by $\mathcal{T}$ a partition of Ω which satisfies the following conditions:

- $\mathcal{T}$ consists of triangles and parallelograms (*affine equivalence*),
- any two elements in $\mathcal{T}$ share at most a common edge or a common vertex (*admissibility*),
- the minimal angle of all elements in $\mathcal{T}$ is bounded away from zero (*shape regularity*).

Note that triangles and parallelograms may be mixed. A single partition is always shape-regular. Shape regularity is relevant for families of partitions which are obtained by subsequent refinements. Then the refinement process must not lead to too small angles.

Denote by $\widehat{K}$ the reference triangle $\{x \in \mathbb{R}^2 : x_1 \geq 0, x_2 \geq 0, x_1 + x_2 \leq 1\}$ or the reference square $[0,1]^2$ and set

$$R_1(\widehat{K}) = \begin{cases} \operatorname{span}\{1, x_1, x_2\} & \text{for the reference triangle,} \\ \operatorname{span}\{1, x_1, x_2, x_1x_2\} & \text{for the reference square.} \end{cases}$$

Then every element $K \in \mathcal{T}$ is the image of the reference element $\widehat{K}$ under an affine diffeomorphism $F_K : \widehat{K} \to K$. Set

$$R_1(K) = \left\{\varphi \circ F_K^{-1} : \varphi \in R_1(\widehat{K})\right\}$$

and define the lowest order conforming finite element space associated with $\mathcal{T}$ by

$$S^{1,0}(\mathcal{T}) = \left\{\varphi \in C(\Omega) : \varphi|_K \in R_1(K) \text{ for all } K \in \mathcal{T}\right\},$$
$$S_D^{1,0}(\mathcal{T}) = \left\{\varphi \in S^{1,0}(\mathcal{T}) : \varphi = 0 \text{ on } \Gamma_D\right\}.$$

We then consider the following finite element discretisation of problem (1.2):

find $u_{\mathcal{T}} \in S_D^{1,0}(\mathcal{T})$ such that

$$\int_\Omega \nabla u_{\mathcal{T}} \cdot \nabla v_{\mathcal{T}} = \int_\Omega f v_{\mathcal{T}} + \int_{\Gamma_N} g v_{\mathcal{T}} \tag{1.3}$$

holds for all $v_{\mathcal{T}} \in S_D^{1,0}(\mathcal{T})$.

The Lax–Milgram lemma [**109**, Theorem 1.1] again implies that problem (1.3) has a unique solution.

1.3 Notations and Auxiliary Results

Local error estimates for quasi-interpolation operators and inverse estimates involving local cut-off functions are fundamental tools in deriving a posteriori error estimates. They require some additional notations.

1.3.1 Vertices and Edges

With a partition $\mathcal{T}$ we associate the sets $\mathcal{N}$ and $\mathcal{E}$ of all vertices and edges, respectively of all elements in $\mathcal{T}$. A subscript K, Ω, Γ, Γ_D, or Γ_N to $\mathcal{N}$ or $\mathcal{E}$ indicates that only those vertices or edges, respectively, are considered which are contained in the corresponding index set. Similarly $\mathcal{N}_E$ denotes the set of the vertices of a given edge E. The union of all edges in $\mathcal{E}$ is denoted by Σ and is called the *skeleton* of $\mathcal{T}$.

For every element or edge $S \in \mathcal{T} \cup \mathcal{E}$ we denote by h_S its *diameter*. Note that the shape regularity of $\mathcal{T}$ implies that for all elements K and K' and all edges E and E' that share at least one vertex, the ratios $\frac{h_K}{h_{K'}}$, $\frac{h_E}{h_{E'}}$, and $\frac{h_K}{h_E}$ are bounded from below and from above by a common constant which only

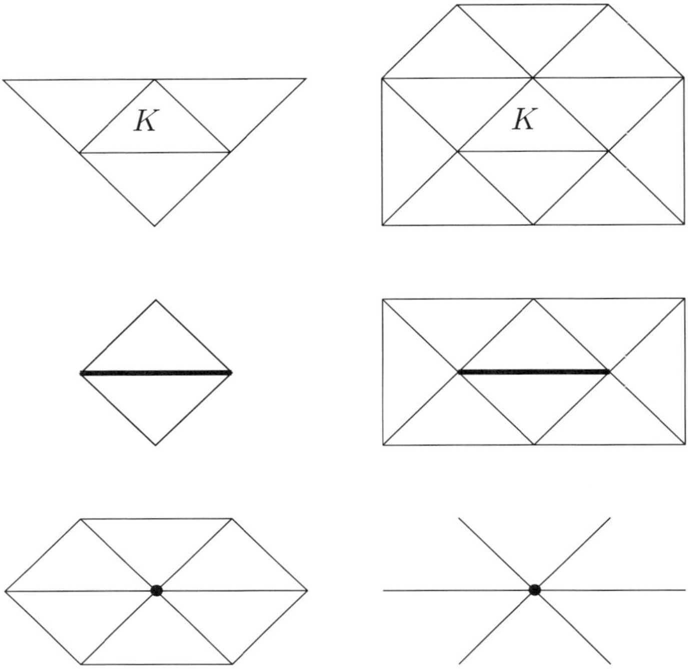

Figure 1.1 Domains ω_K (top left), $\widetilde{\omega}_K$ (top right), ω_E (middle left), $\widetilde{\omega}_E$ (middle right), ω_z (bottom left), and σ_z (bottom right); thick lines: edge E, bullets: vertex z

depends on the sine of the smallest angle in $\mathcal{T}$ and which is referred to as the *shape parameter* of $\mathcal{T}$, cf. equation (3.3) (p. 82).

With every element K, every edge E, and every vertex z we associate the following sets (cf. Figures 1.1 and 1.2)

$$\omega_K = \bigcup_{\mathcal{E}_K \cap \mathcal{E}_{K'} \neq \emptyset} K', \quad \widetilde{\omega}_K = \bigcup_{\mathcal{N}_K \cap \mathcal{N}_{K'} \neq \emptyset} K',$$
$$\omega_E = \bigcup_{E \in \mathcal{E}_{K'}} K', \qquad \widetilde{\omega}_E = \bigcup_{\mathcal{N}_E \cap \mathcal{N}_{K'} \neq \emptyset} K',$$
$$\omega_z = \bigcup_{z \in \mathcal{N}_{K'}} K', \qquad \sigma_z = \bigcup_{z \in \mathcal{N}_E} E.$$

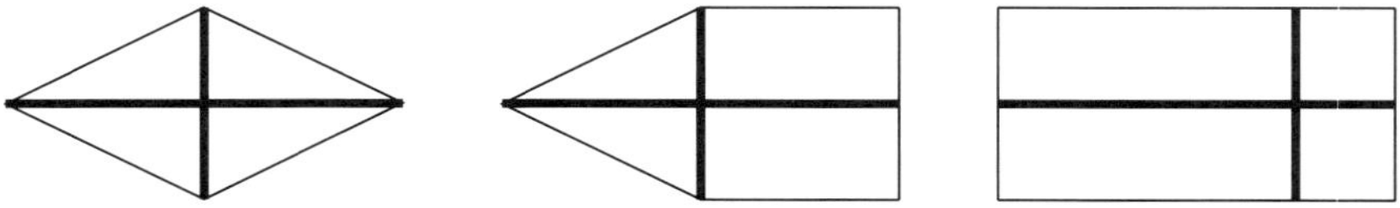

Figure 1.2 Domains ω_z and σ_z (thick lines)

Due to the shape regularity of $\mathcal{T}$, the diameter of any of these sets can be bounded by a multiple of the diameter of any element or edge contained in that set. The constant only depends on the shape parameter of $\mathcal{T}$.

1.3.2 Nodal Shape Functions

With every vertex $z \in \mathcal{N}$ we associate its *nodal shape function* λ_z which is uniquely defined by the conditions

$$\lambda_z \in S^{1,0}(\mathcal{T}), \quad \lambda_z(z) = 1, \quad \lambda_z(x) = 0 \text{ for all } x \in \mathcal{N} \setminus \{z\}.$$

Lemma 1.2 *The nodal shape functions have the following properties:*

$$\begin{aligned} 0 \le \lambda_z \le 1, & \qquad \operatorname{supp} \lambda_z = \omega_z, \\ \sum_{z \in \mathcal{N}_K} \lambda_z = 1 \quad \text{on } K, & \qquad \sum_{z \in \mathcal{N}_E} \lambda_z = 1 \quad \text{on } E. \end{aligned}$$

Proof We leave the straightforward proof to the reader. □

1.3.3 A Quasi-Interpolation Operator

For every measurable set $\omega \subset \mathbb{R}^2$ with positive measure and every integrable function v we define the *average* of v on ω by

$$\overline{v}_\omega = \frac{1}{\mu_2(\omega)} \int_\omega v,$$

where μ_d denotes the standard Lebesgue measure in $\mathbb{R}^d$.

With this notation we define a *quasi-interpolation operator* $I_\mathcal{T} : H^1_D(\Omega) \to S^{1,0}_D(\mathcal{T})$ by

$$I_\mathcal{T} v = \sum_{z \in \mathcal{N}_\Omega \cup \mathcal{N}_{\Gamma_N}} \overline{v}_{\omega_z} \lambda_z. \tag{1.4}$$

Notice that $I_\mathcal{T} v$ is defined for every integrable function v. The operator $I_\mathcal{T}$ has the following approximation properties.

Proposition 1.3 *The following error estimates hold for all elements $K \in \mathcal{T}$, all edges $E \in \mathcal{E}$, and all functions $v \in H^1_D(\Omega)$*

$$\begin{aligned} \|v - I_\mathcal{T} v\|_K &\le c_{A,1}\, h_K\, \|\nabla v\|_{\widetilde{\omega}_K}, \\ \|v - I_\mathcal{T} v\|_E &\le c_{A,2}\, h_E^{\frac{1}{2}}\, \|\nabla v\|_{\widetilde{\omega}_E}. \end{aligned}$$

The constants $c_{A,1}$ and $c_{A,2}$ only depend on the shape parameter of $\mathcal{T}$.

Proof In Section 3.5 (p. 108) we give a detailed proof of Proposition 1.3 using a slight modification of $I_\mathcal{T}$. This proof in particular gives explicit bounds for the constants $c_{A,1}$ and $c_{A,2}$. Here, we only sketch the principal steps of the proof.

For every vertex $z \in \mathcal{N}$ and every function $v \in H^1(\omega_z)$, we have the *Poincaré inequality*

$$\left\| v - \bar{v}_{\omega_z} \right\|_{\omega_z} \leq c_z h_z \left\| \nabla v \right\|_{\omega_z},$$

where h_z is the diameter of ω_z with respect to the Euclidean norm. The standard proof of this inequality is not constructive and proceeds by contradiction. A constructive proof is given in Section 3.4 (p. 91). In particular it shows that the constant c_z equals $\frac{1}{\pi}$ for convex ω_z and that c_z depends on the shape parameter of $\mathcal{T}$ for non-convex ω_z.

For every element K, every edge E thereof, and every function $v \in H^1(K)$, we have the *trace inequality*

$$\|v\|_E \leq c_1 h_K^{-\frac{1}{2}} \|v\|_K + c_2 h_K^{\frac{1}{2}} \|\nabla v\|_K, \tag{1.5}$$

where the constants c_1 and c_2 only depend on the shape parameter of $\mathcal{T}$. The standard proof of the trace inequality again is not constructive. It proceeds by transforming the edge E to an edge $\widehat{E}$ of the reference element $\widehat{K}$, using the continuous embedding of $L^2(\widehat{E})$ in $L^2(\widehat{K})$, i.e. the trace theorem on $\widehat{K}$, and transforming back to the element K. In Section 3.3 (p. 87) we give a constructive proof of the trace inequality which also yields explicit bounds for the constants c_1 and c_2.

Lemma 1.2 and equation (1.4) imply that for every element K and every function $v \in H_D^1(\Omega)$

$$\begin{aligned}\|v - I_{\mathcal{T}} v\|_K &\leq \sum_{z \in \mathcal{N}_K} \left\| \lambda_z (v - \bar{v}_{\omega_z}) \right\|_K + \sum_{z \in \mathcal{N}_{K,\Gamma_D}} \left\| \lambda_z \bar{v}_{\omega_z} \right\|_K \\ &\leq \sum_{z \in \mathcal{N}_K} \left\| v - \bar{v}_{\omega_z} \right\|_K + \sum_{z \in \mathcal{N}_{K,\Gamma_D}} \left\| \bar{v}_{\omega_z} \right\|_K .\end{aligned}$$

The first term on the right-hand side of this estimate can be bounded using the Poincaré inequality.

To estimate the second term, we observe that for every vertex $z \in \mathcal{N}_{K,\Gamma_D}$ there is an edge E_z on the Dirichlet boundary Γ_D and an element K_z such that z is a vertex of E_z and E_z is an edge of K_z. Since v vanishes on E_z, we may therefore use the trace and Poincaré inequalities to bound $\left\| \bar{v}_{\omega_z} \right\|_K$ in terms of $\|\nabla v\|_{\omega_z}$.

Finally, Lemma 1.2 and equation (1.4) imply that for every edge E and every function $v \in H_D^1(\Omega)$

$$\|v - I_{\mathcal{T}} v\|_E \leq \sum_{z \in \mathcal{N}_E} \left\| v - \bar{v}_{\omega_z} \right\|_E + \sum_{z \in \mathcal{N}_{E,\Gamma_D}} \left\| \bar{v}_{\omega_z} \right\|_E .$$

Using the trace inequality, the two terms on the right-hand side of this estimate can be bounded in exactly the same way as the corresponding terms in the estimate for K. □

1.3.4 Local Cut-Off Functions

For every element K and every edge E, we define *local cut-off functions* ψ_K and ψ_E by

$$\psi_K = \beta_K \prod_{z \in \mathcal{N}_K} \lambda_z$$

with

$$\beta_K = \begin{cases} 27 & \text{if } K \text{ is a triangle,} \\ 256 & \text{if } K \text{ is a parallelogram} \end{cases}$$

and

$$\psi_E = 4 \prod_{z \in \mathcal{N}_E} \lambda_z.$$

Proposition 1.4 *The functions ψ_K and ψ_E have the following properties*

$$\begin{aligned} \operatorname{supp} \psi_K &= K, & 0 \le \psi_K &\le 1, & \max_{x \in K} \psi_K(x) &= 1, \\ \operatorname{supp} \psi_E &= \omega_E, & 0 \le \psi_E &\le 1, & \max_{x \in E} \psi_E(x) &= 1. \end{aligned}$$

Moreover the following inverse estimates hold for all linear or bi-linear functions v

$$\begin{aligned} \|v\|_K &\le c_{I,1} \left\| \psi_K^{\frac{1}{2}} v \right\|_K, \\ \left\| \nabla(\psi_K v) \right\|_K &\le c_{I,2}\, h_K^{-1} \|v\|_K, \\ \|v\|_E &\le c_{I,3} \left\| \psi_E^{\frac{1}{2}} v \right\|_E, \\ \left\| \nabla(\psi_E v) \right\|_{\omega_E} &\le c_{I,4}\, h_E^{-\frac{1}{2}} \|v\|_E, \\ \|\psi_E v\|_{\omega_E} &\le c_{I,5}\, h_E^{\frac{1}{2}} \|v\|_E. \end{aligned}$$

The constants $c_{I,1}, \ldots, c_{I,5}$ depend only on the shape parameter of $\mathcal{T}$.

Proof The properties of ψ_K and ψ_E follow from the properties of the functions λ_z listed in Lemma 1.2. The inverse estimates are all proved by transforming the element K or the edge E to the reference element $\widehat{K}$ or the reference edge $\widehat{E} = \widehat{K} \cap \{x_2 = 0\}$, respectively, invoking the equivalence of norms on the finite-dimensional space of linear or bi-linear functions and transforming back to the given element K or edge E. A detailed proof of Proposition 1.4, which yields explicit bounds for the constants $c_{I,1}, \ldots, c_{I,5}$, is given in Section 3.6 (p. 112). □

1.3.5 Jumps

With every edge E, we associate a unit vector $\mathbf{n}_E$. For interior edges $E \in \mathcal{E}_\Omega$ its orientation is arbitrary; for edges $E \in \mathcal{E}_\Gamma$ on the boundary the orientation is chosen such that $\mathbf{n}_E$ points to the exterior of Ω. For any piecewise continuous function v and any interior edge E, we denote by $\mathbb{J}_E(v)$ the jump of v across E in the direction of $\mathbf{n}_E$:

$$\mathbb{J}_E(v)(x) = \lim_{t \to 0+} v(x - t\mathbf{n}_E) - \lim_{t \to 0+} v(x + t\mathbf{n}_E) \quad \text{for all } x \in E. \tag{1.6}$$

Note that $\mathbb{J}_E(v)$ depends on the orientation of $\mathbf{n}_E$ but that expressions of the form $\mathbb{J}_E(\mathbf{n}_E \cdot \nabla v)$ are independent thereof.

1.4 Residual Estimates

Denote by $u \in H^1_D(\Omega)$ and $u_{\mathcal{T}} \in S^{1,0}_D(\mathcal{T})$ the exact solutions of problems (1.2) (p. 4) and (1.3) (p. 5), respectively. In what follows we want to bound the error $\|\nabla(u - u_{\mathcal{T}})\|$ from above and from below by computable quantities which only involve $u_{\mathcal{T}}$, the solution of the discrete problem, and f and g, the data of the differential equation.

1.4.1 The Equivalence of Error and Residual

For all $v \in H^1_D(\Omega)$ the functions u and $u_{\mathcal{T}}$ fulfil the identity

$$\int_\Omega \nabla(u - u_{\mathcal{T}}) \cdot \nabla v = \int_\Omega fv + \int_{\Gamma_N} gv - \int_\Omega \nabla u_{\mathcal{T}} \cdot \nabla v. \tag{1.7}$$

The right-hand side of equation (1.7) implicitly defines the *residual* R of $u_{\mathcal{T}}$ as an element of the dual space of $H^1_D(\Omega)$ by setting for all $v \in H^1_D(\Omega)$

$$\langle R, v \rangle = \int_\Omega fv + \int_{\Gamma_N} gv - \int_\Omega \nabla u_{\mathcal{T}} \cdot \nabla v. \tag{1.8}$$

The Cauchy–Schwarz inequality implies for all $v \in H^1_D(\Omega)$

$$\|\nabla v\| = \sup_{w \in H^1_D(\Omega)\setminus\{0\}} \frac{1}{\|\nabla w\|} \int_\Omega \nabla v \cdot \nabla w. \tag{1.9}$$

This corresponds to the fact that the bi-linear form $H^1_D(\Omega) \ni v, w \mapsto \int_\Omega \nabla v \cdot \nabla w$ defines an isometry of $H^1_D(\Omega)$ onto its dual space and that $\|\nabla \cdot\|$ is the energy norm corresponding to problems (1.2) and (1.3).

Equations (1.7)–(1.9) imply the identity

$$\|\nabla(u - u_{\mathcal{T}})\| = \sup_{w \in H^1_D(\Omega)\setminus\{0\}} \frac{\langle R, w \rangle}{\|\nabla w\|}. \tag{1.10}$$

Since the sup term in equation (1.10) is the norm of the residual in the dual space of $H^1_D(\Omega)$, we have proved:

the norm of the error in $H^1_D(\Omega)$ is equal to the norm of the residual in the dual space of $H^1_D(\Omega)$.

Most a posteriori error indicators try to estimate this dual norm of the residual by quantities that can more easily be computed from f, g, and $u_{\mathcal{T}}$.

1.4.2 Galerkin Orthogonality

We first observe that, since $S^{1,0}_D(\mathcal{T}) \subset H^1_D(\Omega)$, the error is orthogonal to $S^{1,0}_D(\mathcal{T})$, i.e. for all $w_{\mathcal{T}} \in S^{1,0}_D(\mathcal{T})$ there holds

$$\int_\Omega \nabla(u - u_{\mathcal{T}}) \cdot \nabla w_{\mathcal{T}} = 0. \tag{1.11}$$

Using equations (1.7) and (1.8) this can be written as

$$\langle R, w_{\mathcal{T}} \rangle = 0 \tag{1.12}$$

for all $w_{\mathcal{T}} \in S_D^{1,0}(\mathcal{T})$, i.e. the discrete space $S_D^{1,0}(\mathcal{T})$ is contained in the kernel of R. Equation (1.12) reflects the fact that the discretisation (1.3) is consistent and that no additional errors are introduced by numerical integration or by inexact solution of the discrete problem. It is often referred to as *Galerkin orthogonality*.

1.4.3 An L^2-Representation of the Residual

For all $w \in H_D^1(\Omega)$ integration by parts element-wise yields

$$\begin{aligned}\langle R, w \rangle &= \int_\Omega fw + \int_{\Gamma_N} gw - \sum_{K \in \mathcal{T}} \int_K \nabla u_{\mathcal{T}} \cdot \nabla w \\ &= \int_\Omega fw + \int_{\Gamma_N} gw + \sum_{K \in \mathcal{T}} \left\{ \int_K \Delta u_{\mathcal{T}} w - \int_{\partial K} \mathbf{n}_K \cdot \nabla u_{\mathcal{T}} w \right\} \\ &= \sum_{K \in \mathcal{T}} \int_K (f + \Delta u_{\mathcal{T}}) w + \sum_{E \in \mathcal{E}_{\Gamma_N}} \int_E (g - \mathbf{n}_E \cdot \nabla u_{\mathcal{T}}) w - \sum_{E \in \mathcal{E}_\Omega} \int_E \mathbb{J}_E(\mathbf{n}_E \cdot \nabla u_{\mathcal{T}}) w.\end{aligned}$$

Here, $\mathbf{n}_K$ denotes the unit exterior normal to the element K. Note that $\Delta u_{\mathcal{T}}$ vanishes on all triangles.

For brevity, we define *element* and *edge residuals* $r : \Omega \to \mathbb{R}$ and $j : \Sigma \to \mathbb{R}$ by

$$r|_K = f + \Delta u_{\mathcal{T}} \quad \text{on every } K \in \mathcal{T}, \tag{1.13}$$

$$j|_E = \begin{cases} -\mathbb{J}_E(\mathbf{n}_E \cdot \nabla u_{\mathcal{T}}) & \text{if } E \in \mathcal{E}_\Omega, \\ g - \mathbf{n}_E \cdot \nabla u_{\mathcal{T}} & \text{if } E \in \mathcal{E}_{\Gamma_N}, \\ 0 & \text{if } E \in \mathcal{E}_{\Gamma_D}. \end{cases} \tag{1.14}$$

Then, we obtain the following *L^2-representation of the residual*

$$\langle R, w \rangle = \int_\Omega rw + \int_\Sigma jw. \tag{1.15}$$

Together with equation (1.12) this implies that for all $w \in H_D^1(\Omega)$ and all $w_{\mathcal{T}} \in S_D^{1,0}(\mathcal{T})$

$$\langle R, w \rangle = \int_\Omega r(w - w_{\mathcal{T}}) + \int_\Sigma j(w - w_{\mathcal{T}}). \tag{1.16}$$

1.4.4 An Upper Bound on the Error

We fix an arbitrary function $w \in H_D^1(\Omega)$ and choose $w_{\mathcal{T}} = I_{\mathcal{T}} w$ with the quasi-interpolation operator of equation (1.4) (p. 7). The Cauchy–Schwarz inequality for integrals and Proposition 1.3 (p. 7) then yield

$$\begin{aligned}\langle R, w\rangle &= \sum_{K\in\mathcal{T}} \int_K r(w - I_{\mathcal{T}}w) + \sum_{E\in\mathcal{E}} \int_E j(w - I_{\mathcal{T}}w) \\ &\le \sum_{K\in\mathcal{T}} \|r\|_K \, \|w - I_{\mathcal{T}}w\|_K + \sum_{E\in\mathcal{E}} \|j\|_E \, \|w - I_{\mathcal{T}}w\|_E \\ &\le \sum_{K\in\mathcal{T}} \|r\|_K \, c_{A,1} \, h_K \, \|\nabla w\|_{\tilde{\omega}_K} + \sum_{E\in\mathcal{E}} \|j\|_E \, c_{A,2} \, h_E^{\frac{1}{2}} \, \|\nabla w\|_{\tilde{\omega}_E} .\end{aligned}$$

Invoking the Cauchy–Schwarz inequality for sums this gives

$$\begin{aligned}\langle R, w\rangle \le \max\{c_{A,1}, c_{A,2}\} &\left\{\sum_{K\in\mathcal{T}} h_K^2 \, \|r\|_K^2 + \sum_{E\in\mathcal{E}} h_E \, \|j\|_E^2\right\}^{\frac{1}{2}} \\ &\cdot \left\{\sum_{K\in\mathcal{T}} \|\nabla w\|_{\tilde{\omega}_K}^2 + \sum_{E\in\mathcal{E}} \|\nabla w\|_{\tilde{\omega}_E}^2\right\}^{\frac{1}{2}} .\end{aligned}$$

In the final step we observe that the shape regularity of $\mathcal{T}$ implies

$$\left\{\sum_{K\in\mathcal{T}} \|\nabla w\|_{\tilde{\omega}_K}^2 + \sum_{E\in\mathcal{E}} \|\nabla w\|_{\tilde{\omega}_E}^2\right\}^{\frac{1}{2}} \le c_{\mathcal{T}} \, \|\nabla w\|$$

with a constant $c_{\mathcal{T}}$ which only depends on the shape parameter of $\mathcal{T}$ and which takes into account that every element is counted several times on the left-hand side of this inequality.

Combining these estimates with equality (1.10) we obtain the following upper bound for the error

$$\left\|\nabla(u - u_{\mathcal{T}})\right\| \le c^* \left\{\sum_{K\in\mathcal{T}} h_K^2 \, \|r\|_K^2 + \sum_{E\in\mathcal{E}} h_E \, \|j\|_E^2\right\}^{\frac{1}{2}} \tag{1.17}$$

with $c^* = \max\{c_{A,1}, c_{A,2}\} \, c_{\mathcal{T}}$.

The right-hand side of equation (1.17) can be used as an *a posteriori error indicator* since it only involves the known data f and g, the solution $u_{\mathcal{T}}$ of the discrete problem, and the geometrical data of the partition. Inequality (1.17) implies that the a posteriori error indicator is *reliable* in the sense that an inequality of the form 'error indicator $\le$ tolerance' implies that the true error is also less than the tolerance up to the multiplicative constant c^*. We want to prove that the error indicator is also *efficient* in the sense that an inequality of the form 'error indicator $\ge$ tolerance' implies that the true error is also greater than the tolerance possibly up to another multiplicative constant.

For general functions f and g, the exact evaluation of the integrals on the right-hand side of equation (1.17) may be prohibitively expensive or even impossible. The integrals must then be approximated by suitable quadrature formulae. Alternatively, the functions f and g may be approximated by simpler functions, e.g. piecewise polynomial functions, and the resulting integrals can be evaluated exactly. Often, both approaches are equivalent.

1.4.5 A Lower Bound on the Error

In order to prove the announced efficiency, we replace the functions f and g by their averages on the elements and edges, respectively. For every element K and for every edge E on the Neumann boundary we set

$$\bar{f}_K = \frac{1}{\mu_2(K)}\int_K f, \quad \bar{g}_E = \frac{1}{\mu_1(E)}\int_E g \tag{1.18}$$

and define

$$f_{\mathcal{T}} = \sum_{K\in\mathcal{T}} \bar{f}_K \chi_K, \quad g_{\mathcal{E}} = \sum_{E\in\mathcal{E}_{\Gamma_N}} \bar{g}_E \chi_E, \tag{1.19}$$

where χ_A denotes the characteristic function of the set A.

We fix an arbitrary element K and insert the function

$$w_K = (\bar{f}_K + \Delta u_{\mathcal{T}})\psi_K \tag{1.20}$$

in equation (1.15). Taking into account equations (1.7) and (1.8) and that supp $w_K \subset K$, we obtain

$$\int_K r w_K = \int_K \nabla(u - u_{\mathcal{T}}) \cdot \nabla w_K.$$

We add $\int_K (\bar{f}_K - f) w_K$ on both sides of this equation and get

$$\begin{aligned}
\int_K (\bar{f}_K + \Delta u_{\mathcal{T}})^2 \psi_K &= \int_K (\bar{f}_K + \Delta u_{\mathcal{T}}) w_K \\
&= \int_K r w_K + \int_K (\bar{f}_K - f) w_K \\
&= \int_K \nabla(u - u_{\mathcal{T}}) \cdot \nabla w_K + \int_K (\bar{f}_K - f) w_K.
\end{aligned}$$

Proposition 1.4 (p. 9) implies for the left hand-side of this equation

$$\int_K (\bar{f}_K + \Delta u_{\mathcal{T}})^2 \psi_K \geq c_{I,1}^{-2} \left\| \bar{f}_K + \Delta u_{\mathcal{T}} \right\|_K^2$$

and for the two terms on its right-hand side

$$\begin{aligned}
\int_K \nabla(u - u_{\mathcal{T}}) \cdot \nabla w_K &\leq \left\| \nabla(u - u_{\mathcal{T}}) \right\|_K \|\nabla w_K\|_K \\
&\leq \left\| \nabla(u - u_{\mathcal{T}}) \right\|_K c_{I,2}\, h_K^{-1} \left\| \bar{f}_K + \Delta u_{\mathcal{T}} \right\|_K, \\
\int_K (f - \bar{f}_K) w_K &\leq \left\| f - \bar{f}_K \right\|_K \|w_K\|_K \\
&\leq \left\| f - \bar{f}_K \right\|_K \left\| \bar{f}_K + \Delta u_{\mathcal{T}} \right\|_K.
\end{aligned}$$

This proves that

$$h_K \left\| \bar{f}_K + \Delta u_{\mathcal{T}} \right\|_K \leq c_{I,1}^2\, c_{I,2} \left\| \nabla(u - u_{\mathcal{T}}) \right\|_K + c_{I,1}^2\, h_K \left\| f - \bar{f}_K \right\|_K. \tag{1.21}$$

Next, we consider an arbitrary interior edge $E \in \mathcal{E}_\Omega$ and insert the function

$$w_E = -\mathbb{J}_E(\mathbf{n}_E \cdot \nabla u_\mathcal{T})\psi_E \tag{1.22}$$

in equation (1.15). Taking into account equations (1.7) and (1.8), this gives

$$\begin{aligned}\int_E \mathbb{J}_E(\mathbf{n}_E \cdot \nabla u_\mathcal{T})^2 \psi_E &= \int_E j w_E \\ &= \int_{\omega_E} \nabla(u - u_\mathcal{T}) \cdot \nabla w_E - \int_{\omega_E} r w_E \\ &= \int_{\omega_E} \nabla(u - u_\mathcal{T}) \cdot \nabla w_E - \sum_{K \subset \omega_E} \int_K (\bar{f}_K + \Delta u_\mathcal{T}) w_E - \sum_{K \subset \omega_E} \int_K (f - \bar{f}_K) w_E.\end{aligned}$$

Proposition 1.4 (p. 9) implies for the left-hand side of this equation

$$\int_E \mathbb{J}_E(\mathbf{n}_E \cdot \nabla u_\mathcal{T})^2 \psi_E \geq c_{I,3}^{-2} \left\| \mathbb{J}_E(\mathbf{n}_E \cdot \nabla u_\mathcal{T}) \right\|_E^2$$

and for the three terms on its right-hand side

$$\begin{aligned}\int_{\omega_E} \nabla(u - u_\mathcal{T}) \cdot \nabla w_E &\leq \left\| \nabla(u - u_\mathcal{T}) \right\|_{\omega_E} \|\nabla w_E\|_{\omega_E} \\ &\leq \left\| \nabla(u - u_\mathcal{T}) \right\|_{\omega_E} c_{I,4}\, h_E^{-\frac{1}{2}} \left\| \mathbb{J}_E(\mathbf{n}_E \cdot \nabla u_\mathcal{T}) \right\|_E, \\ \sum_{K \subset \omega_E} \int_K (\bar{f}_K + \Delta u_\mathcal{T}) w_E &\leq \sum_{K \subset \omega_E} \left\| \bar{f}_K + \Delta u_\mathcal{T} \right\|_K \|w_E\|_K \\ &\leq \sum_{K \subset \omega_E} \left\| \bar{f}_K + \Delta u_\mathcal{T} \right\|_K \cdot c_{I,5}\, h_E^{\frac{1}{2}} \left\| \mathbb{J}_E(\mathbf{n}_E \cdot \nabla u_\mathcal{T}) \right\|_E, \\ \sum_{K \subset \omega_E} \int_K (f - \bar{f}_K) w_E &\leq \sum_{K \subset \omega_E} \left\| f - \bar{f}_K \right\|_K \|w_E\|_K \\ &\leq \sum_{K \subset \omega_E} \left\| f - \bar{f}_K \right\|_K c_{I,5}\, h_E^{\frac{1}{2}} \left\| \mathbb{J}_E(\mathbf{n}_E \cdot \nabla u_\mathcal{T}) \right\|_E.\end{aligned}$$

This yields

$$\begin{aligned}c_{I,3}^{-2} \left\| \mathbb{J}_E(\mathbf{n}_E \cdot \nabla u_\mathcal{T}) \right\|_E &\leq c_{I,4}\, h_E^{-\frac{1}{2}} \left\| \nabla(u - u_\mathcal{T}) \right\|_{\omega_E} \\ &\quad + \sum_{K \subset \omega_E} c_{I,5}\, h_E^{\frac{1}{2}} \left\| \bar{f}_K + \Delta u_\mathcal{T} \right\|_K + \sum_{K \subset \omega_E} c_{I,5}\, h_E^{\frac{1}{2}} \left\| f - \bar{f}_K \right\|_K.\end{aligned}$$

Combining this estimate with inequality (1.21), we obtain

$$
\begin{aligned}
&h_E^{\frac{1}{2}} \left\| \mathbb{J}_E(\mathbf{n}_E \cdot \nabla u_{\mathcal{T}}) \right\|_E \\
&\quad \leq c_{I,3}^2 \left[c_{I,4} + c_{I,1}^2\, c_{I,2}\, c_{I,5} \right] \left\| \nabla(u - u_{\mathcal{T}}) \right\|_{\omega_E} + c_{I,3}^2\, c_{I,5} \left[1 + c_{I,1}^2 \right] \sum_{K \subset \omega_E} h_K \left\| f - \bar{f}_K \right\|_K .
\end{aligned}
\tag{1.23}
$$

Finally, we fix an edge E on the Neumann boundary, denote by K the adjacent element, and insert the function

$$
w_E = (\bar{g}_E - \mathbf{n}_E \cdot \nabla u_{\mathcal{T}}) \psi_E \tag{1.24}
$$

in equation (1.15). Taking into account equations (1.7) and (1.8), this gives

$$
\int_E j w_E = \int_K \nabla(u - u_{\mathcal{T}}) \cdot \nabla w_E - \int_K r w_E .
$$

We add $\int_E (\bar{g}_E - g) w_E$ on both sides of this equation and obtain

$$
\begin{aligned}
\int_E (\bar{g}_E - \mathbf{n}_E \cdot \nabla u_{\mathcal{T}})^2 \psi_E &= \int_E (\bar{g}_E - \mathbf{n}_E \cdot \nabla u_{\mathcal{T}}) w_E \\
&= \int_E j w_E + \int_E (\bar{g}_E - g) w_E \\
&= \int_K \nabla(u - u_{\mathcal{T}}) \cdot \nabla w_E - \int_K (\bar{f}_K + \Delta u_{\mathcal{T}}) w_E - \int_K (f - \bar{f}_K) w_E + \int_E (\bar{g}_E - g) w_E .
\end{aligned}
$$

Invoking once again Proposition 1.4 (p. 9) and using the same arguments as above, this implies that

$$
\begin{aligned}
&h_E^{\frac{1}{2}} \left\| \bar{g}_E - \mathbf{n}_E \cdot \nabla u_{\mathcal{T}} \right\|_E \\
&\quad \leq c_{I,3}^2 \left[c_{I,4} + c_{I,1}^2\, c_{I,2}\, c_{I,5} \right] \left\| \nabla(u - u_{\mathcal{T}}) \right\|_K + c_{I,3}^2\, c_{I,5} \left[1 + c_{I,1}^2 \right] h_K \left\| f - \bar{f}_K \right\|_K + c_{I,3}^2\, h_E^{\frac{1}{2}} \left\| g - \bar{g}_E \right\|_E .
\end{aligned}
\tag{1.25}
$$

Estimates (1.21), (1.23), and (1.25) prove the announced efficiency of the a posteriori error estimate.

1.4.6 A Posteriori Error Estimates

The results of this section can be summarised as follows.

Theorem 1.5 *Denote by u and $u_{\mathcal{T}}$ the unique solutions of problems (1.2) (p. 4) and (1.3) (p. 5), respectively. For every element $K \in \mathcal{T}$ define the* residual a posteriori error indicator *$\eta_{R,K}$ by*

$$
\eta_{R,K} = \left\{ h_K^2 \left\| \bar{f}_K + \Delta u_{\mathcal{T}} \right\|_K^2 + \frac{1}{2} \sum_{E \in \mathcal{E}_{K,\Omega}} h_E \left\| \mathbb{J}_E(\mathbf{n}_E \cdot \nabla u_{\mathcal{T}}) \right\|_E^2 + \sum_{E \in \mathcal{E}_{K,\Gamma_N}} h_E \left\| \bar{g}_E - \mathbf{n}_E \cdot \nabla u_{\mathcal{T}} \right\|_E^2 \right\}^{\frac{1}{2}} ,
$$

where $\overline{f}_K$ and $\overline{g}_E$ are the averages of f and g on K and E, respectively, defined in equation (1.18) (p. 13). There are two constants c^* and c_*, which only depend on the shape parameter of $\mathcal{T}$, such that the estimates

$$\left\| \nabla (u - u_{\mathcal{T}}) \right\| \le c^* \left\{ \sum_{K \in \mathcal{T}} \eta_{R,K}^2 + \sum_{K \in \mathcal{T}} h_K^2 \left\| f - \overline{f}_K \right\|_K^2 + \sum_{E \in \mathcal{E}_{\Gamma_N}} h_E \left\| g - \overline{g}_E \right\|_E^2 \right\}^{\frac{1}{2}}$$

and

$$\eta_{R,K} \le c_* \left\{ \left\| \nabla (u - u_{\mathcal{T}}) \right\|_{\omega_K}^2 + \sum_{K' \subset \omega_K} h_{K'}^2 \left\| f - \overline{f}_{K'} \right\|_{K'}^2 + \sum_{E \in \mathcal{E}_{K,\Gamma_N}} h_E \left\| g - \overline{g}_E \right\|_E^2 \right\}^{\frac{1}{2}}$$

hold for all elements $K \in \mathcal{T}$.

Remark 1.6 The factor $\frac{1}{2}$ multiplying the second term in $\eta_{R,K}$ takes into account that each interior edge is counted twice when adding all $\eta_{R,K}^2$. Note that $\Delta u_{\mathcal{T}} = 0$ on all triangles.

Remark 1.7 The first term in $\eta_{R,K}$ is related to the residual of $u_{\mathcal{T}}$ with respect to the strong form of the differential equation. The second and third term in $\eta_{R,K}$ are related to the boundary operator which is canonically associated with the strong and weak form of the differential equation. These boundary terms are crucial when considering low order finite element discretisations as done here. To see this, consider problem (1.1) (p. 4) in the unit square $(0, 1)^2$ with Dirichlet boundary conditions on the left and bottom part and exact solution $u(x) = x_1 x_2$. When using a triangulation consisting of right-angled isosceles triangles and evaluating the line integrals by the trapezoidal rule, the solution of problem (1.3) (p. 5) satisfies $u_{\mathcal{T}}(x) = u(x)$ for all $x \in \mathcal{N}$ but $u_{\mathcal{T}} \neq u$. The second and third terms in $\eta_{R,K}$ reflect the fact that $u_{\mathcal{T}} \notin H^2(\Omega)$ and that $u_{\mathcal{T}}$ does not exactly satisfy the Neumann boundary condition.

Remark 1.8 The terms $h_K \left\| f - \overline{f}_K \right\|_K$ and $h_E^{\frac{1}{2}} \left\| g - \overline{g}_E \right\|_E$ in the estimates of Theorem 1.5 are often referred to as *data oscillation*. When defining the error indicator with the original data f and g, they appear in the lower bound for the error only. We have opted to base the error indicator on the discrete data $\overline{f}_K$ and $\overline{g}_E$, since this facilitates its practical computation and its comparison with the error indicators that will be analysed in subsequent sections. The data oscillation terms are in general higher order perturbations of the other terms. In special situations, however, they can be dominant. To see this, assume that $\mathcal{T}$ contains at least one triangle, choose a triangle $K_0 \in \mathcal{T}$, and a non-zero C^∞-function ϱ_0 with support in K_0, and consider problem (1.1) (p. 4) with $f = -\Delta\varrho_0$ and $\Gamma_D = \Gamma$. Since $\int_{K_0} f = -\int_{K_0} \Delta\varrho_0 = 0$ and $f = 0$ outside K_0, we have $\overline{f}_K = 0$ for all $K \in \mathcal{T}$. Since $\int_\Omega f v_{\mathcal{T}} = -\int_{K_0} \Delta\varrho_0 v_{\mathcal{T}} = -\int_{K_0} \varrho_0 \Delta v_{\mathcal{T}} = 0$ for all $v_{\mathcal{T}} \in S_D^{1,0}(\mathcal{T})$, the exact solution of problem (1.3) (p. 5) is $u_{\mathcal{T}} = 0$. Hence, we have $\eta_{R,K} = 0$ for all $K \in \mathcal{T}$, but $\left\| \nabla (u - u_{\mathcal{T}}) \right\| \neq 0$. This effect is not restricted to the particular approximation of f considered here. Since ϱ_0 is completely arbitrary, we will always encounter similar difficulties as long as we do not evaluate $\left\| f \right\|_K$ exactly, which in general is impossible. Obviously, this problem is cured when further refining the mesh.

Remark 1.9 The error indicator $\eta_{R,K}$ was first proposed and analysed for problem (1.1) (p. 4) in one dimension in [**31, 32**]. The idea of using suitable local test functions for establishing local lower bounds was introduced in [**251**]. A different analysis is given in [**12, 13, 30**].

1.5 A Vertex-Oriented Residual Error Indicator

Our starting point is the equivalence of error and residual expressed by equation (1.10) (p. 10). Inspired by this equation we fix an arbitrary function $w \in H^1_D(\Omega)$ with $\|\nabla w\| = 1$ and try to bound $\langle R, w\rangle$. Contrary to the previous section we will not resort to a quasi-interpolation operator for estimating $\langle R, w\rangle$. Instead, our arguments will be based on:

- the observation that the nodal shape functions $\lambda_z, z \in \mathcal{N}$, form a partition of unity,
- the Galerkin orthogonality (1.12) (p. 11),
- the L^2-representation (1.15) (p. 11) of R,
- weighted Poincaré-type inequalities with weight function λ_z.

1.5.1 *Localisation*

As in Lemma 1.2 (p. 7) one easily checks that the nodal shape functions λ_z, $z \in \mathcal{N}$, form a partition of unity, i.e. $\sum_{z\in\mathcal{N}} \lambda_z = 1$. This immediately implies

$$\langle R, w\rangle = \sum_{z\in\mathcal{N}} \langle R, \lambda_z w\rangle \tag{1.26}$$

and

$$\|\nabla w\|^2 = \sum_{z\in\mathcal{N}} \left\| \lambda_z^{\frac{1}{2}} \nabla w \right\|^2_{\omega_z}. \tag{1.27}$$

These two identities suggest bounding $\langle R, \lambda_z w\rangle$ in terms of the corresponding norm $\left\| \lambda_z^{\frac{1}{2}} \nabla w \right\|_{\omega_z}$. To this end we fix a vertex $z \in \mathcal{N}$ and associate with it a real number w_z which is arbitrary subject to the condition that $z \in \mathcal{N}_{\Gamma_D}$ implies $w_z = 0$. We then have $w_z\lambda_z \in S_D^{1,0}(\mathcal{T})$. The Galerkin orthogonality (1.12) (p. 11) therefore implies

$$\langle R, \lambda_z w\rangle = \big\langle R, \lambda_z(w - w_z)\big\rangle. \tag{1.28}$$

The L^2-representation (1.15) (p. 11) of R on the other hand yields

$$\big\langle R, \lambda_z(w - w_z)\big\rangle = \int_{\omega_z} r\lambda_z(w - w_z) + \int_{\sigma_z} j\lambda_z(w - w_z). \tag{1.29}$$

Thus we are left with bounding the right-hand side of this estimate appropriately.

1.5.2 *Weighted Poincaré-Type Inequalities*

The estimation of the right-hand side of equation (1.29) is based on the following auxiliary result.

Lemma 1.10 *For every vertex $z \in \mathcal{N}$ there are constants $c_2(\omega_z)$ and $c_2(\sigma_z)$ such that, for a suitable choice $w_z \in \mathbb{R}$ with $w_z\lambda_z \in S_D^{1,0}(\mathcal{T})$, there hold the following* Poincaré-type inequalities

$$\left\| \lambda_z^{\frac{1}{2}}(w - w_z) \right\|_{\omega_z} \le c_2(\omega_z) h_z \left\| \lambda_z^{\frac{1}{2}} \nabla w \right\|_{\omega_z},$$

$$\left\{ \sum_{E \subset \sigma_z} h_E^{\perp} \left\| \lambda_z^{\frac{1}{2}}(w - w_z) \right\|_E^2 \right\}^{\frac{1}{2}} \le c_2(\sigma_z) h_z \left\| \lambda_z^{\frac{1}{2}} \nabla w \right\|_{\omega_z},$$

where h_z denotes the diameter of ω_z and where $h_E^{\perp} = \int_{\omega_E} \lambda_z / \int_E \lambda_z$.

We will be prove Lemma 1.10 for general L^p-spaces in Section 3.4.8 (p. 107), cf. Lemma 3.29 (p. 107). The proof is based on weighted trace inequalities, cf. Section 3.3 (p. 87), weighted Poincaré inequalities, cf. Section 3.4.5 (p. 99), and weighted Friedrichs' inequalities, cf. Section 3.4.7 (p. 105), with explicit constants. The results of Section 3.4.8 in particular yield $c_2(\omega_z) \le \frac{1}{\pi}$ and $c_2(\sigma_z) \le \frac{2}{\sqrt{\pi}}$ for a convex domain ω_z.

1.5.3 *Upper Bounds*

The Cauchy–Schwarz inequality for integrals and Lemma 1.10 yield for the first term on the right-hand side of equation (1.29)

$$\int_{\omega_z} r\lambda_z(w - w_z) \le \left\| \lambda_z^{\frac{1}{2}} r \right\|_{\omega_z} c_2(\omega_z) h_z \left\| \lambda_z^{\frac{1}{2}} \nabla w \right\|_{\omega_z}.$$

The Cauchy–Schwarz inequality for integrals and sums and Lemma 1.10 imply for the second term

$$\begin{aligned} \int_{\sigma_z} j\lambda_z(w - w_z) &\le \sum_{E \subset \sigma_z} \left\| \lambda_z^{\frac{1}{2}} j \right\|_E \left\| \lambda_z^{\frac{1}{2}}(w - w_z) \right\|_E \\ &\le \left\{ \sum_{E \subset \sigma_z} (h_E^{\perp})^{-1} \left\| \lambda_z^{\frac{1}{2}} j \right\|_E^2 \right\}^{\frac{1}{2}} \left\{ \sum_{E \subset \sigma_z} h_E^{\perp} \left\| \lambda_z^{\frac{1}{2}}(w - w_z) \right\|_E^2 \right\}^{\frac{1}{2}} \\ &\le \left\{ \sum_{E \subset \sigma_z} (h_E^{\perp})^{-1} \left\| \lambda_z^{\frac{1}{2}} j \right\|_E^2 \right\}^{\frac{1}{2}} c_2(\sigma_z) h_z \left\| \lambda_z^{\frac{1}{2}} \nabla w \right\|_{\omega_z}. \end{aligned}$$

These two estimates and equations (1.28) and (1.29) prove

$$\langle R, \lambda_z w \rangle \le \left\{ h_z \left[c_2(\omega_z) \left\| \lambda_z^{\frac{1}{2}} r \right\|_{\omega_z} + c_2(\sigma_z) \left\{ \sum_{E \subset \sigma_z} (h_E^{\perp})^{-1} \left\| \lambda_z^{\frac{1}{2}} j \right\|_E^2 \right\}^{\frac{1}{2}} \right] \right\} \left\| \lambda_z^{\frac{1}{2}} \nabla w \right\|_{\omega_z}.$$

In a last step we insert this estimate in equation (1.26), use the Cauchy–Schwarz inequality for sums, and take into account equation (1.27) and $\|\nabla w\| = 1$. We thus obtain

$$\langle R, w \rangle \le \left\{ \sum_{z \in \mathcal{N}} \eta_{R,z}^2 \right\}^{\frac{1}{2}} \tag{1.30}$$

with

$$\eta_{R,z} = h_z \left[c_2(\omega_z) \left\| \lambda_z^{\frac{1}{2}} r \right\|_{\omega_z} + c_2(\sigma_z) \left\{ \sum_{E \subset \sigma_z} (h_E^{\perp})^{-1} \left\| \lambda_z^{\frac{1}{2}} j \right\|_E^2 \right\}^{\frac{1}{2}} \right]. \tag{1.31}$$

Since $w \in H_D^1(\Omega)$ was arbitrary subject to the condition $\|\nabla w\| = 1$, the equivalence of error and residual (1.10) (p. 10) implies that the right-hand side of inequality (1.30) is an upper bound for the error $\|\nabla(u - u_{\mathcal{T}})\|$ and that $\eta_{R,z}$ is a reliable error indicator.

1.5.4 Lower Bounds

In order to prove the efficiency of $\eta_{R,z}$, we compare it with the residual error indicator $\eta_{R,K}$ of the previous section. To this end we first observe that the shape regularity of $\mathcal{T}$ implies for all vertices $z \in \mathcal{N}$, all elements $K \subset \omega_z$, and all edges $E \subset \sigma_z$

$$h_z \leq c_1 h_K \quad \text{and} \quad h_z^2 (h_E^{\perp})^{-1} \leq c_2 h_E$$

with constants c_1 and c_2 which only depend on the shape parameter of $\mathcal{T}$. Next, the property $0 \leq \lambda_z \leq 1$ of Lemma 1.2 (p. 7) and the triangle inequality yield

$$\left\| \lambda_z^{\frac{1}{2}} r \right\|_{\omega_z} \leq \left\{ 2 \sum_{K \subset \omega_z} \left[\left\| \bar{f}_K + \Delta u_{\mathcal{T}} \right\|_K^2 + \left\| f - \bar{f}_K \right\|_K^2 \right] \right\}^{\frac{1}{2}}.$$

Similarly, we obtain for every edge $E \subset \sigma_z \setminus \Gamma_N$

$$\left\| \lambda_z^{\frac{1}{2}} j \right\|_E \leq \left\| \mathbb{J}_E(\mathbf{n}_E \cdot \nabla u_{\mathcal{T}}) \right\|_E$$

and for every edge $E \subset \sigma_z \cap \Gamma_N$

$$\left\| \lambda_z^{\frac{1}{2}} j \right\|_E \leq \left\| \bar{g}_E - \mathbf{n}_E \cdot \nabla u_{\mathcal{T}} \right\|_E + \left\| g - \bar{g}_E \right\|_E.$$

These estimates show that $\eta_{R,z}$ can be bounded by the indicators $\eta_{R,K}$ and the data errors $h_K \left\| f - \bar{f}_K \right\|_K$ and $h_E^{\frac{1}{2}} \left\| g - \bar{g}_E \right\|_E$ corresponding to the elements and edges contained in ω_z and $\sigma_z \cap \Gamma_N$. Together with the estimates of Section 1.4.5 (p. 12) this proves the efficiency of the indicators $\eta_{R,z}$.

1.5.5 A Posteriori Error Estimates

The results of this section can be summarised as follows.

Theorem 1.11 *Denote by u and $u_{\mathcal{T}}$ the unique solutions of problems (1.2) (p. 4) and (1.3) (p. 5), respectively. For every vertex $z \in \mathcal{N}$ define the* residual a posteriori error indicator $\eta_{R,z}$ *by equation (1.31), where the residuals r and j are given by equations (1.13) (p. 11) and (1.14) (p. 11),* *respectively. There is a constant $\widehat{c}_*$, which only depends on the shape parameter of $\mathcal{T}$, such that the estimates*

$$\|\nabla(u - u_{\mathcal{T}})\| \leq \left\{ \sum_{z \in \mathcal{N}} \eta_{R,z}^2 \right\}^{\frac{1}{2}}$$

and

$$\eta_{R,z} \leq \widehat{c}_* \left\{ \sum_{\substack{K \in \mathcal{T} \\ \omega_z \cap \omega_K \neq \emptyset}} \|\nabla(u - u_{\mathcal{T}})\|_K^2 + \sum_{\substack{K \in \mathcal{T} \\ \omega_z \cap \omega_K \neq \emptyset}} h_K^2 \left\| f - \bar{f}_K \right\|_K^2 + \sum_{E \subset \sigma_z \cap \Gamma_N} h_E \|g - \bar{g}_E\|_E^2 \right\}^{\frac{1}{2}}$$

hold for all vertices $z \in \mathcal{N}$.

Remark 1.12 The results of this section are based on [**247**]. The proof of the lower bounds can be considerably improved by taking into account that the nodal shape functions λ_z are a partition of unity [**247**, Corollary 3.3]. A similar approach based on partitions of unity and weighted trace and Poincaré inequalities is presented in [**95**]. There, contrary to Lemma 1.10 and Theorem 1.11, the weight functions are present only on the left-hand sides of the trace and Poincaré inequalities and are missing in the error indicators.

1.6 Edge Residuals

Theorem 1.5 (p. 15) shows that the edge residuals in $\eta_{R,K}$ yield a lower bound for the error. We want to prove that they also yield an upper bound provided the partition $\mathcal{T}$ exclusively consists of triangles and satisfies an extra condition which is slightly more restrictive than shape regularity but allows local refinement.

1.6.1 Galerkin Orthogonality and the L^2-Projection

We denote by $P_{\mathcal{T}} : H_D^1(\Omega) \to S_D^{1,0}(\mathcal{T})$ the L^2-projection which, for all $w \in H_D^1(\Omega)$ and all $v_{\mathcal{T}} \in S_D^{1,0}(\mathcal{T})$, is defined by

$$\int_\Omega P_{\mathcal{T}} w v_{\mathcal{T}} = \int_\Omega w v_{\mathcal{T}}.$$

We insert $w_{\mathcal{T}} = P_{\mathcal{T}} w$ in equation (1.16) (p. 11). Observing that $\Delta u_{\mathcal{T}} = 0$ on all triangles and taking into account the L^2-orthogonality of $P_{\mathcal{T}} w$, we then obtain for all $w \in H_D^1(\Omega)$

$$\begin{aligned}
\langle R, w \rangle &= \int_\Omega r(w - P_{\mathcal{T}} w) + \int_\Sigma j(w - P_{\mathcal{T}} w) \\
&= \int_\Omega f(w - P_{\mathcal{T}} w) + \int_\Sigma j(w - P_{\mathcal{T}} w) \\
&= \int_\Omega (f - I_{\mathcal{T}} f)(w - P_{\mathcal{T}} w) + \int_\Sigma j(w - P_{\mathcal{T}} w).
\end{aligned}$$

The Cauchy–Schwarz inequality for integrals and a weighted Cauchy–Schwarz inequality for sums then yield for all $w \in H_D^1(\Omega)$

$$\begin{aligned}\langle R, w\rangle &\leq \sum_{K\in\mathcal{T}} \left\|f - I_{\mathcal{T}}f\right\|_K \|w - P_{\mathcal{T}}w\|_K + \sum_{E\in\mathcal{E}} \left\|j\right\|_E \|w - P_{\mathcal{T}}w\|_E \\ &\leq \left\{\sum_{K\in\mathcal{T}} h_K^2 \left\|f - I_{\mathcal{T}}f\right\|_K^2 + \sum_{E\in\mathcal{E}} h_E \left\|j\right\|_E^2\right\}^{\frac{1}{2}} \left\{\sum_{K\in\mathcal{T}} h_K^{-2} \|w - P_{\mathcal{T}}w\|_K^2 + \sum_{E\in\mathcal{E}} h_E^{-1} \|w - P_{\mathcal{T}}w\|_E^2\right\}^{\frac{1}{2}}.\end{aligned}$$

This estimate and inequality (1.10) (p. 10) show that we have achieved our goal if we can prove an estimate of the form

$$\left\{\sum_{K\in\mathcal{T}} h_K^{-2} \|w - P_{\mathcal{T}}w\|_K^2 + \sum_{E\in\mathcal{E}} h_E^{-1} \|w - P_{\mathcal{T}}w\|_E^2\right\}^{\frac{1}{2}} \leq c\,\|\nabla w\| \tag{1.32}$$

for all $w \in H_D^1(\Omega)$.

1.6.2 Some Stability Results for the L^2-Projection

For the proof of inequality (1.32), we need some additional notation and an extra condition on $\mathcal{T}$.

Given an arbitrary element $K_0 \in \mathcal{T}$, define $\mathcal{U}_0(K_0) = K_0$ and then, recursively, for $j \geq 1$, $\mathcal{U}_j(K_0)$ as the union of all $K \in \mathcal{T}$ which are not in $\bigcup_{i<j}\mathcal{U}_i(K_0)$ but which have at least one vertex in $\mathcal{U}_{j-1}(K_0)$. Thus $\mathcal{U}_j(K_0)$ is the union of all triangles K which may be reached by a connected path $P_1, \ldots, P_j$ with P_1 a vertex of K_0, P_j a vertex of K, and P_iP_{i+1} an edge in $\mathcal{E}$ for $1 \leq i < j$, and not by any shorter such path. Note that $\mathcal{U}_1(K_0) \cup K_0 = \widetilde{\omega}_{K_0}$.

For $K \in \mathcal{U}_j(K_0)$ we set $\ell(K_0, K) = j$. It follows, in particular, that $\ell(K_0, K)$ is symmetric in K_0 and K. Denote by $n_j(K_0)$ the number of triangles in $\mathcal{U}_j(K_0)$.

We now assume that $\mathcal{T}$ also satisfies the following

growth condition: there are constants $c_1, c_2, \alpha, \beta, r$ with $\alpha \geq 1, \beta \geq 1$ and

$$\alpha^{\frac{1}{2}}\beta < \sqrt{3} + \sqrt{2} \approx 3.146$$

such that

$$\begin{aligned}\frac{\mu_2(K)}{\mu_2(K_0)} &\leq c_1\alpha^{\ell(K_0,K)} \quad \text{for all } K_0, K \in \mathcal{T}, \\ n_j(K) &\leq c_2 j^r \beta^j \quad \text{for all } K \in \mathcal{T}.\end{aligned}$$

Note that, similar to the shape regularity, the growth condition is relevant for families of partitions which are obtained by some refinement process. The growth condition was introduced by M. Crouzeix and V. Thomée in [115] in order to prove the stability in L^p and $W^{1,p}$ of L^2-projections onto finite element spaces. It was pointed out in [115] that the growth condition is much weaker than quasi-uniformity. The following example, however, shows that it is stronger than shape regularity.

Example 1.13 Consider the unit square $\Omega = (0, 1)^2$. The triangulation $\mathcal{T}_1$ is obtained by dividing Ω into 16 equal squares and by connecting the upper left and lower right corner of each of these

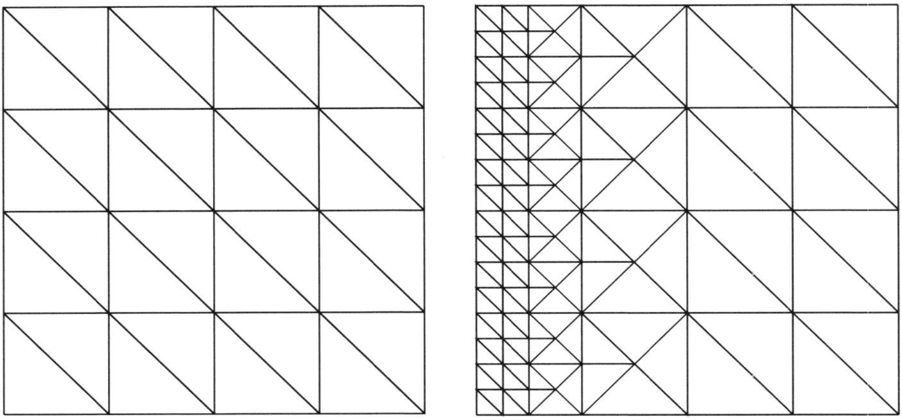

Figure 1.3 Triangulations $\mathcal{T}_1$ (left) and $\mathcal{T}_3$ (right) of Example 1.13

squares. The triangulation $\mathcal{T}_k$, $k \geq 2$, is recursively obtained from $\mathcal{T}_{k-1}$ according to the following rules (cf. Figure 1.3).

- Each $K \in \mathcal{T}_{k-1}$ having a vertex on the left edge of Ω is cut into four new triangles by connecting the midpoints of its edges.
- Each $K \in \mathcal{T}_{k-1}$ having exactly one hanging node on an edge which is not its longest one is cut into three new triangles by connecting the midpoint of its longest edge to the vertex opposite to it and to the hanging node.
- Each $K \in \mathcal{T}_{k-1}$ having exactly one hanging node on its longest edge is bisected by connecting the midpoint of its longest edge to the vertex opposite to it.

One easily checks that all triangulations $\mathcal{T}_k$ consist of right-angled isosceles triangles. Hence, they are all shape regular. In order to verify that the family $\mathcal{T}_k$ does not satisfy the growth condition, consider the triangle $K_k \in \mathcal{T}_k$, which has an edge on the left edge of Ω and the lower left corner of Ω as a vertex, and the triangle $K'_k \in \mathcal{T}_k$, which has an edge on the right edge of Ω and the lower right corner of Ω as a vertex. From the construction of $\mathcal{T}_k$ one concludes that

$$\frac{\mu_2(K'_k)}{\mu_2(K_k)} = 4^k, \quad \ell(K_k, K'_k) = 3 + k, \quad n_{4+k}(K'_{k+1}) = 2n_{3+k}(K'_k).$$

Hence, we have at least $\alpha \geq 4$ and $\beta \geq 2$. Thus the growth condition is violated. Note that things change if there is a $k_0 \geq 2$ such that $\mathcal{T}_{k+1}$ is obtained from $\mathcal{T}_k$ by uniform refinement whenever k is a multiple of k_0. Then, depending on k_0, the constants c_1 and c_2 can be chosen such that the growth condition is satisfied.

To prove estimate (1.32) we define a mesh-dependent norm on $H^1(\Omega)$ by

$$|||w|||_{\mathcal{T}} = \left\{ \sum_{K\in\mathcal{T}} \left[\mu_2(K)^{-1} \|w\|_K^2 + \|\nabla w\|_K^2 \right] \right\}^{\frac{1}{2}}.$$

The definition of h_K implies $\mu_2(K) \le \frac{\pi}{4} h_K^2$ for all triangles K. Since for every triangle K and every edge E of K, the quotient $\frac{h_K}{h_E}$ only depends on the smallest angle of K, this inequality and the trace inequality (1.5) (p. 8) yield for all $w \in H_D^1(\Omega)$

$$\left\{ \sum_{K\in\mathcal{T}} h_K^{-2} \|w - P_\mathcal{T} w\|_K^2 + \sum_{E\in\mathcal{E}} h_E^{-1} \|w - P_\mathcal{T} w\|_E^2 \right\}^{\frac{1}{2}} \le c_3 \, |||w - P_\mathcal{T} w|||_\mathcal{T} \tag{1.33}$$

with a constant c_3 which only depends on the shape parameter of $\mathcal{T}$.

Next we prove the stability of $P_\mathcal{T}$ with respect to $|||\cdot|||_\mathcal{T}$.

Lemma 1.14 *There is a constant c_4, which only depends on the constants in the growth condition, such that*

$$|||P_\mathcal{T} w|||_\mathcal{T} \le c_4 \, |||w|||_\mathcal{T}$$

holds for all $w \in H_D^1(\Omega)$.

Proof Fix a function $w \in H_D^1(\Omega)$. Theorem 4 in [**115**] implies that

$$\left\{ \sum_{K\in\mathcal{T}} \left\| \nabla(P_\mathcal{T} w) \right\|_K^2 \right\}^{\frac{1}{2}} = \left\| \nabla(P_\mathcal{T} w) \right\| \le c_4' \, \|\nabla w\| = c_4' \left\{ \sum_{K\in\mathcal{T}} \|\nabla w\|_K^2 \right\}^{\frac{1}{2}}$$

with a constant c_4' which only depends on the constants in the growth condition. On the other hand, we obviously have $w = \sum_{K\in\mathcal{T}} w|_K \chi_K$, where χ_K denotes the characteristic function of K. From [**115**, Lemma 3] we know that

$$\left\| P_\mathcal{T}(w|_{K'} \chi_{K'}) \right\|_K \le c_4'' \gamma^{\ell(K,K')} \, \|w|_{K'} \chi_{K'}\|_{K'}$$

holds for all $K, K' \in \mathcal{T}$ with $\gamma = \frac{1}{\sqrt{3}+\sqrt{2}}$ and a universal constant c_4''. Hence, we obtain for all $K \in \mathcal{T}$

$$\begin{aligned}
\mu_2(K)^{-\frac{1}{2}} \, \|P_\mathcal{T} w\|_K &\le \mu_2(K)^{-\frac{1}{2}} \sum_{K'\in\mathcal{T}} \left\| P_\mathcal{T}(w|_{K'} \chi_{K'}) \right\|_K \\
&\le c_4'' \mu_2(K)^{-\frac{1}{2}} \sum_{K'\in\mathcal{T}} \gamma^{\ell(K,K')} \, \|w|_{K'} \chi_{K'}\|_{K'} \\
&= c_4'' \sum_{K'\in\mathcal{T}} \gamma^{\ell(K,K')} \mu_2(K)^{-\frac{1}{2}} \mu_2(K')^{\frac{1}{2}} \mu_2(K')^{-\frac{1}{2}} \, \|w\|_{K'} \\
&\le c_4'' \sum_{K'\in\mathcal{T}} [\alpha^{\frac{1}{2}} \gamma]^{\ell(K,K')} \mu_2(K')^{-\frac{1}{2}} \, \|w\|_{K'} .
\end{aligned}$$

This estimate and the arguments used in the proof of Theorem 3 in [**115**] imply that

$$\left\{ \sum_{K\in\mathcal{T}} \mu_2(K)^{-1} \, \|P_\mathcal{T} w\|_K^2 \right\}^{\frac{1}{2}} \le c_4'' \left\{ \sum_{j=0}^{\infty} j^r (\beta \alpha^{\frac{1}{2}} \gamma)^j \right\} \left\{ \sum_{K\in\mathcal{T}} \mu_2(K)^{-1} \, \|w\|_K^2 \right\}^{\frac{1}{2}} .$$

Note that the series on the right-hand side of the last inequality is finite due to the growth condition and the definition of γ. The last estimate and the bound for $\|\nabla(P_{\mathcal{T}}w)\|$ obviously establish the assertion of the proposition. □

In a last step, we combine the L^2-projection $P_{\mathcal{T}}$ and the quasi-interpolation operator $I_{\mathcal{T}}$ of equation (1.4) (p. 7). Since $P_{\mathcal{T}}$ is the identity on $S_D^{1,0}(\mathcal{T})$, we have for all $w \in H_D^1(\Omega)$

$$w - P_{\mathcal{T}}w = w - I_{\mathcal{T}}w + I_{\mathcal{T}}w - P_{\mathcal{T}}w = w - I_{\mathcal{T}}w - P_{\mathcal{T}}(w - I_{\mathcal{T}}w).$$

Lemma 1.14 therefore implies

$$|||w - P_{\mathcal{T}}w|||_{\mathcal{T}} \leq (1 + c_4)\, |||w - I_{\mathcal{T}}w|||_{\mathcal{T}}.$$

Proposition 1.3 (p. 7) on the other hand yields

$$|||w - I_{\mathcal{T}}w|||_{\mathcal{T}} \leq c_5\, \|\nabla w\| \tag{1.34}$$

with a constant c_5 which only depends on the constants $c_{A,1}$ and $c_{A,2}$ and the shape parameter of $\mathcal{T}$. Inequalities (1.33)–(1.34) obviously prove the desired estimate (1.32).

1.6.3 A Posteriori Error Estimates

We summarise the results of this section.

Theorem 1.15 *Assume that the partition $\mathcal{T}$ exclusively consists of triangles and that it satisfies the growth condition in addition to the conditions of Section 1.2 (p. 4). Denote by u and $u_{\mathcal{T}}$ the unique solutions of problems (1.2) (p. 4) and (1.3) (p. 5), respectively. For every edge $E \in \mathcal{E}$ define the* edge residual a posteriori error indicator *$\eta_{R,E}$ by*

$$\eta_{R,E} = \begin{cases} h_E^{\frac{1}{2}} \left\| \mathbb{J}_E(\mathbf{n}_E \cdot \nabla u_{\mathcal{T}}) \right\|_E & \text{if } E \in \mathcal{E}_{\Omega}, \\ h_E^{\frac{1}{2}} \left\| \overline{g}_E - \mathbf{n}_E \cdot \nabla u_{\mathcal{T}} \right\|_E & \text{if } E \in \mathcal{E}_{\Gamma_N}, \\ 0 & \text{if } E \in \mathcal{E}_{\Gamma_D}, \end{cases}$$

where $\overline{g}_E$ denotes the average of g on E defined in equation (1.18) (p. 13). There are two constants $\widetilde{c}^*$ and $\widetilde{c}_*$, which only depend on the shape parameter of $\mathcal{T}$ and the constants in the growth condition, such that the estimates

$$\|\nabla(u - u_{\mathcal{T}})\| \leq \widetilde{c}^* \left\{ \sum_{E \in \mathcal{E}} \eta_{R,E}^2 + \sum_{K \in \mathcal{T}} h_K^2 \left\| f - I_{\mathcal{T}} f \right\|_K^2 + \sum_{E \in \mathcal{E}_{\Gamma_N}} h_E \left\| g - \overline{g}_E \right\|_E^2 \right\}^{\frac{1}{2}}$$

and

$$\eta_{R,E} \leq \widetilde{c}_* \left\{ \left\| \nabla(u - u_{\mathcal{T}}) \right\|_{\omega_E}^2 + \sum_{K' \subset \omega_E} h_{K'}^2 \left\| f - I_{\mathcal{T}} f \right\|_{K'}^2 \right\}^{\frac{1}{2}}$$

and

$$\eta_{R,E} \le \widetilde{c}_* \left\{ \left\| \nabla (u - u_{\mathcal{T}}) \right\|_{\omega_E}^2 + \sum_{K' \subset \omega_E} h_{K'}^2 \left\| f - I_{\mathcal{T}} f \right\|_{K'}^2 + h_E \left\| g - \overline{g}_E \right\|_E^2 \right\}^{\frac{1}{2}}$$

hold for all $E \in \mathcal{E}_\Omega$ *and all* $E \in \mathcal{E}_{\Gamma_N}$*, respectively, where* $I_{\mathcal{T}}$ *is the quasi-interpolation operator of equation* (1.4) *(p. 7).*

Remark 1.16 The $I_{\mathcal{T}} f$-terms in Theorem 1.15 can be replaced by *any* approximation of f in $S_D^{1,0}(\mathcal{T})$.

Remark 1.17 Theorem 1.15 was first established in [**101**] together with a similar result for the L^2-norm of the error. There, it was also shown that one can dispense with the growth condition when replacing the L^2-projection $P_{\mathcal{T}}$ by a weighted L^2-projection involving suitable local cut-off functions as weights. This, however, has to be paid for by an additional correction term $\left\{ \sum_{K \cap \Gamma_D \neq \emptyset} h_K^2 \left\| f \right\|_K^2 \right\}^{\frac{1}{2}}$ which is larger than the corresponding correction terms in Theorem 1.15. Theorem 1.15 was extended to anisotropic triangular and tetrahedral meshes in [**170**], cf. Section 4.5.5 (p. 185). In [**283, 284**] D. Yu proved that on a uniform square grid the error of the finite element method is controlled either by the element residuals or the edge residuals depending on whether the polynomial degree of the finite element functions is even or odd.

1.7 Auxiliary Local Problems

The results of Sections 1.4, 1.5, and 1.6 show that we must reliably estimate the norm of the residual as an element of the dual space of $H_D^1(\Omega)$. This could be achieved by lifting the residual to a suitable subspace of $H_D^1(\Omega)$ by solving auxiliary problems similar to, but simpler than the original discrete problem (1.3) (p. 5). Practical considerations and the results of Sections 1.4, 1.5, and 1.6 suggest that the auxiliary problems should satisfy the following conditions.

- In order to get information on the local behaviour of the error, they should involve only small subdomains of Ω.
- In order to yield accurate information on the error, they should be based on finite element spaces which are more accurate than the original one.
- In order to keep the computational work to a minimum, they should involve as few degrees of freedom as possible.
- To each edge and, if need be, to each element there should correspond at least one degree of freedom in at least one of the auxiliary problems.
- The solution of all auxiliary problems should not cost more than the assembly of the stiffness matrix of problem (1.3) (p. 5).

There are many possible ways to satisfy these conditions. Here, we present three of them. To this end we denote by $\mathbb{P}_1 = \text{span}\{1, x_1, x_2\}$ the space of linear polynomials in two variables.

1.7.1 Dirichlet Problems Associated with Vertices

First, we decide to impose Dirichlet boundary conditions on the auxiliary problems. The fourth condition then implies that the corresponding subdomains must consist of more than one element. A reasonable choice is to consider all vertices $z \in \mathcal{N}_\Omega \cup \mathcal{N}_{\Gamma_N}$ and the corresponding domains ω_z, cf. Figure 1.1 (p. 6). The above conditions then lead to the following definition.

Set for all $z \in \mathcal{N}_\Omega \cup \mathcal{N}_{\Gamma_N}$

$$V_z = \operatorname{span}\{\varphi\psi_K, \rho\psi_E, \sigma\psi_{E'} : K \in \mathcal{T}, z \in \mathcal{N}_K, E \in \mathcal{E}_\Omega, z \in \mathcal{N}_E, \\ E' \in \mathcal{E}_{\Gamma_N}, E' \subset \partial\omega_z, \varphi, \rho, \sigma \in \mathbb{P}_1\} \tag{1.35}$$

and

$$\eta_{D,z} = \|\nabla v_z\|_{\omega_z} \tag{1.36}$$

where $v_z \in V_z$ is the unique solution of

$$\int_{\omega_z} \nabla v_z \cdot \nabla w = \int_{\omega_z} f_\mathcal{T} w + \int_{\partial\omega_z \cap \Gamma_N} g_\mathcal{E} w - \int_{\omega_z} \nabla u_\mathcal{T} \cdot \nabla w \tag{1.37}$$

for all $w \in V_z$. Here, the functions $f_\mathcal{T}$ and $g_\mathcal{E}$ are defined in equation (1.19) (p. 13).

In order to get a different interpretation of problem (1.37), set $u_z = u_\mathcal{T} + v_z$. Then $\eta_{D,z} = \|\nabla(u_z - u_\mathcal{T})\|_{\omega_z}$ and $u_z \in u_\mathcal{T} + V_z$ is the unique solution of

$$\int_{\omega_z} \nabla u_z \cdot \nabla w = \int_{\omega_z} f_\mathcal{T} w + \int_{\partial\omega_z \cap \Gamma_N} g_\mathcal{E} w$$

for all $w \in V_z$. This is a discrete analogue of the Dirichlet problem

$$\begin{aligned} -\Delta\varphi &= f && \text{in } \omega_z \\ \varphi &= u_\mathcal{T} && \text{on } \partial\omega_z \backslash \Gamma_N \\ \frac{\partial\varphi}{\partial n} &= g && \text{on } \partial\omega_z \cap \Gamma_N. \end{aligned}$$

Hence, we can interpret the error indicator $\eta_{D,z}$ in two ways:

- we solve a local analogue of the residual equation (1.7) (p. 10) using a higher order finite element approximation and use a suitable norm of the solution as error indicator;
- we solve a local discrete analogue of the original problem (1.1) (p. 4) using a higher order finite element space and compare the solution of this problem to the one of problem (1.3) (p. 5).

Thus, in a certain sense, $\eta_{D,z}$ is based on an extrapolation technique.

The following proposition shows that $\eta_{D,z}$ yields upper and lower bounds for the error $\|\nabla(u - u_\mathcal{T})\|$ and that it is comparable to the indicator $\eta_{R,K}$.

Theorem 1.18 *Denote by u and $u_{\mathcal{T}}$ the unique solutions of problems (1.2) (p. 4) and (1.3) (p. 5). There are constants $c_{\mathcal{N},1}, \ldots, c_{\mathcal{N},4}$, which only depend on the shape parameter of $\mathcal{T}$, such that the estimates*

$$\eta_{D,z} \leq c_{\mathcal{N},1} \left\{ \sum_{K \subset \omega_z} \eta_{R,K}^2 \right\}^{\frac{1}{2}},$$

$$\eta_{R,K} \leq c_{\mathcal{N},2} \left\{ \sum_{z \in \mathcal{N}_K \setminus \mathcal{N}_{\Gamma_D}} \eta_{D,z}^2 \right\}^{\frac{1}{2}},$$

$$\eta_{D,z} \leq c_{\mathcal{N},3} \left\{ \left\| \nabla (u - u_{\mathcal{T}}) \right\|_{\omega_z}^2 + \sum_{K \subset \omega_z} h_K^2 \left\| f - \bar{f}_K \right\|_K^2 + \sum_{E \subset \Gamma_N \cap \partial \omega_z} h_E \left\| g - \bar{g}_E \right\|_E^2 \right\}^{\frac{1}{2}},$$

and

$$\left\| \nabla (u - u_{\mathcal{T}}) \right\| \leq c_{\mathcal{N},4} \left\{ \sum_{z \in \mathcal{N}_\Omega \cup \mathcal{N}_{\Gamma_N}} \eta_{D,z}^2 + \sum_{K \in \mathcal{T}} h_K^2 \left\| f - \bar{f}_K \right\|_K^2 + \sum_{E \in \mathcal{E}_{\Gamma_N}} h_E \left\| g - \bar{g}_E \right\|_E^2 \right\}^{\frac{1}{2}}$$

hold for all $z \in \mathcal{N}_\Omega \cup \mathcal{N}_{\Gamma_N}$ and all $K \in \mathcal{T}$. Here, $\bar{f}_K$, $\bar{g}_E$, and $\eta_{R,K}$ are as in Theorem 1.5 (p. 15) and $\eta_{D,z}$ is given by equation (1.36).

Proof The third and fourth estimate of Theorem 1.18 follow from the first and second one and Theorem 1.5 (p. 15).

To prove the first estimate, we integrate the right-hand side of equation (1.37) by parts element-wise and obtain for all $w \in V_z$

$$\begin{aligned} \int_{\omega_z} \nabla v_z \cdot \nabla w = \sum_{K \subset \omega_z} \int_K (\bar{f}_K + \Delta u_{\mathcal{T}}) w - \sum_{E \subset \sigma_z \cap \Omega} \int_E \mathbb{J}_E(\mathbf{n}_E \cdot \nabla u_{\mathcal{T}}) w \\ + \sum_{E \subset \Gamma_N \cap \partial \omega_z} \int_E (\bar{g}_E - \mathbf{n}_E \cdot \nabla u_{\mathcal{T}}) w. \end{aligned} \tag{1.38}$$

Since the functions in V_z vanish at all vertices in ω_z, standard scaling arguments for finite elements, cf. Proposition 1.4 (p. 9) and Section 3.6 (p. 112), imply

$$\|w\|_K \leq \tilde{c}_{I1} h_K \|\nabla w\|_K, \quad \|w\|_E \leq \tilde{c}_{I2} h_E^{\frac{1}{2}} \|\nabla w\|_K \tag{1.39}$$

for all $w \in V_z$, all elements K in ω_z, and all edges $E \in \mathcal{E}_K \setminus \Gamma_D$. The constants $\tilde{c}_{I1}$ and $\tilde{c}_{I2}$ only depend on the smallest angle in $\mathcal{T}$.

Now, we insert the solution v_z of problem (1.37) in equation (1.38). The Cauchy–Schwarz inequalities for integrals and sums and the above estimates then imply

$$\begin{aligned}
\eta_{D,z}^2 &= \int_{\omega_z} \nabla v_z \cdot \nabla v_z \\
&\le \sum_{K \subset \omega_z} \left\| \bar{f}_K + \Delta u_{\mathcal{T}} \right\|_K \|v_z\|_K + \sum_{E \subset \sigma_z \cap \Omega} \left\| \mathbb{J}_E(\mathbf{n}_E \cdot \nabla u_{\mathcal{T}}) \right\|_E \|v_z\|_E \\
&\quad + \sum_{E \subset \Gamma_N \cap \partial\omega_z} \left\| \bar{g}_E - \mathbf{n}_E \cdot \nabla u_{\mathcal{T}} \right\|_E \|v_z\|_E \\
&\le c_{\mathcal{N},1} \eta_{D,z} \left\{ \sum_{K \subset \omega_z} \eta_{R,K}^2 \right\}^{\frac{1}{2}}.
\end{aligned}$$

This proves the first estimate of Theorem 1.18.

Next, we observe that the functions w_K and w_E defined in (1.20) (p. 13), (1.22) (p. 14), and (1.24) (p. 15) are all elements of V_z for all elements K and all edges $E \in \mathcal{E}_\Omega \cup \mathcal{E}_{\Gamma_N}$ which are contained in ω_z. Inserting these functions in equation (1.38) and repeating the proofs of estimates (1.21) (p. 13), (1.23) (p. 15), and (1.25) (p. 15) proves the second estimate of Theorem 1.18. □

1.7.2 *Dirichlet Problems Associated with Elements*

We now consider an indicator which is a slight variation of the preceding one. Instead of all $z \in \mathcal{N}_\Omega \cup \mathcal{N}_{\Gamma_N}$ and the corresponding domains ω_z, we consider all $K \in \mathcal{T}$ and the corresponding sets ω_K, cf. Figure 1.1 (p. 6). The considerations given at the beginning of this section then lead to the following definition.

Set for all $K \in \mathcal{T}$

$$\begin{aligned}
\widetilde{V}_K = \operatorname{span}\{\varphi\psi_{K'}, \rho\psi_E, \sigma\psi_{E'} :\ & K' \in \mathcal{T}, \mathcal{E}_{K'} \cap \mathcal{E}_K \neq \emptyset, \\
& E \in \mathcal{E}_K \cap \mathcal{E}_\Omega, \\
& E' \in \mathcal{E}_{\Gamma_N}, E' \subset \partial\omega_K, \\
& \varphi, \rho, \sigma \in \mathbb{P}_1\}
\end{aligned} \tag{1.40}$$

and

$$\eta_{D,K} = \|\nabla \widetilde{v}_K\|_{\omega_K} \tag{1.41}$$

where $\widetilde{v}_K \in \widetilde{V}_K$ is the unique solution of

$$\int_{\omega_K} \nabla \widetilde{v}_K \cdot \nabla w = \int_{\omega_K} f_{\mathcal{T}} w + \int_{\partial\omega_K \cap \Gamma_N} g_{\mathcal{E}} w - \int_{\omega_K} \nabla u_{\mathcal{T}} \cdot \nabla w \tag{1.42}$$

for all $w \in \widetilde{V}_K$.

As before we can interpret $u_{\mathcal{T}} + \widetilde{v}_K$ as an approximate solution of the Dirichlet problem

$$\begin{aligned}
-\Delta\varphi &= f && \text{in } \omega_K \\
\varphi &= u_{\mathcal{T}} && \text{on } \partial\omega_K \backslash \Gamma_N \\
\frac{\partial\varphi}{\partial n} &= g && \text{on } \partial\omega_K \cap \Gamma_N.
\end{aligned}$$

The following proposition shows that $\eta_{D,K}$ also yields upper and lower bounds on the error $\|\nabla(u-u_{\mathcal{T}})\|$ and that it is comparable to $\eta_{D,z}$ and $\eta_{R,K}$. Its proof is completely similar to that of Theorem 1.18 and is therefore left to the reader.

Theorem 1.19 *Denote by u and $u_{\mathcal{T}}$ the unique solutions of problem (1.2) (p. 4) and (1.3) (p. 5). There are constants $c_{\mathcal{T},1}, \ldots, c_{\mathcal{T},4}$, which only depend on the shape parameter of $\mathcal{T}$, such that the estimates*

$$\eta_{D,K} \le c_{\mathcal{T},1} \left\{ \sum_{K' \subset \omega_K} \eta_{R,K'}^2 \right\}^{\frac{1}{2}},$$

$$\eta_{R,K} \le c_{\mathcal{T},2} \left\{ \sum_{K' \subset \omega_K} \eta_{D,K'}^2 \right\}^{\frac{1}{2}},$$

$$\eta_{D,K} \le c_{\mathcal{T},3} \left\{ \left\|\nabla(u-u_{\mathcal{T}})\right\|_{\omega_K}^2 + \sum_{K' \subset \omega_K} h_{K'}^2 \left\|f - \bar{f}_{K'}\right\|_{K'}^2 + \sum_{E' \subset \Gamma_N \cap \partial\omega_K} h_{E'} \left\|g - \bar{g}_{E'}\right\|_{E'}^2 \right\}^{\frac{1}{2}},$$

$$\left\|\nabla(u-u_{\mathcal{T}})\right\| \le c_{\mathcal{T},4} \left\{ \sum_{K \in \mathcal{T}} \eta_{D,K}^2 + \sum_{K \in \mathcal{T}} h_K^2 \left\|f - \bar{f}_K\right\|_K^2 + \sum_{E \in \mathcal{E}_{\Gamma_N}} h_E \left\|g - \bar{g}_E\right\|_E^2 \right\}^{\frac{1}{2}}$$

hold for all $K \in \mathcal{T}$. Here, $\bar{f}_K$, $\bar{g}_E$, and $\eta_{R,K}$ are as in Theorem 1.5 (p. 15), and $\eta_{D,K}$ is defined by equation (1.41).

1.7.3 Neumann Problems

For the third indicator we decide to impose Neumann boundary conditions on the auxiliary problems. Now, it is possible to choose the elements in $\mathcal{T}$ as the corresponding subdomains. This leads to the following definition.

Set for all $K \in \mathcal{T}$

$$V_K = \operatorname{span}\left\{\varphi\psi_K, \rho\psi_E : E \in \mathcal{E}_K \backslash \mathcal{E}_{\Gamma_D}, \varphi, \rho \in \mathbb{P}_1\right\} \tag{1.43}$$

and

$$\eta_{N,K} = \|\nabla v_K\|_K \tag{1.44}$$

where v_K is the unique solution of

$$\int_K \nabla v_K \cdot \nabla w = \int_K (\bar{f}_K + \Delta u_{\mathcal{T}})w - \frac{1}{2}\sum_{E \in \mathcal{E}_{K,\Omega}} \int_E \mathbb{J}_E(\mathbf{n}_E \cdot \nabla u_{\mathcal{T}})w + \sum_{E \in \mathcal{E}_{K,\Gamma_N}} \int_E (\bar{g}_E - \mathbf{n}_E \cdot \nabla u_{\mathcal{T}})w \tag{1.45}$$

for all $w \in V_K$.

Note that the factor $\frac{1}{2}$ multiplying the residuals on interior edges takes into account that interior edges are counted twice when summing the contributions of all elements.

Problem (1.45) can be interpreted as a discrete analogue of the Neumann problem

$$\begin{aligned} -\Delta\varphi &= r && \text{in } K \\ \frac{\partial\varphi}{\partial n} &= \frac{1}{2}j && \text{on } \partial K \cap \Omega \\ \frac{\partial\varphi}{\partial n} &= j && \text{on } \partial K \cap \Gamma_N \\ \varphi &= 0 && \text{on } \partial K \cap \Gamma_D \end{aligned}$$

with the element and edge residuals of equations (1.13) (p. 11) and (1.14) (p. 11).

The following proposition shows that $\eta_{N,K}$ also yields upper and lower bounds on the error $\|\nabla(u-u_{\mathcal{T}})\|$ and that it is comparable to $\eta_{R,K}$. Its proof is completely similar to that of Theorem 1.18 and is left to the reader.

Theorem 1.20 *Denote by u and $u_{\mathcal{T}}$ the unique solutions of problem (1.2) (p. 4) and (1.3) (p. 5). There are constants $c_{\mathcal{T},5}, \ldots, c_{\mathcal{T},8}$, which only depend on the shape parameter of $\mathcal{T}$, such that the estimates*

$$\eta_{N,K} \le c_{\mathcal{T},5}\,\eta_{R,K},$$

$$\eta_{R,K} \le c_{\mathcal{T},6}\left\{\sum_{K'\subset\omega_K} \eta_{N,K'}^2\right\}^{\frac{1}{2}},$$

$$\eta_{N,K} \le c_{\mathcal{T},7}\left\{\left\|\nabla(u-u_{\mathcal{T}})\right\|_{\omega_K}^2 + \sum_{K'\subset\omega_K} h_{K'}^2 \left\|f-\bar{f}_{K'}\right\|_{K'}^2 + \sum_{E\in\mathcal{E}_{K,\Gamma_N}} h_E \left\|g-\bar{g}_E\right\|_E^2\right\}^{\frac{1}{2}},$$

$$\left\|\nabla(u-u_{\mathcal{T}})\right\| \le c_{\mathcal{T},8}\left\{\sum_{K\in\mathcal{T}} \eta_{N,K}^2 + \sum_{K\in\mathcal{T}} h_K^2 \left\|f-\bar{f}_K\right\|_K^2 + \sum_{E\in\mathcal{E}_{\Gamma_N}} h_E \left\|g-\bar{g}_E\right\|_E^2\right\}^{\frac{1}{2}}$$

hold for all $K\in\mathcal{T}$. Here, $\bar{f}_K$, $\bar{g}_E$, and $\eta_{R,K}$ are as in Theorem 1.5 (p. 15), and $\eta_{N,K}$ is defined by equation (1.44).

Remark 1.21 When $\mathcal{T}$ exclusively consists of triangles, $\Delta u_{\mathcal{T}}$ vanishes element-wise and the normal derivatives $\mathbf{n}_E\cdot\nabla u_{\mathcal{T}}$ are edge-wise constant. In this case the functions φ, ρ, and σ can be dropped in equations (1.35), (1.40), and (1.43). This considerably reduces the dimension of the spaces V_z, $\widetilde{V}_K$, and V_K and thus of the discrete auxiliary problems. Figures 1.1 (p. 6) and 1.2 (p. 6) show typical examples of domains ω_z and ω_K. From this it is obvious that, in general, problems (1.37), (1.42), and (1.45) have at least dimensions 12, 7, and 4, respectively. In any case, the computation of $\eta_{D,z}$, $\eta_{D,K}$, and $\eta_{N,K}$ is more expensive than that of $\eta_{R,K}$. This is sometimes paid off by an improved accuracy of the error estimate.

Remark 1.22 The indicator $\eta_{D,z}$ was first introduced in [**31**]. Apparently, the indicator $\eta_{D,K}$ was first presented in [**64**]. The indicator $\eta_{N,K}$, finally, is a slight modification of an indicator which was first introduced and analysed in [**42**]. In its original form presented in [**42**], the indicator $\eta_{N,K}$ is based on the solution of the auxiliary problem (1.45) incorporating only the edge-bubble functions ψ_E. In [**42**], the proof that $\eta_{N,K}$ yields upper and lower bounds on the error is based on a saturation assumption. In [**207**] it was shown that this extra condition can be dispensed with. The present analysis is a slight modification of the one given in [**255**]. A quite different analysis is given in [**13**, **128**, **130**]. Numerical comparisons of these error indicators may be found in [**33**, **194**, **198**].

1.8 A Hierarchical Approach

The key idea of the hierarchical approach is to solve problem (1.2) (p. 4) approximately using a more accurate finite element space and to compare this solution with the solution of problem (1.3) (p. 5). In order to reduce the computational cost of the new problem, the new finite element space is decomposed into the original one and a nearly orthogonal higher order complement. Then only the contribution corresponding to the complement is computed. To further reduce the computational cost, the original bi-linear form is replaced by an equivalent one which leads to a diagonal stiffness matrix.

1.8.1 Higher Order Finite Element Spaces

To describe this idea in detail, we consider a finite element space $Y_{\mathcal{T}}$ which satisfies $S_D^{1,0}(\mathcal{T}) \subset Y_{\mathcal{T}} \subset H_D^1(\Omega)$ and which either consists of higher order elements or corresponds to a refinement of $\mathcal{T}$. We then denote by $w_{\mathcal{T}} \in Y_{\mathcal{T}}$ the unique solution of

$$\int_\Omega \nabla w_{\mathcal{T}} \cdot \nabla v_{\mathcal{T}} = \int_\Omega f v_{\mathcal{T}} + \int_{\Gamma_N} g v_{\mathcal{T}} \tag{1.46}$$

for all $v_{\mathcal{T}} \in Y_{\mathcal{T}}$.

To compare the solutions $w_{\mathcal{T}}$ of problem (1.46) and $u_{\mathcal{T}}$ of problem (1.3) (p. 5), we subtract $\int_\Omega \nabla u_{\mathcal{T}} \cdot \nabla v_{\mathcal{T}}$ from both sides of equation (1.46) and take into account equation (1.7) (p. 10). We thus obtain for all $v_{\mathcal{T}} \in Y_{\mathcal{T}}$

$$\begin{aligned} \int_\Omega \nabla(w_{\mathcal{T}} - u_{\mathcal{T}}) \cdot \nabla v_{\mathcal{T}} &= \int_\Omega f v_{\mathcal{T}} + \int_{\Gamma_N} g v_{\mathcal{T}} - \int_\Omega \nabla u_{\mathcal{T}} \cdot \nabla v_{\mathcal{T}} \\ &= \int_\Omega \nabla(u - u_{\mathcal{T}}) \cdot \nabla v_{\mathcal{T}}, \end{aligned}$$

where $u \in H_D^1(\Omega)$ is the unique solution of problem (1.2) (p. 4). Since $S_D^{1,0}(\mathcal{T}) \subset Y_{\mathcal{T}}$, we may insert $v_{\mathcal{T}} = w_{\mathcal{T}} - u_{\mathcal{T}}$ as a test function in this equation. The Cauchy–Schwarz inequality for integrals then implies

$$\|\nabla(w_{\mathcal{T}} - u_{\mathcal{T}})\| \le \|\nabla(u - u_{\mathcal{T}})\|.$$

To prove the converse estimate, we assume that the space $Y_{\mathcal{T}}$ satisfies a *saturation assumption*: there is a constant β with $0 \le \beta < 1$ such that

$$\|\nabla(u - w_{\mathcal{T}})\| \le \beta \|\nabla(u - u_{\mathcal{T}})\|. \tag{1.47}$$

From the saturation assumption (1.47) and the triangle inequality we immediately conclude that

$$\begin{aligned} \|\nabla(u - u_{\mathcal{T}})\| &\le \|\nabla(u - w_{\mathcal{T}})\| + \|\nabla(w_{\mathcal{T}} - u_{\mathcal{T}})\| \\ &\le \beta \|\nabla(u - u_{\mathcal{T}})\| + \|\nabla(w_{\mathcal{T}} - u_{\mathcal{T}})\| \end{aligned}$$

and therefore

$$\|\nabla(u - u_{\mathcal{T}})\| \leq \frac{1}{1-\beta} \|\nabla(w_{\mathcal{T}} - u_{\mathcal{T}})\| .$$

Thus, we have proved the two-sided error bound

$$\|\nabla(w_{\mathcal{T}} - u_{\mathcal{T}})\| \leq \|\nabla(u - u_{\mathcal{T}})\| \leq \frac{1}{1-\beta} \|\nabla(w_{\mathcal{T}} - u_{\mathcal{T}})\| .$$

Hence, we may use $\|\nabla(w_{\mathcal{T}} - u_{\mathcal{T}})\|$ as an a posteriori error indicator. This device, however, is not efficient since the computation of $w_{\mathcal{T}}$ is at least as costly as that of $u_{\mathcal{T}}$.

1.8.2 Hierarchical Splitting of the Finite Element Spaces

In order to obtain a more efficient error estimation, we use a hierarchical splitting $Y_{\mathcal{T}} = S_D^{1,0}(\mathcal{T}) \oplus Z_{\mathcal{T}}$ and assume that the spaces $S_D^{1,0}(\mathcal{T})$ and $Z_{\mathcal{T}}$ are nearly orthogonal in that they satisfy a *strengthened Cauchy–Schwarz inequality*: there is a constant γ with $0 \leq \gamma < 1$ such that

$$\left| \int_{\Omega} \nabla v_{\mathcal{T}} \cdot \nabla z_{\mathcal{T}} \right| \leq \gamma \, \|\nabla v_{\mathcal{T}}\| \, \|\nabla z_{\mathcal{T}}\| \tag{1.48}$$

holds for all $v_{\mathcal{T}} \in S_D^{1,0}(\mathcal{T}), z_{\mathcal{T}} \in Z_{\mathcal{T}}$.

We want to replace $\|\nabla(w_{\mathcal{T}} - u_{\mathcal{T}})\|$ by $\|\nabla z_{\mathcal{T}}\|$ with a suitable function $z_{\mathcal{T}} \in Z_{\mathcal{T}}$. To this end we denote by $z_{\mathcal{T}} \in Z_{\mathcal{T}}$ the unique solution of

$$\int_{\Omega} \nabla z_{\mathcal{T}} \cdot \nabla \zeta_{\mathcal{T}} = \int_{\Omega} f \zeta_{\mathcal{T}} + \int_{\Gamma_N} g \zeta_{\mathcal{T}} - \int_{\Omega} \nabla u_{\mathcal{T}} \cdot \nabla \zeta_{\mathcal{T}} \tag{1.49}$$

for all $\zeta_{\mathcal{T}} \in Z_{\mathcal{T}}$.

From the definitions (1.2) (p. 4), (1.3) (p. 5), (1.46), and (1.49) of $u, u_{\mathcal{T}}, w_{\mathcal{T}}$, and $z_{\mathcal{T}}$, we infer for all $\zeta_{\mathcal{T}} \in Z_{\mathcal{T}}$ that

$$\begin{aligned} \int_{\Omega} \nabla z_{\mathcal{T}} \cdot \nabla \zeta_{\mathcal{T}} &= \int_{\Omega} \nabla(u - u_{\mathcal{T}}) \cdot \nabla \zeta_{\mathcal{T}} \\ &= \int_{\Omega} \nabla(w_{\mathcal{T}} - u_{\mathcal{T}}) \cdot \nabla \zeta_{\mathcal{T}} \end{aligned} \tag{1.50}$$

and for all $v_{\mathcal{T}} \in S_D^{1,0}(\mathcal{T})$ that

$$\int_{\Omega} \nabla(w_{\mathcal{T}} - u_{\mathcal{T}}) \cdot \nabla v_{\mathcal{T}} = 0. \tag{1.51}$$

We now insert $\zeta_{\mathcal{T}} = z_{\mathcal{T}}$ in equation (1.50). The Cauchy–Schwarz inequality for integrals then yields

$$\|\nabla z_{\mathcal{T}}\| \leq \|\nabla(u - u_{\mathcal{T}})\| .$$

Next, we write $w_\mathcal{T} - u_\mathcal{T}$ in the form $\bar{v}_\mathcal{T} + \bar{z}_\mathcal{T}$ with $\bar{v}_\mathcal{T} \in S_D^{1,0}(\mathcal{T})$ and $\bar{z}_\mathcal{T} \in Z_\mathcal{T}$. From the strengthened Cauchy–Schwarz inequality we then deduce that

$$\begin{aligned}(1-\gamma)\left\{\|\nabla\bar{v}_\mathcal{T}\|^2 + \|\nabla\bar{z}_\mathcal{T}\|^2\right\} &\le \left\|\nabla(w_\mathcal{T} - u_\mathcal{T})\right\|^2 \\ &\le (1+\gamma)\left\{\|\nabla\bar{v}_\mathcal{T}\|^2 + \|\nabla\bar{z}_\mathcal{T}\|^2\right\}\end{aligned}$$

and in particular

$$\|\nabla\bar{z}_\mathcal{T}\| \le \frac{1}{\sqrt{1-\gamma}}\left\|\nabla(w_\mathcal{T} - u_\mathcal{T})\right\|. \tag{1.52}$$

From inequality (1.52) and equations (1.50) and (1.51) with $\zeta_\mathcal{T} = \bar{z}_\mathcal{T}$ we conclude that

$$\begin{aligned}\left\|\nabla(w_\mathcal{T} - u_\mathcal{T})\right\|^2 &= \int_\Omega \nabla(w_\mathcal{T} - u_\mathcal{T})\cdot\nabla(w_\mathcal{T} - u_\mathcal{T}) \\ &= \int_\Omega \nabla(w_\mathcal{T} - u_\mathcal{T})\cdot\nabla(\bar{v}_\mathcal{T} + \bar{z}_\mathcal{T}) \\ &= \int_\Omega \nabla(w_\mathcal{T} - u_\mathcal{T})\cdot\nabla\bar{z}_\mathcal{T} \\ &= \int_\Omega \nabla z_\mathcal{T}\cdot\nabla\bar{z}_\mathcal{T} \\ &\le \|\nabla z_\mathcal{T}\|\,\|\nabla\bar{z}_\mathcal{T}\| \\ &\le \frac{1}{\sqrt{1-\gamma}}\|\nabla z_\mathcal{T}\|\,\left\|\nabla(w_\mathcal{T} - u_\mathcal{T})\right\|\end{aligned}$$

and hence

$$\begin{aligned}\left\|\nabla(u - u_\mathcal{T})\right\| &\le \frac{1}{1-\beta}\left\|\nabla(w_\mathcal{T} - u_\mathcal{T})\right\| \\ &\le \frac{1}{(1-\beta)\sqrt{1-\gamma}}\|\nabla z_\mathcal{T}\|.\end{aligned}$$

Thus, we have established the two-sided error bound

$$\|\nabla z_\mathcal{T}\| \le \left\|\nabla(u - u_\mathcal{T})\right\| \le \frac{1}{(1-\beta)\sqrt{1-\gamma}}\|\nabla z_\mathcal{T}\|.$$

Therefore, $\|\nabla z_\mathcal{T}\|$ can be used as an error indicator.

1.8.3 Lumping of the Stiffness Matrix

At first sight, the computation of $z_\mathcal{T}$ seems to be cheaper than that of $w_\mathcal{T}$ since the dimension of $Z_\mathcal{T}$ is smaller than that of $Y_\mathcal{T}$. The computation of $z_\mathcal{T}$, however, still requires the solution of a global system and is therefore as expensive as the calculation of $u_\mathcal{T}$ and $w_\mathcal{T}$. Yet, in most applications the functions in $Z_\mathcal{T}$ vanish at the vertices of $\mathcal{N}$ since $Z_\mathcal{T}$ is the hierarchical complement of $S_D^{1,0}(\mathcal{T})$ in $Y_\mathcal{T}$. This in particular implies that the stiffness matrix corresponding to $Z_\mathcal{T}$ is spectrally equivalent to a suitably scaled lumped mass matrix. Therefore, $z_\mathcal{T}$ can be replaced by a quantity $z_\mathcal{T}^*$ which can be computed by solving a diagonal linear system of equations.

More precisely, we assume that there is a bi-linear form b on $Z_\mathcal{T} \times Z_\mathcal{T}$ which has a diagonal stiffness matrix and which defines an equivalent norm to $\|\nabla \cdot\|$ on $Z_\mathcal{T}$, i.e.

$$\lambda \, \|\nabla \zeta_\mathcal{T}\|^2 \leq b(\zeta_\mathcal{T}, \zeta_\mathcal{T}) \leq \Lambda \, \|\nabla \zeta_\mathcal{T}\|^2 \tag{1.53}$$

holds for all $\zeta_\mathcal{T} \in Z_\mathcal{T}$ with constants $0 < \lambda \leq \Lambda$.

The conditions on b imply that there is a unique function $z_\mathcal{T}^* \in Z_\mathcal{T}$ which satisfies

$$b(z_\mathcal{T}^*, \zeta_\mathcal{T}) = \int_\Omega f \zeta_\mathcal{T} + \int_{\Gamma_N} g \zeta_\mathcal{T} - \int_\Omega \nabla u_\mathcal{T} \cdot \nabla \zeta_\mathcal{T} \tag{1.54}$$

for all $\zeta_\mathcal{T} \in Z_\mathcal{T}$.

Equations (1.7) (p. 10) and (1.49) yield for all $\zeta_\mathcal{T} \in Z_\mathcal{T}$

$$\begin{aligned} b(z_\mathcal{T}^*, \zeta_\mathcal{T}) &= \int_\Omega \nabla (u - u_\mathcal{T}) \cdot \nabla \zeta_\mathcal{T} \\ &= \int_\Omega \nabla z_\mathcal{T} \cdot \nabla \zeta_\mathcal{T}. \end{aligned}$$

Inserting $\zeta_\mathcal{T} = z_\mathcal{T}$ and $\zeta_\mathcal{T} = z_\mathcal{T}^*$ in this identity and using estimate (1.53), we infer that

$$\begin{aligned} b(z_\mathcal{T}^*, z_\mathcal{T}^*) &= \int_\Omega \nabla (u - u_\mathcal{T}) \cdot \nabla z_\mathcal{T}^* \\ &\leq \|\nabla (u - u_\mathcal{T})\| \, \|\nabla z_\mathcal{T}^*\| \\ &\leq \|\nabla (u - u_\mathcal{T})\| \, \frac{1}{\sqrt{\lambda}} b(z_\mathcal{T}^*, z_\mathcal{T}^*)^{\frac{1}{2}} \end{aligned}$$

and

$$\begin{aligned} \|\nabla z_\mathcal{T}\|^2 &= b(z_\mathcal{T}^*, z_\mathcal{T}) \\ &\leq b(z_\mathcal{T}^*, z_\mathcal{T}^*)^{\frac{1}{2}} b(z_\mathcal{T}, z_\mathcal{T})^{\frac{1}{2}} \\ &\leq b(z_\mathcal{T}^*, z_\mathcal{T}^*)^{\frac{1}{2}} \sqrt{\Lambda} \, \|\nabla z_\mathcal{T}\|. \end{aligned}$$

This proves the final two-sided error bound

$$\sqrt{\lambda} b(z_\mathcal{T}^*, z_\mathcal{T}^*)^{\frac{1}{2}} \leq \|\nabla (u - u_\mathcal{T})\| \leq \frac{\sqrt{\Lambda}}{(1-\beta)\sqrt{1-\gamma}} b(z_\mathcal{T}^*, z_\mathcal{T}^*)^{\frac{1}{2}}.$$

1.8.4 A Posteriori Error Estimates

We may summarise the results of this section as follows.

Theorem 1.23 *Denote by u and $u_\mathcal{T}$ the unique solutions of problems (1.2) (p. 4) and (1.3) (p. 5), respectively. Assume that the space $Y_\mathcal{T} = S_D^{1,0}(\mathcal{T}) \oplus Z_\mathcal{T}$ satisfies the saturation assumption (1.47) (p. 31) and the strengthened Cauchy–Schwarz inequality (1.48) (p. 32) and admits a bi-linear form b on $Z_\mathcal{T} \times Z_\mathcal{T}$ which has a diagonal stiffness matrix and which satisfies estimate (1.53) (p. 34).*

Denote by $z_{\mathcal{T}}^{*} \in Z_{\mathcal{T}}$ *the unique solution of problem* (1.54) *(p. 34) and define the* hierarchical a posteriori error indicator η_H *by*

$$\eta_H = b(z_{\mathcal{T}}^{*}, z_{\mathcal{T}}^{*})^{\frac{1}{2}}.$$

Then the a posteriori error estimates

$$\left\| \nabla (u - u_{\mathcal{T}}) \right\| \leq \frac{\sqrt{\Lambda}}{(1-\beta)\sqrt{1-\gamma}} \eta_H$$

and

$$\eta_H \leq \frac{1}{\sqrt{\lambda}} \left\| \nabla (u - u_{\mathcal{T}}) \right\|$$

are valid.

Remark 1.24 Notice that, contrary to previous a posteriori error estimates, the upper and lower bounds of Theorem 1.23 are *both* global ones.

Example 1.25 For a simple example of a hierarchical error indicator choose

$$Z_{\mathcal{T}} = \operatorname{span}\left\{ \psi_S : S \in \mathcal{T} \cup \mathcal{E}_\Omega \cup \mathcal{E}_{\Gamma_N} \right\}$$

and

$$b\left(\sum_S \alpha_S \psi_S, \sum_S \alpha_S' \psi_S \right) = \sum_S \alpha_S \alpha_S' \int_\Omega \nabla \psi_S \cdot \nabla \psi_S$$

where the sums extend over all elements, all interior edges, and all edges on the Neumann boundary. The corresponding space $Y_{\mathcal{T}}$ contains the space of continuous piecewise quadratic polynomials and is contained in the space of continuous piecewise cubic polynomials. A straightforward calculation yields

$$\eta_H = \left\{ \sum_{S \in \mathcal{T} \cup \mathcal{E}_\Omega \cup \mathcal{E}_{\Gamma_N}} \left(\frac{\langle R, \psi_S \rangle}{\|\nabla \psi_S\|} \right)^2 \right\}^{\frac{1}{2}}.$$

One may check that $Z_{\mathcal{T}}$ and b satisfy the assumptions of Theorem 1.23. Yet, it is much simpler to prove the efficiency and reliability of the error indicator η_H directly. Equations (1.7) (p. 10) and (1.8) (p. 10) and the Cauchy–Schwarz inequality imply

$$\frac{\langle R, \psi_S \rangle}{\|\nabla \psi_S\|} \leq \left\| \nabla (u - u_{\mathcal{T}}) \right\|_{\operatorname{supp} \psi_S}$$

for every $S \in \mathcal{T} \cup \mathcal{E}_\Omega \cup \mathcal{E}_{\Gamma_N}$ and consequently

$$\eta_H \leq c_1 \left\| \nabla (u - u_{\mathcal{T}}) \right\|$$

where the constant c_1 only depends on the shape parameter $C_{\mathcal{T}}$ of $\mathcal{T}$. This proves the efficiency of η_H. Since $Z_{\mathcal{T}}$ contains the local cut-off functions ψ_K and ψ_E for all elements, all interior edges, and all edges on the Neumann boundary, the arguments of Section 1.4.5 (p. 12) yield the estimate

$$\left\{ \sum_{K \in \mathcal{T}} \eta_{R,K}^2 \right\}^{\frac{1}{2}} \le c_2 \left\{ \eta_H^2 + \sum_{K \in \mathcal{T}} h_K^2 \left\| f - \bar{f}_K \right\|_K^2 + \sum_{E \in \mathcal{E}_{\Gamma_N}} h_E \left\| g - \bar{g}_E \right\|_E^2 \right\}^{\frac{1}{2}}$$

with a constant c_2 which only depends on the shape parameter $C_{\mathcal{T}}$ of $\mathcal{T}$. Together with Theorem 1.5 (p. 15) this proves the reliability of η_H.

Remark 1.26 When considering families of partitions obtained by successive refinement, the constants β and γ in the saturation assumption and the strengthened Cauchy–Schwarz inequality should be uniformly less than 1. Similarly, the quotient $\frac{\Lambda}{\lambda}$ should be uniformly bounded.

Remark 1.27 The bi-linear form b can often be constructed as follows. The hierarchical complement $Z_{\mathcal{T}}$ can be chosen such that its elements vanish at the element vertices $\mathcal{N}$. Standard scaling arguments then imply that, on $Z_{\mathcal{T}}$, the H^1-semi-norm $\|\nabla \cdot\|$ is equivalent to a scaled L^2-norm. Similarly, one can then prove that the mass matrix corresponding to this norm is spectrally equivalent to a lumped mass matrix. The lumping process in turn corresponds to a suitable numerical quadrature. The bi-linear form b then is given by the inner product corresponding to the weighted L^2-norm evaluated with the quadrature rule.

Remark 1.28 The strengthened Cauchy–Schwarz inequality, e.g. holds if $Y_{\mathcal{T}}$ consists of continuous piecewise quadratic or bi-quadratic functions. Often it can be established by transforming to the reference element and solving a small eigenvalue problem there [**78**, **131**, **187**].

Remark 1.29 The saturation assumption (1.47) is used to establish the reliability of the error indicator η_H, i.e. the first error estimate of Theorem 1.23. One can prove that the reliability of η_H in turn implies the saturation assumption (1.47) [**74**, Theorem 2.1]. If the space $Y_{\mathcal{T}}$ contains the functions w_K and w_E of equations (1.20) (p. 13), (1.22) (p. 14), and (1.24) (p. 15), one may repeat the proofs of estimates (1.21) (p. 13), (1.23) (p. 15), and (1.25) (p. 15) and obtain that, up to perturbation terms of the form $h_K \left\| f - \bar{f}_K \right\|_K$ and $h_E^{\frac{1}{2}} \left\| g - \bar{g}_E \right\|_E$, the quantity $\left\| \nabla z_{\mathcal{T}}^* \right\|_{\omega_K}$ is bounded from below by $\eta_{R,K}$ for every element K. Together with Theorem 1.5 (p. 15) and inequality (1.54) this proves, up to the perturbation terms, the reliability of η_H without resorting to the saturation assumption. In fact, this result may be used to prove that the saturation assumption holds if the right-hand sides f and g of problem (1.1) (p. 4) are piecewise constant on $\mathcal{T}$ and $\mathcal{E}_{\Gamma_N}$, respectively [**74**, Section 4].

Remark 1.30 Theorem 1.23 together with some extensions may be found in [**13**, **40**, **74**, **123**].

1.9 Gradient Recovery

To avoid unnecessary technical difficulties and to simplify the presentation, we consider in this section problem (1.1) (p. 4) with pure Dirichlet boundary conditions, i.e. $\Gamma_N = \emptyset$, and assume that the partition $\mathcal{T}$ exclusively consists of triangles.

1.9.1 The Basic Idea

The error indicator of this chapter is based on the following ideas. Denote by u and $u_{\mathcal{T}}$ the unique solutions of problems (1.2) (p. 4) and (1.3) (p. 5). As a motivation, suppose that we dispose of an easily computable approximation $Gu_{\mathcal{T}}$ of $\nabla u_{\mathcal{T}}$ such that

$$\|\nabla u - Gu_{\mathcal{T}}\| \leq \beta \, \|\nabla u - \nabla u_{\mathcal{T}}\| \tag{1.55}$$

holds with a constant $0 \leq \beta < 1$. We then have

$$\frac{1}{1+\beta} \, \|Gu_{\mathcal{T}} - \nabla u_{\mathcal{T}}\| \leq \|\nabla u - \nabla u_{\mathcal{T}}\| \leq \frac{1}{1-\beta} \, \|Gu_{\mathcal{T}} - \nabla u_{\mathcal{T}}\|$$

and may therefore choose $\|Gu_{\mathcal{T}} - \nabla u_{\mathcal{T}}\|$ as an error indicator. Since $\nabla u_{\mathcal{T}}$ is a piecewise constant vector field, we may hope that its L^2-projection onto the continuous, piecewise linear vector fields satisfies inequality (1.55). The computation of this projection, however, is as expensive as the solution of problem (1.3) (p. 5). We therefore replace the L^2-scalar product by an approximation which leads to a more tractable auxiliary problem.

1.9.2 Averaging

In order to make these ideas more precise, we denote by $W_{\mathcal{T}}$ the space of all piecewise linear vector fields and set $V_{\mathcal{T}} = W_{\mathcal{T}} \cap C(\Omega, \mathbb{R}^2)$. Note that $\nabla S_D^{1,0}(\mathcal{T}) \subset W_{\mathcal{T}}$ and $V_{\mathcal{T}} = S^{1,0}(\mathcal{T})^2$. We define a mesh-dependent scalar product $(\cdot\,,\,\cdot)_{\mathcal{T}}$ on $W_{\mathcal{T}}$ by

$$(\mathbf{v}\,,\,\mathbf{w})_{\mathcal{T}} = \sum_{K\in\mathcal{T}} \frac{\mu_2(K)}{3} \left\{ \sum_{z\in\mathcal{N}_K} \mathbf{v}|_K(z) \cdot \mathbf{w}|_K(z) \right\},$$

where $\mathbf{v}|_K(z) = \lim_{K\ni y\to z} \mathbf{v}(y)$ for all $\mathbf{v} \in W_{\mathcal{T}}, K \in \mathcal{T}, z \in \mathcal{N}_K$.

Denote by $Gu_{\mathcal{T}} \in V_{\mathcal{T}}$ the $(\cdot\,,\,\cdot)_{\mathcal{T}}$-projection of $\nabla u_{\mathcal{T}}$ onto $V_{\mathcal{T}}$, i.e.

$$(Gu_{\mathcal{T}}\,,\,\mathbf{v}_{\mathcal{T}})_{\mathcal{T}} = (\nabla u_{\mathcal{T}}\,,\,\mathbf{v}_{\mathcal{T}})_{\mathcal{T}} \tag{1.56}$$

for all $\mathbf{v}_{\mathcal{T}} \in V_{\mathcal{T}}$. We then define an error indicator by

$$\eta_{Z,K} = \|Gu_{\mathcal{T}} - \nabla u_{\mathcal{T}}\|_K$$

for all $K \in \mathcal{T}$ and

$$\eta_Z = \left\{ \sum_{K\in\mathcal{T}} \eta_{Z,K}^2 \right\}^{\frac{1}{2}}. \tag{1.57}$$

For the practical computation of $\eta_{Z,K}$, we observe that for all $\mathbf{v}, \mathbf{w} \in V_{\mathcal{T}}$

$$(\mathbf{v}\,,\,\mathbf{w})_{\mathcal{T}} = \frac{1}{3} \sum_{z\in\mathcal{N}} \mu_2(\omega_z)\mathbf{v}(z) \cdot \mathbf{w}(z). \tag{1.58}$$

This identity and the definition (1.56) of $Gu_{\mathcal{T}}$ immediately yield for all $z \in \mathcal{N}$ the representation

$$Gu_{\mathcal{T}}(z) = \sum_{K \subset \omega_z} \frac{\mu_2(K)}{\mu_2(\omega_z)} \nabla u_{\mathcal{T}}|_K. \tag{1.59}$$

1.9.3 A Posteriori Error Estimates

In order to prove that η_Z is an error indicator, we will not establish inequality (1.55). Instead, we will compare $\eta_{Z,K}$ with the edge residual error indicator $\eta_{R,E}$ of Theorem 1.15 (p. 24). The following proposition in combination with Theorem 1.15 shows that, up to perturbation terms of the form $h_K \left\| f - I_{\mathcal{T}} f \right\|_K$, the error indicator η_Z yields upper and lower error bounds provided the partition $\mathcal{T}$ satisfies the growth condition of Section 1.6. According to Remark 1.17 (p. 25) this extra condition may be dropped at the expense of slightly larger perturbation terms close to the boundary. Note that, since $\Gamma_N = \emptyset$, only interior edges yield contributions to $\eta_{R,E}$ and that the perturbation terms in Theorem 1.15 involving g vanish.

Theorem 1.31 *Assume that the partition $\mathcal{T}$ exclusively consists of triangles and that the Neumann boundary Γ_N is the empty set. Denote by $u_{\mathcal{T}}$ the unique solution of problem (1.3) (p. 5). Recall the definition of the error indicator of Theorem 1.15 (p. 24)*

$$\eta_{R,E} = \left\{ \sum_{E \in \mathcal{E}_\Omega} h_E \left\| \mathbb{J}_E(\mathbf{n}_E \cdot \nabla u_{\mathcal{T}}) \right\|_E^2 \right\}^{\frac{1}{2}}$$

and define the error indicator η_Z by equation (1.57). Then there are two constants c_{Z1} and c_{Z2} which only depend on the shape parameter of $\mathcal{T}$ such that the estimates

$$c_{Z1}\eta_{R,E} \le \eta_Z \le c_{Z2}\eta_{R,E}$$

are valid.

Proof Since $\mathcal{T}$ exclusively consists of triangles, the normal derivatives $\mathbf{n}_E \cdot \nabla u_{\mathcal{T}}$ are edge-wise constant. We therefore have

$$\eta_{R,E}^2 = \sum_{E \in \mathcal{E}_\Omega} h_E^2 \mathbb{J}_E(\mathbf{n}_E \cdot \nabla u_{\mathcal{T}})^2.$$

Due to the shape regularity, the right-hand side of this equation is bounded from above and from below by multiples of

$$\sum_{z \in \mathcal{N}} \mu_2(\omega_z) \widetilde{\eta}_{R,z}^2$$

where

$$\widetilde{\eta}_{R,z} = \left\{ \sum_{E \subset \sigma_z \cap \Omega} \mathbb{J}_E(\mathbf{n}_E \cdot \nabla u_{\mathcal{T}})^2 \right\}^{\frac{1}{2}}.$$

One easily checks that for all $\mathbf{v} \in W_{\mathcal{T}}$

$$\frac{1}{4}\|\mathbf{v}\|^2 \le (\mathbf{v}, \mathbf{v})_{\mathcal{T}} \le \|\mathbf{v}\|^2.$$

This inequality implies that η_Z is bounded from above and from below by multiples of $(Gu_{\mathcal{T}} - \nabla u_{\mathcal{T}}, Gu_{\mathcal{T}} - \nabla u_{\mathcal{T}})_{\mathcal{T}}$. Since the quadrature formula

$$\int_K \varphi \approx \frac{\mu_2(K)}{3} \sum_{z \in \mathcal{N}_K} \varphi(z)$$

is exact for all affine functions, we conclude that

$$(\mathbf{v}, \mathbf{w})_{\mathcal{T}} = \int_\Omega \mathbf{v} \cdot \mathbf{w}$$

holds for all vector fields $\mathbf{v}, \mathbf{w} \in W_{\mathcal{T}}$ provided that at least one of them is piecewise constant. This identity and equations (1.56), (1.58), and (1.59) imply that

$$\begin{aligned}
&(Gu_{\mathcal{T}} - \nabla u_{\mathcal{T}}, Gu_{\mathcal{T}} - \nabla u_{\mathcal{T}})_{\mathcal{T}} \\
&= (\nabla u_{\mathcal{T}}, \nabla u_{\mathcal{T}})_{\mathcal{T}} - (Gu_{\mathcal{T}}, Gu_{\mathcal{T}})_{\mathcal{T}} \\
&= \sum_{K \in \mathcal{T}} \mu_2(K) |\nabla u_{\mathcal{T}}|_K|^2 - \sum_{z \in \mathcal{N}} \frac{1}{3} \mu_2(\omega_z) \left| Gu_{\mathcal{T}}(z) \right|^2 \\
&= \sum_{z \in \mathcal{N}} \left\{ \sum_{K \subset \omega_z} \frac{1}{3} \mu_2(K) |\nabla u_{\mathcal{T}}|_K|^2 - \frac{1}{3}\mu_2(\omega_z) \left| \sum_{K \subset \omega_z} \frac{\mu_2(K)}{\mu_2(\omega_z)} \nabla u_{\mathcal{T}}|_K \right|^2 \right\} \\
&= \frac{1}{3} \sum_{z \in \mathcal{N}} \mu_2(\omega_z) \eta_{Z,z}^2,
\end{aligned}$$

where

$$\eta_{Z,z} = \left\{ \sum_{K \subset \omega_z} \frac{\mu_2(K)}{\mu_2(\omega_z)} |\nabla u_{\mathcal{T}}|_K|^2 - \left| \sum_{K \subset \omega_z} \frac{\mu_2(K)}{\mu_2(\omega_z)} \nabla u_{\mathcal{T}}|_K \right|^2 \right\}^{\frac{1}{2}}.$$

Therefore, it is sufficient to prove that

$$c_{Z3} \widetilde{\eta}_{R,z}^2 \le \eta_{Z,z}^2 \le c_{Z4} \widetilde{\eta}_{R,z}^2 \tag{1.60}$$

holds for all $z \in \mathcal{N}$ with constants c_{Z3} and c_{Z4} that only depend on the smallest angle in $\mathcal{T}$.

In order to establish inequality (1.60), we consider an interior node $z \in \mathcal{N}_\Omega$. The case of a boundary node is completely similar. Denote by N the number of triangles in ω_z and enumerate the triangles in counter-clockwise order starting with 0 as depicted in Figure 1.4. Set $K_N = K_0$ and for $0 \le i \le N$

$$v_i = \frac{\partial}{\partial x} u_{\mathcal{T}}|_{K_i}, \quad w_i = \frac{\partial}{\partial y} u_{\mathcal{T}}|_{K_i}, \quad \mu_i = \frac{\mu_2(K_i)}{\mu_2(\omega_z)},$$

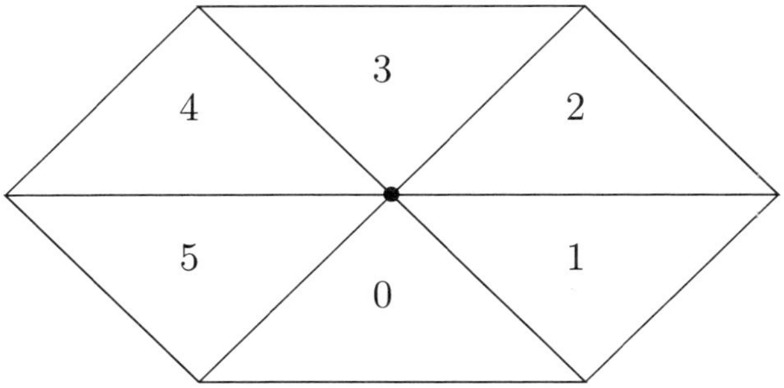

Figure 1.4 Enumeration of triangles in ω_z

as well as

$$v = (v_1, \ldots, v_N)^t, \quad w = (w_1, \ldots, w_N)^t.$$

Since the tangential components of $\nabla u_{\mathcal{T}}$ are continuous across edges, we have

$$\widetilde{\eta}_{R,z}^2 = \sum_{i=1}^{N} \left\{ (v_i - v_{i-1})^2 + (w_i - w_{i-1})^2 \right\} = v^t B v + w^t B w$$

and

$$\eta_{Z,z}^2 = \sum_{i=1}^{N} \mu_i v_i^2 - \left(\sum_{i=1}^{N} \mu_i v_i \right)^2 + \sum_{i=1}^{N} \mu_i w_i^2 - \left(\sum_{i=1}^{N} \mu_i w_i \right)^2 = v^t A v + w^t A w$$

with symmetric, positive semi-definite $N \times N$ matrices A and B. Since $\sum_{i=1}^{N} \mu_i = 1$, we have $e = (1, \ldots, 1)^t \in \ker(A) \cap \ker(B)$.

For any $x \in \mathbb{R}^N$, set

$$\widetilde{x} = (x_1 - x_N, \ldots, x_{N-1} - x_N)^t \in \mathbb{R}^{N-1}$$

and denote by $\widetilde{A}$ and $\widetilde{B}$ the $(N-1) \times (N-1)$ matrices which are obtained by deleting the last column and the last row of A and B. We then have for all $x \in \mathbb{R}^N$

$$x^t A x = (x - x_N e)^t A (x - x_N e) = \widetilde{x}^t \widetilde{A} \widetilde{x}$$

and

$$x^t B x = (x - x_N e)^t B (x - x_N e) = \widetilde{x}^t \widetilde{B} \widetilde{x}.$$

Since

$$\widetilde{B} = \begin{pmatrix} 2 & -1 & & & \\ -1 & 2 & -1 & & \\ & \ddots & \ddots & \ddots & \\ & & \ddots & \ddots & -1 \\ & & & -1 & 2 \end{pmatrix},$$

its eigenvalues are given by

$$\lambda_k(\widetilde{B}) = 4\sin^2\left(\frac{k\pi}{2N}\right), \quad 1 \le k \le N-1.$$

Gerschgorin's theorem [**222**, Theorem 5.2] on the other hand implies that the eigenvalues $\lambda_k(\widetilde{A})$ of $\widetilde{A}$ are bounded by

$$\min_{1\le i\le N} \mu_i^2 \le \lambda_k(\widetilde{A}) \le 2 \quad \text{for all } 1 \le k \le N-1.$$

Hence, we have for all $x \in \mathbb{R}^N$

$$\frac{1}{4}\min_{1\le i\le N} \mu_i^2\, x^t Bx \le x^t Ax, \quad 2\sin^2\left(\frac{\pi}{2N}\right) x^t Ax \le x^t Bx.$$

This proves inequality (1.60). □

Remark 1.32 The indicator η_Z was first proposed by J. Z. Zhu and O. C. Zienkiewicz in [**286**]. The present analysis follows the lines of [**231**]. Similar indicators based on a local averaging or extrapolation of the gradient are analysed in [**13, 16, 128, 137, 138, 150, 169, 281**]. In [**49, 93**] it is proved that each averaging technique leads to an a posteriori error estimate.

1.10 Equilibrated Residuals

In [**172**] P. Ladevèze and D. Leguillon proposed a technique for a posteriori error estimation which is based on a dual variational principle. In what follows we will briefly sketch the underlying idea. For a more detailed presentation we refer to [**11**] and [**172**].

1.10.1 Characterisation of the Error as the Minimum of a Quadratic Functional

We define a quadratic functional J on $H_D^1(\Omega)$ by

$$J(v) = \frac{1}{2}\int_\Omega \nabla v \cdot \nabla v - \int_\Omega fv - \int_{\Gamma_N} gv + \int_\Omega \nabla u_{\mathcal{T}} \cdot \nabla v,$$

where $u_{\mathcal{T}} \in S_D^{1,0}(\mathcal{T})$ is the unique solution of problem (1.3) (p. 5). Equation (1.7) (p. 10) implies for all $v \in H_D^1(\Omega)$ that

$$J(v) = \frac{1}{2}\int_\Omega \nabla v \cdot \nabla v - \int_\Omega \nabla(u - u_{\mathcal{T}}) \cdot \nabla v.$$

Hence, J attains its unique minimum at $u - u_{\mathcal{T}}$. Inserting $v = u - u_{\mathcal{T}}$ in the definition of J therefore yields for all $v \in H_D^1(\Omega)$

$$-\frac{1}{2}\left\|\nabla(u - u_{\mathcal{T}})\right\|^2 = J(u - u_{\mathcal{T}}) \le J(v).$$

Thus, the energy norm of the error can be computed by solving the variational problem

$$\text{minimise } J(v) \text{ in } H_D^1(\Omega).$$

1.10.2 *Extension of the Residual to a Larger Space*

Unfortunately, the above problem is an infinite-dimensional minimisation problem. In order to obtain a more tractable problem we want to replace $H_D^1(\Omega)$ by the broken space

$$H_{\mathcal{T}} = \left\{ v \in L^2(\Omega) : v|_K \in H^1(K) \text{ for all } K \in \mathcal{T},\ v = 0 \text{ on } \Gamma_D \right\}.$$

Obviously, we have

$$H_D^1(\Omega) = \left\{ v \in H_{\mathcal{T}} : \mathbb{J}_E(v) = 0 \text{ for all } E \in \mathcal{E}_\Omega \right\}, \tag{1.61}$$

where $\mathbb{J}_E(v)$ denotes the jump of v across E in direction $\mathbf{n}_E$, cf. equation (1.6) (p. 9).

Proposition III.1.1 in [**90**] implies that a continuous linear functional τ on $H_{\mathcal{T}}$ vanishes on $H_D^1(\Omega)$ if and only if there is a vector field $\sigma \in L^2(\Omega)^2$ with $\operatorname{div}\sigma \in L^2(\Omega)$ and $\mathbf{n}\cdot\sigma = 0$ on Γ_N such that for all $v \in H_{\mathcal{T}}$

$$\tau(v) = \sum_{K\in\mathcal{T}} \int_{\partial K} \mathbf{n}_K \cdot \sigma v.$$

Hence, the polar of $H_D^1(\Omega)$ in $H_{\mathcal{T}}$ can be identified with the space

$$M = \left\{ \sigma \in L^2(\Omega)^2 : \operatorname{div}\sigma \in L^2(\Omega), \mathbf{n}\cdot\sigma = 0 \text{ on } \Gamma_N \right\}.$$

Recall the definition (1.8) (p. 10) of the residual R of $u_{\mathcal{T}}$ as a continuous linear functional on $H_D^1(\Omega)$. We want to extend R to a continuous linear functional on the larger space $H_{\mathcal{T}}$. To this end, we associate with every edge $E \in \mathcal{E}$ a smooth function γ_E. The choice of γ_E is arbitrary subject to the constraint that $\gamma_E = g|_E$ for all $E \in \mathcal{E}_{\Gamma_N}$. The particular choice of the fluxes γ_E for the inter-element boundaries will later on determine the error estimation method; for $E \subset \Gamma_D$ the value of γ_E is completely irrelevant. Once we have chosen the fluxes γ_E, we can associate with every element $K \in \mathcal{T}$ a function γ_K defined on ∂K such that for all $v \in H_{\mathcal{T}}$

$$\sum_{K\in\mathcal{T}} \int_{\partial K} \gamma_K v = \sum_{E\in\mathcal{E}} \int_E \gamma_E \mathbb{J}_E(v).$$

Here, we use the convention that $\mathbb{J}_E(v) = v$ if $E \subset \Gamma$. We then define the extension $\widetilde{R}$ of R to $H_{\mathcal{T}}$ by

$$\langle \widetilde{R}, v \rangle = \sum_{K\in\mathcal{T}} \left\{ \int_K fv - \int_K \nabla u_{\mathcal{T}} \cdot \nabla v + \int_{\partial K} \gamma_K v \right\} - \sum_{E\in\mathcal{E}_\Omega} \int_E \gamma_E \mathbb{J}_E(v)$$

for all $v \in H_{\mathcal{T}}$.

1.10.3 *Characterisation of the Extended Residual as the Saddle-Point of a Lagrange Functional*

Next, we define a functional μ_* by

$$\mu_*(v) = \sum_{E \in \mathcal{E}_\Omega} \int_E \gamma_E \mathbb{J}_E(v)$$

for all $v \in H_{\mathcal{T}}$. Since for all $v \in H^1_D(\Omega)$

$$\sum_{E \in \mathcal{E}_\Omega} \int_E \gamma_E \mathbb{J}_E(v) = 0,$$

we have $\mu_* \in M$. Using this functional, the extension $\widetilde{R}$ of the residual R can be written as

$$\langle \widetilde{R}, v \rangle = \sum_{K \in \mathcal{T}} \left\{ \int_K fv - \int_K \nabla u_{\mathcal{T}} \cdot \nabla v + \int_{\partial K} \gamma_K v \right\} - \mu_*(v).$$

Now, we define a Lagrange functional $\mathcal{L}$ on $H_{\mathcal{T}} \times M$ by

$$\mathcal{L}(v, \mu) = \frac{1}{2} \sum_{K \in \mathcal{T}} \int_K \nabla v \cdot \nabla v - \langle \widetilde{R}, v \rangle - \mu(v).$$

Due to equation (1.61), M is the space of Lagrange multipliers for the constraint

$$\mathbb{J}_E(v) = 0 \text{ for all } E \in \mathcal{E}_\Omega.$$

This implies that

$$-\frac{1}{2} \left\| \nabla(u - u_{\mathcal{T}}) \right\|^2 = \inf_{v \in H^1_D(\Omega)} J(v) = \inf_{w \in H_{\mathcal{T}}} \sup_{\mu \in M} \mathcal{L}(w, \mu) = \sup_{\mu \in M} \inf_{w \in H_{\mathcal{T}}} \mathcal{L}(w, \mu).$$

Hence, we get for all $\mu \in M$

$$\begin{aligned} -\frac{1}{2} \left\| \nabla(u - u_{\mathcal{T}}) \right\|^2 &\geq \inf_{w \in H_{\mathcal{T}}} \mathcal{L}(w, \mu) \\ &= \inf_{w \in H_{\mathcal{T}}} \left\{ \sum_{K \in \mathcal{T}} \left[\frac{1}{2} \int_K \nabla w \cdot \nabla w - \int_K fw + \int_K \nabla u_{\mathcal{T}} \cdot \nabla w - \int_{\partial K} \gamma_K w \right] \right. \\ &\qquad \left. + \mu_*(w) - \mu(w) \right\}. \end{aligned}$$

The particular choice $\mu = \mu_*$ therefore yields

$$\left\| \nabla(u - u_{\mathcal{T}}) \right\|^2 \leq -2 \inf_{w \in H_{\mathcal{T}}} \sum_{K \in \mathcal{T}} \left\{ \frac{1}{2} \int_K \nabla w \cdot \nabla w - \int_K fw + \int_K \nabla u_{\mathcal{T}} \cdot \nabla w - \int_{\partial K} \gamma_K w \right\}.$$

1.10.4 Localisation of the Quadratic Functional

In order to write the last identity in a more compact form, we denote by H_K the restriction of $H_D^1(\Omega)$ to a single element $K \in \mathcal{T}$ and set for all $w \in H_K$ and all $K \in \mathcal{T}$

$$J_K(w) = \frac{1}{2}\int_K \nabla w \cdot \nabla w - \int_K fw + \int_K \nabla u_{\mathcal{T}} \cdot \nabla w - \int_{\partial K} \gamma_K w.$$

We then have $H_{\mathcal{T}} \cong \prod_{K\in\mathcal{T}} H_K$ and therefore

$$\left\|\nabla(u - u_{\mathcal{T}})\right\|^2 \leq -2\sum_{K\in\mathcal{T}} \inf_{w_K\in H_K} J_K(w_K). \tag{1.62}$$

Estimate (1.62) reduces the computation of the energy norm of the error to a family of minimisation problems on the elements in $\mathcal{T}$. Yet, for each $K \in \mathcal{T}$, the corresponding variational problem is still infinite-dimensional. In order to overcome this difficulty, we first rewrite J_K and set for every $K \in \mathcal{T}$

$$r_K = f + \Delta u_{\mathcal{T}}, \quad j_K = \gamma_K - \mathbf{n}_K \cdot \nabla u_{\mathcal{T}}.$$

Using integration by parts, we obtain for every $K \in \mathcal{T}$

$$J_K(w) = \frac{1}{2}\int_K \nabla w \cdot \nabla w - \int_K r_K w - \int_{\partial K} j_K w.$$

Next, for every element $K \in \mathcal{T}$ we define

$$Y_K = \Big\{\tau \in L^2(K)^2 : \operatorname{div}\tau \in L^2(K), \\ \operatorname{div}\tau = r_K, \mathbf{n}_K \cdot \tau = j_K\Big\}. \tag{1.63}$$

The complementary energy principle [**90**, Example I.3.4] then tells us that

$$\inf_{w\in H_K} J_K(w) = \sup_{\tau\in Y_K}\left(-\frac{1}{2}\int_K \tau\cdot\tau\right).$$

Together with inequality (1.62) this implies that

$$\begin{aligned}\left\|\nabla(u - u_{\mathcal{T}})\right\|^2 &\leq -2\sum_{K\in\mathcal{T}} \sup_{\tau\in Y_K}\left(-\frac{1}{2}\int_K \tau\cdot\tau\right)\\ &= \sum_{K\in\mathcal{T}} \inf_{\tau\in Y_K}\left(\int_K \tau\cdot\tau\right).\end{aligned}$$

This is the announced dual variational principle.

1.10.5 A Posteriori Error Estimates

The dual variational principle implies the following a posteriori error estimate.

Theorem 1.33 *Denote by u and $u_{\mathcal{T}}$ the unique solutions of problems* (1.2) *(p. 4) and* (1.3) *(p. 5), respectively. For every element $K \in \mathcal{T}$ define the space Y_K by equation* (1.63). *Then the a posteriori error estimate*

$$\left\|\nabla(u - u_{\mathcal{T}})\right\|^2 \leq \sum_{K\in\mathcal{T}} \int_K \tau_K \cdot \tau_K$$

holds for all $\tau_K \in Y_K$ and all $K \in \mathcal{T}$.

Remark 1.34 Notice that, contrary to all previous results, Theorem 1.33 only provides an upper bound for the error. The concrete realisation of the equilibrated residual method depends on the choice of the τ_K and on the definition of the γ_K. P. Ladevèze and D. Leguillon [**172**] choose γ_K to be the average of $\mathbf{n}_K \cdot \nabla u_{\mathcal{T}}$ from the neighbouring elements plus a suitable piecewise linear function on ∂K. The functions τ_K are often chosen as the solution of a minimisation problem on a finite-dimensional subspace of Y_K corresponding to higher order finite elements. With a proper choice of γ_K and τ_K one may thus recover the error indicator $\eta_{N,K}$ of Theorem 1.20 (p. 30) and the indicator $\rho_{\mathcal{T}}$ of Theorems 1.41 (p. 52) and 3.63 (p. 144). Further results can be found in [**9, 10, 11, 13, 122, 160, 186, 208**].

1.11 Dual Weighted Residuals

1.11.1 Motivation

In this section we consider the following problem:

> given the solutions u and $u_{\mathcal{T}}$ of problems (1.2) (p. 4) and (1.3) (p. 5) and a continuous linear functional $\ell : H_D^1(\Omega) \to \mathbb{R}$ compute the error
>
> $$|\langle \ell\, , u\rangle - \langle \ell\, , u_{\mathcal{T}}\rangle| \,. \tag{1.64}$$

Denoting by $\|\ell\|_*$ the norm of ℓ, we have

$$|\langle \ell\, , u\rangle - \langle \ell\, , u_{\mathcal{T}}\rangle| \leq \|\ell\|_* \left\|\nabla(u - u_{\mathcal{T}})\right\|$$

so that all reliable error indicators of the previous sections yield upper bounds for the quantity (1.64). The purpose of the *dual weighted residual method* is to bound the error (1.64) directly without estimating the energy error $\left\|\nabla(u - u_{\mathcal{T}})\right\|$. The hope is to thus obtain a mesh and a discrete solution which is better adapted to $\langle \ell\, , u\rangle$, the *quantity of interest*.

Here, we want to stress an important difference from the preceding sections:

> *for a general linear functional ℓ we cannot expect lower bounds for the quantity* (1.64) *since $u - u_{\mathcal{T}}$ may be in the kernel of ℓ although $u \neq u_{\mathcal{T}}$.*

1.11.2 Some Examples of Quantities of Interest

Before describing the basic idea of the dual weighted residual method, we present three typical examples for linear functionals ℓ.

Example 1.35 Given the solutions u and $u_{\mathcal{T}}$ of problems (1.2) and (1.3), set for all $v \in H_D^1(\Omega)$

$$\langle \ell , v\rangle = \frac{1}{\left\| \nabla(u - u_{\mathcal{T}})\right\|} \int_\Omega \nabla(u - u_{\mathcal{T}}) \cdot \nabla v.$$

Obviously, the quantity (1.64) is the energy error $\left\| \nabla(u - u_{\mathcal{T}})\right\|$.

Example 1.36 Given the solutions u and $u_{\mathcal{T}}$ of problems (1.2) and (1.3), set for all $v \in L^2(\Omega)$

$$\langle \ell , v\rangle = \frac{1}{\| u - u_{\mathcal{T}}\|} \int_\Omega (u - u_{\mathcal{T}})v.$$

Obviously, the quantity (1.64) now is the L^2-error $\| u - u_{\mathcal{T}}\|$.

Example 1.37 Fix a point x^* in Ω and a positive constant ε such that $B(x^*, \varepsilon)$, the circle with radius ε centred at x^*, is contained in Ω. Set for all $v \in L^1(\Omega)$

$$\langle \ell , v\rangle = \frac{1}{\mu_2(B(x^*, \varepsilon))} \int_{B(x^*,\varepsilon)} v.$$

The quantity $\langle \ell , v\rangle$ is a regularised approximation of the point value $v(x^*)$.

1.11.3 *The Associated Dual Variational Problem*

Due to the Lax–Milgram lemma [**109**, Theorem 1.1], the following *dual problem* admits a unique solution:

find $w \in H_D^1(\Omega)$ such that

$$\int_\Omega \nabla v \cdot \nabla w = \langle \ell , v\rangle \tag{1.65}$$

holds for all $v \in H_D^1(\Omega)$.

Now, we insert the error $u - u_{\mathcal{T}}$ in equation (1.65) and obtain

$$\langle \ell , u\rangle - \langle \ell , u_{\mathcal{T}}\rangle = \int_\Omega \nabla(u - u_{\mathcal{T}}) \cdot \nabla w.$$

The definition (1.8) (p. 10) of the residual, the Galerkin orthogonality (1.12) (p. 11), and the L^2-representation (1.15) (p. 11) of the residual then imply for *every* function $w_{\mathcal{T}} \in S_D^{1,0}(\mathcal{T})$

$$\langle \ell , u\rangle - \langle \ell , u_{\mathcal{T}}\rangle = \int_\Omega r(w - w_{\mathcal{T}}) + \int_\Sigma j(w - w_{\mathcal{T}}).$$

1.11.4 *A Posteriori Error Estimates*

Applying a weighted Cauchy–Schwarz inequality to the last identity yields the basic error estimate of the dual weighted residual method.

Theorem 1.38 *Denote by u and $u_{\mathcal{T}}$ the unique solutions of problems* (1.2) *(p. 4) and* (1.3) *(p. 5), respectively, by w the unique solution of the dual problem* (1.65)*, and by $w_{\mathcal{T}} \in S_D^{1,0}(\mathcal{T})$ any finite element approximation of w. Then the following error estimate holds*

$$|\langle \ell, u\rangle - \langle \ell, u_{\mathcal{T}}\rangle| \leq \sum_{K\in\mathcal{T}} \rho_K \gamma_K$$

with

$$\rho_K = \left\{\|r\|_K^2 + h_K^{-1}\,\|j\|_{\partial K}^2\right\}^{\frac{1}{2}},$$

$$\gamma_K = \left\{\|w - w_{\mathcal{T}}\|_K^2 + h_K\,\|w - w_{\mathcal{T}}\|_{\partial K}^2\right\}^{\frac{1}{2}},$$

where the residuals r and j are given by equations (1.13) *(p. 11) and* (1.14) *(p. 11), respectively.*

1.11.5 Realisation of the Method

In order to make the dual weighted residual method operational, we must explicitly compute or bound the weights γ_K in Theorem 1.38. There are two main approaches to this task.

- Compute an approximation $\widetilde{w}$ to w by solving a discrete analogue of the dual problem (1.65) with either continuous piecewise (bi-)quadratic finite element functions corresponding to $\mathcal{T}$ or continuous piecewise (bi-)linear finite element functions corresponding to a refinement of $\mathcal{T}$. For the computation of the weights γ_K, we then replace w by $\widetilde{w}$ and choose $w_{\mathcal{T}}$ as the linear interpolate of $\widetilde{w}$ in the vertices $\mathcal{N}$ of $\mathcal{T}$.
- Choose $w_{\mathcal{T}}$ as a suitable interpolate of w, use an interpolation error estimate to bound the weights γ_K in terms of a suitable higher order Sobolev norm of w, and invoke a regularity estimate for the dual problem to bound the Sobolev norm of w by the norm of ℓ.

To make this more precise, consider Examples 1.35–1.37.

For Example 1.35 we have $\|\nabla w\| \leq 1$ and choose $w_{\mathcal{T}} = I_{\mathcal{T}} w$ with the quasi-interpolation operator $I_{\mathcal{T}}$ of equation (1.4) (p. 7). Proposition 1.3 (p. 7) then implies

$$\gamma_K \leq \max\{c_{A,1}, c_{A,2}\}\, h_K\, \|\nabla w\|_{\widetilde{\omega}_K}.$$

The Cauchy–Schwarz inequality therefore yields

$$\sum_{K\in\mathcal{T}} \rho_K\gamma_K \leq c\left\{\sum_{K\in\mathcal{T}} h_K^2 \rho_K^2\right\}^{\frac{1}{2}}$$

so that we recover the residual a posteriori error estimate of Theorem 1.5 (p. 15).

In Example 1.36 the norm of ℓ equals the constant c_F in *Friedrichs' inequality*, cf. [**3**] and Section 3.4.7 (p. 105)

$$\|v\| \leq c_F\,\|\nabla v\| \quad \text{for all } v \in H_D^1(\Omega). \tag{1.66}$$

If Ω is convex, the second-order weak derivatives of w are all contained in $L^2(\Omega)$ and fulfil the regularity estimate $\left\|\nabla^2 w\right\| \leq c_\Omega c_F$ [**147**]. We choose $w_{\mathcal{T}}$ as the piecewise linear interpolate of w in the vertices $\mathcal{N}$, i.e. $w_{\mathcal{T}}(x) = w(x)$ for all $x \in \mathcal{N}$. A standard interpolation error estimate [**109**, Theorem 16.2] then implies

$$\gamma_K \leq ch_K^2 \left\| \nabla^2 w \right\|_K .$$

The Cauchy–Schwarz inequality therefore yields

$$\sum_{K \in \mathcal{T}} \rho_K \gamma_K \leq c \left\{ \sum_{K \in \mathcal{T}} h_K^4 \rho_K^2 \right\}^{\frac{1}{2}}$$

so that we gain an additional factor h_K element-wise when compared with the residual a posteriori error estimate of Theorem 1.5 (p. 15).

In Example 1.37 finally, the solution w of the dual problem (1.65) behaves like $\ln(\sqrt{|x - x^*|^2 + \varepsilon^2})$ [**35**, Example 3.3]. We choose $w_{\mathcal{T}}$ as for Example 1.36 and now obtain from [**109**, Theorem 16.2]

$$\gamma_K \leq c \frac{h_K^3}{\varepsilon^2 + \max_{x \in K} |x - x^*|^2}.$$

Thus elements 'far away' from x^* have a considerably smaller weight.

Remark 1.39 The first references to the dual weighted residual method probably are [**53**] and [**158**]. An overview of this approach may be found in [**13, 54, 224**] and in particular [**35**].

1.12 The Hyper-Circle Method

In this section we will frequently use the space $H(\text{div}; \Omega)$ of all vector fields in $L^2(\Omega)^2$ which have their divergence in $L^2(\Omega)$. The spaces $H_D^1(\Omega)$ and $H(\text{div}; \Omega)$ are closely related since, due to the Gauss theorem, the divergence is the adjoint operator of the gradient. We refer to Section 4.8 (p. 208) and [**90**, Section III.1] for properties of the space $H(\text{div}; \Omega)$. Here, we will only use the following result which is proved in the usual way using the definition of weak derivatives and integration by parts element-wise [**90**, Proposition III.1.2]: a vector field ρ, which is element-wise continuously differentiable, is in $H(\text{div}; \Omega)$ if and only if its normal components $\rho \cdot \mathbf{n}_E$ are continuous across interior edges.

1.12.1 *The Prager–Synge Theorem*

The hyper-circle method is based on the following result of W. Prager and J. L. Synge [**221**].

Proposition 1.40 *Denote by $u \in H_D^1(\Omega)$ the unique solution of problem (1.2) (p. 4) and by $\rho \in H(\text{div}; \Omega)$ any vector field with*

$$\begin{aligned} -\operatorname{div} \rho &= f \quad \text{in } \Omega \\ \rho \cdot \mathbf{n} &= g \quad \text{on } \Gamma_N. \end{aligned} \tag{1.67}$$

Then the identity

$$\|\nabla u - \nabla v\|^2 + \|\nabla u - \rho\|^2 = \|\nabla v - \rho\|^2$$

holds for all functions $v \in H_D^1(\Omega)$.

Proof Using integration by parts, equation (1.67) implies for every function $w \in H_D^1(\Omega)$

$$\begin{aligned}\int_\Omega fw + \int_{\Gamma_N} gw &= -\int_\Omega \operatorname{div} \rho w + \int_{\Gamma_N} \rho \cdot \mathbf{n} w \\ &= \int_\Omega \rho \cdot \nabla w.\end{aligned}$$

Combining this identity with equation (1.2) (p. 4) we obtain for every $w \in H_D^1(\Omega)$

$$\int_\Omega \nabla u \cdot \nabla w = \int_\Omega \rho \cdot \nabla w.$$

Now we consider an arbitrary function $v \in H_D^1(\Omega)$ and apply this identity to $w = u$ and $w = v$. We thus obtain

$$\begin{aligned}&\|\nabla u - \nabla v\|^2 + \|\nabla u - \rho\|^2 \\ &= 2\|\nabla u\|^2 - 2\int_\Omega \nabla u \cdot \nabla v + \|\nabla v\|^2 - 2\int_\Omega \rho \cdot \nabla u + \|\rho\|^2 \\ &= 2\int_\Omega \rho \cdot \nabla u - 2\int_\Omega \rho \cdot \nabla v + \|\nabla v\|^2 - 2\int_\Omega \rho \cdot \nabla u + \|\rho\|^2 \\ &= \|\nabla v\|^2 - 2\int_\Omega \rho \cdot \nabla v + \|\rho\|^2 \\ &= \|\nabla v - \rho\|^2.\end{aligned}$$

□

1.12.2 A General A Posteriori Error Estimate

If we apply Proposition 1.40 to the solution $u_{\mathcal{T}}$ of the discrete problem (1.3) (p. 5), we immediately obtain

$$\left\|\nabla(u - u_{\mathcal{T}})\right\| \leq \|\nabla u_{\mathcal{T}} - \rho\| \tag{1.68}$$

for *every* vector field $\rho \in H(\mathrm{div};\Omega)$ which satisfies equation (1.67). Therefore, for every such vector field, the quantity $\|\nabla u_{\mathcal{T}} - \rho\|$ is a reliable error indicator. Moreover, neither the error indicator nor the a posteriori error estimate (1.68) contain any hidden or unknown constant.

This is the main attraction of the hyper-circle method. Yet, to make it operational, we must find a vector field ρ which satisfies equation (1.67), which is easy to compute, and which makes the right-hand side of inequality (1.68) as small as possible. Apparently this is a non-trivial task. There are many more or less complex approaches to solve this problem. Here, we only present the simplest one. As a side-effect it reveals relations to the methods of Sections 1.7 (p. 25) and 1.10 (p. 41).

1.12.3 Construction of a Suitable Vector Field

The basic idea is to construct a piecewise linear vector field $\rho_{\mathcal{T}}$ such that

$$\begin{aligned}-\operatorname{div} \rho_{\mathcal{T}} &= \bar{f}_K && \text{on every } K \in \mathcal{T} \\ \mathbb{J}_E(\mathbf{n}_E \cdot \rho_{\mathcal{T}}) &= -\mathbb{J}_E(\mathbf{n}_E \cdot \nabla u_{\mathcal{T}}) && \text{on every } E \in \mathcal{E}_\Omega \\ \mathbf{n} \cdot \rho_{\mathcal{T}} &= \bar{g}_E - \mathbf{n} \cdot \nabla u_{\mathcal{T}} && \text{on every } E \in \mathcal{E}_{\Gamma_N}\end{aligned} \tag{1.69}$$

where $\bar{f}_K$ and $\bar{g}_E$ are given by equation (1.18) (p. 13). Then the vector field $\rho = \rho_{\mathcal{T}} + \nabla u_{\mathcal{T}}$ is contained in $H(\text{div}; \Omega)$. It satisfies equation (1.67) provided that

- $\mathcal{T}$ exclusively consists of triangles,
- f is piecewise constant, i.e. $f = f_{\mathcal{T}}$ with $f_{\mathcal{T}}$ given by equation (1.19) (p. 13),
- g is piecewise constant, i.e. $g = g_{\mathcal{E}}$ with $g_{\mathcal{E}}$ given by equation (1.19) (p. 13).

In order to simplify the presentation, we will from now on assume that these conditions are satisfied. Parallelograms could be treated by changing the definition (1.70) of the vector fields $\gamma_{K,E}$, cf. Lemma 3.1 (p. 87). General functions f and g introduce additional data errors as in Theorem 1.5 (p. 15).

For every triangle K and every edge E thereof, we denote by $a_{K,E}$ the vertex of K which is not contained in E and set

$$\gamma_{K,E}(x) = \frac{\mu_1(E)}{2\mu_2(K)}(x - a_{K,E}). \tag{1.70}$$

The vector fields $\gamma_{K,E}$ are the shape functions of the lowest order Raviart–Thomas space, cf. Section 4.8 (p. 208) and [**90**, Section III.3], and have the following properties

$$\begin{aligned} \text{div}\,\gamma_{K,E} &= \frac{\mu_1(E)}{\mu_2(K)} && \text{on } K, \\ \mathbf{n}_K \cdot \gamma_{K,E} &= 0 && \text{on } \partial K \setminus E, \\ \mathbf{n}_K \cdot \gamma_{K,E} &= 1 && \text{on } E, \\ \left\| \gamma_{K,E} \right\|_K &\leq c h_K, \end{aligned} \tag{1.71}$$

where $\mathbf{n}_K$ denotes the unit exterior normal of K and where the constant c only depends on the shape parameter of $\mathcal{T}$.

Before proceeding we want to check these properties.

The first property is obvious.

The second one follows from the fact that, for every point $x \in \partial K \setminus E$, the vector $x - a_{K,E}$ is tangential to the corresponding edge.

To verify the third property, we observe that $\mathbf{n}_K \cdot \gamma_{K,E}$ is constant on E and use the Gauss theorem and the first and second property to obtain

$$\mu_1(E) = \int_K \text{div}\,\gamma_{K,E} = \int_{\partial K} \mathbf{n}_K \cdot \gamma_{K,E} = \mu_1(E)(\mathbf{n}_K \cdot \gamma_{K,E})|_E.$$

To prove the last property, we denote by m_1, m_2, and m_3 the mid-points of the edges of K and recall that $\int_K \varphi = \frac{\mu_2(K)}{3} \sum_{i=1}^3 \varphi(m_i)$ holds for every quadratic function φ. The property of $\gamma_{K,E}$ then follows from the observation that $\left|\gamma_{K,E}(m_i)\right| \leq c$ holds for every i.

Now, we consider an arbitrary interior vertex $z \in \mathcal{N}_\Omega$. We enumerate the triangles in ω_z from 1 to n and the edges emanating from z from 0 to n such that (cf. Figure 1.5)

- $E_0 = E_n$,
- E_{i-1} and E_i are edges of K_i for every i.

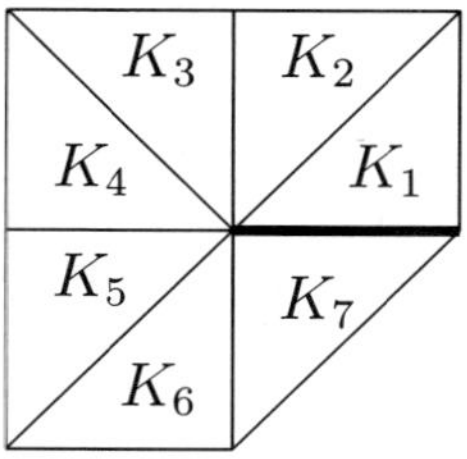

Figure 1.5 Enumeration of elements in ω_z with the edge E_n indicated by a bold line

We set

$$\alpha_0 = 0$$

and recursively for $i = 1, \ldots, n$

$$\alpha_i = -\frac{\mu_2(K_i)}{3\mu_1(E_i)} f|_{K_i} + \frac{\mu_1(E_{i-1})}{2\mu_1(E_i)} \mathbb{J}_{E_{i-1}}(\mathbf{n}_{E_{i-1}} \cdot \nabla u_{\mathcal{T}}) + \frac{\mu_1(E_{i-1})}{\mu_1(E_i)} \alpha_{i-1}.$$

By induction we obtain

$$\mu_1(E_n)\alpha_n = -\sum_{i=1}^{n} \frac{\mu_2(K_i)}{3} f|_{K_i} + \sum_{j=0}^{n-1} \frac{\mu_1(E_j)}{2} \mathbb{J}_{E_j}(\mathbf{n}_{E_j} \cdot \nabla u_{\mathcal{T}}).$$

Since

$$\int_{K_i} \lambda_z = \frac{\mu_2(K_i)}{3} \quad \text{and} \quad \int_{E_j} \lambda_z = \frac{\mu_1(E_j)}{2}$$

for every $i \in \{1, \ldots, n\}$ and for every $j \in \{0, \ldots, n-1\}$, we conclude from equations (1.12)–(1.15), using the assumption that f and g are piecewise constant and taking into account that $\Delta u_{\mathcal{T}} = 0$ element-wise, that

$$-\sum_{i=1}^{n} \frac{\mu_2(K_i)}{3} f|_{K_i} + \sum_{j=0}^{n-1} \frac{\mu_1(E_j)}{2} \mathbb{J}_{E_j}(\mathbf{n}_{E_j} \cdot \nabla u_{\mathcal{T}}) = -\int_{\Omega} r\lambda_z - \int_{\Sigma} j\lambda_z = 0.$$

Hence we have $\alpha_n = 0 = \alpha_0$. Therefore we can define a vector field ρ_z by setting for every $i \in \{1, \ldots, n\}$

$$\rho_z|_{K_i} = \alpha_i \gamma_{K_i,E_i} - \left(\frac{1}{2} \mathbb{J}_{E_{i-1}}(\mathbf{n}_{E_{i-1}} \cdot \nabla u_{\mathcal{T}}) + \alpha_{i-1} \right) \gamma_{K_i,E_{i-1}}. \tag{1.72}$$

Equations (1.71) and the definition of the α_i imply that

$$\begin{aligned} -\operatorname{div} \rho_z &= \frac{1}{3} f && \text{on } K_i \\ \mathbb{J}_{E_i}(\rho_z \cdot \mathbf{n}_{E_i}) &= -\frac{1}{2} \mathbb{J}_E(\mathbf{n}_{E_i} \cdot \nabla u_{\mathcal{T}}) && \text{on } E_i \end{aligned} \tag{1.73}$$

holds for every $i \in \{1, \ldots, n\}$.

For a vertex on the boundary Γ, the construction of ρ_z must be modified as follows.

- For every edge on the Neumann boundary Γ_N we must replace $-\mathbb{J}_E(\mathbf{n}_E \cdot \nabla u_{\mathcal{T}})$ by $g - \mathbf{n} \cdot \nabla u_{\mathcal{T}}$ as in equation (1.14) (p. 11).
- If z is a vertex on the Dirichlet boundary, there is at least one edge emanating from z which is contained in Γ_D. We must choose the enumeration of the edges such that E_n is one of these edges.

With these modifications, equations (1.72) and (1.73) carry over although in general $\alpha_n \neq 0$ for vertices on the boundary Γ.

In a final step, we extend the vector fields ρ_z by zero outside ω_z and set

$$\rho_{\mathcal{T}} = \sum_{z \in \mathcal{N}} \rho_z. \tag{1.74}$$

Since every triangle has three vertices and every edge has two vertices, we conclude from equations (1.73) that $\rho_{\mathcal{T}}$ has the desired properties (1.69) so that we may apply Proposition 1.40 to $\rho = \rho_{\mathcal{T}} + \nabla u_{\mathcal{T}}$.

The last inequality in (1.71), the definition of the α_i, and the observation that $\Delta u_{\mathcal{T}}$ vanishes element-wise imply that

$$\|\rho_z\|_{\omega_z} \leq c \left\{ \sum_{K \subset \omega_z} h_K^2 \|f + \Delta u_{\mathcal{T}}\|_K^2 + \sum_{E \subset \sigma_z \cap \Omega} h_E \|\mathbb{J}_E(\mathbf{n}_E \cdot \nabla u_{\mathcal{T}})\|_E^2 \right.$$
$$\left. + \sum_{E \subset \sigma_z \cap \Gamma_N} h_E \|g - \mathbf{n}_E \cdot \nabla u_{\mathcal{T}}\|_E^2 \right\}^{\frac{1}{2}}$$

holds for every vertex $z \in \mathcal{N}$ with a constant which only depends on the shape parameter of $\mathcal{T}$.

1.12.4 A Posteriori Error Estimates

The last estimate, Theorem 1.5 (p. 15), and inequality (1.68) prove the following a posteriori error estimates.

Theorem 1.41 *Assume that $\mathcal{T}$ exclusively consists of triangles and that the functions f and g are piecewise constant on $\mathcal{T}$ and $\mathcal{E}$, respectively. Denote by u and $u_{\mathcal{T}}$ the unique solutions of problems* (1.2) *(p. 4) and* (1.3) *(p. 5), respectively, and define the vector field $\rho_{\mathcal{T}}$ by equations* (1.72) *and* (1.74). *Then there is a constant c_* which only depends the shape parameter of $\mathcal{T}$ such that the estimates*

$$\|\nabla(u - u_{\mathcal{T}})\| \leq \|\rho_{\mathcal{T}}\|$$

and

$$\|\rho_{\mathcal{T}}\| \leq c_* \|\nabla(u - u_{\mathcal{T}})\|$$

are valid.

Remark 1.42 Notice that Theorem 1.41 provides a *global* lower bound for the error. There is a vast literature on a posteriori error estimates based on the hyper-circle method, e.g. [**83**, **186**, **202**,

227]. Further references may be found in [202]. The presentation of this section follows the approach of [79, Section 9] and [83]. It suggests that the hyper-circle method is structurally different from the methods of the preceding sections. This impression, however, is wrong. In Section 3.8.3 (p. 139) we will present an approach to the hyper-circle method which has the same structure as the methods of the preceding sections, in particular as the residual estimates of Section 1.4 (p. 10). This approach is based on the observation that every continuous linear functional on $H^1_D(\Omega)$ which has an L^2-representation analogous to equation (1.15) (p. 11) admits an explicit lifting to a space of piecewise $H(\mathrm{div})$-vector fields [272].

1.13 Efficiency and Asymptotic Exactness

The quality of an a posteriori error indicator is often measured by its *efficiency index*, i.e. the ratio of the estimated error to the true error. An error indicator is called *efficient* if its efficiency index together with its inverse remain bounded for all mesh-sizes. It is called *asymptotically exact* if its efficiency index tends to one when the mesh-size converges to zero.

1.13.1 Efficiency

In general we have

$$\left\{\sum_{K\in\mathcal{T}} h_K^2 \left\|f - \bar{f}_K\right\|_K^2\right\}^{\frac{1}{2}} = o(h)$$

and

$$\left\{\sum_{E\in\mathcal{E}_{\Gamma_N}} h_E \left\|g - \bar{g}_E\right\|_E^2\right\}^{\frac{1}{2}} = o(h),$$

where $h = \max_{K\in\mathcal{T}} h_K$ denotes the maximal mesh-size. On the other hand, the solutions of problems (1.2) (p. 4) and (1.3) (p. 5) satisfy

$$\left\|\nabla(u - u_{\mathcal{T}})\right\| \geq ch$$

always but in trivial cases. Hence, Theorems 1.5 (p. 15), 1.11 (p. 19), 1.15 (p. 24), 1.18 (p. 27), 1.19 (p. 29), 1.20 (p. 30), 1.23 (p. 34), and 1.31 (p. 38) imply that the corresponding error indicators are efficient. Their efficiency indices can in principle be estimated explicitly since the constants in the above theorems only depend on the constants in the quasi-interpolation error estimate 1.3 (p. 7) and the inverse inequalities 1.4 (p. 9) for which explicit bounds are derived in Sections 3.5 (p. 108) and 3.6 (p. 112).

1.13.2 Asymptotic Exactness and Super-Convergence

Using super-convergence results, one can also prove that on special meshes the error indicators of Sections 1.4–1.9 are asymptotically exact. As an example, we will show this for the indicator $\eta_{N,K}$ defined by equation (1.44) (p. 29). To do this we need the following notations.

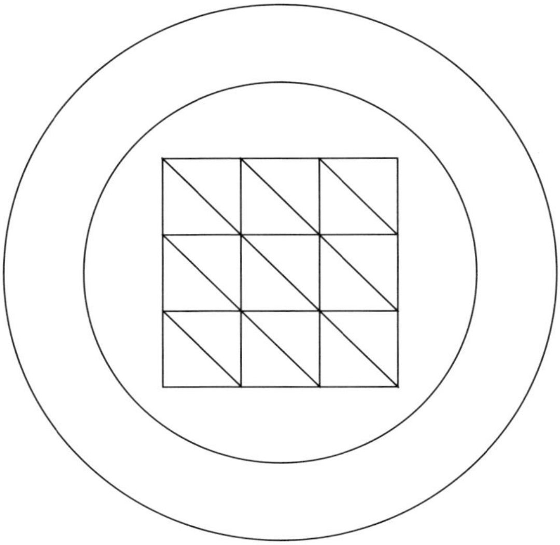

Figure 1.6 Domains Ω, Ω_1, and Ω_0 (from large to small) with parallel mesh in Ω_0

Assume that the partition $\mathcal{T}$ exclusively consists of triangles. Denote by Ω_0 and Ω_1 two open subsets of Ω such that $\overline{\Omega}_0 \subset \Omega_1$ and $\overline{\Omega}_0 = \bigcup_{K\subset\Omega_0} K$. The triangulation $\mathcal{T}$ is called *parallel* in Ω_0 if ω_E is a parallelogram for all edges $E \subset \Omega_0$, cf. Figure 1.6. Set

$$e = u - u_{\mathcal{T}}$$

where u and $u_{\mathcal{T}}$ are the solutions of problems (1.2) (p. 4) and (1.3) (p. 5). Denote by v_K, $K \in \mathcal{T}$, the unique solution of problem (1.45) (p. 29) and define $\varepsilon \in L^2(\Omega)$ and $\nabla_{\mathcal{T}}\varepsilon \in L^2(\Omega)^2$ by

$$\varepsilon|_K = v_K, \quad (\nabla_{\mathcal{T}}\varepsilon)|_K = \nabla v_K$$

for all $K \in \mathcal{T}$. The following super-convergence result is proved in [**280**]. Here, $H^k(\Omega)$ denotes the standard *Sobolev space* of all L^2-functions which have all their weak derivatives up to and including the order k in $L^2(\Omega)$, cf. Section 3.1 (p. 79).

Lemma 1.43 *Assume that $\mathcal{T}$ is parallel in Ω_0 and that $u|_{\Omega_1} \in H^3(\Omega_1)$. Then the following* super-convergence *result holds*

$$\left\|\nabla(i_{\mathcal{T}}u - u_{\mathcal{T}})\right\|_{\Omega_0} \leq c\left\{h^2 \left\|\nabla^3 u\right\|_{\Omega_1} + \|e\|_{\Omega_1}\right\}.$$

Here, $i_{\mathcal{T}}u \in S_D^{1,0}(\mathcal{T})$ denotes the piecewise linear interpolation of u in the nodes $\mathcal{N}$ which is defined by $i_{\mathcal{T}}u(z) = u(z)$ for all $z \in \mathcal{N}$.

With the help of Lemma 1.43 we can prove the following auxiliary result.

Lemma 1.44 *Assume that the conditions of Lemma 1.43 are satisfied. Then the estimate*

$$\|\nabla e - \nabla_{\mathcal{T}}\varepsilon\|_{\Omega_0} \le c\left\{h^2 \left\|\nabla^3 u\right\|_{\Omega_1} + \|e\|_{\Omega_1}\right\}$$

holds for sufficiently small h.

Proof Assume that h is small enough, such that $\omega_K \subset \Omega_1$ for all elements $K \in \mathcal{T}$ that are contained in Ω_0. Given an arbitrary element $K \in \mathcal{T}$ with $K \subset \Omega_0$ and an arbitrary function $\varphi \in H^2(\omega_K)$ we then denote by $v_\varphi \in V_K$ the unique solution of

$$\int_K \nabla v_\varphi \cdot \nabla w = -\int_K (\Delta\varphi)|_K w - \frac{1}{2}\sum_{E\in\mathcal{E}_K}\int_E \mathbb{J}_E(\mathbf{n}_E \cdot \nabla(i_{\mathcal{T}}\varphi))w \tag{1.75}$$

for all $w \in V_K$. Here, V_K is as in equation (1.43) (p. 29). With this notation, we obtain for all elements $K \in \mathcal{T}$ that are contained in Ω_0

$$\begin{aligned}\|\nabla e - \nabla_{\mathcal{T}}\varepsilon\|_K &= \|\nabla u - \nabla u_{\mathcal{T}} - \nabla v_K\|_K \\ &\le \left\|\nabla u - \nabla(i_{\mathcal{T}}u) - \nabla v_u\right\|_K + \left\|\nabla(i_{\mathcal{T}}u) - \nabla u_{\mathcal{T}}\right\|_K + \|\nabla v_u - \nabla v_K\|_K \\ &= S_{1,K} + S_{2,K} + S_{3,K}.\end{aligned} \tag{1.76}$$

Lemma 1.43 implies

$$\left\{\sum_{K\subset\Omega_0} S_{2,K}^2\right\}^{\frac{1}{2}} \le c\left\{h^2\left\|\nabla^3 u\right\|_{\Omega_1} + \|e\|_{\Omega_1}\right\}. \tag{1.77}$$

Consider an arbitrary element $K \in \mathcal{T}$ which is contained in Ω_0 and an arbitrary quadratic polynomial φ on ω_K. We claim that

$$v_\varphi = \varphi - i_{\mathcal{T}}\varphi \quad \text{on } K. \tag{1.78}$$

To prove this, we first observe that $(\varphi - i_{\mathcal{T}}\varphi)|_K \in V_K$ for every element K. Next, we recall from [**128**] that

$$\mathbf{n}_K \cdot \nabla(\varphi - i_{\mathcal{T}}\varphi)(z_E) = -\frac{1}{2}\mathbb{J}_E(\mathbf{n}_E \cdot \nabla(i_{\mathcal{T}}\varphi)) \tag{1.79}$$

holds for all edges $E \in \mathcal{E}_K$, where z_E denotes the midpoint of E. The fact that $\Delta\varphi$ is constant on K, integration by parts, application of Simpson's rule, and equation (1.79) imply for all $w \in V_K$, recalling that w vanishes at the vertices of K,

$$\begin{aligned}\int_K \nabla(\varphi - i_{\mathcal{T}}\varphi)\cdot\nabla w &= -\int_K \Delta(\varphi - i_{\mathcal{T}}\varphi)w + \sum_{E\in\mathcal{E}_K}\int_E \mathbf{n}_K\cdot\nabla(\varphi - i_{\mathcal{T}}\varphi)w \\ &= -\int_K (\Delta\varphi)|_K w + \sum_{E\in\mathcal{E}_K}\frac{2}{3}h_E\mathbf{n}_K\cdot\nabla(\varphi - i_{\mathcal{T}}\varphi)(z_E)w(z_E) \\ &= -\int_K (\Delta\varphi)|_K w - \frac{1}{2}\sum_{E\in\mathcal{E}_K}\int_E \mathbb{J}_E(\mathbf{n}_E\cdot\nabla(i_{\mathcal{T}}\varphi))w.\end{aligned}$$

Since the solution of problem (1.75) is unique, this proves equation (1.78).

Inequalities (1.39) (p. 27) and a standard local inverse estimate imply that

$$\left\| \nabla v_\varphi \right\|_K \le c h_K \left\| \nabla^2 \varphi \right\|_{\omega_K}$$

and thus

$$\left\| \nabla \varphi - \nabla(i_{\mathcal{T}} \varphi) - \nabla v_\varphi \right\|_K \le c h_K \left\| \nabla^2 \varphi \right\|_{\omega_K} \tag{1.80}$$

hold for all $\varphi \in H^2(\omega_K)$. Equation (1.78), inequality (1.80), and the Bramble–Hilbert lemma [**109**, Theorem 28.1] yield for all $\varphi \in H^3(\omega_K)$

$$\left\| \nabla \varphi - \nabla(i_{\mathcal{T}} \varphi) - \nabla v_\varphi \right\|_K \le c h_K^2 \left\| \nabla^3 \varphi \right\|_{\omega_K}. \tag{1.81}$$

Applying estimate (1.81) to u gives

$$\left\{ \sum_{K \subset \Omega_0} S_{1,K}^2 \right\}^{\frac{1}{2}} \le c h^2 \left\| \nabla^3 u \right\|_{\Omega_1}. \tag{1.82}$$

Equations (1.45) (p. 29) and (1.75) finally imply that for all $w \in V_K$

$$\int_K \nabla(v_u - v_K) \cdot \nabla w = \frac{1}{2} \sum_{E \in \mathcal{E}_K} \int_E \mathbb{J}_E(\mathbf{n}_E \cdot \nabla(u_{\mathcal{T}} - i_{\mathcal{T}} u)) w.$$

Together with inequalities (1.39) (p. 27) and a standard local inverse estimate this shows that

$$\| \nabla v_u - \nabla v_K \|_K \le c \left\| \nabla(u_{\mathcal{T}} - i_{\mathcal{T}} u) \right\|_{\omega_K}. \tag{1.83}$$

Estimate (1.83) and Lemma 1.43 yield

$$\left\{ \sum_{K \subset \Omega_0} S_{3,K}^2 \right\}^{\frac{1}{2}} \le c \left\{ h^2 \left\| \nabla^3 u \right\|_{\Omega_1} + \| e \|_{\Omega_1} \right\}. \tag{1.84}$$

Inequalities (1.76), (1.77), (1.82), and (1.84) establish the desired result. □

Under suitable assumptions, Lemma 1.44 implies the asymptotic exactness of $\eta_{N,K}$.

Theorem 1.45 *Assume that the partition $\mathcal{T}$ is parallel in Ω_0, that the solution u of the differential equation satisfies $u|_{\Omega_1} \in H^3(\Omega_1)$, and that the error fulfils $\| \nabla e \|_{\Omega_0} \ge ch$ and $\| e \|_{\Omega_1} \le ch^{1+\varepsilon}$ for some $\varepsilon > 0$. Then the error indicator $\eta_{N,K}$ of equation (1.44) (p. 29) is asymptotically exact, i.e.*

$$\left\{ \sum_{K \subset \Omega_0} \eta_{N,K}^2 \right\}^{\frac{1}{2}} \Big/ \| \nabla e \|_{\Omega_0} \xrightarrow[h \to 0]{} 1.$$

Proof Note that

$$\sum_{K \subset \Omega_0} \eta_{N,K}^2 = \|\nabla_{\mathcal{T}} \varepsilon\|_{\Omega_0}^2 .$$

Moreover, the assumptions and Lemma 1.44 imply that

$$\frac{\|\nabla e - \nabla_{\mathcal{T}} \varepsilon\|_{\Omega_0}}{\|\nabla e\|_{\Omega_0}} \leq c \left\{ h \left\| \nabla^3 u \right\|_{\Omega_1} + h^{\varepsilon} \right\} \underset{h \to 0}{\longrightarrow} 0. \qquad \square$$

Remark 1.46 The third assumption of Theorem 1.45 is satisfied in all but trivial cases. The fourth assumption of Theorem 1.45 follows from standard regularity results if $f \in L^2(\Omega)$ and g is the restriction to Γ_N of an H^1-function.

1.13.3 A Counter-Example

The following example shows that asymptotic exactness may not hold on general meshes even if they are strongly structured.

Example 1.47 Consider problem (1.1) (p. 4) on the square $\Omega = (0, 1)^2$ with $\Gamma_N = ((0, 1) \times \{0\}) \cup ((0, 1) \times \{1\})$, $g = 0$, and $f = 1$. The exact solution is $u(x, y) = \frac{1}{2}x(1 - x)$. The triangulation $\mathcal{T}$ is obtained as follows, cf. Figure 1.7: Ω is divided into n^2 squares with sides of length $h = \frac{1}{n}$ with $n \geq 1$; each square is cut into four triangles by drawing the two diagonals. This triangulation, which is often called a *criss-cross grid*, is obviously not parallel. Since the solution u of problem (1.1) (p. 4) is quadratic and the Neumann boundary conditions are homogeneous, one easily checks that the solution $u_{\mathcal{T}}$ of problem (1.3) (p. 5) is given by

$$u_{\mathcal{T}}(x) = \begin{cases} u(z) & \text{if } z \text{ is a vertex of a square,} \\ u(z) - \frac{h^2}{24} & \text{if } z \text{ is a midpoint of a square.} \end{cases}$$

Using this expression for $u_{\mathcal{T}}$, one can explicitly calculate the error and the error indicator. After some computations one obtains for any square Q, which is disjoint from Γ_N,

$$\left\{ \sum_{K \subset Q} \eta_{N,K}^2 \right\}^{\frac{1}{2}} \Big/ \, \|\nabla e\|_Q = \sqrt{\frac{17}{6}} \approx 1.68.$$

Hence, the error indicator cannot be asymptotically exact.

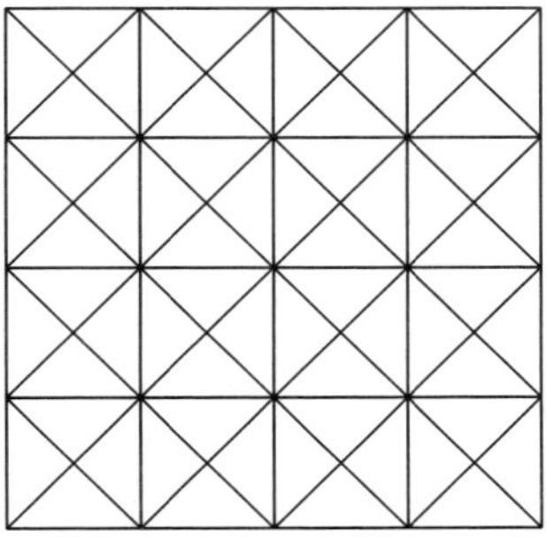

Figure 1.7 Triangulation of Example 1.47 corresponding to $n = 4$

Remark 1.48 The present analysis follows the lines of [**130**]. The asymptotic exactness of other error indicators is proved in [**45, 46, 128, 231**] using similar super-convergence results. An analogue of Example 1.47 is also established there. The efficiency of the residual error indicator $\eta_{R,K}$ is established in [**30**] by comparing the linear finite element solution to solutions obtained by the so-called p-version of the finite element method. This analysis also gives realistic bounds on the efficiency index of $\eta_{R,K}$. Different error indicators are compared in [**33, 194, 198**]. Further results on super-convergence phenomena can be found in [**164, 165, 276**].

1.14 Convergence of the Adaptive Process I

In this section we want to prove the convergence of Algorithm 1.1 (p. 2) based on the residual error indicator $\eta_{R,K}$ of Theorem 1.5 (p. 15) and a suitable marking strategy in step (4). To simplify the exposition, we consider problem 1.1 (p. 4) with pure Dirichlet conditions, i.e. we assume that the Neumann boundary Γ_N is empty. The present analysis will be refined and extended in Sections 4.13 (p. 264) and 6.8 (p. 362).

1.14.1 Piecewise Constant Right-Hand Sides

In a first step, we assume that the right-hand side f is piecewise constant on all partitions. This restriction will be overcome in the second step.

Consider a partition $\mathcal{T}_1$ of Ω and a *refinement* $\mathcal{T}_2$ of $\mathcal{T}_1$, i.e. every element in $\mathcal{T}_1$ is the union of elements in $\mathcal{T}_2$. The corresponding finite element spaces are then *nested*, i.e.

$$S_D^{1,0}(\mathcal{T}_1) \subset S_D^{1,0}(\mathcal{T}_2). \tag{1.85}$$

Denote by u and u_1, u_2 the unique solutions of problem (1.2) (p. 4) and of problem (1.3) (p. 5) corresponding to the partitions $\mathcal{T}_1$ and $\mathcal{T}_2$, respectively. Relation (1.85) and the Galerkin orthogonality (1.11) (p. 10) imply

$$\left\|\nabla(u-u_2)\right\|^2 = \left\|\nabla(u-u_1)\right\|^2 - \left\|\nabla(u_1-u_2)\right\|^2. \tag{1.86}$$

Since f is piecewise constant on $\mathcal{T}_1$, Theorem 1.5 (p. 15) yields

$$\left\|\nabla(u-u_1)\right\|^2 \le c^{*2} \sum_{K\in\mathcal{T}_1} \eta_{R,K}^2. \tag{1.87}$$

Now, we fix a parameter $\theta \in (0,1)$ and determine with Algorithm 2.2 (p. 65) a subset $\widetilde{\mathcal{T}}_1$ of $\mathcal{T}_1$ such that

$$\sum_{K\in\widetilde{\mathcal{T}}_1} \eta_{R,K}^2 \ge \theta \sum_{K\in\mathcal{T}_1} \eta_{R,K}^2. \tag{1.88}$$

Combining estimates (1.87) and (1.88), we obtain

$$\left\|\nabla(u-u_1)\right\|^2 \le \frac{c^{*2}}{\theta} \sum_{K\in\widetilde{\mathcal{T}}_1} \eta_{R,K}^2. \tag{1.89}$$

Next, we assume that the refined partition $\mathcal{T}_2$ satisfies the following conditions:

- the midpoint of every edge of every element in $\widetilde{\mathcal{T}}_1$ is a vertex of an element in $\mathcal{T}_2$;
- for every element in $\widetilde{\mathcal{T}}_1$ there is a point in its interior which is a vertex of an element in $\mathcal{T}_2$.

These conditions can be fulfilled by applying to every element in $\widetilde{\mathcal{T}}_1$ either two steps of the red refinement of Section 2.1.2 (p. 65), cf. the top left element in Figure 2.2 (p. 66), or three steps of the marked edge bisection of Section 2.1.4, cf. the bottom right element in Figure 2.5 (p. 69).

Thanks to these conditions, for every element K in $\widetilde{\mathcal{T}}_1$ and every edge thereof, we may replace the local cut-off functions ψ_E and ψ_K by the nodal shape functions of $S_D^{1,0}(\mathcal{T}_2)$ which correspond to the midpoint of E and the vertex of $\mathcal{T}_2$ in the interior of K, respectively. One easily checks that the proof of Proposition 1.4 (p. 9) carries over to these functions. Now, we replace in equations (1.20) (p. 13) and (1.22) (p. 14) the functions ψ_K and ψ_E by these nodal shape functions of $S_D^{1,0}(\mathcal{T}_2)$. The corresponding functions w_K and w_E are then contained in $S_D^{1,0}(\mathcal{T}_2)$. Therefore, we may replace u by u_2 in estimates (1.21) (p. 13) and (1.23) (p. 15). Moreover, the terms $f - \bar{f}_K$ on the right-hand sides of these estimates vanish since the function f is assumed to be piecewise constant on $\mathcal{T}_1$. The arguments of Section 1.4 therefore yield

$$\sum_{K\in\widetilde{\mathcal{T}}_1} \eta_{R,K}^2 \le c_*^2 \left\|\nabla(u_2 - u_1)\right\|^2 . \tag{1.90}$$

Estimates (1.89) and (1.90) imply

$$-\left\|\nabla(u_2 - u_1)\right\|^2 \le -\frac{1}{c_*^2} \sum_{K\in\widetilde{\mathcal{T}}_1} \eta_{R,K}^2 \le -\frac{\theta}{c_*^2 c^{*2}} \left\|\nabla(u - u_1)\right\|^2 .$$

Inserting this in estimate (1.86), we finally arrive at

$$\left\|\nabla(u - u_2)\right\|^2 \le \left(1 - \frac{\theta}{c_*^2 c^{*2}}\right) \left\|\nabla(u - u_1)\right\|^2 .$$

This proves the convergence of Algorithm 1.1 (p. 2) provided that the right-hand side f is piecewise constant on the coarsest partition $\mathcal{T}_0$. Every iteration of Algorithm 1.1 reduces the error at least by the factor $\sqrt{1 - \frac{\theta}{c_*^2 c^{*2}}}$ which only depends on the parameter θ and the shape parameter of $\mathcal{T}_0$.

1.14.2 General Right-Hand Sides

Now, we want to overcome the restriction to piecewise constant right-hand sides. Consider an arbitrary function $f \in L^2(\Omega)$ and an arbitrary partition $\mathcal{T}$ and define the piecewise constant function $f_{\mathcal{T}}$ by equation (1.19) (p. 13). Denote by u the unique solution of problem (1.2) (p. 4) with right-hand side f, by $\widetilde{u}$ the unique solution of problem (1.2) with right-hand side $f_{\mathcal{T}}$, and by $\widetilde{u}_{\mathcal{T}}$ the unique solution of the discrete problem (1.3) (p. 5) with right-hand side $f_{\mathcal{T}}$. Since $u_{\mathcal{T}}$, the solution of problem (1.3) (p. 5) with right-hand side f, is the best approximation to u in $S_D^{1,0}(\mathcal{T})$ with respect to the energy norm $\|\nabla\cdot\|$, we have

$$\left\|\nabla(u - u_{\mathcal{T}})\right\| \le \left\|\nabla(u - \widetilde{u}_{\mathcal{T}})\right\| .$$

Moreover, the triangle inequality yields

$$\left\| \nabla(u - \widetilde{u}_{\mathcal{T}}) \right\| \le \left\| \nabla(u - \widetilde{u}) \right\| + \left\| \nabla(\widetilde{u} - \widetilde{u}_{\mathcal{T}}) \right\|.$$

The arguments of the first part of this section can be applied to the term $\left\| \nabla(\widetilde{u} - \widetilde{u}_{\mathcal{T}}) \right\|$. Hence, we only have to control the term $\left\| \nabla(u - \widetilde{u}) \right\|$.

To do this, we insert $v = u - \widetilde{u}$ into equation (1.9) (p. 10) and obtain

$$\left\| \nabla(u - \widetilde{u}) \right\| = \sup_{w \in H_D^1(\Omega)\setminus\{0\}} \frac{1}{\|\nabla w\|} \int_\Omega \nabla(u - \widetilde{u}) \cdot \nabla w.$$

Taking into account that u and $\widetilde{u}$ solve problem (1.2) with right-hand sides f and $f_{\mathcal{T}}$, respectively, we conclude for every $w \in H_D^1(\Omega)$ that

$$\int_\Omega \nabla(u - \widetilde{u}) \cdot \nabla w = \int_\Omega (f - f_{\mathcal{T}}) w.$$

Since $f_{\mathcal{T}}$ is the L^2-projection of f onto the space of piecewise constant functions corresponding to $\mathcal{T}$, we have

$$\int_\Omega (f - f_{\mathcal{T}}) w = \int_\Omega (f - f_{\mathcal{T}})(w - w_{\mathcal{T}}) = \sum_{K \in \mathcal{T}} \int_K (f - \bar{f}_K)(w - \overline{w}_K).$$

The Cauchy–Schwarz inequality therefore yields

$$\int_\Omega \nabla(u - \widetilde{u}) \cdot \nabla w \le \sum_{K \in \mathcal{T}} \left\| f - \bar{f}_K \right\|_K \|w - \overline{w}_K\|_K .$$

Since every element K is convex, we know from Section 3.4 (p. 91) and [**50, 107, 217**] that

$$\|w - \overline{w}_K\|_K \le \frac{h_K}{\pi} \|\nabla w\|_K$$

holds for all elements. A weighted Cauchy–Schwarz inequality therefore implies

$$\left\| \nabla(u - \widetilde{u}) \right\| \le \frac{1}{\pi} \left\{ \sum_{K \in \mathcal{T}} h_K^2 \left\| f - \bar{f}_K \right\|_K^2 \right\}^{\frac{1}{2}}.$$

Hence, for general right-hand sides f, we have to control the right-hand side of this estimate which is usually referred to as *data oscillation*. This can be done with a modification of Algorithms 1.1 (p. 2) and 2.2 (p. 65) as follows.

As in the first part of this section, we consider a partition $\mathcal{T}_1$ and a refinement $\mathcal{T}_2$ of $\mathcal{T}_1$. Given a parameter $\theta \in (0, 1)$, we apply Algorithm 2.2 to $\mathcal{T}_1$ with η_K^2 replaced by $h_K^2 \left\| f - \bar{f}_K \right\|_K^2$ and thus determine a subset $\widetilde{\mathcal{T}}_1$ of $\mathcal{T}_1$ such that

$$\sum_{K \in \widetilde{\mathcal{T}}_1} h_K^2 \left\| f - \bar{f}_K \right\|^2 \ge \theta \sum_{K \in \mathcal{T}_1} h_K^2 \left\| f - \bar{f}_K \right\|_K^2 .$$

We now assume that $\mathcal{T}_2$ satisfies the following condition:

> every element K in $\widetilde{\mathcal{T}}_1$ is the union of elements in $\mathcal{T}_2$ such that each of these elements has a diameter $\frac{1}{2}h_K$ at most.

Again, this condition can be fulfilled by applying to every element K in $\widetilde{\mathcal{T}}_1$ either two steps of the red refinement of Section 2.1.2 (p. 65) or three steps of the marked edge bisection of Section 2.1.4 (p. 68).

Now, we split $\mathcal{T}_2$ into two disjoint subsets $\mathcal{T}_{2,R}$ and $\mathcal{T}_{2,U}$ such that $\bigcup_{K\in\mathcal{T}_{2,R}} K = \bigcup_{K\in\widetilde{\mathcal{T}}_1} K$. Then we have

$$\sum_{K\in\mathcal{T}_{2,U}} h_K^2 \left\| f - \bar{f}_K \right\|_K^2 \le \sum_{K\in\mathcal{T}_1\setminus\widetilde{\mathcal{T}}_1} h_K^2 \left\| f - \bar{f}_K \right\|_K^2 = \sum_{K\in\mathcal{T}_1} h_K^2 \left\| f - \bar{f}_K \right\|_K^2 - \sum_{K\in\widetilde{\mathcal{T}}_1} h_K^2 \left\| f - \bar{f}_K \right\|_K^2$$

and

$$\sum_{K\in\mathcal{T}_{2,R}} h_K^2 \left\| f - \bar{f}_K \right\|_K^2 \le \frac{1}{4} \sum_{K\in\widetilde{\mathcal{T}}_1} h_K^2 \left\| f - \bar{f}_K \right\|_K^2 .$$

This implies

$$\begin{aligned}
\sum_{K\in\mathcal{T}_2} h_K^2 \left\| f - \bar{f}_K \right\|_K^2 &= \sum_{K\in\mathcal{T}_{2,R}} h_K^2 \left\| f - \bar{f}_K \right\|_K^2 + \sum_{K\in\mathcal{T}_{2,U}} h_K^2 \left\| f - \bar{f}_K \right\|_K^2 \\
&\le \sum_{K\in\mathcal{T}_1} h_K^2 \left\| f - \bar{f}_K \right\|_K^2 - \frac{3}{4} \sum_{K\in\widetilde{\mathcal{T}}_1} h_K^2 \left\| f - \bar{f}_K \right\|_K^2 \\
&\le \left(1 - \frac{3\theta}{4}\right) \sum_{K\in\mathcal{T}_1} h_K^2 \left\| f - \bar{f}_K \right\|_K^2 .
\end{aligned}$$

We now apply Algorithm 1.1 (p. 2) with the following modifications:

- step (2) is omitted;
- in step (3), the a posteriori error estimate is replaced by the quantities $h_K \left\| f - \bar{f}_K \right\|_K$.

Every iteration then reduces the corresponding quantity $\left\| \nabla(u - \widetilde{u}) \right\|$ at least by the factor $\sqrt{1 - \frac{3\theta}{4}}$.

Given any fixed tolerance ε, after finitely many iterations, this algorithm furnishes a partition $\mathcal{T}$ with $\left\| \nabla(u - \widetilde{u}) \right\| \le \frac{\varepsilon}{2}$. This partition is used as starting point for the adaptive algorithm described in the first part of this section which then, after finitely many iterations, provides a refined partition $\mathcal{T}'$ with $\left\| \nabla(\widetilde{u} - \widetilde{u}_{\mathcal{T}'}) \right\| \le \frac{\varepsilon}{2}$ and therefore $\left\| \nabla(u - u_{\mathcal{T}'}) \right\| \le \varepsilon$.

Remark 1.49 The presentation of this section is based on the arguments of W. Dörfler [**126**] who was the first to prove the convergence of an adaptive finite element method based on a posteriori error estimates. The arguments of W. Dörfler were later generalised and improved. In particular, one nowadays controls the error indicator and the data oscillation simultaneously [**70, 196, 197, 198, 199, 200**].

Remark 1.50 The above results prove the convergence of Algorithm 1.1 (p. 2). Given any positive tolerance ε it provides a finest partition such that the corresponding finite element solution has an energy error at most ε. Yet, the arguments do not establish the *optimality* of the process. Due to general results of nonlinear approximation theory [**124**], this would require us to prove that the final partition produced by Algorithm 1.1 has approximately $\varepsilon^{-\frac{1}{2}}$ vertices. The optimality of the adaptive process can only be ensured by incorporating a suitable *coarsening strategy*, cf. Section 2.2 (p. 69) and [**70**]. Further optimality and convergence results for adaptive finite element methods can be found in [**70, 196, 197, 198, 199, 200**].

1.15 Summary and Outlook

In this chapter we have presented several a posteriori error indicators. All are *reliable*, i.e. they yield upper bounds for the error of the discrete solution. Most of them are also *efficient*, i.e. they yield lower bounds for the error.

The upper bounds were always global bounds, whereas the lower bounds were often local bounds. This is not by chance but has a structural reason. The upper bounds are based on the stability of the variational problem and use the inverse of the differential operator which is a global operator. The lower bounds, on the other hand, are based on the continuity of the variational problem and use the differential operator itself which is a local operator.

The crucial steps in deriving the a posteriori error estimates were:

- the equivalence of the error and a suitable dual norm of the residual, cf. equation (1.10) (p. 10);
- the Galerkin orthogonality of the error, cf. equations (1.11) and (1.12) (p. 11);
- an L^2-representation of the residual, cf. equations (1.13), (1.14), and (1.15) (p. 11);
- error estimates for a suitable quasi-interpolation operator, cf. Proposition 1.3 (p. 7);
- inverse estimates for suitable local cut-off functions, cf. Proposition 1.4 (p. 9).

The first step reveals a fundamental difference between a priori and a posteriori error estimates: a priori error estimates are based on the stability of the discrete problem; a posteriori error estimates require the stability of the infinite-dimensional variational problem.

The analysis of this chapter was performed for the simplest model problem. Some generalisations are obvious: general linear elliptic equations, cf. Chapter 4 (p. 151), nonlinear elliptic problems, cf. Chapter 5 (p. 281), and time-dependent linear and nonlinear parabolic equations, cf. Chapter 6 (p. 309). Here, we want to stress an important phenomenon which does not appear in the model problem: *robustness*. To understand this, we add a large reaction term to the left-hand side of problem (1.1) (p. 4) and consider the reaction–diffusion equation

$$\begin{aligned} -\Delta u + \kappa^2 u &= f && \text{in } \Omega \\ u &= 0 && \text{on } \Gamma_D \\ \frac{\partial u}{\partial n} &= g && \text{on } \Gamma_N \end{aligned} \tag{1.91}$$

with $\kappa \gg 1$. The variational formulation of problem (1.91) and its discretisation are obtained by adding the terms $\kappa^2 \int_\Omega uv$ and $\kappa^2 \int_\Omega u_\mathcal{T} v_\mathcal{T}$ to the left-hand sides of equations (1.2) (p. 4) and (1.3) (p. 5), respectively.

If we still measure the error with the norm $\|\nabla \cdot\|$, equation (1.10) (p. 10) now takes the form

$$\left\|\nabla(u-u_{\mathcal{T}})\right\| \leq \sup_{w\in H_D^1(\Omega)\setminus\{0\}} \frac{\langle R, w\rangle}{\|\nabla w\|} \leq (1+c_F^2\kappa^2)\left\|\nabla(u-u_{\mathcal{T}})\right\|,$$

where c_F is the constant in Friedrichs' inequality (1.66) (p. 47). Thus the relation between the norm of the error and the dual norm of the residual depends on the parameter κ. The remaining estimates of Section 1.4 (p. 10) immediately carry over to the present problem provided that we add $\kappa^2 u_{\mathcal{T}}$ to the residual r and to the element residual in $\eta_{R,K}$. Yet, due to the above estimate, the quantity $c^* c_*$ which measures the quality of the error indicator is now proportional to κ^2.

To avoid this undesirable phenomenon, we may decide to measure the error with the energy norm of problem (1.7) which is given by

$$|||v||| = \left\{\|\nabla v\|^2 + \kappa^2 \|v\|^2\right\}^{\frac{1}{2}}.$$

Equation (1.10) (p. 10) then takes the form

$$|||u-u_{\mathcal{T}}||| = \sup_{w\in H_D^1(\Omega)\setminus\{0\}} \frac{\langle R, w\rangle}{|||w|||}.$$

Hence, the energy norm of the error still equals the corresponding dual norm of the residual. Yet, now we have to prove Propositions 1.3 (p. 7) and 1.4 (p. 9) with the norm $\|\nabla\cdot\|$ replaced by the energy norm $|||\cdot|||$. If we do this, we observe that the constants $c_{A,1}$ and $c_{A,2}$ and $c_{I,1}, \ldots, c_{I,5}$ now depend on κ. Correspondingly, the product $c^* c_*$ of the constants in Theorem 1.5 (p. 15) is again proportional to κ^2.

Thus, both approaches yield upper and lower bounds for the residual error indicator which deteriorate with increasing κ, i.e. the residual error indicator is *not robust*. In Section 4.3 (p. 159) we will show that one may obtain a robust residual error indicator by suitably modifying the weighting factors of the element and edge residuals.

2

Implementation

In this chapter we show how to use a posteriori error indicators for devising self-adaptive discretisations. The main techniques are mesh-refinement, mesh-coarsening, and mesh-smoothing. We also present some sample data structures and a few examples demonstrating the benefits of an adaptive discretisation.

2.1 Mesh-Refinement

For step (4) of Algorithm 1.1 (p. 2), we must provide a device that constructs the next mesh $\mathcal{T}_{k+1}$ from the current mesh $\mathcal{T}_k$ disposing of an error indicator η_K for every element $K \in \mathcal{T}_k$. This requires two key ingredients:

- a *marking strategy* that decides which elements should be refined and
- *refinement rules* which determine the actual subdivision of a single element.

Since we want to ensure the admissibility of the next partition, we have to avoid *hanging nodes*, cf. Figure 2.1. Therefore, the refinement process will proceed in two stages.

- In the first stage we determine a subset $\widetilde{\mathcal{T}}_k$ of $\mathcal{T}_k$ consisting of all those elements that must be refined due to a large value of η_K. The refinement of these elements is usually called *regular*.
- In the second stage additional elements are refined in order to eliminate the hanging nodes which may have been created in the first stage. The refinement of these additional elements is sometimes referred to as *irregular*.

2.1.1 *Marking Strategies*

There are two popular marking strategies for determining the set $\widetilde{\mathcal{T}}_k$: the *maximum strategy* and the *equilibration strategy*, which is sometimes also called the *Dörfler strategy* or *bulk chasing*.

Algorithm 2.1 (Maximum strategy). *Given: a partition $\mathcal{T}$, error indicators η_K for the elements $K \in \mathcal{T}$, and a threshold $\theta \in (0, 1)$.*

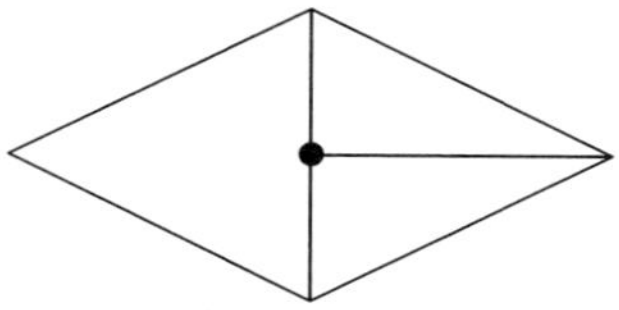

Figure 2.1 Hanging node •

Sought: a subset $\widetilde{\mathcal{T}}$ *of* marked *elements that should be refined.*

(1) *Compute* $\eta_{\mathcal{T},\max} = \max_{K\in\mathcal{T}} \eta_K$.

(2) *If* $\eta_K \geq \theta\eta_{\mathcal{T},\max}$ *mark K for refinement and put it into the set* $\widetilde{\mathcal{T}}$.

Algorithm 2.2 (Equilibration strategy). *Given: a partition* $\mathcal{T}$, *error indicators* η_K *for the elements* $K \in \mathcal{T}$, *and a threshold* $\theta \in (0, 1)$.

Sought: a subset $\widetilde{\mathcal{T}}$ *of* marked *elements that should be refined.*

(1) *Compute* $\Theta_{\mathcal{T}} = \sum_{K\in\mathcal{T}} \eta_K^2$. *Set* $\Sigma_{\mathcal{T}} = 0$ *and* $\widetilde{\mathcal{T}} = \emptyset$.

(2) *If* $\Sigma_{\mathcal{T}} \geq \theta\Theta_{\mathcal{T}}$ *return* $\widetilde{\mathcal{T}}$; stop. *Otherwise go to step* (3).

(3) *Compute* $\widetilde{\eta}_{\mathcal{T},\max} = \max_{K\in\mathcal{T}\setminus\widetilde{\mathcal{T}}} \eta_K$.

(4) *For all elements* $K \in \mathcal{T}\setminus\widetilde{\mathcal{T}}$ *check whether* $\eta_K = \widetilde{\eta}_{\mathcal{T},\max}$. *If this is the case, put K in* $\widetilde{\mathcal{T}}$ *and add* η_K^2 *to* $\Sigma_{\mathcal{T}}$. *Otherwise skip K. When all elements have been checked, return to step* (2).

At the end of this algorithm the set $\widetilde{\mathcal{T}}$ satisfies

$$\sum_{K\in\widetilde{\mathcal{T}}} \eta_K^2 \geq \theta \sum_{K\in\mathcal{T}} \eta_K^2.$$

Both marking strategies yield comparable results. The maximum strategy obviously is cheaper than the equilibration strategy. In the maximum strategy, a large value of θ leads to small sets $\widetilde{\mathcal{T}}$, i.e. very few elements are marked; a small value of θ leads to large sets $\widetilde{\mathcal{T}}$, i.e. nearly all elements are marked. For the equilibration strategy the effect is reversed: a small value of θ leads to a small set $\widetilde{\mathcal{T}}$; a large value of θ leads to a large set $\widetilde{\mathcal{T}}$. A popular and well-established choice for both strategies is $\theta \approx 0.5$.

In many applications, e.g. convection dominated problems as in Example 2.9 (p. 77), one encounters the difficulty that very few elements have an extremely large estimated error, whereas the remaining ones split into the vast majority with an extremely small estimated error and a third group of medium size consisting of elements which have an estimated error much less than the error of the elements in the first group and much larger than the error of the elements in the second group. In this situation, Algorithms 2.1 and 2.2 will only refine the elements of the first group. This deteriorates the performance of the adaptive algorithm. It can be enhanced substantially by the following simple modification.

> Given a small percentage ε, e.g. 0.05 or 0.1, we first mark the ε% elements with largest estimated error for refinement and then apply Algorithms 2.1 and 2.2 only to the remaining elements.

2.1.2 *Regular Refinement*

Elements that are marked for refinement are often refined by connecting their midpoints of edges. The resulting elements are called *red*.

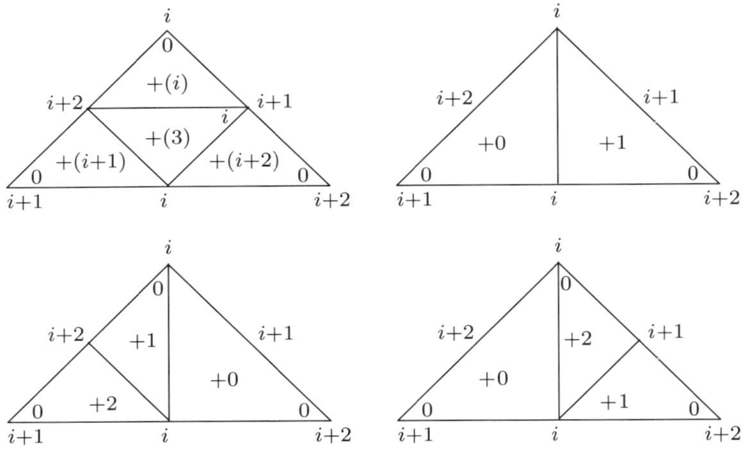

Figure 2.2 Refinement of triangles (red: top-left, green: top-right, blue: bottom) and enumeration of vertices, edges, and descendants

Triangles and quadrilaterals are thus subdivided into four smaller triangles and quadrilaterals that are similar to the parent element and have the same angles. Thus the shape parameter $\frac{h_K}{\rho_K}$ of the elements does not change, cf. Sections 1.3.1 (p. 5) and 3.2 (p. 81).

This refinement is illustrated by the top-left triangle of Figure 2.2 and by the top square of Figure 2.3. The numbers outside the elements indicate the local enumeration of edges and vertices of the parent element. The numbers inside the elements close to the vertices indicate the local enumeration of the vertices of the descendants. The numbers +0, +1, etc. inside the elements give the enumeration of the children.

Note that the enumeration of new elements and new vertices is chosen in such a way that triangles and quadrilaterals may be treated simultaneously without case selections.

Parallelepipeds are also subdivided into eight smaller similar parallelepipeds by joining the midpoints of edges.

For tetrahedrons, the situation is more complicated. Joining the midpoints of edges introduces four smaller similar tetrahedrons at the vertices of the parent tetrahedron plus a double pyramid in its interior. The latter one is subdivided into four small tetrahedrons by cutting it along two orthogonal planes. These tetrahedrons, however, are not similar to the parent tetrahedron. Yet, there are rules which determine the cutting planes such that a repeated refinement according to these rules leads to at most four similarity classes of elements originating from a parent element [47]. Thus these rules guarantee that the shape parameter of the partition does not deteriorate during a repeated adaptive refinement procedure.

2.1.3 *Irregular Refinement*

Since not all elements are refined regularly, we need additional refinement rules in order to avoid *hanging nodes*, cf. Figure 2.1 (p. 65), and to ensure the admissibility of the refined partition. These rules are illustrated in Figures 2.2 and 2.3.

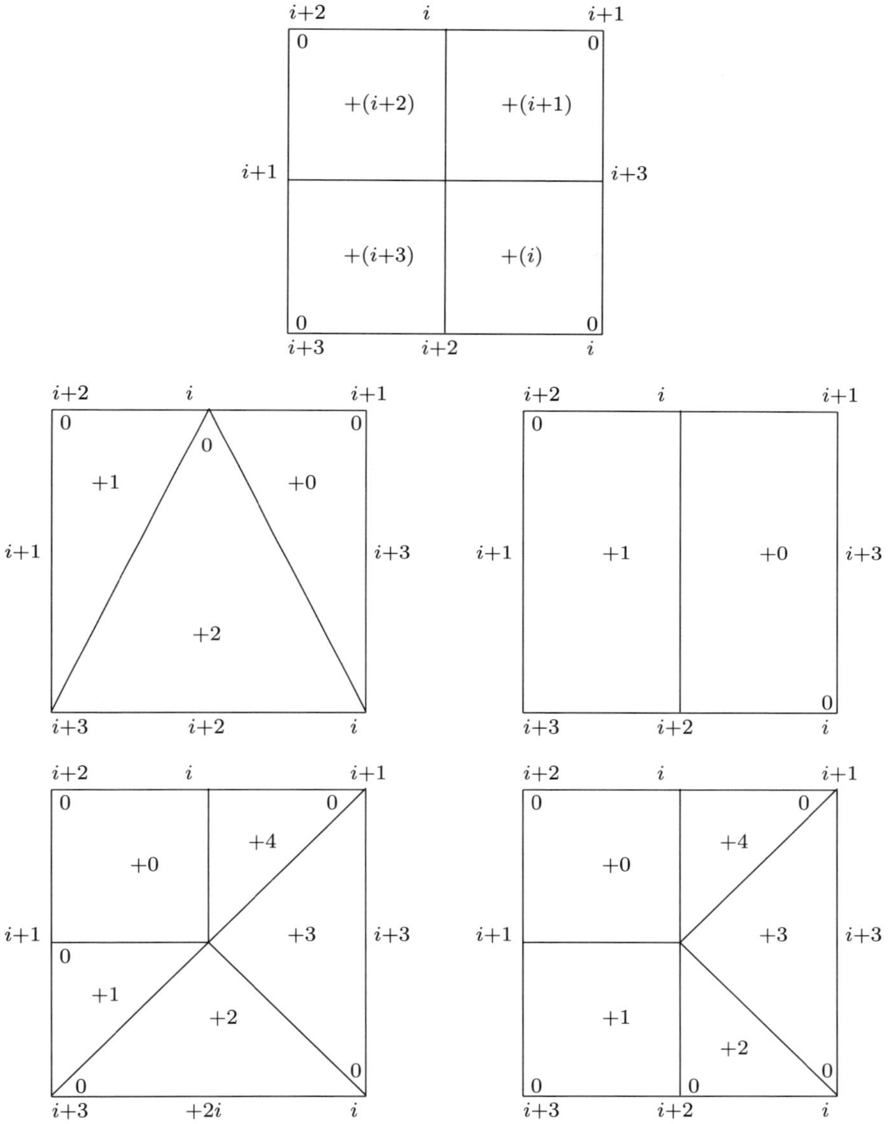

Figure 2.3 Refinement of quadrilaterals (red: top, green: middle-left, blue: middle-right and bottom-left, purple: bottom-right) and enumeration of vertices, edges, and descendants

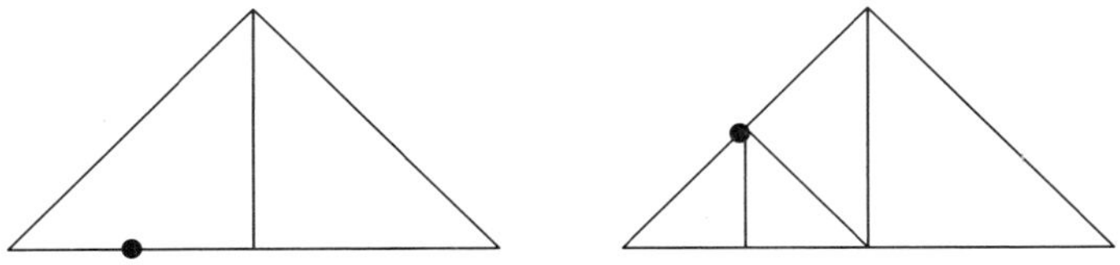

Figure 2.4 Forbidden green refinement and substituting blue refinement

For brevity we call the resulting elements *green, blue,* and *purple*. They are obtained as follows:

- a green element by bisecting exactly one edge;
- a blue element by bisecting exactly two edges;
- a purple quadrilateral by bisecting exactly three edges.

In order to avoid too acute or too obtuse triangles, the blue and green refinement of triangles obey the following two rules:

- In a blue refinement of a triangle, the longest one of the refinement edges is bisected first.
- Before performing a green refinement of a triangle it is checked whether the refinement edge is part of an edge which has been bisected during the last `ng`, e.g. 1 or 2, generations. If this is the case, a blue refinement is performed instead.

The second rule is illustrated in Figure 2.4. The bullet in the left part represents a hanging node which should be eliminated by a green refinement. The right part shows the blue refinement which is performed instead. Here, the bullet represents the new hanging node which is created by the blue refinement. Numerical experiments indicate that a reasonable value of `ng` is 1. Larger values result in an excessive blow-up of the refinement zone.

2.1.4 Marked Edge Bisection

The *marked edge bisection* is an alternative to the described regular red refinement which does not require additional refinement rules for avoiding hanging nodes. It is performed according to the following rules.

- The coarsest mesh $\mathcal{T}_0$ is constructed such that the longest edge of any element is also the longest edge of the adjacent element unless it is a boundary edge.
- The longest edges of the elements in $\mathcal{T}_0$ are marked.
- Given a partition $\mathcal{T}_k$ and an element thereof which should be refined, it is bisected by joining the midpoint of its marked edge with the vertex opposite to this edge.
- When bisecting the edge of an element, its two remaining edges become the marked edges of the two resulting new triangles.

This process is illustrated in Figure 2.5. The marked edges are labelled by •. The marked edge bisection is described in detail in [**191, 228, 229**].

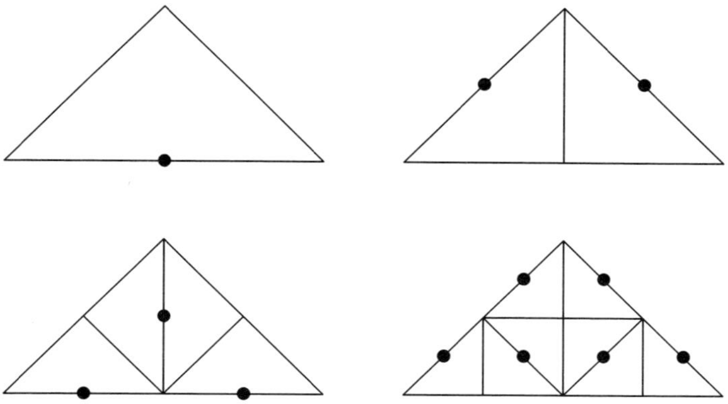

Figure 2.5 Subsequent marked edge bisection, the marked edges are labelled by •

2.2 Mesh-Coarsening

Algorithm 1.1 (p. 2) in combination with the marking strategies of Algorithms 2.1 and 2.2 (p. 65) produces a sequence of increasingly refined partitions. In many situations, however, some elements must be coarsened in the course of the adaptive process. For time-dependent problems this is obvious: a critical region, e.g. an interior layer, may move through the spatial domain in the course of time. For stationary problems this is less obvious. Yet, for elliptic problems one can prove that a possible coarsening is mandatory to ensure the optimal complexity of the adaptive process [**70**].

The basic idea of the coarsening process is to go back in the hierarchy of partitions and to cluster elements with too small an error. The following algorithm goes m generations backwards, accumulates the error indicators, and then advances $n > m$ generations using the marking strategies of Algorithms 2.1 and 2.2 (p. 65). For stationary problems, typical values are $m = 1$ and $n = 2$. For time-dependent problems one may choose $m > 1$ and $n > m + 1$ to enhance the temporal movement of the refinement zone.

Algorithm 2.3 *Given: a hierarchy $\mathcal{T}_0, \ldots, \mathcal{T}_k$ of adaptively refined partitions, error indicators η_K for the elements K of $\mathcal{T}_k$, and parameters $1 \le m \le k$ and $n > m$.*

Sought: a new partition $\mathcal{T}_{k-m+n}$.

(1) *For every element $K \in \mathcal{T}_{k-m}$ set $\widetilde{\eta}_K = 0$.*

(2) *For every element $K \in \mathcal{T}_k$ determine its ancestor $K' \in \mathcal{T}_{k-m}$ and add η_K^2 to $\widetilde{\eta}_{K'}^2$.*

(3) *Successively apply Algorithms 2.1 or 2.2 (p. 65) n times with $\widetilde{\eta}$ as error indicator. In this process, equally distribute $\widetilde{\eta}_K$ over the children of K once an element K is subdivided.*

The following algorithm is particularly suited for the marked edge bisection of Section 2.1.4. In the framework of Algorithm 2.3 its parameters are $m = 1$ and $n = 2$, i.e. it constructs the partition of the next level simultaneously refining and coarsening elements of the current partition. For its description we need some notation.

- An element K of the current partition $\mathcal{T}$ has *refinement level ℓ* if it is obtained by subdividing ℓ times an element of the coarsest partition.

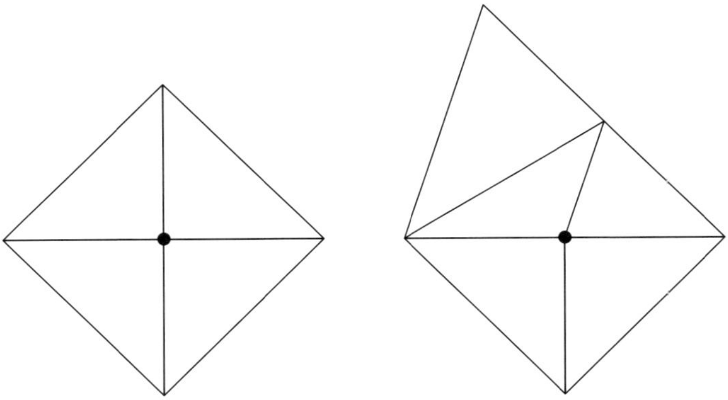

Figure 2.6 The vertex marked • is resolvable in the left patch but not in the right one

- Given a triangle K of the current partition $\mathcal{T}$ which is obtained by bisecting a parent triangle K', the vertex of K which is not a vertex of K' is called the *refinement vertex* of K.
- A vertex $z \in \mathcal{N}$ of the current partition $\mathcal{T}$ and the corresponding patch ω_z are called *resolvable*, cf. Figure 2.6, if
  - z is the refinement vertex of all elements contained in ω_z and
  - all elements contained in ω_z have the same refinement level.

Algorithm 2.4 *Given: a partition $\mathcal{T}$, error indicators η_K for all elements K of $\mathcal{T}$, and parameters $0 < \theta_1 < \theta_2 < 1$.*
Sought: subsets $\mathcal{T}_c$ and $\mathcal{T}_r$ of elements that should be coarsened and refined, respectively.

(1) *Set $\mathcal{T}_c = \emptyset$, $\mathcal{T}_r = \emptyset$ and compute $\eta_{\mathcal{T},\max} = \max_{K\in\mathcal{T}} \eta_K$.*

(2) *For all $K \in \mathcal{T}$ check whether $\eta_K \geq \theta_2 \eta_{\mathcal{T},\max}$. If this is the case, put K into $\mathcal{T}_r$.*

(3) *For all vertices $z \in \mathcal{N}$ check whether z is resolvable. If this is the case and if $\max_{K\subset\omega_z} \eta_K \leq \theta_1 \eta_{\mathcal{T},\max}$, put all elements contained in ω_z into $\mathcal{T}_c$.*

Remark 2.5 Algorithm 2.4 is obviously a modification of the maximum strategy of Algorithm 2.1 (p. 64). A coarsening of elements can also be incorporated in the equilibration strategy of Algorithm 2.2 (p. 65). We refer to [235] for a more detailed description.

2.3 Mesh-Smoothing

In this section we describe mesh-smoothing strategies which try to improve the quality of a partition while retaining its topological structure. The vertices of the partition are moved, but the number of elements and their adjacency remain unchanged. All strategies use a process similar to the well-known Gauss–Seidel algorithm to optimise a suitable *quality function q* over the class of all partitions having the same topological structure. They differ in the choice of the quality function. The strategies of this section do not replace the mesh-refinement methods of the previous sections, they complement them. In particular an improved partition may thus be obtained when a further refinement is impossible due to exhausted storage.

In order to simplify the presentation, we assume throughout this section that all partitions exclusively consist of triangles.

2.3.1 *The Optimisation Process*

We first describe the optimisation process. To this end we assume that we dispose of a quality function q which associates with every element a non-negative number such that a larger value of q indicates a better quality. Given a partition $\mathcal{T}$ we want to find an improved partition $\widetilde{\mathcal{T}}$ with the same number of elements and the same adjacency such that

$$\min_{\widetilde{K}\in\widetilde{\mathcal{T}}} q(\widetilde{K}) > \min_{K\in\mathcal{T}} q(K).$$

To this end we perform several iterations of the following *smoothing procedure* similar to Gauss–Seidel iteration:

> for every vertex z in the current partition $\mathcal{T}$, fix the vertices of $\partial\omega_z$ and find a new vertex $\widetilde{z}$ inside ω_z such that

$$\min_{\widetilde{K}\subset\omega_{\widetilde{z}}} q(\widetilde{K}) > \min_{K\subset\omega_z} q(K). \tag{2.1}$$

The practical solution of the local optimisation problem (2.1) depends on the choice of the quality function q. In what follows we will present three possible choices for q.

2.3.2 *A Quality Function Based on Geometrical Criteria*

The first choice is purely based on the geometry of the partitions and tries to obtain a partition which consists of equilateral triangles. To describe this approach, we enumerate the vertices and edges of a given triangle consecutively in counter-clockwise order from 0 to 2 such that edge i is opposite to vertex i, cf. Figure 2.2 (p. 66) and Section 2.4 (p. 74). Then edge i has the vertices $i+1$ and $i+2$ as its endpoints where all expressions have to be taken modulo 3. With these notations we define the *geometric quality function* q_G by

$$q_G(K) = \frac{4\sqrt{3}\mu_2(K)}{\mu_1(E_0)^2 + \mu_1(E_1)^2 + \mu_1(E_2)^2}.$$

The function q_G is normalised such that it attains its maximal value 1 for an equilateral triangle.

To obtain a more explicit representation of q_G and to solve problem (2.1), we denote by $x_0 = (x_{0,1}, x_{0,2})$, $x_1 = (x_{1,1}, x_{1,2})$, and $x_2 = (x_{2,1}, x_{2,2})$ the coordinates of the vertices. Then we have

$$\mu_2(K) = \frac{1}{2}\left\{(x_{1,1} - x_{0,1})(x_{2,2} - x_{0,2}) - (x_{2,1} - x_{0,1})(x_{1,2} - x_{0,2})\right\}$$

and

$$\mu_1(E_i)^2 = (x_{i+2,1} - x_{i+1,1})^2 + (x_{i+2,2} - x_{i+1,2})^2$$

for $i = 0, 1, 2$.

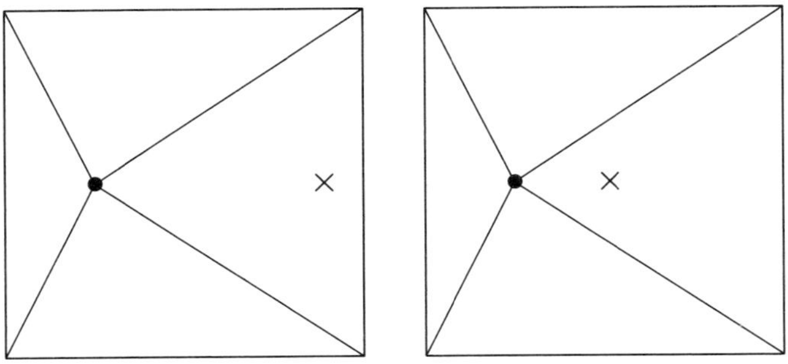

Figure 2.7 Points z (•) and z' (×) for solving the minimisation problem (2.1); left: first strategy, right: second strategy

There are two main possibilities to solve problem (2.1) for q_G. In the first approach, we determine a triangle K_1 in ω_z such that

$$q_G(K_1) = \min_{K \subset \omega_z} q_G(K)$$

and start the enumeration of its vertices at the vertex z. Then we determine a point z' such that the points z', x_1, and x_2 are the vertices of an equilateral triangle and that this enumeration of vertices is in counter-clockwise order. Now, we try to find a point $\widetilde{z}$ satisfying inequality (2.1) by a line search on the straight-line segment connecting z and z', cf. the left part of Figure 2.7.

In the second approach, we determine two triangles K_1 and K_2 in ω_z such that

$$q_G(K_1) = \min_{K \subset \omega_z} q_G(K) \quad \text{and} \quad q_G(K_2) = \min_{K \subset \omega_z \setminus K_1} q_G(K).$$

Then, we determine the unique point z' such that the two triangles corresponding to K_1 and K_2 with z replaced by z' have equal qualities q_G. This point can be computed explicitly from the coordinates of K_1 and K_2 which remain unchanged. If z' is within ω_z, we accept it as an approximation for the optimal solution $\widetilde{z}$ of problem (2.1). Otherwise we again try to find $\widetilde{z}$ by a line search on the straight-line segment connecting z and z', cf. the right part of Figure 2.7.

2.3.3 *A Quality Function Based on Interpolation*

Our second candidate for a quality function is given by

$$q_I(K) = \left\| \nabla(u_Q - u_L) \right\|_K^2,$$

where u_Q and u_L denote the quadratic and linear interpolation, respectively, of an unknown function u. To render the optimisation process operative, we must find a good approximation to $q_I(K)$ which can explicitly be computed without referring to the unknown function u. To this end we recall the functions ψ_E of Section 1.3.4 (p. 8). Then we have

$$u_Q - u_L = \sum_{i=0}^{2} d_i \psi_{E_i}$$

with

$$d_i = u\left(\frac{1}{2}(x_{i+1} + x_{i+2})\right) - \frac{1}{2}u(x_{i+1}) - \frac{1}{2}u(x_{i+2})$$

for $i = 0, 1, 2$ where again all indices have to be taken modulo 3. Hence, we obtain

$$q_I(K) = v^t B v \quad \text{with} \quad v = \begin{pmatrix} d_0 \\ d_1 \\ d_2 \end{pmatrix} \quad \text{and} \quad B_{ij} = \int_K \nabla\psi_{E_i} \cdot \nabla\psi_{E_j}$$

for $i, j = 0, 1, 2$. A straightforward calculation yields

$$B_{ii} = \frac{\mu_1(E_0)^2 + \mu_1(E_1)^2 + \mu_1(E_2)^2}{3\mu_2(K)} = \frac{4}{\sqrt{3}} \frac{1}{q_G(K)}$$

for all i and

$$B_{ij} = \frac{2(x_{i+2} - x_{i+1}) \cdot (x_{j+2} - x_{j+1})}{3\mu_2(K)}$$

for $i \neq j$. Since B is spectrally equivalent to its diagonal, we approximate $q_I(K)$ by

$$\widetilde{q}_I(K) = \frac{1}{q_G(K)} \sum_{i=0}^{2} d_i^2.$$

To obtain an explicit representation of $\widetilde{q}_I$ in terms of the geometrical data of K, we assume that the second derivatives of u are constant on K. Denoting by H_K the Hessian matrix of u on K, Taylor's formula then yields

$$d_i = -\frac{1}{8}(x_{i+2} - x_{i+1})^t H_K (x_{i+2} - x_{i+1})$$

for $i = 0, 1, 2$. Hence, with this assumption, $\widetilde{q}_I$ is a rational function with quadratic polynomials in the numerator and denominator. Problem (2.1) can therefore be solved approximately with a few steps of a damped Newton iteration. Alternatively we may adopt our previous geometrical reasoning with q_G replaced by $\widetilde{q}_I$.

2.3.4 A Quality Function Based on an Error Indicator

The third choice of a quality function is given by

$$q_E(K) = \int_K \left| \sum_{i=0}^{2} e_i \nabla\psi_{E_i} \right|^2,$$

where the coefficients e_0, e_1, and e_2 are computed from an error indicator η_K. Once we dispose of these coefficients, the optimisation problem (2.1) for the function q_E may be solved in the same way as for q_I.

The computation of the coefficients e_0, e_1, and e_2 is particularly simple for the error indicator $\eta_{N,K}$ of Section 1.7.3 (p. 29) which is based on the solution of local Neumann problems on the elements. Denoting by v_K the solution of the auxiliary problem (1.45) (p. 29), we compute the e_i by solving the least-squares problem

$$\text{minimise} \int_K \left| \nabla v_K - \sum_{i=0}^{2} e_i \nabla \psi_{E_i} \right|^2 . \tag{2.2}$$

For the error indicator $\eta_{D,K}$ of Section 1.7.2 (p. 28) which is based on the solution of an auxiliary discrete Dirichlet problem on the patch ω_K, we may proceed in a similar way and simply replace v_K by the restriction to K of $\widetilde{v}_K$, the solution of problem (1.42) (p. 28).

For the error indicator $\eta_{R,E}$ of Section 1.6 (p. 20) and Theorem 1.15 (p. 24), we simply set

$$e_i = -h_{E_i}^{\frac{1}{2}} \mathbb{J}_E(\mathbf{n}_{E_i} \cdot \nabla u_{\mathcal{T}})$$

for $i = 0, 1, 2$ with the obvious modification for edges on the boundary Γ. The scaling factor $h_{E_i}^{\frac{1}{2}}$ takes account of the weighting factors of Theorem 1.15 (p. 24) and the inverse estimates of Proposition 1.4 (p. 9).

For the residual error indicator $\eta_{R,K}$ of Section 1.4 (p. 10) and Theorem 1.5 (p. 15), finally, we replace in problem (2.2) the function v_K by

$$h_K \left(\bar{f}_K + \Delta u_{\mathcal{T}} \right) \psi_K - \sum_{i=0}^{2} h_{E_i}^{\frac{1}{2}} \mathbb{J}_E(\mathbf{n}_{E_i} \cdot \nabla u_{\mathcal{T}}) \psi_{E_i}$$

with the obvious modifications for edges on the boundary Γ. The scaling factors again take into account the weighting factors of Theorem 1.5 (p. 15) and the inverse estimates of Proposition 1.4 (p. 9).

Remark 2.6 The presentation of this section follows [**41**]. Quality criteria for partitions are discussed in [**171**]. For further references we refer to [**142, 143**] and the literature cited there.

2.4 Data Structures

In this section we briefly describe the required data structures for a `Java` or `C++` implementation of an adaptive finite element algorithm. For simplicity we consider only the two-dimensional case. Note that the data structures are independent of the particular differential equation and apply to all engineering problems which require the approximate solution of partial differential equations. The described data structures are realised in the `Java` demonstration applet `ALF` (Adaptive Linear Finite elements) and the `Scilab` function library `AFEM` (Adative Finite Element Methods). Notice that `ALF` uses triangular and quadrilateral elements simultaneously with the regular and irregular refinement of Sections 2.1.2 (p. 65) and 2.1.3 (p. 66), while `AFEM` exclusively uses triangular elements with the marked edge bisection of Section 2.1.4 (p. 68). Both are available at the address `www.rub.de/num1/softwareE.html` together with short user guides in pdf format. We stress that `ALF` and `AFEM` are both devised for demonstration purposes only. Much more sophisticated and general codes are available with `ALBERTA` [**235**], `DEALII` [**36, 37**], `FEAST` [**51**], `FreeFem++` [**149**], `OOFEM` [**215, 216**], `PLTMG` [**38, 39**] to name only a few.

The class NODE realises the concept of a node, i.e. of a vertex of a grid. It has three members c, t, and d.

- The member c stores the coordinates in Euclidean 2-space. It is a double array of length 2.
- The member t stores the type of the node. It equals 0 if it is an interior point of the computational domain. It is k, $k > 0$, if the node belongs to the k-th component of the Dirichlet boundary. It equals $-k$, $k > 0$, if the node is on the k-th component of the Neumann boundary.
- The member d gives the address of the corresponding degree of freedom. It equals -1 if the corresponding node is not a degree of freedom since, e.g. it lies on the Dirichlet boundary. This member takes into account that not every node actually is a degree of freedom.

The class ELEMENT realises the concept of an element. It has six members nv, v, e, p, c, and t.

- The member nv determines the element type, i.e. triangle or quadrilateral.
- The members v and e realise the vertex and edge information, respectively. Both are integer arrays of length 4.
  The vertices are enumerated consecutively in counter-clockwise order; v$[i]$ gives the global number of the i-th vertex. It is assumed that v$[3] = -1$ if nv$= 3$.
  The edges are also enumerated consecutively in counter-clockwise order such that the i-th edge has the vertices $i + 1$ mod nv and $i + 2$ mod nv as its endpoints. Thus, in a triangle, edge i is opposite vertex i.
  A value e$[i] = -1$ indicates that the corresponding edge is on a straight part of the boundary. Similarly e$[i] = -k - 2$, $k \geq 0$, indicates that the endpoints of the corresponding edge are on the k-th curved part of the boundary. A value e$[i] = j \geq 0$ indicates that edge i of the current element is adjacent to element number j. Thus the member e describes the neighbourhood relation of elements.
- The members p, c, and t realise the grid hierarchy and give the number of the parent, the number of the first child, and the refinement type, respectively. In particular we have

$$\mathtt{t} \in \begin{cases} \{0\} & \text{if the element is not refined,} \\ \{1,\ldots,4\} & \text{if the element is refined green,} \\ \{5\} & \text{if the element is refined red,} \\ \{6,\ldots,24\} & \text{if the element is refined blue,} \\ \{25,\ldots,100\} & \text{if the element is refined purple.} \end{cases}$$

At first sight it may seem strange to keep the information about nodes and elements in different classes. Yet, this approach has several advantages.

- It minimises the storage requirement. The coordinates of a node must be stored only once. If nodes and elements are represented by a common structure, these coordinates are stored 4–6 times.
- The elements represent the topology of the grid which is independent of the particular position of the nodes. If nodes and elements are represented by different structures it is much easier to implement mesh-smoothing algorithms which affect the position of the nodes but do not change the mesh topology.

When creating a hierarchy of adaptively refined grids, the nodes are completely hierarchical, i.e. a node of grid $\mathcal{T}_i$ is also a node of any grid $\mathcal{T}_j$ with $j > i$. Since in general the grids are only partly refined, the elements are not completely hierarchical. Therefore, all elements of all grids are stored.

The information about the different grids is implemented by the class `LEVEL`. Its members `nn`, `nt`, `nq`, and `ne` give the number of nodes, triangles, quadrilaterals, and edges, respectively of a given grid. The members `first` and `last` give the addresses of the first element of the current grid and of the first element of the next grid, respectively. The member `dof` finally yields the number of degrees of freedom of the corresponding discrete finite element problems.

2.5 Numerical Examples

The examples of this section are computed with the `Scilab` function library `AFEM` (Adative Finite Element Methods) using `Scilab` Version 5.3.3 on a 2 GHz Intel Core i7 MacBook Pro with 4 GB RAM. `AFEM` is available at the address `www.rub.de/num1/softwareE.html` together with a short user guide in pdf format. The discretisation is based on linear triangular elements, i.e. the space $S_D^{1,0}(\mathcal{T})$ associated with an admissible, shape-regular triangulation $\mathcal{T}$. The assembly of the discrete problem is based on exact integration using piecewise linear approximations of the data of the differential equation. The linear systems are solved exactly using `Scilab`'s built-in sparse-matrix solver. Uniform and adaptive refinement is based on the marked edge bisection of Section 2.1.4 (p. 68).

Tables 2.1–2.3 give for all examples the following quantities:

`L` the number of refinement levels,

`NT` the number of triangles,

`NV` the number of vertices,

`NN` the number of unknowns,

ε the true relative error $\frac{\|\nabla(u-u_{\mathcal{T}})\|}{\|\nabla u\|}$, if the exact solution is known, and the estimated relative error $\frac{\eta}{\|\nabla u_{\mathcal{T}}\|}$, if the exact solution is unknown,

`q` the efficiency index $\frac{\eta}{\|\nabla(u-u_{\mathcal{T}})\|}$ of the error estimator provided the exact solution is known.

Example 2.7 We consider the Poisson equation

$$\begin{aligned} -\Delta u &= 0 \quad \text{in } \Omega \\ u &= g \quad \text{on } \Gamma \end{aligned}$$

in the L-shaped domain $\Omega = (-1,1)^2 \setminus ((0,1) \times (-1,0))$. The boundary data g are chosen such that the exact solution in polar coordinates is $u = r^{\frac{2}{3}} \sin\left(\frac{3}{2}\pi\varphi\right)$. The coarsest mesh consists of six isosceles right-angled triangles with longest sides oriented from north-west to south-east. We first apply a uniform refinement until the storage capacity is exhausted. Then we apply an adaptive refinement strategy based on the edge-residual error estimator $\eta_{R,E}$ of Theorem 1.15 (p. 24) and the maximum strategy of Algorithm 2.1 (p. 64). The refinement process is stopped as soon as we obtain a solution with about the same relative error as the solution on the finest uniform mesh. The corresponding numbers are given in Table 2.1. Plate 2 in the colour plate section shows the finest mesh obtained by the adaptive process and the corresponding discrete solution.

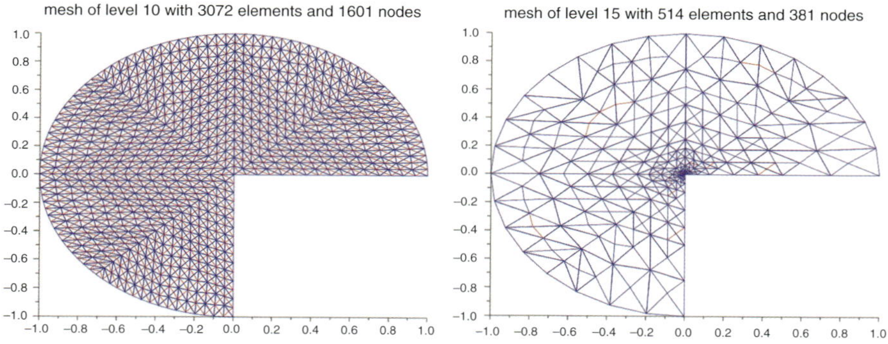

Plate 1 Triangulation obtained by uniform (left) and adaptive (right) refinement for the example of Section 1.1, cf. Table 1.1

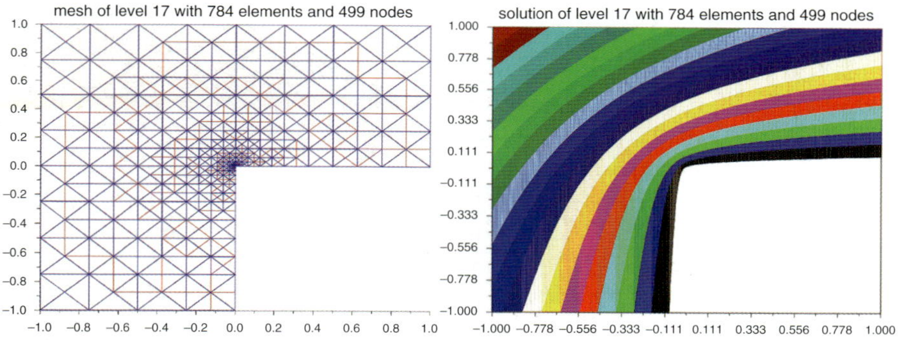

Plate 2 Adaptively refined triangulation of level 17 and corresponding discrete solution for Example 2.7

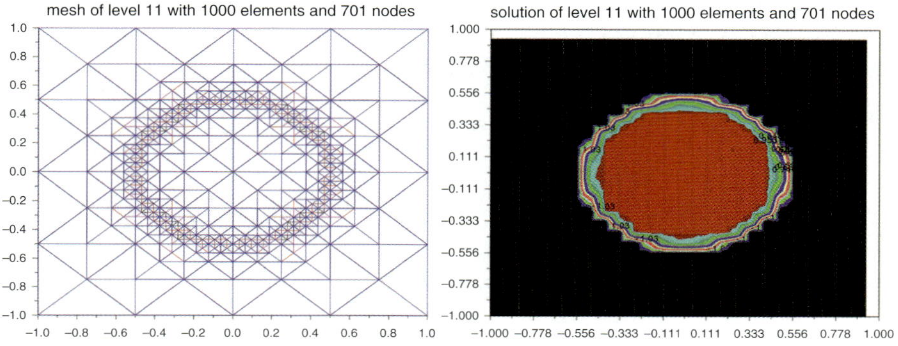

Plate 3 Adaptively refined triangulation of level 11 and the corresponding discrete solution for Example 2.8

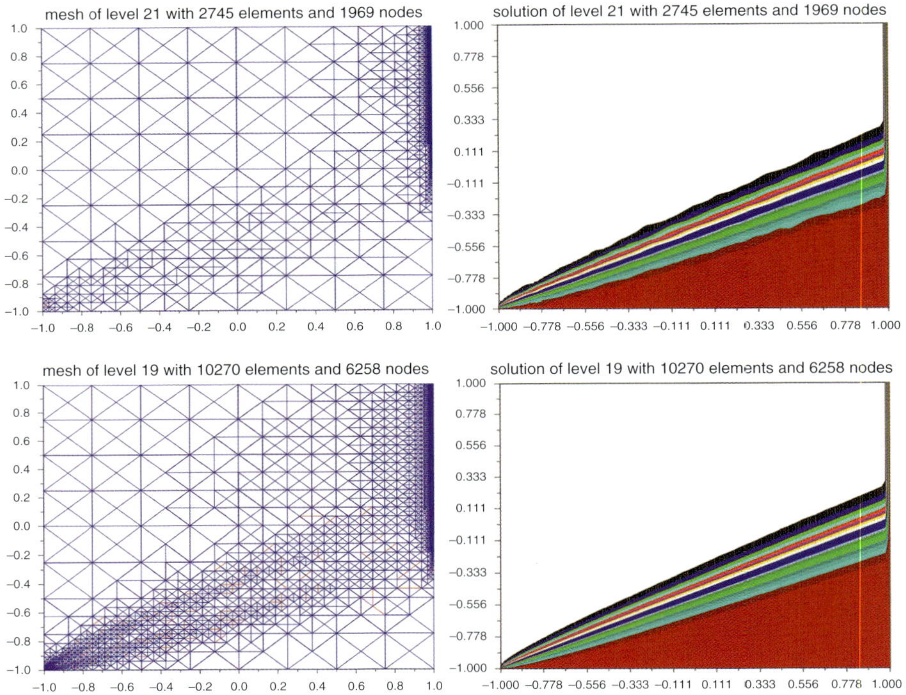

Plate 4 Adaptively refined triangulations and corresponding discrete solutions for Example 2.9 with refinement based on the maximum strategy with parameters $\theta = 0.5$ (top) and $\theta = 0.1$ (bottom)

Table 2.1 Comparison of uniform and adaptive refinement for Example 2.7

	uniform	adaptive
L	12	17
NT	12288	784
NV	6273	499
NN	6017	365
$\varepsilon(\%)$	0.13	0.2
q	–	19.58

Example 2.8 Now we consider the reaction–diffusion equation

$$\begin{aligned} -\Delta u + \kappa^2 u &= f \quad \text{in } \Omega \\ u &= 0 \quad \text{on } \Gamma \end{aligned}$$

in the square $\Omega = (-1, 1)^2$. The reaction parameter κ is chosen equal to 100. The right-hand side f is such that the exact solution is $u = \tanh\left(\kappa\left(x^2 + y^2 - \frac{1}{4}\right)\right)$. It exhibits an interior layer along the boundary of the circle of radius $\frac{1}{2}$ centred at the origin. The coarsest mesh consists of eight isosceles right-angled triangles with longest sides oriented from north-west to south-east. For the comparison of adaptive and uniform refinement we proceed as in the previous example. In order to take account of the reaction term, the error estimator is now the modified residual estimator $\eta_{R,K}$ of Theorem 4.10 (p. 160). Table 2.2 gives the corresponding numbers and Plate 3 in the colour plate section shows the finest mesh obtained by the adaptive process and the corresponding discrete solution.

Example 2.9 Finally we consider the convection–diffusion equation

$$\begin{aligned} -\varepsilon \Delta u + \mathbf{a} \cdot \nabla u &= 0 \quad \text{in } \Omega \\ u &= g \quad \text{on } \Gamma \end{aligned}$$

Table 2.2 Comparison of uniform and adaptive refinement for Example 2.8

	uniform	adaptive
L	11	11
NT	8192	1000
NV	4225	701
NN	3969	493
$\varepsilon(\%)$	2.13	2.05
q	–	8.33

Table 2.3 Comparison of adaptive refinement for Example 2.9 with parameters $\theta = 0.5$ and $\theta = 0.1$ in the maximum strategy of Algorithm 2.1

	$\theta = 0.5$	$\theta = 0.1$
L	21	19
NT	2745	10270
NV	1969	6258
NN	1224	4000
$\varepsilon(\%)$	41.7	18.1

in the square $\Omega = (-1, 1)^2$. The diffusion parameter is $\varepsilon = \frac{1}{100}$, the convection is $\mathbf{a} = \binom{2}{1}$, and the boundary condition is

$$g = \begin{cases} 0 & \text{on the left and top boundary,} \\ 100 & \text{on the bottom and right boundary.} \end{cases}$$

The exact solution of this problem is unknown, but it is known that it exhibits an exponential boundary layer at the boundary $x = 1, y > 0$ and a parabolic interior layer along the line connecting the points $(-1, -1)$ and $(1, 0)$. The coarsest mesh is as in Example 2.8. Since the exact solution is unknown, we cannot give the efficiency index $\mathfrak{q}$ and perform only an adaptive refinement. The error estimator is the one of Theorem 4.19 (p. 168). Since the exponential layer is far stronger than the parabolic one, the maximum strategy of Algorithm 2.1 (p. 64) leads to a refinement preferably close to the boundary $x = 1, y > 0$ and has difficulties in catching the parabolic interior layer. This is demonstrated by Plate 4 in the colour plate section. It shows the finest mesh obtained by the adaptive process and the corresponding discrete solution for the two parameters $\theta = 0.5$ and $\theta = 0.1$ in the maximum strategy. Table 2.3 gives the corresponding numbers for both values of θ.

3

Auxiliary Results

In this chapter we collect the technical prerequisites for the a posteriori error estimates of Chapters 4–6. After introducing the required notation for function spaces and finite element meshes and spaces, we provide explicit bounds for the constants in the trace, Poincaré, and Friedrichs' inequalities as well as the interpolation error and inverse estimates and prove a strengthened Cauchy–Schwarz inequality for the decomposition of affine functions with values in suitable dual spaces. While these results are of a more or less technical nature, the final section on the estimation of dual norms of certain continuous linear functionals, labelled residuals, is at the heart of all subsequent results and is of independent interest.

3.1 Function Spaces

In this section we extend the notation of Section 1.2 (p. 4) and collect all results concerning Lebesgue and Sobolev spaces that will be needed in what follows.

3.1.1 Lebesgue Spaces

We denote by $\Omega \subset \mathbb{R}^d$, $d \geq 2$, an open, connected, bounded, polyhedral domain with Lipschitz boundary Γ consisting of two disjoint parts Γ_D and Γ_N. We assume that the *Dirichlet boundary* Γ_D is closed relative to Γ and has a positive $(d-1)$-dimensional Hausdorff measure. The *Neumann boundary* Γ_N may be empty.

The *Lebesgue measure* and the $(d-1)$-dimensional *Hausdorff measure* in $\mathbb{R}^d$, $d \geq 2$, are denoted by μ_d and μ_{d-1}, respectively.

In what follows p, q, etc. are *Lebesgue exponents* with $1 \leq p, q \leq \infty$. The dual exponent is denoted by p', q', etc. and is defined by $\frac{1}{p} + \frac{1}{p'} = 1$ with the obvious modification for the case $p = \infty$.

For every open polyhedral subset ω of Ω with Lipschitz boundary γ and every Lebesgue exponent p with $1 \leq p \leq \infty$, we denote by $L^p(\omega)$ and $L^p(\gamma)$ the standard *Lebesgue spaces* equipped with the norms [3], [109, Section I.2]

$$\|\varphi\|_{p;\omega} = \begin{cases} \left\{ \int_\omega |\varphi|^p \, d\mu_d \right\}^{\frac{1}{p}}, & \text{if } p < \infty, \\ \operatorname*{ess.\,sup}_{x \in \omega} |\varphi(x)| & \text{if } p = \infty, \end{cases}$$

and

$$\|\varphi\|_{p;\gamma} = \begin{cases} \left\{ \int_\gamma |\varphi|^p \, \mathrm{d}\mu_{d-1} \right\}^{\frac{1}{p}}, & \text{if } p < \infty, \\ \underset{x\in\gamma}{\text{ess. sup}} \, |\varphi(x)| & \text{if } p = \infty. \end{cases}$$

For vector fields $\mathbf{v} : \omega \to \mathbb{R}^d$ we use the same definition with the convention that $|\cdot|$ is the Euclidean norm in $\mathbb{R}^d$.

3.1.2 Sobolev Spaces

For $k \geq 1$ and $1 \leq p \leq \infty$, we denote by $W^{k,p}(\omega)$ the standard *Sobolev spaces* equipped with the norm

$$\|\varphi\|_{k,p;\omega} = \begin{cases} \left\{ \sum_{\ell=0}^{k} \left\|\nabla^\ell \varphi\right\|_{p;\omega}^p \right\}^{\frac{1}{p}}, & \text{if } p < \infty, \\ \max_{0\leq \ell \leq k} \left\|\nabla^\ell \varphi\right\|_{\infty;\omega}, & \text{if } p = \infty. \end{cases}$$

As usual, we write $H^k(\omega)$ and $\|\cdot\|_{k;\omega}$ instead of $W^{k,2}(\omega)$ and $\|\cdot\|_{k,2;\omega}$, respectively. In analogy to Section 1.2 (p. 4) we further set for $1 \leq p < \infty$

$$W_D^{1,p}(\Omega) = \left\{ \varphi \in W^{1,p}(\Omega) : \varphi = 0 \text{ on } \Gamma_D \right\}.$$

Given a Lebesgue exponent p with $1 \leq p < \infty$, a positive number ε and a non-negative number β, we will frequently equip the space $W^{1,p}(\omega)$ with the (semi-) norm

$$|||\varphi|||_\omega = \left\{ \varepsilon \, \|\nabla\varphi\|_{p;\omega}^p + \beta \, \|\varphi\|_{p;\omega}^p \right\}^{\frac{1}{p}}. \tag{3.1}$$

The following cases will be of particular interest:

$\varepsilon = 1,$	$\beta = 0$	the standard $W^{1,p}$-semi-norm,
$\varepsilon = 1,$	$\beta = 1$	the standard $W^{1,p}$-norm,
$\varepsilon \ll 1,$	$\beta \approx 1$	for singularly perturbed problems.

We do not add a subscript p, ε, or β to $|||\cdot|||$ since the actual values of these parameters will be evident from the context.

If $p = 2$ or $\omega = \Omega$ we omit the corresponding subscript in the norms, i.e. $\|\cdot\|_p = \|\cdot\|_{p;\Omega}$, $\|\cdot\|_\omega = \|\cdot\|_{2;\omega}$, $\|\cdot\| = \|\cdot\|_{2;\Omega}$, $|||\cdot||| = |||\cdot|||_\Omega$, etc.

For every Lebesgue exponent p with $1 < p < \infty$, the dual space of $W_D^{1,p}(\Omega)$ is denoted by $W^{-1,p'}(\Omega)$ and equipped with the norm

$$|||\varphi|||_* = \sup_{v\in W_D^{1,p}(\Omega)\setminus\{0\}} \frac{\langle \varphi, v \rangle}{|||v|||}, \tag{3.2}$$

where $\langle\cdot\,,\,\cdot\rangle$ denotes the standard duality pairing. Thus, the dual norm $|||\cdot|||_*$ depends on the parameters p, ε, and β which characterise the primal norm $|||\cdot|||$.

For the lifting of residuals in Section 3.8.3 (p. 139) and the mixed variational problems of Sections 4.8 (p. 208) and 4.9 (p. 223), we need the space $H(\text{div};\omega)$ of all vector fields $\mathbf{v}$ in $L^2(\omega)^d$ which have their divergence $\operatorname{div}\mathbf{v}$ in $L^2(\omega)$, cf. Section 1.12 (p. 48). This space is equipped with its graph norm

$$\|\mathbf{v}\|_{H(\text{div};\omega)} = \left\{\|\mathbf{v}\|_\omega^2 + \|\operatorname{div}\mathbf{v}\|_\omega^2\right\}^{\frac{1}{2}}.$$

3.1.3 *Spaces for Parabolic Problems*

For the analysis of parabolic equations in Chapter 6, we need spaces of functions of one real variable with values in suitable Sobolev spaces. Given a separable Banach space V with norm $\|\cdot\|_V$, two real numbers a and b with $a < b$, and a real number p with $1 \le p \le \infty$, we denote by $L^p(a,b;V)$ the space of all measurable functions u defined on (a,b) with values in V such that the mapping $t \mapsto \left\|u(\cdot,t)\right\|_V$ is in $L^p((a,b))$. $L^p(a,b;V)$ is a Banach space equipped with the norm [**119**, Chapter XVIII, Section 1]

$$\|u\|_{L^p(a,b;V)} = \begin{cases} \left\{\displaystyle\int_a^b \left\|u(\cdot,t)\right\|_V^p \mathrm{d}t\right\}^{\frac{1}{p}}, & \text{if } p < \infty, \\ \operatorname*{ess.\,sup}_{t\in(a,b)} \left\|u(\cdot,t)\right\|_V, & \text{if } p = \infty. \end{cases}$$

For any $u \in L^p(a,b;V)$ we denote by $\partial_t u$ its partial derivative with respect to time in the distributional sense [**119**, loc. cit.].

Given a second separable Banach space W with norm $\|\cdot\|_W$ such that V is continuously embedded in W, we introduce the Banach space

$$W^p(a,b;V,W) = \left\{u \in L^p(a,b;V) : \partial_t u \in L^p(a,b;W)\right\}$$

and equip it with the norm

$$\|u\|_{W^p(a,b;V,W)} = \begin{cases} \left\{\displaystyle\int_a^b \left\|u(\cdot,t)\right\|_V^p \mathrm{d}t + \int_a^b \left\|\partial_t u(\cdot,t)\right\|_W^p \mathrm{d}t\right\}^{\frac{1}{p}}, & \text{if } p < \infty, \\ \operatorname*{ess.\,sup}_{t\in(a,b)} \max\left\{\left\|u(\cdot,t)\right\|_V, \left\|\partial_t u(\cdot,t)\right\|_W\right\}, & \text{if } p = \infty. \end{cases}$$

If $p > 1$, we know from [**119**, Chapter XVIII, Section 1, Proposition 9] that for any $u \in W^p(a,b;V,W)$ the traces $u(\cdot,a)$ and $u(\cdot,b)$ are defined as elements of W.

3.2 Finite Element Meshes and Spaces

In this section we extend the notation of Sections 1.2 (p. 4) and 1.3 (p. 5) and collect all results concerning finite element meshes and functions that will be needed in what follows.

3.2.1 *Finite Element Meshes*

Given a domain $\Omega \subset \mathbb{R}^d$ as in the previous section, we denote by $\mathcal{T}$ a partition of Ω which satisfies the following conditions.

- The closure of Ω is the union of all elements in $\mathcal{T}$.
- The Dirichlet boundary Γ_D is the union of $(d-1)$-dimensional faces of elements in $\mathcal{T}$.
- Every element has at least one vertex in $\Omega \cup \Gamma_N$.
- Every element in $\mathcal{T}$ is either a *simplex* or a *parallelepiped*, i.e. it is the image of the d-dimensional *reference simplex* $\widehat{K}_d = \left\{x \in \mathbb{R}^d : x_1 \geq 0, \ldots, x_d \geq 0, x_1 + \ldots + x_d \leq 1\right\}$ or of the d-dimensional *reference cube* $\widehat{K}_d = [0,1]^d$ (reference element in short) under an affine mapping (*affine-equivalence*).
- Any two elements in $\mathcal{T}$ are either disjoint or share a complete lower dimensional face of their boundaries (*admissibility*).
- For any element K, the ratio of its diameter h_K to the diameter ρ_K of the largest ball inscribed into K is bounded independently of K (*shape regularity*).

With every partition $\mathcal{T}$ we associate its *shape parameter*

$$C_{\mathcal{T}} = \max_{K \in \mathcal{T}} \frac{h_K}{\rho_K}. \tag{3.3}$$

When considering families of partitions obtained by successive local or global refinement, the shape parameter $C_{\mathcal{T}}$ must be bounded *uniformly with respect to all partitions*. In two dimensions, $C_{\mathcal{T}}$ is proportional to the inverse of the sine of the smallest angle and shape regularity means that the smallest angles of all elements stay bounded away from zero.

With every partition $\mathcal{T}$, we associate the sets $\mathcal{N}$ and $\mathcal{E}$ of all vertices and $(d-1)$-dimensional faces (faces for short), respectively of all elements in $\mathcal{T}$. A subscript K, Ω, Γ, Γ_D, or Γ_N on $\mathcal{N}$ or $\mathcal{E}$ indicates that only those vertices or faces, respectively, are considered which are contained in the corresponding set. Similarly $\mathcal{N}_E$ denotes the set of the vertices of a given face E. The union of all faces in $\mathcal{E}$ is denoted by Σ and is called the *skeleton* of $\mathcal{T}$.

The sets ω_K, $\widetilde{\omega}_K$, ω_E, $\widetilde{\omega}_E$, ω_z, and σ_z are defined as in Section 1.3 (p. 5), cf. Figures 1.1 (p. 6) and 1.2 (p. 6), with the modification that the notion 'edge' must be replaced by 'face'.

For every element or face $S \in \mathcal{T} \cup \mathcal{E}$, we denote by $h_S = \operatorname{diam}(S)$ its diameter and set $h_z = h_{\omega_z} = \operatorname{diam}(\omega_z)$ for every vertex $z \in \mathcal{N}$. As in Section 1.3 (p. 5), the shape regularity of $\mathcal{T}$ implies that for all elements K and K' and all faces E and E' that share at least one vertex, the ratios $\frac{h_K}{h_{K'}}$, $\frac{h_E}{h_{E'}}$, and $\frac{h_K}{h_E}$ are proportional to the shape parameter $C_{\mathcal{T}}$ of $\mathcal{T}$.

For every Lebesgue exponent p with $1 \leq p < \infty$, every positive number ε and every non-negative number β, we further define the *modified diameter* of any bounded set ω by

$$\hbar_\omega = \begin{cases} \min\left\{\varepsilon^{-\frac{1}{p}} \operatorname{diam}(\omega), \beta^{-\frac{1}{p}}\right\} & \text{if } \beta > 0, \\ \varepsilon^{-\frac{1}{p}} \operatorname{diam}(\omega) & \text{if } \beta = 0. \end{cases} \tag{3.4}$$

We do not add a subscript p, ε, or β to $\hbar$ since the actual values of these parameters will be evident from the context.

With every element K of $\mathcal{T}$, we finally associate the parameter

$$\nu_K = \begin{cases} d & \text{if } K \text{ is a simplex}, \\ 1 & \text{if } K \text{ is a parallelepiped}. \end{cases} \tag{3.5}$$

3.2.2 *Finite Element Spaces*

For every multi-index $\alpha \in \mathbb{N}^d$, we set for brevity

$$|\alpha|_1 = \alpha_1 + \cdots + \alpha_d, \ |\alpha|_\infty = \max\{\alpha_i : 1 \le i \le d\}, \ x^\alpha = x_1^{\alpha_1} \cdot \ldots \cdot x_d^{\alpha_d}.$$

With every integer k we then associate a space $R_k(\widehat{K}_d)$ of polynomials by

$$R_k(\widehat{K}_d) = \begin{cases} \operatorname{span}\{x^\alpha : |\alpha|_1 \le k\} & \text{for the reference simplex,} \\ \operatorname{span}\{x^\alpha : |\alpha|_\infty \le k\} & \text{for the reference cube} \end{cases}$$

and set for every element K in any partition $\mathcal{T}$

$$R_k(K) = \left\{\varphi \circ F_K^{-1} : \varphi \in R_k(\widehat{K}_d)\right\},$$

where F_K is an affine diffeomorphism mapping $\widehat{K}_d$ onto K.

Using this notation, we define finite element spaces by

$$\begin{aligned} S^{k,-1}(\mathcal{T}) &= \left\{\varphi : \Omega \to \mathbb{R} : \varphi|_K \in R_k(K) \ \text{ for all } K \in \mathcal{T}\right\}, \\ S^{k,0}(\mathcal{T}) &= S^{k,-1}(\mathcal{T}) \cap C(\overline{\Omega}), \\ S_D^{k,0}(\mathcal{T}) &= S^{k,0}(\mathcal{T}) \cap H_D^1(\Omega) = \left\{\varphi \in S^{k,0}(\mathcal{T}) : \varphi = 0 \ \text{ on } \Gamma_D\right\}, \\ S_0^{k,0}(\mathcal{T}) &= S^{k,0}(\mathcal{T}) \cap H_0^1(\Omega) = \left\{\varphi \in S^{k,0}(\mathcal{T}) : \varphi = 0 \ \text{ on } \Gamma\right\}. \end{aligned}$$

The number k may be 0 for the first space, but must be at least 1 for the other spaces.

3.2.3 *Shape and Cut-Off Functions*

We associate with every vertex $z \in \mathcal{N}$ its *nodal shape function* λ_z which is uniquely defined by the conditions, cf. Section 1.3.2 (p. 7)

$$\lambda_z \in S^{1,0}(\mathcal{T}), \quad \lambda_z(z) = 1, \quad \lambda_z(x) = 0 \text{ for all } x \in \mathcal{N}_\mathcal{T} \setminus \{z\}.$$

Lemma 1.2 (p. 7) immediately carries over to arbitrary space dimensions. As in two dimensions, the set ω_z is the support of the function λ_z and $\int_\Sigma \lambda_z = \int_{\sigma_z} \lambda_z$.

The definition of the local cut-off functions ψ_K and ψ_E can be extended to arbitrary space dimensions by

$$\psi_K = \beta_K \prod_{z \in \mathcal{N}_K} \lambda_z, \quad \psi_E = \beta_E \prod_{z \in \mathcal{N}_E} \lambda_z$$

with

$$\beta_K = \begin{cases} (d+1)^{d+1} & \text{if } K \text{ is a simplex,} \\ (2^d)^{2^d} & \text{if } K \text{ is a parallelepiped,} \end{cases}$$

$$\beta_E = \begin{cases} 2^2 & \text{if } E \text{ is an edge,} \\ d^d & \text{if } E \text{ is a } (d-1)\text{-dimensional simplex,} \\ (2^{d-1})^{2^{d-1}} & \text{if } E \text{ is a } (d-1)\text{-dimensional parallelepiped.} \end{cases}$$

Notice that the factors β_K and β_E are chosen such that the functions ψ_K and ψ_E attain the value 1 at the barycentre of K and E, respectively.

With this normalisation, the first part of Proposition 1.4 (p. 9) immediately carries over to arbitrary space dimensions. Its second part, i.e. the inverse estimates, will be proved in Section 3.6 (p. 112).

3.2.4 Modified Cut-Off Functions

For some applications, in particular singularly perturbed differential equations, we need a modification of the cut-off functions ψ_E. To describe it, we associate with every positive parameter ϑ the transformation

$$\begin{aligned}\Phi_\vartheta : \mathbb{R}^d &\to \mathbb{R}^d \\ (x_1, \ldots, x_d) &\mapsto (x_1, \ldots, x_{d-1}, \vartheta x_d).\end{aligned}$$

Given a face E and an element K adjacent to E, we denote by $F_{E,K} : \mathbb{R}^d \to \mathbb{R}^d$ an orientation-preserving, affine transformation which maps the reference element $\widehat{K}_d$ to the element K and the intersection $\widehat{E}_d$ of $\widehat{K}_d$ with the d-th coordinate plane $\{x \in \mathbb{R}^d : x_d = 0\}$ to the face E. With these notations we define the 'squeezed' element K_ϑ by, cf. Figure 3.1

$$K_\vartheta = F_{E,K} \circ \Phi_\vartheta \circ F_{E,K}^{-1}(K)$$

and denote by λ_{z,K_ϑ}, $z \in \mathcal{N}_{K_\vartheta}$, its nodal shape functions. The modified cut-off function is then defined by

$$\psi_{E,\vartheta} = \begin{cases} \beta_E \prod_{z \in \mathcal{N}_E} \lambda_{z,K_\vartheta} & \text{on all } K \subset \omega_E, \\ 0 & \text{on } \Omega \setminus \omega_E, \end{cases}$$

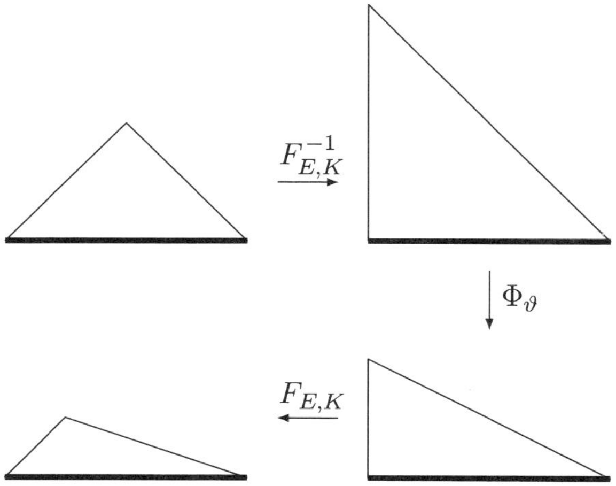

Figure 3.1 Original element K (top-left), reference element $\widehat{K}_d$ (top-right), squeezed reference element $\Phi_\vartheta(\widehat{K}_d)$ (bottom-right), and squeezed element K_ϑ (bottom-left); thick lines indicate the faces E and $\widehat{E}_d$

where β_E is as above and ascertains that $\psi_{E,\vartheta}$ attains the value 1 at the barycentre of E. Note that, for $\vartheta = 1$, the functions $\psi_{E,\vartheta}$ and ψ_E are equal.

3.2.5 Smooth Cut-Off Functions

The cut-off functions ψ_K and ψ_E are piecewise polynomials and globally continuous. Hence they are contained in $W^{1,\infty}(\Omega)$ but not in $W^{2,1}(\Omega)$. In some places we will need local cut-off functions which are of class C^1 or even smoother.

To achieve this goal, we fix an integer $m \geq 1$. For every element $K \in \mathcal{T}$ we set

$$\psi_{K,m} = (\psi_K)^{m+1}.$$

Then, $\psi_{K,m}$, together with all its derivatives up to order m, vanishes on the boundary of K. Hence, $\psi_{K,m}$ is contained in $C^m(\Omega)$ and consequently in $W^{m+1,p}(\Omega)$ for every Lebesgue exponent p.

For the functions ψ_E this simple device does not suffice since they extend over two adjacent elements and since the nodal shape functions are not differentiable across inter-element boundaries. To overcome this difficulty, we denote for every element $K \in \mathcal{T}$ and every vertex $z \in \mathcal{N}_K$ thereof by $\lambda_{K,z}$ the extension to $\mathbb{R}^d$ of the nodal shape function of z associated with K, i.e.

$$\lambda_{K,z}|_K \in R_1(K), \quad \lambda_{K,z}(z) = 1, \quad \lambda_{K,z}(x) = 0 \text{ for all } x \in \mathcal{N}_K \setminus \{z\}.$$

Note that the functions $\lambda_{K,z}$ are *global* polynomials whereas the nodal shape functions λ_z are *piecewise* polynomials, cf. Figure 3.2 for the one-dimensional situation.

We now define

$$\psi_{E,m} = \beta_{E,m} \left(\prod_{K \subset \omega_E} \prod_{z \in \mathcal{N}_E} \lambda_{K,z} \chi_{\omega_E} \right)^{m+1},$$

where χ_{ω_E} denotes the characteristic function of ω_E and where $\beta_{E,m}$ is determined by the condition $\max_{x \in E} \psi_{E,m}(x) = 1$. Since the functions $\psi_{E,m}$ are built upon global polynomials and vanish on the boundary of ω_E together with all their derivatives up to order m, they are contained in $C^m(\Omega)$ and consequently in $W^{m+1,p}(\Omega)$ for every Lebesgue exponent p. Figure 3.3 shows a sketch of the functions $(\psi_E)^2$ and $\psi_{E,1}$ for a domain ω_E consisting of two isosceles right-angled triangles sharing their common longest edge; depicted are the approximated graphs along a cross-section perpendicular to E passing through the midpoint of E.

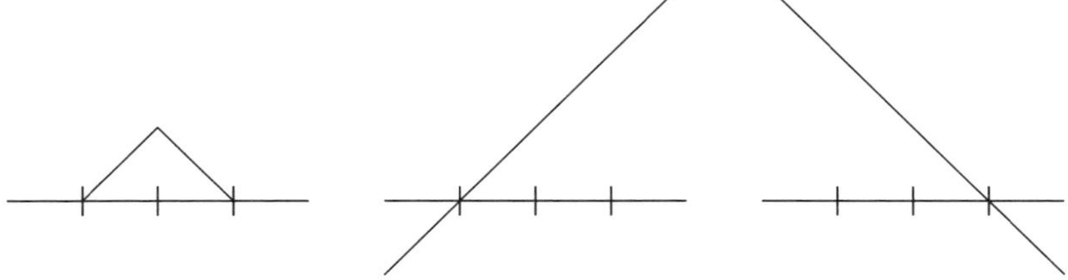

Figure 3.2 Functions λ_z (left) and $\lambda_{K,z}$ for the intervals K having z as right endpoint (middle) and left endpoint (right)

Figure 3.3 Sketch of functions $(\psi_E)^2$ (left) and $\psi_{E,1}$ (right) for a domain ω_E consisting of two right-angled isosceles triangles sharing their common longest edge; depicted are the approximated graphs along a cross-section perpendicular to E passing through the midpoint of E

In Section 4.11 (p. 248) we will need yet another type of smooth cut-off function $\Psi_{E,1}$ associated with the interior faces $E \in \mathcal{E}_\Omega$. For its definition we denote by $K_{E,1}$ and $K_{E,2}$ the two elements adjacent to E and choose the enumeration such that $\mathbf{n}_E$ is the unit outward normal of $K_{E,1}$. Recalling the parameter ν_K of equation (3.5) we then set

$$\Psi_{E,1} = \psi_{E,1} \left(\frac{\mu_d(K_2)}{\nu_{K_2}\mu_{d-1}(E)} \sum_{z \in \mathcal{N}_{K_2} \setminus \mathcal{N}_E} \lambda_{K_2,z} - \frac{\mu_d(K_1)}{\nu_{K_1}\mu_{d-1}(E)} \sum_{z \in \mathcal{N}_{K_1} \setminus \mathcal{N}_E} \lambda_{K_1,z} \right),$$

where $\psi_{E,1}$ is the C^1-cut-off function corresponding to E defined above. Obviously, the function $\Psi_{E,1}$ vanishes together with its first-order derivatives on $\partial\omega_E$. Taking into account that $\frac{\mu_d(K_i)}{\nu_{K_i}\mu_{d-1}(E)}$ is the height of K_i above E and that $\lambda_{K_i,z}$ vanishes on E for $z \in \mathcal{N}_{K_i} \setminus \mathcal{N}_E$ and $i = 1, 2$, we conclude that $\Psi_{E,1}$ vanishes on E and that its normal derivative $\mathbf{n}_E \cdot \nabla\Psi_{E,1}$ is proportional to $\psi_{E,1}$ on E.

3.2.6 *Space–Time Finite Elements*

The parabolic equations of Chapter 6 are all posed on bounded space–time cylinders of the form $\Omega \times (0, T]$ with an open bounded polyhedral cross-section $\Omega \subset \mathbb{R}^d$ and an arbitrary, but fixed finite final time T. They will be discretised using space–time finite elements. For these we consider partitions $\mathcal{I} = \{[t_{n-1}, t_n] : 1 \le n \le N_\mathcal{I}\}$ of the time interval $[0, T]$ into sub-intervals satisfying $0 = t_0 < \cdots < t_{N_\mathcal{I}} = T$. For every n with $1 \le n \le N_\mathcal{I}$, we denote by $I_n = [t_{n-1}, t_n]$ the n-th sub-interval and by $\tau_n = t_n - t_{n-1}$ its length.

With every intermediate time t_n, $0 \le n \le N_\mathcal{I}$, we associate an admissible, affine-equivalent, shape-regular partition $\mathcal{T}_n$ of Ω, cf. Figure 3.4, and a corresponding finite element space X_n. In addition to the conditions of Section 3.2.1, the partitions $\mathcal{I}$ and $\mathcal{T}_n$ and the spaces X_n must satisfy the following assumptions.

- For every n with $1 \le n \le N_\mathcal{I}$ there is an affine-equivalent, admissible, and shape-regular partition $\widetilde{\mathcal{T}}_n$ such that it is a refinement of both $\mathcal{T}_n$ and $\mathcal{T}_{n-1}$ and such that

  $$\sup_{1 \le n \le N_\mathcal{I}} \sup_{K \in \widetilde{\mathcal{T}}_n} \sup_{\substack{K' \in \mathcal{T}_n \\ K \subset K'}} \frac{h_{K'}}{h_K} < \infty$$

  uniformly with respect to all partitions $\mathcal{I}$ which are obtained by adaptive or uniform refinement of any initial partition of $[0, T]$ (*transition condition*).
- Each X_n consists of continuous functions which are piecewise polynomials, the degrees being at least one and being bounded uniformly with respect to all partitions $\mathcal{T}_n$ and $\mathcal{I}$ (*degree condition*).

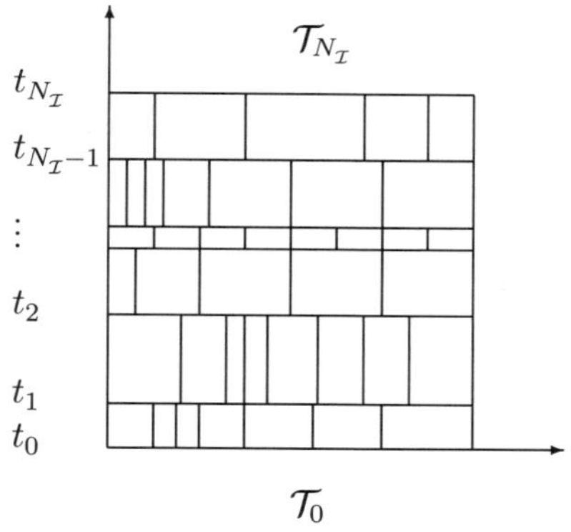

Figure 3.4 Space–time partition

The transition condition is due to the simultaneous presence of finite element functions defined on different grids. Usually the partition $\mathcal{T}_n$ is obtained from $\mathcal{T}_{n-1}$ by a combination of refinement and of coarsening. In this case the transition condition only restricts the coarsening: it should not be too abrupt nor too strong.

The lower bound on the polynomial degrees is needed for the construction of suitable quasi-interpolation operators. The upper bound ensures that the constants in inverse estimates are uniformly bounded.

Notice that we do not impose any shape condition of the form $\max_n \tau_n \leq c \min_n \tau_n$.

3.3 Trace Inequalities

It is well known [3] that the *trace inequality*

$$\|w\|_{p;E} \leq \left\{c_1(K)\, \|w\|_{p;K}^p + c_2(K)\, \|\nabla w\|_{p;K}^p\right\}^{\frac{1}{p}}$$

holds for all Lebesgue exponents p with $1 \leq p < \infty$, all elements K, all faces E of K, and all functions $w \in W^{1,p}(K)$. Transforming to the reference element, invoking the trace inequality there and transforming back, one easily checks that the constants $c_1(K)$ and $c_2(K)$ are proportional to h_K^{-1} and h_K^{p-1}, respectively. For our purposes, however, this form of the trace inequality is not sharp enough. Moreover, we need more explicit expressions for the involved constants.

To achieve this goal, we first construct a suitable vector field.

Lemma 3.1 *For every element $K \in \mathcal{T}$ and every face E of K, there exists a vector field $\gamma_{K,E}$ with the following properties*

$$\begin{aligned}
\operatorname{div} \gamma_{K,E} &= \nu_K && \text{in } K,\\
\gamma_{K,E} \cdot \mathbf{n}_K &= 0 && \text{on } \partial K \setminus E,\\
\gamma_{K,E} \cdot \mathbf{n}_K &= \frac{\nu_K \mu_d(K)}{\mu_{d-1}(E)} && \text{on } E,\\
\left\|\gamma_{K,E}\right\|_{\infty;K} &\leq h_K,
\end{aligned}$$

where ν_K is given by equation (3.5) (p. 82).

Proof First, we consider the case that K is a simplex and set

$$\gamma_{K,E}(x) = x - a_{K,E},$$

where $a_{K,E}$ denotes the vertex of K opposite to E. One easily verifies the first property. The restriction of $\gamma_{K,E}$ to any face of K different from E is tangential to this face, whence $\gamma_{K,E} \cdot \mathbf{n}_K = 0$ on $\partial K \setminus E$. Since $\gamma_{K,E}$ is affine, $\gamma_{K,E} \cdot \mathbf{n}_K$ is constant on every face. Consequently, the divergence theorem implies the third property. The last estimate is obvious.

Next, we consider the case that K is a parallelepiped, which partially mimics the simplex case. Choose a vertex $a_{K,E}$ of K that is not contained in E and denote by $b_{K,E}$ the vertex of E which is connected with $a_{K,E}$ by an edge of K. Setting

$$\mathbf{m}_{K,E} = \frac{1}{|b_{K,E} - a_{K,E}|}(b_{K,E} - a_{K,E}) \quad \text{and} \quad \mathbf{n}_{K,E} = \mathbf{n}_K\big|_E,$$

we now define

$$\gamma_{K,E}(x) = \frac{(x - a_{K,E}) \cdot \mathbf{n}_{K,E}}{\mathbf{m}_{K,E} \cdot \mathbf{n}_{K,E}} \mathbf{m}_{K,E}.$$

Note that $\mathbf{m}_{K,E} \cdot \mathbf{n}_{K,E} > 0$ because K is affine-equivalent to the d-cube. The first property of $\gamma_{K,E}$ follows by a short calculation. Since $\mathbf{m}_{K,E}$ is tangential to all faces that are not parallel to E, we have the first part of the second property except for the face parallel to E. For the rest of this property and the third one, we observe that $(x - a_{K,E}) \cdot \mathbf{n}_{K,E}$ is constant on faces parallel to E. Consequently, $\gamma_{K,E}$ vanishes on the face through $a_{K,E}$ parallel to E and the divergence theorem completes the proof of the second and third identity. It remains to verify the estimate of $\|\gamma_{K,E}\|_{\infty;K}$. To this end, observe that $\gamma_{K,E}$ is affine, vanishes on the face parallel to E, and is constant on E. The convex function $|\gamma_{K,E}|$ on K thus assumes its maximum in the vertices of K or, more precisely, in $b_{K,E}$. This yields

$$\|\gamma_{K,E}\|_{\infty;K} = \frac{(b_{K,E} - a_{K,E}) \cdot \mathbf{n}_{K,E}}{\mathbf{m}_{K,E} \cdot \mathbf{n}_{K,E}} = |b_{K,E} - a_{K,E}| \leq h_K. \qquad \square$$

With the help of Lemma 3.1 and the Gauss theorem we can establish the following *trace equality*.

Proposition 3.2 *For every face E of an element K, we have for all $w \in W^{1,1}(K)$*

$$\frac{1}{\mu_{d-1}(E)} \int_E w - \frac{1}{\mu_d(K)} \int_K w = \frac{1}{\nu_K \mu_d(K)} \int_K \gamma_{K,E} \cdot \nabla w.$$

Proof Apply the Gauss theorem to $w\gamma_{K,E}$, use Lemma 3.1, and divide by $\nu_K \mu_d(K)$. $\square$

Next, we prove a variant of Proposition 3.2 with weight function λ_z. For brevity, we set for every element K, every face E of K, and every vertex z of E

$$\mu_{d,z}(K) = \int_K \lambda_z d\mu_d, \qquad \mu_{d-1,z}(E) = \int_E \lambda_z d\mu_{d-1}. \tag{3.6}$$

Proposition 3.3 *For every element K, every face E thereof, every vertex z of E, and every function $w \in W^{1,1}(K)$ we have*

$$\frac{1}{\mu_{d-1,z}(E)} \int_E \lambda_z w - \frac{1}{\mu_{d,z}(K)} \int_K \lambda_z w = \frac{1}{(\nu_K + 1)\mu_{d,z}(K)} \int_K \lambda_z \gamma_{K,E} \cdot \nabla w.$$

Proof We start by applying Proposition 3.2 to the function $\lambda_z w$ and obtain

$$\begin{aligned}\frac{1}{\mu_{d-1}(E)} \int_E \lambda_z w - \frac{1}{\mu_d(K)} \int_K \lambda_z w &= \frac{1}{\nu_K \mu_d(K)} \int_K \gamma_{K,E} \cdot \nabla (\lambda_z w) \\ &= \frac{1}{\nu_K \mu_d(K)} \int_K \lambda_z \gamma_{K,E} \cdot \nabla w + \frac{1}{\nu_K \mu_d(K)} \int_K w \gamma_{K,E} \cdot \nabla \lambda_z.\end{aligned}$$

Since

$$\frac{\mu_{d-1}(E)\mu_{d,z}(K)}{\mu_{d-1,z}(E)\mu_d(K)} = \frac{\nu_K}{\nu_K + 1},$$

multiplication of this identity with $\frac{\mu_{d-1}(E)}{\mu_{d-1,z}(E)}$ yields

$$\begin{aligned}\frac{1}{\mu_{d-1,z}(E)} \int_E \lambda_z w - \frac{\nu_K}{(\nu_K + 1)\mu_{d,z}(K)} \int_K \lambda_z w &= \frac{1}{(\nu_K + 1)\mu_{d,z}(K)} \int_K \lambda_z \gamma_{K,E} \cdot \nabla w \\ &\quad + \frac{1}{(\nu_K + 1)\mu_{d,z}(K)} \int_K w \gamma_{K,E} \cdot \nabla \lambda_z.\end{aligned} \tag{3.7}$$

We claim that $\gamma_{K,E} \cdot \nabla \lambda_z$ is proportional to λ_z, i.e.

$$\gamma_{K,E} \cdot \nabla \lambda_z = \beta \lambda_z \tag{3.8}$$

for some factor $\beta \in \mathbb{R}$. To see this, we first observe that both sides are polynomials which are determined by their values in the vertices of K, irrespective of whether K is a simplex or parallelepiped. Next, let E' be a face that does not contain z. Then, E' belongs to the zero level set of λ_z. Consequently, $\nabla \lambda_z$ is parallel to $\mathbf{n}_K$ on E' and, since $E' \neq E$ thanks to $z \in E$, Lemma 3.1 yields that $\gamma_{K,E} \cdot \nabla \lambda_z$ vanishes on E', too. In other words, λ_z and $\gamma_{K,E} \cdot \nabla \lambda_z$ vanish in all vertices of K except z and hence (3.8) holds. To determine β, we apply Proposition 3.2 to λ_z and obtain

$$\begin{aligned}\frac{1}{\mu_{d-1}(E)} \int_E \lambda_z - \frac{1}{\mu_d(K)} \int_K \lambda_z &= \frac{1}{\nu_K \mu_d(K)} \int_K \gamma_{K,E} \cdot \nabla \lambda_z \\ &= \beta \frac{1}{\nu_K \mu_d(K)} \int_K \lambda_z.\end{aligned}$$

Evaluating the integrals with the midpoint rule and solving for β yields $\beta = 1$. Now, inserting (3.8) into (3.7) completes the proof. □

Remark 3.4 An inspection of the proof of Proposition 3.3 reveals that similar statements can be derived for weighted integrals if the weight function ρ satisfies (3.8) with ρ in place of λ_z.

Propositions 3.3 allows us to prove the following weighted trace inequality. For its unweighted counterpart, cf. Remark 3.6.

Proposition 3.5 *For every Lebesgue exponent p with $1 \le p < \infty$, every element K, every face E thereof, every vertex z of E, and every function $u \in W^{1,p}(K)$, we have*

$$\frac{1}{\mu_{d-1,z}(E)} \left\| \lambda_z^{\frac{1}{p}} u \right\|_{p;E}^p \le \frac{1}{\mu_{d,z}(K)} \left\| \lambda_z^{\frac{1}{p}} u \right\|_{p;K}^p + \frac{ph_K}{(\nu_K + 1)\mu_{d,z}(K)} \left\| \lambda_z^{\frac{1}{p}} u \right\|_{p;K}^{p-1} \left\| \lambda_z^{\frac{1}{p}} \nabla u \right\|_{p;K},$$

where the quantities $\mu_{d,z}(K)$ and $\mu_{d-1,z}(E)$ are given by equation (3.6).

Proof We apply Proposition 3.3 to the function $w = |u|^p$. Taking into account the last estimate of Lemma 3.1 we obtain

$$\begin{aligned}\frac{1}{\mu_{d-1,z}(E)} \left\| \lambda_z^{\frac{1}{p}} u \right\|_{p;E}^p - \frac{1}{\mu_{d,z}(K)} \left\| \lambda_z^{\frac{1}{p}} u \right\|_{p;K}^p &= \frac{1}{\mu_{d-1,z}(E)} \int_E \lambda_z w - \frac{1}{\mu_{d,z}(K)} \int_K \lambda_z w \\ &\le \frac{1}{(\nu_K + 1)\mu_{d,z}(K)} \int_K \left| \lambda_z \gamma_{K,E} \cdot \nabla w \right| \\ &\le \frac{ph_K}{(\nu_K + 1)\mu_{d,z}(K)} \int_K \lambda_z |\nabla u| \, |u|^{p-1}.\end{aligned}$$

This proves the assertion for the case $p = 1$. If $p > 1$ we conclude with the weighted Hölder inequality

$$\int_K \lambda_z |\nabla u| \, |u|^{p-1} \le \left\| \lambda_z^{\frac{1}{p}} u \right\|_{p;K}^{p-1} \left\| \lambda_z^{\frac{1}{p}} \nabla u \right\|_{p;K}. \qquad \square$$

Remark 3.6 Proposition 3.2 shows that Proposition 3.5 holds without the weights $\lambda_z^{1/p}$ with $\nu_K + 1$ replaced by ν_K and $\mu_{d,z}(K)$ and $\mu_{d-1,z}(E)$ replaced by $\mu_d(K)$ and $\mu_{d-1}(E)$, respectively.

Remark 3.7 Inserting $w = \left|\gamma_{K,E}\right|$ in Propositions 3.2 and 3.3 shows that, for $p = 1$, Proposition 3.5 and its unweighted counterpart are sharp.

Finally, by summing applications of Proposition 3.5, we obtain a trace inequality that bounds scaled and weighted Sobolev norms on the skeleton within a star ω_z.

Corollary 3.8 *For every vertex $z \in \mathcal{N}$ and every function $u \in W^{1,p}(\omega_z)$ with $1 \le p < \infty$, we have*

$$\sum_{E \subset \sigma_z} h_E^{\perp} \left\| \lambda_z^{\frac{1}{p}} u \right\|_{p;E}^p \le d \left\| \lambda_z^{\frac{1}{p}} u \right\|_{p;\omega_z}^p + d \max_{K \subset \omega_z} \frac{ph_K}{\nu_K + 1} \left\| \lambda_z^{\frac{1}{p}} u \right\|_{p;\omega_z}^{p-1} \left\| \lambda_z^{\frac{1}{p}} \nabla u \right\|_{p;\omega_z},$$

where $h_E^{\perp} = \mu_{d,z}(\omega_E)/\mu_{d-1,z}(E)$.

Proof We first consider an arbitrary face E contained in σ_z and apply Proposition 3.5 to E and all elements K contained in ω_E. Taking the sum with respect to the elements and using Hölder's inequality for sums, we get

$$h_E^{\perp} \left\| \lambda_z^{\frac{1}{p}} u \right\|_{p;E}^p \le \left\| \lambda_z^{\frac{1}{p}} u \right\|_{p;\omega_E}^p + \max_{K \subset \omega_E} \frac{ph_K}{\nu_K + 1} \left\| \lambda_z^{\frac{1}{p}} u \right\|_{p;\omega_E}^{p-1} \left\| \lambda_z^{\frac{1}{p}} \nabla u \right\|_{p;\omega_E}. \tag{3.9}$$

Now we sum these estimates with respect to all faces E contained in σ_z. Since every element has d faces sharing the fixed vertex z, every element in ω_z is hit at most d times in the sums on the right-hand side. This proves the assertion if $p = 1$. If $p > 1$, we additionally use the following Hölder inequality for sums

$$\sum_{E \subset \sigma_z} \left\| \lambda_z^{\frac{1}{p}} u \right\|_{p;\omega_E}^{p-1} \left\| \lambda_z^{\frac{1}{p}} \nabla u \right\|_{p;\omega_E} \leq \left\{ \sum_{E \subset \sigma_z} \left\| \lambda_z^{\frac{1}{p}} u \right\|_{p;\omega_E}^{p} \right\}^{\frac{p-1}{p}} \left\{ \sum_{E \subset \sigma_z} \left\| \lambda_z^{\frac{1}{p}} \nabla u \right\|_{p;\omega_E}^{p} \right\}^{\frac{1}{p}}. \qquad \square$$

Remark 3.9 The trace equalities and inequalities of this section were first established in [247]. For triangles, the vector field of Lemma 3.1 was used in the trace theorem [7, Lemma 10].

3.4 Poincaré and Friedrichs' Inequalities

Throughout this section, we denote by ω, ζ, and μ_ζ an open, connected, bounded, polyhedral domain in $\mathbb{R}^d$ with Lipschitz boundary, a measurable, non-negative weight function or density with $\int_\omega \zeta > 0$, and the corresponding measure, i.e. $\mu_\zeta(\omega) = \int_\omega \zeta$. In the applications, the domain ω will usually be the support ω_z of a nodal shape function λ_z and the density ζ will either be 1 or the shape function λ_z.

It is well-known [3] that the *Poincaré inequality*

$$\inf_{c \in \mathbb{R}} \left\| \zeta^{\frac{1}{p}} (v - c) \right\|_{p;\omega} \leq c(\omega) \operatorname{diam}(\omega) \left\| \zeta^{\frac{1}{p}} \nabla v \right\|_{p;\omega}$$

holds for every $v \in W^{1,p}(\omega)$. The optimal constant in this inequality is given by

$$C_{P,p,\zeta}(\omega) = \sup_{v \in W^{1,p}(\omega) \setminus \mathbb{R}} \frac{\inf_{c \in \mathbb{R}} \left\| \zeta^{\frac{1}{p}} (v - c) \right\|_{p;\omega}}{\operatorname{diam}(\omega) \left\| \zeta^{\frac{1}{p}} \nabla v \right\|_{p;\omega}}.$$

It only depends on ω, p, and ζ and is often referred to as the *Poincaré constant* of ω. Here, as usual, $\operatorname{diam}(\omega)$ denotes the diameter of ω with respect to the Euclidean norm. Notice that sometimes the quantity

$$\overline{C}_{P,p,\zeta}(\omega) = \sup_{v \in W^{1,p}(\omega) \setminus \mathbb{R}} \frac{\left\| \zeta^{\frac{1}{p}} (v - \overline{v}_\omega) \right\|_{p;\omega}}{\operatorname{diam}(\omega) \left\| \zeta^{\frac{1}{p}} \nabla v \right\|_{p;\omega}}$$

is also called the Poincaré constant of ω where $\overline{v}_\omega = \int_\omega \zeta v / \int_\omega \zeta$ is the average of v with respect to the measure μ_ζ. Since

$$\inf_{c \in \mathbb{R}} \left\| \zeta^{\frac{1}{p}} (v - c) \right\|_{p;\omega} \leq \left\| \zeta^{\frac{1}{p}} (v - \overline{v}_\omega) \right\|_{p;\omega} \leq 2 \inf_{c \in \mathbb{R}} \left\| \zeta^{\frac{1}{p}} (v - c) \right\|_{p;\omega}$$

both quantities are related by

$$C_{P,p,\zeta}(\omega) \leq \overline{C}_{P,p,\zeta}(\omega) \leq 2\, C_{P,p,\zeta}(\omega).$$

The Poincaré inequality is usually proved by a contradiction and scaling argument. This, however, gives no information on the size of the Poincaré constant. In what follows we will present several possibilities to derive explicit and reasonably sharp bounds for the Poincaré constant $C_{P,p,\zeta}(\omega)$ from geometric properties of the set ω. In doing so we always have in mind the application to the sets ω_z associated with the nodal shape functions λ_z.

In the case $p = \infty$, for all densities ζ, the mean value theorem implies $C_{P,p,\zeta}(\omega) \leq 1$ for all convex domains and $C_{P,p,\zeta}(\omega) \leq 2$ for all domains which are star-shaped with respect to a point. Therefore, we will focus in what follows on Lebesgue exponents in the interval $[1, \infty)$.

3.4.1 Convex Domains

For convex domains the following result follows from [**2**], [**107**, Theorem 1.2 (3)] for $p = 1$, [**50**], [**107**, Theorem 1.2 (4)], [**217**] for $p = 2$ and [**107**, Theorem 1.2 (1)] for arbitrary $p \in (1, \infty) \setminus \{2\}$.

Proposition 3.10 *Assume that ω is* convex *and the density ζ is* admissible, *i.e. a positive power of a positive concave function. Then for all Lebesgue exponents $p \in [1, \infty)$ the Poincaré constant of ω is bounded by*

$$\begin{aligned} C_{P,1,\zeta}(\omega) &\leq \overline{C}_{P,1,\zeta}(\omega) \leq \overline{C}_{P,1} = \frac{1}{2}, \\ C_{P,2,\zeta}(\omega) &\leq \overline{C}_{P,2,\zeta}(\omega) \leq \overline{C}_{P,2} = \frac{1}{\pi}, \\ C_{P,p,\zeta}(\omega) &\leq \overline{C}_{P,p,\zeta}(\omega) \leq \overline{C}_{P,p} = 2\left(\frac{p}{2}\right)^{\frac{1}{p}}. \end{aligned} \tag{3.10}$$

These bounds are sharp for the cases $p = 1$ and $p = 2$.

If ω is *not convex*, we are not aware of a similarly nice and general result. In fact, Examples 3.23 (p. 101) and 3.24 (p. 102) show that for non-convex, star-shaped domains the Poincaré constant may be arbitrarily large. These examples indicate that bounds for the Poincaré constants of non-convex domains must incorporate some parameter which measures the lack of convexity. In the following subsections we propose several approaches in this direction. In doing so we do not aim at a general result for arbitrary non-convex domains. Instead we always have in mind the application to the sets ω_z. These are in general not convex, but are star-shaped with respect to the point z. Contrary to general star-shaped domains, they have additional structures which allow one to bound their Poincaré constants in terms of a few easily computable geometric quantities.

3.4.2 Transformation

The following result describes the behaviour of Poincaré constants under transformations.

Proposition 3.11 *For every $p \in [1, \infty)$ and every bijective Lipschitz continuous transformation $F : \omega \to \widehat{\omega}$ with Lipschitz continuous inverse we have*

$$C_{P,p,\zeta}(\omega) \leq C_{P,p,\zeta\circ F^{-1}|\det DF^{-1}|}(\widehat{\omega}) \max_{\omega} \left|\left|\left|DF^{-1}\right|\right|\right|_2 \frac{\operatorname{diam}(\widehat{\omega})}{\operatorname{diam}(\omega)}$$

and

$$C_{P,p,\zeta}(\omega) \leq C_{P,p,\zeta\circ F^{-1}}(\widehat{\omega}) \left(\frac{\max_{\omega} |\det DF|}{\min_{\omega} |\det DF|}\right)^{\frac{1}{p}} \max_{\omega} \left|\left|\left|DF^{-1}\right|\right|\right|_2 \frac{\operatorname{diam}(\widehat{\omega})}{\operatorname{diam}(\omega)}$$

where $|||\cdot|||_2$ denotes the spectral norm of $d \times d$ matrices.

Proof Since F and F^{-1} are Lipschitz functions, the transformation rule yields for every function $v \in L^p(\omega)$ and every constant c

$$\left\| \zeta^{\frac{1}{p}} (v - c) \right\|_{p;\omega}^{p} = \int_{\widehat{\omega}} \zeta \circ F^{-1} \left| v \circ F^{-1} - c \right|^p \left| \det DF^{-1} \right| \\ \leq \frac{1}{\min_\omega |\det DF|} \int_{\widehat{\omega}} \zeta \circ F^{-1} \left| v \circ F^{-1} - c \right|^p .$$

We choose c as the best L^p-approximation by constants of $v \circ F^{-1}$ with respect to either the density $\zeta \circ F^{-1} \left| \det DF^{-1} \right|$ or the density $\zeta \circ F^{-1}$ and obtain

$$\int_{\widehat{\omega}} \zeta \circ F^{-1} \left| v \circ F^{-1} - c \right|^p \left| \det DF^{-1} \right| \\ \leq C_{P,p,\zeta \circ F^{-1} |\det DF^{-1}|}(\widehat{\omega}) \operatorname{diam}(\widehat{\omega})^p \int_{\widehat{\omega}} \zeta \circ F^{-1} \left| \nabla (v \circ F^{-1}) \right|^p \left| \det DF^{-1} \right|$$

and

$$\int_{\widehat{\omega}} \zeta \circ F^{-1} \left| v \circ F^{-1} - c \right|^p \leq C_{P,p,\zeta \circ F^{-1}}(\widehat{\omega}) \operatorname{diam}(\widehat{\omega})^p \int_{\widehat{\omega}} \zeta \circ F^{-1} \left| \nabla (v \circ F^{-1}) \right|^p .$$

Transforming back to ω proves the assertion. □

3.4.3 Reduction

The following result serves to describe the behaviour of the Poincaré constant when passing to a subdomain.

Lemma 3.12 *Assume that the domains ω and $\widehat{\omega} \subset \omega$ are star-shaped with respect to the common point z and that there are a subset $\mathbb{S}$ of the unit sphere S^{d-1} in $\mathbb{R}^d$ and two continuous, strictly positive functions $r, \rho : \mathbb{S} \to \mathbb{R}$ such that*

$$\omega = \left\{ x \in \mathbb{R}^d : x = z + s\sigma \,,\, \sigma \in \mathbb{S} \,,\, 0 \leq s < r(\sigma) \right\} , \\ \widehat{\omega} = \left\{ x \in \mathbb{R}^d : x = z + s\sigma \,,\, \sigma \in \mathbb{S} \,,\, 0 \leq s < \rho(\sigma) \right\} .$$

Moreover assume that for every $\sigma \in \mathbb{S}$ the function $[0, r(\sigma)] \ni s \mapsto \zeta(z + s\sigma)$ is decreasing. Set

$$\kappa = \max_{\sigma \in \mathbb{S}} \frac{r(\sigma)}{\rho(\sigma)} .$$

Then the estimate

$$\left\| \zeta^{\frac{1}{p}} v \right\|_{p;\omega}^{p} \leq \left[1 + 2^{p-1} (\kappa^d - 1) \right] \left\| \zeta^{\frac{1}{p}} v \right\|_{p;\widehat{\omega}}^{p} \\ + 2^{p-1} K_{p,d}(\kappa) \operatorname{diam}(\widehat{\omega})^p \left\| \zeta^{\frac{1}{p}} \nabla v \right\|_{p;\omega \setminus \widehat{\omega}}^{p} \\ + 2^{p-1} \frac{p}{d} (\kappa^d - 1) \operatorname{diam}(\widehat{\omega}) \left\| \zeta^{\frac{1}{p}} v \right\|_{p;\widehat{\omega}}^{p-1} \left\| \zeta^{\frac{1}{p}} \nabla v \right\|_{p;\widehat{\omega}}$$

holds for all $p \in [1,\infty)$ and all $v \in W^{1,p}(\omega)$. Here, the function $K_{p,d}$ is defined by

$$K_{p,d}(x) = \begin{cases} \frac{1}{d}(x^d - 1) & \text{if } p = 1, \\ \int_1^x s^{d-1}\left\{\int_1^s t^{-\frac{d-1}{p-1}}\mathrm{d}t\right\}^{p-1}\mathrm{d}s & \text{if } p > 1, \end{cases} \tag{3.11}$$

and satisfies

$$K_{p,d}(x) \le \begin{cases} \frac{1}{d}\left(\frac{p-1}{d-p}\right)^{p-1} x^d\left(1 - x^{-\frac{d-p}{p-1}}\right)^{p-1} & \text{if } p < d, \\ \frac{1}{d}x^d(\ln x)^{d-1} & \text{if } p = d, \\ \frac{1}{p}\left(\frac{p-1}{p-d}\right)^{p-1} x^p\left(1 - x^{-\frac{p-d}{p-1}}\right)^{p-1} & \text{if } p > d. \end{cases} \tag{3.12}$$

Remark 3.13 The assumptions on the functions r and ρ are regularity conditions for the boundaries of the sets ω and $\widehat{\omega}$.

Proof of Lemma 3.12 We may assume that $z = 0$ since the Lebesgue integral is translation invariant and that v is Lipschitz continuous in view of the definition of $W^{1,p}(\omega)$, cf. the first step of the proof of Theorem 1.1 in [**107**]. Since the Lebesgue integral is additive, we have

$$\left\|\zeta^{\frac{1}{p}}v\right\|_{p;\omega}^p = \left\|\zeta^{\frac{1}{p}}v\right\|_{p;\omega\backslash\widehat{\omega}}^p + \left\|\zeta^{\frac{1}{p}}v\right\|_{p;\widehat{\omega}}^p. \tag{3.13}$$

For the first term on the right-hand side of this equation we obtain

$$\begin{aligned} \left\|\zeta^{\frac{1}{p}}v\right\|_{p;\omega\backslash\widehat{\omega}}^p \le{}& 2^{p-1}\int_{\mathbb{S}}\int_{\rho(\sigma)}^{r(\sigma)}\left|v(s\sigma) - v(\rho(\sigma)\sigma)\right|^p s^{d-1}\zeta(s\sigma)\mathrm{d}s\mathrm{d}\sigma \\ &+ 2^{p-1}\int_{\mathbb{S}}\int_{\rho(\sigma)}^{r(\sigma)}\left|v(\rho(\sigma)\sigma)\right|^p s^{d-1}\zeta(s\sigma)\mathrm{d}s\mathrm{d}\sigma. \end{aligned} \tag{3.14}$$

The mean value theorem implies

$$\left|v(s\sigma) - v(\rho(\sigma)\sigma)\right|^p = \left|\int_{\rho(\sigma)}^s \nabla v(t\sigma)\cdot\sigma\,\mathrm{d}t\right|^p \le \left\{\int_{\rho(\sigma)}^s \left|\nabla v(t\sigma)\right.\mathrm{d}t\right\}^p.$$

Since the function $s \mapsto \zeta(s\sigma)$ is decreasing for all $\sigma \in \mathbb{S}$, we have in the case $p = 1$

$$\begin{aligned} &\int_{\mathbb{S}}\int_{\rho(\sigma)}^{r(\sigma)}\left\{\int_{\rho(\sigma)}^s\left|\nabla v(t\sigma)\right|\mathrm{d}t\right\}s^{d-1}\zeta(s\sigma)\mathrm{d}s\mathrm{d}\sigma \\ &\le \int_{\mathbb{S}}\int_{\rho(\sigma)}^{r(\sigma)}\int_{\rho(\sigma)}^s\left|\nabla v(t\sigma)\right|t^{d-1}\mathrm{d}t\left(\frac{s}{\rho(\sigma)}\right)^{d-1}\zeta(s\sigma)\mathrm{d}s\mathrm{d}\sigma \\ &\le \int_{\mathbb{S}}\int_{\rho(\sigma)}^{r(\sigma)}\int_{\rho(\sigma)}^s\left|\nabla v(t\sigma)\right|t^{d-1}\zeta(t\sigma)\mathrm{d}t\left(\frac{s}{\rho(\sigma)}\right)^{d-1}\mathrm{d}s\mathrm{d}\sigma \\ &\le \|\zeta\nabla v\|_{1;\omega\backslash\widehat{\omega}}\max_{\sigma\in\mathbb{S}}\rho(\sigma)K_{1,d}\left(\frac{r(\sigma)}{\rho(\sigma)}\right) \\ &\le K_{1,d}(\kappa)\operatorname{diam}(\widehat{\omega})\,\|\zeta\nabla v\omega\|_{1;\omega\backslash\widehat{\omega}}. \end{aligned}$$

If $p > 1$, Hölder's inequality and the monotonicity of the function $s \mapsto \zeta(s\sigma)$ imply

$$\begin{aligned}
&\int_{\mathbb{S}} \int_{\rho(\sigma)}^{r(\sigma)} \left\{ \int_{\rho(\sigma)}^{s} |\nabla v(t\sigma)| \, \mathrm{d}t \right\}^p s^{d-1} \zeta(s\sigma) \mathrm{d}s \mathrm{d}\sigma \\
&\leq \int_{\mathbb{S}} \int_{\rho(\sigma)}^{r(\sigma)} \int_{\rho(\sigma)}^{s} |\nabla v(t\sigma)|^p \, t^{d-1} \mathrm{d}t \left\{ \int_{\rho(\sigma)}^{s} t^{-\frac{d-1}{p-1}} \mathrm{d}t \right\}^{p-1} s^{d-1} \zeta(s\sigma) \mathrm{d}s \mathrm{d}\sigma \\
&\leq \int_{\mathbb{S}} \int_{\rho(\sigma)}^{r(\sigma)} \int_{\rho(\sigma)}^{s} |\nabla v(t\sigma)|^p \, t^{d-1} \zeta(t\sigma) \mathrm{d}t \left\{ \int_{\rho(\sigma)}^{s} t^{-\frac{d-1}{p-1}} \mathrm{d}t \right\}^{p-1} s^{d-1} \mathrm{d}s \mathrm{d}\sigma \\
&\leq \left\| \zeta^{\frac{1}{p}} \nabla v \right\|_{p;\omega \setminus \widehat{\omega}}^p \max_{\sigma \in \mathbb{S}} \rho(\sigma)^p K_{p,d} \left(\frac{r(\sigma)}{\rho(\sigma)} \right) \\
&\leq K_{p,d}(\kappa) \operatorname{diam}(\widehat{\omega})^p \left\| \zeta^{\frac{1}{p}} \nabla v \right\|_{p;\omega \setminus \widehat{\omega}}^p .
\end{aligned}$$

Thus we have for all p

$$\int_{\mathbb{S}} \int_{\rho(\sigma)}^{r(\sigma)} |v(s\sigma) - v(\rho(\sigma)\sigma)|^p s^{d-1} \zeta(s\sigma) \mathrm{d}s \mathrm{d}\sigma \leq K_{p,d}(\kappa) \operatorname{diam}(\widehat{\omega})^p \left\| \zeta^{\frac{1}{p}} \nabla v \right\|_{p;\omega \setminus \widehat{\omega}}^p . \tag{3.15}$$

Using once again the monotonicity of the function $s \mapsto \zeta(s\sigma)$ and the boundedness of v, we obtain for the second term on the right-hand side of (3.14)

$$\begin{aligned}
\int_{\rho(\sigma)}^{r(\sigma)} |v(\rho(\sigma)\sigma)|^p s^{d-1} \zeta(s\sigma) \mathrm{d}s &\leq \int_{\rho(\sigma)}^{r(\sigma)} |v(\rho(\sigma)\sigma)|^p s^{d-1} \zeta(\rho(\sigma)\sigma) \mathrm{d}s \\
&= \frac{1}{d} \left[\left(\frac{r(\sigma)}{\rho(\sigma)} \right)^d - 1 \right] \rho(\sigma)^d |v(\rho(\sigma)\sigma)|^p \zeta(\rho(\sigma)\sigma) \\
&= \frac{1}{d} \left[\left(\frac{r(\sigma)}{\rho(\sigma)} \right)^d - 1 \right] \int_0^{\rho(\sigma)} \frac{\mathrm{d}}{\mathrm{d}t} \left[t^d |v(t\sigma)|^p \right] \mathrm{d}t \, \zeta(\rho(\sigma)\sigma) \\
&\leq \frac{1}{d} \left[\left(\frac{r(\sigma)}{\rho(\sigma)} \right)^d - 1 \right] \int_0^{\rho(\sigma)} \left| \frac{\mathrm{d}}{\mathrm{d}t} \left[t^d |v(t\sigma)|^p \right] \right| \zeta(t\sigma) \mathrm{d}t.
\end{aligned}$$

Since

$$\begin{aligned}
\int_0^{\rho(\sigma)} \left| \frac{\mathrm{d}}{\mathrm{d}t} \left[t^d |v(t\sigma)|^p \right] \right| \zeta(t\sigma) \mathrm{d}t &\leq d \int_0^{\rho(\sigma)} t^{d-1} |v(t\sigma)|^p \zeta(t\sigma) \mathrm{d}t \\
&\quad + p\rho(\sigma) \int_0^{\rho(\sigma)} t^{d-1} |v(t\sigma)|^{p-1} |\nabla v(t\sigma)| \, \zeta(t\sigma) \mathrm{d}t,
\end{aligned}$$

we conclude that

$$\begin{aligned}
\int_{\mathbb{S}} \int_{\rho(\sigma)}^{r(\sigma)} |v(\rho(\sigma)\sigma)|^p s^{d-1} \zeta(s\sigma) \mathrm{d}s \mathrm{d}\sigma &\leq \left(\kappa^d - 1 \right) \left\| \zeta^{\frac{1}{p}} v \right\|_{p;\widehat{\omega}}^p \\
&\quad + \frac{p}{d} \left(\kappa^d - 1 \right) \operatorname{diam}(\widehat{\omega}) \left\| \zeta^{\frac{1}{p}} v \right\|_{p;\widehat{\omega}}^{p-1} \left\| \zeta^{\frac{1}{p}} \nabla v \right\|_{p;\widehat{\omega}} .
\end{aligned} \tag{3.16}$$

Combining equation (3.13) and inequalities (3.14), (3.15), and (3.16) proves the assertion. □

Lemma 3.12 implies the following estimate for Poincaré constants.

Proposition 3.14 *Assume that the conditions of Lemma 3.12 are fulfilled. Then, for every Lebesgue exponent $p \in [1, \infty)$, the Poincaré constants of the domains ω and $\widehat{\omega} \subset \omega$ are related by*

$$C_{P,p,\zeta}(\omega) \le \frac{\operatorname{diam}(\widehat{\omega})}{\operatorname{diam}(\omega)} \Big\{ C_{P,p,\zeta}(\widehat{\omega})^p \left[2^{p-1}(\kappa^d - 1) + 1\right] + 2^{p-1} \max\left\{K_{p,d}(\kappa), C_{P,p,\zeta}(\widehat{\omega})^{p-1} \frac{p}{d}(\kappa^d - 1)\right\}\Big\}^{\frac{1}{p}}$$

where the parameter κ and the function $K_{p,d}$ are as in Lemma 3.12.

Remark 3.15 Proposition 3.14 slightly generalises older results of [**264, 265**] where the case $p = 2$ and a constant function ρ are considered.

3.4.4 Decomposition

For brevity we say that two open disjoint subsets ω_1 and ω_2 of ω are *neighbours*, if the closure of every set ω_i contains a closed simplex or parallelepiped K_i such that K_1 and K_2 share a complete $(d-1)$-dimensional face. This notion is motivated by the trace inequality of Proposition 3.5 (p. 90) and its unweighted counterpart, cf. Remark 3.6 (p. 90).

Assume that ω_1 and ω_2 are neighbours and that the density ζ is either the constant function 1 or the nodal shape function λ_z of a common vertex of the elements K_1 and K_2. Proposition 3.5 and Remark 3.6 and the observation that $\nu_K \ge 1$ for all elements then yield for all constants v_1 and v_2

$$\begin{aligned} |v_1 - v_2|^p = \frac{1}{\mu_\zeta(E)} \left\| \zeta^{\frac{1}{p}}(v_1 - v_2) \right\|_{p;E}^p &\le \frac{2^{p-1}}{\mu_\zeta(K_1)} \left\| \zeta^{\frac{1}{p}}(v - v_1) \right\|_{p;K_1}^p + \frac{2^{p-1}}{\mu_\zeta(K_2)} \left\| \zeta^{\frac{1}{p}}(v - v_2) \right\|_{p;K_2}^p \\ &+ \frac{2^{p-1} p h_{K_1}}{\mu_\zeta(K_1)} \left\| \zeta^{\frac{1}{p}}(v - v_1) \right\|_{p;K_1}^{p-1} \left\| \zeta^{\frac{1}{p}} \nabla v \right\|_{p;K_1} + \frac{2^{p-1} p h_{K_2}}{\mu_\zeta(K_2)} \left\| \zeta^{\frac{1}{p}}(v - v_2) \right\|_{p;K_2}^{p-1} \left\| \zeta^{\frac{1}{p}} \nabla v \right\|_{p;K_2}, \end{aligned}$$

where E is the common face of K_1 and K_2. With the help of this estimate we can prove the following auxiliary result.

Lemma 3.16 *Assume that there are open subsets ω_i, $1 \le i \le n$, of ω with the following properties: the sets ω_i are pairwise disjoint, ω is the interior of the union of the closures of the ω_i, and for every pair i, j of different indices there is a sequence $i = k_0, \dots, k_\ell = j$ of indices such that, for every m, the sets $\omega_{k_{m-1}}$ and ω_{k_m} are neighbours. Furthermore suppose that the density either equals 1 or is a nodal shape function for all K_i. Then the estimate*

$$\begin{aligned} \left\| \zeta^{\frac{1}{p}} \left(v - \sum_{j=1}^n \alpha_j v_j \right) \right\|_{p;\omega}^p &\le \sum_{i=1}^n \left[\alpha_i + 2^{p-1}(1 - \alpha_i)\right] \left\| \zeta^{\frac{1}{p}}(v - v_i) \right\|_{p;\omega_i}^p \\ &+ \sum_{i=1}^n \left[2(n-1)\right]^{p-1} A_i W_i(v)^p \end{aligned}$$

holds for all functions $v \in W^{1,p}(\omega)$, all constants $v_i \in \mathbb{R}$, and all weights $\alpha_i \geq 0$ with $\sum_i \alpha_i = 1$ where

$$W_i(v)^p = \frac{2^{p-1}}{\mu_\zeta(K_i)} \left\| \zeta^{\frac{1}{p}} (v - v_i) \right\|_{p;K_i}^p + \frac{2^{p-1} p h_{K_i}}{\mu_\zeta(K_i)} \left\| \zeta^{\frac{1}{p}} (v - v_i) \right\|_{p;K_i}^{p-1} \left\| \zeta^{\frac{1}{p}} \nabla v \right\|_{p;K_i}$$

$$A_i = 2 \sum_{\substack{j=1 \\ j \neq i}}^{n} (1 - \alpha_j) \mu_\zeta(\omega_j) - \alpha_i \mu_\zeta(\omega) + \mu_\zeta(\omega_i).$$

Proof Since the sets ω_i are pairwise disjoint, we have for every L^p-function

$$\left\| \zeta^{\frac{1}{p}} v \right\|_{p;\omega}^p = \sum_{i=1}^{n} \left\| \zeta^{\frac{1}{p}} v \right\|_{p;\omega_i}^p .$$

Since the p-th power of the L^p-norm is convex, we further have for every index i

$$\left\| \zeta^{\frac{1}{p}} \left(v - \sum_{j=1}^{n} \alpha_j v_j \right) \right\|_{p;\omega_i}^p \leq \sum_{j=1}^{n} \alpha_j \left\| \zeta^{\frac{1}{p}} (v - v_j) \right\|_{p;\omega_i}^p .$$

For every pair of different indices we obtain

$$\left\| \zeta^{\frac{1}{p}} (v - v_j) \right\|_{p;\omega_i}^p \leq 2^{p-1} \left\| \zeta^{\frac{1}{p}} (v - v_i) \right\|_{p;\omega_i}^p + 2^{p-1} \mu_\zeta(\omega_i) \left| v_i - v_j \right|^p$$

and

$$\begin{aligned} \left| v_i - v_j \right|^p &\leq \ell^{p-1} \sum_{m=1}^{\ell} \left| v_{k_m} - v_{k_{m-1}} \right|^p \\ &\leq (n-1)^{p-1} \sum_{m=1}^{\ell} \left[W_{k_m}(v)^p + W_{k_{m-1}}(v)^p \right] \\ &\leq (n-1)^{p-1} \left[2 \sum_{k=1}^{n} W_k(v)^p - W_i(v)^p - W_j(v)^p \right]. \end{aligned}$$

Since ω is the union of the pairwise disjoint sets ω_i, we get for every index k

$$\begin{aligned} &2 \sum_{i=1}^{n} \sum_{\substack{j=1 \\ j \neq i}}^{n} \alpha_j \mu_\zeta(\omega_i) - \sum_{\substack{j=1 \\ j \neq k}}^{n} \alpha_j \mu_\zeta(\omega_k) - \sum_{\substack{i=1 \\ i \neq k}}^{n} \alpha_k \mu_\zeta(\omega_i) \\ &= 2 \sum_{i=1}^{n} (1 - \alpha_i) \mu_\zeta(\omega_i) - (1 - \alpha_k) \mu_\zeta(\omega_k) - \alpha_k \left[\mu_\zeta(\omega) - \mu_\zeta(\omega_k) \right] \\ &= 2 \sum_{\substack{i=1 \\ i \neq k}}^{n} (1 - \alpha_i) \mu_\zeta(\omega_i) + (1 - \alpha_k) \mu_\zeta(\omega_k) - \alpha_k \left[\mu_\zeta(\omega) - \mu_\zeta(\omega_k) \right] \\ &= A_k. \end{aligned}$$

Combining all these results proves the assertion. □

Lemma 3.16 implies the following bounds for Poincaré constants.

Proposition 3.17 *Suppose that the assumptions of Lemma 3.16 are fulfilled. Then the Poincaré constant of ω is bounded by*

$$\begin{aligned}
C_{P,p,\zeta}(\omega) \le \min_{1\le k\le n} \Bigg\{ \max \Bigg\{ \Bigg(1 + \big[4(n-1)\big]^{p-1} \frac{\mu_\zeta(\omega) - \mu_\zeta(\omega_k)}{\mu_\zeta(K_k)} \\
\cdot \Bigg[1 + \frac{p}{C_{P,p,\zeta}(\omega_k)} \Bigg] \Bigg) C_{P,p,\zeta}(\omega_k)^p \frac{\operatorname{diam}(\omega_k)^p}{\operatorname{diam}(\omega)^p}, \\
\max_{i\ne k} \Bigg(2^{p-1} + \big[4(n-1)\big]^{p-1} \frac{2\mu_\zeta(\omega) - 2\mu_\zeta(\omega_k) - \mu_\zeta(\omega_i)}{\mu_\zeta(K_i)} \\
\cdot \Bigg[1 + \frac{p}{C_{P,p,\zeta}(\omega_i)} \Bigg] \Bigg) C_{P,p,\zeta}(\omega_i)^p \frac{\operatorname{diam}(\omega_i)^p}{\operatorname{diam}(\omega)^p} \Bigg\} \Bigg\}^{\frac{1}{p}} \\
\le \max_{1\le i\le n} \Bigg\{ 2\big[4(n-1)\big]^{p-1} \Bigg(1 - \min_k \frac{\mu_\zeta(\omega_k)}{\mu_\zeta(\omega)} \Bigg) \\
\cdot \Bigg[1 + \frac{p}{C_{P,p,\zeta}(\omega_i)} \Bigg] C_{P,p,\zeta}(\omega_i)^p \frac{\mu_\zeta(\omega)\operatorname{diam}(\omega_i)^p}{\mu_\zeta(K_i)\operatorname{diam}(\omega)^p} \Bigg\}^{\frac{1}{p}}.
\end{aligned}$$

Proof In Lemma 3.16 we choose every v_i as a best L^p-approximation of v on ω_i. Since $h_{K_i} \le \operatorname{diam}(\omega_i)$ for every index i, we thus obtain

$$\begin{aligned}
C_{P,p,\zeta}(\omega) \le \min_{\alpha_i} \max_{1\le i\le n} \Bigg\{ \Bigg[\alpha_i + 2^{p-1}(1-\alpha_i) + (4(n-1))^{p-1} \frac{A_i}{\mu_\zeta(K_i)} \Bigg] \\
\cdot C_{P,p,\zeta}(\omega_i)^p \Bigg[1 + \frac{p}{C_{P,p,\zeta}(\omega_i)} \Bigg] \frac{\operatorname{diam}(\omega_i)^p}{\operatorname{diam}(\omega)^p} \Bigg\}^{\frac{1}{p}}
\end{aligned}$$

where the minimum has to be taken with respect to all sets of convex weights. Since the right-hand side of this estimate is an affine function of the weights, it is minimised by choosing an appropriate index k and setting $\alpha_k = 1$ and $\alpha_i = 0$ for the other indices. Obviously we have

$$\alpha_k + 2^{p-1}(1-\alpha_k) = 1, \quad A_k = \mu_\zeta(\omega) - \mu_\zeta(\omega_k)$$

and for all $i \ne k$

$$\alpha_i + 2^{p-1}(1-\alpha_i) = 2^{p-1}, \quad A_i = 2\mu_\zeta(\omega) - 2\mu_\zeta(\omega_k) - \mu_\zeta(\omega_i).$$

This proves the first estimate of the proposition. The second one follows from the first one taking into account that

$$\mu_\zeta(\omega) - \mu_\zeta(\omega_k) \le 2\mu_\zeta(\omega) - 2\min_i \mu_\zeta(\omega_i) - \mu_\zeta(\omega_k), \quad \mu_\zeta(\omega_k) \ge \mu_\zeta(K_k)$$

for all k and that $[4(n-1)]^{p-1} \ge 2^{p-1}$. □

3.4.5 Application to Finite Element Stars

Using the auxiliary results of the preceding subsections we are now able to give explicit bounds for the Poincaré constants of *finite element stars* ω_z. We first consider the case of *convex* stars.

Proposition 3.18 *Suppose that the star ω_z consists either of simplices or of parallelepipeds and is* convex. *Moreover assume that the density ζ either equals 1 or is the nodal shape function λ_z. Then the Poincaré constant of ω_z satisfies $C_{P,p,\zeta}(\omega_z) \le \overline{C}_{P,p}$ where $\overline{C}_{P,p}$ is given by (3.10).*

Proof It suffices to prove that the basis function λ_z is an admissible weight.

We first consider the simplex case. Since the minimum of concave functions is again a concave function, we may prove the admissibility of λ_z by verifying the following identity

$$\lambda_z = \min_{K \subset \omega_z} \lambda_{K,z} \qquad \text{in } \omega_z, \tag{3.17}$$

where $\lambda_{K,z}$ denotes the global affine function in $\mathbb{R}^d$ that coincides with λ_z on K, cf. Section 3.2.5 and Figure 3.2 (p. 85). To this end note first that ω_z is convex, $\{\lambda_{K,z} = 0\}$ is a supporting plane for ω_z, and $\lambda_{K,z}(z) = 1$. Hence there holds $\omega_z \subset \{\lambda_{K,z} \ge 0\}$ for all $K \subset \omega_z$. Next, fix an element K of ω_z and consider, in K, the function $\lambda_{K,z} - \lambda_{K',z}$ where K' is another element of ω_z. This function is affine and therefore attains its maximum in the vertices of K. Checking the nodal values with the help of $K \subset \left\{\lambda_{K',z} \ge 0\right\}$ yields $\lambda_{K,z} \le \lambda_{K',z}$ in K. Consequently, (3.17) holds and λ_z is an admissible weight in the simplex case.

Next, we consider the parallelepiped case. Note that here the restriction of λ_z to an element is not concave. Yet, we claim that $\sqrt[d]{\lambda_z}$ is concave on its support ω_z and that consequently λ_z is an admissible weight also in this case. To this end, map by an affine transformation the parallelepiped star onto a reference star consisting of bricks, i.e. parallelepipeds with faces parallel to the coordinate planes. The transformed shape function is then the product of d non-negative concave one-dimensional hat functions. Consequently, its d-th root is concave and so the d-th root of the original shape function λ_z. □

Next we consider the case of *non-convex* stars. To formulate our results we need a few notations. For every star ω_z we denote by n_z the number of its elements and by γ_z the union of all faces in the boundary of ω_z which do not contain the vertex z. Without loss of generality we may assume that every star ω_z satisfies an *exterior cone condition*. Therefore we can associate with every ω_z the following geometric quantities

$$R_z = \max_{y \in \gamma_z} |y - z|, \qquad \rho_z = \min_{y \in \gamma_z} |y - z|, \qquad \kappa_z = \frac{R_z}{\rho_z}. \tag{3.18}$$

Notice for practical computations that R_z is the maximal distance from z of any vertex of ω_z different from z and that $\rho_z = \min_{E \subset \gamma_z} \nu_{K_E} \frac{\mu_d(K_E)}{\mu_{d-1}(E)}$ where K_E is the element adjacent to E and where ν_{K_E} is given by equation (3.5) (p. 82).

Our first result estimates the Poincaré constant of ω_z in terms of the parameter κ_z. Here, $\overline{C}_{P,p}$ is given by (3.10) and a vertex of a polyhedron is called a *re-entrant corner* if the intersection of the polyhedron with every ball centred at the vertex is non-convex.

Proposition 3.19 *Suppose that the density ζ either equals 1 or is the nodal shape function λ_z. Then the Poincaré constant of any star ω_z is bounded by*

$$C_{P,p,\zeta}(\omega_z) \le 4\overline{C}_{P,p}\kappa_z^{\frac{d}{p}-1}\left(\frac{1}{2}+\frac{1}{2}\max\left\{\kappa_z^{-d}K_{p,d}(\kappa_z), \frac{p}{d\overline{C}_{P,p}}\right\}\right)^{\frac{1}{p}}$$

if z is not a re-entrant corner of ω_z and by

$$C_{P,p,\zeta}(\omega_z) \le 8\overline{C}_{P,p}k_d\,(k_d\kappa_z)^{\frac{d}{p}-1}\left(\frac{1}{2}+\frac{1}{2}\max\left\{(k_d\kappa_z)^{-d}K_{p,d}(k_d\kappa_z), \frac{p}{d\overline{C}_{P,p}}\right\}\right)^{\frac{1}{p}}$$

if z is a re-entrant corner of ω_z. Here, the parameter κ_z is as in (3.18), the quantity k_d is given by

$$k_d = \begin{cases} 1+\sqrt{2} & \text{if } d=2, \\ 2+\sqrt{3} & \text{if } d=3, \\ \dfrac{3d}{2} & \text{if } d>3, \end{cases} \tag{3.19}$$

and the function $K_{p,d}$ is defined by equation (3.11) (p. 94) and can be estimated as in inequality (3.12) (p. 94).

Remark 3.20 Estimate (3.12) shows that, for large κ_z, the asymptotic behaviour of the bound for the Poincaré constant is of order $\kappa_z^{d/p-1}$ for $p < d$, $(\ln \kappa_z)^{d-1}$ for $p = d$ and 1 for $p > d$.

Proof of Proposition 3.19 Due to the definition of ρ_z we can apply Lemma 3.12 to $\omega = \omega_z$ with the constant function $\rho = \rho_z$. We then have $\kappa = \kappa_z$ and $\operatorname{diam}(\widehat{\omega})/\operatorname{diam}(\omega) \le 2/\kappa_z$. If z is an interior point of ω_z, the set $\widehat{\omega}$ is the ball centred at z with radius ρ_z, otherwise it is a strict subset of this ball. Still, the domain $\widehat{\omega}$ is convex, if z is not a re-entrant corner. Thus Proposition 3.14, inequality (3.10), and the estimates $2^{p-1}(\kappa^d - 1) + 1 \le 2^{p-1}\kappa^d$ and $\kappa^d - 1 \le \kappa^d$ prove Proposition 3.19 in this case.

For the case of a re-entrant corner, we need an additional argument: we first transform ω_z to a reference polyhedron $\widetilde{\omega}_z$ with vertex z such that z is not a re-entrant corner of $\widetilde{\omega}_z$, then apply Proposition 3.14 to $\omega = \widetilde{\omega}_z$, and express the corresponding geometric quantities in terms of R_z, ρ_z, and κ_z. The transformation F is the composition of $d-1$ transformations F_i. The i-th transformation F_i only affects the coordinates x_i and x_{i+1} and is piecewise affine on the four quadrants of the (x_i, x_{i+1})-plane. In this plane, up to rotation, the restriction of F_i to any of the quadrants has the derivative $\begin{pmatrix} 1 & 1/\sqrt{2} \\ 0 & 1/\sqrt{2} \end{pmatrix}$. Hence we have $\max \det DF / \min \det DF = 1$. Since $\left|\left|\left|DF^{-1}\right|\right|\right|_2 \operatorname{diam}(\widetilde{\omega}_z) \le 2\,|||DF|||_2 \left|\left|\left|DF^{-1}\right|\right|\right|_2 \operatorname{diam}(\omega_z)$, Proposition 3.11 yields

$$C_{P,p,\zeta}(\omega_z) \le 2\,|||DF|||_2 \left|\left|\left|DF^{-1}\right|\right|\right|_2 C_{P,p,\zeta}(\widetilde{\omega}_z).$$

Since F is the composition of piecewise affine transformations, the function λ_z is transformed into the nodal shape function corresponding to z and $\widetilde{\omega}_z$. Therefore, we can apply Proposition 3.14 to $\omega = \widetilde{\omega}_z$ with the constant function $\rho = \widetilde{\rho}_z$ where $\widetilde{\rho}_z$ is defined as ρ_z with ω_z replaced by $\widetilde{\omega}_z$. We then have $\kappa \le |||DF|||_2 \left|\left|\left|DF^{-1}\right|\right|\right|_2 \kappa_z$. Since F maps $\mathbb{R}^d$ to $\mathbb{R} \times \mathbb{R}_+^{d-1}$, the vertex z is not a re-entrant corner of $\widetilde{\omega}_z$. Therefore, the set $\widehat{\omega}$ is now convex and Proposition 3.14 and estimate (3.10) prove Proposition 3.19 if we can show that $|||DF|||_2 \left|\left|\left|DF^{-1}\right|\right|\right|_2 \le k_d$ with k_d as in (3.19). To this end we observe that the 2^d different values of DF only differ by multiplicative rotations. Therefore we only have to consider the restriction of DF to $\mathbb{R}_+^d$. There, we obtain

$$DF = \begin{pmatrix} 1 & \frac{1}{\sqrt{2}} & 0 & 0 & \dots \\ 0 & \frac{1}{\sqrt{2}} & \frac{1}{\sqrt{2}} & 0 & \dots \\ 0 & 0 & \ddots & \ddots & \\ \vdots & \vdots & & \ddots & \end{pmatrix}, \quad DF^{-1} = \begin{pmatrix} 1 & -1 & 1 & -1 & \dots \\ 0 & \sqrt{2} & -\sqrt{2} & \sqrt{2} & \dots \\ 0 & 0 & \sqrt{2} & -\sqrt{2} & \\ \vdots & \vdots & & \ddots & \end{pmatrix}.$$

For $d = 2$ and $d = 3$ an elementary calculation yields the desired estimate. For general $d > 3$ it follows from Gerschgorin's theorem [**222**, Theorem 5.2] applied to $DF^t DF$ and $DF^{-t}DF^{-1}$. □

Our next result is complementary to the first one in that it estimates the Poincaré constant in terms of the number n_z of simplices and parallelepipeds and is independent of the parameter κ_z. As in Proposition 3.19 the quantities $\overline{C}_{P,p}$ are given by (3.10).

Proposition 3.21 *Suppose that the density ζ either equals 1 or is the nodal shape function λ_z. Then the Poincaré constant of any star ω_z is bounded by*

$$\begin{aligned} C_{P,p,\zeta}(\omega_z) &\le \overline{C}_{P,p} \max_{1\le i\le n_z} \frac{h_{K_i}\mu_\zeta(\omega_z)^{\frac{1}{p}}}{\mu_\zeta(K_i)^{\frac{1}{p}} h_z} \cdot \left\{2\left[4(n_z-1)\right]^{p-1}\left[1 - \min_{1\le i\le n_z}\frac{\mu_\zeta(K_i)}{\mu_\zeta(\omega_z)}\right]\left[1+\frac{p}{\overline{C}_{P,p}}\right]\right\}^{\frac{1}{p}} \\ &\le 4\overline{C}_{P,p}\,(n_z-1)^{1-\frac{1}{p}}\left[\frac{1}{2}+\frac{p}{2\overline{C}_{P,p}}\right]^{\frac{1}{p}} \max_{1\le i\le n_z}\frac{h_{K_i}\mu_\zeta(\omega_z)^{\frac{1}{p}}}{\mu_\zeta(K_i)^{\frac{1}{p}} h_z}, \end{aligned}$$

where h_z denotes the diameter of ω_z.

Proof In Lemma 3.16 and Proposition 3.17 we choose the elements as the subdomains ω_i. Since the set ω_z is assumed to satisfy an exterior cone condition, the assumptions of Lemma 3.16 are fulfilled. Proposition 3.21 therefore follows from Proposition 3.17. □

Remark 3.22 The bounds of Propositions 3.19 and 3.21 can easily be computed from geometric data of the sets ω_z. The examples of the next subsection show that Propositions 3.19 and 3.21 are complementary in that in some cases Proposition 3.19 yields the better result, in others Proposition 3.21. The examples moreover show that the minimum of the bounds of Propositions 3.19 and 3.21 gives an estimate for the Poincaré constants of the ω_z which is sharp when taking the supremum of the Poincaré constants with respect to all possible such sets.

3.4.6 Examples

In the following examples the density ζ always equals either 1 or the nodal shape function λ_z associated with the corresponding star.

Our first example shows that the Poincaré constants of stars may be arbitrarily large and that Proposition 3.19 can capture the correct asymptotic growth.

Example 3.23 For $0 < \varepsilon \ll 1$ we consider stars $\omega_{z,\varepsilon}$ with vertices $(\pm 1, \pm 1)$, $(\pm 1, 0)$, and $(0, \pm\varepsilon)$ that consist of $n_z = 8$ triangles which all share the vertex $z = (0,0)$, cf. Figure 3.5. They are obtained from a standard union-jack partition of the square $(-1,1)^2$ by contracting by a factor ε the edges on the x_2-axes. On $\omega_{z,\varepsilon}$ we consider the functions

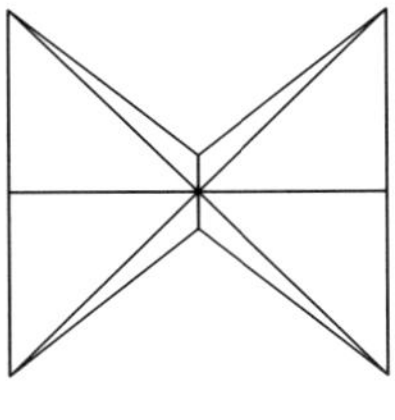

Figure 3.5 Domain of Example 3.23

$$u_\varepsilon(x) = \begin{cases} -1 & \text{if } x_1 \leq -\varepsilon \\ \dfrac{x_1}{\varepsilon} & \text{if } -\varepsilon < x_1 < \varepsilon \\ 1 & \text{if } x_1 \geq \varepsilon. \end{cases}$$

Obviously we have $\operatorname{diam}(\omega_{z,\varepsilon}) = 2\sqrt{2}$ and

$$\left\|\zeta^{\frac{1}{p}} u_\varepsilon\right\|^p_{p;\omega_{z,\varepsilon}} \geq \mu_\zeta(\omega_{z,\varepsilon} \cap \{|x_1| \geq \varepsilon\})$$
$$\left\|\zeta^{\frac{1}{p}} \nabla u_\varepsilon\right\|^p_{p;\omega_{z,\varepsilon}} = \varepsilon^{-p}\mu_\zeta(\omega_{z,\varepsilon} \cap \{|x_1| \leq \varepsilon\}).$$

Since u_ε has vanishing average, this implies

$$C_{P,p,\zeta}(\omega_{z,\varepsilon}) \geq \frac{1}{2}\overline{C}_{P,p,\zeta}(\omega_{z,\varepsilon}) \gtrsim \frac{\sqrt{2}}{8} 3^{-\frac{1}{p}} \varepsilon^{1-\frac{2}{p}}.$$

For Proposition 3.19 we have $\kappa_z = \sqrt{2}\varepsilon^{-1}$ and thus obtain

$$C_{P,p,\zeta}(\omega_{z,\varepsilon}) \lesssim 4\overline{C}_{P,p}\varepsilon^{1-\frac{2}{p}}.$$

Thus Proposition 3.19 yields the correct asymptotic behaviour of the Poincaré constant. Since $\max_i \frac{h^p_{K_i}\mu_\zeta(\omega_{z,\varepsilon})}{\mu_\zeta(K_i)\operatorname{diam}(\omega_{z,\varepsilon})^p} \approx \varepsilon^{-1}$, Proposition 3.21 yields for this example the incorrect asymptotic growth $C_{P,p,\zeta}(\omega_{z,\varepsilon}) \lesssim \varepsilon^{-\frac{1}{p}}$.

Our second example is a modification of the first one. It shows that, at least for $p = 1$, Proposition 3.21 can capture the correct asymptotic growth of the Poincaré constants of a family of degenerating stars.

Example 3.24 For integers $m \geq 2$ we consider stars $\omega_{z,m}$ with vertices $(0, \pm\frac{1}{m^2})$, $(\pm 1, 0)$, and $(\pm 1, \pm\tan(\frac{\pi\ell}{4m}))$, $r_\ell(\pm 1, \pm\tan(\frac{\pi(2\ell-1)}{8m}))$, $1 \leq \ell \leq m$, which consist of $n_z = 8m + 4$ triangles that all share the vertex $z = (0, 0)$, cf. Figure 3.6. The r_ℓ are chosen such that all triangles have the same area $\frac{1}{2m^2}$. Since $r_\ell \approx \frac{1}{m}$ for all ℓ and since $\mu_\zeta(\omega_{z,m} \cap \{|x_1| \geq m^{-2}\}) \gtrsim 8m^{-1} - 2m^{-2} \geq 6m^{-1}$, the arguments of Example 3.23 with $\varepsilon = \frac{1}{m^2}$ imply that

$$C_{P,p,\zeta}(\omega_{z,m}) \gtrsim m^{\frac{4}{p}-3}.$$

Since all triangles have the same area and a diameter of at most $\sqrt{2}$, Proposition 3.21 yields

$$C_{P,p,\zeta}(\omega_{z,m}) \lesssim m$$

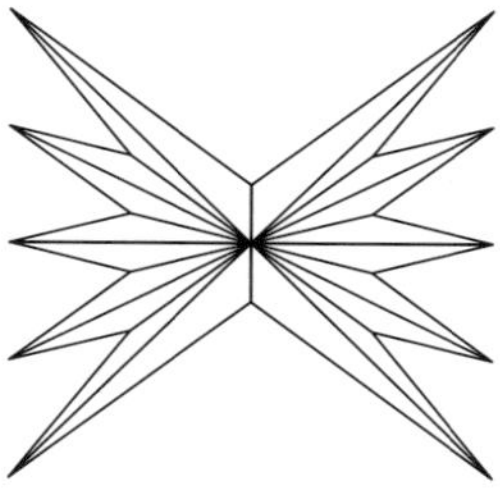

Figure 3.6 Domain of Example 3.24

which is the correct asymptotic behaviour for $p = 1$. Since $\kappa_z = \sqrt{2}m^2$, Proposition 3.19 gives for $1 \leq p < 2$ the bound $C_{P,p,\zeta}(\omega_{z,m}) \lesssim m^{\frac{4}{p}-2}$ which is too large by a factor m.

Our next example is complementary to the first and second one. It shows that the Poincaré constants of a family of degenerating stars may be uniformly bounded and that Proposition 3.21 can capture the correct asymptotic behaviour.

Example 3.25 For $0 < \varepsilon \ll 1$ we consider stars $\omega_{z,\varepsilon}$ with vertices $(\pm\varepsilon^{-1}, \pm\varepsilon^{-1})$, $(\pm\varepsilon, 0)$, and $(0, \pm\varepsilon)$ that consist of $n_z = 8$ triangles sharing the vertex $z = (0, 0)$, cf. Figure 3.7. They are obtained from a standard union-jack partition of the square $(-1, 1)^2$ by contracting by a factor ε the edges on the coordinate axes and by stretching by a factor ε^{-1} the edges on the bisectors of the plane. Obviously we have $\operatorname{diam}(\omega_{z,\varepsilon}) = 2\varepsilon^{-1}$, $\operatorname{diam}(K_i) = \varepsilon^{-1}$, and $\frac{\mu_\zeta(\omega_{z,\varepsilon})}{\mu_\zeta(K_i)} = 8$ for all triangles. Hence, for all Lebesgue exponents p, Proposition 3.21 yields the uniform bound

$$C_{P,p,\zeta}(\omega_{z,\varepsilon}) \leq 14\overline{C}_{P,p}\left[\frac{1}{2} + \frac{p}{2\overline{C}_{P,p}}\right]^{\frac{1}{p}}$$

for the Poincaré constants. Since $\kappa_z = \varepsilon^{-2}$, Proposition 3.19 yields a uniform bound for the Poincaré constants only for Lebesgue exponents $p > d$.

The next example is typical for a locally uniform finite element mesh near a re-entrant corner.

Example 3.26 We consider stars $\omega_{z,\alpha}$ which are obtained from a union-jack partition of the square $(-1, 1)^2$ by removing a circular sector centred at the origin with angle $\alpha \in (0, \pi)$, cf. Figure 3.8. Obviously we have $R_z = \sqrt{2}$, $\rho_z = 1$, $\kappa_z = \sqrt{2}$. Hence, for all Lebesgue exponents p and all angles α, Proposition 3.19 yields the uniform bound

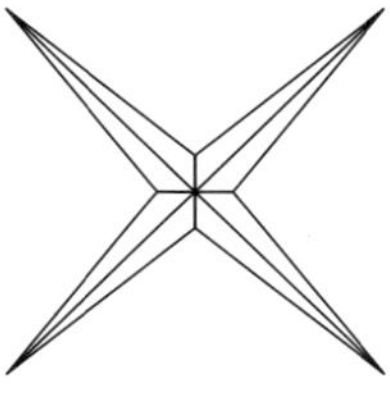

Figure 3.7 Domain of Example 3.25

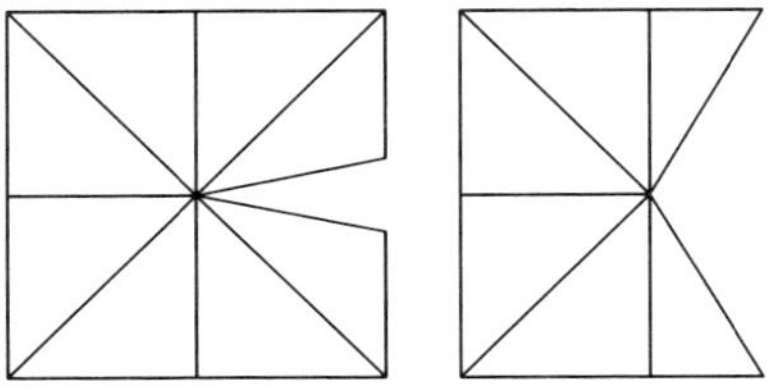

Figure 3.8 Domains of Example 3.26 for $\alpha \leq \frac{\pi}{2}$ (left) and $\frac{\pi}{2} \leq \alpha < \pi$ (right)

$$C_{P,p,\zeta}(\omega_{z,\alpha}) \leq 4\overline{C}_{P,p}\sqrt{2}(\sqrt{2}+1)^{\frac{2}{p}} \cdot \left(1 + \max\left\{\left(2+\sqrt{2}\right)^{-2} K_{p,2}(2+\sqrt{2}), \frac{p}{2\overline{C}_{P,p}}\right\}\right)^{\frac{1}{p}}.$$

Proposition 3.21 does not yield a uniform bound for the Poincaré constants since $\max_i \frac{h_{K_i}^p \mu_\zeta(\omega_{z,\alpha})}{\mu_\zeta(K_i)\,\mathrm{diam}(\omega_{z,\alpha})^p}$ tends to infinity when the angle α approaches $\frac{\pi}{2}$ or π. For this class of stars, Proposition 3.11 combined with the piecewise affine transformation used in the proof of Proposition 3.19 yields the even better bound

$$C_{P,p,\zeta}(\omega_{z,\alpha}) \leq \left(1+\sqrt{2}\right)\overline{C}_{P,p}$$

for all values of α.

Our final example is typical for a boundary-layer adapted anisotropic finite element mesh close to a re-entrant corner.

Example 3.27 For $0 < \varepsilon \ll 1$ we consider stars $\omega_{z,\varepsilon}$ which consist of two rectangles with sides of length 1 and ε and a square with sides of length ε sharing the vertex $z = (0,0)$, cf. Figure 3.9. Obviously we have $n_z = 3$, $\mathrm{diam}(\omega_{z,\varepsilon}) = \sqrt{2}(1+\varepsilon)$, and

$$\max_i \frac{h_{K_i}^p \mu_\zeta(\omega_{z,\varepsilon})}{\mu_\zeta(K_i)\,\mathrm{diam}(\omega_{z,\varepsilon})^p} = \max\left\{2^{-\frac{p}{2}}\,3,\ \varepsilon^{p-1}\,3\right\} \leq 3.$$

Hence, for all Lebesgue exponents p, Proposition 3.21 yields the uniform bound

$$C_{P,p,\zeta}(\omega_{z,\varepsilon}) \leq 8\overline{C}_{P,p}\left[\frac{3}{4} + \frac{3p}{4\overline{C}_{P,p}}\right]^{\frac{1}{p}}$$

for the Poincaré constants. Since $\kappa_z = \varepsilon^{-1}$, Proposition 3.19 yields a uniform bound for the Poincaré constants only for Lebesgue exponents $p > d$. As in the previous example, we may

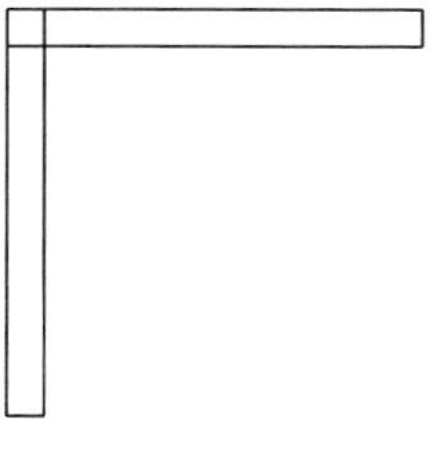

Figure 3.9 Domain of Example 3.27

obtain an even better bound by applying Proposition 3.11 with a suitable transformation. More precisely, we subdivide the three quadrilaterals of $\omega_{z,\varepsilon}$ into six triangles by drawing the diagonals emanating from the point z and enumerating these in counter-clockwise order. To triangles 1–3 we then apply a transformation with Jacobian $\left(\begin{smallmatrix}1&1\\0&1\end{smallmatrix}\right)$ and to triangles 4–6 a transformation with Jacobian $\left(\begin{smallmatrix}1&1\\-1&0\end{smallmatrix}\right)$. The star $\omega_{z,\varepsilon}$ is thus transformed into a symmetric trapezoid with height ε and parallel sides of length 2 and $2+2\varepsilon$, which is decomposed into two parallelograms and one triangle. The transformed shape function is given by $\frac{1}{\varepsilon}(\varepsilon - x_2)(1-|x_1|+x_2)$. Since it is the product of two non-negative concave functions, its square root is concave. Hence it is an admissible weight. Proposition 3.11 therefore yields

$$C_{P,p,\zeta}(\omega_{z,\varepsilon}) \le \sqrt{3+\sqrt{5}}\,\overline{C}_{P,p}$$

for all values of ε.

3.4.7 Friedrichs' Inequalities

If $\partial\omega \cap \Gamma_D$ has a positive $(d-1)$-dimensional Hausdorff measure, it is well known [3] that *Friedrichs' inequality*

$$\left\|\zeta^{\frac{1}{p}} v\right\|_{p;\omega} \le c(\omega)\,\mathrm{diam}(\omega)\left\|\zeta^{\frac{1}{p}}\nabla v\right\|_{p;\omega}$$

holds for every $v \in W_D^{1,p}(\Omega)$. The optimal constant in this inequality is given by

$$C_{F,p,\zeta}(\omega) = \sup_{v\in W_D^{1,p}(\Omega)\setminus\{0\}} \frac{\left\|\zeta^{\frac{1}{p}} v\right\|_{p;\omega}}{\mathrm{diam}(\omega)\left\|\zeta^{\frac{1}{p}}\nabla v\right\|_{p;\omega}}.$$

It only depends on ω, p, ζ, and the size of $\partial\omega\cap\Gamma_D$ and is often referred to as *Friedrichs' constant* of ω. Friedrichs' inequality is usually proved by a contradiction and scaling argument. This, however, gives no information on the size of Friedrichs' constant. For finite element stars ω_z and densities 1 and λ_z, the following result gives explicit bounds for Friedrichs' constant of ω_z in terms of the corresponding Poincaré constant and easily computable geometric parameters of the star.

Proposition 3.28 *Suppose that the boundary of the star ω_z shares at least one face with the Dirichlet boundary, i.e. $\partial\omega_z \cap \mathcal{E}_{\Gamma_D} \neq \emptyset$, and that the density equals either 1 or the nodal shape function λ_z. Then Friedrichs' constant of ω_z satisfies*

$$C_{F,p,\zeta}(\omega_z) \le \left[1 + M_z \max_{E\subset\Gamma_D\cap\partial\omega_z}\left(\frac{h_z^\perp}{h_E^\perp}\right)^{\frac{1}{p}}\right] C_{P,p,\zeta}(\omega_z) + M_z \max_{E\subset\Gamma_D\cap\partial\omega_z}\left[\frac{h_{\omega_E}}{h_z\widetilde{\nu}_E}\left(\frac{h_z^\perp}{h_E^\perp}\right)^{\frac{1}{p}}\right],$$

where $M_z \le \max_{K\subset\omega_z}(d+1-\nu_K)$ is the maximal number of Dirichlet faces per element,

$$h_z^\perp = \frac{\mu_d(\omega_z)}{\mu_{d-1}(\Gamma_D\cap\partial\omega_z)}, \quad h_E^\perp = \frac{\mu_d(\omega_E)}{\mu_{d-1}(E)}, \quad \widetilde{\nu}_E = \nu_{\omega_E} \quad \text{if } \zeta = 1,$$

$$h_z^\perp = \frac{\mu_{d,z}(\omega_z)}{\mu_{d-1,z}(\Gamma_D\cap\partial\omega_z)}, \quad h_E^\perp = \frac{\mu_{d,z}(\omega_E)}{\mu_{d-1,z}(E)}, \quad \widetilde{\nu}_E = \nu_{\omega_E}+1 \quad \text{if } \zeta = \lambda_z,$$

and where ν_K and $\mu_{d,z}$, $\mu_{d-1,z}$ are given by equations (3.5) (p. 82) and (3.6) (p. 88), respectively.

Proof We only consider the case $\zeta = \lambda_z$. The other case follows with the same arguments by replacing the trace inequality of Proposition 3.5 (p. 90) by its unweighted counterpart and taking into account Remark 3.6 (p. 90).

Since the boundary $\partial\omega_z$ contains at least one Dirichlet face, we have $\mu_{d-1,z}(\Gamma_{D,z}) > 0$. Consider an arbitrary function $v \in W_D^{1,p}(\Omega)$. Then we have $\int_{\Gamma_{D,z}} v\lambda_z = 0$. Consequently, for any real number $c \in \mathbb{R}$ we may estimate

$$\left\|\lambda_z^{\frac{1}{p}} v\right\|_{p;\omega_z} \leq \left\|\lambda_z^{\frac{1}{p}}(v-c)\right\|_{p;\omega_z} + \mu_{d,z}(\omega_z)^{\frac{1}{p}}\,|c|$$
$$\leq \left\|\lambda_z^{\frac{1}{p}}(v-c)\right\|_{p;\omega_z} + \frac{\mu_{d,z}(\omega_z)^{\frac{1}{p}}}{\mu_{d-1,z}(\Gamma_{D,z})}\left|\int_{\Gamma_{D,z}} \lambda_z(v-c)\right|.$$

In order to estimate the last term of the right-hand side, we first consider the corresponding term for any Dirichlet face E in $\partial\omega_z$. Since ω_E denotes the element containing E, Proposition 3.5 (p. 90) with $p = 1$ and Hölder's inequality for integrals yield

$$\int_E \lambda_z\,|v-c| \leq \frac{\mu_{d-1,z}(E)}{\mu_{d,z}(\omega_E)^{\frac{1}{p}}}\left(\left\|\lambda_z^{\frac{1}{p}}(v-c)\right\|_{p;\omega_E} + \frac{h_{\omega_E}}{\nu_{\omega_E}+1}\left\|\lambda_z^{\frac{1}{p}}\nabla v\right\|_{p;\omega_E}\right).$$

We next sum over all Dirichlet faces E of $\partial\omega_z$. In doing so, elements in ω_z may be counted several times. Yet, since every element K has at least one vertex in $\Omega \cup \Gamma_N$ the corresponding multiplicity of K is at most $d + 1 - \nu_K$. Hölder's inequality for sums therefore gives

$$\sum_{E\subset\partial\omega_z\cap\Gamma_D} \frac{\mu_{d-1,z}(E)^{\frac{1}{p'}+\frac{1}{p}}}{\mu_{d,z}(\omega_E)^{\frac{1}{p}}}\left\|\lambda_z^{\frac{1}{p}}(v-c)\right\|_{p;\omega_E} \leq M_z \max_{E\subset\partial\omega_z\cap\Gamma_D} (h_E^{\perp})^{-\frac{1}{p}}\mu_{d-1,z}(\partial\omega_z\cap\Gamma_D)^{\frac{1}{p'}}\left\|\lambda_z^{\frac{1}{p}}(v-c)\right\|_{p;\omega_z}$$

for the first product of the right-hand side and

$$\sum_{E\subset\partial\omega_z\cap\Gamma_D} \frac{\mu_{d-1,z}(E)^{\frac{1}{p'}+\frac{1}{p}}}{\mu_{d,z}(\omega_E)^{\frac{1}{p}}}\frac{h_{\omega_E}}{\nu_{\omega_E}+1}\left\|\lambda_z^{\frac{1}{p}}\nabla v\right\|_{p;\omega_E}$$
$$\leq M_z \max_{E\subset\partial\omega_z\cap\Gamma_D}\left[\frac{h_{\omega_E}}{(\nu_{\omega_E}+1)(h_E^{\perp})^{\frac{1}{p}}}\right]\mu_{d-1,z}(\partial\omega_z\cap\Gamma_D)^{\frac{1}{p'}}\left\|\lambda_z^{\frac{1}{p}}\nabla v\right\|_{p;\omega_z}$$

for the second one. Hence, we have

$$\frac{\mu_{d,z}(\omega_z)^{\frac{1}{p}}}{\mu_{d-1,z}(\omega_z\cap\Gamma_D)}\left|\int_{\omega_z\cap\Gamma_D}\lambda_z(v-c)\right| \leq M_z \max_{E\subset\omega_z\cap\Gamma_D}\left(\frac{h_z^{\perp}}{h_E^{\perp}}\right)^{\frac{1}{p}}\left\|\lambda_z^{\frac{1}{p}}(v-c)\right\|_{p;\omega_z}$$
$$+ h_z M_z \max_{E\subset\partial\omega_z\cap\Gamma_D}\left[\frac{h_{\omega_E}}{h_z(\nu_{\omega_E}+1)}\left(\frac{h_z^{\perp}}{h_E^{\perp}}\right)^{\frac{1}{p}}\right]\left\|\lambda_z^{\frac{1}{p}}\nabla v\right\|_{p;\omega_z}.$$

We conclude the proof by combining all estimates and invoking the Poincaré inequality. □

3.4.8 *Weighted Poincaré-Type Inequalities for Stars and Skeletons*

With the help of the results of the preceding subsections we can now prove a generalisation of Lemma 1.10 (p. 18) to $W^{1,p}$-spaces equipped with either the standard semi-norm $\|\nabla\cdot\|_{p;\omega}$ or the general norm $|||\cdot|||_{\omega}$. We start with the standard semi-norm.

Lemma 3.29 *For every vertex $z \in \mathcal{N}$ and every $v \in W_D^{1,p}(\Omega)$ there is a $v_z \in \mathbb{R}$ with $\lambda_z v_z \in S_D^{1,0}(\mathcal{T})$ such that the weighted Poincaré-type inequalities*

$$\left\|\lambda_z^{\frac{1}{p}}(v - v_z)\right\|_{p;\omega_z} \leq c_p(\omega_z) h_z \left\|\lambda_z^{\frac{1}{p}} \nabla v\right\|_{p;\omega_z}$$

$$\left\{\sum_{E \subset \sigma_z} h_E^{\perp} \left\|\lambda_z^{\frac{1}{p}}(v - v_z)\right\|_{p;E}^{p}\right\}^{\frac{1}{p}} \leq c_p(\sigma_z) h_z \left\|\lambda_z^{\frac{1}{p}} \nabla v\right\|_{p;\omega_z}$$

hold with $h_E^{\perp} = \mu_{d,z}(\omega_E)/\mu_{d-1,z}(E)$,

$$c_p(\omega_z) = \begin{cases} C_{P,p,\lambda_z}(\omega_z) & \text{if } z \notin \mathcal{N}_{\Gamma_D}, \\ C_{F,p,\lambda_z}(\omega_z) & \text{if } z \in \mathcal{N}_{\Gamma_D}, \end{cases}$$

$$c_p(\sigma_z) \leq c_p(\omega_z) \left\{ d \left[1 + \frac{p}{c_p(\omega_z)} \max_{K \subset \omega_z} \frac{h_K}{(\nu_K + 1) h_z} \right] \right\}^{\frac{1}{p}}.$$

Proof The first estimate follows from Sections 3.4.5 and 3.4.7. Corollary 3.8 (p. 90) implies the second estimate. □

Next, we consider the general norm $|||\cdot|||_{\omega}$ defined in equation (3.1) (p. 80). For brevity, we introduce its weighted counterpart and set for every vertex $z \in \mathcal{N}$

$$|||v|||_z = \left\{ \varepsilon \left\|\lambda_z^{\frac{1}{p}} \nabla v\right\|_{p;\omega_z}^{p} + \beta \left\|\lambda_z^{\frac{1}{p}} v\right\|_{p;\omega_z}^{p} \right\}^{\frac{1}{p}}. \tag{3.20}$$

Lemma 3.30 *For every vertex $z \in \mathcal{N}$ and every $v \in W_D^{1,p}(\Omega)$ there is a $v_z \in \mathbb{R}$ with $\lambda_z v_z \in S_D^{1,0}(\mathcal{T})$ such that the weighted Poincaré-type inequalities*

$$\left\|\lambda_z^{\frac{1}{p}}(v - v_z)\right\|_{p;\omega_z} \leq \widetilde{c}_p(\omega_z) \hbar_{\omega_z} |||v|||_z$$

$$\left\{\sum_{E \subset \sigma_z} h_E^{\perp} \left\|\lambda_z^{\frac{1}{p}}(v - v_z)\right\|_{p;E}^{p}\right\}^{\frac{1}{p}} \leq \widetilde{c}_p(\sigma_z) \varepsilon^{-\frac{1}{p^2}} \hbar_{\omega_z}^{1-\frac{1}{p}} h_z^{\frac{1}{p}} |||v|||_z$$

hold with $h_E^{\perp} = \mu_{d,z}(\omega_E)/\mu_{d-1,z}(E)$,

$$\widetilde{c}_p(\omega_z) = \begin{cases} \max\{c_p(\omega_z), 1\} & \text{if } \beta > 0, \\ c_p(\omega_z) & \text{if } \beta = 0, \end{cases}$$

$$\widetilde{c}_p(\sigma_z) \leq \widetilde{c}_p(\omega_z) \left\{ d \left[1 + \frac{p}{\widetilde{c}_p(\omega_z)} \max_{K \subset \omega_z} \frac{h_K}{(\nu_K + 1) h_z} \right] \right\}^{\frac{1}{p}},$$

where $\hbar_{\omega_z}$ and $c_p(\omega_z)$ are as in (3.4) (p. 82) and Lemma 3.29, respectively.

Proof If $\beta = 0$, Lemma 3.30 is a reformulation of Lemma 3.29. Therefore, we only have to consider the case $\beta > 0$. The first estimate of Lemma 3.29 and the definition of $|||\cdot|||_z$ imply that there is a constant v_z such that

$$\left\| \lambda_z^{\frac{1}{p}} (v - v_z) \right\|_{p;\omega_z} \leq c_p(\omega_z) \varepsilon^{-\frac{1}{p}} h_z \, |||v|||_z .$$

On the other hand, we obviously have

$$\|v\|_{p;\omega_z} \leq \beta^{-\frac{1}{p}} \, |||v|||_z .$$

Taking the minimum of these two estimates proves the first estimate of Lemma 3.30. The second one follows from Corollary 3.8 (p. 90), the first estimate, and $\varepsilon^{1/p} h_z^{-1} \hbar_{\omega_z} \leq 1$. □

Remark 3.31 Since $\nu_K \geq 1$ and $h_K \leq h_z$ hold for all elements K, the constants $c_p(\sigma_z)$ and $\widetilde{c}_p(\sigma_z)$ can be bounded by

$$c_p(\sigma_z) \leq \left\{ d\left[1 + \frac{p}{2}\right]\right\}^{\frac{1}{p}} \max\left\{ c_p(\omega_z), c_p(\omega_z)^{1-\frac{1}{p}} \right\},$$

$$\widetilde{c}_p(\sigma_z) \leq \left\{ d\left[1 + \frac{p}{2}\right]\right\}^{\frac{1}{p}} \max\left\{ \widetilde{c}_p(\omega_z), \widetilde{c}_p(\omega_z)^{1-\frac{1}{p}} \right\}.$$

Remark 3.32 The presentation of this section follows [**248**]. Proposition 3.28 was first implicitly proved in [**247**].

3.5 Interpolation Error Estimates

3.5.1 First-Order Interpolation

We first consider a modification of the quasi-interpolation operator $I_{\mathcal{T}}$ of Section 1.3.3 (p. 7) which is based on *weighted averages*. To this end we set for every vertex $z \in \mathcal{N}$ and every function $v \in L^1(\omega_z)$

$$\overline{v}_z = \begin{cases} \dfrac{\int_{\omega_z} \lambda_z v}{\int_{\omega_z} \lambda_z} & \text{if } z \in \mathcal{N}_\Omega \cup \mathcal{N}_{\Gamma_N} \\ 0 & \text{if } z \in \mathcal{N}_{\Gamma_D} \end{cases} \tag{3.21}$$

and define the *quasi-interpolation operator* $I_{\mathcal{T}} : L^1(\Omega) \to S_D^{1,0}(\mathcal{T})$ by

$$I_{\mathcal{T}} v = \sum_{z \in \mathcal{N}} \overline{v}_z \lambda_z. \tag{3.22}$$

Recall that $\widetilde{\omega}_K$ and $\widetilde{\omega}_E$ denote the union of all elements that share at least a vertex with a given element K or face E, respectively, cf. Figure 1.1 (p. 6), that ρ_K is the diameter of the largest ball contained in K, that $h_E^\perp = \mu_{d,z}(\omega_E)/\mu_{d-1,z}(E)$, that ν_K is given by equation (3.5) (p. 82), and that $\delta_{p,q}$ denotes the Kronecker symbol. With this notation we want to prove the following generalisation of Proposition 1.3 (p. 7).

Proposition 3.33 *For every Lebesgue exponent p, every element $K \in \mathcal{T}$, and every face $E \in \mathcal{E}_K$ thereof, the local error estimates*

$$\begin{aligned}
\|v - I_{\mathcal{T}}v\|_{p;K} &\leq C_{A,1,p}(K)\, \|v\|_{p;\widetilde{\omega}_K} \\
\|v - I_{\mathcal{T}}v\|_{p;K} &\leq C_{A,2,p}(K) h_K\, \|\nabla v\|_{p;\widetilde{\omega}_K} \\
\left\|\nabla(v - I_{\mathcal{T}}v)\right\|_{p;K} &\leq C_{A,3,p}(K)\, \|\nabla v\|_{p;\widetilde{\omega}_K} \\
\|v - I_{\mathcal{T}}v\|_{p;E} &\leq C_{A,4,p}(E) h_E^{1-\frac{1}{p}}\, \|\nabla v\|_{p;\widetilde{\omega}_E}
\end{aligned}$$

hold for all functions $v \in W_D^{1,p}(\Omega)$ with

$$C_{A,1,p}(K) = 2 - \delta_{p,2},$$

$$C_{A,2,p}(K) = \max\left\{\max_{z\in\mathcal{N}_{K,\Omega}\cup\mathcal{N}_{K,\Gamma_N}} \overline{C}_{P,p,\lambda_z}(\omega_z)\frac{h_z}{h_K}, \max_{z\in\mathcal{N}_{K,\Gamma_D}} C_{F,p,\lambda_z}(\omega_z)\frac{h_z}{h_K}\right\},$$

$$C_{A,3,p}(K) = 1 + \sum_{z\in\mathcal{N}_{K,\Omega}\cup\mathcal{N}_{K,\Gamma_N}} C_{P,p,1}(\omega_z)\frac{h_z}{\rho_K}\left[1 + \left(\frac{\mu_d(K)}{\mu_{d,z}(\omega_z)}\right)^{\frac{1}{p}}\right] + \sum_{z\in\mathcal{N}_{K,\Gamma_D}} C_{F,p,1}(\omega_z)\frac{h_z}{\rho_K},$$

$$\begin{aligned}
C_{A,4,p}(E) = \max\Bigg\{&\max_{z\in\mathcal{N}_{K,\Omega}\cup\mathcal{N}_{K,\Gamma_N}} \overline{C}^p_{P,p,\lambda_z}(\omega_z)\frac{h_z^p}{h_E^{\perp} h_E^{p-1}}\cdot\left[1 + \max_{K\subset\omega_E}\frac{ph_K}{\overline{C}_{P,p,\lambda_z}(\omega_z)(\nu_K+1)h_z}\right], \\
&\max_{z\in\mathcal{N}_{K,\Gamma_D}} C^p_{F,p,\lambda_z}(\omega_z)\frac{h_z^p}{h_E^{\perp} h_E^{p-1}}\cdot\left[1 + \max_{K\subset\omega_E}\frac{ph_K}{C_{F,p,\lambda_z}(\omega_z)(\nu_K+1)h_z}\right]\Bigg\}^{\frac{1}{p}}
\end{aligned}$$

with the obvious modifications for $p = \infty$.

Proof We first consider an arbitrary element K. Since $(\lambda_z)_{z\in\mathcal{N}_K}$ form a partition of unity on K, we obtain for every function $w \in L^{p'}(K)$

$$\int_K (v - I_{\mathcal{T}}v)w = \sum_{z\in\mathcal{N}_K}\int_K \lambda_z(v - \overline{v}_z)w.$$

Hölder's inequality for integrals and sums gives

$$\begin{aligned}
\sum_{z\in\mathcal{N}_K}\int_K \lambda_z(v - \overline{v}_z)w &\leq \left\{\sum_{z\in\mathcal{N}_K}\left\|\lambda_z^{\frac{1}{p}}(v - \overline{v}_z)\right\|_{p;K}^p\right\}^{\frac{1}{p}}\left\{\sum_{z\in\mathcal{N}_K}\left\|\lambda_z^{\frac{1}{p'}} w\right\|_{p';K}^{p'}\right\}^{\frac{1}{p'}} \\
&\leq \left\{\sum_{z\in\mathcal{N}_K}\left\|\lambda_z^{\frac{1}{p}}(v - \overline{v}_z)\right\|_{p;K}^p\right\}^{\frac{1}{p}} \|w\|_{p';K}.
\end{aligned}$$

Since $w \in L^{p'}(K)$ was arbitrary and since

$$\|v - I_{\mathcal{T}}v\|_{p;K} = \sup_{w\in L^{p'}(K)\setminus\{0\}} \frac{\int_K (v - I_{\mathcal{T}}v)w}{\|w\|_{p';K}},$$

this proves that

$$\|v - I_{\mathcal{T}} v\|_{p;K} \leq \left\{ \sum_{z \in \mathcal{N}_K} \left\| \lambda_z^{\frac{1}{p}} (v - \overline{v}_z) \right\|_{p;K}^p \right\}^{\frac{1}{p}} . \tag{3.23}$$

If z is a vertex on the Dirichlet boundary, we obviously have

$$\left\| \lambda_z^{\frac{1}{p}} (v - \overline{v}_z) \right\|_{p;K} \leq \left\| \lambda_z^{\frac{1}{p}} v \right\|_{p;\omega_z} \leq C_{F,p,\lambda_z}(\omega_z) h_z \left\| \lambda_z^{\frac{1}{p}} \nabla v \right\|_{p;\omega_z} . \tag{3.24}$$

If, on the other hand, z is not on the Dirichlet boundary, we take into account that $\overline{v}_z$ is the best weighted L^2-approximation of v by constants and obtain

$$\left\| \lambda_z^{\frac{1}{p}} (v - \overline{v}_z) \right\|_{p;K} \leq (2 - \delta_{p,2}) \inf_{c \in \mathbb{R}} \left\| \lambda_z^{\frac{1}{p}} (v - c) \right\|_{p;\omega_z} \leq (2 - \delta_{p,2}) \left\| \lambda_z^{\frac{1}{p}} v \right\|_{p;\omega_z}$$

and

$$\left\| \lambda_z^{\frac{1}{p}} (v - \overline{v}_z) \right\|_{p;K} \leq \overline{C}_{P,p,\lambda_z}(\omega_z) h_z \left\| \lambda_z^{\frac{1}{p}} \nabla v \right\|_{p;\omega_z} . \tag{3.25}$$

Since $(\lambda_z)_{z \in \mathcal{N}_K}$ form a partition of unity on K, we have

$$\left\{ \sum_{z \in \mathcal{N}_K} \left\| \lambda_z^{\frac{1}{p}} v \right\|_{p;\omega_z}^p \right\}^{\frac{1}{p}} \leq \|v\|_{p;\widetilde{\omega}_K} , \quad \left\{ \sum_{z \in \mathcal{N}_K} \left\| \lambda_z^{\frac{1}{p}} \nabla v \right\|_{p;\omega_z}^p \right\}^{\frac{1}{p}} \leq \|\nabla v\|_{p;\widetilde{\omega}_K} .$$

Inserting these estimates in (3.23) proves the first two estimates of Proposition 3.33 and the expressions for $C_{A,1,p}(K)$ and $C_{A,2,p}(K)$.

For the proof of the third estimate, we consider an arbitrary vector field $\varphi \in L^{p'}(K)^d$ and obtain

$$\begin{aligned}
\int_K \nabla(v - I_{\mathcal{T}} v) \cdot \varphi &= \sum_{z \in \mathcal{N}_K} \int_K \nabla \left(\lambda_z (v - \overline{v}_z) \right) \cdot \varphi \\
&= \sum_{z \in \mathcal{N}_K} \int_K \lambda_z \nabla v \cdot \varphi + \sum_{z \in \mathcal{N}_K} \int_K (v - \overline{v}_z) \nabla \lambda_z \cdot \varphi .
\end{aligned}$$

Hölder's inequality for integrals and sums yields for the first term on the right-hand side

$$\begin{aligned}
\sum_{z \in \mathcal{N}_K} \int_K \lambda_z \nabla v \cdot \varphi &\leq \left\{ \sum_{z \in \mathcal{N}_K} \left\| \lambda_z^{\frac{1}{p}} \nabla v \right\|_{p;K}^p \right\}^{\frac{1}{p}} \left\{ \sum_{z \in \mathcal{N}_K} \left\| \lambda_z^{\frac{1}{p'}} \varphi \right\|_{p';K} \right\}^{\frac{1}{p'}} \\
&\leq \|\nabla v\|_{p;K} \, \|\varphi\|_{p';K} .
\end{aligned}$$

For the second term on the right-hand side, on the other hand, we get

$$\sum_{z\in\mathcal{N}_K}\int_K (v-\bar{v}_z)\nabla\lambda_z\cdot\varphi \leq \sum_{z\in\mathcal{N}_K} \|v-\bar{v}_z\|_{p;K}\,\|\nabla\lambda_z\|_{\infty;K}\,\|\varphi\|_{p';K}\,.$$

Since $\varphi \in L^{p'}(K)^d$ was arbitrary and since

$$\left\|\nabla(v-I_{\mathcal{T}}v)\right\|_{p;K} = \sup_{\varphi\in L^{p'}(K)^d\setminus\{0\}} \frac{\int_K \nabla(v-I_{\mathcal{T}}v)\cdot\varphi}{\|\varphi\|_{p';K}},$$

this proves that

$$\left\|\nabla(v-I_{\mathcal{T}}v)\right\|_{p;K} \leq \|\nabla v\|_{p;K} + \sum_{z\in\mathcal{N}_K}\|v-\bar{v}_z\|_{p;K}\,\|\nabla\lambda_z\|_{\infty;K}\,. \tag{3.26}$$

An elementary calculation yields $\|\nabla\lambda_z\|_{\infty;K} \leq \frac{1}{\rho_K}$. If z is a vertex on the Dirichlet boundary, we have

$$\|v-\bar{v}_z\|_{p;K} \leq \|v\|_{p;\omega_z} \leq C_{F,p,1}(\omega_z)h_z\,\|\nabla v\|_{p;\omega_z}\,.$$

If, on the other hand, z is not on the Dirichlet boundary, we obtain for every $c \in \mathbb{R}$ using Hölder's inequality and $0 \leq \lambda_z \leq 1$

$$\begin{aligned}\|v-\bar{v}_z\|_{p;K} &\leq \|v-c\|_{p;K} + \|c-\bar{v}_z\|_{p;K}\\ &\leq \|v-c\|_{p;K} + \left(\frac{\mu_d(K)}{\mu_{d,z}(\omega_z)}\right)^{\frac{1}{p}}\|v-c\|_{p;\omega_z}\end{aligned}$$

and thus

$$\|v-\bar{v}_z\|_{p;K} \leq C_{P,p,1}(\omega_z)h_z\left[1+\left(\frac{\mu_d(K)}{\mu_{d,z}(\omega_z)}\right)^{\frac{1}{p}}\right]\|\nabla v\|_{p;\omega_z}\,.$$

Inserting these estimates in (3.26) proves the third estimate of Proposition 3.33 and the expression for $C_{A,3,p}(K)$.

We finally consider an arbitrary face E. With the same arguments as in the first part of the proof we now obtain

$$\|v-I_{\mathcal{T}}v\|_{p;E} \leq \left\{\sum_{z\in\mathcal{N}_E}\left\|\lambda_z^{\frac{1}{p}}(v-\bar{v}_z)\right\|_{p;E}^p\right\}^{\frac{1}{p}}. \tag{3.27}$$

Estimate (3.9) (p. 90) yields for every $z \in \mathcal{N}_E$

$$\left\|\lambda_z^{\frac{1}{p}}(v-\bar{v}_z)\right\|_{p;E}^p \leq \frac{1}{h_E^{\perp}}\left\|\lambda_z^{\frac{1}{p}}(v-\bar{v}_z)\right\|_{p;\omega_E}^p + \max_{K\subset\omega_E}\frac{ph_K}{(\nu_K+1)h_E^{\perp}}\left\|\lambda_z^{\frac{1}{p}}(v-\bar{v}_z)\right\|_{p;\omega_E}^{p-1}\left\|\lambda_z^{\frac{1}{p}}\nabla v\right\|_{p;\omega_E}.$$

Combining this estimate with inequalities (3.24) and (3.25) and inserting the result in inequality (3.27) proves the last estimate of Proposition 3.33 and the expression for $C_{A,4,p}(E)$. □

Remark 3.34 For later use we note that equations (3.23) and (3.27) hold for *every* quasi-interpolation operator of the form $v \mapsto \sum_{z\in\mathcal{N}} v_z\lambda_z$ with $v_z \in \mathbb{R}$ for all vertices $z \in \mathcal{N}$. Thus, for all these operators, local error estimates can be derived from the corresponding Poincaré and Friedrichs' inequalities.

3.5.2 Higher Order Interpolation

For some applications, we need interpolation error estimates with respect to higher order Sobolev norms. Since $W^{2,1}(\Omega)$ is continuously embedded in $C(\overline{\Omega})$ we can then use the standard *nodal interpolation operator* $i_{\mathcal{T}} : C(\overline{\Omega}) \to S_D^{1,0}(\mathcal{T})$ which is defined by

$$i_{\mathcal{T}}v(z) = v(z)$$

for all $z \in \mathcal{N}$. It has the following well-known local approximation properties [**109**, Theorem 16.2].

Proposition 3.35 *The following error estimates hold for all elements $K \in \mathcal{T}$, all faces $E \in \mathcal{E}_K$ thereof, all Lebesgue exponents p, all integers $m \in \{0, 1\}$, and all functions $v \in W^{2,p}(\Omega) \cap W_D^{1,p}(\Omega)$*

$$\left\|\nabla^m(v - i_{\mathcal{T}}v)\right\|_{p;K} \le C_{A,1,p,m}(K)h_K^{2-m}\,\|v\|_{2,p;K}\,,$$

$$\left\|\nabla^m(v - i_{\mathcal{T}}v)\right\|_{p;E} \le C_{A,2,p,m}(E)h_E^{2-m-\frac{1}{p}}\,\|v\|_{2,p;\omega_E}\,.$$

The constants $C_{A,1,p,m}(K)$ and $C_{A,2,p,m}(E)$ only depend on p, m, and the shape parameter $C_{\mathcal{T}}$ of $\mathcal{T}$.

Remark 3.36 A result similar to Proposition 3.33 without the weight functions λ_z was first proved in [**264**]. Other quasi-interpolation operators such as the operators of Clément [**110**], of Scott-Zhang [**237**], and of Section (1.3.3) (p. 7) can be analysed in the same way.

3.6 Inverse Estimates

In this section we generalise Proposition 1.4 (p. 9) to higher space dimensions, arbitrary polynomial degrees, and general L^p-spaces. Recall that, for $\vartheta = 1$ and every face E, the modified cut-off function $\psi_{E,\vartheta}$ equals the standard cut-off function ψ_E.

Proposition 3.37 *The following inverse estimates hold for all elements K, all faces E of K, all polynomial degrees k, all positive parameters ϑ, and all polynomials $v \in R_k(K)$ and $w \in R_k(E)$*

$$\|v\|_K \le C_{I,1,k}(K)\left\|\psi_K^{\frac{1}{2}}v\right\|_K,$$

$$\left\|\nabla(\psi_K v)\right\|_K \le C_{I,2,k}(K)h_K^{-1}\,\|v\|_K\,,$$

$$\|w\|_E \le C_{I,3,k}(E)\left\|\psi_{E,\vartheta}^{\frac{1}{2}}w\right\|_E,$$

$$\left\|\nabla(\psi_{E,\vartheta}w)\right\|_K \le C_{I,4,k}(E,K)\vartheta^{-\frac{1}{2}}h_E^{-\frac{1}{2}}\,\|w\|_E\,,$$

$$\left\|\psi_{E,\vartheta}w\right\|_K \le C_{I,5,k}(E,K)\vartheta^{\frac{1}{2}}h_E^{\frac{1}{2}}\,\|w\|_E\,,$$

and, for all Lebesgue exponents p,

$$\|v\|_{p;K} = C_{I,1,k,p}(K) \sup_{u\in R_k(K)\setminus\{0\}} \frac{\int_K \psi_K v u}{\|u\|_{p';K}},$$

$$\left\|\nabla(\psi_K v)\right\|_{p;K} \le C_{I,2,k,p}(K) h_K^{-1} \|v\|_{p;K},$$

$$\|w\|_{p;E} = C_{I,3,k,p}(E) \sup_{u\in R_k(E)\setminus\{0\}} \frac{\int_E \psi_{E,\vartheta} w u}{\|u\|_{p';E}},$$

$$\left\|\nabla(\psi_{E,\vartheta} w)\right\|_{p;K} \le C_{I,4,k,p}(E,K) \vartheta^{\frac{1}{p}-1} h_E^{\frac{1}{p}-1} \|w\|_{p;E},$$

$$\left\|\psi_{E,\vartheta} w\right\|_{p;K} \le C_{I,5,k,p}(E,K) \vartheta^{\frac{1}{p}} h_E^{\frac{1}{p}} \|w\|_{p;E}.$$

The constants $C_{I,1,k}(K), \ldots, C_{I,5,k,p}(E,K)$ only depend on the shape parameter $C_{\mathcal{T}}$ of $\mathcal{T}$, the polynomial degree k, and the Lebesgue exponent p.

One easily checks that the expressions on the left-hand and right-hand sides of theses inequalities all define norms on the corresponding spaces of polynomials. Therefore, Proposition 3.37 can be proved in the standard way by transforming to the reference element, using the equivalence of norms on finite-dimensional spaces there, and transforming back [**254, 255**].

This proof, however, gives no information on the size of the constants and their dependence on the polynomial degree. To achieve this, we have to perform a more detailed analysis which is the object of the following subsections. We will first consider the L^2-case. It plays a particular role since it allows a dimension reduction argument and relies on properties of Legendre polynomials. Then, we will consider general L^p-spaces. Here, the results are weaker than for the L^2-case since some of the arguments used there are restricted to the Hilbert setting and do not carry over to the general situation.

3.6.1 The L^2-Setting

For every element K, every face E of K, and every integer $k \ge 1$ the optimal constants in the first part of Proposition 3.37 are given by

$$C_{I,1,k}(K) = \sup_{v\in R_k(K)\setminus\{0\}} \frac{\|v\|_K}{\left\|\psi_K^{\frac{1}{2}} v\right\|_K},$$

$$C_{I,2,k}(K) = \sup_{v\in R_k(K)\setminus\{0\}} \frac{h_K \left\|\nabla(\psi_K v)\right\|_K}{\|v\|_K},$$

$$C_{I,3,k}(E) = \sup_{v\in R_k(E)\setminus\{0\}} \frac{\|v\|_E}{\left\|\psi_{E,\vartheta}^{\frac{1}{2}} v\right\|_K},$$

$$C_{I,4,k}(E,K) = \sup_{v\in R_k(E)\setminus\{0\}} \frac{\vartheta^{\frac{1}{2}} h_E^{\frac{1}{2}} \left\|\nabla(\psi_{E,\vartheta} v)\right\|_K}{\|v\|_E},$$

$$C_{I,5,k}(E,K) = \sup_{v\in R_k(E)\setminus\{0\}} \frac{\left\|\psi_{E,\vartheta} v\right\|_K}{\vartheta^{\frac{1}{2}} h_E^{\frac{1}{2}} \|v\|_E}.$$

When compared with Proposition 1.4 (p. 9) we have

$$c_{I,1} = C_{I,1,1}(K), \quad c_{I,2} = C_{I,2,1}(K),$$
$$c_{I,3} = C_{I,3,1}(E), \quad c_{I,4} = C_{I,4,1}(E,K), \quad c_{I,5} = C_{I,5,1}(E,K).$$

In order to obtain explicit bounds for the constants $C_{I,1,k}(K), \ldots, C_{I,5,k}(E,K)$ we will proceed as follows.

- We first investigate the behaviour of the constants under affine transformations. This allows us to reduce the case of arbitrary elements and faces to the particular case of suitable reference elements and faces.
- As an intermediate step, we then consider the one-dimensional situation.
- Next, we treat the case of the d-cube as a reference element. Its tensor product structure allows for a dimension reduction argument to the one-dimensional situation.
- Finally, we consider the case of the standard d-simplex as a reference element. Again, a dimension reduction argument is used. Yet, it is technically more involved since the intersection of the standard d-simplex with a hyperplane parallel to a coordinate plane is a $(d-1)$-dimensional simplex of varying size.

All these steps are performed for the particular case $\vartheta = 1$ of the standard cut-off functions ψ_E. The general case of the modified cut-off functions $\psi_{E,\vartheta}$ can be reduced to this special case by applying the results of the next subsection to the affine transformation $F_{E,K} \circ \Phi_\vartheta \circ F_{E,K}^{-1}$ of Section 3.2.4 (p. 84) and by observing that the restriction of the standard and modified cut-off functions to the corresponding face E are identical.

3.6.2 Transformation to the Reference Element

The following proposition describes the behaviour of the constants $C_{I,1,k}(K), \ldots, C_{I,5,k}(E,K)$ under affine transformations.

Proposition 3.38 *Assume that the element K and its face E are the image of the element K' and of its face E', respectively, under an affine transformation. Then for every polynomial degree k, the quantities $C_{I,1,k}(K), \ldots, C_{I,5,k}(E,K)$ corresponding to K and E on one hand and the quantities $C_{I,1,k}(K'), \ldots, C_{I,5,k}(E',K')$ corresponding to K' and E' on the other hand are related by*

$$\begin{aligned}
C_{I,1,k}(K) &= C_{I,1,k}(K'),\\
C_{I,2,k}(K) &\le \frac{h_K}{\rho_K}\, C_{I,2,k}(K'),\\
C_{I,3,k}(E) &= C_{I,3,k}(E'),\\
C_{I,4,k}(E,K) &\le \left(\frac{\mu_d(K) h_E}{\mu_{d-1}(E)\rho_K^2} \, \frac{\mu_{d-1}(E') h_{K'}^2}{\mu_d(K') h_{E'}} \right)^{\frac{1}{2}} C_{I,4,k}(E',K'),\\
C_{I,5,k}(E,K) &= \left(\frac{\mu_d(K)}{\mu_{d-1}(E) h_E} \, \frac{\mu_{d-1}(E') h_{E'}}{\mu_d(K')} \right)^{\frac{1}{2}} C_{I,5,k}(E',K').
\end{aligned}$$

Proof Denote by $F : x' \mapsto x = b + Bx'$ the affine transformation which maps K' and E' onto K and E, respectively. Since its Jacobian DF is constant, it induces via $v \mapsto v' = v \circ F$ an isomorphism of

$R_k(K)$ and $R_k(E)$ onto $R_k(K')$ and $R_k(E')$, respectively. Moreover, we have $\psi_{K'} = \psi_K \circ F$ and $\psi_{E'} = \psi_E \circ F$.

The transformation rule for integrals yields for every smooth function v defined on K or E

$$\|v\|_K = \left(\frac{\mu_d(K)}{\mu_d(K')}\right)^{\frac{1}{2}} \|v'\|_{K'} \quad \text{and} \quad \|v\|_E = \left(\frac{\mu_{d-1}(E)}{\mu_{d-1}(E')}\right)^{\frac{1}{2}} \|v'\|_{E'}.$$

This proves the first and third identity of Proposition 3.38.

Denote by ∇_x and $\nabla_{x'}$ the gradient with respect to the variable x and x', respectively, and by $|||\cdot|||_2$ the spectral norm on $\mathbb{R}^{d\times d}$ which is the matrix norm associated with the Euclidean norm $|\cdot|$. The chain rule for differentiation then yields for every smooth function v

$$\nabla_x v = B^{-t}\nabla_{x'}v' \quad \text{and} \quad |\nabla_x v| \leq \left|\left|\left|B^{-t}\right|\right|\right|_2 \left|\nabla_{x'}v'\right|.$$

Since K contains a ball with diameter ρ_K and since K' is contained in a ball with diameter $h_{K'}$, we conclude that $\left|\left|\left|B^{-t}\right|\right|\right|_2 \leq \frac{h_{K'}}{\rho_K}$. We therefore have for every $v \in R_k(K) \setminus \{0\}$

$$\begin{aligned}
\frac{h_K \left\|\nabla_x(\psi_K v)\right\|_K}{\|v\|_K} &= \frac{h_K \left\|B^{-t}\nabla_{x'}(\psi_{K'}v')\right\|_{K'}}{\|v'\|_{K'}} \\
&\leq \frac{h_{K'}}{\rho_K} \frac{h_K \left\|\nabla_{x'}(\psi_{K'}v')\right\|_{K'}}{\|v'\|_{K'}} \\
&= \frac{h_K}{\rho_K} \frac{h_{K'} \left\|\nabla_{x'}(\psi_{K'}v')\right\|_{K'}}{\|v'\|_{K'}}.
\end{aligned}$$

This proves the second estimate of Proposition 3.38.

Similarly, we obtain for every function $w \in R_k(E) \setminus \{0\}$

$$\begin{aligned}
\frac{h_E^{\frac{1}{2}} \left\|\nabla_x(\psi_E w)\right\|_K}{\|w\|_E} &= \left(\frac{\mu_d(K)}{\mu_d(K')} \frac{\mu_{d-1}(E')}{\mu_{d-1}(E)}\right)^{\frac{1}{2}} \frac{h_E^{\frac{1}{2}} \left\|B^{-t}\nabla_{x'}(\psi_{E'}w')\right\|_{K'}}{\|w'\|_{E'}} \\
&\leq \left(\frac{\mu_d(K)}{\mu_d(K')} \frac{\mu_{d-1}(E')}{\mu_{d-1}(E)}\right)^{\frac{1}{2}} \frac{h_{K'}}{\rho_K} \frac{h_E^{\frac{1}{2}} \left\|\nabla_{x'}(\psi_{E'}w')\right\|_{K'}}{\|w'\|_{E'}} \\
&= \left(\frac{\mu_d(K)}{\mu_d(K')} \frac{\mu_{d-1}(E')}{\mu_{d-1}(E)} \frac{h_E}{h_{E'}} \frac{h_{K'}^2}{\rho_K^2}\right)^{\frac{1}{2}} \frac{h_{E'}^{\frac{1}{2}} \left\|\nabla_{x'}(\psi_{E'}w')\right\|_{K'}}{\|w'\|_{E'}}.
\end{aligned}$$

This proves the fourth estimate of Proposition 3.38.

Finally, we get for every function $w \in R_k(E) \setminus \{0\}$

$$\begin{aligned}
\frac{\|\psi_E w\|_K}{h_E^{\frac{1}{2}} \|w\|_E} &= \left(\frac{\mu_d(K)}{\mu_d(K')} \frac{\mu_{d-1}(E')}{\mu_{d-1}(E)}\right)^{\frac{1}{2}} \frac{\|\psi_{E'}w'\|_{K'}}{h_E^{\frac{1}{2}} \|w'\|_{E'}} \\
&= \left(\frac{\mu_d(K)}{\mu_d(K')} \frac{\mu_{d-1}(E')}{\mu_{d-1}(E)} \frac{h_{E'}}{h_E}\right)^{\frac{1}{2}} \frac{\|\psi_{E'}w'\|_{K'}}{h_{E'}^{\frac{1}{2}} \|w'\|_{E'}}.
\end{aligned}$$

This proves the fifth estimate of Proposition 3.38. □

Remark 3.39 The above proof implies that the inequalities in Proposition 3.38 are equalities whenever the affine transformation is a combination of translations, rotations, and scalings. Moreover, in one space dimension, the proof of Proposition 3.38 yields

$$C_{I,1,k}((a,b)) = C_{I,1,k}((0,1)), \quad C_{I,2,k}((a,b)) = C_{I,2,k}((0,1)),$$

for every interval (a, b).

Denote by $\widehat{K}_d$ the *reference d-simplex* $\{x \in \mathbb{R}^d : x_1 \geq 0, \ldots, x_d \geq 0, x_1 + \cdots + x_d \leq 1\}$ or the *reference d-cube* $[0,1]^d$ and by $\widehat{E}_d$ its intersection with the d-th coordinate plane $\{x_d = 0\}$. One easily checks that

$$h_{\widehat{K}_d} = \begin{cases} \sqrt{2} & \text{if } \widehat{K}_d \text{ is the reference } d \text{ simplex,} \\ \sqrt{d} & \text{if } \widehat{K}_d \text{ is the reference } d \text{ cube,} \end{cases}$$

$$h_{\widehat{E}_d} = \begin{cases} 1 & \text{if } d = 2, \\ \sqrt{2} & \text{if } \widehat{K}_d \text{ is the reference } d \text{ simplex and } d \geq 3, \\ \sqrt{d-1} & \text{if } \widehat{K}_d \text{ is the reference } d \text{ cube and } d \geq 3. \end{cases}$$

Proposition 3.38 with $K' = \widehat{K}_d$ and $E' = \widehat{E}_d$ therefore yields:

Corollary 3.40 *Assume that the element K and its face E are the image of the reference d-simplex or reference d-cube $\widehat{K}_d$ and of its face $\widehat{E}_d$, respectively, under an affine transformation. Then for every polynomial degree k, the quantities $C_{I,1,k}(K), \ldots, C_{I,5,k}(E,K)$ corresponding to K and E on one hand and the quantities $C_{I,1,k}(\widehat{K}_d), \ldots, C_{I,5,k}(\widehat{E}_d, \widehat{K}_d)$ corresponding to $\widehat{K}_d$ and $\widehat{E}_d$ on the other hand are related by*

$$\begin{aligned} C_{I,1,k}(K) &= C_{I,1,k}(\widehat{K}_d), \\ C_{I,2,k}(K) &\leq \frac{h_K}{\rho_K} C_{I,2,k}(\widehat{K}_d), \\ C_{I,3,k}(E) &= C_{I,3,k}(\widehat{E}_d), \\ C_{I,4,k}(E,K) &\leq \sqrt{2d}\left(\frac{\mu_d(K) h_E}{\mu_{d-1}(E)\rho_K^2}\right)^{\frac{1}{2}} C_{I,4,k}(\widehat{E}_d, \widehat{K}_d), \\ C_{I,5,k}(E,K) &\leq \sqrt{\sqrt{2d}}\left(\frac{\mu_d(K)}{\mu_{d-1}(E) h_E}\right)^{\frac{1}{2}} C_{I,5,k}(\widehat{E}, \widehat{K}_d). \end{aligned}$$

Remark 3.41 If K is the image of $\widehat{K}_d$ under an affine transformation and if E is any $(d-1)$-dimensional face of K, the affine transformation can always be chosen such that $\widehat{E}_d$ is mapped onto E.

3.6.3 *Inverse Estimates on Intervals*

Now, we consider the one-dimensional situation.

Lemma 3.42 *The following inverse estimates hold for all univariate polynomials p of degree k and all integers k*

$$\begin{aligned} \left\| (1-x^2)^{\frac{1}{2}} p' \right\|_{(-1,1)} &\leq \sqrt{k(k+1)} \, \|p\|_{(-1,1)}, \\ \|p\|_{(-1,1)} &\leq (k+2) \, \|(1-x^2)p\|_{(-1,1)}. \end{aligned}$$

Proof Denote by L_k the k-th Legendre polynomial with leading coefficient 1. Consider two integers $0 < \ell \le k$. Since $(1-x^2)L'_\ell(x)$ vanishes at $x = \pm 1$, integration by parts yields

$$\int_{-1}^{1} (1-x^2)L'_k(x)L'_\ell(x)\mathrm{d}x = -\int_{-1}^{1} L_k(x)\left[(1-x^2)L'_\ell(x)\right]'\mathrm{d}x.$$

Since $\left[(1-x^2)L'_\ell(x)\right]'$ is a polynomial of degree ℓ with leading coefficient $-\ell(\ell+1)$, the orthogonality of the Legendre polynomials implies that

$$\int_{-1}^{1} (1-x^2)L'_k(x)L'_\ell(x)\mathrm{d}x = \begin{cases} k(k+1)\,\|L_k\|^2_{(-1,1)} & \text{if } \ell = k, \\ 0 & \text{if } \ell < k. \end{cases} \tag{3.28}$$

Now, consider a polynomial p of degree k. It may be written in the form $p = \sum_{\ell=0}^{k} \alpha_\ell L_\ell$. The orthogonality of the Legendre polynomials and equation (3.28) imply that

$$\|p\|^2_{(-1,1)} = \sum_{\ell=0}^{k} \alpha_\ell^2\,\|L_\ell\|^2_{(-1,1)}$$

and

$$\begin{aligned} \left\|(1-x^2)^{\frac{1}{2}}p'\right\|^2_{(-1,1)} &= \int_{-1}^{1}(1-x^2)p'(x)^2\mathrm{d}x \\ &= \sum_{\ell=0}^{k} \alpha_\ell^2\ell(\ell+1)\,\|L_\ell\|^2_{(-1,1)} \\ &\le k(k+1)\,\|p\|^2_{(-1,1)}. \end{aligned}$$

This proves the first estimate.

Denote by $1 > x_{1,k+2} > \cdots > x_{k+2,k+2} > -1$ the zeros of L_{k+2} and by $\omega_{1,k+2}, \ldots, \omega_{k+2,k+2}$ the weights of the corresponding Gaussian quadrature formula. Consider an arbitrary polynomial p of degree k. Since p^2 and $(1-x^2)p^2$ are polynomials of degree at most $2k+2$, we conclude that

$$\int_{-1}^{1} p(x)^2\mathrm{d}x = \sum_{i=1}^{k+2} \omega_{i,k+2}p(x_{i,k+2})^2,$$

and

$$\int_{-1}^{1} (1-x^2)p(x)^2\mathrm{d}x = \sum_{i=1}^{k+2} \omega_{i,k+2}\,(1-x_{i,k+2}^2)p(x_{i,k+2})^2$$

which implies

$$\begin{aligned} \int_{-1}^{1}(1-x^2)p(x)^2\mathrm{d}x &\ge (1-x_{1,k+2}^2)\sum_{i=1}^{k+2}\omega_{i,k+2}p(x_{i,k+2})^2 \\ &= (1-x_{1,k+2}^2)\int_{-1}^{1}p(x)^2\mathrm{d}x \end{aligned}$$

or, equivalently,

$$\int_{-1}^{1} p(x)^2 \mathrm{d}x \le \frac{1}{1 - x_{1,k+2}^2} \int_{-1}^{1} (1 - x^2) p(x)^2 \mathrm{d}x.$$

Since $x_{1,k+2} \le \cos\left(\frac{\pi}{2(k+2)}\right)$ [**242**, Theorem VI.6.21.3] and $\sin z \ge \frac{2}{\pi} z$ for all $z \in [0, \frac{\pi}{2}]$, this establishes the second estimate of Lemma 3.42. □

Remark 3.43 The proof of the second estimate of Lemma 3.42 and Remark 3.39 imply that

$$\int_a^b q(x)\mathrm{d}x \le (k+2)^2 \int_a^b \psi_{(a,b)}(x) q(x) \mathrm{d}x$$

holds for every integer k, every interval (a, b), and every non-negative polynomial q of degree at most $2k + 1$.

Since $1 - x^2 = \psi_{(-1,1)}$ and $\operatorname{diam}((-1,1)) = 2$ and

$$\begin{aligned} \left\| ((1 - x^2)p)' \right\|_{(-1,1)} &\le \|2xp\|_{(-1,1)} + \|(1 - x^2)p'\|_{(-1,1)} \\ &\le 2\|p\|_{(-1,1)} + \|(1 - x^2)p'\|_{(-1,1)}, \end{aligned}$$

Lemma 3.42 and Remark 3.39 imply the following estimates.

Proposition 3.44 *For every interval (a, b) and every polynomial degree k, the constants $\mathrm{C}_{I,1,k}((a,b))$ and $\mathrm{C}_{I,2,k}((a,b))$ can be estimated by*

$$\mathrm{C}_{I,1,k}((a,b)) \le k + 2, \quad \mathrm{C}_{I,2,k}((a,b)) \le 4 + 2\sqrt{k(k+1)}.$$

3.6.4 Inverse Estimates on the Reference Cube

Since the reference cube is the tensor product of intervals, a dimension reduction argument can be used to express the constants $\mathrm{C}_{I,1,k}(\widehat{K}_d), \ldots, \mathrm{C}_{I,5,k}(\widehat{E}_d, \widehat{K}_d)$ in terms of the corresponding constants for intervals.

Proposition 3.45 *The constants $\mathrm{C}_{I,1,k}(\widehat{K}_d), \ldots, \mathrm{C}_{I,5,k}(\widehat{E}_d, \widehat{K}_d)$ corresponding to the reference d-cube $\widehat{K}_d$ and its $(d-1)$-face $\widehat{E}_d$ can be bounded by*

$$\begin{aligned} \mathrm{C}_{I,1,k}(\widehat{K}_d) &\le (k+2)^d, \\ \mathrm{C}_{I,2,k}(\widehat{K}_d) &\le d\left(4 + 2\sqrt{k(k+1)}\right), \\ \mathrm{C}_{I,3,k}(\widehat{E}_d) &\le (k+2)^{d-1}, \\ \mathrm{C}_{I,4,k}(\widehat{E}_d, \widehat{K}_d) &\le d^{\frac{1}{4}} \left[1 + \frac{d-1}{3}\left(4 + 2\sqrt{k(k+1)}\right)^2\right]^{\frac{1}{2}}, \\ \mathrm{C}_{I,5,k}(\widehat{E}_d, \widehat{K}_d) &\le \frac{1}{\sqrt{3}(d-1)^{\frac{1}{4}}}. \end{aligned}$$

Proof We prove Proposition 3.45 by induction on the dimension d. To this end we write every $x \in \mathbb{R}^d$ in the form $x = (x', x_d)$ with $x' \in \mathbb{R}^{d-1}$ and note that

$$\psi_{\widehat{K}_d}(x) = \psi_{(0,1)}(x_d)\psi_{\widehat{K}_{d-1}}(x') \quad \text{and} \quad \psi_{\widehat{E}_d}(x) = (1 - x_d)\psi_{\widehat{K}_{d-1}}(x')$$

holds for all $x \in \widehat{K}_d$.

To prove the first and second estimate of Proposition 3.45, consider an arbitrary polynomial $v \in R_k(\widehat{K}_d)$. Fubini's theorem implies that

$$\|v\|_{\widehat{K}_d}^2 = \int_0^1 \left\{ \int_{\widehat{K}_d \cap \{x_d = t\}} v(x', t)^2 \mathrm{d}x' \right\} \mathrm{d}t.$$

Since $\widehat{K}_d \cap \{x_d = t\}$ is, for every $t \in [0, 1]$, a translate of $\widehat{K}_{d-1}$, the induction hypotheses and Proposition 3.38 yield for every $t \in [0, 1]$

$$\int_{\widehat{K}_d \cap \{x_d = t\}} v(x', t)^2 \mathrm{d}x' \leq C_{I,1,k}(\widehat{K}_{d-1})^2 \int_{\widehat{K}_d \cap \{x_d = t\}} \psi_{\widehat{K}_{d-1}}(x') v(x', t)^2 \mathrm{d}x'.$$

Since $t \mapsto \int_{\widehat{K}_d \cap \{x_d = t\}} \psi_{\widehat{K}_{d-1}}(x') v(x', t)^2 \mathrm{d}x'$ is a non-negative polynomial of degree at most $2k$, we conclude from Remark 3.43 that

$$\begin{aligned}
&\int_0^1 \left\{ \int_{\widehat{K}_d \cap \{x_d = t\}} \psi_{\widehat{K}_{d-1}}(x') v(x', t)^2 \mathrm{d}x' \right\} \mathrm{d}t \\
&\quad \leq (k+2)^2 \int_0^1 \psi_{(0,1)}(t) \left\{ \int_{\widehat{K}_d \cap \{x_d = t\}} \psi_{\widehat{K}_{d-1}}(x') v(x', t)^2 \mathrm{d}x' \right\} \mathrm{d}t \\
&\quad = (k+2)^2 \int_{\widehat{K}_d} \psi_{\widehat{K}_d}(x) v(x)^2 \mathrm{d}x.
\end{aligned}$$

Combining these estimates we obtain

$$C_{I,1,k}(\widehat{K}_d) \leq (k+2)\, C_{I,1,k}(\widehat{K}_{d-1}).$$

Together with Proposition 3.44 this proves the first estimate of Proposition 3.45.

To prove the second estimate, we first consider the partial derivative with respect to x_d. Fubini's theorem implies that

$$\begin{aligned}
h_{\widehat{K}_d}^2 \left\| \frac{\partial}{\partial x_d} (\psi_{\widehat{K}_d} v) \right\|_{\widehat{K}_d}^2 &= d \int_{\widehat{K}_{d-1}} \left\{ \int_0^1 \left[\frac{\partial}{\partial x_d} (\psi_{\widehat{K}_d}(x', x_d) v(x', x_d)) \right]^2 \mathrm{d}x_d \right\} \mathrm{d}x' \\
&= d \int_{\widehat{K}_{d-1}} \psi_{\widehat{K}_{d-1}}(x')^2 \left\{ \int_0^1 \left[\frac{\partial}{\partial x_d} (\psi_{(0,1)}(x_d) v(x', x_d)) \right]^2 \mathrm{d}x_d \right\} \mathrm{d}x'.
\end{aligned}$$

From Proposition 3.44 we obtain for every $x' \in \widehat{K}_{d-1}$

$$\begin{aligned}
\int_0^1 \left[\frac{\partial}{\partial x_d} (\psi_{(0,1)}(x_d) v(x', x_d)) \right]^2 \mathrm{d}x_d &= \left\| (\psi_{(0,1)}(\cdot) v(x', \cdot))' \right\|_{(0,1)}^2 \\
&\leq C_{I,2,k}((0,1))^2 \left\| v(x', \cdot) \right\|_{(0,1)}^2 \\
&\leq \left(4 + 2\sqrt{k(k+1)}\right)^2 \left\| v(x', \cdot) \right\|_{(0,1)}^2.
\end{aligned}$$

Since, due to Fubini's theorem,

$$\int_{\widehat{K}_{d-1}} \psi_{\widehat{K}_{d-1}}(x')^2 \left\| v(x',\cdot) \right\|_{(0,1)}^2 \,\mathrm{d}x' \le \int_{\widehat{K}_{d-1}} \left\| v(x',\cdot) \right\|_{(0,1)}^2 \,\mathrm{d}x' = \|v\|_{\widehat{K}_d}^2 ,$$

this proves

$$h_{\widehat{K}_d} \left\| \frac{\partial}{\partial x_d} \left(\psi_{\widehat{K}_d} v\right) \right\|_{\widehat{K}_d} \le \sqrt{d}\left(4 + 2\sqrt{k(k+1)}\right) \|v\|_{\widehat{K}_d} .$$

Since $\widehat{K}_d$, $\psi_{\widehat{K}_d}$, and the Lebesgue integral are invariant under rotations, the same estimate holds for the other partial derivatives. This proves the second estimate of Proposition 3.45.

The third estimate of Proposition 3.45 follows from the first one and Remark 3.39, since $\widehat{E}_d$ is isometric to $\widehat{K}_{d-1}$.

For the proof of the last two estimates of Proposition 3.45, consider an arbitrary polynomial $w \in R_k(\widehat{E}_d)$. Since w is constant with respect to x_d, we conclude from Fubini's theorem that

$$\frac{\partial}{\partial x_d} \left(\psi_{\widehat{E}_d} w\right) = -\psi_{\widehat{K}_{d-1}} w$$

and thus

$$\begin{aligned} h_{\widehat{E}_d} \left\| \frac{\partial}{\partial x_d} \left(\psi_{\widehat{E}_d} w\right) \right\|_{\widehat{K}_d}^2 &= \sqrt{d} \int_{\widehat{E}_d} \left\{ \int_0^1 \left(\psi_{\widehat{K}_{d-1}}(x') w(x')\right)^2 \mathrm{d}x_d \right\} \mathrm{d}x' \\ &\le \sqrt{d}\, \|w\|_{\widehat{E}_d}^2 . \end{aligned}$$

For the derivatives with respect to x', we similarly obtain

$$\begin{aligned} h_{\widehat{E}_d} \left\| \nabla_{x'} \left(\psi_{\widehat{E}_d} w\right) \right\|_{\widehat{K}_d}^2 &= \sqrt{d} \int_{\widehat{E}_d} \left\{ \int_0^1 \left[(1-x_d)\nabla_{x'} \left(\psi_{\widehat{K}_{d-1}}(x') w(x')\right)\right]^2 \mathrm{d}x_d \right\} \mathrm{d}x' \\ &= \frac{1}{3}\sqrt{d} \int_{\widehat{E}_d} \left[\nabla_{x'} \left(\psi_{\widehat{K}_{d-1}}(x') w(x')\right)\right]^2 \mathrm{d}x' \\ &= \frac{1}{3}\frac{\sqrt{d}}{d-1} h_{\widehat{K}_{d-1}}^2 \left\| \nabla_{x'} \left(\psi_{\widehat{K}_{d-1}} w\right) \right\|_{\widehat{K}_{d-1}}^2 . \end{aligned}$$

Together with the induction hypotheses this implies

$$\begin{aligned} h_{\widehat{E}_d} \left\| \nabla_{x'} \left(\psi_{\widehat{E}_d} w\right) \right\|_{\widehat{K}_d}^2 &\le \frac{1}{3}\frac{\sqrt{d}}{d-1} C_{I,2,k}(\widehat{K}_{d-1})^2 \|w\|_{\widehat{K}_{d-1}}^2 \\ &= \frac{1}{3}\frac{\sqrt{d}}{d-1} C_{I,2,k}(\widehat{K}_{d-1})^2 \|w\|_{\widehat{E}_d}^2 \\ &\le \frac{1}{3}\frac{\sqrt{d}}{d-1} \left[(d-1)(4 + 2\sqrt{k(k+1)})\right]^2 \|w\|_{\widehat{E}_d}^2 . \end{aligned}$$

Combining these estimates we obtain the fourth estimate of Proposition 3.45.

Finally, Fubini's theorem yields

$$\begin{aligned}\|\psi_{\widehat{E}_d} w\|_{\widehat{K}_d}^2 &= \int_{\widehat{E}_d} \left\{ \int_0^1 (1-x_d)^2 \left(\psi_{\widehat{K}_{d-1}}(x')w(x')\right)^2 \mathrm{d}x_d \right\} \mathrm{d}x' \\ &= \frac{1}{3} \int_{\widehat{E}_d} \left(\psi_{\widehat{K}_{d-1}}(x')w(x')\right)^2 \mathrm{d}x' \\ &\le \frac{1}{3} \|w\|_{\widehat{E}_d}^2 \\ &\le \frac{1}{3\sqrt{d-1}} h_{\widehat{E}_d} \|w\|_{\widehat{E}_d}^2 .\end{aligned}$$

This proves the fifth estimate of Proposition 3.45. □

3.6.5 *Inverse Estimates on the Reference Simplex*

We again use a dimension reduction argument. The details are now slightly more involved since the reference simplex does not have the nice tensor product form of the reference cube.

Proposition 3.46 *The constants $C_{I,1,k}(\widehat{K}_d), \ldots, C_{I,5,k}(\widehat{E}_d, \widehat{K}_d)$ corresponding to the reference d-simplex $\widehat{K}_d$ and its $(d-1)$-face $\widehat{E}_d$ can be bounded by*

$$C_{I,1,k}(\widehat{K}_d) \le [2(k+2)]^d \left[\frac{1}{(d-1)!} \left(\frac{d}{d+1} \right)^{d+1} \right]^{\frac{1}{2}},$$

$$C_{I,2,k}(\widehat{K}_d) \le \min\{d, 2\}\, d^{\frac{3}{2}} \left(\frac{d+1}{d} \right)^{d+1} \left(1 + \frac{1}{2}\sqrt{k(k+1)} \right),$$

$$C_{I,3,k}(\widehat{E}_d) \le [2(k+2)]^{d-1} \left[\frac{1}{(d-2)!} \left(\frac{d-1}{d} \right)^{d} \right]^{\frac{1}{2}},$$

$$\begin{aligned}C_{I,4,k}(\widehat{E}_d, \widehat{K}_d) \le \min\left\{(d-1)^{\frac{1}{4}}, 2^{\frac{1}{4}}\right\} \Bigg[&\frac{2d-1}{4} \left(\frac{2d}{2d-1} \right)^{2d} \\ &+ \frac{1}{3} \min\left\{(d-1)^2, 4\right\} (d-1)^3 \left(\frac{d}{d-1} \right)^{2d} \cdot \left(1 + \frac{1}{2}\sqrt{k(k+1)} \right)^2 \Bigg]^{\frac{1}{2}},\end{aligned}$$

$$C_{I,5,k}(\widehat{E}_d, \widehat{K}_d) \le \frac{3}{2\min\left\{(d-1)^{\frac{1}{4}}, 2^{\frac{1}{4}}\right\} \sqrt{2d+1}} \left(\frac{2d}{2d+1} \right)^d .$$

Remark 3.47 At the expense of slightly larger right-hand sides the estimates of Proposition 3.46 can be simplified to

$$C_{I,1,k}(\widehat{K}_d) \le \frac{[2(k+2)]^d}{\sqrt{(d-1)!}},$$

$$C_{I,2,k}(\widehat{K}_d) \le 9 d^{\frac{3}{2}} \left(1 + \frac{1}{2}\sqrt{k(k+1)} \right),$$

$$C_{I,3,k}(\widehat{E}_d) \le \frac{[2(k+2)]^{d-1}}{\sqrt{(d-2)!}},$$

$$C_{I,4,k}(\widehat{E}_d, \widehat{K}_d) \le 2^{\frac{1}{4}} \left[\frac{3d}{2} + 4d^3 \left(1 + \frac{1}{2}\sqrt{k(k+1)} \right)^2 \right]^{\frac{1}{2}},$$
$$C_{I,5,k}(\widehat{E}_d, \widehat{K}_d) \le \frac{3}{2\sqrt{2d+1}}.$$

Proof of Proposition 3.46 The proof is similar to that of Proposition 3.45. Some modifications, however, arise from the missing tensor-product structure of the simplex $\widehat{K}_d$ and the cut-off function $\psi_{\widehat{K}_d}$. In particular, the intersection of $\widehat{K}_d$ with a hyperplane $\{x_d = t\}$ is now a simplex with varying size and the functions $\psi_{\widehat{K}_d}$ and $\psi_{\widehat{E}_d}$ now satisfy

$$\begin{aligned}\psi_{\widehat{K}_d}(x', t) &= \frac{(d+1)^{d+1}}{d^d} t(1-t)^d \psi_{\widehat{K}_d \cap \{x_d = t\}}(x') \\ &= \frac{(d+1)^{d+1}}{4d^d} \psi_{(0,1-|x'|_1)}(t)(1 - |x'|_1)^{d-1} \psi_{\widehat{K}_{d-1}}(x')\end{aligned}$$

and

$$\psi_{\widehat{E}_d}(x', t) = \frac{1 - |x'|_1 - t}{1 - |x'|_1} \psi_{\widehat{K}_{d-1}}(x').$$

To prove the first and second estimate of Proposition 3.46, consider an arbitrary polynomial $v \in R_k(\widehat{K}_d)$.

For every $x \in \widehat{K}_d$, we have $1 \ge |x|_1 \ge d \min_{1 \le i \le d} x_i$ and therefore $\widehat{K}_d = \bigcup_{1 \le i \le d} \widehat{K}_{d,i}$ with $\widehat{K}_{d,i} = \left\{ x \in \widehat{K}_d : x_i \le \frac{1}{d} \right\}$. Fubini's theorem implies that

$$\|v\|^2_{\widehat{K}_{d,d}} = \int_0^{\frac{1}{d}} \left\{ \int_{\widehat{K}_d \cap \{x_d = t\}} v(x', t)^2 \mathrm{d}x' \right\} \mathrm{d}t.$$

Since $\widehat{K}_d \cap \{x_d = t\}$ is for every $t \in [0, 1]$ the image of $\widehat{K}_{d-1}$ under a translation and a scaling, the induction hypotheses and Proposition 3.38 yield for every $t \in [0, 1]$

$$\int_{\widehat{K}_d \cap \{x_d = t\}} v(x', t)^2 \mathrm{d}x' \le C_{I,1,k}(\widehat{K}_{d-1})^2 \int_{\widehat{K}_d \cap \{x_d = t\}} \psi_{\widehat{K}_d \cap \{x_d = t\}}(x') v(x', t)^2 \mathrm{d}x'.$$

Since $t \mapsto \int_{\widehat{K}_d \cap \{x_d = t\}} \psi_{\widehat{K}_d \cap \{x_d = t\}}(x') v(x', t)^2 \mathrm{d}x'$ is a non-negative polynomial of degree at most $2k + 1$, we conclude from Remark 3.43 that

$$\begin{aligned}&\int_0^{\frac{1}{d}} \left\{ \int_{\widehat{K}_d \cap \{x_d = t\}} \psi_{\widehat{K}_d \cap \{x_d = t\}}(x') v(x', t)^2 \mathrm{d}x' \right\} \mathrm{d}t \\ &\le (k+2)^2 \int_0^{\frac{1}{d}} \psi_{(0,\frac{1}{d})}(t) \left\{ \int_{\widehat{K}_d \cap \{x_d = t\}} \psi_{\widehat{K}_d \cap \{x_d = t\}}(x') v(x', t)^2 \mathrm{d}x' \right\} \mathrm{d}t \\ &= (k+2)^2 \int_0^{\frac{1}{d}} \left\{ \int_{\widehat{K}_d \cap \{x_d = t\}} \frac{4d^{d+2}}{(d+1)^{d+1}(d-1)^2} (1-t)^{-d} (1 - \frac{1}{d} - t) \psi_{\widehat{K}_d}(x', t) v(x', t)^2 \mathrm{d}x' \right\} \mathrm{d}t \\ &\le (k+2)^2 \frac{4d^{2d+1}}{(d+1)^{d+1}(d-1)^{d+1}} \int_{\widehat{K}_{d,d}} \psi_{\widehat{K}_d}(x) v(x)^2 \mathrm{d}x.\end{aligned}$$

Combining these estimates we obtain

$$\|v\|_{\widehat{K}_{d,d}}^2 \le C_{I,1,k}(\widehat{K}_{d-1})^2(k+2)^2 \frac{4d^{2d+1}}{(d+1)^{d+1}(d-1)^{d+1}} \left\|\psi_{\widehat{K}_d} v\right\|_{\widehat{K}_{d,d}}^2 .$$

Interchanging the roles of x_d and x_i yields corresponding estimates for the remaining sets $\widehat{K}_{d,i}$, $1 \le i \le d-1$. Hence we have

$$C_{I,1,k}(\widehat{K}_d) \le \frac{2}{\sqrt{d}} \left(\frac{d^2}{d^2-1}\right)^{\frac{d+1}{2}} (k+2)\, C_{I,1,k}(\widehat{K}_{d-1}).$$

Together with Proposition 3.44 this proves the first estimate of Proposition 3.46.

To prove the second estimate, we first consider the partial derivative with respect to x_d. Fubini's theorem implies that

$$\begin{aligned} h_{\widehat{K}_d}^2 \left\| \frac{\partial}{\partial x_d}\left(\psi_{\widehat{K}_d} v\right)\right\|_{\widehat{K}_d}^2 &= \min\{d,2\} \int_{\widehat{K}_{d-1}} \left\{ \int_0^{1-|x'|_1} \left[\frac{\partial}{\partial x_d}\left(\psi_{\widehat{K}_d}(x',x_d)v(x',x_d)\right)\right]^2 \mathrm{d}x_d \right\} \mathrm{d}x' \\ &= \min\{d,2\} \frac{(d+1)^{2d+2}}{16d^{2d}} \int_{\widehat{K}_{d-1}} \psi_{\widehat{K}_{d-1}}(x')^2 \\ &\quad \cdot \left(1-\left|x'\right|_1\right)^2 \left\{ \int_0^{1-|x'|_1} \left[\frac{\partial}{\partial x_d}\left(\psi_{(0,1-|x'|_1)}(x_d)v(x',x_d)\right)\right]^2 \mathrm{d}x_d \right\} \mathrm{d}x'. \end{aligned}$$

From Proposition 3.44 we obtain for every $x' \in \widehat{K}_{d-1}$

$$\begin{aligned} &\left(1-\left|x'\right|_1\right)^2 \int_0^{1-|x'|_1} \left[\frac{\partial}{\partial x_d}\left(\psi_{(0,1-|x'|_1)}(x_d)v(x',x_d)\right)\right]^2 \mathrm{d}x_d \\ &\quad = h_{(0,1-|x'|_1)}^2 \left\| \left(\psi_{(0,1-|x'|_1)}(\cdot)v(x',\cdot)\right)'\right\|_{(0,1-|x'|_1)}^2 \\ &\quad \le C_{I,2,k}((0,1-\left|x'\right|_1))^2 \left\|v(x',\cdot)\right\|_{(0,1-|x'|_1)}^2 \\ &\quad \le \left(4+2\sqrt{k(k+1)}\right)^2 \left\|v(x',\cdot)\right\|_{(0,1-|x'|_1)}^2 . \end{aligned}$$

Since due to Fubini's theorem

$$\begin{aligned} \int_{\widehat{K}_{d-1}} \psi_{\widehat{K}_{d-1}}(x')^2 \left\|v(x',\cdot)\right\|_{(0,1-|x'|_1)}^2 \mathrm{d}x' &\le \int_{\widehat{K}_{d-1}} \left\|v(x',\cdot)\right\|_{(0,1-|x'|_1)}^2 \mathrm{d}x' \\ &= \|v\|_{\widehat{K}_d}^2 , \end{aligned}$$

this proves

$$h_{\widehat{K}_d} \left\| \frac{\partial}{\partial x_d}\left(\psi_{\widehat{K}_d} v\right)\right\|_{\widehat{K}_d} \le \min\{d,2\} \frac{(d+1)^{d+1}}{4d^d} \left(4+2\sqrt{k(k+1)}\right) \|v\|_{\widehat{K}_d} .$$

Since $\widehat{K}_d$, $\psi_{\widehat{K}_d}$, and the Lebesgue integral are invariant under permutations of the coordinates, the same estimate holds for the other partial derivatives. This proves the second estimate of Proposition 3.46.

The third estimate of Proposition 3.46 follows from the first one and Remark 3.39, since $\widehat{E}_d$ is isometric to $\widehat{K}_{d-1}$.

For the proof of the last two estimates of Proposition 3.45, consider an arbitrary polynomial $w \in R_k(\widehat{E}_d)$. Since w is constant with respect to x_d, we conclude that

$$\frac{\partial}{\partial x_d}\left(\psi_{\widehat{E}_d}(x',x_d)w(x',x_d)\right) = \frac{-1}{1-|x'|_1}\psi_{\widehat{K}_{d-1}}(x')w(x')$$

holds for all $(x',x_d) \in \widehat{K}_d$. Fubini's theorem therefore implies

$$\begin{aligned} h_{\widehat{E}_d}\left\|\frac{\partial}{\partial x_d}\left(\psi_{\widehat{E}_d}w\right)\right\|^2_{\widehat{K}_d} &= \min\left\{\sqrt{d-1},\sqrt{2}\right\} \\ &\quad\cdot \int_{\widehat{K}_{d-1}}\left\{\int_0^{1-|x'|_1}\left(\frac{-1}{1-|x'|_1}\psi_{\widehat{K}_{d-1}}(x')w(x')\right)^2 \mathrm{d}x_d\right\}\mathrm{d}x' \\ &= \min\left\{\sqrt{d-1},\sqrt{2}\right\}\int_{\widehat{K}_{d-1}}\frac{1}{1-|x'|_1}\psi_{\widehat{K}_{d-1}}(x')^2w(x')^2\mathrm{d}x' \\ &\le \max_{x'\in\widehat{K}_{d-1}}\left|\frac{1}{1-|x'|_1}\psi_{\widehat{K}_{d-1}}(x')^2\right| \min\left\{\sqrt{d-1},\sqrt{2}\right\}\|w\|^2_{\widehat{E}_d}. \end{aligned}$$

Since

$$\frac{1}{1-|x'|_1}\psi_{\widehat{K}_{d-1}}(x')^2 = d^{2d}\left(1-\sum_{i=1}^{d-1}x_i\right)\prod_{i=1}^{d-1}x_i^2,$$

a straightforward calculation yields

$$\max_{x'\in\widehat{K}_{d-1}}\left|\frac{1}{1-|x'|_1}\psi_{\widehat{K}_{d-1}}(x')^2\right| = \frac{2d-1}{4}\left(\frac{2d}{2d-1}\right)^{2d}$$

and thus proves that

$$h_{\widehat{E}_d}\left\|\frac{\partial}{\partial x_d}\left(\psi_{\widehat{E}_d}w\right)\right\|^2_{\widehat{K}_d} \le \min\left\{\sqrt{d-1},\sqrt{2}\right\}\frac{2d-1}{4}\left(\frac{2d}{2d-1}\right)^{2d}\|w\|^2_{\widehat{E}_d}.$$

For the derivatives with respect to x', we observe that for every $x' \in \widehat{E}_d$ the mapping $t \mapsto \nabla_{x'}\left(\psi_{\widehat{E}_d}(x',t)w(x')\right)$ is a linear polynomial in t which attains the value $\nabla_{x'}\left(\psi_{\widehat{E}_d}(x',0)w(x')\right) = \nabla_{x'}\left(\psi_{\widehat{K}_{d-1}}(x')w(x')\right)$ for $t=0$ and the value 0 for $t = 1-|x'|_1$. The Simpson rule therefore yields for every $x' \in \widehat{E}_d$

$$\int_0^{1-|x'|_1}\left(\nabla_{x'}\left(\psi_{\widehat{E}_d}(x',t)w(x')\right)\right)^2\mathrm{d}t = \frac{1}{3}\left(1-|x'|_1\right)\left(\nabla_{x'}\left(\psi_{\widehat{K}_{d-1}}(x')w(x')\right)\right)^2.$$

Combining this result with Fubini's theorem and using the induction hypotheses, we obtain

$$\begin{aligned} h_{\widehat{E}_d}\left\|\nabla_{x'}\left(\psi_{\widehat{E}_d}w\right)\right\|^2_{\widehat{K}_d} &= \min\left\{\sqrt{d-1},\sqrt{2}\right\}\int_{\widehat{E}_d}\left\{\int_0^{1-|x'|_1}\left(\nabla_{x'}\left(\psi_{\widehat{E}_d}(x',t)w(x')\right)\right)^2\mathrm{d}t\right\}\mathrm{d}x' \\ &= \frac{1}{3}\min\left\{\sqrt{d-1},\sqrt{2}\right\}\int_{\widehat{E}_d}\left(1-|x'|_1\right)\left(\nabla_{x'}\left(\psi_{\widehat{K}_{d-1}}(x')w(x')\right)\right)^2\mathrm{d}x' \end{aligned}$$

$$
\begin{aligned}
&\leq \frac{1}{3}\min\left\{\sqrt{d-1},\sqrt{2}\right\}\int_{\widehat{E}_d}\left(\nabla_{x'}\left(\psi_{\widehat{K}_{d-1}}(x')w(x')\right)\right)^2 \mathrm{d}x' \\
&= \frac{1}{3}\min\left\{\sqrt{d-1},\sqrt{2}\right\}\left\|\nabla\left(\psi_{\widehat{K}_{d-1}}w\right)\right\|^2_{\widehat{K}_{d-1}} \\
&\leq \frac{1}{3}\min\left\{\sqrt{d-1},\sqrt{2}\right\} C_{I,2,k}(\widehat{K}_{d-1})^2 \left\|w\right\|^2_{\widehat{K}_{d-1}} \\
&= \frac{1}{3}\min\left\{\sqrt{d-1},\sqrt{2}\right\} C_{I,2,k}(\widehat{K}_{d-1})^2 \left\|w\right\|^2_{\widehat{E}_d}.
\end{aligned}
$$

Inserting the bound for $C_{I,2,k}(\widehat{K}_{d-1})$ and adding the estimates for both types of partial derivatives, establishes the fourth estimate of Proposition 3.46.

Since, for every $x' \in \widehat{E}_d$, the mapping $t \mapsto \psi_{\widehat{E}_d}(x',t)w(x')$ is a linear polynomial in t which attains the value

$$
\psi_{\widehat{E}_d}(x',0)w(x') = \psi_{\widehat{K}_{d-1}}(x',t)w(x')
$$

at $t = 0$ and the value 0 at $t = 1 - |x'|_1$, Fubini's theorem and the Simpson rule finally yield

$$
\begin{aligned}
\left\|\psi_{\widehat{E}_d}w\right\|^2_{\widehat{K}_d} &= \int_{\widehat{E}_d}\left\{\int_0^{1-|x'|_1}\left(\psi_{\widehat{E}_d}(x',x_d)w(x')\right)^2 \mathrm{d}x_d\right\}\mathrm{d}x' \\
&= \frac{1}{3}\int_{\widehat{E}_d}(1-|x'|_1)\left(\psi_{\widehat{K}_{d-1}}(x')w(x')\right)^2 \mathrm{d}x' \\
&\leq \frac{1}{3}\max_{x'\in\widehat{K}_{d-1}}\left|(1-|x'|_1)\psi_{\widehat{K}_{d-1}}(x')^2\right| \left\|w\right\|^2_{\widehat{E}_d} \\
&= \frac{1}{3\min\left\{\sqrt{d-1},\sqrt{2}\right\}}\max_{x'\in\widehat{K}_{d-1}}\left|(1-|x'|_1)\psi_{\widehat{K}_{d-1}}(x')^2\right| h_{\widehat{E}_d}\left\|w\right\|^2_{\widehat{E}_d}.
\end{aligned}
$$

A straightforward calculation yields

$$
\max_{x'\in\widehat{K}_{d-1}}\left|(1-|x'|_1)\psi_{\widehat{K}_{d-1}}(x')^2\right| = \frac{27}{4(2d+1)}\left(\frac{2d}{2d+1}\right)^{2d}
$$

and thus proves the fifth estimate of Proposition 3.46. □

3.6.6 L^p-Estimates

Now, we extend the previous inverse estimates to general L^p-spaces. The estimates are less sharp with respect to their dependence on the polynomial degree, since we use an appropriate inverse estimate to reduce the L^p-case to the L^2-setting.

The optimal constants in the second part of Proposition 3.37 (p. 112) are given by

$$
\begin{aligned}
C_{I,1,k,p}(K) &= \sup_{v\in R_k(K)\setminus\{0\}}\ \inf_{u\in R_k(K)\setminus\{0\}}\frac{\|v\|_{p;K}\|u\|_{p';K}}{\left|\int_K \psi_K vu\right|}, \\
C_{I,2,k,p}(K) &= \sup_{v\in R_k(K)\setminus\{0\}}\frac{h_K\left\|\nabla(\psi_K v)\right\|_{p;K}}{\|v\|_{p;K}}, \\
C_{I,3,k,p}(E) &= \sup_{v\in R_k(E)\setminus\{0\}}\ \inf_{u\in R_k(E)\setminus\{0\}}\frac{\|v\|_{p;E}\|u\|_{p';E}}{\left|\int_E \psi_{E,\vartheta} vu\right|},
\end{aligned}
$$

$$C_{I,4,k,p}(E,K) = \sup_{v\in R_k(E)\setminus\{0\}} \frac{\vartheta^{1-\frac{1}{p}} h_E^{1-\frac{1}{p}} \left\| \nabla(\psi_{E,\vartheta} v) \right\|_{p;K}}{\|v\|_{p;E}},$$

$$C_{I,5,k,p}(E,K) = \sup_{v\in R_k(E)\setminus\{0\}} \frac{\left\| \psi_{E,\vartheta} v \right\|_{p;K}}{\vartheta^{\frac{1}{p}} h_E^{\frac{1}{p}} \|v\|_{p;E}}.$$

Notice that

$$C_{I,1,k,2}(K) = C_{I,1,k}(K)^2, \qquad C_{I,2,k,2}(K) = C_{I,2,k}(K),$$
$$C_{I,3,k,2}(E) = C_{I,3,k}(E)^2, \qquad C_{I,4,k,2}(E,K) = C_{I,4,k}(E,K),$$
$$C_{I,5,k,2}(E,K) = C_{I,5,k}(E,K).$$

As in the L^2-case we only have to consider the particular case $\vartheta = 1$ of the standard cut-off functions ψ_E. The general case is again reduced to this special one by applying the results of Subsection 3.6.1 to the affine transformation $F_{E,K} \circ \Phi_\vartheta \circ F_{E,K}^{-1}$ of Section 3.2.4 (p. 84).

As in Subsections 3.6.4 and 3.6.5, a dimension reduction argument can be used to express the constants $C_{I,2,k,\infty}(K)$, $C_{I,4,k,\infty}(E,K)$, and $C_{I,5,k,\infty}(E,K)$ in terms of their one-dimensional analogues. These in turn correspond to well-known quantities used in the proof of the classical Bernstein and Markoff inequalities. By interpolation one can then express the constants $C_{I,2,k,p}(K)$, $C_{I,4,k,p}(E,K)$, and $C_{I,5,k,p}(E,K)$ in terms of the corresponding constants for the cases $p = 2$ and $p = \infty$. A duality argument then yields the constants $C_{I,2,k,p}(K)$ and $C_{I,5,k,p}(E,K)$ for the case $1 \le p < 2$. For the constant $C_{I,4,k,p}(E,K)$, however, this duality argument unfortunately does not work. For the constants $C_{I,1,k,\infty}(K)$ and $C_{I,3,k,\infty}(K)$, unfortunately the dimension reduction argument is not applicable or leads to one-dimensional analogues which are not known from classical theory.

To avoid these difficulties, for general p, we directly express the constants $C_{I,1,k,p}(K), \ldots,$ $C_{I,5,k,p}(E,K)$ in terms of the corresponding quantities for $p = 2$. This is achieved using appropriate inverse estimates for L^p-norms of polynomials by their L^2-norms. To this end, we set

$$\overline{C}_{I,k,p}(K) = \sup_{v\in R_k(K)\setminus\{0\}} \frac{\mu_d(K)^{\frac{1}{2}-\frac{1}{p}} \|v\|_{p;K}}{\|v\|_K},$$

$$\underline{C}_{I,k,p}(K) = \sup_{v\in R_k(K)\setminus\{0\}} \frac{\|v\|_K}{\mu_d(K)^{\frac{1}{2}-\frac{1}{p}} \|v\|_{p;K}}.$$

Proposition 3.48 *The constants $\overline{C}_{I,k,p}(K)$ and $\underline{C}_{I,k,p}(K)$ are invariant under affine transformations of the set K. They are monotonically increasing functions of the polynomial degree k. For every Lebesgue exponent p with $1 \le p \le \infty$ they satisfy the relations*

$$\overline{C}_{I,k,p}(K) = \underline{C}_{I,k,p'}(K),$$

$$\overline{C}_{I,k,p}(K) \le \begin{cases} 1 & \text{if } 1 \le p \le 2, \\ \overline{C}_{I,k,\infty}(K)^{1-\frac{2}{p}} & \text{if } 2 < p \le \infty, \end{cases}$$

$$\underline{C}_{I,k,p}(K) \le \begin{cases} \overline{C}_{I,k,\infty}(K)^{\frac{2}{p}-1} & \text{if } 1 \le p < 2, \\ 1 & \text{if } 2 \le p \le \infty, \end{cases}$$

$$\overline{C}_{I,k,p}(K)\underline{C}_{I,k,p}(K) \le \overline{C}_{I,k,\infty}(K)^{\left|1-\frac{2}{p}\right|}.$$

Proof The affine-invariance of $\overline{C}_{I,k,p}(K)$ and $\underline{C}_{I,k,p}(K)$ and their monotone dependence on the polynomial degree are obvious. The relation $\overline{C}_{I,k,p}(K) = \underline{C}_{I,k,p'}(K)$ follows by duality. The bound of $\overline{C}_{I,k,p}(K)$ follows from Hölder's inequality for $1 \le p \le 2$ and by interpolation for $2 < p \le \infty$. The last two estimates follow from the first two relations. □

Proposition 3.49 *For every interval (a, b) and every polynomial degree k, we have*

$$\overline{C}_{I,k,\infty}((a,b)) \le \sqrt{2k+2}.$$

For every element K which is the affine image of either the reference d-simplex or of the reference d-cube and every polynomial degree k, we have

$$\overline{C}_{I,k,\infty}(K) \le (2k+2)^{\frac{1}{2}}(4k+2)^{\frac{d-1}{2}}.$$

Proof From [**242**, Section 7.21 and Theorem 7.3.1] we know that

$$\|v\|_{\infty;(-1,1)} \le \sqrt{k+1}\,\|v\|_{(-1,1)}$$

holds for all univariate polynomials v of degree at most k. Together with Proposition 3.48 this proves the first assertion of Proposition 3.49.

To prove the second assertion, we first consider the reference d-cube $\widehat{K}_d$. Choose an arbitrary polynomial $v \in R_k(\widehat{K}_d)$ and a point $x^* \in \widehat{K}_d$ with $|v(x^*)| = \|v\|_{\infty;\widehat{K}_d}$. Since $\mu_d(\widehat{K}_d) = \mu_{d-1}(\widehat{K}_{d-1}) = 1$ and since $\widehat{K}_d \cap \{x_d = x_d^*\}$ is the image of $\widehat{K}_{d-1}$ under a translation, we have

$$\begin{aligned}
\mu_d(\widehat{K}_d)\,\|v\|^2_{\infty;\widehat{K}_d} &= |v(x^*)|^2 \\
&\le \|v\|^2_{\infty;\widehat{K}_d\cap\{x_d=x_d^*\}} \\
&= \mu_{d-1}\left(\widehat{K}_d \cap \{x_d = x_d^*\}\right)\|v\|^2_{\infty;\widehat{K}_d\cap\{x_d=x_d^*\}} \\
&\le \overline{C}_{I,k,\infty}(\widehat{K}_{d-1})^2\,\|v\|^2_{\widehat{K}_d\cap\{x_d=x_d^*\}}.
\end{aligned}$$

Since the mapping $t \mapsto \|v\|^2_{\widehat{K}_d\cap\{x_d=t\}}$ is a univariate polynomial of degree at most $2k$, the first assertion of Proposition 3.49 implies that

$$\begin{aligned}
\|v\|^2_{\widehat{K}_d\cap\{x_d=x_d^*\}} &\le \sup_{0\le t\le 1}\|v\|^2_{\widehat{K}_d\cap\{x_d=t\}} \\
&\le (4k+2)\int_0^1 \|v\|^2_{\widehat{K}_d\cap\{x_d=t\}}\,dt \\
&= (4k+2)\,\|v\|^2_{\widehat{K}_d}.
\end{aligned}$$

Combining these estimates we obtain for the reference d-cube

$$\overline{C}_{I,k,\infty}(\widehat{K}_d) \le \sqrt{4k+2}\,\overline{C}_{I,k,\infty}(\widehat{K}_{d-1}).$$

By induction this proves the second estimate of Proposition 3.49 if K is the affine image of the reference d-cube.

We now consider the reference d-simplex $\widehat{K}_d$ and again choose an arbitrary polynomial $v \in R_k(\widehat{K}_d)$ and a point $x^* \in \widehat{K}_d$ with $|v(x^*)| = \|v\|_{\infty;\widehat{K}_d}$. Since $|x^*|_1 \leq 1$ there is a component x_i^* of x^* with $x_i^* \leq \frac{1}{d}$. By applying a suitable rotation if need be, we may assume that $i = d$. Since $\mu_d(\widehat{K}_d) = \frac{1}{d!}$ and $\mu_{d-1}(\widehat{K}_{d-1}) = \frac{1}{(d-1)!}$ and since $\widehat{K}_d \cap \{x_d = x_d^*\}$ is the image of $\widehat{K}_{d-1}$ under a translation and a scaling by the factor $1 - x_d^*$, we now obtain

$$\begin{aligned}
\mu_d(\widehat{K}_d)\,\|v\|^2_{\infty;\widehat{K}_d} &= \frac{1}{d!}\,|v(x^*)|^2 \\
&\leq \frac{1}{d!}\,\|v\|^2_{\infty;\widehat{K}_d\cap\{x_d=x_d^*\}} \\
&\leq \frac{1}{d!}\overline{C}_{I,k,\infty}(\widehat{K}_{d-1})^2\mu_{d-1}\left(\widehat{K}_d \cap \{x_d = x_d^*\}\right)^{-1}\|v\|^2_{\widehat{K}_d\cap\{x_d=x_d^*\}} \\
&= \frac{(d-1)!}{d!}\overline{C}_{I,k,\infty}(\widehat{K}_{d-1})^2(1-x_d^*)^{1-d}\,\|v\|^2_{\widehat{K}_d\cap\{x_d=x_d^*\}} \\
&\leq \frac{1}{d}\left(\frac{d}{d-1}\right)^{d-1}\overline{C}_{I,k,\infty}(\widehat{K}_{d-1})^2\,\|v\|^2_{\widehat{K}_d\cap\{x_d=x_d^*\}}.
\end{aligned}$$

Since the mapping $t \mapsto \|v\|^2_{\widehat{K}_d\cap\{x_d=t\}}$ is a univariate polynomial of degree at most $2k$, we conclude that

$$\begin{aligned}
\|v\|^2_{\widehat{K}_d\cap\{x_d=x_d^*\}} &\leq \sup_{0\leq t\leq 1}\|v\|^2_{\widehat{K}_d\cap\{x_d=t\}} \\
&\leq (4k+2)\int_0^1 \|v\|^2_{\widehat{K}_d\cap\{x_d=t\}}\,dt \\
&= (4k+2)\,\|v\|^2_{\widehat{K}_d}.
\end{aligned}$$

Since $\frac{1}{d}\left(\frac{d}{d-1}\right)^{d-1} \leq 1$, induction proves the second estimate of Proposition 3.49 if K is the affine image of the reference d-simplex. □

Proposition 3.50 *For every element K, every face E of K, every Lebesgue exponent p with $1 \leq p \leq \infty$, and every polynomial degree k, we have*

$$\begin{aligned}
C_{I,1,k,p}(K) &\leq \overline{C}_{I,k,\infty}(\widehat{K}_d)^{\left|1-\frac{2}{p}\right|}\,C_{I,1,k}(K)^2, \\
C_{I,2,k,p}(K) &\leq \overline{C}_{I,k+\nu_K+1,\infty}(\widehat{K}_d)^{\left|1-\frac{2}{p}\right|}\,C_{I,2,k}(K), \\
C_{I,3,k,p}(E) &\leq \overline{C}_{I,k,\infty}(\widehat{K}_{d-1})^{\left|1-\frac{2}{p}\right|}\,C_{I,3,k}(K)^2, \\
C_{I,4,k,p}(E,K) &\leq \left(\frac{\mu_{d-1}(E)h_E}{\mu_d(K)}\right)^{\frac{1}{2}-\frac{1}{p}}\overline{C}_{I,k+\nu_K+1,p}(\widehat{K}_d)\underline{C}_{I,k,p}(\widehat{K}_{d-1})\cdot C_{I,4,k}(E,K), \\
C_{I,5,k,p}(E,K) &\leq \left(\frac{\mu_{d-1}(E)h_E}{\mu_d(K)}\right)^{\frac{1}{2}-\frac{1}{p}}\overline{C}_{I,k+\nu_K+1,p}(\widehat{K}_d)\underline{C}_{I,k,p}(\widehat{K}_{d-1})\cdot C_{I,5,k}(E,K).
\end{aligned}$$

Here, the parameter ν_K is as in (3.5) (p. 82), the constants $\overline{C}_{I,k,p}(K)$ and $\underline{C}_{I,k,p}(K)$ are given by Propositions 3.48 and 3.49, and the constants $C_{I,1,k}(K), \ldots, C_{I,5,k}(E,K)$ are as in Corollary 3.40 and Propositions 3.45 and 3.46.

Proof From Proposition 3.48 we obtain for all polynomials $v, u \in R_k(K)$

$$\begin{aligned}\|v\|_{p;K}\,\|u\|_{p';K} &\le \overline{C}_{I,k,p}(K)\overline{C}_{I,k,p'}(K)\,\|v\|_K\,\|u\|_K\\ &\le \overline{C}_{I,k,\infty}(K)^{\left|\frac{1}{2}-\frac{1}{p}\right|}\,\|v\|_K\,\|u\|_K\,.\end{aligned}$$

This proves that

$$\begin{aligned}C_{I,1,k,p}(K) &\le \overline{C}_{I,k,\infty}(K)^{\left|\frac{1}{2}-\frac{1}{p}\right|}\,C_{I,1,k,2}(K)\\ &= \overline{C}_{I,k,\infty}(K)^{\left|\frac{1}{2}-\frac{1}{p}\right|}\,C_{I,1,k}(K)^2\end{aligned}$$

and thus establishes the first estimate of the proposition.

Since $\psi_K \in R_{\nu_K+1}(K)$, we obtain for every polynomial $v \in R_k(K)$

$$\begin{aligned}h_K\left\|\nabla(\psi_K v)\right\|_{p;K} &\le \overline{C}_{I,k+\nu_K+1,p}(K)\mu_d(K)^{\frac{1}{p}-\frac{1}{2}}h_K\left\|\nabla(\psi_K v)\right\|_K\\ &\le \overline{C}_{I,k+\nu_K+1,p}(K)\,C_{I,2,k}(K)\mu_d(K)^{\frac{1}{p}-\frac{1}{2}}\,\|v\|_K\\ &\le \overline{C}_{I,k+\nu_K+1,p}(K)\underline{C}_{I,k,p}(K)\,C_{I,2,k}(K)\,\|v\|_{p;K}\,.\end{aligned}$$

Taking into account Proposition 3.48 this proves the second estimate of Proposition 3.50.

The bound for $C_{I,3,k,p}(E)$ is proved in exactly the same way as the bound for $C_{I,1,k,p}(K)$ taking into account that E is the affine image of $\widehat{K}_{d-1}$.

Since $\psi_E \in R_{\nu_K+1}(E)$, we obtain for every polynomial $w \in R_k(E)$

$$\begin{aligned}h_E^{1-\frac{1}{p}}\left\|\nabla(\psi_E w)\right\|_{p;K} &\le h_E^{\frac{1}{2}-\frac{1}{p}}\overline{C}_{I,k+\nu_K+1,p}(\widehat{K}_d)\mu_d(K)^{\frac{1}{p}-\frac{1}{2}}h_E^{\frac{1}{2}}\left\|\nabla(\psi_E w)\right\|_K\\ &\le h_E^{\frac{1}{2}-\frac{1}{p}}\overline{C}_{I,k+\nu_K+1,p}(\widehat{K}_d)\mu_d(K)^{\frac{1}{p}-\frac{1}{2}}\,C_{I,4,k}(E,K)\,\|w\|_E\\ &\le h_E^{\frac{1}{2}-\frac{1}{p}}\,\mu_d(K)^{\frac{1}{p}-\frac{1}{2}}\,\mu_{d-1}(E)^{\frac{1}{2}-\frac{1}{p}}\,\overline{C}_{I,k+\nu_K+1,p}(\widehat{K}_d)\underline{C}_{I,k,p}(\widehat{K}_{d-1})\\ &\quad\cdot C_{I,4,k}(E,K)\,\|w\|_{p;E}\,.\end{aligned}$$

This proves the fourth estimate of Proposition 3.50.

The last estimate is proved in exactly the same way as the fourth one. □

3.6.7 *Smooth Cut-Off Functions*

For some applications, we need inverse estimates with respect to higher order Sobolev norms and the smooth cut-off functions of Section 3.2.5 (p. 85). They are proved in exactly the same way as Proposition 3.37.

Proposition 3.51 *The following inverse estimates hold for all Lebesgue exponents p with $1 \le p \le \infty$, all polynomials v of degree at most k, and all integers $m \ge 1$, $1 \le \ell \le m+1$, and $1 \le \ell' \le 2$*

$$\begin{aligned}\|v\|_{p;K} &\le C_{I,1,k,m,p}(K)\sup_{u\in R_k(K)\setminus\{0\}}\frac{\int_K \psi_{K,m}vu}{\|u\|_{p';K}},\\ \left\|\nabla^{\ell}(\psi_{K,m}v)\right\|_{p;K} &\le C_{I,2,k,m,p}(K)h_K^{-\ell}\,\|v\|_{p;K}\,,\\ \|v\|_{p;E} &\le C_{I,3,k,m,p}(E)\sup_{u\in R_k(E)\setminus\{0\}}\frac{\int_E \psi_{E,m}vu}{\|u\|_{p';E}},\end{aligned}$$

$$\left\|\nabla^{\ell}(\psi_{E,m}v)\right\|_{p;\omega_E} \le C_{I,4,k,m,p}(E,K)h_E^{-\ell+\frac{1}{p}} \|v\|_{p;E},$$

$$\left\|\psi_{E,m}v\right\|_{p;\omega_E} \le C_{I,5,k,m,p}(E,K)h_E^{\frac{1}{p}} \|v\|_{p;E},$$

$$\|v\|_{p;E} \le C_{I,6,k,m,p}(E) \sup_{u\in R_k(E)\setminus\{0\}} \frac{\int_E \mathbf{n}_E \cdot \nabla(\Psi_{E,1}u)v}{\|u\|_{p';E}},$$

$$\left\|\nabla^{\ell'}(\Psi_{E,1}v)\right\|_{p;\omega_E} \le C_{I,7,k,m,p}(E,K)h_E^{-\ell'+\frac{1}{p}} \|v\|_{p;E},$$

$$\left\|\Psi_{E,1}v\right\|_{p;\omega_E} \le C_{I,8,k,m,p}(E,K)h_E^{\frac{1}{p}} \|v\|_{p;E}.$$

The constants $C_{I,1,k,m,p}(K), \ldots, C_{I,8,k,m,p}(E,K)$ only depend on the shape parameter $C_{\mathcal{T}}$ of $\mathcal{T}$, the polynomial degree k, the Lebesgue exponent p, and and the integer m.

Remark 3.52 Inverse estimates for local cut-off functions were first proved in [**251**] and then extended in [**253, 254, 255, 256, 258**]. The modified local cut-off functions were introduced in [**261, 263**]. The exposition of this section concerning the L^2-case follows [**266**]; the part concerning general L^p-norms is new.

3.7 Decomposition of Affine Functions in $L^p(0, 1; Y^*)$

In this section we prove a technical result concerning the decomposition of affine functions of one real variable with values in dual spaces. This result is needed in Chapter 6 for the estimation of residuals associated with finite element approximations of parabolic differential equations. Basically, it is a strengthened Cauchy–Schwarz inequality. Yet, it is proved in a different way which is better suited for L^p-norms and functions with values in dual spaces.

In what follows Y is a Banach space with norm $\|\cdot\|_Y$ and Y^* denotes the dual space of Y equipped with the standard dual norm

$$\|\varphi\|_* = \sup_{v\in Y\setminus\{0\}} \frac{\langle\varphi, v\rangle}{\|v\|_Y}.$$

Lemma 3.53 *For every Lebesgue exponent $p \in (1,\infty)$ and every parameter $\theta \in [\frac{1}{2}, 1]$ there is a constant $\beta_{p,\theta} > 0$ such that the inequalities*

$$\begin{aligned}\beta_{p,\theta}\left\{\|\varphi\|^p_{L^p(0,1;Y^*)} + \left\|(\theta-t)\psi\right\|^p_{L^p(0,1;Y^*)}\right\}^{\frac{1}{p}} &\le \left\|\varphi + (\theta-t)\psi\right\|_{L^p(0,1;Y^*)} \\ &\le \|\varphi\|_{L^p(0,1;Y^*)} + \left\|(\theta-t)\psi\right\|_{L^p(0,1;Y^*)}\end{aligned}$$

hold for all $\varphi, \psi \in Y^*$. In particular we have

$$\beta_{p,\theta} \ge \begin{cases}\sqrt{\dfrac{5}{14}}\left(1-\dfrac{\sqrt{3}}{2}\right) & \textit{for } p=2, \theta\in[\frac{1}{2},1],\\ 2^{-\frac{1}{p}}[1-p^{-\frac{2}{p}}(1-\frac{1}{p^2})^{1-\frac{1}{p}}] & \textit{for } p\in(1,\infty), \theta=\frac{1}{2}.\end{cases}$$

Proof The upper bound is the triangle inequality. To prove the lower bound, we set for brevity $\gamma_{q,\theta} = \frac{1}{q}[\theta^q + (1-\theta)^q]$ for $q \ge 1$, $\frac{1}{2} \le \theta \le 1$. A straightforward calculation then yields

$$\|\varphi\|_{L^p(0,1;Y^*)} = \|\varphi\|_*, \qquad \left\|(\theta - t)\psi\right\|_{L^p(0,1;Y^*)} = \gamma_{p+1,\theta}^{\frac{1}{p}} \|\psi\|_*.$$

Due to the Riesz isomorphism between Y and Y^* there are functions $v, w \in Y$ with

$$\langle\varphi, v\rangle = \|\varphi\|_*^p, \qquad \|v\|_Y = \|\varphi\|_*^{p-1},$$
$$\langle\psi, w\rangle = \|\psi\|_*^p, \qquad \|w\|_Y = \|\psi\|_*^{p-1}.$$

Denote by $\alpha \geq 0$ an arbitrary non-negative parameter which will be fixed at the end of this proof. The triangle inequality and Hölder's inequality in $\mathbb{R}^2$ then yield

$$\begin{aligned}
\left\|(\alpha+1)t^\alpha v + (\theta - t)^{p-1} w\right\|_{L^{p'}(0,1;Y)} &\leq \left\|(\alpha+1)t^\alpha v\right\|_{L^{p'}(0,1;Y)} + \left\|(\theta - t)^{p-1} w\right\|_{L^{p'}(0,1;Y)} \\
&= \frac{\alpha+1}{(\alpha p' + 1)^{\frac{1}{p'}}} \|v\|_Y + \gamma_{p+1,\theta}^{\frac{1}{p'}} \|w\|_Y \\
&= \frac{\alpha+1}{(\alpha p' + 1)^{\frac{1}{p'}}} \|\varphi\|_{L^p(0,1;Y^*)}^{p-1} + \left\|(\theta - t)\psi\right\|_{L^p(0,1;Y^*)}^{p-1} \\
&\leq \left\{\frac{(\alpha+1)^p}{(\alpha p' + 1)^{\frac{p}{p'}}} + 1\right\}^{\frac{1}{p}} \Big\{\|\varphi\|_{L^p(0,1;Y^*)}^p \\
&\qquad + \left\|(\theta - t)\psi\right\|_{L^p(0,1;Y^*)}^p\Big\}^{\frac{p-1}{p}}.
\end{aligned}$$

The definition of the dual norm $\|\cdot\|_*$, on the other hand, implies that

$$\begin{aligned}
&\int_0^1 \left\langle\varphi + (\theta - t)\psi, (\alpha+1)t^\alpha v + (\theta - t)^{p-1} w\right\rangle \mathrm{d}t \\
&\geq \langle\varphi, v\rangle - \gamma_{p,\theta} |\langle\varphi, w\rangle| - \left|\theta - \frac{\alpha+1}{\alpha+2}\right| |\langle\psi, v\rangle| + \gamma_{p+1,\theta} \langle\psi, w\rangle \\
&\geq \|\varphi\|_*^p - \gamma_{p,\theta} \|\varphi\|_* \|\psi\|_*^{p-1} - \left|\theta - \frac{\alpha+1}{\alpha+2}\right| \|\psi\|_* \|\varphi\|_*^{p-1} + \gamma_{p+1,\theta} \|\psi\|_*^p \\
&= \|\varphi\|_{L^p(0,1;Y^*)}^p - \gamma_{p,\theta} \gamma_{p+1,\theta}^{\frac{1-p}{p}} \|\varphi\|_{L^p(0,1;Y^*)} \left\|(\theta - t)\psi\right\|_{L^p(0,1;Y^*)}^{p-1} \\
&\quad - \left|\theta - \frac{\alpha+1}{\alpha+2}\right| \gamma_{p+1,\theta}^{-\frac{1}{p}} \left\|(\theta - t)\psi\right\|_{L^p(0,1;Y^*)} \|\varphi\|_{L^p(0,1;Y^*)}^{p-1} + \left\|(\theta - t)\psi\right\|_{L^p(0,1;Y^*)}^p.
\end{aligned}$$

To this estimate we apply Young's inequality $-ab \geq -\frac{\varepsilon^{p'}}{p'} a^{p'} - \frac{1}{p\varepsilon^p} b^p$ with $\varepsilon = (p-1)^{-\frac{p-1}{p^2}}$ twice and obtain

$$\begin{aligned}
&\int_0^1 \left\langle\varphi + (\theta - t)\psi, (\alpha+1)t^\alpha v + (\theta - t)^{p-1} w\right\rangle \mathrm{d}t \\
&\geq \left\{1 - \frac{1}{p}(p-1)^{1-\frac{1}{p}} \gamma_{p+1,\theta}^{-\frac{1}{p}} \left|\frac{\alpha+1}{\alpha+2} - \theta\right| - \frac{1}{p}(p-1)^{1-\frac{1}{p}} \gamma_{p,\theta} \gamma_{p+1,\theta}^{\frac{1}{p}-1}\right\} \\
&\quad \cdot \left\{\|\varphi\|_{L^p(0,1;Y^*)}^p + \left\|(\theta - t)\psi\right\|_{L^p(0,1;Y^*)}^p\right\}^{\frac{p-1}{p}}.
\end{aligned}$$

Combined with the upper bound for the norm of $(\alpha + 1)t^\alpha v + (\theta - t)^{p-1}w$ this proves the lower bound of the lemma with

$$\beta_{p,\theta} = \sup_{\alpha \geq 0} \left\{ \frac{(\alpha + 1)^p}{(\alpha p' + 1)^{\frac{p}{p'}}} + 1 \right\}^{-\frac{1}{p}} \left\{ 1 - \frac{1}{p}(p-1)^{1-\frac{1}{p}} \gamma_{p+1,\theta}^{-\frac{1}{p}} \left| \frac{\alpha + 1}{\alpha + 2} - \theta \right| - \frac{1}{p}(p-1)^{1-\frac{1}{p}} \gamma_{p,\theta} \gamma_{p+1,\theta}^{\frac{1}{p}-1} \right\}.$$

An elementary calculation yields the lower bounds for $\beta_{2,\theta}$ and $\beta_{p,\frac{1}{2}}$ by choosing $\alpha = 2$ in the first case and $\alpha = 0$ in the second. □

Remark 3.54 The estimates of Lemma 3.53 are invariant under affine transformations of the interval $(0, 1)$.

Remark 3.55 Lemma 3.53 is implicitly established in [**267**, Section 7], [**270**, Section 7], and [**63**, Lemma 5.1].

3.8 Estimation of Residuals

Motivated by the results of Sections 1.4–1.6, we derive in this section upper and lower bounds for the dual norms of certain continuous linear functionals. The bounds can easily be evaluated explicitly and are close to optimal. Motivated by Section 1.12, we moreover construct a representation of these functionals by piecewise $H(\text{div})$-functions. The results of this section are at the heart of all subsequent a posteriori error estimates since these are always based on the equivalence of a primal norm of the error to the corresponding dual norm of the associated residual.

To make things more precise, we fix a Lebesgue exponent p with $1 < p < \infty$, parameters $\varepsilon > 0$ and $\beta \geq 0$, and an admissible, affine-equivalent, shape-regular partition $\mathcal{T}$ together with its skeleton Σ. We equip $W_D^{1,p}(\Omega)$ with the norm $|||\cdot|||$ of (3.1) (p. 80) and its dual space with the corresponding dual norm $|||\cdot|||_*$ of (3.2) (p. 80). For every vertex $z \in \mathcal{N}$ we moreover use the weighted counterpart $|||\cdot|||_z$ of the norm $|||\cdot|||$ defined in (3.20) (p. 107).

With these notations, we consider a continuous linear functional R on $W_D^{1,p}(\Omega)$, briefly labelled the *residual*, which has the following two properties:

- there are two functions $r \in L^{p'}(\Omega)$ and $j \in L^{p'}(\Sigma)$ such that

$$\langle R, v \rangle = \int_\Omega rv + \int_\Sigma jv \tag{3.29}$$

  holds for all $v \in W_D^{1,p}(\Omega)$ (*L^p-representation*);
- the space $S_D^{1,0}(\mathcal{T})$ is contained in the kernel of R, i.e.

$$\langle R, v_\mathcal{T} \rangle = 0 \tag{3.30}$$

  holds for all $v_\mathcal{T} \in S_D^{1,0}(\mathcal{T})$ (*Galerkin orthogonality*).

The L^p-representation is crucial for the subsequent analysis. The Galerkin orthogonality, on the other hand, is primarily a technical assumption which simplifies the exposition and which will be dropped in Subsection 3.8.4.

3.8.1 Upper Bounds

In what follows v is an arbitrary function in $W_D^{1,p}(\Omega)$ which is kept fixed. Since the nodal shape functions λ_z, $z \in \mathcal{N}$, form a partition of unity, i.e. $\sum_{z\in\mathcal{N}} \lambda_z = 1$, we have

$$\sum_{z\in\mathcal{N}} |||v|||_z^p = |||v|||^p \quad \text{and} \quad \langle R, v\rangle = \sum_{z\in\mathcal{N}} \langle R, \lambda_z v\rangle .$$

Next we fix a vertex $z \in \mathcal{N}$ and associate with it a real number v_z which is arbitrary subject to the condition that $z \in \mathcal{N}_{\Gamma_D}$ implies $v_z = 0$. We then have $v_z\lambda_z \in S_D^{1,0}(\mathcal{T})$. The Galerkin orthogonality therefore implies

$$\langle R, \lambda_z v\rangle = \big\langle R, \lambda_z(v - v_z)\big\rangle . \tag{3.31}$$

The L^p-representation of R on the other hand yields

$$\big\langle R, \lambda_z(v - v_z)\big\rangle = \int_{\omega_z} r\lambda_z(v - v_z) + \int_{\sigma_z} j\lambda_z(v - v_z).$$

Lemma 3.30 (p. 107) implies that we can choose the v_z such that

$$\begin{aligned} \left\| \lambda_z^{\frac{1}{p}}(v - v_z)\right\|_{p;\omega_z} &\le \widetilde{c}_p(\omega_z)\hbar_{\omega_z}\, |||v|||_z , \\ \left\{\sum_{E\subset\sigma_z} h_E^{\perp} \left\| \lambda_z^{\frac{1}{p}}(v - v_z)\right\|_{p;E}^p\right\}^{\frac{1}{p}} &\le \widetilde{c}_p(\sigma_z)\varepsilon^{-\frac{1}{p^2}}\hbar_{\omega_z}^{1-\frac{1}{p}} h_z^{\frac{1}{p}}\, |||v|||_z \end{aligned}$$

where $h_E^{\perp} = \mu_{d,z}(\omega_E)/\mu_{d-1,z}(E)$. We now apply Hölder's inequality to the two terms on the right-hand side of the above equation and use these estimates to obtain

$$\int_{\omega_z} r\lambda_z(v - v_z) \le \left\| \lambda_z^{\frac{1}{p'}} r\right\|_{p';\omega_z} \widetilde{c}_p(\omega_z)\hbar_{\omega_z}\, |||v|||_z$$

and

$$\begin{aligned} \int_{\sigma_z} j\lambda_z(v - v_z) &\le \sum_{E\subset\sigma_z} \left\| \lambda_z^{\frac{1}{p'}} j\right\|_{p';E} \left\| \lambda_z^{\frac{1}{p}}(v - v_z)\right\|_{p;E} \\ &\le \left\{\sum_{E\subset\sigma_z} \left(h_E^{\perp}\right)^{-\frac{p'}{p}} \left\| \lambda_z^{\frac{1}{p'}} j\right\|_{p';E}^{p'}\right\}^{\frac{1}{p'}} \left\{\sum_{E\subset\sigma_z} h_E^{\perp} \left\| \lambda_z^{\frac{1}{p}}(v - v_z)\right\|_{p;E}^{p}\right\}^{\frac{1}{p}} \\ &\le \left\{\sum_{E\subset\sigma_z} \left(h_E^{\perp}\right)^{1-p'} \left\| \lambda_z^{\frac{1}{p'}} j\right\|_{p';E}^{p'}\right\}^{\frac{1}{p'}} \widetilde{c}_p(\sigma_z)\varepsilon^{-\frac{1}{p^2}}\hbar_{\omega_z}^{1-\frac{1}{p}} h_z^{\frac{1}{p}}\, |||v|||_z . \end{aligned}$$

This gives

$$\langle R, \lambda_z v\rangle \le |||v|||_z \left[\widetilde{c}_p(\omega_z)\hbar_{\omega_z} \left\| \lambda_z^{\frac{1}{p'}} r \right\|_{p';\omega_z} + \widetilde{c}_p(\sigma_z) \left\{ \sum_{E\subset\sigma_z} \left(\frac{h_z}{h_E^{\perp}}\right)^{p'-1} \varepsilon^{-\frac{(p'-1)^2}{p'}} \hbar_{\omega_z} \left\| \lambda_z^{\frac{1}{p'}} j \right\|_{p';E}^{p'} \right\}^{\frac{1}{p'}} \right].$$

Since $v \in W_D^{1,p}(\Omega)$ was arbitrary, this estimate together with Hölder's inequality for sums and the first two identities of the subsection yields the following upper bound for $|||R|||_*$.

Theorem 3.56 *The dual norm of the residual R is bounded from above by*

$$|||R|||_* \le \left\{ \sum_{z\in\mathcal{N}} \left[\widetilde{c}_p(\omega_z)\hbar_{\omega_z} \left\| \lambda_z^{\frac{1}{p'}} r \right\|_{p';\omega_z} + \widetilde{c}_p(\sigma_z) \left\{ \sum_{E\subset\sigma_z} \left(\frac{h_z}{h_E^{\perp}}\right)^{p'-1} \varepsilon^{-\frac{(p'-1)^2}{p'}} \hbar_{\omega_z} \left\| \lambda_z^{\frac{1}{p'}} j \right\|_{p';E}^{p'} \right\}^{\frac{1}{p'}} \right]^{p'} \right\}^{\frac{1}{p'}}$$

where $h_E^{\perp} = \mu_{d,z}(\omega_E)/\mu_{d-1,z}(E)$ and where the constants $\widetilde{c}_p(\omega_z)$ and $\widetilde{c}_p(\sigma_z)$ are as in Lemma 3.30 (p. 107).

When applied to a finite element solution of a partial differential equation, Theorem 3.56 yields a vertex oriented residual error indicator which generalises (1.31) (p. 19). In order to obtain an element oriented error indicator, we regroup the terms on the right-hand side of the above estimate. The triangle inequality in $\ell^{p'}$ yields

$$\left\{ \sum_{z\in\mathcal{N}} \left[\widetilde{c}_p(\omega_z)\hbar_{\omega_z} \left\| \lambda_z^{\frac{1}{p'}} r \right\|_{p';\omega_z} + \widetilde{c}_p(\sigma_z) \left\{ \sum_{E\subset\sigma_z} \left(\frac{h_z}{h_E^{\perp}}\right)^{p'-1} \varepsilon^{-\frac{(p'-1)^2}{p'}} \hbar_{\omega_z} \left\| \lambda_z^{\frac{1}{p'}} j \right\|_{p';E}^{p'} \right\}^{\frac{1}{p'}} \right]^{p'} \right\}^{\frac{1}{p'}}$$

$$\le \left\{ \sum_{z\in\mathcal{N}} \widetilde{c}_p(\omega_z)^{p'} \hbar_{\omega_z}^{p'} \left\| \lambda_z^{\frac{1}{p'}} r \right\|_{p';\omega_z}^{p'} \right\}^{\frac{1}{p'}} + \left\{ \sum_{z\in\mathcal{N}} \sum_{E\subset\sigma_z} \widetilde{c}_p(\sigma_z)^{p'} \left(\frac{h_z}{h_E^{\perp}}\right)^{p'-1} \varepsilon^{-\frac{(p'-1)^2}{p'}} \hbar_{\omega_z} \left\| \lambda_z^{\frac{1}{p'}} j \right\|_{p';E}^{p'} \right\}^{\frac{1}{p'}}.$$

Taking into account once more that the λ_z form a partition of unity, we obtain

$$\begin{aligned}
\sum_{z\in\mathcal{N}} \widetilde{c}_p(\omega_z)^{p'} \hbar_{\omega_z}^{p'} \left\| \lambda_z^{\frac{1}{p'}} r \right\|_{p';\omega_z}^{p'} &= \sum_{z\in\mathcal{N}} \widetilde{c}_p(\omega_z)^{p'} \hbar_{\omega_z}^{p'} \left\{ \sum_{K\subset\omega_z} \left\| \lambda_z^{\frac{1}{p'}} r \right\|_{p';K}^{p'} \right\} \\
&= \sum_{K\in\mathcal{T}} \left\{ \sum_{z\in\mathcal{N}_K} \widetilde{c}_p(\omega_z)^{p'} \hbar_{\omega_z}^{p'} \left\| \lambda_z^{\frac{1}{p'}} r \right\|_{p';K}^{p'} \right\} \\
&\le \sum_{K\in\mathcal{T}} \max_{z\in\mathcal{N}_K} \left\{ \widetilde{c}_p(\omega_z)^{p'} \hbar_{\omega_z}^{p'} \right\} \left\{ \sum_{z\in\mathcal{N}_K} \left\| \lambda_z^{\frac{1}{p'}} r \right\|_{p';K}^{p'} \right\} \\
&= \sum_{K\in\mathcal{T}} \max_{z\in\mathcal{N}_K} \left\{ \widetilde{c}_p(\omega_z)^{p'} \hbar_{\omega_z}^{p'} \right\} \|r\|_{p';K}^{p'}
\end{aligned}$$

and, using the same arguments,

$$\sum_{z\in\mathcal{N}}\sum_{E\subset\sigma_z}\widetilde{c}_p(\sigma_z)^{p'}\left(\frac{h_z}{h_E^{\perp}}\right)^{p'-1}\varepsilon^{-\frac{(p'-1)^2}{p'}}\hbar_{\omega_z}\left\|\lambda_z^{\frac{1}{p'}}j\right\|_{p';E}^{p'}$$

$$\leq\sum_{E\in\mathcal{E}}\max_{z\in\mathcal{N}_E}\left\{\widetilde{c}_p(\sigma_z)^{p'}\left(\frac{h_z}{h_E^{\perp}}\right)^{p'-1}\varepsilon^{-\frac{(p'-1)^2}{p'}}\hbar_{\omega_z}\right\}\left\|j\right\|_{p';E}^{p'}.$$

This establishes the following upper bound for $|||R|||_*$.

Theorem 3.57 *For all $K\in\mathcal{T}$ and all $E\in\mathcal{E}$ set*

$$c_p(K)=\max_{z\in\mathcal{N}_K}\left\{\widetilde{c}_p(\omega_z)\frac{\hbar_{\omega_z}}{\hbar_K}\right\},$$

$$c_p(E)=\max_{z\in\mathcal{N}_E}\left\{\widetilde{c}_p(\sigma_z)\left(\frac{h_z}{h_E^{\perp}}\right)^{1-\frac{1}{p'}}\left(\frac{\hbar_{\omega_z}}{\hbar_E}\right)^{\frac{1}{p'}}\right\}.$$

Then the dual norm of the residual R is bounded from above by

$$|||R|||_*\leq\left\{\sum_{K\in\mathcal{T}}c_p(K)^{p'}\hbar_K^{p'}\left\|r\right\|_{p';K}^{p'}\right\}^{\frac{1}{p'}}+\left\{\sum_{E\in\mathcal{E}}c_p(E)^{p'}\varepsilon^{-\frac{(p'-1)^2}{p'}}\hbar_E\left\|j\right\|_{p';E}^{p'}\right\}^{\frac{1}{p'}}.$$

3.8.2 *Lower Bounds*

We now establish lower bounds for the dual norm of the residual R. Since the arguments are based on the inverse inequalities of Section 3.6 (p. 112), we must assume that the functions r and j are piecewise polynomials. In applications, this assumption will be satisfied by a suitable approximation of the original functions r and j. This approximation then introduces additional data errors which will be bounded separately and which can often be proved to be of higher order.

We first establish lower bounds locally on the elements and faces.

Theorem 3.58 *Assume that the functions r and j are piecewise polynomials of degree at most k. For every element K and every face E, there are functions w_K and w_E with the following properties*

$$\begin{aligned}\operatorname{supp} w_K&\subset K,\\ \|w_K\|_{p;K}&\leq 1,\\ \|r\|_{p';K}&\leq C_{I,1,k,p'}(K)\int_K rw_K,\\ |||w_K|||_K&\leq 2^{\frac{1}{p}}\max\left\{C_{I,2,k,p}(K),1\right\}\hbar_K^{-1}\end{aligned}$$

and

$$\begin{aligned}\operatorname{supp} w_E&\subset\omega_E,\\ \|w_E\|_{p;E}&\leq 1,\\ \|j\|_{p';E}&\leq C_{I,3,k,p'}(E)\int_E jw_E,\end{aligned}$$

$$|||w_E|||_{\omega_E} \le 2^{\frac{1}{p}} \max_{K\subset\omega_E} \max\left\{C_{I,4,k,p}(E,K), C_{I,5,k,p}(E,K)\right\} \varepsilon^{\frac{(p'-1)^2}{p'^2}} \hbar_E^{-\frac{1}{p'}},$$

$$\|w_E\|_{p;\omega_E} \le 2^{\frac{1}{p}} \max_{K\subset\omega_E} C_{I,5,k,p}(E,K)\varepsilon^{\frac{1}{p^2}} \hbar_E^{\frac{1}{p}}.$$

Proof Consider first an arbitrary element K. Due to Proposition 3.37 (p. 112) there is a polynomial $\rho_K \in R_k(K)$ with $\|\rho_K\|_{p;K} = 1$ and

$$\|r\|_{p';K} \le C_{I,1,k,p'}(K) \int_K r\psi_K\rho_K.$$

Set $w_K = \psi_K\rho_K$. Then the first three assertions of the theorem are obviously satisfied.

Taking into account the definition (3.4) (p. 82) of $\hbar_K$, Proposition 3.37 (p. 112) implies

$$\begin{aligned}|||w_K|||_K &\le \left\{\varepsilon\, C_{I,2,k,p}(K)^p h_K^{-p} + \beta\right\}^{\frac{1}{p}} \\ &\le \max\left\{C_{I,2,k,p}(K), 1\right\} 2^{\frac{1}{p}} \max\left\{\varepsilon^{\frac{1}{p}} h_K^{-1}, \beta^{\frac{1}{p}}\right\} \\ &= 2^{\frac{1}{p}} \max\left\{C_{I,2,k,p}(K), 1\right\} \hbar_K^{-1}.\end{aligned}$$

Next, we consider an arbitrary face E and set

$$\vartheta = \varepsilon^{\frac{1}{p}} h_E^{-1}\hbar_E = \min\left\{1, \varepsilon^{\frac{1}{p}} h_E^{-1}\beta^{-\frac{1}{p}}\right\}. \tag{3.32}$$

Due to Proposition 3.37 (p. 112), there is a polynomial $\tau_E \in R_K(E)$ with $\|\tau_E\|_{p;E} = 1$ and

$$\|j\|_{p';E} \le C_{I,3,k,p'}(E) \int_E j\psi_{E,\vartheta}\tau_E.$$

We identify τ_E with its natural extension to $\mathbb{R}^d$ and set $w_E = \psi_{E,\vartheta}\tau_E$. This function obviously satisfies assertions five to seven of the Proposition.

The last one follows from the last estimate of Proposition 3.37 (p. 112) and the definition of ϑ.

For the second last one, we obtain from Proposition 3.37

$$\begin{aligned}|||w_E|||_{\omega_E} &\le \left\{\sum_{K\subset\omega_E}\left[\varepsilon\, C_{I,4,k,p}(E,K)^p\vartheta^{1-p}h_E^{1-p} + \beta\, C_{I,5,k,p}(E,K)^p\vartheta h_E\right]\right\}^{\frac{1}{p}} \\ &\le 2^{\frac{1}{p}} \max_{K\subset\omega_E}\max\left\{C_{I,4,k,p}(E,K), C_{I,5,k,p}(E,K)\right\}\cdot\max\left\{\varepsilon^{\frac{1}{p}}\vartheta^{\frac{1}{p}-1}h_E^{\frac{1}{p}-1}, \beta^{\frac{1}{p}}\vartheta^{\frac{1}{p}}h_E^{\frac{1}{p}}\right\}.\end{aligned}$$

Since

$$\begin{aligned}\max\left\{\varepsilon^{\frac{1}{p}}\vartheta^{\frac{1}{p}-1}h_E^{\frac{1}{p}-1}, \beta^{\frac{1}{p}}\vartheta^{\frac{1}{p}}h_E^{\frac{1}{p}}\right\} &= \max\left\{\varepsilon^{\frac{1}{p}}\varepsilon^{\frac{1}{p^2}-\frac{1}{p}}\hbar_E^{\frac{1}{p}-1}, \beta^{\frac{1}{p}}\varepsilon^{\frac{1}{p^2}}\hbar_E^{\frac{1}{p}}\right\} \\ &= \varepsilon^{\frac{1}{p^2}}\hbar_E^{\frac{1}{p}-1}\max\left\{1, \beta^{\frac{1}{p}}\hbar_E\right\} \\ &\le \varepsilon^{\frac{1}{p^2}}\hbar_E^{\frac{1}{p}-1} \\ &= \varepsilon^{\frac{(p'-1)^2}{p'^2}}\hbar_E^{-\frac{1}{p'}},\end{aligned}$$

this proves the theorem. □

Next, we establish a lower bound which is global with respect to a subset $\mathcal{T}'$ of $\mathcal{T}$. The choice $\mathcal{T}' = \mathcal{T}$ then in particular yields a global lower bound for the dual norm $|||R|||_*$ of R.

Theorem 3.59 *Assume that the functions r and j are piecewise polynomials of degree at most k and that $\mathcal{T}'$ is a non-empty subset of $\mathcal{T}$. Denote by $\mathcal{E}'$ the collection of all faces of all elements in $\mathcal{T}'$ and set $\omega_{\mathcal{T}'} = \bigcup_{K\in\mathcal{T}'} K \cup \bigcup_{E\in\mathcal{E}'} \omega_E$ and*

$$\eta_{\mathcal{T}'} = \left\{ \sum_{K\in\mathcal{T}'} \hbar_K^{p'} \|r\|_{p';K}^{p'} \right\}^{\frac{1}{p'}} + \left\{ \sum_{E\in\mathcal{E}'} \varepsilon^{-\frac{(p'-1)^2}{p'}} \hbar_E \|j\|_{p';E}^{p'} \right\}^{\frac{1}{p'}}.$$

Then there is a function $w \in L^p(\omega_{\mathcal{T}'})$ with the following properties

$$\langle R, w\rangle \geq \eta_{\mathcal{T}'}^{p'} \quad \text{and} \quad |||w|||_{\omega_{\mathcal{T}'}} \leq c\eta_{\mathcal{T}'}^{p'-1}.$$

The constant c only depends on the constants $C_{I,1,k,p'}(K)$, $C_{I,2,k,p}(K)$, $C_{I,3,k,p'}(E)$, $C_{I,4,k,p}(E,K)$, and $C_{I,5,k,p}(E,K)$ associated with the elements and faces contained in $\mathcal{T}'$ and $\mathcal{E}'$ and on $\max_{K\in\mathcal{T}'} \#\mathcal{E}_K$, the maximal number of faces per element.

Proof For brevity, we set for every element K in $\mathcal{T}'$ and every face E in $\mathcal{E}'$

$$\alpha_K = \hbar_K^{p'} \|r\|_{p';K}^{p'-1}, \quad \beta_E = \varepsilon^{-\frac{(p'-1)^2}{p'}} \hbar_E \|j\|_{p';E}^{p'-1},$$

so that

$$\eta_{\mathcal{T}'} = \left\{ \sum_{K\in\mathcal{T}'} \alpha_K \|r\|_{p';K} \right\}^{\frac{1}{p'}} + \left\{ \sum_{E\in\mathcal{E}'} \beta_E \|j\|_{p';E} \right\}^{\frac{1}{p'}}.$$

For the function w, we make the ansatz $w = \gamma_1 w_1 + \gamma_2 w_2$ with

$$w_1 = \sum_{K\in\mathcal{T}'} \alpha_K w_K, \quad w_2 = \sum_{E\in\mathcal{E}'} \beta_E w_E,$$

where the functions w_K and w_E are as in Theorem 3.58 and where the positive parameters γ_1 and γ_2 will be determined later. Since the function w_1 vanishes on all faces, we obtain from equation (3.29)

$$\begin{aligned}
\langle R, w\rangle &= \int_\Omega rw + \int_\Sigma jw \\
&= \gamma_1 \int_\Omega rw_1 + \gamma_2 \int_\Omega rw_2 + \gamma_2 \int_\Sigma jw_2 \\
&= \gamma_1 \sum_{K\in\mathcal{T}'} \alpha_K \int_K rw_K + \gamma_2 \sum_{E\in\mathcal{E}'} \beta_E \int_{\omega_E} rw_E + \gamma_2 \sum_{E\in\mathcal{E}'} \beta_E \int_E jw_E.
\end{aligned}$$

Proposition 3.37 (p. 112) and Theorem 3.58 yield for the three terms on the right-hand side of this estimate

$$\sum_{K\in\mathcal{T}'}\alpha_K\int_K rw_K \geq \left[\max_{K\in\mathcal{T}'} C_{I,1,k,p'}(K)\right]^{-1}\sum_{K\in\mathcal{T}'}\alpha_K\|r\|_{p';K}$$

$$= \left[\max_{K\in\mathcal{T}'} C_{I,1,k,p'}(K)\right]^{-1}\sum_{K\in\mathcal{T}'}\hbar_K^{p'}\|r\|_{p';K}^{p'},$$

$$\sum_{E\in\mathcal{E}'}\beta_E\int_E jw_E \geq \left[\max_{E\in\mathcal{E}'} C_{I,3,k,p'}(E)\right]^{-1}\sum_{E\in\mathcal{E}'}\beta_E\left\|j\right\|_{p';E}$$

$$= \left[\max_{E\in\mathcal{E}'} C_{I,3,k,p'}(E)\right]^{-1}\sum_{E\in\mathcal{E}'}\varepsilon^{-\frac{(p'-1)^2}{p'}}\hbar_E\left\|j\right\|_{p';E}^{p'},$$

$$\sum_{E\in\mathcal{E}'}\beta_E\int_{\omega_E} rw_E \leq \sum_{E\in\mathcal{E}'}\sum_{K\subset\omega_E}\beta_E\|r\|_{p';K}C_{I,5,k,p}(E,K)\varepsilon^{\frac{1}{p^2}}\hbar_E^{\frac{1}{p}}$$

$$\leq \max_{\substack{K\in\mathcal{T}'\\E\in\mathcal{E}_K}} C_{I,5,k,p}(E,K)\left\{\sum_{K\in\mathcal{T}'}\hbar_K^{p'}\|r\|_{p';K}^{p'}\right\}^{\frac{1}{p'}}\cdot\left\{\sum_{E\in\mathcal{E}'}\hbar_K^{-p}\varepsilon^{\frac{1}{p}}\hbar_E\beta_E^p\right\}^{\frac{1}{p}}$$

$$\leq \max_{\substack{K\in\mathcal{T}'\\E\in\mathcal{E}_K}} C_{I,5,k,p}(E,K)\left\{\sum_{K\in\mathcal{T}'}\hbar_K^{p'}\|r\|_{p';K}^{p'}\right\}^{\frac{1}{p'}}\cdot\left\{\sum_{E\in\mathcal{E}'}\varepsilon^{-\frac{(p'-1)^2}{p'}}\hbar_E\left\|j\right\|_{p';E}^{p'}\right\}^{\frac{1}{p}}.$$

Combining these estimates and using Young's inequality $ab \leq \frac{1}{p}a^p + \frac{1}{p'}b^{p'}$ with

$$a = \left[\max_{E\in\mathcal{E}'} C_{I,3,k,p'}(E)\right]^{-\frac{1}{p}}\left\{\sum_{E\in\mathcal{E}'}\varepsilon^{-\frac{(p'-1)^2}{p'}}\hbar_E\left\|j\right\|_{p';E}^{p'}\right\}^{\frac{1}{p}},$$

$$b = \left[\max_{E\in\mathcal{E}'} C_{I,3,k,p'}(E)\right]^{\frac{1}{p}}\max_{\substack{K\in\mathcal{T}'\\E\in\mathcal{E}_K}} C_{I,5,k,p}(E,K)\left\{\sum_{K\in\mathcal{T}'}\hbar_K^{p'}\|r\|_{p';K}^{p'}\right\}^{\frac{1}{p'}}$$

reveals that a suitable choice of γ_1 and γ_2, depending on the constants $C_{I,1,k,p'}(K)$, $C_{I,3,k,p'}(E)$, and $C_{I,5,k,p}(E,K)$, yields the first estimate of the proposition.

On the other hand, we obviously have

$$|||w|||_{\omega_{\mathcal{T}'}} \leq \gamma_1\,|||w_1|||_{\omega_{\mathcal{T}'}} + \gamma_2\,|||w_2|||_{\omega_{\mathcal{T}'}}.$$

Since the supports of the functions w_K are mutually disjoint, we conclude from Theorem 3.58 that

$$|||w_1|||_{\omega_{\mathcal{T}'}}^p = \sum_{K\in\mathcal{T}'}\alpha_K^p\,|||w_K|||_K^p$$

$$\leq 2\max_{K\in\mathcal{T}'}\max\left\{C_{I,1,k,p'}(K)^p, 1\right\}\sum_{K\in\mathcal{T}'}\hbar_K^{p'}\|r\|_{p';K}^{p'}.$$

Since the support of a fixed function w_E intersects the support of at most $\max_{K\in\mathcal{T}'}\#\mathcal{E}_K$ other functions of this type, we obtain with the same arguments

$$
\begin{aligned}
|||w_2|||^p_{\omega_{\mathcal{T}'}} &\le \max_{K\in\mathcal{T}'} \#\mathcal{E}_K \sum_{E\in\mathcal{E}'} \beta_E^p \, |||w_E|||^p_{\omega_E} \\
&\le 2 \max_{K\in\mathcal{T}'} \#\mathcal{E}_K \max_{\substack{K\in\mathcal{T}'\\ E\in\mathcal{E}_K}} \max\left\{C_{I,4,k,p}(E,K)^p, C_{I,5,k,p}(E,K)^p\right\} \cdot \sum_{E\in\mathcal{E}'} \varepsilon^{-\frac{(p'-1)^2}{p'}} \hbar_E \left\|j\right\|^{p'}_{p';E}.
\end{aligned}
$$

Combining these two estimates proves the second estimate of the proposition. □

3.8.3 *An* $H(\mathrm{div})$*-Lifting of Residuals*

The results of this subsection generalise those of Section 1.12 (p. 48) concerning the hyper-circle method. In what follows we assume that $p = 2$ and denote by

$$
H_{\mathcal{T}}(\mathrm{div}) = \left\{\tau \in L^2(\Omega)^d : \tau|_K \in H(\mathrm{div}; K) \text{ for all } K \in \mathcal{T}\right\}
$$

the space of all piecewise $H(\mathrm{div})$-vector fields.

We start with an auxiliary result.

Proposition 3.60 *For every element K and every pair of functions $f \in L^2(K)$ and $g \in L^2(\partial K)$ which satisfy the compatibility condition*

$$
\int_K f + \int_{\partial K} g = 0, \tag{3.33}
$$

there is a vector field $\tau_K \in H(\mathrm{div}; K)$ with

$$
\begin{aligned}
-\operatorname{div} \tau_K &= f \quad \text{in } K, \\
\tau_K \cdot \mathbf{n}_K &= g \quad \text{on } \partial K.
\end{aligned} \tag{3.34}
$$

Here, $\mathbf{n}_K$ denotes the unit exterior normal of K. The vector field τ_K can be chosen such that the stability estimate

$$
\|\tau_K\|_K \le c_{K,1} h_K \left\|f\right\|_K + c_{K,2} h_K^{\frac{1}{2}} \left\|g\right\|_{\partial K} \tag{3.35}
$$

holds with

$$
c_{K,1} = \frac{1}{\pi}, \qquad c_{K,2} = \frac{\sqrt{2\pi+1}}{\pi} \left(\frac{h_K \mu_{d-1}(\partial K)}{\mu_d(K)}\right)^{\frac{1}{2}}.
$$

If the functions f and g are piecewise polynomials of degree k, the vector field τ_K can be chosen from the Raviart–Thomas space RT_k or the Brezzi–Douglas–Marini space BDM_{k+1} such that properties (3.34) and (3.35) remain valid. In this case both constants depend on the polynomial degree k and the shape parameter $C_{\mathcal{T}}$.

Proof The existence of τ_K is well known and follows from [**90**, Lemmas III.1.1 and III.1.2] and [**145**, Lemma 2.1] or [**72**].

The stability result (3.35) should also be well known, but we could not find a reference. Therefore we prove it in what follows.

We first consider the analytical case. Due to the compatibility condition (3.33) there is a unique function $v_K \in H^2(K)$ which solves the Neumann problem

$$-\Delta v_K = f \text{ in } K, \quad \mathbf{n}_K \cdot \nabla v_K = g \text{ on } \partial K, \quad \int_K v_K = 0.$$

For every $w \in H^1(K)$ it satisfies

$$\int_K \nabla v_K \cdot \nabla w = \int_K fw + \int_{\partial K} gw. \tag{3.36}$$

The vector field $\tau_K = \nabla v_K$ obviously fulfils conditions (3.34). In order to prove the stability result we insert $w = v_K$ in equation (3.36). This yields

$$\|\tau_K\|_K^2 = \int_K f v_K + \int_{\partial K} g v_K \leq \|f\|_K \|v_K\|_K + \|g\|_{\partial K} \|v_K\|_{\partial K}.$$

Since v_K has vanishing mean value on K, we know from the Poincaré inequality, Proposition 3.10 (p. 92), that

$$\|v_K\|_K \leq \frac{1}{\pi} h_K \|\nabla v_K\|_K = \frac{1}{\pi} h_K \|\tau_K\|_K.$$

Together with the trace inequality, Proposition 3.5 (p. 90), and Remark 3.6 (p. 90), this gives for every face E contained in ∂K

$$\begin{aligned}\|v_K\|_E^2 &\leq \frac{\mu_{d-1}(E)}{\mu_d(K)} \|v_K\|_K^2 + \frac{2h_K\mu_{d-1}(E)}{\mu_d(K)} \|\nabla v_K\|_K \|v_K\|_K \\ &\leq \frac{2\pi + 1}{\pi^2} \frac{h_K\mu_{d-1}(E)}{\mu_d(K)} h_K \|\tau_K\|_K.\end{aligned}$$

Combining these estimates proves the stability result in this case.

Next we consider the discrete case where f and g are piecewise polynomials of degree k. We now set $\tau_K = J_K \nabla v_K$ where J_K is the interpolation operator of [**90**] which maps $H^1(K)^d$ into the corresponding space RT_k or BDM_{k+1}. (Notice that J_K is labelled ρ_K in [**90**].) Its properties imply that τ_K fulfils conditions (3.34). Since J_K uses moments on the boundary of K, we cannot expect to bound $\|\tau_K\|_K$ in terms of $\|\nabla v_K\|_K$. Yet, from [**90**, Proposition III.3.6] we know that

$$\|\tau_K\|_K \leq \|\nabla v_K\|_K + \|\nabla v_K - \tau_K\|_K \leq \|\nabla v_K\|_K + ch_K \left\|\nabla^2 v_K\right\|_K$$

with a constant c which depends on the polynomial degree k and the shape parameter $C_{\mathcal{T}}$. Hence, it remains to prove that the second term on the right-hand side of this estimate can be bounded by the right-hand side of (3.35) with appropriate constants depending on k and $C_{\mathcal{T}}$. We do this with the help of a scaling argument. Denote by $F_K : \widehat{x} \mapsto a_K + B_K\widehat{x}$ an orientation-preserving affine transformation of the reference simplex or reference cube $\widehat{K}$ onto K and set $\widehat{f} = \frac{\mu_d(K)}{\mu_d(\widehat{K})} f \circ F_K$ on $\widehat{K}$ and $\widehat{g} = \frac{\mu_{d-1}(F_K(\widehat{E}))}{\mu_{d-1}(\widehat{E})} g|_{F_K(\widehat{E})} \circ F_K$ on every face $\widehat{E}$ of $\widehat{K}$. The compatibility condition (3.33) implies that $\int_{\widehat{K}} \widehat{f} + \int_{\partial\widehat{K}} \widehat{g} = 0$. Hence, there is a unique function $\widehat{v} \in H^2(\widehat{K})$ which solves the Neumann problem

$$-\Delta\widehat{v} = \widehat{f} \text{ on } \widehat{K}, \quad \widehat{\mathbf{n}} \cdot \nabla\widehat{v} = \widehat{g} \text{ on } \partial\widehat{K}, \quad \int_{\widehat{K}} \widehat{v} = 0,$$

where $\widehat{\mathbf{n}}$ is the unit exterior normal of $\widehat{K}$. The properties of the Piola transformation $\widehat{\tau} \mapsto \frac{\mu_d(\widehat{K})}{\mu_d(K)} B_K \widehat{\tau}$ imply that $\nabla v_K = \frac{\mu_d(\widehat{K})}{\mu_d(K)} B_K \nabla \widehat{v}$ and thanks to [**90**, Lemma III.1.7]

$$\left\| \nabla^2 v_K \right\|_K \leq \left(\frac{\mu_d(\widehat{K})}{\mu_d(K)} \right)^{\frac{1}{2}} |||B_K|||_2 \left|\left|\left|B_K^{-1}\right|\right|\right|_2 \left\| \nabla^2 \widehat{v} \right\|_{\widehat{K}},$$

where $|||\cdot|||_2$ denotes the spectral norm of matrices. Since $\widehat{K}$ is convex, the H^2-norm of $\widehat{v}$ can be bounded by the L^2-norm of $\widehat{f}$ and the $H^{\frac{1}{2}}$-norm of $\widehat{g}$. Since $\widehat{g}$ is a piecewise polynomial, the latter can be bounded by the corresponding L^2-norm. Hence, we have

$$\begin{aligned} \left\| \nabla^2 \widehat{v} \right\|_{\widehat{K}} &\leq \widehat{c}_1 \left\{ \left\| \widehat{f} \right\|_{\widehat{K}} + \widehat{c}_2 \left\| \widehat{g} \right\|_{\partial \widehat{K}} \right\} \\ &\leq \widehat{c}_1 \left\{ \left(\frac{\mu_d(K)}{\mu_d(\widehat{K})} \right)^{\frac{1}{2}} \left\| f \right\|_K + \widehat{c}_2 \max_{\widehat{E} \in \mathcal{E}_{\widehat{K}}} \left(\frac{\mu_{d-1}(F_K(\widehat{E}))}{\mu_{d-1}(\widehat{E})} \right)^{\frac{1}{2}} \left\| g \right\|_{\partial K} \right\} \end{aligned}$$

with a constant $\widehat{c}_2$ which depends on the polynomial degree k. Combining all these estimates completes the proof. □

Remark 3.61 We stress that in the case of polynomial data f and g, problem (3.34) can directly be solved on the discrete level by considering the corresponding Galerkin formulation in the appropriate Raviart–Thomas or Brezzi–Douglas–Marini spaces. The Neumann problems and their solutions v_K and $\widehat{v}$ are needed to prove the stability result (3.35) but not to compute the discrete version of τ_K.

The following proposition shows that the residual R can be lifted to the space of piecewise $H(\text{div})$-vector fields.

Proposition 3.62 *There is a vector field $\rho_{\mathcal{T}} \in H_{\mathcal{T}}(\text{div})$ such that*

$$\langle R, v \rangle = \int_\Omega \rho_{\mathcal{T}} \cdot \nabla v \tag{3.37}$$

holds for all $v \in H^1_D(\Omega)$. The vector field $\rho_{\mathcal{T}}$ can explicitly be computed by sweeping through the elements of $\mathcal{T}$. Moreover, $\rho_{\mathcal{T}}$ can be chosen from the Raviart–Thomas space RT_{k+1} or the Brezzi–Douglas–Marini space BDM_{k+2} if the functions r and j are piecewise polynomials of degree k.

Proof We construct $\rho_{\mathcal{T}}$ piecewise on the patches ω_z by generalising the construction of Section 1.12 (p. 48) and the ideas of [**79**, Algorithm 9.3] and [**83**, **122**]. To this end we recall the definition (3.21) (p. 108) of the weighted average $\overline{v}_z$ of a function $v \in L^1(\omega_z)$.

Since Ω has a Lipschitz boundary, the same applies to ω_z. Hence, any pair of elements in ω_z can be connected by a path of adjacent elements in ω_z. Therefore, we can enumerate the elements in ω_z from 1 to $n_z = \#\{K \subset \omega_z\}$ such that, cf. Figure 3.10 for the two-dimensional situation:

- for every $i \in \{1, \ldots, n_z - 1\}$ the elements K_i and K_{i+1} share a face E_i which is contained in σ_z;
- either K_{n_z} and K_1 share a face E_{n_z} which is contained in σ_z and which is different from the faces $E_1, \ldots, E_{n_z-1}$, or K_{n_z} has a face E_{n_z} which is contained in $\sigma_z \cap \Gamma$;
- if z is on the Dirichlet boundary, the face E_{n_z} is also contained in the Dirichlet boundary.

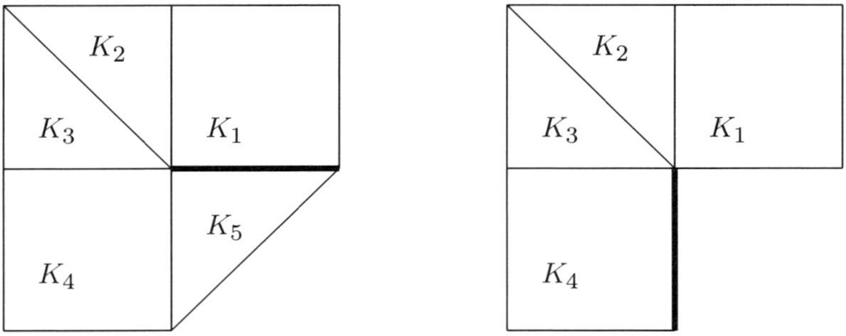

Figure 3.10 Enumeration of the elements in ω_z for a vertex z in Ω (left) and on Γ (right); the face E_{n_z} is indicated by a bold line

Set

$$\alpha_1 = \frac{1}{\mu_{d-1}(E_1)} \left\{ \int_{K_1} \lambda_z r + \int_{\partial K_1 \cap \sigma_z} \lambda_z j \right\}$$

and, for $2 \leq i \leq n_z$,

$$\alpha_i = \frac{1}{\mu_{d-1}(E_i)} \left\{ \int_{K_i} \lambda_z r + \int_{(\partial K_i \cap \sigma_z) \setminus E_{i-1}} \lambda_z j + \mu_{d-1}(E_{i-1}) \alpha_{i-1} \right\}.$$

Obviously we have

$$\mu_{d-1}(E_{n_z}) \alpha_{n_z} = \int_{\omega_z} \lambda_z r + \int_{\sigma_z} \lambda_z j.$$

If z is not on the Dirichlet boundary, the Galerkin orthogonality (3.30) therefore implies $\alpha_{n_z} = 0$. If, on the other hand, z is on the Dirichlet boundary, α_{n_z} in general is not zero.

Due to the definition of $\alpha_1, \ldots, \alpha_{n_z}$, we may apply Proposition 3.60 to the sets $K_1, \ldots, K_{n_z}$ with

$$f = \lambda_z r \text{ in } K_1, \quad g = \begin{cases} \lambda_z j & \text{on } (\partial K_1 \cap \sigma_z) \setminus E_1, \\ \lambda_z j - \alpha_1 & \text{on } E_1, \\ 0 & \text{on } \partial K_1 \setminus \sigma_z \end{cases}$$

and, for $i \in \{2, \ldots, n_z\}$,

$$f = \lambda_z r \text{ in } K_i, \quad g = \begin{cases} \lambda_z j & \text{on } (\partial K_i \cap \sigma_z) \setminus (E_{i-1} \cup E_i), \\ \alpha_{i-1} & \text{on } E_{i-1}, \\ \lambda_z j - \alpha_i & \text{on } E_i, \\ 0 & \text{on } \partial K_i \setminus \sigma_z \end{cases}$$

and thus obtain vector fields $\rho_i \in H(\text{div}; K_i)$ for $1 \leq i \leq n_z$. We extend every ρ_i by zero outside K_i and set

$$\rho_z = \sum_{i=1}^{n_z} \rho_i.$$

Consider an arbitrary function v in $H^1_D(\Omega)$. Integration by parts on K_1 yields

$$\begin{aligned}\int_{K_1} \rho_1 \cdot \nabla v &= \int_{K_1} \rho_1 \cdot \nabla(v - \overline{v}_z) \\ &= -\int_{K_1} \operatorname{div} \rho_1 (v - \overline{v}_z) + \int_{\partial K_1} \mathbf{n}_{K_1} \cdot \rho_1 (v - \overline{v}_z) \\ &= \int_{K_1} \lambda_z r(v - \overline{v}_z) + \int_{\partial K_1 \cap \sigma_z} \lambda_z j(v - \overline{v}_z) - \int_{E_1} \alpha_1 (v - \overline{v}_z).\end{aligned}$$

Similarly we obtain for every $i \in \{2, \ldots, n_z\}$

$$\begin{aligned}\int_{K_i} \rho_i \cdot \nabla v &= \int_{K_i} \rho_i \cdot \nabla(v - \overline{v}_z) \\ &= -\int_{K_i} \operatorname{div} \rho_i (v - \overline{v}_z) + \int_{\partial K_i} \mathbf{n}_{K_i} \cdot \rho_i (v - \overline{v}_z) \\ &= \int_{K_i} \lambda_z r(v - \overline{v}_z) + \int_{(\partial K_i \cap \sigma_z) \setminus E_{i-1}} \lambda_z j(v - \overline{v}_z) + \int_{E_{i-1}} \alpha_{i-1} (v - \overline{v}_z) - \int_{E_i} \alpha_i (v - \overline{v}_z).\end{aligned}$$

Summation with respect to i yields

$$\int_{\omega_z} \rho_z \cdot \nabla v = \int_{\omega_z} \lambda_z r(v - \overline{v}_z) + \int_{\sigma_z} \lambda_z j(v - \overline{v}_z) - \int_{E_{n_z}} \alpha_{n_z} (v - \overline{v}_z).$$

If z is not on the Dirichlet boundary, we have $\alpha_{n_z} = 0$. If, on the other hand, z is on the Dirichlet boundary, E_{n_z} is contained in Γ_D and both v and $\overline{v}_z$ vanish on E_{n_z}. Hence, the contribution of E_{n_z} vanishes in both cases and we obtain for every vertex $z \in \mathcal{N}$

$$\int_{\omega_z} \rho_z \cdot \nabla v = \int_{\omega_z} \lambda_z r(v - \overline{v}_z) + \int_{\sigma_z} \lambda_z j(v - \overline{v}_z).$$

Now, we extend every ρ_z by zero outside ω_z and set

$$\rho_{\mathcal{T}} = \sum_{z \in \mathcal{N}} \rho_z.$$

Taking into account that the functions λ_z form a partition of unity, summation with respect to all vertices yields

$$\begin{aligned}\int_{\Omega} \rho_{\mathcal{T}} \cdot \nabla v &= \sum_{z \in \mathcal{N}} \int_{\Omega} \rho_z \cdot \nabla v \\ &= \sum_{z \in \mathcal{N}} \int_{\omega_z} \rho_z \cdot \nabla v \\ &= \sum_{z \in \mathcal{N}} \left\{ \int_{\omega_z} \lambda_z r(v - \overline{v}_z) + \int_{\sigma_z} \lambda_z j(v - \overline{v}_z) \right\}\end{aligned}$$

$$= \sum_{z\in\mathcal{N}} \left\{ \int_\Omega \lambda_z r(v - \bar{v}_z) + \int_\Sigma \lambda_z j(v - \bar{v}_z) \right\}$$

$$= \langle R, v\rangle - \left\langle R, \sum_{z\in\mathcal{N}} \bar{v}_z \lambda_z \right\rangle$$

for all $v \in H^1_D(\Omega)$. Taking into account the Galerkin orthogonality (3.30), this proves Proposition 3.62. □

The following theorem shows that the L^2-norm of $\rho_\mathcal{T}$ yields upper and lower bounds for the dual norm $|||R|||_*$ of the residual. Moreover, the upper bound does not contain any unknown constant.

Theorem 3.63 *The dual norm of the residual can be bounded from above by*

$$|||R|||_* \leq \varepsilon^{-\frac{1}{2}} \|\rho_\mathcal{T}\| . \tag{3.38}$$

Here, $\rho_\mathcal{T}$ is as in Proposition 3.62. The vector field $\rho_\mathcal{T}$ also satisfies the stability estimate

$$\varepsilon^{-\frac{1}{2}} \|\rho_\mathcal{T}\| \leq \left\{ \sum_{z\in\mathcal{N}} \left[c_\mathcal{T}^2 \varepsilon^{-1} h_z^2 \|r\|^2_{\omega_z} + c_\mathcal{E}^2 \varepsilon^{-1} h_z \|j\|^2_{\sigma_z} \right] \right\}^{\frac{1}{2}} . \tag{3.39}$$

The constant $c_\mathcal{E}$ depends on the shape parameter $C_\mathcal{T}$. If the vector field $\rho_\mathcal{T}$ is chosen from a Raviart–Thomas or Brezzi–Douglas–Marini space, both constants $c_\mathcal{T}$ and $c_\mathcal{E}$ depend on the shape parameter and on the polynomial degree of $\rho_\mathcal{T}$.

Proof The reliability, inequality (3.38), is an obvious consequence of the representation (3.37) and the definition of the dual norm.

For the proof of the efficiency, inequality (3.39), we apply Proposition 3.60 and the stability estimate (3.35) at the different stages of the construction of $\rho_\mathcal{T}$.

Consider an arbitrary vertex $z \in \mathcal{N}$. Inequality (3.35) implies for the vector field ρ_1 and the element K_1

$$\|\rho_1\|^2_{K_1} \leq 2c^2_{K_1,1} h^2_{K_1} \|\lambda_z r\|^2_{K_1} + 4c^2_{K_1,2} h_{K_1} \left\|\lambda_z j\right\|^2_{\partial K_1 \cap \sigma_z} + 4c^2_{K_1,2} h_{K_1} \mu_{d-1}(E_1) \left|\alpha_1\right|^2 .$$

The definition of α_1 and the Cauchy–Schwarz inequality, on the other hand, yield

$$\begin{aligned} h_{K_1} \mu_{d-1}(E_1) \left|\alpha_1\right|^2 \leq {} & 2 \frac{\mu_d(K_1)}{\mu_{d-1}(E_1) h_{K_1}} h^2_{K_1} \|\lambda_z r\|^2_{K_1} \\ & + 2 \frac{\mu_{d-1}(\partial K_1 \cap \sigma_z)}{\mu_{d-1}(E_1)} h_{K_1} \left\|\lambda_z j\right\|^2_{\partial K_1 \cap \sigma_z} . \end{aligned}$$

Similarly, we obtain for all $i \in \{2, \ldots, n_z\}$

$$\begin{aligned} \|\rho_i\|^2_{K_i} \leq {} & 2c^2_{K_i,1} h^2_{K_i} \|\lambda_z r\|^2_{K_i} + 4c^2_{K_i,2} h_{K_i} \left\|\lambda_z j\right\|^2_{(\partial K_i \cap \sigma_z)\setminus E_{i-1}} \\ & + 4c^2_{K_i,2} h_{K_i} \mu_{d-1}(E_{i-1}) \left|\alpha_{i-1}\right|^2 + 4c^2_{K_i,2} h_{K_i} \mu_{d-1}(E_i) \left|\alpha_i\right|^2 \end{aligned}$$

and

$$
\begin{aligned}
h_{K_i}\mu_{d-1}(E_i)\,|\alpha_i|^2 \le\ & 3\frac{\mu_d(K_i)}{\mu_{d-1}(E_i)h_{K_i}}h_{K_i}^2\,\|\lambda_z r\|_{K_i}^2 \\
&+ 3\frac{\mu_{d-1}((\partial K_i \cap \sigma_z)\setminus E_{i-1})}{\mu_{d-1}(E_i)}h_{K_i}\,\left\|\lambda_z j\right\|_{(\partial K_i\cap\sigma_z)\setminus E_{i-1}}^2 \\
&+ 3\frac{h_{K_i}\mu_{d-1}(E_{i-1})}{h_{K_{i-1}}\mu_{d-1}(E_i)}h_{K_{i-1}}\mu_{d-1}(E_i)\,|\alpha_{i-1}|^2 .
\end{aligned}
$$

Adding all estimates we arrive at

$$
\|\rho_z\|_{\omega_z}^2 \le c_{z,1}^2 h_z^2\,\|\lambda_z r\|_{\omega_z}^2 + c_{z,2}^2 h_z\,\left\|\lambda_z j\right\|_{\sigma_z}^2 , \tag{3.40}
$$

where the constants $c_{z,1}$ and $c_{z,2}$ depend on n_z and the quantities

$$
\begin{aligned}
&\max_{1\le i\le n_z} c_{K_i,1}, && \max_{1\le i\le n_z} c_{K_i,2}, \\
&\max_{1\le i\le n_z} \frac{\mu_d(K_i)}{\mu_{d-1}(E_i)h_{K_i}}, && \max_{1\le i\le n_z} \frac{\mu_{d-1}(\partial K_i\cap\sigma_z)}{\mu_{d-1}(E_i)}, \\
&\max_{1\le i<j\le n_z} \frac{h_{K_i}\mu_{d-1}(E_j)}{h_{K_j}\mu_{d-1}(E_i)}.
\end{aligned}
$$

Since every vector field ρ_z vanishes outside ω_z and since every element K intersects $\#\mathcal{N}_K$ patches ω_z, we have

$$
\begin{aligned}
\|\rho_{\mathcal{T}}\|^2 &= \sum_{K\in\mathcal{T}} \|\rho_{\mathcal{T}}\|_K^2 \\
&= \sum_{K\in\mathcal{T}} \left\| \sum_{z\in\mathcal{N}_K} \rho_z \right\|_K^2 \\
&\le \left(\max_{K\in\mathcal{T}} \#\mathcal{N}_K\right) \sum_{K\in\mathcal{T}}\sum_{z\in\mathcal{N}_K} \|\rho_z\|_K^2 \\
&= \left(\max_{K\in\mathcal{T}} \#\mathcal{N}_K\right) \sum_{z\in\mathcal{N}} \|\rho_z\|_{\omega_z}^2 .
\end{aligned}
$$

Combining these estimates proves the stability result (3.39) and Theorem 3.63. □

Remark 3.64 Theorem 3.63 shows that $\varepsilon^{-\frac{1}{2}}\,\|\rho_{\mathcal{T}}\|$ is a reliable and efficient error indicator. A comparison with Theorems 3.57 and 3.59 with $\mathcal{T}' = \mathcal{T}$, however, reveals that it is not a robust one. In fact its quality depends on the mesh Peclét number. This defect is not a technical but a structural one. It is due to the fact that the right-hand side of equation (3.37) only gives control of the gradient of the test function v and not of the function itself. This defect can be remedied by taking $\sup_v \frac{\int_\Omega \rho_{\mathcal{T}}\nabla v}{\|v\|}$ as error indicator where the supremum is taken with respect to all functions in $H_D^1(\Omega)$. This gives a robust error indicator. Yet, we do not know of any practical and efficient way to compute this quantity. Another possible remedy is to use

$\min\left\{\varepsilon^{-\frac{1}{2}} \|\rho_{\mathcal{T}}\|, \sum_K \left[\beta_K \|\operatorname{div}\rho_{\mathcal{T}}\|_K + \beta_{\partial K} \|\rho_{\mathcal{T}} \cdot \mathbf{n}\|_{\partial K}\right]\right\}$ as an error indicator. Here, the sum refers to the elements of $\mathcal{T}$ and β_K and $\beta_{\partial K}$ are suitable weights similar to those in Theorem 3.57. A variant of this approach is used in [**105, 134, 148**].

3.8.4 *Coping with Missing Galerkin Orthogonality*

We now drop the assumption (3.30) of Galerkin orthogonality. Since the lower bounds of Section 3.8.2 do not require Galerkin orthogonality, we only have to review the upper bounds of Section 3.8.1 and the $H(\operatorname{div})$-lifting of Section 3.8.3.

We now replace equation (3.31) by

$$\langle R, \lambda_z v\rangle = \langle R, \lambda_z(v - \overline{v}_z)\rangle + \langle R, \lambda_z \overline{v}_z\rangle$$

for every vertex $z \in \mathcal{N}$ and every function $v \in W_D^{1,p}(\Omega)$. Here, the constants $\overline{v}_z$ are given by equation (3.21) (p. 108). The first term on the right-hand side of the above identity can be handled as before. Summing with respect to all vertices, the second term gives rise to $\langle R, I_{\mathcal{T}} v\rangle = \langle I_{\mathcal{T}}^* R, v\rangle$. Here, $I_{\mathcal{T}}^*$ denotes the adjoint of $I_{\mathcal{T}}$, the quasi-interpolation operator defined in (3.22) (p. 108). Therefore, in the absence of Galerkin orthogonality, Theorems 3.56 and 3.57 remain valid provided that the term $\left|\left|\left| I_{\mathcal{T}}^* R \right|\right|\right|_*$ is added to the right-hand sides of the upper bounds for $|||R|||_*$. Obviously this term vanishes in the presence of Galerkin orthogonality. Note that $I_{\mathcal{T}}$ can be replaced by any quasi-interpolation operator $Q_{\mathcal{T}} : L^1(\Omega) \to S_D^{1,0}(\mathcal{T})$ of the form $Q_{\mathcal{T}} v = \sum_{z\in\mathcal{N}} v_z \lambda_z$ provided that $v_z = 0$ holds for all $z \in \mathcal{N}_{\Gamma_D}$ and that the estimates of Lemma 3.30 (p. 107) remain valid.

Next we consider the $H(\operatorname{div})$-lifting of Section 3.8.3. Obviously we have

$$\langle R, v\rangle = \langle R, v - I_{\mathcal{T}} v\rangle + \langle R, I_{\mathcal{T}} v\rangle$$

for all $v \in W_D^{1,p}(\Omega)$. For every $z \in \mathcal{N}$ set

$$\overline{R}_z = \begin{cases} \dfrac{\langle R, \lambda_z\rangle}{\int_{\omega_z} \lambda_z} = \dfrac{\int_{\omega_z} \lambda_z r + \int_{\sigma_z} \lambda_z j}{\int_{\omega_z} \lambda_z} & \text{if } z \in \mathcal{N}_\Omega \cup \mathcal{N}_{\Gamma_N}, \\ 0 & \text{if } z \in \mathcal{N}_{\Gamma_D}. \end{cases}$$

The definition (3.21) of $\overline{v}_z$ implies for all $v \in W_D^{1,p}(\Omega)$ and all $z \in \mathcal{N}_\Omega \cup \mathcal{N}_{\Gamma_N}$

$$\begin{aligned}\langle R, v - I_{\mathcal{T}} v\rangle &= \sum_{z\in\mathcal{N}} \int_{\omega_z} r\lambda_z(v - \overline{v}_z) + \sum_{z\in\mathcal{N}} \int_{\sigma_z} j\lambda_z(v - \overline{v}_z) \\ &= \sum_{z\in\mathcal{N}} \int_{\omega_z} (r - \overline{R}_z)\lambda_z(v - \overline{v}_z) + \sum_{z\in\mathcal{N}} \int_{\sigma_z} j\lambda_z(v - \overline{v}_z)\end{aligned}$$

and

$$\int_{\omega_z} (r - \overline{R}_z)\lambda_z + \int_{\sigma_z} j\lambda_z = 0.$$

Therefore, in the absence of Galerkin orthogonality, we must replace r by $r - \overline{R}_z$ in the definition of the vector fields ρ_z. With this modification in the construction of $\rho_{\mathcal{T}}$, we must add the term $\langle R, I_{\mathcal{T}} v\rangle$

on the right-hand side of equation (3.37) and the term $\left|\left|\left|I_{\mathcal{T}}^{*}R\right|\right|\right|_{*}$ on the right-hand side of estimate (3.38). For every vertex $z \in \mathcal{N}$, estimate (3.40) now takes the form

$$\begin{aligned}\|\rho_z\|_{\omega_z}^2 &\le c_{z,1}^2 h_z^2 \left\|\lambda_z(r-\overline{R}_z)\right\|_{\omega_z}^2 + c_{z,2}^2 h_z \left\|\lambda_z j\right\|_{\sigma_z}^2 \\ &\le 2c_{z,1}^2 h_z^2 \left\|\lambda_z \overline{R}_z\right\|_{\omega_z}^2 + 2c_{z,1}^2 h_z^2 \|\lambda_z r\|_{\omega_z}^2 + c_{z,2}^2 h_z \left\|\lambda_z j\right\|_{\sigma_z}^2 .\end{aligned}$$

From the definition of $\overline{R}_z$ and the Cauchy–Schwarz inequality we conclude that

$$\begin{aligned}\left\|\lambda_z \overline{R}_z\right\|_{\omega_z}^2 &= \frac{\left(\int_{\omega_z}\lambda_z^2\right)}{\left(\int_{\omega_z}\lambda_z\right)^2}\left|\int_{\omega_z}\lambda_z r + \int_{\sigma_z}\lambda_z j\right|^2 \\ &\le 2\frac{\mu_d(\omega_z)}{\left(\int_{\omega_z}\lambda_z\right)}\|\lambda_z r\|_{\omega_z}^2 + 2\frac{h_z\mu_{d-1}(\sigma_z)}{\left(\int_{\omega_z}\lambda_z\right)}h_z^{-1}\left\|\lambda_z j\right\|_{\sigma_z}^2 .\end{aligned}$$

Therefore, in the absence of Galerkin orthogonality, estimate (3.39) remains valid with different constants $c_{\mathcal{T}}$ and $c_{\mathcal{E}}$ depending on the shape parameter $C_{\mathcal{T}}$. Note that the arguments for the $H(\mathrm{div})$-lifting take advantage of the particular form of the operator $I_{\mathcal{T}}$.

In summary the results of Sections 3.8.1 and 3.8.3 extend to the case of missing Galerkin orthogonality provided that the upper bounds for $|||R|||_{*}$ are augmented by the term $\left|\left|\left|I_{\mathcal{T}}^{*}R\right|\right|\right|_{*}$ which measures the consistency error. Thus we are left with the task of establishing computable upper bounds for this term. This of course depends on the cause of the consistency error. Two of the most important examples are SUPG-discretisations, cf. Sections 4.4.2 (p. 166), 4.10.2 (p. 240), 5.4.2 (p. 302), 6.2.2 (p. 318), and 6.5.2 (p. 336), and nested iterative solvers. In what follows we briefly indicate how to bound $\left|\left|\left|I_{\mathcal{T}}^{*}R\right|\right|\right|_{*}$ in these cases.

3.8.4.1 SUPG-discretisations

These discretisations usually give rise to a consistency error of the form

$$\langle R, I_{\mathcal{T}}v\rangle = \sum_{K\in\mathcal{T}} \delta_K \int_K r\mathbf{a}\cdot\nabla(I_{\mathcal{T}}v),$$

cf. [**139, 153, 271**] and Sections 4.4.2 (p. 166), 4.10.2 (p. 240), 5.4.2 (p. 302), 6.2.2 (p. 318), and 6.5.2 (p. 336). Here, $\mathbf{a} \in L^\infty(\Omega)^d$ is a given vector field modelling a convection and δ_K are non-negative stabilisation parameters.

A standard inverse inequality and the definitions (3.1) (p. 80) of $|||\cdot|||$ and (3.4) (p. 82) of $\hbar_K$ yield for every element K

$$\begin{aligned}\left|\int_K r\mathbf{a}\cdot\nabla(I_{\mathcal{T}}v)\right| &\le \min\left\{\varepsilon^{-\frac{1}{p}}, c_I h_K^{-1}\beta^{-\frac{1}{p}}\right\}\|\mathbf{a}\|_{\infty;K}\|r\|_{p';K}|||I_{\mathcal{T}}v|||_K \\ &\le \max\{c_I,1\}h_K^{-1}\|\mathbf{a}\|_{\infty;K}\hbar_K\|r\|_{p';K}|||I_{\mathcal{T}}v|||_K .\end{aligned}$$

Hence, we have

$$\left|\left|\left|I_{\mathcal{T}}^{*}R\right|\right|\right|_{*} \le \max\{c_I,1\}\max_{K\in\mathcal{T}}(\delta_K h_K^{-1}\|\mathbf{a}\|_{\infty;K})\,\|I_{\mathcal{T}}\|_{\mathcal{L}}\left\{\sum_{K\in\mathcal{T}}\hbar_K^{p'}\|r\|_{p';K}^{p'}\right\}^{\frac{1}{p'}} .$$

Here, $\|I_{\mathcal{T}}\|_{\mathcal{L}}$ denotes the operator norm of $I_{\mathcal{T}}$ which can be bounded using Proposition 3.33 (p. 109). Since $\max_{K\in\mathcal{T}}(\delta_K h_K^{-1} \|\mathbf{a}\|_{\infty;K})$ is of order 1 for all SUPG-discretisations used in practice, this shows that $\left|\left|\left|I_{\mathcal{T}}^* R\right|\right|\right|_*$ can be absorbed by the right-hand side of the upper bound in Theorem 3.57.

Since the shape functions λ_z form a partition of unity, we have

$$\begin{aligned}\sum_{K\in\mathcal{T}} \delta_K \int_K r\mathbf{a}\cdot\nabla(I_{\mathcal{T}}v) &= \sum_{K\in\mathcal{T}}\sum_{z\in\mathcal{N}_K} \delta_K \int_K \lambda_z r\mathbf{a}\cdot\nabla(I_{\mathcal{T}}v)\\ &= \sum_{z\in\mathcal{N}}\sum_{K\subset\omega_z} \delta_K \int_K \lambda_z r\mathbf{a}\cdot\nabla(I_{\mathcal{T}}v).\end{aligned}$$

Therefore, the previous arguments show that $\left|\left|\left|I_{\mathcal{T}}^* R\right|\right|\right|_*$ can be absorbed by the right-hand side of the upper bound in Theorem 3.56 too.

For the $H(\mathrm{div})$-lifting of Section 3.8.3 we choose an arbitrary function $v \in H_D^1(\Omega)$ and define the function $w \in L^2(\Omega)$ by $w|_K = \delta_K \mathbf{a}\cdot\nabla(I_{\mathcal{T}}v)$ for all $K \in \mathcal{T}$. The property $\sum_z \lambda_z = 1$, the definitions of $\rho_{\mathcal{T}}$, $\overline{R}_z$, and $I_{\mathcal{T}}$, and a weighted Cauchy–Schwarz inequality imply that

$$\begin{aligned}\langle R, I_{\mathcal{T}}v\rangle &= \sum_{K\in\mathcal{T}}\sum_{z\in\mathcal{N}_K}\int_K \lambda_z(r-\overline{R}_z)w + \sum_{K\in\mathcal{T}}\sum_{z\in\mathcal{N}_K}\int_K \lambda_z\overline{R}_z w\\ &= \int_\Omega (-\operatorname{div}\rho_{\mathcal{T}})w + \langle R, I_{\mathcal{T}}w\rangle\\ &\le \left\{\sum_{K\in\mathcal{T}} \varepsilon^{-1}h_K^2 \|\operatorname{div}\rho_{\mathcal{T}}\|_K^2\right\}^{\frac12} \left\{\sum_{K\in\mathcal{T}} \varepsilon h_K^{-2} \|w\|_K^2\right\}^{\frac12} + |||R|||_*\, |||I_{\mathcal{T}}w|||\,.\end{aligned}$$

Standard inverse estimates yield for every element K

$$\begin{aligned}\|\operatorname{div}\rho_{\mathcal{T}}\|_K &\le c_I h_K^{-1} \|\rho_{\mathcal{T}}\|_K,\\ \|w\|_K &\le \delta_K \|\mathbf{a}\|_{\infty;K} \min\left\{\varepsilon^{-\frac12}, \widetilde{c}_I h_K^{-1}\right\} |||I_{\mathcal{T}}v|||_K,\\ |||I_{\mathcal{T}}w|||_K &\le \left\{\varepsilon\widetilde{c}_I^{\,2} h_K^{-2} + 1\right\}^{\frac12} \|I_{\mathcal{T}}w\|_K,\end{aligned}$$

where c_I and $\widetilde{c}_I$ both depend on the shape parameter $C_{\mathcal{T}}$ and where c_I in addition depends on the polynomial degree of $\rho_{\mathcal{T}}$. The definition of $I_{\mathcal{T}}$ implies that

$$\|I_{\mathcal{T}}w\|_K \le \sum_{z\in\mathcal{N}_K} \|\lambda_z\|_K |\overline{w}_z| \le \sum_{z\in\mathcal{N}_K} \frac{\|\lambda_z\|_K^2}{\int_{\omega_z}\lambda_z} \|w\|_{\omega_z} \le \sum_{z\in\mathcal{N}_K} \|w\|_{\omega_z}$$

holds for every element K. Moreover there is a constant $\widehat{c}_I$ which only depends on the shape parameter $C_{\mathcal{T}}$ such that the inequality

$$\left\{\varepsilon\widetilde{c}_I^{\,2} h_{K'}^{-2} + 1\right\} \min\left\{\varepsilon^{-1}h_K^2, \widetilde{c}_I^{\,2}\right\} \le 2\widetilde{c}_I^{\,2} \frac{\min\left\{\varepsilon^{-1}h_K^2, \widetilde{c}_I^{\,2}\right\}}{\min\left\{\varepsilon^{-1}h_{K'}^2, \widetilde{c}_I^{\,2}\right\}} \le \widehat{c}_I^{\,2}$$

holds for every pair of elements K, K' which share a vertex. Finally, every element K intersects at most $\#\mathcal{N}_K$ patches ω_z. Combining all these estimates we obtain the upper bound

$$\left\|\left|I_{\mathcal{T}}^* R\right|\right\|_* \leq \varepsilon^{-\frac{1}{2}} \|\rho_{\mathcal{T}}\| \max_{K\in\mathcal{T}}(h_K^{-1}\delta_K)c_I \|\mathbf{a}\|_\infty \|I_{\mathcal{T}}\|_{\mathcal{L}} + |||R|||_* \max_{K\in\mathcal{T}}(\delta_K h_K^{-1} \#\mathcal{N}_K)\widehat{c_I} \|\mathbf{a}\|_\infty \|I_{\mathcal{T}}\|_{\mathcal{L}} .$$

Since, according to [139, 153], the stability parameters δ_K should be less than the local mesh-size h_K, we may assume that

$$\max_{K\in\mathcal{T}}(\delta_K h_K^{-1}) \leq \frac{1}{2}\max\left\{c_I \|\mathbf{a}\|_\infty \|I_{\mathcal{T}}\|_{\mathcal{L}}, \max_{K\in\mathcal{T}}(\#\mathcal{N}_K)\widehat{c_I}\|\mathbf{a}\|_\infty \|I_{\mathcal{T}}\|_{\mathcal{L}}\right\}^{-1}.$$

The above estimate of $\left\|\left|I_{\mathcal{T}}^* R\right|\right\|_*$ and the first part of this subsection then imply that

$$|||R|||_* \leq \varepsilon^{-\frac{1}{2}} \|\rho_{\mathcal{T}}\| + \left\|\left|I_{\mathcal{T}}^* R\right|\right\|_* \leq \frac{3}{2}\varepsilon^{-\frac{1}{2}} \|\rho_{\mathcal{T}}\| + \frac{1}{2}|||R|||_*$$

and consequently

$$|||R|||_* \leq 3\varepsilon^{-\frac{1}{2}} \|\rho_{\mathcal{T}}\| .$$

Thus, in the case of a SUPG-discretisation, we have to pay for the missing Galerkin orthogonality by a factor of 3 in the upper bound (3.38) of Theorem 3.63.

3.8.4.2 Nested iterative solvers

We finally consider the case of a nested iterative solver as it is often used within an adaptive finite element code [79, Section V.4]. Such an algorithm is associated with a sequence $\mathcal{T}_0, \ldots, \mathcal{T}_\ell$ of partitions obtained by repeated adaptive refinement with $\mathcal{T}_0$ and $\mathcal{T}_\ell$ being the coarsest and finest partition, respectively. It usually has the following properties.

- The discrete problem corresponding to $\mathcal{T}_0$ is solved exactly.
- The discrete problem corresponding to $\mathcal{T}_k$, $k \geq 1$, is solved approximately with an iterative solver. The computed approximate solution corresponding to $\mathcal{T}_{k-1}$ is used as starting value and the iteration is stopped once the initial residual is reduced by a factor θ_k.

For every $k \in \{0, \ldots, \ell\}$, we denote by R_k the residual of the computed approximate solution of the discrete problem corresponding to $\mathcal{T}_k$. We assume that all R_k admit an L^p-representation (3.29) and that R_0 satisfies the Galerkin orthogonality (3.30). The second assumption can be relaxed but it simplifies the exposition. Denote by η_k, $0 \leq k \leq \ell$, any of the right-hand sides of the upper bounds for $|||R_k|||_*$ in Theorems 3.56, 3.57, or 3.63 which will serve as error indicators. With these notations and assumptions we have $\left\|\left|I_{\mathcal{T}_0}^* R_0\right|\right\|_* = 0$ and for $k \geq 1$

$$\begin{aligned}
\left\|\left|I_{\mathcal{T}_k}^* R_k\right|\right\|_* &\leq \left\|I_{\mathcal{T}_k}\right\|_{\mathcal{L}} \sup_{v_k\in S_D^{1,0}(\mathcal{T}_k)} \frac{\langle R_k, v_k\rangle}{|||v_k|||} \\
&\leq \theta_k \left\|I_{\mathcal{T}_k}\right\|_{\mathcal{L}} \sup_{v_k\in S_D^{1,0}(\mathcal{T}_k)} \frac{\langle R_{k-1}, v_k\rangle}{|||v_k|||} \\
&\leq \theta_k \left\|I_{\mathcal{T}_k}\right\|_{\mathcal{L}} |||R_{k-1}|||_* \\
&\leq \theta_k \left\|I_{\mathcal{T}_k}\right\|_{\mathcal{L}} \left\{\eta_{k-1} + \left\|\left|I_{\mathcal{T}_{k-1}}^* R_{k-1}\right|\right\|_*\right\}.
\end{aligned}$$

By induction this estimate and the arguments of the first part of this subsection prove that

$$|||R_\ell|||_* \le \eta_\ell + \sum_{k=0}^{\ell-1} \eta_k \left(\prod_{m=k+1}^{\ell} \theta_m \left\| I_{\mathcal{T}_m} \right\|_{\mathcal{L}} \right).$$

Thus, in the case of a nested iterative solver, the residual R_ℓ can be controlled by the error indicators $\eta_0, \ldots, \eta_\ell$ and the stopping tolerances $\theta_1, \ldots, \theta_\ell$.

Remark 3.65 The results of Section 3.8.1 are based on [**247**], those of Sections 3.8.3 and 3.8.4 on [**272**]. The techniques of Section 3.8.2 were first developed in [**251**, **254**]; the current presentation follows [**267**, **270**].

4

Linear Elliptic Equations

In this chapter we establish a posteriori error estimates for the discretisation of linear elliptic equations departing from a general abstract framework. This approach in particular accentuates the generic equivalence of the error to the associated residual and highlights that this is a structural property of the variational problem which is completely independent of the discretisation. Thus, the derivation of a posteriori error estimates amounts to the task of establishing computable bounds for dual norms of residuals. This is achieved using the general tools provided in Chapter 3. The concrete realisation of this task depends on the given differential equation and discretisation. In particular we address the following topics: robustness of the a posteriori error estimates with respect to parameters inherent in the variational problem and with respect to anisotropy of the meshes or of the differential equation, mixed finite element methods for saddle-point problems, differential equations of higher order, and non-conforming methods. Finally, we take up the results of Section 1.14 and give a general convergence proof for a generic adaptive algorithm. This establishes the convergence of the adaptive algorithm for all problems and all error indicators considered in this chapter using any marking and refinement strategy of Chapter 2.

4.1 Abstract Linear Problems

In this section we present the abstract framework which is the common basis for the results of this chapter.

4.1.1 *Spaces and Norms*

Given two Banach spaces X and Y with norms $\|\cdot\|_X$ and $\|\cdot\|_Y$, respectively, we denote by

- $\mathcal{L}(X,Y)$ the space of continuous linear mappings from X into Y equipped with the norm

$$\|L\|_{\mathcal{L}(X,Y)} = \sup_{\varphi\in X\setminus\{0\}} \frac{\|L\varphi\|_Y}{\|\varphi\|_X},$$

- $Y^* = \mathcal{L}(Y,\mathbb{R})$ the space of continuous linear functionals on Y,

- $\langle \cdot , \cdot \rangle_Y$ or $\langle \cdot , \cdot \rangle$ in short the dual pairing of Y^* and Y, i.e. $\langle \ell , \psi \rangle$ is the value of the functional $\ell \in Y^*$ applied to the element ψ of Y,
- $\mathcal{L}^2(X, Y, \mathbb{R})$ the space of continuous bi-linear mappings from $X \times Y$ into $\mathbb{R}$ equipped with the norm

$$\|B\|_{\mathcal{L}^2(X,Y,\mathbb{R})} = \sup_{\varphi \in X \setminus \{0\}} \sup_{\psi \in Y \setminus \{0\}} \frac{|B(\varphi, \psi)|}{\|\varphi\|_X \, \|\psi\|_Y}.$$

The spaces $\mathcal{L}^2(X, Y, \mathbb{R})$ and $\mathcal{L}(X, Y^*)$ are isomorphic, the canonical isometry is given by

$$\begin{aligned} \mathcal{L}^2(X, Y, \mathbb{R}) \ni B &\leftrightarrow L \in \mathcal{L}(X, Y^*) \\ B(\varphi, \psi) &= \langle L\varphi , \psi \rangle_Y \end{aligned} \tag{4.1}$$

for all $\varphi \in X$ and all $\psi \in Y$.

4.1.2 Variational Problem

Given a bi-linear mapping $B \in \mathcal{L}^2(X, Y, \mathbb{R})$ and a linear functional $\ell \in Y^*$, we consider in this chapter linear problems of the form:

find $\varphi \in X$ such that

$$B(\varphi, \psi) = \langle \ell , \psi \rangle_Y \tag{4.2}$$

holds for all $\psi \in Y$.

Using the isomorphism (4.1), problem (4.2) can be written in the equivalent form

$$L\varphi = \ell. \tag{4.3}$$

The following proposition gives criteria for the well-posedness of problems (4.2) and (4.3).

Proposition 4.1 *Assume that the space Y is reflexive, that*

$$\sup_{\varphi \in X} B(\varphi, \psi) > 0 \tag{4.4}$$

holds for all $\psi \in Y \setminus \{0\}$, and that

$$\inf_{\varphi \in X \setminus \{0\}} \sup_{\psi \in Y \setminus \{0\}} \frac{B(\varphi, \psi)}{\|\varphi\|_X \, \|\psi\|_Y} = \beta > 0. \tag{4.5}$$

Then problem (4.2) admits a unique solution for every right-hand side $\ell \in Y^*$ and the solution depends continuously on the right-hand side.

Proof Denote by $L \in \mathcal{L}(X, Y^*)$ the linear mapping which corresponds to B via the isomorphism (4.1). Condition (4.5) then implies that L is injective. Due to the closed range theorem [**282**, Section VII.5.1] the range of L is a closed subspace of Y^*. If L were not surjective, the Hahn–Banach theorem [**282**, Section IV.1] and the reflexivity of Y would imply the existence of a $\psi_0 \in Y \setminus \{0\}$ with

$$\langle L\varphi , \psi_0 \rangle_Y = 0$$

for all $\varphi \in X$. Due to relation (4.1) this contradicts condition (4.4). Hence, L is surjective, too. Condition (4.5) then implies the continuity of L^{-1}. Since problems (4.2) and (4.3) are equivalent, this proves the assertion. □

Remark 4.2 Proposition 4.1 was first established independently by I. Babuška in [**29**] and F. Brezzi in [**88**]. Condition (4.5) is often referred to as *inf–sup condition, Babuška–Brezzi condition, Ladyzhenskaja–Babuška–Brezzi condition,* or *LBB condition.* Note that

$$\left\| L^{-1} \right\|_{\mathcal{L}(Y^*,X)} = \frac{1}{\beta}.$$

The conditions of Proposition 4.1 are in particular satisfied when B is coercive and symmetric, i.e.

- $X = Y$,
- $B(\varphi, \psi) = B(\psi, \varphi)$ for all $\varphi, \psi \in X$, and
- there is a constant $\alpha > 0$ such that for all $\varphi \in X$

$$B(\varphi, \varphi) \geq \alpha \left\| \varphi \right\|_X^2.$$

In this special case the reflexivity of Y can be dropped; the constant β equals α.

Remark 4.3 The proof of Proposition 4.1 also reveals that, for reflexive spaces Y, conditions (4.4) and (4.5) are necessary for the well-posedness of problems (4.2) and (4.3).

Remark 4.4 Since a continuous linear operator and its adjoint have the same norm, the proof of Proposition 4.1 and Remark 4.2 imply that

$$\inf_{\varphi \in X\setminus\{0\}} \sup_{\psi \in Y\setminus\{0\}} \frac{B(\varphi, \psi)}{\left\| \varphi \right\|_X \left\| \psi \right\|_Y} = \inf_{\psi \in Y\setminus\{0\}} \sup_{\varphi \in X\setminus\{0\}} \frac{B(\varphi, \psi)}{\left\| \varphi \right\|_X \left\| \psi \right\|_Y}$$

cf. Lemma 4.86 (p. 270), too.

4.1.3 Discrete Problem

For the discretisation of problem (4.2), we choose finite-dimensional subspaces $X_\mathcal{T}$ of X and $Y_\mathcal{T}$ of Y, a bi-linear form $B_\mathcal{T}$ on $X_\mathcal{T} \times Y_\mathcal{T}$, and a linear functional $\ell_\mathcal{T}$ on $Y_\mathcal{T}$. The discrete analogue of problem (4.2) is then:

find $\varphi_\mathcal{T} \in X_\mathcal{T}$ such that

$$B_\mathcal{T}(\varphi_\mathcal{T}, \psi_\mathcal{T}) = \langle \ell_\mathcal{T}, \psi_\mathcal{T} \rangle_{Y_\mathcal{T}} \tag{4.6}$$

holds for all $\psi_\mathcal{T} \in Y_\mathcal{T}$.

Since the spaces $X_\mathcal{T}$ and $Y_\mathcal{T}$ are finite-dimensional, $B_\mathcal{T}$ and $\ell_\mathcal{T}$ are automatically continuous. Denoting by $L_\mathcal{T} \in \mathcal{L}(X_\mathcal{T}, Y_\mathcal{T}^*)$ the linear map associated with $B_\mathcal{T}$ via the isomorphism (4.1), problem (4.6) takes the equivalent form

$$L_\mathcal{T} \varphi_\mathcal{T} = \ell_\mathcal{T}. \tag{4.7}$$

Since finite-dimensional spaces are always reflexive, Proposition 4.1 applied to problem (4.6) gives the following criteria for the well-posedness of the discrete problems (4.6) and (4.7).

Proposition 4.5 *Assume that*

$$\sup_{\varphi_{\mathcal{T}} \in X_{\mathcal{T}}} B_{\mathcal{T}}(\varphi_{\mathcal{T}}, \psi_{\mathcal{T}}) > 0$$

holds for all $\psi_{\mathcal{T}} \in Y_{\mathcal{T}} \setminus \{0\}$ *and that*

$$\inf_{\varphi_{\mathcal{T}} \in X_{\mathcal{T}} \setminus \{0\}} \sup_{\psi_{\mathcal{T}} \in Y_{\mathcal{T}} \setminus \{0\}} \frac{B_{\mathcal{T}}(\varphi_{\mathcal{T}}, \psi_{\mathcal{T}})}{\|\varphi_{\mathcal{T}}\|_{X_{\mathcal{T}}} \, \|\psi_{\mathcal{T}}\|_{Y_{\mathcal{T}}}} = \beta_{\mathcal{T}} > 0.$$

Then problem (4.6) admits a unique solution for every right-hand side $\ell_{\mathcal{T}} \in Y_{\mathcal{T}}^*$.

Remark 4.6 Proposition 4.5 is a complicated way to state that, after choosing bases for $X_{\mathcal{T}}$ and $Y_{\mathcal{T}}$, the matrix corresponding to $B_{\mathcal{T}}$ must have maximal rank. When deriving a priori error estimates, the discretisation must be *stable*, i.e. $\beta_{\mathcal{T}}$ must uniformly be positive. When deriving a posteriori error estimates, we do not need this condition. We only need the unique solvability of problem (4.6) and the stability of the analytical problem (4.2) which is reflected in the inf–sup condition (4.5). Since we assume that $X_{\mathcal{T}} \subset X$ and $Y_{\mathcal{T}} \subset Y$, our discretisations are always *conforming*. Whenever this condition is violated, a posteriori error estimates are still possible, but the theory is less elegant and more involved, cf. Section 4.12 (p. 261).

4.1.4 A Posteriori Error Estimation

Assume that the conditions of Propositions 4.1 and 4.5 are satisfied and denote by $\varphi \in X$ and $\varphi_{\mathcal{T}} \in X_{\mathcal{T}}$ the unique solutions of problems (4.2) and (4.6), respectively. Using the operator L associated with B via the isomorphism (4.1) we have

$$L(\varphi - \varphi_{\mathcal{T}}) = \ell - L\varphi_{\mathcal{T}}.$$

This immediately implies

$$\|L\|_{\mathcal{L}(X,Y^*)}^{-1} \|\ell - L\varphi_{\mathcal{T}}\|_{Y^*} \le \|\varphi - \varphi_{\mathcal{T}}\|_X \le \left\|L^{-1}\right\|_{\mathcal{L}(Y^*,X)} \|\ell - L\varphi_{\mathcal{T}}\|_{Y^*} \tag{4.8}$$

and for all $\psi \in Y$ with $\|\psi\|_Y = 1$

$$\langle \ell - L\varphi_{\mathcal{T}}, \psi \rangle_Y = \langle L(\varphi - \varphi_{\mathcal{T}}), \psi \rangle_Y \le \|L\|_{\mathcal{L}(X,Y^*)} \|\varphi - \varphi_{\mathcal{T}}\|_X .$$

This is the generic equivalence of error and residual which we established in estimate (1.10) (p. 10) for a special situation.

Unfortunately, the dual norm of the residual $\|\ell - L\varphi_{\mathcal{T}}\|_{Y^*}$ cannot be used for practical purposes since its exact evaluation requires the solution of an infinite-dimensional variational problem which is as difficult a task as the solution of the original variational problem (4.2). To obtain error bounds which are better amenable to practical computations, we have to find an approximation which is close to exact and easy to compute at the same time. This is the place where the results of Section 3.8 (p. 132) come into play.

To apply the results of Sections 3.8.1 (p. 133) and 3.8.3 (p. 139), we have to check the following points:

- Y is a suitable subspace of a suitable $W^{1,p}$-space;
- the residual $R = \ell - L\varphi_{\mathcal{T}}$ admits an L^p-representation as in equation (3.29) (p. 132);
- the residual $R = \ell - L\varphi_{\mathcal{T}}$ satisfies the Galerkin orthogonality of equation (3.30) (p. 132).

The first point usually presents no difficulty. The value of p is typically determined by the differential equation and its variational formulation. For the second point, we may usually use integration by parts element-wise. For the third point, we need a closer look.

We first consider the important special case that the discrete forms $B_{\mathcal{T}}$ and $\ell_{\mathcal{T}}$ are the restrictions of the forms B and ℓ to the finite-dimensional spaces, i.e.

$$B_{\mathcal{T}}(\varphi_{\mathcal{T}}, \psi_{\mathcal{T}}) = B(\varphi_{\mathcal{T}}, \psi_{\mathcal{T}}), \quad \langle \ell_{\mathcal{T}}, \psi_{\mathcal{T}} \rangle_{Y_{\mathcal{T}}} = \langle \ell, \psi_{\mathcal{T}} \rangle_Y$$

holds for all $\varphi_{\mathcal{T}} \in X_{\mathcal{T}}$ and all $\psi_{\mathcal{T}} \in Y_{\mathcal{T}}$. Then we have *Galerkin orthogonality*, i.e. for all $\psi_{\mathcal{T}} \in Y_{\mathcal{T}}$ there holds

$$B(\varphi - \varphi_{\mathcal{T}}, \psi_{\mathcal{T}}) = 0$$

or equivalently

$$\langle \ell - L\varphi_{\mathcal{T}}, \psi_{\mathcal{T}} \rangle_Y = \langle L(\varphi - \varphi_{\mathcal{T}}), \psi_{\mathcal{T}} \rangle_Y = 0.$$

Hence, in this case, we may apply the results of Sections 3.8.1 (p. 133) and 3.8.3 (p. 139) to the residual $R = \ell - L\varphi_{\mathcal{T}}$ and immediately obtain upper bounds for $\|\ell - L\varphi_{\mathcal{T}}\|_{Y^*}$.

For the general case, Section 3.8.4 (p. 146) suggests that we may introduce a suitable restriction or projection operator $Q_{\mathcal{T}} : Y \to Y_{\mathcal{T}}$ and split the residual into the form

$$\ell - L\varphi_{\mathcal{T}} = (Id_Y - Q_{\mathcal{T}})^*(\ell - L\varphi_{\mathcal{T}}) + Q_{\mathcal{T}}^*(\ell - L\varphi_{\mathcal{T}}).$$

To the first term on the right-hand side of this equation we may then apply the techniques of Sections 3.8.1 and 3.8.3. The second term represents a consistency error. Section 3.8.4 (p. 146) provides methods for controlling this term in two important particular cases. In the general situation, it may be estimated using methods similar to the second Strang lemma [**109**, Theorem 31.1].

Both cases can be summarised as

$$\|\ell - L\varphi_{\mathcal{T}}\|_{Y^*} \le \left\|(Id_Y - Q_{\mathcal{T}})^*(\ell - L\varphi_{\mathcal{T}})\right\|_{Y^*} + \left\|Q_{\mathcal{T}}^*(\ell - L\varphi_{\mathcal{T}})\right\|_{Y^*}$$

and

$$\|\varphi - \varphi_{\mathcal{T}}\|_X \le \left\|L^{-1}\right\|_{\mathcal{L}(Y^*,X)} \left\{ \left\|(Id_Y - Q_{\mathcal{T}})^*(\ell - L\varphi_{\mathcal{T}})\right\|_{Y^*} + \left\|Q_{\mathcal{T}}^*(\ell - L\varphi_{\mathcal{T}})\right\|_{Y^*} \right\},$$

where the terms involving $Q_{\mathcal{T}}^*$ vanish in case of Galerkin orthogonality.

To obtain lower bounds for the residual, we apply the techniques of Section 3.8.2 (p. 135). In the present abstract framework this amounts in choosing a finite-dimensional subspace $\widetilde{Y}_{\mathcal{T}}$ of Y with $Y_{\mathcal{T}} \subset \widetilde{Y}_{\mathcal{T}} \subset Y$ and replacing $\|\ell - L\varphi_{\mathcal{T}}\|_{Y^*}$ by $\|\ell - L\varphi_{\mathcal{T}}\|_{\widetilde{Y}_{\mathcal{T}}^*}$. We then have

$$\|\ell - L\varphi_{\mathcal{T}}\|_{\widetilde{Y}_{\mathcal{T}}^*} \le \|\ell - L\varphi_{\mathcal{T}}\|_{Y^*}$$

and, using estimates (4.8),

$$\|\ell - L\varphi_{\mathcal{T}}\|_{\widetilde{Y}_{\mathcal{T}}^*} \le \|L\|_{\mathcal{L}(X,Y^*)} \|\varphi - \varphi_{\mathcal{T}}\|_X .$$

These estimates show that ideally the restriction operator $Q_{\mathcal{T}}$ and the space $\widetilde{Y}_{\mathcal{T}}$ fit together such that

$$\left\|(Id_Y - Q_{\mathcal{T}})^*(\ell - L\varphi_{\mathcal{T}})\right\|_{Y^*} + \left\|Q_{\mathcal{T}}^*(\ell - L\varphi_{\mathcal{T}})\right\|_{Y^*} \le c\,\|\ell - L\varphi_{\mathcal{T}}\|_{\widetilde{Y}_{\mathcal{T}}^*}$$

holds with a constant c of moderate size. This requirement is not a trivial one since the left-hand side of the estimate involves the supremum with respect to an infinite-dimensional space whereas the right-hand side involves the supremum with respect to a finite-dimensional space. Yet, if this estimate can be established, we have reduced the problem of a posteriori error estimation to the task of finding an error indicator $\eta_{\mathcal{T}}$ which only depends on the discrete solution $\varphi_{\mathcal{T}}$ and the given data of the variational problem, which is easy to evaluate, and which yields lower bounds for $\|\ell - L\varphi_{\mathcal{T}}\|_{\widetilde{Y}_{\mathcal{T}}^*}$ and upper bounds for $\left\|(Id_Y - Q_{\mathcal{T}})^*(\ell - L\varphi_{\mathcal{T}})\right\|_{Y^*}$ and $\left\|Q_{\mathcal{T}}^*(\ell - L\varphi_{\mathcal{T}})\right\|_{Y^*}$.

The results of Chapter 1 and Section 3.8.2 (p. 135) show that in addition we may have to approximate given data. This introduces an additional data error $\theta_{\mathcal{T}}$ which we have to take into account.

The abstract framework for a posteriori error estimation may therefore be summarised as follows.

Theorem 4.7 *Assume that the conditions of Propositions 4.1 and 4.5 are satisfied and denote by φ and $\varphi_{\mathcal{T}}$ the unique solutions of problems (4.2) and (4.6), respectively. Assume that there are a restriction operator $Q_{\mathcal{T}} \in \mathcal{L}(Y, Y_{\mathcal{T}})$, a finite-dimensional subspace $\widetilde{Y}_{\mathcal{T}}$ of Y with $Y_{\mathcal{T}} \subset \widetilde{Y}_{\mathcal{T}} \subset Y$, an error indicator $\eta_{\mathcal{T}}$, which only depends on the discrete solution $\varphi_{\mathcal{T}}$ and the given data of the variational problem (4.2), and a data error $\theta_{\mathcal{T}}$, which only depends on the data of the variational problem, such that the estimates*

$$\left\|(Id_Y - Q_{\mathcal{T}})^*(\ell - L\varphi_{\mathcal{T}})\right\|_{Y^*} \le c_A\left(\eta_{\mathcal{T}} + \theta_{\mathcal{T}}\right), \quad \left\|Q_{\mathcal{T}}^*(\ell - L\varphi_{\mathcal{T}})\right\|_{Y^*} \le c_C\left(\eta_{\mathcal{T}} + \theta_{\mathcal{T}}\right) \tag{4.9}$$

and

$$\eta_{\mathcal{T}} \le c_I\left(\|\ell - L\varphi_{\mathcal{T}}\|_{\widetilde{Y}_{\mathcal{T}}^*} + \theta_{\mathcal{T}}\right) \tag{4.10}$$

are fulfilled. Then the error $\varphi - \varphi_{\mathcal{T}}$ can be estimated from above by

$$\|\varphi - \varphi_{\mathcal{T}}\|_X \le \left\|L^{-1}\right\|_{\mathcal{L}(Y^*,X)} (c_A + c_C)\left(\eta_{\mathcal{T}} + \theta_{\mathcal{T}}\right)$$

and from below by

$$\eta_{\mathcal{T}} \le c_I\left(\|L\|_{\mathcal{L}(X,Y^*)} \|\varphi - \varphi_{\mathcal{T}}\|_X + \theta_{\mathcal{T}}\right).$$

Remark 4.8 The quantity

$$\|L\|_{\mathcal{L}(X,Y^*)} \left\|L^{-1}\right\|_{\mathcal{L}(Y^*,X)} (c_A + c_C)c_I$$

is a measure for the quality of the error indicator $\eta_{\mathcal{T}}$. It corresponds to the condition number of a linear operator. It should be uniformly bounded with respect to families of discretisations of the same variational problem and with respect to parameters such as, e.g. the size of reaction or diffusion terms or mesh-Péclet-numbers, which are inherent in the differential equation. This uniformity with respect to parameters is often referred to as *robustness*.

Remark 4.9 When applied to linear elliptic equations in the following sections, the present abstract framework yields a posteriori error estimates with respect to an energy norm which is usually a $W^{1,p}$-norm. It can be modified to also obtain error estimates with respect to a weaker norm which is usually an L^p-norm. To this end we must replace the space X by a larger space X_- with a weaker norm and the space Y by a smaller space Y_+ with a stronger norm. Moreover, we must assume that the space $\widetilde{Y}_{\mathcal{T}}$ is contained in Y_+ while retaining the condition $Y_{\mathcal{T}} \subset Y$. Note that this only affects the choice of the local cut-off functions but not the discretisation. We will present this modification in more detail in the context of nonlinear elliptic equations, cf. Remarks 5.4 (p. 284) and 5.8 (p. 290) and the results of Sections 5.2.5 (p. 298) and 5.4.5 (p. 307).

4.2 The Model Problem Revisited

Within the abstract framework of Section 4.1, the model problem (1.1) (p. 4) and its discretisation (1.3) (p. 5) correspond to the following data

$$
\begin{aligned}
X = Y &= H_D^1(\Omega), \\
B(\varphi, \psi) &= \int_\Omega \nabla\varphi \cdot \nabla\psi, \\
\langle \ell, \psi \rangle_Y &= \int_\Omega f\psi + \int_{\Gamma_N} g\psi, \\
X_{\mathcal{T}} = Y_{\mathcal{T}} &= S_D^{k,0}(\mathcal{T}), \\
B_{\mathcal{T}}(\varphi_{\mathcal{T}}, \psi_{\mathcal{T}}) &= B(\varphi_{\mathcal{T}}, \psi_{\mathcal{T}}), \\
\langle \ell_{\mathcal{T}}, \psi_{\mathcal{T}} \rangle_{Y_{\mathcal{T}}} &= \langle \ell, \psi_{\mathcal{T}} \rangle_Y,
\end{aligned}
$$

where in Chapter 1 the space dimension d and the polynomial degree k always equal 2 and 1, respectively.

Since the space $H_D^1(\Omega)$ is reflexive and the bi-linear form B is symmetric and coercive, the conditions of Propositions 4.1 (p. 152) and 4.5 (p. 154) are satisfied. Since the discrete forms $B_{\mathcal{T}}$ and $\ell_{\mathcal{T}}$ are the restrictions of B and ℓ, respectively, we have Galerkin orthogonality.

Therefore, we do not need the restriction operator $Q_{\mathcal{T}}$ and can directly apply Theorems 3.56 (p. 134) and 3.57 (p. 135) to obtain upper bounds for the error. The functions r and j are given by equations (1.13) (p. 11) and (1.14) (p. 11), respectively.

Theorem 3.59 (p. 137) immediately yields lower bounds for the error upon approximating the functions f and g by piecewise polynomials $f_{\mathcal{T}}$ and $g_{\mathcal{E}}$, respectively. This approximation introduces an additional data error. Theorem 3.59 fits into the abstract framework with

$$
\widetilde{Y}_{\mathcal{T}} = \operatorname{span}\left\{\psi_K\varphi, \psi_E\sigma : \varphi \in R_{\max\{k-2,0\}}(K), \sigma \in R_{k-1}(E), K \in \mathcal{T}, E \in \mathcal{E}\right\},
$$

where ψ_K, ψ_E are the local cut-off functions of Section 3.2 (p. 81).

The data error $\theta_{\mathcal{T}}$ of Theorem 4.7 (p. 156) is given by

$$\theta_{\mathcal{T}} = \left\{ \sum_{K \in \mathcal{T}} h_K^2 \left\| f - \bar{f}_K \right\|_K^2 + \sum_{E \in \mathcal{E}_{\Gamma_N}} h_E \left\| g - \bar{g}_E \right\|_E^2 \right\}^{\frac{1}{2}} .$$

The error estimator $\eta_{\mathcal{T}}$ is one of the quantities

$$\left\{ \sum_{K \in \mathcal{T}} \eta_{R,K}^2 \right\}^{\frac{1}{2}}, \quad \left\{ \sum_{z \in \mathcal{N}_\Omega \cup \mathcal{N}_{\Gamma_N}} \eta_{D,z}^2 \right\}^{\frac{1}{2}},$$

$$\left\{ \sum_{K \in \mathcal{T}} \eta_{D,K}^2 \right\}^{\frac{1}{2}}, \quad \left\{ \sum_{K \in \mathcal{T}} \eta_{N,K}^2 \right\}^{\frac{1}{2}}$$

of Theorem 1.5 (p. 15) and equations (1.36) (p. 26), (1.41) (p. 28), and (1.44) (p. 29).

The constant c_C in Theorem 4.7 (p. 156) equals 0; the constants c_A and c_I are rational functions of the constants $\widetilde{c}_2(\omega_z)$ and $\widetilde{c}_2(\sigma_z)$ of Theorem 3.57 (p. 135) and $C_{I,1,k}(K), \ldots, C_{I,S,k}(E,K)$ of Proposition 3.37 (p. 112), respectively.

This in particular shows that the results of Sections 1.4 (p. 10) and 1.7 (p. 25) immediately carry over to higher space dimensions and higher polynomial degrees.

When $k = 1$, the results of Section 1.6 (p. 20) fit into the abstract framework of Section 4.1 (p. 151) with the L^2-projection $P_{\mathcal{T}}$ as restriction operator $Q_{\mathcal{T}}$. The estimator $\eta_{\mathcal{T}}$ is then given by

$$\eta_{\mathcal{T}} = \left\{ \sum_{E \in \mathcal{E}_\Omega \cup \mathcal{E}_{\Gamma_N}} \eta_{R,E}^2 \right\}^{\frac{1}{2}}$$

with $\eta_{R,E}$ as in Theorem 1.15 (p. 24).

After approximating the functions f and g by piecewise polynomials $f_{\mathcal{T}}$ and $g_{\mathcal{E}}$, respectively, we may apply Proposition 3.62 (p. 141) and Theorem 3.63 (p. 144) to the model problem. We thus obtain an $H(\text{div})$-lifting $\rho_{\mathcal{T}}$ of the residual. Estimate (3.38) (p. 144) of Theorem 3.63 (p. 144) immediately yields the upper bound

$$\left\| \nabla(u - u_{\mathcal{T}}) \right\| \le \|\rho_{\mathcal{T}}\| + \theta_{\mathcal{T}}.$$

Estimate (3.39) (p. 144) of Theorem 3.63 combined with the lower bounds of Theorem 3.59 (p. 137) gives the lower bound

$$\|\rho_{\mathcal{T}}\| \le c_* \left\{ \left\| \nabla(u - u_{\mathcal{T}}) \right\| + \theta_{\mathcal{T}} \right\}$$

with a constant c_* which only depends on the shape parameter $C_{\mathcal{T}}$ of $\mathcal{T}$ and the polynomial degree k. This generalises Theorem 1.41 (p. 52) concerning the hyper-circle method.

The model problem is a very special case of the quasilinear elliptic equations discussed in Section 5.2 (p. 290). Consequently the results of Section 5.2.5 (p. 298) immediately yield L^2-error estimates for the model problem. When estimating the L^2-norm of the error, the element and face residuals in $\eta_{R,K}$ and $\eta_{R,E}$ must be multiplied by additional factors h_K and h_E, respectively, and the solutions of the auxiliary local discrete problems which are at the base of the estimators $\eta_{D,z}$, $\eta_{D,K}$, and $\eta_{N,K}$ must be evaluated with respect to an appropriate local L^2-norm.

4.3 Reaction–Diffusion Equations

We consider the singularly perturbed reaction–diffusion equation

$$\begin{aligned} -\Delta u + \kappa^2 u &= f \quad \text{in } \Omega \\ u &= 0 \quad \text{on } \Gamma \end{aligned} \tag{4.11}$$

where $\kappa \gg 1$ and $\Omega \subset \mathbb{R}^d$, $d \geq 2$, is a bounded polygonal domain with Lipschitz boundary Γ. Problems of this type arise, e.g. in the modelling of thin plates, in the numerical treatment of reaction processes, and from the time-discretisation of the heat equation.

4.3.1 *Variational Problem and Discretisation*

The variational formulation of problem (4.11) and its finite element discretisation fit into the the framework of Section 4.1 (p. 151) with the choices

$$\begin{aligned} X = Y &= H_0^1(\Omega), \\ B(\varphi, \psi) &= \int_\Omega \{\nabla\varphi \cdot \nabla\psi + \kappa^2 \varphi\psi\}, \\ \langle \ell, \psi \rangle_Y &= \int_\Omega f\psi, \\ X_\mathcal{T} = Y_\mathcal{T} &= S_0^{k,0}(\mathcal{T}), \\ B_\mathcal{T}(\varphi_\mathcal{T}, \psi_\mathcal{T}) &= B(\varphi_\mathcal{T}, \psi_\mathcal{T}), \\ \langle \ell_\mathcal{T}, \psi_\mathcal{T} \rangle_{Y_\mathcal{T}} &= \langle \ell, \psi_\mathcal{T} \rangle_Y . \end{aligned}$$

At first sight, we have two natural choices for the norms of X and Y:

- the standard H^1-norm $\|\cdot\|_1$ and
- the energy norm $|||\cdot||| = \left\{\|\nabla\cdot\|^2 + \kappa^2 \|\cdot\|^2\right\}^{\frac{1}{2}}$ associated with problem (4.11).

The standard H^1-norm, however, is not a good choice since we then have for all $\varphi, \psi \in X$

$$B(\varphi, \varphi) = |||\varphi|||^2 \geq \|\varphi\|_1^2$$

and

$$B(\varphi, \psi) \leq |||\varphi|||\ |||\psi||| \leq \kappa^2 \|\varphi\|_1 \|\psi\|_1$$

which gives

$$\|L\|_{\mathcal{L}(X,Y^*)} \left\|L^{-1}\right\|_{\mathcal{L}(Y^*,X)} = \kappa^2.$$

Therefore, Theorem 4.7 (p. 156) shows that any a posteriori error estimate based on the H^1-norm will lead to a gap of at least κ^2 between upper and lower error bounds and thus will be worthless. The energy norm on the other hand immediately yields

$$\|L\|_{\mathcal{L}(X,Y^*)} \left\|L^{-1}\right\|_{\mathcal{L}(Y^*,X)} = 1.$$

Hence, we will henceforth equip the spaces X and Y with this norm.

Since the space $H_0^1(\Omega)$ is reflexive and the bi-linear form B is symmetric and coercive, the conditions of Propositions 4.1 (p. 152) and 4.5 (p. 154) are satisfied.

4.3.2 A Posteriori Error Estimates

Since the discrete forms $B_{\mathcal{T}}$ and $\ell_{\mathcal{T}}$ are the restrictions of B and ℓ, respectively, we have Galerkin orthogonality. Therefore, we do not need the restriction operator $Q_{\mathcal{T}}$ and can directly apply Theorems 3.56 (p. 134) and 3.57 (p. 135) to obtain upper bounds on the error. The functions r and j are given by

$$r|_K = f + \Delta u_{\mathcal{T}} - \kappa^2 u_{\mathcal{T}} \quad \text{for all } K,$$

$$j|_E = \begin{cases} -\mathbb{J}_E(\mathbf{n}_E \cdot \nabla u_{\mathcal{T}}) & \text{if } E \in \mathcal{E}_\Omega, \\ 0 & \text{if } E \in \mathcal{E}_\Gamma. \end{cases}$$

Theorem 3.59 (p. 137) immediately yields lower bounds for the error upon replacing the function f by a piecewise polynomial approximation $f_{\mathcal{T}}$ which introduces an additional data error. Theorems 3.59 and 3.63 fit into the abstract framework with

$$\widetilde{Y}_{\mathcal{T}} = \operatorname{span}\left\{\psi_K \varphi, \psi_{E,\vartheta}\sigma : \varphi \in R_{\max\{k-2,0\}}(K), \sigma \in R_{k-1}(E), K \in \mathcal{T}, E \in \mathcal{E}\right\},$$

where ψ_K, $\psi_{E,\vartheta}$ are the local cut-off functions of Section 3.2 (p. 81) and where according to equation (3.32) (p. 136) $\vartheta = \min\{1, h_E^{-1}\kappa^{-1}\}$.

Combining these results, we arrive at the following a posteriori error estimates.

Theorem 4.10 *Denote by u the unique weak solution of problem (4.11) and by $u_{\mathcal{T}}$ its finite element approximation, respectively. For every element $K \in \mathcal{T}$ define the residual a posteriori error estimator $\eta_{R,K}$ by*

$$\eta_{R,K} = \left\{ \hbar_K^2 \left\|f_{\mathcal{T}} + \Delta u_{\mathcal{T}} - \kappa^2 u_{\mathcal{T}}\right\|_K^2 + \frac{1}{2} \sum_{E \in \mathcal{E}_{K,\Omega}} \hbar_E \left\|\mathbb{J}_E(\mathbf{n}_E \cdot \nabla u_{\mathcal{T}})\right\|_E^2 \right\}^{\frac{1}{2}},$$

where $\hbar_S = \min\left\{\operatorname{diam}(S), \kappa^{-1}\right\}$ is as in equation (3.4) (p. 82) and where $f_{\mathcal{T}} \in S^{k,-1}(\mathcal{T})$ is any approximation of f such as, e.g. its L^2-projection onto the piecewise constant functions $S^{0,-1}(\mathcal{T})$. Then there are two constants c^* and c_*, which only depend on the shape parameter $C_{\mathcal{T}}$ of $\mathcal{T}$ defined in *equation (3.3) (p. 82) and the polynomial degree k, but which are independent of κ, such that the error measured in the energy norm $|||\cdot||| = \left\{\|\nabla\cdot\|^2 + \kappa^2 \|\cdot\|^2\right\}^{\frac{1}{2}}$ can be bounded from above by*

$$|||u - u_{\mathcal{T}}||| \leq c^* \left\{ \sum_{K\in\mathcal{T}} \eta_{R,K}^2 + \sum_{K\in\mathcal{T}} \hbar_K^2 \left\|f - f_{\mathcal{T}}\right\|_K^2 \right\}^{\frac{1}{2}}$$

and, for every element $K \in \mathcal{T}$, bounded from below by

$$\eta_{R,K} \leq c_* \left\{ |||u - u_{\mathcal{T}}|||_{\omega_K}^2 + \sum_{K'\subset\omega_K} \hbar_{K'}^2 \left\|f - f_{\mathcal{T}}\right\|_{K'}^2 \right\}^{\frac{1}{2}},$$

where $|||\cdot|||_{\omega_K}$ denotes the restriction of $|||\cdot|||$ to ω_K.

Remark 4.11 Notice that the residual error indicators $\eta_{R,K}$ for the model problem and for the present reaction–diffusion equation only differ in the scaling factors of the element and edge residuals. For the model problem these are $h_K = \text{diam}(K)$ and $h_E = \text{diam}(E)$; here they are $\hbar_K = \min\{\text{diam}(K), \kappa^{-1}\}$ and $\hbar_E = \min\{\text{diam}(E), \kappa^{-1}\}$.

Remark 4.12 The error estimators of Section 1.7 (p. 25), which are based on the solution of auxiliary local discrete problems, can easily be adapted to the present reaction–diffusion equation. One only has to perform the following modifications.

- In the definitions (1.35) (p. 26), (1.40) (p. 28), and (1.43) (p. 29) of the spaces V_z, $\widetilde{V}_K$, and V_K, the bubble functions ψ_E must be replaced by their modified counterparts $\psi_{E,\vartheta}$ with $\vartheta = \min\{1, h_E^{-1}\kappa^{-1}\}$.
- The auxiliary problems must also be reaction–diffusion equations, i.e. the integrals on the left-hand sides of equations (1.37) (p. 26), (1.42) (p. 28), and (1.45) (p. 29) must be changed from $\int_\omega \nabla\varphi \cdot \nabla\psi$ to $\int_\omega \{\nabla\varphi \cdot \nabla\psi + \kappa^2\varphi\psi\}$ with the corresponding domains ω and functions φ and ψ.
- In the definitions (1.36) (p. 26), (1.41) (p. 28), and (1.44) (p. 29) of the error estimators, the H^1-norm $\|\nabla\cdot\|_\omega$ must be replaced by the energy norm $|||\cdot|||_\omega$ with the corresponding domain ω.

Remark 4.13 The results of this section can easily be extended to reaction–diffusion equations with mixed Dirichlet and Neumann boundary conditions similar to problem (1.1) (p. 4). One only has to perform the following modifications.

- The spaces $H_0^1(\Omega)$ and $S_0^{k,0}(\mathcal{T})$ must be replaced by $H_D^1(\Omega)$ and $S_D^{k,0}(\mathcal{T})$, respectively, and the terms $\int_{\Gamma_N} gv$ and $\int_{\Gamma_N} gv_{\mathcal{T}}$ must be added in the definitions of ℓ and $\ell_{\mathcal{T}}$.
- The boundary residuals

$$\sum_{E\in\mathcal{E}_{\Gamma_N}} \hbar_E \left\|g_{\mathcal{E}} - \mathbf{n}_E \cdot \nabla u_{\mathcal{T}}\right\|_E^2$$

must be added to $\eta_{R,K}$, where $g_{\mathcal{E}}$ is any approximation of g which is piecewise polynomial on the partition of Γ_N induced by $\mathcal{T}$.

- The data error

$$\sum_{E \in \mathcal{E}_{\Gamma_N}} \hbar_E \left\| g - g_{\mathcal{E}} \right\|_E^2$$

must be added on the right-hand sides of the error estimates in Theorem 4.10.

Remark 4.14 Consider the singularly perturbed reaction–diffusion equation

$$\begin{aligned} -\varepsilon \Delta u + u &= f \quad \text{in } \Omega \\ u &= 0 \quad \text{on } \Gamma \end{aligned} \tag{4.12}$$

where $0 < \varepsilon \ll 1$. Its variational formulation and finite element discretisation corresponds to the data

$$\begin{aligned} X = Y &= H_0^1(\Omega), \\ B(\varphi, \psi) &= \int_\Omega \{\varepsilon \nabla\varphi \cdot \nabla\psi + \varphi\psi\}, \\ \langle \ell, \psi \rangle_Y &= \int_\Omega f\psi, \\ X_{\mathcal{T}} = Y_{\mathcal{T}} &= S_0^{k,0}(\mathcal{T}), \\ B_{\mathcal{T}}(\varphi_{\mathcal{T}}, \psi_{\mathcal{T}}) &= B(\varphi_{\mathcal{T}}, \psi_{\mathcal{T}}), \\ \langle \ell_{\mathcal{T}}, \psi_{\mathcal{T}} \rangle_{Y_{\mathcal{T}}} &= \langle \ell, \psi_{\mathcal{T}} \rangle_Y . \end{aligned}$$

The natural energy norm is given by

$$|||\varphi||| = \left\{ \varepsilon \left\| \nabla\varphi \right\|^2 + \left\| \varphi \right\|^2 \right\}^{\frac{1}{2}} .$$

With the same arguments as above we conclude that

$$\eta_{R,K,\varepsilon}^2 = \hbar_K^2 \left\| f_{\mathcal{T}} + \varepsilon \Delta v_{\mathcal{T}} - v_{\mathcal{T}} \right\|_K^2 + \frac{1}{2} \sum_{E \in \mathcal{E}_{K,\Omega}} \varepsilon^{-\frac{1}{2}} \hbar_E \left\| \mathbb{J}_E(\varepsilon \mathbf{n}_E \cdot \nabla v_{\mathcal{T}}) \right\|_E^2$$

with $\hbar_S = \min\left\{ \varepsilon^{-\frac{1}{2}} \operatorname{diam}(S), 1 \right\}$ as in (3.4) (p. 82) is a robust error indicator for this problem.

Remark 4.15 After approximating the function f by piecewise polynomials $f_{\mathcal{T}}$, we may apply Proposition 3.62 (p. 141) and Theorem 3.63 (p. 144) to problem (4.12). We thus obtain an $H(\operatorname{div})$-lifting $\rho_{\mathcal{T}}$ of the residual. Estimate (3.38) (p. 144) of Theorem 3.63 (p. 144) immediately yields the upper bound

$$|||u - u_{\mathcal{T}}||| \leq \varepsilon^{-\frac{1}{2}} \left\| \rho_{\mathcal{T}} \right\| + c^* \theta_{\mathcal{T}}$$

with

$$\theta_{\mathcal{T}} = \left\{ \sum_{K \in \mathcal{T}} \hbar_K^2 \left\| f - f_{\mathcal{T}} \right\|_K^2 \right\}^{\frac{1}{2}} ,$$

where the constant c^* only depends on the shape parameter $C_{\mathcal{T}}$ of $\mathcal{T}$ and the polynomial degree of $f_{\mathcal{T}}$.

Estimate (3.39) (p. 144) of Theorem 3.63 combined with the lower bounds of Theorem 3.59 (p. 137) on the other hand gives the lower bound

$$\varepsilon^{-\frac{1}{2}} \|\rho_{\mathcal{T}}\| \leq c_* \max_{z\in\mathcal{N}} \max\left\{\varepsilon^{-\frac{1}{2}} \operatorname{diam}(\omega_z), 1\right\} \{|||u - u_{\mathcal{T}}||| + \theta_{\mathcal{T}}\}$$

with a constant c_* which only depends on the shape parameter $C_{\mathcal{T}}$ of $\mathcal{T}$ and the polynomial degree k. This generalises Theorem 1.41 (p. 52) concerning the hyper-circle method and shows that it is not fully robust. This lack of robustness is not a technical but a structural drawback. It is due to the fact that the lifting (3.37) (p. 141) only incorporates the principal part of the differential operator of problem (4.12) and gives no control on the dominant lower order terms. The same observation applies to problem (4.11).

Remark 4.16 Theorem 4.10 and Remarks 4.12–4.14 were first presented in [**263**]. Remark 4.15 is due to [**272**]. Robust error indicators based on the equilibration of residuals and on the solution of auxiliary local discrete problems may be found in [**8, 15**].

4.4 Convection–Diffusion Equations

We consider the stationary convection–diffusion equation

$$\begin{aligned} -\varepsilon\Delta u + \mathbf{a}\cdot\nabla u + bu &= f && \text{in } \Omega \\ u &= 0 && \text{on } \Gamma_D \\ \varepsilon\frac{\partial u}{\partial n} &= g && \text{on } \Gamma_N \end{aligned} \tag{4.13}$$

in a polygonal domain Ω in $\mathbb{R}^d$, $d \geq 2$, with Lipschitz boundary Γ consisting of two disjoint components Γ_D and Γ_N. The data have to satisfy the following conditions:

(A1) $0 < \varepsilon \ll 1$;

(A2) $\mathbf{a} \in W^{1,\infty}(\Omega)^d$, $b \in L^\infty(\Omega)$;

(A3) there are two constants $\beta \geq 0$ and $c_b \geq 0$, which do not depend on ε, such that

$$-\frac{1}{2}\operatorname{div}\mathbf{a} + b \geq \beta \quad \text{and} \quad \|b\|_\infty \leq c_b\beta;$$

(A4) the Dirichlet boundary Γ_D has positive $(d-1)$-dimensional Hausdorff measure and includes the inflow boundary, i.e.

$$\left\{x \in \Gamma : \mathbf{a}(x)\cdot\mathbf{n}(x) < 0\right\} \subset \Gamma_D.$$

Assumption (A3) allows us to handle simultaneously the case of a non-vanishing zero-order reaction term and that of absent reaction, the latter one corresponding to $\beta = 0$. If $\beta = 0$ we set $c_b = 0$. Assumption (A1) of course means that we are interested in the convection-dominated regime.

4.4.1 Variational Formulation

The variational formulation of problem (4.13) fits into the abstract framework of Section 4.1 (p. 151) with

$$X = Y = H^1_D(\Omega),$$
$$B(\varphi,\psi) = \int_\Omega \{\varepsilon\nabla\varphi\cdot\nabla\psi + \mathbf{a}\cdot\nabla\varphi\psi + b\varphi\psi\},$$
$$\langle \ell\,,\,\psi\rangle_Y = \int_\Omega f\psi + \int_{\Gamma_N} g\psi.$$

Integration by parts of the convection term yields

$$\begin{aligned} B(\varphi,\varphi) &= \int_\Omega \left\{\varepsilon\,|\nabla\varphi|^2 + \frac{1}{2}\mathbf{a}\cdot\nabla(\varphi^2) + b\varphi^2\right\} \\ &= \int_\Omega \left\{\varepsilon\,|\nabla\varphi|^2 + \left(b - \frac{1}{2}\operatorname{div}\mathbf{a}\right)\varphi^2\right\} + \int_\Gamma \mathbf{n}\cdot\mathbf{a}\varphi^2 \end{aligned} \tag{4.14}$$

for all $\varphi \in X$. Hence, due to Assumptions (A3) and (A4), the natural energy norm for problem (4.13) is given by

$$|||\varphi||| = \left\{\varepsilon\,\|\nabla\varphi\|^2 + \beta\,\|\varphi\|^2\right\}^{\frac{1}{2}}.$$

The corresponding dual norm on X^* is as usual defined by

$$|||\varphi|||_* = \sup_{\psi\in X\setminus\{0\}} \frac{\langle\varphi\,,\,\psi\rangle_X}{|||\psi|||}.$$

We equip the space Y with the energy norm $|||\cdot|||$ and the space X with the norm $|||\cdot||| + |||\mathbf{a}\cdot\nabla\cdot|||_*$, i.e.

$$\|\varphi\|_X = |||\varphi||| + |||\mathbf{a}\cdot\nabla\varphi|||_*, \quad \|\psi\|_Y = |||\psi|||.$$

This definition is possible since we have $\mathbf{a}\cdot\nabla\varphi \in L^2(\Omega)$ for all $\varphi \in X$ and since

$$(X, |||\cdot|||) \hookrightarrow (L^2(\Omega), \|\cdot\|) \hookrightarrow (X^*, |||\cdot|||_*)$$

with continuous embeddings.

This, at first sight unusual, choice of norms is motivated by the results of Section 6.2 (p. 317) for the corresponding in-stationary convection–diffusion equation and the observation that the convection term $\mathbf{a}\cdot\nabla\varphi$ is the stationary counterpart of the material derivative $\partial_t\varphi + \mathbf{a}\cdot\nabla\varphi$ in the time-dependent regime.

The following proposition shows that, with this choice of norms, the bi-linear form B fulfils the conditions of Proposition 4.1 (p. 152) with an inf–sup constant which is independent of ε. In particular it implies that problem (4.13) admits a unique solution, that the operator L corresponding to the bi-linear form B via the isometry (4.1) (p. 152) is an isomorphism of X onto Y^*, and that the corresponding condition number $\|L\|_{\mathcal{L}(X,Y^*)}\left\|L^{-1}\right\|_{\mathcal{L}(Y^*,X)}$ is uniformly bounded with respect to ε.

Notice that the, at first sight, more natural choice $\|\cdot\|_X = \|\cdot\|_Y = |||\cdot|||$ leads, due to the convection, to a condition number $\|L\|_{\mathcal{L}(X,Y^*)} \left\|L^{-1}\right\|_{\mathcal{L}(Y^*,X)}$ of order $\varepsilon^{-\frac{1}{2}}$. Hence, this choice of norms could not yield robust a posteriori error estimates.

Proposition 4.17 *The bi-linear form B is coercive, i.e.*

$$B(\varphi, \varphi) \geq |||\varphi|||^2$$

for all $\varphi \in X$, and fulfils the upper bound

$$B(\varphi, \psi) \leq \max\{c_b, 1\}\{|||\varphi||| + |||\mathbf{a}\cdot\nabla\varphi|||_*\}\,|||\psi|||$$

for all $\varphi \in X$, $\psi \in Y$ and the inf–sup condition

$$\inf_{\varphi\in X\setminus\{0\}} \sup_{\psi\in Y\setminus\{0\}} \frac{B(\varphi,\psi)}{\{|||\varphi||| + |||\mathbf{a}\cdot\nabla\varphi|||_*\}\,|||\psi|||} \geq \frac{1}{2+\max\{c_b,1\}}.$$

Here, the constant c_b is that of Assumption (A3).

Proof The coercivity of B follows from equation (4.14) and assumptions (A3) and (A4).

The upper bound for B follows from Assumption (A3) and the definitions of the norms $|||\cdot|||$ and $|||\cdot|||_*$.

To prove the inf–sup condition, we fix an arbitrary function $\varphi \in X$. Due to the definition of the dual norm, there is a function $\psi \in X = Y$ with

$$|||\psi||| = 1, \qquad \int_\Omega \mathbf{a}\cdot\nabla\varphi\psi = \langle \mathbf{a}\cdot\nabla\varphi\,,\,\psi\rangle_X \geq |||\mathbf{a}\cdot\nabla\varphi|||_*\,.$$

Set

$$\Psi = \varphi + \frac{1}{1+\max\{c_b,1\}}\,|||\varphi|||\,\psi.$$

The bi-linearity and coercivity of B then yield

$$B(\varphi,\Psi) = B(\varphi,\varphi) + \frac{1}{1+\max\{c_b,1\}}\,|||\varphi|||\,B(\varphi,\psi) \geq |||\varphi|||^2 + \frac{1}{1+\max\{c_b,1\}}\,|||\varphi|||\,B(\varphi,\psi).$$

The definition of ψ and Assumption (A3) on the other hand imply that

$$B(\varphi,\psi) = \int_\Omega \mathbf{a}\cdot\nabla\varphi\psi + \varepsilon\int_\Omega \nabla\varphi\cdot\nabla\psi + \int_\Omega b\varphi\psi \geq |||\mathbf{a}\cdot\nabla\varphi|||_* - \max\{c_b,1\}\,|||\varphi|||$$

and therefore

$$B(\varphi,\Psi) \geq \frac{1}{1+\max\{c_b,1\}}\left\{|||\varphi|||^2 + |||\mathbf{a}\cdot\nabla\varphi|||_*\,|||\varphi|||\right\}.$$

Since

$$|||\Psi||| \leq \frac{2+\max\{c_b,1\}}{1+\max\{c_b,1\}}\,|||\varphi|||\,,$$

this estimate yields

$$\sup_{\psi \in Y \setminus \{0\}} \frac{B(\varphi, \psi)}{|||\psi|||} \geq \frac{B(\varphi, \Psi)}{|||\Psi|||} \geq \frac{1}{2 + \max\{c_b, 1\}} \{|||\varphi||| + |||\mathbf{a} \cdot \nabla u|||_*\}.$$

Since $\varphi \in X$ was arbitrary, this proves the inf–sup condition. □

Remark 4.18 A similar result is established in [**233**]. There, however, the spaces X and Y are equipped with the same norm which is an interpolation norm between $|||\cdot|||$ and $|||\cdot||| + |||\mathbf{a} \cdot \nabla \cdot|||_*$. The present result is better suited for our purposes, since we thus have to estimate quasi-interpolates and local cut-off functions with respect to the energy norm and can use the results of Chapter 3.

4.4.2 Discrete Problem

The finite element discretisation of problem (4.13) fits into the framework of Section 4.1 (p. 151) with

$$X_{\mathcal{T}} = Y_{\mathcal{T}} = S_D^{k,0}(\mathcal{T}),$$
$$B_{\mathcal{T}}(\varphi_{\mathcal{T}}, \psi_{\mathcal{T}}) = B(\varphi_{\mathcal{T}}, \psi_{\mathcal{T}}) + \sum_{K \in \mathcal{T}} \vartheta_K \int_K \{-\varepsilon \Delta \varphi_{\mathcal{T}} + \mathbf{a} \cdot \nabla \varphi_{\mathcal{T}} + b \varphi_{\mathcal{T}}\} \mathbf{a} \cdot \nabla \psi_{\mathcal{T}},$$
$$\langle \ell_{\mathcal{T}}, \psi_{\mathcal{T}} \rangle_{Y_{\mathcal{T}}} = \langle \ell, \psi_{\mathcal{T}} \rangle_Y + \sum_{K \in \mathcal{T}} \vartheta_K \int_K f \mathbf{a} \cdot \nabla \psi_{\mathcal{T}}.$$

The ϑ_K are non-negative stabilisation parameters. We will always assume that

$$\vartheta_K \|\mathbf{a}\|_{\infty;K} \leq c_S h_K \tag{4.15}$$

holds for all elements $K \in \mathcal{T}$ with a constant c_S which is independent of ε and any mesh-size. The choice $\vartheta_K = 0$ for all K yields the standard Galerkin discretisation; the choice $\vartheta_K > 0$ for all K corresponds to the SUPG discretisations [**139, 153, 154**]. Condition (4.15) is satisfied for all choices of ϑ_K used in practice.

Assumptions (A3), (A4), and (4.15) and standard arguments for SUPG discretisations imply that $B_{\mathcal{T}}$ fulfils the conditions of Proposition 4.5 (p. 154) and that, correspondingly, the discrete problem admits a unique solution.

4.4.3 Residual A Posteriori Error Estimates

The residual $R = f - Lu_{\mathcal{T}}$ admits an L^2-representation (3.29) (p. 132) with

$$r|_K = f + \varepsilon \Delta u_{\mathcal{T}} - \mathbf{a} \cdot \nabla u_{\mathcal{T}} - b u_{\mathcal{T}},$$
$$j|_E = \begin{cases} -\mathbb{J}_E(\varepsilon \mathbf{n}_E \cdot \nabla u_{\mathcal{T}}) & \text{if } E \in \mathcal{E}_\Omega, \\ g - \varepsilon \mathbf{n}_E \cdot \nabla u_{\mathcal{T}} & \text{if } E \in \mathcal{E}_{\Gamma_N}, \\ 0 & \text{if } E \in \mathcal{E}_{\Gamma_D} \end{cases}$$

for all elements K and faces E. For the lower error bounds, we introduce piecewise polynomial approximations $f_{\mathcal{T}}$, $\mathbf{a}_{\mathcal{T}}$, $b_{\mathcal{T}}$, and $g_{\mathcal{E}}$ of the data f, $\mathbf{a}$, b, and g, respectively, and define $r_{\mathcal{T}}$ and $j_{\mathcal{E}}$ as r and j with f,

a, b, and g replaced by these approximations. As usual, this approximation will introduce an additional data error.

Due to the stabilisation terms, the Galerkin orthogonality (3.30) (p. 132) does not hold. Hence, we proceed as in Section 3.8.4 (p. 146) and choose the quasi-interpolation operator $I_{\mathcal{T}}$ of Proposition 3.33 (p. 109) as restriction operator $Q_{\mathcal{T}}$. Observing that

$$L_{\mathcal{T}} u_{\mathcal{T}} = \ell_{\mathcal{T}}$$

and taking into account the definitions of $B_{\mathcal{T}}$ and $\ell_{\mathcal{T}}$, we then have for every $w \in Y$

$$\begin{aligned}\left\langle I_{\mathcal{T}}^{*}(\ell - Lu_{\mathcal{T}}), w \right\rangle_{Y} &= \langle \ell - Lu_{\mathcal{T}}, I_{\mathcal{T}} w \rangle_{Y} \\ &= \sum_{K \in \mathcal{T}} \vartheta_K \int_K \{ f + \varepsilon \Delta u_{\mathcal{T}} - \mathbf{a} \cdot \nabla u_{\mathcal{T}} - b u_{\mathcal{T}} \} \, \mathbf{a} \cdot \nabla (I_{\mathcal{T}} w) \\ &= \sum_{K \in \mathcal{T}} \vartheta_K \int_K r \mathbf{a} \cdot \nabla (I_{\mathcal{T}} w).\end{aligned}$$

From the Cauchy–Schwarz inequality we conclude that

$$\int_K r\mathbf{a} \cdot \nabla (I_{\mathcal{T}} w) \leq \|r\|_K \, \|\mathbf{a}\|_{\infty;K} \left\| \nabla (I_{\mathcal{T}} w) \right\|_K$$

holds for every element K. For the estimation of $\left\| \nabla (I_{\mathcal{T}} w) \right\|_K$, we recall the definition (3.4) (p. 82): $\hbar_\omega = \min\left\{ \varepsilon^{-\frac{1}{2}} \operatorname{diam}(\omega), \beta^{-\frac{1}{2}} \right\}$ with the obvious modification if $\beta = 0$. A standard inverse estimate, the definition of the energy norm, Proposition 3.33 (p. 109), and Remark 3.36 (p. 112) then yield for every element K

$$\begin{aligned}\left\| \nabla (I_{\mathcal{T}} w) \right\|_K &\leq c_I \min\left\{ \varepsilon^{-\frac{1}{2}}, h_K^{-1} \beta^{-\frac{1}{2}} \right\} |||I_{\mathcal{T}} w|||_K \\ &= c_I h_K^{-1} \hbar_K \, |||I_{\mathcal{T}} w|||_K \\ &\leq c_I c_A h_K^{-1} \hbar_K \, |||w|||_{\tilde{\omega}_K}\end{aligned}$$

with constants c_I and c_A which only depend on the shape parameter $C_{\mathcal{T}}$ of $\mathcal{T}$. These estimates and assumption (4.15) imply that

$$\left\| I_{\mathcal{T}}^{*}(\ell - Lu_{\mathcal{T}}) \right\|_{Y^{*}} \leq c_C \left\{ \sum_{K \in \mathcal{T}} \hbar_K^2 \, \|r\|_K^2 \right\}^{\frac{1}{2}}.$$

The constant c_C only depends on the shape parameter $C_{\mathcal{T}}$, the constant in assumption (4.15), and $\|\mathbf{a}\|_\infty$.

This estimate, Theorem 3.57 (p. 135), Theorem 3.59 (p. 137), Section 3.8.4 (p. 146), and Proposition 4.17 verify the assumptions of Theorem 4.7 (p. 156). It therefore yields the following a posteriori error estimates.

Theorem 4.19 *Denote by u the unique weak solution of problem (4.13) and by $u_{\mathcal{T}}$ its finite element approximation. Define for every element $K \in \mathcal{T}$ the* residual a posteriori error indicator *$\eta_{R,K}$ by*

$$\eta_{R,K} = \Bigg\{ \hbar_K^2 \left\| f_{\mathcal{T}} + \varepsilon \Delta u_{\mathcal{T}} - \mathbf{a} \cdot \nabla_{\mathcal{T}} \cdot \nabla u_{\mathcal{T}} - b_{\mathcal{T}} u_{\mathcal{T}} \right\|_K^2 + \frac{1}{2} \sum_{E \in \mathcal{E}_{K,\Omega}} \varepsilon^{-\frac{1}{2}} \hbar_E \left\| \mathbb{J}_E(\varepsilon \mathbf{n}_E \cdot \nabla u_{\mathcal{T}}) \right\|_E^2$$

$$+ \sum_{E \in \mathcal{E}_{K,\Gamma_N}} \varepsilon^{-\frac{1}{2}} \hbar_E \left\| g_{\mathcal{E}} - \varepsilon \mathbf{n}_E \cdot \nabla u_{\mathcal{T}} \right\|_E^2 \Bigg\}^{\frac{1}{2}}$$

and the data error indicator *θ_K by*

$$\theta_K = \left\{ \hbar_K^2 \left\| f - f_{\mathcal{T}} + (\mathbf{a}_{\mathcal{T}} - \mathbf{a}) \cdot \nabla u_{\mathcal{T}} + (b_{\mathcal{T}} - b) u_{\mathcal{T}} \right\|_K^2 + \sum_{E \in \mathcal{E}_K \cap \mathcal{E}_{\Gamma_N}} \varepsilon^{-\frac{1}{2}} \hbar_E \left\| g - g_{\mathcal{E}} \right\|_E^2 \right\}^{\frac{1}{2}},$$

where $\hbar_S$, $S \in \mathcal{T} \cup \mathcal{E}$, is given by equation (3.4) (p. 82) and where $f_{\mathcal{T}} \in S^{k,-1}(\mathcal{T})$, $\mathbf{a}_{\mathcal{T}} \in S^{k,-1}(\mathcal{T})^d$, $b_{\mathcal{T}} \in S^{k,-1}(\mathcal{T})$, and $g_{\mathcal{E}} \in S^{k,-1}(\mathcal{E})$ are approximations of the data f, $\mathbf{a}$, b, and g, respectively. Then the error $u - u_{\mathcal{T}}$ can be bounded from above by

$$|||u - u_{\mathcal{T}}||| + \left\| \left| \mathbf{a} \cdot \nabla (u - u_{\mathcal{T}}) \right| \right\|_* \leq c^* \left\{ \sum_{K \in \mathcal{T}} \left[\eta_{R,K}^2 + \theta_K^2 \right] \right\}^{\frac{1}{2}}$$

and from below by

$$\left\{ \sum_{K \in \mathcal{T}} \eta_{R,K}^2 \right\}^{\frac{1}{2}} \leq c_* \left[|||u - u_{\mathcal{T}}||| + \left\| \left| \mathbf{a} \cdot \nabla (u - u_{\mathcal{T}}) \right| \right\|_* + \left\{ \sum_{K \in \mathcal{T}} \theta_K^2 \right\}^{\frac{1}{2}} \right].$$

The constant c^* only depends on the shape parameter $C_{\mathcal{T}}$ of $\mathcal{T}$; the constant c_* depends on $C_{\mathcal{T}}$ and the polynomial degree k. Both constants are independent of ε.

Theorem 4.19 in particular implies that the error indicator $\eta_{R,K}$ yields upper bounds for the energy norm of the error. The following result shows that it also yields lower bounds for $|||u - u_{\mathcal{T}}|||$ which are even local ones. This, however, has to be paid for by a factor which depends on the *mesh-Péclet-number* $\varepsilon^{-1} h_K \|\mathbf{a}\|_{\infty;K}$.

Theorem 4.20 *With the conditions and notation of Theorem 4.19, the energy norm of the error can be bounded from above by*

$$|||u - u_{\mathcal{T}}||| \leq \tilde{c}^* \left\{ \sum_{K \in \mathcal{T}} \left[\eta_{R,K}^2 + \theta_K^2 \right] \right\}^{\frac{1}{2}}$$

and for every element $K \in \mathcal{T}$ from below by

$$\eta_{R,K} \leq \tilde{c}_* \left[\left(1 + \varepsilon^{-\frac{1}{2}} \hbar_K \|\mathbf{a}\|_{\infty;\omega_K} \right) |||u - u_{\mathcal{T}}||| + \theta_K \right].$$

Again, the constant $\tilde{c}^*$ only depends on the shape parameter $C_{\mathcal{T}}$; the constant $\tilde{c}_*$ depends on $C_{\mathcal{T}}$ and the polynomial degree k. Both constants are independent of ε.

Proof Due to Theorem 4.19, we only have to prove the lower error bound. To do this we apply Theorem 3.59 (p. 137) with $\mathcal{T}' = \{K\}$. It yields a function $w \in L^2(\omega_K)$ with

$$|||w|||_{\omega_K} \leq c\eta_{R,K}$$

and

$$\eta_{R,K}^2 \leq \int_{\omega_K} r_{\mathcal{T}} w + \int_{\partial K} j_{\mathcal{E}} w = \langle R, w\rangle + \int_{\omega_K} (r_{\mathcal{T}} - r)w + \int_{\partial K} (j_{\mathcal{E}} - j)w.$$

Retracing the proof of Theorem 3.59 and taking into account Theorem 3.58 (p. 135), we conclude that

$$\|w\|_{\omega_K} \leq c\hbar_K \, |||w|||_{\omega_K} .$$

We therefore obtain for the data errors

$$\int_{\omega_K} (r_{\mathcal{T}} - r)w + \int_{\partial K} (j_{\mathcal{E}} - j)w \leq c\eta_{R,K}\theta_K.$$

The definition of the residual R, of the bi-linear form B, and of the energy norm $|||\cdot|||$, on the other hand, yield

$$\begin{aligned}
\langle R, w\rangle &= B(u - u_{\mathcal{T}}, w) \\
&\leq |||u - u_{\mathcal{T}}|||_{\omega_K} \, |||w|||_{\omega_K} + \|\mathbf{a}\|_{\infty;\omega_K} \left\|\nabla(u - u_{\mathcal{T}})\right\|_{\omega_K} \|w\|_{\omega_K} \\
&\leq \left(1 + \|\mathbf{a}\|_{\infty;\omega_K} \varepsilon^{-\frac{1}{2}} \hbar_K\right) |||u - u_{\mathcal{T}}|||_{\omega_K} \, |||w|||_{\omega_K} .
\end{aligned}$$

This proves the lower bound for the error. □

Remark 4.21 A comparison of Theorems 4.19 and 4.20 reveals that the indicator $\eta_{R,K}$ may overestimate the energy norm of the error. If this happens, the dual norm of the error's convective derivative must be large. This indicates an interior or boundary layer which is not sufficiently well resolved.

4.4.4 The Hyper-Circle Method

After approximating the data f, $\mathbf{a}$, b, and g by piecewise polynomials $f_{\mathcal{T}}$, $\mathbf{a}_{\mathcal{T}}$, $b_{\mathcal{T}}$, and $g_{\mathcal{E}}$, respectively, we may apply Proposition 3.62 (p. 141) and Theorem 3.63 (p. 144) to problem (4.13). We thus obtain an $H(\mathrm{div})$-lifting $\rho_{\mathcal{T}}$ of the residual. Estimate (3.38) (p. 144) of Theorem 3.63 (p. 144) immediately yields the upper bound

$$|||u - u_{\mathcal{T}}||| + \left\|\mathbf{a} \cdot \nabla(u - u_{\mathcal{T}})\right\|_* \leq \varepsilon^{-\frac{1}{2}} \|\rho_{\mathcal{T}}\| + c^*\theta_{\mathcal{T}}$$

where the data error $\theta_{\mathcal{T}}$ is as in Theorem 4.19 and where the constant c^* only depends on the shape parameter $C_{\mathcal{T}}$ of $\mathcal{T}$ and the polynomial degree of $f_{\mathcal{T}}$, $\mathbf{a}_{\mathcal{T}}$, $b_{\mathcal{T}}$, and $g_{\mathcal{E}}$.

Estimate (3.39) (p. 144) of Theorem 3.63 combined with the lower bounds of Theorem 3.59 (p. 137) on the other hand gives the lower bound

$$\varepsilon^{-\frac{1}{2}} \|\rho_{\mathcal{T}}\| \le c_* \max_{z\in\mathcal{N}} \max\left\{\varepsilon^{-\frac{1}{2}} \operatorname{diam}(\omega_z), 1\right\} \left\{|||u - u_{\mathcal{T}}||| + \left|\left|\left|\mathbf{a} \cdot \nabla(u - u_{\mathcal{T}})\right|\right|\right|_* + \theta_{\mathcal{T}}\right\}$$

with a constant c_* which only depends on the shape parameter $C_{\mathcal{T}}$ of $\mathcal{T}$ and the polynomial degree k. This generalises Theorem 1.41 (p. 52) concerning the hyper-circle method and shows that it is not fully robust. As for reaction–diffusion problems, this lack of robustness is not a technical but a structural drawback. It is due to the fact that the lifting (3.37) (p. 141) only incorporates the principal part of the differential operator of problem (4.13) and gives no control on the dominant lower order terms.

4.4.5 Auxiliary Local Discrete Problems

Due to the dominant convection, special attention must be paid to the well-posedness of the auxiliary problems when deriving error indicators which are based on the solution of auxiliary local discrete problems. As an example we extend to problem (4.13) the indicators $\eta_{D,K}$ and $\eta_{N,K}$ of Section 1.7 (p. 25) which are based on the solution of local Dirichlet and Neumann problems, respectively.

4.4.5.1 Local Dirichlet problems

We start with the indicator $\eta_{D,K}$. To this end we fix an arbitrary element $K \in \mathcal{T}$ and set

$$\widetilde{V}_K = \operatorname{span}\left\{\psi_{K'} v, \psi_{E,\vartheta}\sigma : K' \subset \omega_K, E \subset \partial K \setminus \Gamma_D, v \in R_k(K'), \sigma \in R_k(E)\right\},$$

where $\vartheta = \varepsilon^{\frac{1}{2}} h_E^{-1} \hbar_E$ is as in equation (3.32) (p. 136). Then we consider the problem:

find $v \in \widetilde{V}_K$ such that

$$\begin{aligned} \int_{\omega_K} \{\varepsilon \nabla v \cdot \nabla w + \mathbf{a} \cdot \nabla v w + b v w\} &= \int_{\omega_K} f_{\mathcal{T}} w + \int_{\partial K \cap \Gamma_N} g_{\mathcal{E}} w \\ &\quad - \int_{\omega_K} \{\varepsilon \nabla u_{\mathcal{T}} \cdot \nabla w + \mathbf{a}_{\mathcal{T}} \cdot \nabla u_{\mathcal{T}} w + b_{\mathcal{T}} u_{\mathcal{T}} w\} \end{aligned} \tag{4.16}$$

for all $w \in \widetilde{V}_K$.

The following proposition is a discrete analogue of Proposition 4.17. In particular it implies the unique solvability of problem (4.16).

Lemma 4.22 *Denote by $\pi_{\widetilde{V}_K}$ the L^2-projection onto $\widetilde{V}_K$. Then the following estimates are valid*

$$\sup_{v\in\widetilde{V}_K\setminus\{0\}} \sup_{w\in\widetilde{V}_K\setminus\{0\}} \frac{\int_{\omega_K} \{\varepsilon \nabla v \cdot \nabla w + \mathbf{a} \cdot \nabla v w + b v w\}}{\left\{|||v|||_{\omega_K}^2 + \hbar_K^2 \left\|\pi_{\widetilde{V}_K}(\mathbf{a} \cdot \nabla v)\right\|_{\omega_K}^2\right\}^{\frac{1}{2}} |||w|||_{\omega_K}} \le \sqrt{2} \max\{c_a, c_b, 1\}$$

and

$$\inf_{v\in\widetilde{V}_K\setminus\{0\}} \sup_{w\in\widetilde{V}_K\setminus\{0\}} \frac{\displaystyle\int_{\omega_K}\{\varepsilon\nabla v\cdot\nabla w+\mathbf{a}\cdot\nabla v w+bvw\}}{\left\{|||v|||_{\omega_K}^2+\hbar_K^2\left\|\pi_{\widetilde{V}_K}(\mathbf{a}\cdot\nabla v)\right\|_{\omega_K}^2\right\}^{\frac{1}{2}}|||w|||_{\omega_K}} \geq \frac{1}{1+3c_a^2\max\{c_b,1\}^2}.$$

The constant c_b is that of Assumption (A3). The constant c_a only depends on the polynomial degree k and the shape parameter $C_{\mathcal{T}}$, cf. (4.17) below.

Proof The definition of $\pi_{\widetilde{V}_K}$ implies, for all $v, w \in \widetilde{V}_K$,

$$\int_{\omega_K}(\mathbf{a}\cdot\nabla v)w=\int_{\omega_K}\pi_{\widetilde{V}_K}(\mathbf{a}\cdot\nabla v)w.$$

This identity, Assumption (A3), and the definition of the energy norm yield, for all $v, w \in \widetilde{V}_K$,

$$\int_{\omega_K}\{\varepsilon\nabla v\cdot\nabla w+\mathbf{a}\cdot\nabla v w+bvw\}\leq\max\{c_b,1\}\,|||v|||_{\omega_K}\,|||w|||_{\omega_K}+\left\|\pi_{\widetilde{V}_K}(\mathbf{a}\cdot\nabla v)\right\|_{\omega_K}\|w\|_{\omega_K}.$$

Since the functions in $\widetilde{V}_K$ vanish at the vertices of K, the norms $h_K\|\nabla\cdot\|_{\omega_K}$ and $\|\cdot\|_{\omega_K}$ are equivalent on $\widetilde{V}_K$. Therefore $\hbar_K\,|||\cdot|||_{\omega_K}$ and $\|\cdot\|_{\omega_K}$ are also equivalent norms on $\widetilde{V}_K$, i.e. there is a constant $c_a \geq 1$ which only depends on the polynomial degree k and the shape parameter $C_{\mathcal{T}}$ such that

$$\frac{1}{c_a}\hbar_K\,|||w|||_{\omega_K}\leq\|w\|_{\omega_K}\leq c_a\hbar_K\,|||w|||_{\omega_K} \tag{4.17}$$

holds for all $w \in \widetilde{V}_K$. These inequalities prove the upper bound of Lemma 4.22.

For the proof of the lower bound, we proceed as in the proof of Proposition 4.17. We consider an arbitrary function $v \in \widetilde{V}_K$ and set

$$w_\gamma=v+\gamma\hbar_K^2\pi_{\widetilde{V}_K}(\mathbf{a}\cdot\nabla v).$$

The constant γ is arbitrary at present and will be determined below. The norm-equivalence (4.17) implies

$$\begin{aligned}|||w_\gamma|||_{\omega_K}&\leq|||v|||_{\omega_K}+c_a\gamma\hbar_K\left\|\pi_{\widetilde{V}_K}(\mathbf{a}\cdot\nabla v)\right\|_{\omega_K}\\&\leq\left\{1+c_a^2\gamma^2\right\}^{\frac{1}{2}}\left\{|||v|||_{\omega_K}^2+\hbar_K^2\left\|\pi_{\widetilde{V}_K}(\mathbf{a}\cdot\nabla v)\right\|_{\omega_K}^2\right\}^{\frac{1}{2}}.\end{aligned}$$

Assumptions (A3) and (A4), integration by parts, and the definition of the energy norm, on the other hand, imply that the bi-linear form on the left-hand side of problem (4.16) is coercive on $\widetilde{V}_K$ with constant 1. Inserting w_γ as a test function in this bi-linear form therefore yields

$$\begin{aligned}
&\int_{\omega_K} \left\{\varepsilon \nabla v \cdot \nabla w_\gamma + \mathbf{a} \cdot \nabla v w_\gamma + b v w_\gamma\right\} \\
&\quad = \int_{\omega_K} \{\varepsilon \nabla v \cdot \nabla v + \mathbf{a} \cdot \nabla v v + b v v\} \\
&\qquad + \gamma \hbar_K^2 \int_{\omega_K} \left\{\varepsilon \nabla v \cdot \nabla(\pi_{\widetilde{V}_K}(\mathbf{a} \cdot \nabla v)) + (\mathbf{a} \cdot \nabla v)\, \pi_{\widetilde{V}_K}(\mathbf{a} \cdot \nabla v) + b v\, \pi_{\widetilde{V}_K}(\mathbf{a} \cdot \nabla v)\right\} \\
&\quad \geq |||v|||_{\omega_K}^2 + \gamma \hbar_K^2 \left\|\pi_{\widetilde{V}_K}(\mathbf{a} \cdot \nabla v)\right\|_{\omega_K}^2 - c_a \max\{c_b, 1\}\, \gamma \hbar_K\, |||v|||_{\omega_K} \left\|\pi_{\widetilde{V}_K}(\mathbf{a} \cdot \nabla v)\right\|_{\omega_K} \\
&\quad \geq \left(1 - \frac{1}{2} c_a^2 \max\{c_b, 1\}^2 \gamma\right) |||v|||_{\omega_K}^2 + \frac{1}{2} \gamma \hbar_K^2 \left\|\pi_{\widetilde{V}_K}(\mathbf{a} \cdot \nabla v)\right\|_{\omega_K}^2 .
\end{aligned}$$

Now, we choose

$$\gamma = \frac{2}{1 + c_a^2 \max\{c_b, 1\}^2}$$

and obtain

$$\begin{aligned}
&\sup_{w \in \widetilde{V}_K \setminus \{0\}} \frac{\int_{\omega_K} \{\varepsilon \nabla v \cdot \nabla w + \mathbf{a} \cdot \nabla v w + b v w\}}{\left\{|||v|||_{\omega_K}^2 + \hbar_K^2 \left\|\pi_{\widetilde{V}_K}(\mathbf{a} \cdot \nabla v)\right\|_{\omega_K}^2\right\}^{\frac{1}{2}} |||w|||_{\omega_K}} \\
&\quad \geq \frac{\int_{\omega_K} \left\{\varepsilon \nabla v \cdot \nabla w_\gamma + \mathbf{a} \cdot \nabla v w_\gamma + b v w_\gamma\right\}}{\left\{|||v|||_{\omega_K}^2 + \hbar_K^2 \left\|\pi_{\widetilde{V}_K}(\mathbf{a} \cdot \nabla v)\right\|_{\omega_K}^2\right\}^{\frac{1}{2}} |||w_\gamma|||_{\omega_K}} \\
&\quad \geq \frac{1}{\left(1 + c_a^2 \max\{c_b, 1\}^2\right) \left\{1 + \dfrac{4 c_a^2}{\left(1 + c_a^2 \max\{c_b, 1\}^2\right)^2}\right\}^{\frac{1}{2}}} .
\end{aligned}$$

Since

$$\begin{aligned}
\left(1 + c_a^2 \max\{c_b, 1\}^2\right) \left\{1 + \frac{4 c_a^2}{\left(1 + c_a^2 \max\{c_b, 1\}^2\right)^2}\right\}^{\frac{1}{2}} &= \left\{\left(1 + c_a^2 \max\{c_b, 1\}^2\right)^2 + 4 c_a^2\right\}^{\frac{1}{2}} \\
&\leq 1 + 3 c_a^2 \max\{c_b, 1\}^2
\end{aligned}$$

this proves the inf–sup condition. □

We denote by $\widetilde{v}_K \in \widetilde{V}_K$ the unique solution of problem (4.16) and define the error indicator $\eta_{D,K}$ by

$$\eta_{D,K} = \left\{|||\widetilde{v}_K|||_{\omega_K}^2 + \hbar_K^2 \left\|\pi_{\widetilde{V}_K}(\mathbf{a} \cdot \nabla \widetilde{v}_K)\right\|_{\omega_K}^2\right\}^{\frac{1}{2}} .$$

Theorem 4.23 *With the notation and assumptions of Theorem 4.19, there are two constants $c_\dagger$ and $c^\dagger$ which only depend on the polynomial degree k and on the shape parameter $C_{\mathcal{T}}$ such that the estimate*

$$\frac{1}{c_\dagger}\eta_{R,K} \le \eta_{D,K} \le c^\dagger \left\{ \sum_{K' \subset \omega_K} \eta_{R,K'}^2 \right\}^{\frac{1}{2}}$$

holds for all elements $K \in \mathcal{T}$. Moreover $\eta_{D,K}$ yields the upper bound for the error

$$|||u - u_{\mathcal{T}}||| + \left\|| \mathbf{a} \cdot \nabla(u - u_{\mathcal{T}}) \right\||_* \le \widetilde{c}^* \left\{ \sum_{K \in \mathcal{T}} \left[\eta_{D,K}^2 + \theta_K^2 \right] \right\}^{\frac{1}{2}}$$

and the lower bound for the error

$$\left\{ \sum_{K \in \mathcal{T}} \eta_{D,K}^2 \right\}^{\frac{1}{2}} \le \widetilde{c}_* \left[|||u - u_{\mathcal{T}}||| + \left\|| \mathbf{a} \cdot \nabla(u - u_{\mathcal{T}}) \right\||_* + \left\{ \sum_{K \in \mathcal{T}} \theta_K^2 \right\}^{\frac{1}{2}} \right],$$

where θ_K is as in Theorem 4.19. The constants $\widetilde{c}_*$ and $\widetilde{c}^*$ only depend on the polynomial degree k and shape parameter $C_{\mathcal{T}}$, but are independent of ε.

Proof In view of Theorem 4.19, we only have to prove the estimates comparing $\eta_{R,K}$ and $\eta_{D,K}$. Integration by parts of the right-hand side of problem (4.16) yields, for all $w \in \widetilde{V}_K$,

$$\int_{\omega_K} f_{\mathcal{T}} w + \int_{\partial K \cap \Gamma_N} g_{\mathcal{E}} w - \int_{\omega_K} \{\varepsilon \nabla u_{\mathcal{T}} \cdot \nabla w + \mathbf{a}_{\mathcal{T}} \cdot \nabla u_{\mathcal{T}} w + b_{\mathcal{T}} u_{\mathcal{T}} w\} = \int_{\omega_K} r_{\mathcal{T}} w + \int_{\partial K} j_{\mathcal{E}} w.$$

Hence, we have

$$\sup_{w \in \widetilde{V}_K} \frac{1}{|||w|||_{\omega_K}} \int_{\omega_K} \{\varepsilon \nabla \widetilde{v}_K \cdot \nabla w + \mathbf{a} \cdot \nabla \widetilde{v}_K w + b \widetilde{v}_K w\} = \sup_{w \in \widetilde{V}_K} \frac{1}{|||w|||_{\omega_K}} \left\{ \int_{\omega_K} r_{\mathcal{T}} w + \int_{\partial K} j_{\mathcal{E}} w \right\}. \tag{4.18}$$

Lemma 4.22 implies that the left-hand side of equation (4.18) is bounded from above and from below by constant multiples of

$$\left\{ |||\widetilde{v}_K|||_{\omega_K}^2 + \hbar_K^2 \left\| \pi_{\widetilde{V}_K} (\mathbf{a} \cdot \nabla \widetilde{v}_K) \right\|_{\omega_K}^2 \right\}^{\frac{1}{2}}.$$

The norm equivalence (4.17) yields that the right-hand side of equation (4.18) is bounded from above by a constant multiple of

$$\left\{ \sum_{K' \subset \omega_K} \eta_{R,K'}^2 \right\}^{\frac{1}{2}}.$$

Theorem 3.59 (p. 137), on the other hand, shows that the right-hand side of equation (4.18) can be bounded from below by a constant multiple of

$$\left\{ \sum_{K' \subset \omega_K} \hbar_{K'}^2 \, \|r_{\mathcal{T}}\|_{K'}^2 + \sum_{E \in \mathcal{E}_K \setminus \mathcal{E}_{\Gamma_D}} \varepsilon^{-\frac{1}{2}} \hbar_E \, \|j_{\mathcal{E}}\|_E^2 \right\}^{\frac{1}{2}}$$

which in turn is an upper bound for $\eta_{R,K}$. □

4.4.5.2 Local Neumann problems

Next, we extend the indicator $\eta_{N,K}$ which is based on the solution of an auxiliary local discrete Neumann problem. The main difficulty now is to ensure the coercivity of the corresponding bilinear form. To achieve this, we have to approximate the reaction b and the convection $\mathbf{a}$ by discrete quantities such that Assumption (A3) remains valid for this approximation and such that the normal component of the discrete convection is piecewise constant on the faces of $\mathcal{T}$.

To this end, we fix the approximation $b_{\mathcal{T}}$ to be the L^2-projection of b onto the space $S^{0,-1}(\mathcal{T})$ of piecewise constant functions corresponding to $\mathcal{T}$. For the approximation of the convection we choose for $\mathbf{a}_{\mathcal{T}}$ the L^2-projection of $\mathbf{a}$ onto the lowest order *Raviart–Thomas space* $\mathrm{RT}_0(\mathcal{T})$ corresponding to $\mathcal{T}$ which is defined by [**90**, Sections III.3.1 and III 3.2]

$$\mathbf{a}_{\mathcal{T}}|_K \in R_0(K)^d + \begin{pmatrix} x_1 \\ \vdots \\ x_d \end{pmatrix} R_0(K) \quad \text{for all } K \in \mathcal{T},$$

$$\int_E \mathbf{n}_E \cdot \mathbf{a}_{\mathcal{T}} = \int_E \mathbf{n}_E \cdot \mathbf{a} \qquad \text{for all } E \in \mathcal{E}.$$

Note that $\mathbf{n}_E \cdot \mathbf{a}_{\mathcal{T}}$ is piecewise constant on the faces. Therefore, we can associate with each element K the collection of its outflow faces by setting

$$\mathcal{E}_K^+ = \{E \in \mathcal{E}_K : \mathbf{n}_K \cdot \mathbf{a}_{\mathcal{T}} \geq 0\},$$

where $\mathbf{n}_K$ denotes the outward normal to K.

With these definitions, we set

$$V_K = \operatorname{span}\left\{\psi_K v, \psi_{E,\vartheta}\sigma : v \in R_k(K), E \in \mathcal{E}_K^+, \sigma \in R_k(E)\right\}$$

with $\vartheta = \varepsilon^{\frac{1}{2}} h_E^{-1} \hbar_E$ and consider the problem:

find $v \in V_K$ such that

$$\int_K \{\varepsilon \nabla v \cdot \nabla w + \mathbf{a}_{\mathcal{T}} \cdot \nabla v w + b_{\mathcal{T}} v w\} = \int_K r_{\mathcal{T}} w + \sum_{E \in \mathcal{E}_K^+} \int_E j_{\mathcal{E}} w \tag{4.19}$$

for all $w \in V_K$.

The following proposition is an analogue of Lemma 4.22. In particular it implies the unique solvability of problem (4.19).

Lemma 4.24 *For all elements K, we have*

$$\sup_{v\in V_K\setminus\{0\}} \sup_{w\in V_K\setminus\{0\}} \frac{\int_K \{\varepsilon\nabla v\cdot\nabla w + \mathbf{a}_{\mathcal{T}}\cdot\nabla v w + b_{\mathcal{T}} v w\}}{\left\{|||v|||_K^2 + \hbar_K^2\,\|\mathbf{a}_{\mathcal{T}}\cdot\nabla v\|_K^2\right\}^{\frac{1}{2}}\,|||w|||_K} \le \sqrt{2}\max\{\widehat{c_a}, c_b, 1\}$$

and

$$\inf_{v\in V_K\setminus\{0\}} \sup_{w\in V_K\setminus\{0\}} \frac{\int_K \{\varepsilon\nabla v\cdot\nabla w + \mathbf{a}_{\mathcal{T}}\cdot\nabla v w + b_{\mathcal{T}} v w\}}{\left\{|||v|||_K^2 + \hbar_K^2\,\|\mathbf{a}_{\mathcal{T}}\cdot\nabla v_K\|_K^2\right\}^{\frac{1}{2}}\,|||w|||_K} \ge \frac{1}{1+3\widehat{c_a}^2\max\{c_b,1\}^2}.$$

The constant c_b is that of Assumption (A3). The constant $\widehat{c_a}$ only depends on the polynomial degree k and the shape parameter $C_{\mathcal{T}}$, cf. inequality (4.20) below.

Proof Choose an arbitrary element K and keep it fixed in what follows. Since $b_{\mathcal{T}}$ is constant on K, we conclude from Assumption (A3) that

$$\|b_{\mathcal{T}}\|_{\infty;K} = \left|\frac{1}{\mu_d(K)}\int_K b_{\mathcal{T}}\right| = \left|\frac{1}{\mu_d(K)}\int_K b\right| \le c_b\beta.$$

From [**90**, Sections III.3.1 and III 3.2] we know that $\operatorname{div}_K \mathbf{a}_{\mathcal{T}}$, the divergence of the restriction of $\mathbf{a}_{\mathcal{T}}$ to K, is constant on K and satisfies

$$\int_K \operatorname{div}_K \mathbf{a}_{\mathcal{T}} = \int_K \operatorname{div}\mathbf{a}.$$

Hence, we get from Assumption (A3)

$$b_{\mathcal{T}} - \frac{1}{2}\operatorname{div}_K \mathbf{a}_{\mathcal{T}} = \frac{1}{\mu_d(K)}\int_K \left\{b_{\mathcal{T}} - \frac{1}{2}\operatorname{div}_K \mathbf{a}_{\mathcal{T}}\right\} = \frac{1}{\mu_d(K)}\int_K \left\{b - \frac{1}{2}\operatorname{div}\mathbf{a}\right\} \ge \beta.$$

This proves that $b_{\mathcal{T}}$ and $\mathbf{a}_{\mathcal{T}}$ satisfy Assumption (A3) with the same constants β and c_b. Hence, for all $v, w \in V_K$, we obtain as in the proof of Lemma 4.22 the estimate

$$\int_K \{\varepsilon\nabla v\nabla w + \mathbf{a}_{\mathcal{T}}\cdot\nabla v w + b_{\mathcal{T}} v w\} \le \max\{c_b, 1\}\,|||v_K|||_K\,|||w|||_K + \|\mathbf{a}_{\mathcal{T}}\cdot\nabla v\|_K\,\|w\|_K.$$

The same arguments as in the proof of estimate (4.17) imply that there is a constant $\widehat{c_a} \ge 1$ which only depends on the polynomial degree k and on the shape parameter $C_{\mathcal{T}}$ such that

$$\frac{1}{\widehat{c_a}}\hbar_K\,|||w|||_K \le \|w\|_{\omega_K} \le \widehat{c_a}\hbar_K\,|||w|||_{\omega_K} \tag{4.20}$$

holds for all $w \in V_K$. These inequalities prove the upper bound of Lemma 4.24.

For the proof of the lower bound of Lemma 4.24, we only have to check the coercivity of the bi-linear form on the left-hand side of problem (4.19). Once this is done, the inf–sup condition is established with the same arguments as in the proof of Lemma 4.22. Since the functions in V_K vanish on the inflow boundary of K, we have for every $w \in V_K$

$$\begin{aligned}
&\int_K \{\varepsilon \nabla w \cdot \nabla w + \mathbf{a}_{\mathcal{T}} \cdot \nabla w w + b_{\mathcal{T}} w w\} \\
&= \varepsilon \|\nabla w\|_K^2 + \int_K \frac{1}{2} \operatorname{div}_K(\mathbf{a}_{\mathcal{T}} w^2) + \int_K \left\{ b_{\mathcal{T}} - \frac{1}{2} \operatorname{div}_K \mathbf{a}_{\mathcal{T}} \right\} w^2 \\
&= \varepsilon \|\nabla w\|_K^2 + \int_{\partial K} \frac{1}{2} \mathbf{n}_K \cdot \mathbf{a}_{\mathcal{T}} w^2 + \int_K \left\{ b_{\mathcal{T}} - \frac{1}{2} \operatorname{div}_K \mathbf{a}_{\mathcal{T}} \right\} w^2 \\
&\geq \varepsilon \|\nabla w\|_K^2 + \beta \|w\|_K^2 .
\end{aligned}$$

□

We denote by $v_K \in V_K$ the unique solution of problem (4.19) and define the error indicator $\eta_{N,K}$ by

$$\eta_{N,K} = \left\{ |||v_K|||_K^2 + \hbar_K^2 \|\mathbf{a}_{\mathcal{T}} \cdot \nabla v_K\|_K^2 \right\}^{\frac{1}{2}} .$$

Theorem 4.25 *With the notation and assumptions of Theorem 4.19, there are two constants $\widehat{c}_+$ and $\widehat{c}^+$ which only depend on the polynomial degree k and the shape parameter $C_{\mathcal{T}}$ such that the estimate*

$$\frac{1}{\widehat{c}_+} \left\{ \hbar_K^2 \|r_{\mathcal{T}}\|_K^2 + \sum_{E \in \mathcal{E}_K^+} \varepsilon^{-\frac{1}{2}} \hbar_E \left\| j_{\mathcal{E}} \right\|_E^2 \right\}^{\frac{1}{2}} \leq \eta_{N,K} \leq \widehat{c}^+ \eta_{R,K}$$

holds for all elements K. Moreover $\eta_{N,K}$ yields the upper bound for the error

$$|||u - u_{\mathcal{T}}||| + \left\| |\mathbf{a} \cdot \nabla(u - u_{\mathcal{T}}) \right\| |_* \leq \widehat{c}^* \left\{ \sum_{K \in \mathcal{T}} \left[\eta_{N,K}^2 + \theta_K^2 \right] \right\}^{\frac{1}{2}}$$

and the lower bound for the error

$$\left\{ \sum_{K \in \mathcal{T}} \eta_{N,K}^2 \right\}^{\frac{1}{2}} \leq \widehat{c}_* \left[|||u - u_{\mathcal{T}}||| + \left\| |\mathbf{a} \cdot \nabla(u - u_{\mathcal{T}}) \right\| |_* + \left\{ \sum_{K \in \mathcal{T}} \theta_K^2 \right\}^{\frac{1}{2}} \right],$$

where θ_K is as in Theorem 4.19. The constants $\widehat{c}_*$ and $\widehat{c}^*$ only depend on the polynomial degree k and the shape parameter $C_{\mathcal{T}}$, but are independent of ε.

Proof The first estimate is proved with the same arguments as the corresponding estimate of Theorem 4.23. The error bounds follow from the first estimate, Theorem 4.19, and the observation that for every face E, which is not part of the Dirichlet boundary Γ_D, there is at least one element K_E with $E \in \mathcal{E}_{K_E}^+$. □

Remark 4.26 Theorems 4.19, 4.23, and 4.25 were first established in [**271**]. Theorem 4.20 is originally due to [**261**]. Computational comparisons are given in [**213, 214**].

4.5 Anisotropic Meshes

When considering singularly perturbed elliptic equations with dominant reaction or convection terms as in the previous sections, one often encounters solutions with boundary or interior layers. These layers can most efficiently be resolved by so-called *anisotropic elements* which are very small in the direction perpendicular to the layer and relatively large in the directions parallel to the layer. These elements no longer fulfil the requirement of shape-regularity of Sections 1.3 (p. 5) and 3.2 (p. 81), i.e. the ratio $\frac{h_K}{\rho_K}$ may be very large. Nevertheless to obtain reasonable a posteriori error bounds for meshes containing anisotropic elements, we must

- appropriately define the notion of anisotropy,
- identify new geometric quantities which suitably characterise the elements, and
- modify the results concerning quasi-interpolation operators and local cut-off functions such that they take account of the anisotropy.

In what follows we focus on simplicial elements and postpone the discussion of tensorial elements to the final Subsection 4.5.8 (p. 191).

4.5.1 *Anisotropic Triangles and Tetrahedrons*

We enumerate the vertices P_0, P_1, P_2 of a given triangle K such that, cf. Figure 4.1

- P_0P_1 is the longest edge and
- P_0P_2 is the shortest one.

We denote by

- $\mathbf{p}_{1,K}$ the vector $\overrightarrow{P_0P_1}$ and
- $\mathbf{p}_{2,K}$ the vector perpendicular to P_0P_1 pointing to P_2.

With these notations we set

$$h_{i,K} = \left|\mathbf{p}_{i,K}\right| \quad i = 1, 2$$

and

$$h_{\min,K} = \min\left\{h_{1,K}, h_{2,K}\right\} = h_{2,K}.$$

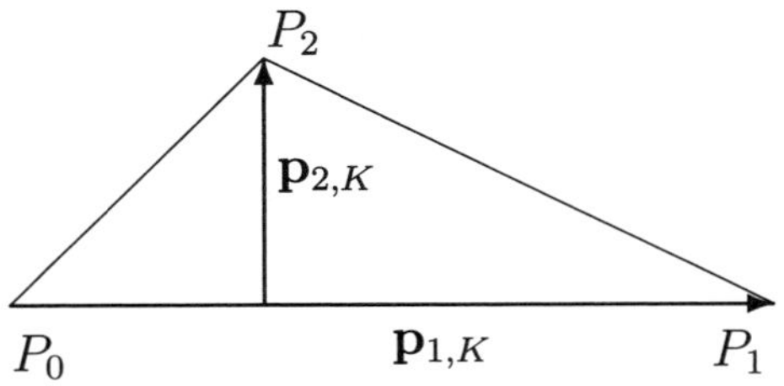

Figure 4.1 Geometric data of an anisotropic triangle

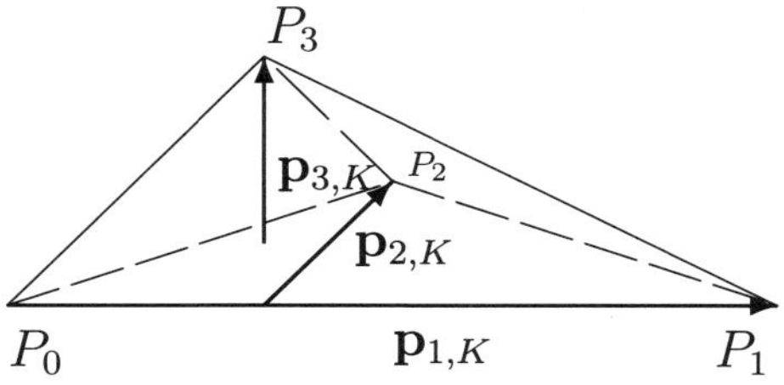

Figure 4.2 Geometric data of an anisotropic tetrahedron

For a tetrahedron K, we enumerate its vertices $P_0, \ldots, P_3$ such that, cf. Figure 4.2

- P_0P_1 is the longest edge,
- the triangle $\Delta P_0P_1P_2$ has largest area among the two triangles adjacent to P_0P_1, and
- P_0P_2 is the shortest edge of $\Delta P_0P_1P_2$.

We then define three vectors $\mathbf{p}_1, \ldots, \mathbf{p}_3$ as follows:

- $\mathbf{p}_{1,K} = \overrightarrow{P_0P_1}$,
- $\mathbf{p}_{2,K}$ is the vector in the plane $P_0P_1P_2$ which is perpendicular to P_0P_1 and which points to P_2,
- $\mathbf{p}_{3,K}$ is the vector which is perpendicular to the plane $P_0P_1P_2$ and which points to P_3.

With these notations we set

$$h_{i,K} = \left|\mathbf{p}_{i,K}\right| \quad i = 1, 2, 3$$

and

$$h_{\min,K} = \min\left\{h_{1,K}, h_{2,K}, h_{3,K}\right\} = h_{3,K}.$$

The condition of shape regularity is now replaced by the following assumptions:

- the number of elements sharing an arbitrary vertex $z \in \mathcal{N}$ is uniformly bounded;
- the ratios $\frac{h_{i,K}}{h_{i,K'}}$ are uniformly bounded for all $1 \le i \le d$ and all elements K and K' which share at least one vertex.

To quantify these assumptions we denote by N_z the number of elements sharing a given vertex $z \in \mathcal{N}$ and define the *anisotropic shape parameter* by

$$C_{\mathcal{T},a} = \max\left\{\max_{z\in\mathcal{N}} N_z, \max_{\substack{K,K'\in\mathcal{T} \\ \mathcal{N}_K\cap\mathcal{N}_{K'}\neq\emptyset}} \max_{1\le i\le d} \frac{h_{i,K}}{h_{i,K'}}\right\}. \tag{4.21}$$

With every triangle or tetrahedron K, we associate three matrices M_K, H_K, and B_K as follows:

- the vectors $\mathbf{p}_{1,K}, \ldots, \mathbf{p}_{d,K}$ are the columns of the matrix M_K;
- $H_K = \operatorname{diag}(h_{1,K}, \ldots, h_{d,K})$;
- the vectors $\overrightarrow{P_0P_1}, \ldots, \overrightarrow{P_0P_d}$ are the columns of the matrix B_K.

Then the transformation

$$F_K : \widehat{x} \mapsto \overrightarrow{0P_0} + B_K\widehat{x}$$

maps the reference triangle or tetrahedron onto K. One easily checks [**168**] that there are two constants c_1 and c_2 such that

$$\begin{aligned} \left|\left|\left|B_K^T M_K^{-T}\right|\right|\right|_2 &= \left|\left|\left|M_K^{-1} B_K\right|\right|\right|_2 \le c_1 \\ \left|\left|\left|B_K^{-1} M_K\right|\right|\right|_2 &= \left|\left|\left|M_K^T B_K^{-T}\right|\right|\right|_2 \le c_2 \end{aligned} \tag{4.22}$$

where $|||\cdot|||_2$ denotes the spectral norm on $\mathbb{R}^{d\times d}$.

To simplify subsequent estimates, we introduce the piecewise constant function $M_{\mathcal{T}} \in S^{0,-1}(\mathcal{T})^{d\times d}$ which takes the value M_K on the element K.

Given any edge or face E, we denote by K_E the element of $\mathcal{T}$ which is adjacent to E and which has minimal $h_{\min,K}$. Set

$$h_{\min,E} = h_{\min,K_E} \quad \text{and} \quad h_E^{\perp} = d\frac{\mu_d(K_E)}{\mu_{d-1}(E)}.$$

Note that neither $h_{\min,E}$ nor $h_E^{\perp}$ are the length or diameter of E and that the quantity $h_E^{\perp}$ is called h_E in [**168, 170**]. Also note that the present definition of $h_E^{\perp}$ differs slightly from that used in Sections 3.3 (p. 87), 3.4.7 (p. 105), and 3.4.8 (p. 107). The following results can also be proved with the form of $h_E^{\perp}$ used there, but the present one is slightly better adapted to the actual situation.

4.5.2 Quasi-Interpolation Error Estimates

Following Remark 3.34 (p. 112) we start with a modification of the Poincaré inequality which is adapted to anisotropic meshes.

Lemma 4.27 *For every vertex $z \in \mathcal{N}$ and every function $\varphi \in H^1(\omega_z)$, the L^2-projection $\overline{\varphi}_{\omega_z} = \frac{1}{\mu_d(\omega_z)} \int_{\omega_z} \varphi$ satisfies the Poincaré inequality*

$$\left\|\varphi - \overline{\varphi}_{\omega_z}\right\|_{\omega_z} \le c \left\|M_{\mathcal{T}} \nabla\varphi\right\|_{\omega_z}.$$

The constant c only depends on the anisotropic shape parameter $C_{\mathcal{T},a}$ defined in equation (4.21).

Proof Lemma 4.27 is proved by a macro-element technique.

Every set ω_z with $z \in \mathcal{N}$ is a polygon, if $d = 2$, or a polyhedron, if $d = 3$. It consists of N_z triangles or tetrahedrons, respectively, and has L_z vertices different from z, which are labelled $a_1, \ldots, a_{L_z}$. Note that z is a vertex of ω_z if and only if z is on the boundary Γ. We say that two vertices z and z' and the corresponding sets ω_z and $\omega_{z'}$ are equivalent if

- $N_z = N_{z'}$,
- $L_z = L_{z'}$,
- the elements and vertices corresponding to ω_z and $\omega_{z'}$ can be enumerated such that: if $K_i \subset \omega_z$ has the vertices $z, a_{j_1}, \ldots, a_{j_d}$ then $K_i' \subset \omega_{z'}$ has the vertices $z', a'_{j_1}, \ldots, a'_{j_d}$.

Denote by $\mathcal{M}$ the resulting set of equivalence classes. From the conditions on $\mathcal{T}$ and a compactness argument, we conclude that the size of $\mathcal{M}$ is finite uniformly with respect to any family of meshes obtained by successive local or global refinement.

With each $z \in \mathcal{N}$ we can associate a reference set $\widehat{\omega}_z \in \mathcal{M}$ and a piecewise affine mapping F_z such that

- $F_z : \widehat{\omega}_z \to \omega_z$ is bijective,
- for every $i \in \{1, \ldots, N_z\}$ the i-th element $\widehat{K}_i$ in $\widehat{\omega}_z$ is mapped by F_z onto the i-th element K_i in ω_z.

We denote by B_i the Jacobian of F_z restricted to $\widehat{K}_i$.

With these notations, we define an averaging operator $I : L^1(\omega_z) \to \mathbb{R}$ by

$$I\varphi = \frac{1}{\mu_d(\widehat{\omega}_z)} \sum_{i=1}^{N_z} \int_{K_i} \varphi \, |\det B_i|^{-1} .$$

Since

$$\int_{K_i} \varphi \, |\det B_i|^{-1} = \int_{\widehat{K}_i} \varphi \circ F_z,$$

we have

$$I\varphi = \overline{(\varphi \circ F_z)}_{\widehat{\omega}_z}.$$

The Poincaré inequality on $\widehat{\omega}_z$, cf. [**50**], [**118**, Proposition IV.7.2], [**217**], and Section 3.4 (p. 91), therefore implies

$$\|\varphi \circ F_z - I\varphi\|_{\widehat{\omega}_z} \leq C_{P,2}(\widehat{\omega}_z) \operatorname{diam}(\widehat{\omega}_z) \|\nabla(\varphi \circ F_z)\|_{\widehat{\omega}_z}.$$

Since $\mathcal{M}$ is finite, the quantity

$$\widehat{c}_P = \max_{z \in \mathcal{N}} C_{P,2}(\widehat{\omega}_z) \operatorname{diam}(\widehat{\omega}_z)$$

is finite, too.

From the transformation rule we obtain for every $z \in \mathcal{N}$

$$\begin{aligned} \|\varphi - I\varphi\|_{\omega_z}^2 &= \sum_{i=1}^{N_z} \int_{K_i} (\varphi - I\varphi)^2 \\ &= \sum_{i=1}^{N_z} \int_{\widehat{K}_i} (\varphi \circ F_z - I\varphi)^2 \, |\det B_i| \\ &\leq \max_{1 \leq i \leq N_z} |\det B_i| \sum_{i=1}^{N_z} \int_{\widehat{K}_i} (\varphi \circ F_z - I\varphi)^2. \end{aligned}$$

The previous Poincaré-type inequality, on the other hand, yields

$$\begin{aligned} \sum_{i=1}^{N_z} \int_{\widehat{K}_i} (\varphi \circ F_z - I\varphi)^2 &= \|\varphi \circ F_z - I\varphi\|_{\widehat{\omega}_z}^2 \\ &\leq \widehat{c}_P^2 \|\nabla(\varphi \circ F_z)\|_{\widehat{\omega}_z}^2 \\ &= \widehat{c}_P^2 \sum_{i=1}^{N_z} \int_{\widehat{K}_i} |\nabla(\varphi \circ F_z)|^2 . \end{aligned}$$

The definition of the anisotropic shape parameter implies that

$$\max_{1 \leq i \leq N_z} |\det B_i| \sum_{i=1}^{N_z} \int_{\widehat{K}_i} |\nabla(\varphi \circ F_z)|^2 \leq C_{\mathcal{T},a} \sum_{i=1}^{N_z} \int_{\widehat{K}_i} |\nabla(\varphi \circ F_z)|^2 \, |\det B_i| \, .$$

Invoking the transformation rule and estimate (4.22), we finally get

$$\begin{aligned} \sum_{i=1}^{N_z} \int_{\widehat{K}_i} |\nabla(\varphi \circ F_z)|^2 \, |\det B_i| &= \sum_{i=1}^{N_z} \int_{K_i} |B_i^T \nabla\varphi|^2 \\ &\leq \sum_{i=1}^{N_z} \int_{K_i} \left\| B_i^T M_i^{-T} \right\|_2 |M_i^T \nabla\varphi|^2 \\ &\leq c_1 \sum_{i=1}^{N_z} \int_{K_i} |M_i^T \nabla\varphi|^2 \\ &= c_1 \, \|M_{\mathcal{T}} \nabla\varphi\|_{\omega_z}^2 \end{aligned}$$

and thus

$$\|\varphi - I\varphi\|_{\omega_z} \leq c \, \|M_{\mathcal{T}} \nabla\varphi\|_{\omega_z} .$$

Since $\overline{\varphi}_{\omega_z}$ is the best L^2-approximation of φ by constant functions, we have

$$\left\|\varphi - \overline{\varphi}_{\omega_z}\right\|_{\omega_z} \leq \|\varphi - I\varphi\|_{\omega_z} .$$

This estimate, together with the above bound for $\|\varphi - I\varphi\|_{\omega_z}$, yields the desired result. □

When proving the reliability of a posteriori error estimates on anisotropic meshes, it turns out that the correct alignment of the grid with the sought solution is crucial. To quantify this condition, we define a so-called *matching function* $m_{\mathcal{T}} : H^1(\Omega) \to \mathbb{R}$ by

$$m_{\mathcal{T}}(\varphi) = \left\{ \sum_{K\in\mathcal{T}} \sum_{i=1}^{d} h_{\min,K}^{-2} \left\| \mathbf{p}_{i,K} \cdot \nabla\varphi \right\|_K^2 \right\}^{\frac{1}{2}} \|\nabla\varphi\|^{-1}.$$

Since the vectors $\mathbf{p}_{1,K}, \ldots, \mathbf{p}_{d,K}$ are mutually orthogonal, the matching function can be expressed in terms of the matrices M_K as

$$m_{\mathcal{T}}(\varphi) = \left\{ \sum_{K\in\mathcal{T}} h_{\min,K}^{-2} \left\| M_K^T \nabla\varphi \right\|_K^2 \right\}^{\frac{1}{2}} \|\nabla\varphi\|_K^{-1}.$$

We then obtain the following analogue of Proposition 1.3 (p. 7).

Lemma 4.28 *The quasi-interpolation operator $I_{\mathcal{T}}$ of equation (1.4) (p. 7) satisfies the following error estimate for all functions $\varphi \in H_D^1(\Omega)$*

$$\left\{ \sum_{K\in\mathcal{T}} h_{\min,K}^{-2} \|\varphi - I_{\mathcal{T}}\|_K^2 + \sum_{E\in\mathcal{E}} h_E^{\perp} h_{\min,E}^{-2} \|\varphi - I_{\mathcal{T}}\varphi\|_E^2 \right\}^{\frac{1}{2}} \le c_{A,a}\, m_{\mathcal{T}}(\varphi)\, \|\nabla\varphi\|.$$

The constant $c_{A,a}$ only depends on the anisotropic shape parameter $C_{\mathcal{T},a}$ defined in equation (4.21).

Proof The proof of Lemma 4.28 is similar to that of Proposition 3.33 (p. 109) taking into account Remark 3.34 (p. 112) and Remark 3.6 (p. 90). The Poincaré inequality is replaced by Lemma 4.27. For the trace inequality, we observe that the classical trace inequality (1.5) (p. 8) on the reference element $\widehat{K}$, estimates (4.22), and the definition of $h_E^{\perp}$ yield

$$\begin{aligned}
\|\varphi\|_E &= \left(\frac{\mu_{d-1}(E)}{\mu_{d-1}(\widehat{E})} \right)^{\frac{1}{2}} \|\varphi \circ F_K\|_{\widehat{E}} \\
&\le \widehat{c} \left(\frac{\mu_{d-1}(E)}{\mu_{d-1}(\widehat{E})} \right)^{\frac{1}{2}} \left\{ \|\varphi \circ F_K\|_{\widehat{K}}^2 + \left\|\nabla(\varphi \circ F_K)\right\|_{\widehat{K}}^2 \right\}^{\frac{1}{2}} \\
&\le \widehat{c} \left(\frac{\mu_{d-1}(E)\mu_d(\widehat{K})}{\mu_{d-1}(\widehat{E})\mu_d(K_E)} \right)^{\frac{1}{2}} \left\{ \|\varphi\|_{K_E}^2 + \left\| B_K^T \nabla\varphi \right\|_{K_E}^2 \right\}^{\frac{1}{2}} \\
&\le c_1 (h_E^{\perp})^{-\frac{1}{2}} \left\{ \|\varphi\|_{K_E}^2 + \left\|B_K^T M_K^{-T}\right\|_2 \left\| M_K^T \nabla\varphi \right\|_{K_E}^2 \right\}^{\frac{1}{2}} \\
&\le c_2 (h_E^{\perp})^{-\frac{1}{2}} \left\{ \|\varphi\|_{K_E}^2 + \left\| M_K^T \nabla\varphi \right\|_{K_E}^2 \right\}^{\frac{1}{2}}.
\end{aligned}$$

□

4.5.3 *Local Cut-Off Functions*

The local cut-off functions $\psi_K, K \in \mathcal{T}$, and $\psi_E, E \in \mathcal{E}$, are defined as in Sections 1.3.4 (p. 8) and 3.2.3 (p. 83). We then have the following analogue of Propositions 1.4 (p. 9) and 3.37 (p. 112).

Lemma 4.29 *The following inverse estimates hold for all polynomials φ of degree at most k, all elements $K \in \mathcal{T}$, and all edges or faces $E \in \mathcal{E}$*

$$\|\varphi\|_K \le c_{I,a,1,k} \left\| \psi_K^{\frac{1}{2}} \varphi \right\|_K ,$$

$$\left\| \nabla(\psi_K \varphi) \right\|_K \le c_{I,a,2,k}\, h_{\min,K}^{-1} \|\varphi\|_K ,$$

$$\|\varphi\|_E \le c_{I,a,3,k} \left\| \psi_E^{\frac{1}{2}} \varphi \right\|_E ,$$

$$\left\| \nabla(\psi_E \varphi) \right\|_{\omega_E} \le c_{I,a,4,k}\, (h_E^{\perp})^{\frac{1}{2}} h_{\min,E}^{-1} \|\varphi\|_E .$$

The constants $c_{I,a,1,k}, \ldots, c_{I,a,4,k}$ only depend on the polynomial degree k and the anisotropic shape parameter $C_{\mathcal{T},a}$ defined in equation (4.21).

Proof Lemma 4.29 is proved in exactly the same way as Propositions 1.4 (p. 9) and 3.37 (p. 112) by transforming to the reference element, using the equivalence of norms on finite-dimensional spaces there and transforming back again. For the second estimate, we in addition have to take into account that the orthogonality of the matrix $H_K^{-1} M_K^T$, the transformation rule, and estimate (4.22) yield for every function $\psi \in H^1(K)$ and every element K

$$\begin{aligned}
\|\nabla \psi\|_K &= \left\| H_K^{-1} M_K^T \nabla \psi \right\|_K \\
&\le c_1 h_{\min,K}^{-1} \left\| M_K^T \nabla \psi \right\|_K \\
&= c_1 h_{\min,K}^{-1} \left(\frac{\mu_d(K)}{\mu_d(\widehat{K})} \right)^{\frac{1}{2}} \left\| M_K^T B_K^{-T} \nabla(\psi \circ F_K) \right\|_{\widehat{K}} \\
&\le c_2 h_{\min,K}^{-1} \left(\frac{\mu_d(K)}{\mu_d(\widehat{K})} \right)^{\frac{1}{2}} \left\| \nabla(\psi \circ F_K) \right\|_{\widehat{K}} .
\end{aligned}$$

For the fourth estimate, we in addition have to take into account that the transformation rule and the definition of $h_E^{\perp}$ give for every function $\psi \in H^1(K)$, every element K, and every edge or face E thereof

$$\|\psi\|_K = \left(\frac{\mu_d(K)}{\mu_d(\widehat{K})} \right)^{\frac{1}{2}} \|\psi \circ F_K\|_{\widehat{K}}$$

$$\|\psi \circ F_K\|_{\widehat{E}} = \left(\frac{\mu_{d-1}(\widehat{E})}{\mu_{d-1}(E)} \right)^{\frac{1}{2}} \|\psi\|_E$$

and

$$\frac{\mu_d(K)\mu_{d-1}(\widehat{E})}{\mu_d(\widehat{K})\mu_{d-1}(E)} \le c h_E^{\perp}.$$

□

4.5.4 Residual Estimates for the Model Problem

Lemmas 4.28 and 4.29 and the results of Section 4.2 (p. 157) immediately yield the following analogue of Theorem 1.5 (p. 15).

Theorem 4.30 *Denote by u the unique solution of problem (1.2) (p. 4) and by $u_{\mathcal{T}}$ its finite element approximation. For every element $K \in \mathcal{T}$ define the* residual a posteriori error indicator $\eta_{R,K}$ *by*

$$\eta_{R,K} = \left\{ h_{\min,K}^2 \left\| f_{\mathcal{T}} + \Delta u_{\mathcal{T}} \right\|_K^2 + \frac{1}{2} \sum_{E \in \mathcal{E}_{K,\Omega}} h_{\min,E}^2 (h_E^{\perp})^{-1} \left\| \mathbb{J}_E(\mathbf{n}_E \cdot \nabla u_{\mathcal{T}}) \right\|_E^2 + \sum_{E \in \mathcal{E}_{K,\Gamma_N}} h_{\min,E}^2 (h_E^{\perp})^{-1} \left\| g_{\mathcal{E}} - \mathbf{n}_E \cdot \nabla u_{\mathcal{T}} \right\|_E^2 \right\}^{\frac{1}{2}},$$

where $f_{\mathcal{T}} \in S^{k,-1}(\mathcal{T})$ and $g_{\mathcal{E}} \in S^{k,-1}(\mathcal{E}_{\Gamma_N})$ are approximations of f and g, respectively. There are two constants c_a^* and c_{a*}, which only depend on the polynomial degree k and the anisotropic shape parameter $C_{\mathcal{T},a}$ defined in equation (4.21), such that the estimates

$$\left\| \nabla(u - u_{\mathcal{T}}) \right\| \le c_a^* m_{\mathcal{T}}(u - u_{\mathcal{T}}) \left\{ \sum_{K \in \mathcal{T}} \eta_{R,K}^2 + \sum_{K \in \mathcal{T}} h_{\min,K}^2 \left\| f - f_{\mathcal{T}} \right\|_K^2 + \sum_{E \in \mathcal{E}_{\Gamma_N}} h_{\min,E}^2 (h_E^{\perp})^{-1} \left\| g - g_{\mathcal{E}} \right\|_E^2 \right\}^{\frac{1}{2}}$$

and

$$\eta_{R,K} \le c_{a*} \left\{ \left\| \nabla(u - u_{\mathcal{T}}) \right\|_{\omega_K}^2 + \sum_{K' \subset \omega_K} h_{\min,K'}^2 \left\| f - f_{\mathcal{T}} \right\|_{K'}^2 + \sum_{E \in \mathcal{E}_{K,\Gamma_N}} h_{\min,E}^2 (h_E^{\perp})^{-1} \left\| g - g_{\mathcal{E}} \right\|_E^2 \right\}^{\frac{1}{2}}$$

hold for all $K \in \mathcal{T}$.

Proof Theorem 4.30 is proved in exactly the same way as Theorem 1.5 (p. 15). We only have to replace the quasi-interpolation error estimates of Proposition 1.3 (p. 7) by that of Lemma 4.28 and the inverse estimates of Proposition 1.3 (p. 7) by those of Lemma 4.29. Moreover, we must take into account that the coercivity of the bi-linear form

$$\varphi, \psi \mapsto \int_{\Omega} \nabla\varphi \cdot \nabla\psi$$

and the Galerkin orthogonality

$$\int_{\Omega} \nabla(u - u_{\mathcal{T}}) \cdot \nabla v_{\mathcal{T}} = 0$$

for all $v_{\mathcal{T}} \in S_D^{k,0}(\mathcal{T})$ imply that, with the notation of Section 4.1 (p. 151),

$$\begin{aligned}
\left\|(Id_Y - Q_{\mathcal{T}})^*(\ell - Lu_{\mathcal{T}})\right\|_{Y^*} &= \sup_{\psi \in Y\setminus\{0\}} \|\psi\|_Y^{-1} \left\langle \ell - Lu_{\mathcal{T}}, (Id_Y - Q_{\mathcal{T}})\psi\right\rangle_Y \\
&= \sup_{\psi \in Y\setminus\{0\}} \|\psi\|_Y^{-1} \left\langle L(u - u_{\mathcal{T}}), (Id_Y - Q_{\mathcal{T}})\psi\right\rangle_Y \\
&= \sup_{\psi \in Y\setminus\{0\}} \|\psi\|_Y^{-1} \left\langle L(u - u_{\mathcal{T}}), \psi\right\rangle_Y \\
&= \left\|\nabla(u - u_{\mathcal{T}})\right\|^{-1} \left\langle L(u - u_{\mathcal{T}}), u - u_{\mathcal{T}}\right\rangle_Y \\
&= \left\|\nabla(u - u_{\mathcal{T}})\right\|^{-1} \left\langle L(u - u_{\mathcal{T}}), (Id_Y - Q_{\mathcal{T}})(u - u_{\mathcal{T}})\right\rangle_Y \\
&= \|u - u_{\mathcal{T}}\|_Y^{-1} \left\langle \ell - Lu_{\mathcal{T}}, (Id_Y - Q_{\mathcal{T}})(u - u_{\mathcal{T}})\right\rangle_Y . \qquad \square
\end{aligned}$$

Remark 4.31 The relation

$$\left\|(Id_Y - Q_{\mathcal{T}})^*(\ell - Lu_{\mathcal{T}})\right\|_{Y^*} = \|u - u_{\mathcal{T}}\|_Y^{-1} \left\langle \ell - Lu_{\mathcal{T}}, (Id_Y - Q_{\mathcal{T}})(u - u_{\mathcal{T}})\right\rangle_Y$$

which we established in the proof of Theorem 4.30 is crucial. Without it the term $m_{\mathcal{T}}(u - u_{\mathcal{T}})$ multiplying the upper bound of the error in Theorem 4.30 must be replaced by $\sup_{v \in H_0^1(\Omega)} m_{\mathcal{T}}(v) = \frac{h_{\max,K}}{h_{\min,K}}$ with $h_{\max,K} = \max_{1 \le i \le d} h_{i,K} = h_{1,K}$. Since $\frac{h_{\max,K}}{h_{\min,K}}$ is proportional to the shape parameter $C_{\mathcal{T}}$ of $\mathcal{T}$ as defined in equation (3.3) (p. 82), the resulting upper bound of the error would be worthless in the case of anisotropic meshes.

Remark 4.32 Taking into account Remark 4.31, a comparison of Theorems 1.5 (p. 15) and 4.30 reveals, that, for shape-regular meshes, Theorem 4.30 reproduces Theorem 1.5.

4.5.5 Edge Residuals for the Model Problem

We want to extend the results of Section 1.6 (p. 20) to anisotropic triangular or tetrahedral meshes. Therefore, we assume throughout this subsection that $\mathcal{T}$ exclusively consists of triangles, if $d = 2$, or tetrahedrons, if $d = 3$, and that $X_{\mathcal{T}} = S_D^{1,0}(\mathcal{T})$, i.e. we use a linear finite element discretisation. In addition to the notation of Section 1.6, we set

$$\gamma_d = \begin{cases} \sqrt{3} + \sqrt{2} \approx 3.146 & \text{if } d = 2 \\ \frac{3+\sqrt{5}}{2} \approx 2.618 & \text{if } d = 3 \end{cases}$$

and assume that $\mathcal{T}$ satisfies the following

- *anisotropic growth condition*: there are constants c_1, c_2, α, β, and r with $\alpha \ge 1$, $\beta \ge 1$, and $\alpha\beta < \gamma_d$ such that

$$\begin{aligned}
\frac{h_{\min,K}}{h_{\min,K'}} &\le c_1 \alpha^{\ell(K,K')} && \text{for all } K, K' \in \mathcal{T}, \\
n_j(K) &\le c_2 j^r \beta^j && \text{for all } K \in \mathcal{T}.
\end{aligned}$$

Here, as in Section 1.6.2 (p. 21), $\ell(K, K')$ is the length of the shortest path connecting K and K' and $n_j(K)$ is the number of elements K' with distance $\ell(K, K')$ at most j from K.

Similarly to Section 1.6, we define two mesh-dependent norms on $H^1(\Omega)$ by

$$|\varphi|_{\mathcal{T}} = \left\{ \sum_{K\in\mathcal{T}} h_{\min,K}^{-2} \, \|\varphi\|_K^2 \right\}^{\frac{1}{2}},$$

$$|||\varphi|||_{\mathcal{T}} = \left\{ |\varphi|_{\mathcal{T}}^2 + \sum_{E\in\mathcal{E}_\Omega} h_E^{\perp} h_{\min,E}^{-2} \, \|\varphi\|_E^2 \right\}^{\frac{1}{2}}.$$

Lemma 4.33 *The estimates*

$$|\varphi|_{\mathcal{T}} \leq |||\varphi|||_{\mathcal{T}} \leq \left\{1 + 2(d+1)^2\right\}^{\frac{1}{2}} |\varphi|_{\mathcal{T}}$$

hold for all $\varphi \in S_D^{1,0}(\mathcal{T})$.

Proof The first estimate is obvious from the definition of the norms.

To prove the second one, consider an arbitrary edge or face $E \in \mathcal{E}$. Denote by $\widehat{K}$ the standard reference d-simplex and by $\widehat{E}$ its horizontal edge or face, respectively. Let $K \in \mathcal{T}$ be the element adjacent to E with minimal $h_{\min,K}$. By construction we have

$$h_E^{\perp} = d \frac{\mu_d(K)}{\mu_{d-1}(E)}.$$

Denote by F_K an affine transformation which maps $\widehat{K}$ onto K and $\widehat{E}$ onto E. Set

$$\widehat{\varphi} = \varphi \circ F_K.$$

From the transformation rule we then get

$$h_E^{\perp} \, \|\varphi\|_E^2 = h_E^{\perp} \frac{\mu_{d-1}(E)}{\mu_{d-1}(\widehat{E})} \, \|\widehat{\varphi}\|_{\widehat{E}}^2 = d \frac{\mu_d(K)}{\mu_{d-1}(\widehat{E})} \, \|\widehat{\varphi}\|_{\widehat{E}}^2$$

and

$$\mu_d(K) \, \|\widehat{\varphi}\|_{\widehat{K}}^2 = \mu_d(\widehat{K}) \, \|\varphi\|_K^2 .$$

Since $\mathbb{P}_1$ is finite-dimensional, we have for all $\widehat{\varphi} \in \mathbb{P}_1$

$$\|\widehat{\varphi}\|_{\widehat{E}}^2 \leq c \, \|\widehat{\varphi}\|_{\widehat{K}}^2 .$$

Solving the corresponding eigenvalue problem yields $c = 2(d+1)$. Since $\mu_{d-1}(\widehat{E}) = d\mu_d(\widehat{K})$, we obtain

$$h_E^{\perp} \, \|\varphi\|_E^2 \leq 2(d+1) \, \|\varphi\|_K^2 .$$

Since every element is counted at most $(d+1)$-times when summing over all $E \in \mathcal{E}_{\mathcal{T},\Omega}$, this establishes the second estimate. □

Recall that $P_{\mathcal{T}} : H^1_D(\Omega) \to S^{1,0}_D(\mathcal{T})$ denotes the L^2-projection.

Lemma 4.34 *For all $K, K_0 \in \mathcal{T}$ and all $\varphi \in L^2(\Omega)$ with $\operatorname{supp}\varphi \subset K_0$, the estimate*

$$\|P_{\mathcal{T}}\varphi\|_K \leq \gamma_d^{-\ell(K,K_0)} \|\varphi\|_{K_0}$$

holds with the constant γ_d from the anisotropic growth condition.

Proof Lemma 4.34 is proved in [**115**, Lemma 3] for two dimensions, cf. also the proof of Lemma 1.14 (p. 23). A close inspection of this proof reveals that it does not depend on the dimension except for the computation of the quantities

$$\kappa_{j,d} = \max_{\varphi \in \mathbb{P}_1} \frac{-1}{\|\varphi\|_K^2} \int_K \varphi Q_j \varphi \quad , 1 \leq j \leq d,$$
$$\kappa_d = \max_{1 \leq j \leq d} \kappa_{j,d}.$$

Here, K is an arbitrary d-simplex and the operator Q_j sets a given φ to zero at j vertices of K and leaves it unchanged at the remaining $d + 1 - j$ vertices. Once these constants are computed, the arguments of [**115**, Lemma 3] yield the desired estimate with

$$\gamma_d = \left[\frac{1 + \kappa_d}{\kappa_d}\right]^{\frac{1}{2}}.$$

In order to determine κ_d, we first observe that the quotient in the definition of $\kappa_{j,d}$ is invariant under affine transformations. Hence, the quantities $\kappa_{1,d}, \ldots, \kappa_{d,d}$ can be computed for the reference element and do not depend on the enumeration of its vertices. Given $j \in \{1, \ldots, d\}$, the quantity $\kappa_{j,d}$ is therefore the largest solution of the generalised eigenvalue problem

$$\det(\lambda A_d + B_{d,j}) = 0.$$

Here, the symmetric $(d+1) \times (d+1)$ matrix A_d has diagonal entries 2 and off-diagonal entries 1. The symmetric matrix $B_{d,j}$ is obtained from A_d by setting to $\frac{1}{2}$ its values in the first $d + 1 - j$ rows of its last j columns and in the first $d + 1 - j$ columns of its last j rows and by setting to 0 its values in the last j rows of its last j columns. A straightforward calculation yields

$$\kappa_{1,2} = \kappa_{2,2} = \frac{\sqrt{6} - 2}{4}, \quad \kappa_{1,3} = \kappa_{3,3} = \frac{2\sqrt{10} - 5}{10}, \quad \kappa_{2,3} = \frac{3\sqrt{5} - 5}{10}.$$

This establishes the desired result. □

The following stability result is proved in exactly the same way as Lemma 1.14 (p. 23). We only have to replace [**115**, Lemma 3] by Lemma 4.34.

Lemma 4.35 *Assume that the anisotropic growth condition is fulfilled. Then there is a constant c, which only depends on the constants in the anisotropic growth condition, such that*

$$|P_{\mathcal{T}}\varphi|_{\mathcal{T}} \leq c\,|\varphi|_{\mathcal{T}}$$

holds for all $\varphi \in H_0^1(\Omega)$.

Now we can prove the following analogue of estimate (1.32) (p. 21).

Lemma 4.36 *Assume that the anisotropic growth condition is fulfilled. Then there is a constant* c'*, which only depends on the constants in the anisotropic growth condition, such that*

$$|||\varphi - P_{\mathcal{T}}\varphi|||_{\mathcal{T}} \leq c' m_{\mathcal{T}}(\varphi)\,\|\nabla\varphi\|$$

holds for all $\varphi \in H_D^1(\Omega)$.

Proof Fix a function $\varphi \in H_D^1(\Omega)$. We then have

$$|||\varphi - P_{\mathcal{T}}\varphi|||_{\mathcal{T}} \leq |||\varphi - I_{\mathcal{T}}\varphi|||_{\mathcal{T}} + |||I_{\mathcal{T}}\varphi - P_{\mathcal{T}}\varphi|||_{\mathcal{T}}\,.$$

Since $I_{\mathcal{T}}\varphi - P_{\mathcal{T}}\varphi \in S_D^{1,0}(\mathcal{T})$, we may apply Lemma 4.33 and obtain

$$|||I_{\mathcal{T}}\varphi - P_{\mathcal{T}}\varphi|||_{\mathcal{T}} \leq \left\{1 + 2(d+1)^2\right\}^{\frac{1}{2}}\,|I_{\mathcal{T}}\varphi - P_{\mathcal{T}}\varphi|_{\mathcal{T}}\,.$$

Since $I_{\mathcal{T}}\varphi = P_{\mathcal{T}}I_{\mathcal{T}}\varphi$, Lemmas 4.35 and 4.33 yield

$$|I_{\mathcal{T}}\varphi - P_{\mathcal{T}}\varphi|_{\mathcal{T}} = \left|P_{\mathcal{T}}(I_{\mathcal{T}}\varphi - \varphi)\right|_{\mathcal{T}} \leq c\,|I_{\mathcal{T}}\varphi - \varphi|_{\mathcal{T}} \leq c\,|||I_{\mathcal{T}}\varphi - \varphi|||_{\mathcal{T}}\,.$$

Combining these estimates with Lemma 4.28 completes the proof. □

Lemmas 4.29 and 4.36 and the arguments of Section 1.6 (p. 20) yield the following anisotropic analogue of Theorem 1.15 (p. 24).

Theorem 4.37 *Assume that the partition* $\mathcal{T}$ *exclusively consists of triangles or tetrahedrons and that it satisfies the anisotropic growth condition. Denote by u and* $u_{\mathcal{T}}$ *the unique solutions of problems* (1.2) *(p. 4) and* (1.3) *(p. 5), respectively. For every edge or face* $E \in \mathcal{E}$*, define the* edge residual a posteriori error indicator $\eta_{R,E}$ *by*

$$\eta_{R,E} = \begin{cases} h_{\min,E}(h_E^{\perp})^{-\frac{1}{2}}\,\left\|\mathbb{J}_E(\mathbf{n}_E\cdot\nabla u_{\mathcal{T}})\right\|_E & \text{if } E \in \mathcal{E}_\Omega, \\ h_{\min,E}(h_E^{\perp})^{-\frac{1}{2}}\,\left\|\bar{g}_E - \mathbf{n}_E\cdot\nabla u_{\mathcal{T}}\right\|_E & \text{if } E \in \mathcal{E}_{\Gamma_N}, \\ 0 & \text{if } E \in \mathcal{E}_{\Gamma_D}, \end{cases}$$

where $\bar{g}_E$ *denotes the mean value of g on E. Then there are two constants* $\widetilde{c}^*$ *and* $\widetilde{c}_*$*, which only depend on the constants in the anisotropic growth condition, such that the estimates*

$$\left\|\nabla(u - u_{\mathcal{T}})\right\| \leq m_{\mathcal{T}}(u - u_{\mathcal{T}})\widetilde{c}^*\left\{\sum_{E\in\mathcal{E}}\eta_{R,E}^2 + \sum_{K\in\mathcal{T}}h_{\min,K}^2\,\|f - I_{\mathcal{T}}f\|_K^2 + \sum_{E\in\mathcal{E}_{\Gamma_N}}h_{\min,E}^2(h_E^{\perp})^{-1}\,\|g - \bar{g}_E\|_E^2\right\}^{\frac{1}{2}}$$

and

$$\eta_{R,E} \leq \widetilde{c}_* \left\{ \left\| \nabla(u - u_{\mathcal{T}}) \right\|_{\omega_E}^2 + \sum_{K' \subset \omega_E} h_{\min,K'}^2 \left\| f - I_{\mathcal{T}} f \right\|_{K'}^2 \right\}^{\frac{1}{2}}$$

and

$$\eta_{R,E} \leq \widetilde{c}_* \left\{ \left\| \nabla(u - u_{\mathcal{T}}) \right\|_{\omega_E}^2 + \sum_{K' \subset \omega_E} h_{\min,K'}^2 \left\| f - I_{\mathcal{T}} f \right\|_{K'}^2 + h_{\min,E}^2 (h_E^{\perp})^{-1} \left\| g - \bar{g}_E \right\|_E^2 \right\}^{\frac{1}{2}}$$

hold for all $E \in \mathcal{E}_\Omega$ *and all* $E \in \mathcal{E}_{\Gamma_N}$*, respectively, where* $I_{\mathcal{T}}$ *is the quasi-interpolation operator of Proposition 1.3 (p. 7).*

4.5.6 *Reaction–Diffusion Equations*

Theorem 4.10 (p. 160) and Remark 4.14 (p. 162) can be extended to anisotropic meshes with the following modifications.

- The weights $\hbar_K$ and $\hbar_E$ must be defined by

$$\begin{aligned} \hbar_K &= \min\left\{h_{\min,K}, \kappa^{-1}\right\}, \\ \hbar_E &= \frac{h_{\min,E}}{h_E^{\perp}} \min\left\{h_{\min,E}, \kappa^{-1}\right\} \end{aligned} \tag{4.23}$$

  for problem (4.11) (p. 159) and

$$\begin{aligned} \hbar_K &= \min\left\{\varepsilon^{-\frac{1}{2}} h_{\min,K}, 1\right\}, \\ \hbar_E &= \frac{h_{\min,E}}{h_E^{\perp}} \min\left\{\varepsilon^{-\frac{1}{2}} h_{\min,E}, 1\right\}, \end{aligned}$$

  for problem (4.12) (p. 162), respectively.
- The upper bounds on the error must be multiplied by the matching function $m_{\mathcal{T}}(u - u_{\mathcal{T}})$ introduced in Subsection 4.5.2.

To see this, we briefly retrace the arguments of Section 4.3 (p. 159) for problem (4.11) and show how to adapt them to the anisotropic regime using the techniques developed in this section. Problem (4.12) (p. 162) is treated similarly with obvious modifications of the corresponding estimates.

We start with the trace inequality of Proposition 3.5 (p. 90) and Remark 3.6 (p. 90). Starting from the standard trace inequality (1.5) (p. 8) on the reference element, transforming to the actual element, and arguing as in the proof of Lemma 4.27, we now obtain the estimate

$$\|\psi\|_E \leq \widehat{c}_{tr} (h_E^{\perp})^{-\frac{1}{2}} \left\{ \|\psi\|_K + \|\psi\|_K^{\frac{1}{2}} \left\| M_K^T \nabla \psi \right\|_K^{\frac{1}{2}} \right\}$$

with a new constant $\widehat{c}_{tr}$ which is independent of the mesh and its geometric properties.

Next, we turn to the quasi-interpolation error estimates. The relevant estimates are now

$$\begin{aligned}\left\|\varphi - \overline{\varphi}_{\omega_z}\right\|_{\omega_z} &\leq \|\varphi\|_{\omega_z}\\ \left\|\varphi - \overline{\varphi}_{\omega_z}\right\|_{\omega_z} &\leq c_1 \left\|M_{\mathcal{T}} \nabla\varphi\right\|_{\omega_z}\\ \left\|M_{\mathcal{T}} \nabla(\varphi - \overline{\varphi}_{\omega_z})\right\|_{\omega_z} &\leq c_2 \left\|M_{\mathcal{T}} \nabla\varphi\right\|_{\omega_z}.\end{aligned}$$

The first one is obvious, the second one is proved in Lemma 4.27, and the third one is established with similar arguments.

These estimates and the arguments of the proofs of Proposition 3.33 (p. 109) and Lemma 4.28 yield the error estimate

$$\begin{aligned}\|\varphi - I_{\mathcal{T}}\varphi\|_K &\leq c m_{\mathcal{T}}(\varphi)\hbar_K \, |||\varphi|||_{\widetilde{\omega}_K}\\ \|\varphi - I_{\mathcal{T}}\varphi\|_E &\leq c m_{\mathcal{T}}(\varphi)\hbar_E^{\frac{1}{2}} \, |||\varphi|||_{\widetilde{\omega}_E}\end{aligned}$$

with $\hbar_K$ and $\hbar_E$ given by (4.23), the matching function $m_{\mathcal{T}}$ of Subsection 4.5.2, and the energy norm $|||\cdot|||$ of Section 4.3.

Finally, we turn to the inverse estimates for the local cut-off functions. Proposition 3.46 (p. 121) carries over without any modification since it is formulated on the reference element and does not use any transformation between elements. The arguments of the proofs of Proposition 3.38 (p. 114), Corollary 3.40 (p. 116), and Lemma 4.27 then yield the following analogue of Proposition 3.37

$$\begin{aligned}\|\sigma\|_E &\leq c^*_{I,3,k} \left\|\psi_{E,\vartheta}^{\frac{1}{2}}\sigma\right\|_E,\\ \left|\left|\left|\psi_{E,\vartheta}\sigma\right|\right|\right|_{\omega_E} &\leq c^*_{I,4,k}(h_E^{\perp})^{\frac{1}{2}}\left\{h_{\min,E}^{-1}\vartheta^{-\frac{1}{2}} + \kappa\vartheta^{\frac{1}{2}}\right\}\|\sigma\|_E,\\ \left\|\psi_{E,\vartheta}\sigma\right\|_{\omega_E} &\leq c^*_{I,5,k}(h_E^{\perp})^{\frac{1}{2}}\vartheta^{\frac{1}{2}}\|\sigma\|_E.\end{aligned}$$

The constants $c^*_{I,3,k}$, $c^*_{I,4,k}$, and $c^*_{I,5,k}$ only depend on the polynomial degree k and the anisotropic shape parameter $C_{\mathcal{T},a}$ defined in equation (4.21). The optimal choice of ϑ is now $\vartheta = h_E^{\perp} h_{\min,E}^{-2}\hbar_E$.

4.5.7 *Convection–Diffusion Equations*

Remark 4.31 implies that Theorem 4.19 (p. 168) cannot be extended to anisotropic meshes since its proof hinges on Proposition 4.17 (p. 165) and thus needs more than coercivity. Theorem 4.20 (p. 168) on the other hand extends to anisotropic meshes since it only needs the coercivity of the bi-linear forms. The arguments given in the previous subsection show that we only have to define the weights $\hbar_K$ and $\hbar_E$ by

$$\hbar_K = \min\left\{\varepsilon^{-\frac{1}{2}}h_{\min,K}, \beta^{-\frac{1}{2}}\right\}, \quad \hbar_E = \frac{h_{\min,E}}{h_E^{\perp}}\min\left\{\varepsilon^{-\frac{1}{2}}h_{\min,E}, \beta^{-\frac{1}{2}}\right\}$$

and that we must multiply the upper bound of the error by $m_{\mathcal{T}}(u - u_{\mathcal{T}})$.

4.5.8 *Anisotropic Tensorial Elements*

Following [**238**] we now assume that the domain Ω and the partition $\mathcal{T}$ have a tensor-product structure, i.e. there are $L \geq 2$ positive numbers $\ell_1, \ldots, \ell_L$ such that $\Omega = \Omega_1 \times \cdots \times \Omega_L$ with $\Omega_i \subset \mathbb{R}^{\ell_i}$ and every element K has the form $K = K_1 \times \cdots \times K_L$ with ℓ_i-dimensional simplices K_i contained in Ω_i. In two dimension this restricts the elements to anisotropic rectangles. In three dimensions the elements may either be anisotropic cuboids or anisotropic prisms corresponding, up to permutation, to the cases $L = 3, \ell_1 = \ell_2 = \ell_3 = 1$, and $L = 2, \ell_1 = 2, \ell_2 = 1$, respectively.

The quantities $h_{\min,K}$, $h_{\min,E}$, $h_E^{\perp}$, and $C_{\mathcal{T},a}$ are defined as before with $h_{i,K}$ replaced by $\text{diam}(K_i)$. The assumption that $C_{\mathcal{T},a}$ is of moderate size now means that the partitions of the Ω_i induced by the K_i are shape-regular, i.e. the anisotropy takes place in a fixed global direction determined by the splitting of Ω.

With these modifications Theorem 4.30 (p. 184) carries over immediately. When comparing Theorem 4.30 with the a posteriori error estimates of [**238**, Lemmas 4.1 and 5.3], one may be startled at first sight since the matching function is not present there. Instead, [**238**, Lemma 4.1] is based on the assumption that the error is consistent with the anisotropy. When looking into the qualitative aspect [**238**, equation (3.1)] of this assumption, however, one realises that it is equivalent to the condition that the matching function $m_{\mathcal{T}}(u - u_{\mathcal{T}})$ is of moderate size. Thus Theorem 4.30 and Lemmas 4.1, 5.3 of [**238**] are equivalent.

In order to gain more flexibility of the partition, one may abandon the tensor-product structure and consider anisotropic parallelograms or parallelepipeds. Similarly to the simplicial case, the affine transformation to the reference element is then best decomposed into a first transformation to an anisotropic rectangle or cuboid and a second transformation of the anisotropic tensorial element to the standard reference square or cube. The first transformation is described by the matrices M_K, the second one by the matrices H_K defined in Subsection 4.5.1 (p. 177). The above tensorial situation then corresponds to $M_K^{-1}B_K = B_K^{-1}M_K = I$. Thus assumption (4.22) (p. 179) is also appropriate for anisotropic parallelograms and parallelepipeds. The effect of the anisotropy is then taken into account by the matrices H_K and can be controlled as in [**238**].

Remark 4.38 The first a posteriori error analysis for anisotropic partitions is due to K. G. Siebert [**238**] where anisotropic tensorial elements are considered. These results were generalised and extended to simplicial meshes in [**22**, **23**]. The concept of a quality function measuring the correct alignment of the anisotropic mesh was introduced in [**168**]. The presentation of this section is based on [**168**, **170**]. A posteriori error estimates based on averaging methods for anisotropic meshes are considered in [**169**].

4.6 Non-Smooth Coefficients

In this section we consider the diffusion equation

$$\begin{aligned} -\operatorname{div}(A\nabla u) &= f \quad \text{in } \Omega \\ u &= 0 \quad \text{on } \Gamma \end{aligned} \tag{4.24}$$

in a bounded two- or three-dimensional polyhedral domain with a Lipschitz boundary. Here, A denotes a function with values in square, symmetric, positive definite matrices of order 2 or 3 according to the space dimension d. We are interested in two rather different situations.

- Either the function A is discontinuous: it is smooth only on a finite number of subdomains and has large jumps across the interfaces between the subdomains. This models for instance several layers of fluids with rather different viscosity which weakly depend on the depth.
- Or the matrix A is constant on the whole domain but has eigenvalues of very different sizes, which results in an anisotropy of equation (4.24). This models for instance elastic materials in thin layers.

Of course both phenomena may occur simultaneously.

The variational formulation of problem (4.24) and its finite element approximation fit into the abstract framework of Section 4.1 (p. 151) with

$$\begin{aligned} X = Y &= H_0^1(\Omega), \\ B(\varphi, \psi) &= \int_\Omega \nabla\varphi \cdot A\nabla\psi, \\ \langle \ell\,,\, \psi \rangle_Y &= \int_\Omega f\psi, \\ X_{\mathcal{T}} = Y_{\mathcal{T}} &= S_0^{k,0}(\mathcal{T}), \\ B_{\mathcal{T}}(\varphi_{\mathcal{T}}, \psi_{\mathcal{T}}) &= B(\varphi_{\mathcal{T}}, \psi_{\mathcal{T}}), \\ \langle \ell_{\mathcal{T}}\,,\, \psi_{\mathcal{T}} \rangle_{Y_{\mathcal{T}}} &= \langle \ell\,,\, \psi_{\mathcal{T}} \rangle_Y\,. \end{aligned}$$

We equip the spaces X and Y with the natural energy norm, i.e.

$$\|\varphi\|_X = \|\varphi\|_Y = B(\varphi, \varphi)^{\frac{1}{2}} = \left\{ \int_\Omega \nabla\varphi \cdot A\nabla\varphi \right\}^{\frac{1}{2}}$$

for all $\varphi \in X = Y$. The results of Sections 4.1 and 4.2 and the arguments of Section 1.4 then immediately yield a residual a posteriori error indicator which gives global upper and local lower bounds for the energy norm of the error. The quality of this indicator in the sense of Remark 4.8 (p. 156), however, depends on the variation of the diffusion A and the ratio of its extremal eigenvalues, both of which can be large. To avoid this undesirable phenomenon, we must refine our arguments.

4.6.1 Piecewise Constant Scalar Diffusion

We first consider equation (4.24) with $A = \alpha I$ where α is a given, scalar, piecewise constant function on Ω and I is the identity matrix. Accordingly we introduce a disjoint partition of Ω into a finite number of open subdomains Ω_ℓ, $1 \le \ell \le L$, such that the function α is equal to the constant α_ℓ on each Ω_ℓ as illustrated in Figure 4.3. Of course we assume that this partition is consistent with $\mathcal{T}$, i.e. α is piecewise constant on the elements of $\mathcal{T}$.

We define the two parameters

$$\alpha_{\min} = \min_{1 \le \ell \le L} \alpha_\ell, \qquad \alpha_{\max} = \max_{1 \le \ell \le L} \alpha_\ell$$

and we assume that $\alpha_{\min}$ is positive. We are particularly interested in the critical case where the ratio $\frac{\alpha_{\max}}{\alpha_{\min}}$ is large. Our goal is to establish estimates which are independent of this ratio.

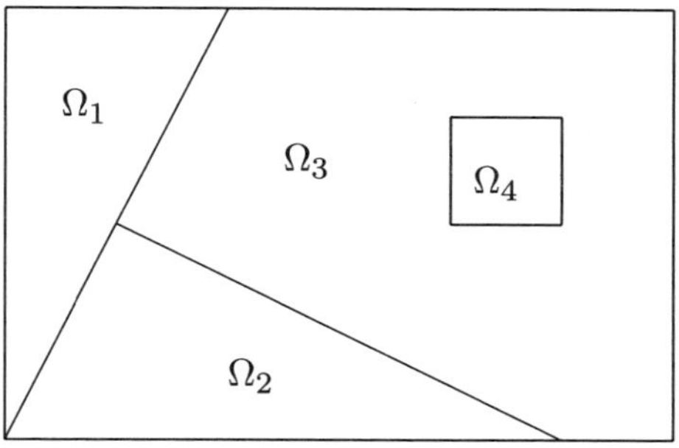

Figure 4.3 Partition of the domain Ω

Note that the energy norm now takes the particular form

$$\|\varphi\|_X = \left\|\alpha^{\frac{1}{2}} \nabla\varphi\right\| = \left\{\sum_{\ell=1}^{L} \alpha_\ell \|\nabla\varphi\|^2_{\Omega_\ell}\right\}^{\frac{1}{2}}.$$

Inspired by the results of the previous sections, we are looking for an error indicator of the form

$$\eta_{\mathcal{T}} = \left\{\sum_{K\in\mathcal{T}} \mu_K^2 \left\|f_{\mathcal{T}} + \operatorname{div}(\alpha\nabla u_{\mathcal{T}})\right\|_K^2 + \sum_{E\in\mathcal{E}_\Omega} \mu_E \left\|\mathbb{J}_E(\alpha\mathbf{n}_E \cdot \nabla u_{\mathcal{T}})\right\|_E^2\right\}^{\frac{1}{2}}$$

with an approximation $f_{\mathcal{T}} \in S^{k,-1}(\mathcal{T})$ of f and appropriate weights μ_K and μ_E.

To get an idea of what a good choice of weights might be, we start with the lower bound of the error.

Integration by parts element-wise gives for all $w \in X$ the standard L^2-representation of the residual

$$\langle \ell - Lu_{\mathcal{T}}, w\rangle_Y = \sum_{K\in\mathcal{T}} \int_K \left\{f_{\mathcal{T}} + \operatorname{div}(\alpha\nabla u_{\mathcal{T}})\right\} w + \sum_{K\in\mathcal{T}} \int_K \{f - f_{\mathcal{T}}\} w - \sum_{E\in\mathcal{E}_\Omega} \int_E \mathbb{J}_E(\alpha\mathbf{n}_E \cdot \nabla u_{\mathcal{T}}) w.$$

We fix an element K and insert

$$w_K = \psi_K \left\{f_{\mathcal{T}} + \operatorname{div}(\alpha\nabla u_{\mathcal{T}})\right\}$$

in this equation to obtain

$$\int_K \psi_K \left\{f_{\mathcal{T}} + \operatorname{div}(\alpha\nabla u_{\mathcal{T}})\right\}^2 = \int_K \nabla(u - u_{\mathcal{T}}) \cdot \alpha\nabla w_K - \int_K \{f - f_{\mathcal{T}}\} w_K.$$

Denoting by α_K the constant value of α on K, Proposition 3.37 (p. 112) yields

$$\left\|f_{\mathcal{T}} + \operatorname{div}(\alpha\nabla u_{\mathcal{T}})\right\|_K^2 \le C_{I,1,k}(K)^2 \int_K \psi_K \left\{f_{\mathcal{T}} + \operatorname{div}(\alpha\nabla u_{\mathcal{T}})\right\}^2,$$

$$\int_K \{f - f_{\mathcal{T}}\} w_K \le \left\|f - f_{\mathcal{T}}\right\|_K \left\|f_{\mathcal{T}} + \operatorname{div}(\alpha\nabla u_{\mathcal{T}})\right\|_K,$$

and

$$\begin{aligned}\int_K \nabla(u-u_{\mathcal{T}})\cdot\alpha\nabla w_K &\le \alpha_K \left\|\nabla(u-u_{\mathcal{T}})\right\|_K \|\nabla w_K\|_K \\ &\le C_{I,2,k}(K)h_K^{-1}\alpha_K \left\|\nabla(u-u_{\mathcal{T}})\right\|_K \cdot \left\|f_{\mathcal{T}}+\operatorname{div}(\alpha\nabla u_{\mathcal{T}})\right\|_K.\end{aligned}$$

Taking into account the particular form of the energy norm, this suggests the choice

$$\mu_K = h_K\alpha_K^{-\frac{1}{2}}. \tag{4.25}$$

Next, we fix a face $E\in\mathcal{E}_\Omega$ and insert the function

$$w_E = -\psi_E \mathbb{J}_E(\alpha\mathbf{n}_E\cdot\nabla u_{\mathcal{T}})$$

in the L^2-representation of the residual. This gives

$$\begin{aligned}\int_E \psi_E\mathbb{J}_E(\alpha\mathbf{n}_E\cdot\nabla u_{\mathcal{T}})^2 = &\int_{\omega_E}\nabla(u-u_{\mathcal{T}})\cdot\alpha\nabla w_E \\ &-\sum_{K\subset\omega_E}\int_K\left\{f_{\mathcal{T}}+\operatorname{div}(\alpha\nabla u_{\mathcal{T}})\right\}w_E \\ &-\sum_{K\subset\omega_E}\int_K\{f-f_{\mathcal{T}}\}\,w_E.\end{aligned}$$

Since α is constant on the elements K, Proposition 3.37 (p. 112) yields for all elements K contained in ω_E

$$\left\|\mathbb{J}_E(\alpha\mathbf{n}_E\cdot\nabla u_{\mathcal{T}})\right\|_E^2 \le C_{I,3,k}(K)^2\int_E\psi_E\mathbb{J}_E(\alpha\mathbf{n}_E\cdot\nabla u_{\mathcal{T}})^2,$$

$$\int_K\left\{f_{\mathcal{T}}+\operatorname{div}(\alpha\nabla u_{\mathcal{T}})\right\}w_E \le C_{I,5,k}(E,K)h_E^{\frac{1}{2}}\left\|f_{\mathcal{T}}+\operatorname{div}(\alpha\nabla u_{\mathcal{T}})\right\|_K\cdot\left\|\mathbb{J}_E(\alpha\mathbf{n}_E\cdot\nabla u_{\mathcal{T}})\right\|_E,$$

$$\int_K\{f-f_{\mathcal{T}}\}\,w_E \le C_{I,5,k}(E,K)h_E^{\frac{1}{2}}\left\|f-f_{\mathcal{T}}\right\|_K\cdot\left\|\mathbb{J}_E(\alpha\mathbf{n}_E\cdot\nabla u_{\mathcal{T}})\right\|_E,$$

and

$$\begin{aligned}\int_K\nabla(u-u_{\mathcal{T}})\cdot\alpha\nabla w_E &\le \alpha_K\left\|\nabla(u-u_{\mathcal{T}})\right\|_K\|\nabla w_E\|_K \\ &\le C_{I,4,k}(E,K)h_E^{-\frac{1}{2}}\alpha_K\left\|\nabla(u-u_{\mathcal{T}})\right\|_K\cdot\left\|\mathbb{J}_E(\alpha\mathbf{n}_E\cdot\nabla u_{\mathcal{T}})\right\|_E.\end{aligned}$$

This suggests the choice

$$\mu_E = h_E\alpha_E^{-1} \quad\text{with}\quad \alpha_E = \max_{K\subset\omega_E}\alpha_K. \tag{4.26}$$

To prove that these weights really yield the desired result, we must now construct a restriction operator $Q_{\mathcal{T}} : X \to X_{\mathcal{T}}$ such that

$$\left\{ \sum_{K \in \mathcal{T}} \mu_K^{-2} \|v - Q_{\mathcal{T}} v\|_K^2 + \sum_{E \in \mathcal{E}_\Omega} \mu_E^{-1} \|v - Q_{\mathcal{T}} v\|_E^2 \right\}^{\frac{1}{2}} \leq c_I \|v\|_X$$

holds for all $v \in X$ with a constant c_I which is independent of any mesh-size and the ratio $\frac{\alpha_{\max}}{\alpha_{\min}}$. To this end, we modify the quasi-interpolation operator $I_{\mathcal{T}}$ of equation (1.4) (p. 7) and Section 3.5 (p. 108) as follows. With every vertex $z \in \mathcal{N}$ we associate a number $\ell(z)$ in $\{1, \ldots, L\}$ such that

- z is contained in the closure of $\Omega_{\ell(z)}$ and
- $\alpha_{\ell(z)}$ is maximal among all α_j such that z is contained in the closure of Ω_j.

For every function $v \in L^1(\Omega)$ we then set

$$\widetilde{v}_z = \begin{cases} \dfrac{1}{\mu_d(\omega_z \cap \Omega_{\ell(z)})} \displaystyle\int_{\omega_z \cap \Omega_{\ell(z)}} v & \text{if } z \in \mathcal{N}_\Omega \\ 0 & \text{if } z \in \mathcal{N}_\Gamma \end{cases}$$

and define the quasi-interpolation operator $\widetilde{I}_{\mathcal{T}} : X \to X_{\mathcal{T}}$ by

$$\widetilde{I}_{\mathcal{T}} v = \sum_{z \in \mathcal{N}} \lambda_z \widetilde{v}_z.$$

Lemma 4.39 *Assume that at most three subdomains $\overline{\Omega}_\ell$ share a common point. Then the following error estimates hold for all elements $K \in \mathcal{T}$, all interior faces $E \in \mathcal{E}_\Omega$, and all functions $v \in X$*

$$\left\| v - \widetilde{I}_{\mathcal{T}} v \right\|_K \leq \widetilde{c}_{A1} \mu_K \left\| \alpha^{\frac{1}{2}} \nabla v \right\|_{\widetilde{\omega}_K},$$

$$\left\| v - \widetilde{I}_{\mathcal{T}} v \right\|_E \leq \widetilde{c}_{A2} \mu_E^{\frac{1}{2}} \left\| \alpha^{\frac{1}{2}} \nabla v \right\|_{\widetilde{\omega}_E}.$$

Here, the weights μ_K and μ_E are given by equations (4.25) and (4.26). The constants $\widetilde{c}_{A1}$ and $\widetilde{c}_{A2}$ only depend on the shape parameter $C_{\mathcal{T}}$ of $\mathcal{T}$.

Proof According to Remark 3.34 (p. 112) we must bound the quantities $\left\| \lambda_z^{\frac{1}{2}} (v - \widetilde{v}_z) \right\|_K$ and $\left\| \lambda_z^{\frac{1}{2}} (v - \widetilde{v}_z) \right\|_E$ for every element K, every face E, and every vertex z of K or E.

Consider first an arbitrary element K and an arbitrary vertex z of K. If z is not contained in the boundary of any subdomain including the boundary of Ω, the Poincaré inequality of Section 3.4 (p. 91) yields

$$\begin{aligned}
\left\| \lambda_z^{\frac{1}{2}} (v - \widetilde{v}_z) \right\|_K &\le \left\| \lambda_z^{\frac{1}{2}} (v - \widetilde{v}_z) \right\|_{\omega_z} \\
&\le \overline{C}_{P,2}(\omega_z) \operatorname{diam}(\omega_z) \|\nabla v\|_{\omega_z} \\
&\le c_2 h_K \alpha_K^{-\frac{1}{2}} \left\| \alpha^{\frac{1}{2}} \nabla v \right\|_{\omega_z} \\
&= c_2 \mu_K \left\| \alpha^{\frac{1}{2}} \nabla v \right\|_{\omega_z} .
\end{aligned}$$

If, on the other hand, z is on the boundary Γ, the same estimate follows from Friedrichs' inequality of Section 3.4.7 (p. 105).

Finally, consider a vertex which is not on the boundary Γ but which is in $\partial\Omega_{\ell(K)}$ where $\ell(K)$ is such that K is contained in the closure of $\Omega_{\ell(K)}$. If $\ell(K) = \ell(z)$ the previous arguments remain valid with ω_z replaced by $\omega_z \cap \Omega_{\ell(K)}$. If $\ell(K) \neq \ell(z)$, we must argue differently. From the definition of $\widetilde{v}_z$ we now obtain

$$\begin{aligned}
\left\| \lambda_z^{\frac{1}{2}} (v - \widetilde{v}_z) \right\|_K &= \left\| \lambda_z^{\frac{1}{2}} \left(v - \frac{1}{\mu_d(\omega_z \cap \Omega_{\ell(z)})} \int_{\omega_z \cap \Omega_{\ell(z)}} v \right) \right\|_K \\
&\le \left\| \lambda_z^{\frac{1}{2}} \left(v - \frac{1}{\mu_d(\omega_z \cap \Omega_{\ell(K)})} \int_{\omega_z \cap \Omega_{\ell(K)}} v \right) \right\|_K \\
&\quad + \left\| \lambda_z^{\frac{1}{2}} \left(\frac{1}{\mu_d(\omega_z \cap \Omega_{\ell(K)})} \int_{\omega_z \cap \Omega_{\ell(K)}} v - \frac{1}{\mu_d(\omega_z \cap \Omega_{\ell(z)})} \int_{\omega_z \cap \Omega_{\ell(z)}} v \right) \right\|_K .
\end{aligned}$$

The first term can be estimated exactly as before. Since $\lambda_z \le 1$, the second term may be estimated as follows

$$\begin{aligned}
&\left\| \lambda_z^{\frac{1}{2}} \left(\frac{1}{\mu_d(\omega_z \cap \Omega_{\ell(K)})} \int_{\omega_z \cap \Omega_{\ell(K)}} v - \frac{1}{\mu_d(\omega_z \cap \Omega_{\ell(z)})} \int_{\omega_z \cap \Omega_{\ell(z)}} v \right) \right\|_K \\
&\le \mu_d(K)^{\frac{1}{2}} \left| \frac{1}{\mu_d(\omega_z \cap \Omega_{\ell(K)})} \int_{\omega_z \cap \Omega_{\ell(K)}} v - \frac{1}{\mu_d(\omega_z \cap \Omega_{\ell(z)})} \int_{\omega_z \cap \Omega_{\ell(z)}} v \right| .
\end{aligned}$$

Since at most three subdomains $\overline{\Omega}_\ell$ share a common point, the subdomains $\Omega_{\ell(K)}$ and $\Omega_{\ell(z)}$ are adjacent and they share a common edge, if $d = 2$, or face, if $d = 3$, which is labelled E. Hence we have

$$\begin{aligned}
&\left| \frac{1}{\mu_d(\omega_z \cap \Omega_{\ell(K)})} \int_{\omega_z \cap \Omega_{\ell(K)}} v - \frac{1}{\mu_d(\omega_z \cap \Omega_{\ell(z)})} \int_{\omega_z \cap \Omega_{\ell(z)}} v \right| \\
&\le \mu_{d-1}(E)^{-\frac{1}{2}} \left\| \frac{1}{\mu_d(\omega_z \cap \Omega_{\ell(K)})} \int_{\omega_z \cap \Omega_{\ell(K)}} v - v \right\|_E \\
&\quad + \mu_{d-1}(E)^{-\frac{1}{2}} \left\| v - \frac{1}{\mu_d(\omega_z \cap \Omega_{\ell(z)})} \int_{\omega_z \cap \Omega_{\ell(z)}} v \right\|_E .
\end{aligned}$$

The first estimate of the proposition now follows by estimating the two terms on the right-hand side of the last inequality with the help of the trace inequalities of Proposition 3.5 (p. 90) and Remark 3.6 (p. 90). The estimation of $\left\| \lambda_z^{\frac{1}{2}} (v - \widetilde{v}_z) \right\|_E$ follows along the same lines using the trace inequalities of Proposition 3.5 and Remark 3.6. Here, the element K adjacent to E must be chosen such that α is maximal. □

Remark 4.40 The assumption of Lemma 4.39 may be weakened as follows [**66**, Hypotheses 2.7]: for any two different subdomains $\overline{\Omega}_\ell$ and $\overline{\Omega}_k$, which share at least one point, there is a connected path passing from $\overline{\Omega}_\ell$ to $\overline{\Omega}_k$ through adjacent subdomains such that the function α is monotonic along this path. Here, as usual, adjacent means that the corresponding subdomains share an edge, if $d = 2$, or a face, if $d = 3$. Figure 4.4 shows a configuration violating this monotonicity assumption.

Standard arguments show that Lemma 4.39 implies the desired error estimate for the restriction operator. We therefore obtain the following a posteriori error estimates.

Theorem 4.41 *Denote by u the unique weak solution of problem (4.24) (p. 191) and by $u_{\mathcal{T}}$ its finite element approximation. For every element $K \in \mathcal{T}$ define the* residual a posteriori error indicator *$\eta_{R,K}$ by*

$$\eta_{R,K} = \left\{ \mu_K^2 \left\| f_{\mathcal{T}} + \operatorname{div}(\alpha \nabla u_{\mathcal{T}}) \right\|_K^2 + \frac{1}{2} \sum_{E \in \mathcal{E}_{K,\Omega}} \mu_E \left\| \mathbb{J}_E(\mathbf{n}_E \cdot \alpha \nabla u_{\mathcal{T}}) \right\|_E^2 \right\}^{\frac{1}{2}},$$

where $f_{\mathcal{T}}$ is any finite element approximation of f and the weights μ_K and μ_E are given by equations (4.25) and (4.26), respectively. Assume that the conditions of Lemma 4.39 or of Remark 4.40 are satisfied. Then there are two constants c^* and c_*, which only depend on the shape parameter $C_{\mathcal{T}}$ of $\mathcal{T}$, such that the estimates

$$\left\| \alpha^{\frac{1}{2}} \nabla (u - u_{\mathcal{T}}) \right\| \le c^* \left\{ \sum_{K \in \mathcal{T}} \eta_{R,K}^2 + \sum_{K \in \mathcal{T}} \mu_K^2 \left\| f - f_{\mathcal{T}} \right\|_K^2 \right\}^{\frac{1}{2}}$$

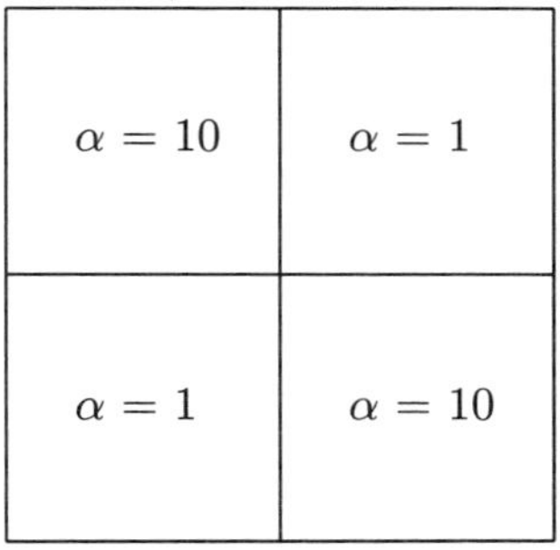

Figure 4.4 Piecewise constant diffusion violating the monotonicity assumption of Remark 4.40

and

$$\eta_{R,K} \leq c_* \left\{ \left\| \alpha^{\frac{1}{2}} \nabla (u - u_{\mathcal{T}}) \right\|_{\omega_K}^2 + \sum_{K' \subset \omega_K} \mu_{K'}^2 \left\| f - f_{\mathcal{T}} \right\|_{K'}^2 \right\}^{\frac{1}{2}}$$

hold for all $K \in \mathcal{T}$.

4.6.2 Piecewise Constant Isotropic Diffusion

Now, we consider problem (4.24) with a piecewise constant function A with values in the space of square, symmetric, positive definite matrices of order d. Denote by A_ℓ the constant value of A on Ω_ℓ and by $\lambda_{\max}(A_\ell)$ and $\lambda_{\min}(A_\ell)$ the extremal eigenvalues of A_ℓ. Set

$$\alpha_{\min} = \min_{1 \leq \ell \leq L} \lambda_{\min}(A_\ell), \ \alpha_{\max} = \max_{1 \leq \ell \leq L} \lambda_{\max}(A_\ell), \ \kappa = \max_{1 \leq \ell \leq L} \frac{\lambda_{\max}(A_\ell)}{\lambda_{\min}(A_\ell)}.$$

We are interested in the case where $\frac{\alpha_{\max}}{\alpha_{\min}}$ is large, but κ is of moderate size. The case of a large κ is treated in Section 4.6.4.

As in the previous subsection, we assume that the partition $\Omega_1, \ldots, \Omega_L$ of Ω is consistent with $\mathcal{T}$. Denote by A_K the constant value of A on the element K and set

$$\mu_K = h_K \lambda_{\max}(A_K)^{-\frac{1}{2}} \quad \text{and} \quad \mu_E = h_E \left\{ \max_{K \subset \omega_E} \lambda_{\max}(A_K) \right\}^{-1}.$$

With this choice of weights Theorem 4.41 directly carries over to the new situation provided, of course, the diffusion αI is replaced by A.

4.6.3 Piecewise Smooth Scalar Diffusion

We now consider problem (4.24) with $A = \alpha I$ where α is a bounded and piecewise twice continuously differentiable function. The relevant parameters are now

$$\alpha_{\ell,\min} = \inf_{x \in \Omega_\ell} \alpha(x), \ \alpha_{\ell,\max} = \sup_{x \in \Omega_\ell} \alpha(x), \ \kappa = \max_{1 \leq \ell \leq L} \frac{\alpha_{\ell,\max}}{\alpha_{\ell,\min}}$$

and

$$\alpha_{\min} = \min_{1 \leq \ell \leq L} \alpha_{\ell,\min}, \quad \alpha_{\max} = \max_{1 \leq \ell \leq L} \alpha_{\ell,\max}.$$

We are interested in the case that the ratio $\frac{\alpha_{\max}}{\alpha_{\min}}$ is large, but that the quantity κ is of moderate size.

As in the previous subsections, we assume that the partition $\Omega_1, \ldots, \Omega_L$ of Ω is consistent with $\mathcal{T}$. We now define the weights μ_K and μ_E by

$$\mu_K = h_K \left\{ \sup_{x \in K} \alpha(x) \right\}^{-\frac{1}{2}} \quad \text{and} \quad \mu_E = h_E \left\{ \sup_{x \in \omega_E} \alpha(x) \right\}^{-1}.$$

Since the diffusion α is no longer constant, we must proceed as in Section 4.4 (p. 163) and approximate it to obtain a computable error indicator. To this end we denote by $\alpha_{\mathcal{T}}$ the L^2-projection of α onto the space $S^{k,-1}(\mathcal{T})$ and replace the diffusion α by $\alpha_{\mathcal{T}}$ in the definition of $\eta_{R,K}$. Similarly to Theorem 4.19 (p. 168), this introduces in the error estimates an additional data error

$$\theta_K = \left\{ \mu_K^2 \left\| \operatorname{div}((\alpha - \alpha_{\mathcal{T}})\nabla u_{\mathcal{T}}) \right\|_K^2 + \frac{1}{2} \sum_{E \in \mathcal{E}_{K,\Omega}} \mu_E \left\| \mathbb{J}_E(\mathbf{n}_E \cdot (\alpha - \alpha_{\mathcal{T}})\nabla u_{\mathcal{T}}) \right\|_E^2 \right\}^{\frac{1}{2}}.$$

With these modifications, Theorem 4.41 directly carries over to the new situation.

The data error θ_K can explicitly be estimated as follows. Using a standard inverse estimate and well-known approximation results for finite element functions and invoking the quantity κ defined above, we obtain for every element

$$\begin{aligned}
\left\| \operatorname{div}((\alpha - \alpha_{\mathcal{T}})\nabla u_{\mathcal{T}}) \right\|_K &\le \left\| \nabla(\alpha - \alpha_{\mathcal{T}}) \right\|_{\infty;K} \|\nabla u_{\mathcal{T}}\|_K + \|\alpha - \alpha_{\mathcal{T}}\|_{\infty;K} \|\Delta u_{\mathcal{T}}\|_K \\
&\le c h_K^k \|\alpha - \alpha_{\mathcal{T}}\|_{k+1,\infty;K} \|\nabla u_{\mathcal{T}}\|_K \\
&\le c h_K^k \|\alpha - \alpha_{\mathcal{T}}\|_{k+1,\infty;K} \cdot \left\{ \inf_{x \in K} \alpha(x) \right\}^{-\frac{1}{2}} \left\| \alpha^{\frac{1}{2}} \nabla u_{\mathcal{T}} \right\|_K \\
&\le c h_K^{k-1} \mu_K \kappa^{\frac{1}{2}} \|\alpha - \alpha_{\mathcal{T}}\|_{k+1,\infty;K} \cdot \left\| \alpha^{\frac{1}{2}} \nabla u_{\mathcal{T}} \right\|_K
\end{aligned}$$

and for every face

$$\begin{aligned}
\left\| \mathbb{J}_E(\mathbf{n}_E \cdot (\alpha - \alpha_{\mathcal{T}})\nabla u_{\mathcal{T}}) \right\|_E &\le \|\alpha - \alpha_{\mathcal{T}}\|_{\infty;E} \left\| \mathbb{J}_E(\mathbf{n}_E \cdot \nabla u_{\mathcal{T}}) \right\|_E \\
&\le c \|\alpha - \alpha_{\mathcal{T}}\|_{\infty;\omega_E} h_E^{-\frac{1}{2}} \|\nabla u_{\mathcal{T}}\|_{\omega_E} \\
&\le c h_E^{k+\frac{1}{2}} \|\alpha - \alpha_{\mathcal{T}}\|_{k+1,\infty;\omega_E} \|\nabla u_{\mathcal{T}}\|_{\omega_E} \\
&\le c h_E^k \mu_E^{\frac{1}{2}} \kappa^{\frac{1}{2}} \|\alpha - \alpha_{\mathcal{T}}\|_{k+1,\infty;\omega_E} \cdot \left\| \alpha^{\frac{1}{2}} \nabla u_{\mathcal{T}} \right\|_{\omega_E}.
\end{aligned}$$

4.6.4 *Constant Anisotropic Diffusion*

We consider problem (4.24) with a constant, symmetric, positive definite matrix A such that the ratio of its largest eigenvalue $\lambda_{\max}$ to its smallest one $\lambda_{\min}$ is large. We want to derive a posteriori error estimates which are independent of this ratio.

To this end, we consider the transformation

$$\begin{aligned}
\Phi : \mathbb{R}^d &\to \mathbb{R}^d \\
x &\mapsto \widetilde{x} = A^{-\frac{1}{2}} x
\end{aligned}$$

and denote by $\widetilde{\Omega}$ the image of Ω under Φ. An elementary calculation shows that u is a weak solution of problem (4.24) if and only if

$$\widetilde{u} = u \circ \Phi^{-1}$$

is a weak solution of the Laplace equation (1.1) (p. 4) on $\widetilde{\Omega}$ with right-hand side

$$\widetilde{f} = f \circ \Phi^{-1}$$

and homogeneous Dirichlet boundary conditions.

The transformation Φ maps the partition $\mathcal{T}$ of Ω into a partition $\widetilde{\mathcal{T}}$ of $\widetilde{\Omega}$ via

$$\widetilde{\mathcal{T}} = \{\Phi(K) : K \in \mathcal{T}\}$$

as illustrated in Figure 4.5. Since the ratio $\frac{\lambda_{\max}}{\lambda_{\min}}$ is large, the shape-regular partition $\mathcal{T}$ is thus transformed into an anisotropic partition $\widetilde{\mathcal{T}}$. The finite element spaces $S_0^{k,0}(\mathcal{T})$ and $S_0^{k,0}(\widetilde{\mathcal{T}})$ corresponding to these partitions are related by

$$v \in S_0^{k,0}(\mathcal{T}) \quad \Longleftrightarrow \quad \widetilde{v} = v \circ \Phi^{-1} \in S_0^{k,0}(\widetilde{\mathcal{T}}).$$

Therefore, $u_{\mathcal{T}} \in S_0^{k,0}(\mathcal{T})$ is a solution of the discretisation of problem (4.24) corresponding to $\mathcal{T}$ if and only if

$$\widetilde{u}_{\mathcal{T}} = u_{\mathcal{T}} \circ \Phi^{-1} \in S_0^{k,0}(\widetilde{\mathcal{T}})$$

is the solution of the discretisation of the transformed problem (1.1) (p. 4) on $\widetilde{\Omega}$ corresponding to $\widetilde{\mathcal{T}}$.

Since $\widetilde{\mathcal{T}}$ in general is anisotropic, Theorem 4.30 (p. 184) yields a good a posteriori error indicator for the transformed problem and its discretisation. Taking into account that we now consider pure Dirichlet boundary conditions, the indicator and the matching function, which enters in the upper bound on the error, take the particular form

$$\widetilde{\eta}_{R,\widetilde{K}} = \left\{ h_{\min,\widetilde{K}}^2 \left\| \widetilde{f}_{\mathcal{T}} + \Delta \widetilde{u}_{\mathcal{T}} \right\|_{\widetilde{K}}^2 + \frac{1}{2} \sum_{\widetilde{E} \in \mathcal{E}_{\widetilde{K},\widetilde{\Omega}}} h_{\min,\widetilde{E}}^2 (h_{\widetilde{E}}^{\perp})^{-1} \left\| \mathbb{J}_E(\mathbf{n}_{\widetilde{E}} \cdot \nabla \widetilde{u}_{\mathcal{T}}) \right\|_{\widetilde{E}}^2 \right\}^{\frac{1}{2}}$$

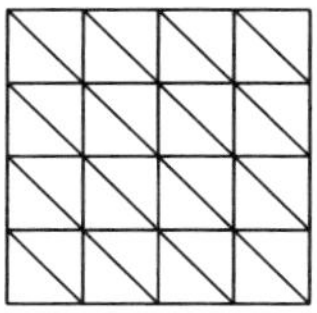

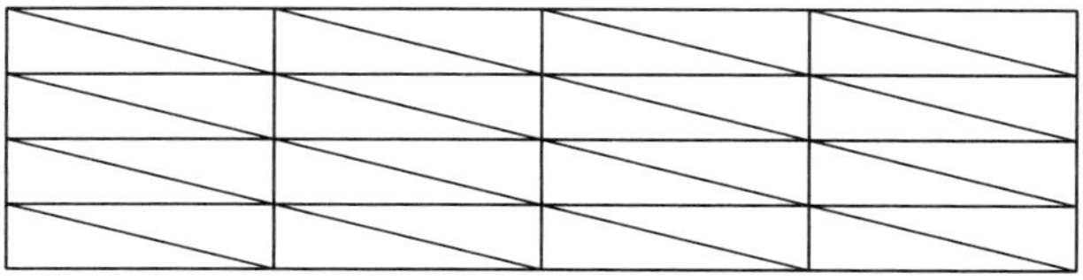

Figure 4.5 Partitions $\mathcal{T}$ (top) and $\widetilde{\mathcal{T}}$ (bottom) of the domains Ω and $\widetilde{\Omega} = \Phi(\Omega)$

and

$$\widetilde{m}_{\widetilde{\mathcal{T}}}(\widetilde{\varphi}) = \left\{ \sum_{\widetilde{K}\in\widetilde{\mathcal{T}}} \sum_{i=1}^{d} h_{\min,\widetilde{K}}^{-2} \left\| \widetilde{\mathbf{p}}_i \cdot \nabla\widetilde{\varphi} \right\|_{\widetilde{K}}^2 \right\}^{\frac{1}{2}} \|\nabla\widetilde{\varphi}\|_{\widetilde{\Omega}}^{-1}.$$

Since

$$\left\| A^{\frac{1}{2}} \nabla(u - u_{\mathcal{T}}) \right\| = \det(A)^{\frac{1}{4}} \left\| \nabla(\widetilde{u} - \widetilde{u}_{\mathcal{T}}) \right\|_{\widetilde{\Omega}}$$

Theorem 4.30 (p. 184) suggests we use

$$\eta_{R,K} = \det(A)^{\frac{1}{4}} \widetilde{\eta}_{R,\Phi(K)}$$

as an error indicator for problem (4.24).

Next, we want to express $\eta_{R,K}$ by quantities which only refer to the element K and which do not resort to the transformation Φ. To this end we define an A-dependent norm $|\cdot|_A$ on $\mathbb{R}^d$ by

$$|x|_A = \left| A^{-\frac{1}{2}} x \right| = |\Phi(x)|,$$

where $|\cdot|$ denotes the standard Euclidean norm on $\mathbb{R}^d$.

We start with the weights. Denote by $P_0, \ldots P_d$ the vertices of a given element K such that they are the pre-images of the vertices $\widetilde{P}_0, \ldots, \widetilde{P}_d$ which correspond to $\widetilde{K} = \Phi(K)$ and which are defined as in Section 4.5. In two dimensions, we immediately conclude that

$$|P_0 - P_1|_A = \max_{0\le i<j\le 2} \left| P_i - P_j \right|_A$$

and

$$h_{\min,\Phi(K)} = h_{A,\min,K} = \inf_{y\in P_0P_1} \left| P_2 - y \right|_A. \tag{4.27}$$

In three dimensions, we observe that, among two faces of a tetrahedron sharing an edge, the face, which has the maximal height above the common edge, has maximal area. Hence, we conclude that

$$|P_0 - P_1|_A = \max_{0\le i<j\le 3} \left| P_i - P_j \right|_A,$$

$$\inf_{y\in P_0P_1} \left| P_2 - y \right|_A = \max_{2\le i\le 3} \inf_{y\in P_0P_1} \left| P_i - y \right|_A,$$

and

$$h_{\min,\Phi(K)} = h_{A,\min,K} = \inf_{y\in \Delta P_0P_1P_2} \left| P_3 - y \right|_A. \tag{4.28}$$

Given a face E in $\mathcal{E}_\Omega$, denote by K_E the element adjacent to E which has minimal $h_{A,\min,K}$ and set

$$h_E^{\perp} = d\frac{\mu_d(K_E)}{\mu_{d-1}(E)}. \tag{4.29}$$

Note that $h_E^\perp$ is the height of K_E above E measured in the *Euclidean* norm and that it is *not* $h_{\widetilde{E}}^\perp$ expressed in quantities referring to E.

Next, we consider the element residuals. From the transformation rule we immediately obtain

$$\det(A)^{\frac{1}{4}} h_{\min,\Phi(K)} \left\| \widetilde{f}_{\mathcal{T}} + \Delta \widetilde{u}_{\mathcal{T}} \right\|_{\Phi(K)} = h_{A,\min,K} \left\| f_{\mathcal{T}} + \operatorname{div}(A \nabla u_{\mathcal{T}}) \right\|_K .$$

Now, we turn to the face residuals. As a preparatory step, for all faces $E \in \mathcal{E}$, we want to prove the identity

$$\mathbf{n}_{\Phi(E)} \cdot \nabla \widetilde{u}_{\mathcal{T}} = \frac{\mu_{d-1}(E)}{\mu_{d-1}(\Phi(E))} \det(A)^{-\frac{1}{2}} \mathbf{n}_E \cdot A \nabla u_{\mathcal{T}}. \tag{4.30}$$

In two dimensions, $d = 2$, we denote by e a vector that has the same length as E, is parallel to E, and satisfies $\det(\mathbf{n}_E, e) > 0$. The vector $\widetilde{e}$ is defined correspondingly with $\Phi(E)$ instead of E. Set

$$P = \begin{pmatrix} 0 & 1 \\ -1 & 0 \end{pmatrix}.$$

Since for any regular, symmetric matrix B of order 2

$$PBP^T = \det(B) B^{-1},$$

we obtain

$$\begin{aligned}
\mathbf{n}_{\Phi(E)} \cdot \nabla \widetilde{u}_{\mathcal{T}} &= \mu_1(\Phi(E))^{-1} P\widetilde{e} \cdot (A^{\frac{1}{2}} \nabla u_{\mathcal{T}}) \\
&= \mu_1(\Phi(E))^{-1} P A^{-\frac{1}{2}} e \cdot (A^{\frac{1}{2}} \nabla u_{\mathcal{T}}) \\
&= \mu_1(\Phi(E))^{-1} P A^{-\frac{1}{2}} P^T P e \cdot (A^{\frac{1}{2}} \nabla u_{\mathcal{T}}) \\
&= \mu_1(\Phi(E))^{-1} \mu_1(E) \det(A)^{-\frac{1}{2}} A^{\frac{1}{2}} \mathbf{n}_E \cdot (A^{\frac{1}{2}} \nabla u_{\mathcal{T}}) \\
&= \frac{\mu_1(E)}{\mu_1(\Phi(E))} \det(A)^{-\frac{1}{2}} \mathbf{n}_E \cdot (A \nabla u_{\mathcal{T}}).
\end{aligned}$$

In the case $d = 3$, we choose two different edges E_a and E_b of the face E. Denote by α and β two vectors that are parallel to E_a and E_b, have the same length as E_a and E_b, and satisfy $\det(\alpha, \beta, \mathbf{n}_E) > 0$. Set $\widetilde{E}_a = \Phi(E_a)$, $\widetilde{E}_b = \Phi(E_b)$, and denote by $\widetilde{\alpha}$ and $\widetilde{\beta}$ the corresponding vectors. Denoting by $\times$ the vector product in $\mathbb{R}^3$, we then get

$$\begin{aligned}
\mathbf{n}_{\Phi(E)} \cdot \nabla \widetilde{u}_{\mathcal{T}} &= \mu_2(\widetilde{\alpha} \times \widetilde{\beta})^{-1} (\widetilde{\alpha} \times \widetilde{\beta}) \cdot (A^{\frac{1}{2}} \nabla u_{\mathcal{T}}) \\
&= \mu_2(\widetilde{\alpha} \times \widetilde{\beta})^{-1} (A^{-\frac{1}{2}} \alpha \times A^{-\frac{1}{2}} \beta) \cdot (A^{-\frac{1}{2}} A \nabla u_{\mathcal{T}}).
\end{aligned}$$

Denote by $a_1, \ldots, a_3$ the columns of $A^{-\frac{1}{2}}$ and by $e_1, \ldots, e_3$ the standard unit vectors of $\mathbb{R}^3$. Since $A^{-\frac{1}{2}}$ is symmetric and since the mapping

$$x, y, z \mapsto x \cdot (y \times z) = \det(x, y, z)$$

is an alternating trilinear form on $\mathbb{R}^3$, we conclude that

$$\begin{aligned} A^{-\frac{1}{2}}(A^{-\frac{1}{2}}\alpha \times A^{-\frac{1}{2}}\beta) &= \sum_{1\le i,j,k\le 3} e_i\alpha_j\beta_k a_i \cdot (a_j \times a_k) \\ &= \det(A)^{-\frac{1}{2}}\alpha \times \beta. \end{aligned}$$

We therefore obtain

$$\begin{aligned} \mathbf{n}_{\Phi(E)} \cdot \nabla\widetilde{u}_{\mathcal{T}} &= \mu_2(\widetilde{\alpha} \times \widetilde{\beta})^{-1} \det(A)^{-\frac{1}{2}}(\alpha \times \beta) \cdot (A\nabla u_{\mathcal{T}}) \\ &= \frac{\mu_2(\alpha \times \beta)}{\mu_2(\widetilde{\alpha} \times \widetilde{\beta})} \det(A)^{-\frac{1}{2}} \mathbf{n}_E \cdot (A\nabla u_{\mathcal{T}}) \\ &= \frac{\mu_2(E)}{\mu_2(\Phi(E))} \det(A)^{-\frac{1}{2}} \mathbf{n}_E \cdot (A\nabla u_{\mathcal{T}}) \end{aligned}$$

which completes the proof of equation (4.30).

With the help of equation (4.30) we may rewrite the face residuals as follows

$$\begin{aligned} &\det(A)^{\frac{1}{4}} h_{\min,\Phi(K)} (h_{\Phi(E)}^{\perp})^{-\frac{1}{2}} \left\| \mathbb{J}_{\Phi(e)}(\mathbf{n}_{\Phi(E)} \cdot \nabla \widetilde{u}_{\mathcal{T}}) \right\|_{\Phi(E)} \\ &= h_{A,\min,K} \left\{ \frac{\mu_{d-1}(E)}{h_{\Phi(E)}^{\perp} \mu_{d-1}(\Phi(E)) \det(A)^{\frac{1}{2}}} \right\}^{\frac{1}{2}} \left\| \mathbb{J}_E(\mathbf{n}_E \cdot A\nabla u_{\mathcal{T}}) \right\|_E . \end{aligned}$$

Since

$$\begin{aligned} d\mu_d(\Phi(K)) &= h_{\Phi(E)}^{\perp} \mu_{d-1}(\Phi(E)), \\ d\mu_d(K) &= h_E^{\perp} \mu_{d-1}(E), \\ \mu_d(K) &= \det(A)^{\frac{1}{2}} \mu_d(\Phi(K)), \end{aligned}$$

this yields the identity

$$\det(A)^{\frac{1}{4}} h_{\min,\Phi(K)} (h_{\Phi(E)}^{\perp})^{-\frac{1}{2}} \left\| \mathbb{J}_{\Phi(e)}(\mathbf{n}_{\Phi(E)} \cdot \nabla\widetilde{u}_{\mathcal{T}}) \right\|_{\Phi(E)} = h_{A,\min,K} (h_E^{\perp})^{-\frac{1}{2}} \left\| \mathbb{J}_E(\mathbf{n}_E \cdot A\nabla u_{\mathcal{T}}) \right\|_E .$$

Finally, we consider the matching function. Denote by $\mathbf{p}_{1,K}, \ldots, \mathbf{p}_{d,K}$ the pre-images of the vectors $\widetilde{\mathbf{p}}_{1,\widetilde{K}}, \ldots, \widetilde{\mathbf{p}}_{d,\widetilde{K}}$ which correspond to $\widetilde{K} = \Phi(K)$ and are defined as in Section 4.5.1 (p. 177). Without resorting to the transformation Φ, these can be computed as follows:

- $\mathbf{p}_{1,K}$ is parallel to P_0P_1 and points to P_1;
- $\mathbf{p}_{2,K,}$ lies in the plane $P_0P_1P_2$, is A^{-1}-orthogonal to $\mathbf{p}_{1,K}$ and points to P_2;
- if $d = 3$, $\mathbf{p}_{3,K}$ is A^{-1}-orthogonal to the plane $P_0P_1P_2$ and points to P_3.

Here, A^{-1}-orthogonality of two vectors x and y means that $x \cdot A^{-1}y = 0$. From these properties we conclude that

$$\widetilde{\mathbf{p}}_{i,\widetilde{K}} \cdot \nabla\widetilde{u}_{\mathcal{T}} = \mathbf{p}_{i,K} \cdot \nabla u_{\mathcal{T}}.$$

This yields

$$\widetilde{m}_{\mathcal{T}}(\widetilde{u}-\widetilde{u}_{\mathcal{T}}) = \left\{\sum_{K\in\mathcal{T}}\sum_{i=1}^{d} h_{A,\min,K}^{-2} \left\|\mathbf{p}_{i,K}\cdot\nabla(u-u_{\mathcal{T}})\right\|_K^2\right\}^{\frac{1}{2}} \left\|A^{\frac{1}{2}}\nabla(u-u_{\mathcal{T}})\right\|^{-1}.$$

Since the vectors $\mathbf{p}_{i,K}$ are mutually A^{-1}-orthogonal and satisfy

$$\left|\mathbf{p}_{i,K}\right|_A \leq h_{A,\max,K}$$

with

$$h_{A,\max,K} = \max_{x,y\in K}\left|x-y\right|_A, \tag{4.31}$$

the matching function can be bounded by

$$\widetilde{m}_{\mathcal{T}}(\widetilde{u}-\widetilde{u}_{\mathcal{T}}) \leq \max_{K\in\mathcal{T}} \frac{h_{A,\max,K}}{h_{A,\min,K}}.$$

Summarising all these results, we arrive at the following a posteriori error estimates.

Theorem 4.42 *Define the quantities $h_{A,\min,K}$ and $h_E^{\perp}$ as in equations (4.27), (4.28), and (4.29) and set*

$$\eta_{R,K} = \left\{ h_{A,\min,K}^2 \left\|f_{\mathcal{T}} + \operatorname{div}(A\nabla u_{\mathcal{T}})\right\|_K^2 + \frac{1}{2}\sum_{E\in\mathcal{E}_{K,\Omega}} h_{A,\min,K}^2 (h_E^{\perp})^{-1} \left\|\mathbb{J}_E(\mathbf{n}_E\cdot A\nabla u_{\mathcal{T}})\right\|_E^2 \right\}^{\frac{1}{2}}.$$

Then the following a posteriori error estimates hold

$$\left\|A^{\frac{1}{2}}\nabla(u-u_{\mathcal{T}})\right\| \leq c_1 m_{\mathcal{T}}(u-u_{\mathcal{T}})\cdot\left\{\sum_{K\in\mathcal{T}}\left[\eta_{R,K}^2 + h_{A,\min,K}^2\left\|f-f_{\mathcal{T}}\right\|_K^2\right]\right\}^{\frac{1}{2}}$$

and

$$\eta_{R,K} \leq c_2\left\{\left\|A^{\frac{1}{2}}\nabla(u-u_{\mathcal{T}})\right\|_{\omega_K}^2 + \sum_{K'\subset\omega_K} h_{A,\min,K'}^2\left\|f-f_{\mathcal{T}}\right\|_{K'}^2\right\}^{\frac{1}{2}}.$$

Here, $f_{\mathcal{T}}$ is any finite element approximation of f corresponding to $\mathcal{T}$. The constants c_1 and c_2 neither depend on h, nor on any shape parameter $C_{\mathcal{T}}$ of $\mathcal{T}$, nor on the ratio $\frac{\lambda_{\max}}{\lambda_{\min}}$. The term $m_{\mathcal{T}}(u-u_{\mathcal{T}})$ is given by

$$m_{\mathcal{T}}(u-u_{\mathcal{T}}) = \left\{\sum_{K\in\mathcal{T}}\sum_{i=1}^{d} h_{A,\min,K}^{-2}\left\|\mathbf{p}_{i,K}\cdot\nabla(u-u_{\mathcal{T}})\right\|_K^2\right\}^{\frac{1}{2}}\left\|A^{\frac{1}{2}}\nabla(u-u_{\mathcal{T}})\right\|^{-1},$$

and can be bounded from above by $\max_{K\in\mathcal{T}}\frac{h_{A,\max,K}}{h_{A,\min,K}}$ with $h_{A,\max,K}$ defined in equation (4.31).

Remark 4.43 In the trivial case $A = \alpha I$ with $\alpha \in \mathbb{R}$, the indicators of Theorems 4.41 and 4.42 are of course identical. Theorem 4.42 can be generalised to the case of a piecewise smooth, anisotropic diffusion along the lines indicated in Sections 4.6.2 and 4.6.3.

Remark 4.44 The results of this section were first published in [66]. Similar results based on the hyper-circle method may be found in [275].

4.7 Eigenvalue Problems

As an example for the treatment of eigenvalue problems, we consider in this section the problem

$$\begin{aligned} -\nabla \cdot (A(x)\nabla u) + b(x)u &= \lambda u \quad \text{in } \Omega \\ u &= 0 \quad \text{on } \Gamma \\ \|u\| &= 1. \end{aligned} \tag{4.32}$$

Here, Ω is a bounded polyhedral domain with Lipschitz boundary, $b \in C(\Omega, \mathbb{R}_+)$ and $A \in C^1(\Omega, \mathbb{R}^{d\times d})$ is symmetric and uniformly positive definite on Ω.

The variational formulation of problem (4.32) is given by:

find $\lambda \in \mathbb{R}$ and $u \in H_0^1(\Omega)$ with $\|u\| = 1$ such that

$$\int_\Omega (\nabla u \cdot A\nabla v + buv) = \lambda \int_\Omega uv \tag{4.33}$$

holds for all $v \in H_0^1(\Omega)$.

Problem (4.33) has an infinite sequence of solutions λ_j, u_j with positive eigenvalues λ_j diverging to ∞. We assume that the eigenvalues are ordered increasingly, i.e. $\lambda_1 \le \lambda_2 \le \cdots$.

The standard finite element discretisation of problem (4.33) takes the form:

find $\lambda_\mathcal{T} \in \mathbb{R}$ and $u_\mathcal{T} \in V_\mathcal{T}$ with $\|u_\mathcal{T}\| = 1$ such that

$$\int_\Omega (\nabla u_\mathcal{T} \cdot A\nabla v_\mathcal{T} + bu_\mathcal{T}v_\mathcal{T}) = \lambda_\mathcal{T} \int_\Omega u_\mathcal{T}v_\mathcal{T} \tag{4.34}$$

holds for all $v_\mathcal{T} \in V_\mathcal{T}$.

Here, $V_\mathcal{T} \subset H_0^1(\Omega)$ is any finite element space associated with an admissible, affine-equivalent, shape-regular partition $\mathcal{T}$ as in Section 3.2.1 (p. 81) which contains the space $S_0^{1,0}(\mathcal{T})$. Problem (4.34) has $N_\mathcal{T} = \dim V_\mathcal{T}$ solutions $\lambda_{\mathcal{T},i}$, $u_{\mathcal{T},i}$ with positive eigenvalues which we also assume increasingly ordered, i.e. $\lambda_{\mathcal{T},1} \le \lambda_{\mathcal{T},2} \le \cdots \le \lambda_{\mathcal{T},N_\mathcal{T}}$.

In what follows we compare solutions λ, u and $\lambda_\mathcal{T}$, $u_\mathcal{T}$ of problems (4.33) and (4.34), respectively which fit together in the sense that $\lambda = \lambda_i$ and $\lambda_\mathcal{T} = \lambda_{\mathcal{T},i}$ for a common index $i \in \{1, \ldots, N_\mathcal{T}\}$. We equip $H_0^1(\Omega)$ with the energy norm

$$|||v||| = \left\{ \int_\Omega (\nabla u \cdot A\nabla u + bu^2) \right\}^{\frac{1}{2}}$$

and its dual space $H^{-1}(\Omega)$ with the associated dual norm

$$|||\varphi|||_* = \sup_{v \in H_0^1(\Omega)\setminus\{0\}} \frac{\langle \varphi, v\rangle}{|||v|||}.$$

With every solution $\lambda_\mathcal{T}, u_\mathcal{T}$ of problem (4.34) we associate a residual $R \in H^{-1}(\Omega)$ by setting for all $v \in H_0^1(\Omega)$

$$\langle R, v\rangle = \lambda_\mathcal{T} \int_\Omega u_\mathcal{T} v - \int_\Omega (\nabla u_\mathcal{T} \cdot A\nabla v + b u_\mathcal{T} v).$$

Since the space $V_\mathcal{T}$ is supposed to contain the space $S_0^{1,0}(\mathcal{T})$, the residual R satisfies the Galerkin orthogonality (3.30) (p. 132). Integration by parts element-wise shows that it satisfies the L^2-representation (3.29) (p. 132) with

$$r|_K = \lambda_\mathcal{T} u_\mathcal{T} + \operatorname{div}(A\nabla u_\mathcal{T}) - b u_\mathcal{T} \quad \text{for all } K,$$

$$j|_E = \begin{cases} -\mathbb{J}_E(\mathbf{n}_E \cdot A\nabla u_\mathcal{T}) & \text{if } E \in \mathcal{E}_\Omega, \\ 0 & \text{if } E \in \mathcal{E}_\Gamma. \end{cases}$$

Hence, $|||R|||_*$ can be bounded from below and from above using the methods of Section 3.8 (p. 132).

Next, we establish a link between the residual R and the error $u - u_\mathcal{T}$. Combining the definition of R and equation (4.33), we obtain for every $v \in H_0^1(\Omega)$

$$\begin{aligned} \langle R, v\rangle &= \lambda_\mathcal{T} \int_\Omega u_\mathcal{T} v - \lambda \int_\Omega uv + \int_\Omega [\nabla(u - u_\mathcal{T}) \cdot A\nabla v + b(u - u_\mathcal{T})v] \\ &= \int_\Omega [\nabla(u - u_\mathcal{T}) \cdot A\nabla v + b(u - u_\mathcal{T})v] \\ &\quad + \frac{1}{2}(\lambda_\mathcal{T} - \lambda)\int_\Omega (u_\mathcal{T} + u)v + \frac{1}{2}(\lambda_\mathcal{T} + \lambda)\int_\Omega (u_\mathcal{T} - u)v. \end{aligned}$$

This identity, the Cauchy–Schwarz inequality, the normalisation $\|u\| = \|u_\mathcal{T}\| = 1$, and Friedrichs' inequality

$$\|v\| \le c_F\, |||v|||$$

imply for every $v \in H_0^1(\Omega)$

$$\frac{\langle R, v\rangle}{|||v|||} \le |||u - u_\mathcal{T}|||_{\operatorname{supp} v} + c_F\,|\lambda - \lambda_\mathcal{T}| + \frac{1}{2} c_F (\lambda_\mathcal{T} + \lambda)\, \|u - u_\mathcal{T}\|_{\operatorname{supp} v}.$$

To obtain an estimate in the reverse direction, we insert the function $v = u - u_\mathcal{T}$ in the above identity for R and take into account that

$$\int_\Omega (u_\mathcal{T} + u)(u - u_\mathcal{T}) = \|u\|^2 - \|u_\mathcal{T}\|^2 = 0.$$

We thus arrive at

$$|||u - u_{\mathcal{T}}|||^2 = \langle R, u - u_{\mathcal{T}} \rangle + \frac{1}{2} (\lambda_{\mathcal{T}} + \lambda) \|u - u_{\mathcal{T}}\|^2$$

and

$$|||u - u_{\mathcal{T}}||| \leq |||R|||_* + \frac{1}{2} c_F (\lambda_{\mathcal{T}} + \lambda) \|u - u_{\mathcal{T}}\| .$$

Hence, the results of Section 3.8 (p. 132) yield global upper and local lower bounds for the energy norm of the error $u - u_{\mathcal{T}}$ up to the perturbation terms $|\lambda - \lambda_{\mathcal{T}}|$ and $\|u - u_{\mathcal{T}}\|_\omega$ with suitable subsets ω of Ω. To control the latter, we recall the a priori error estimates [**226**, Theorem 6.4-3 and Lemma 6.4-4]

$$\begin{aligned} |\lambda - \lambda_{\mathcal{T}}| &\leq c_1 \, |||u - u_{\mathcal{T}}|||^2 , \\ \|u - u_{\mathcal{T}}\| &\leq c_2 h_{\mathcal{T}}^r \, |||u - u_{\mathcal{T}}||| , \end{aligned} \tag{4.35}$$

where $r = 1$ if Ω is convex and $r = \frac{1}{2}$ if Ω is not convex.

Combining the above results yields the following a posteriori error estimates.

Theorem 4.45 *Assume that λ, u and $\lambda_{\mathcal{T}}$, $u_{\mathcal{T}}$ are solutions of problems (4.33) and (4.34), respectively, such that λ and $\lambda_{\mathcal{T}}$ are both at the same position in the increasingly ordered spectrum of the corresponding problem. Define the* residual error indicator *$\eta_{R,K}$ and the data error θ_K by*

$$\eta_{R,K} = \left\{ \sum_{K \in \mathcal{T}} h_K^2 \left\| \lambda_{\mathcal{T}} u_{\mathcal{T}} + \operatorname{div}(A_{\mathcal{T}} \nabla u_{\mathcal{T}}) - b_{\mathcal{T}} u_{\mathcal{T}} \right\|_K^2 + \frac{1}{2} \sum_{E \in \mathcal{E}_{K,\Omega}} h_E \left\| \mathbb{J}_E(\mathbf{n}_E \cdot A_{\mathcal{T}} \nabla u_{\mathcal{T}}) \right\|_E^2 \right\}^{\frac{1}{2}}$$

and

$$\begin{aligned} \theta_K = \Bigg\{ & h_K^2 \left\| -\nabla \cdot ((A - A_{\mathcal{T}}) \nabla u_{\mathcal{T}}) + (b - b_{\mathcal{T}}) u_{\mathcal{T}} \right\|_K^2 \\ & + \sum_{E \in \mathcal{E}_{K,\Omega}} h_E \left\| \mathbb{J}_E(\mathbf{n}_E \cdot ((A - A_{\mathcal{T}}) \nabla u_{\mathcal{T}})) \right\|_E \Bigg\}^{\frac{1}{2}} , \end{aligned}$$

where $A_{\mathcal{T}}$ and $b_{\mathcal{T}}$ are piecewise polynomial approximations of A and b, respectively. Then there are two constants c^* and c_* such that the error is bounded from above by

$$|||u - u_{\mathcal{T}}||| \leq c^* \left\{ \sum_{K \in \mathcal{T}} \eta_{R,K}^2 \right\}^{\frac{1}{2}} + c^* \left\{ \sum_{K \in \mathcal{T}} \theta_K^2 \right\}^{\frac{1}{2}} + c^* (\lambda_{\mathcal{T}} + \lambda) \|u - u_{\mathcal{T}}\|$$

and, for every element $K \in \mathcal{T}$, bounded from below by

$$\eta_{R,K} \leq c_* \, |||u - u_{\mathcal{T}}|||_{\omega_K} + c^* \left\{ \sum_{K' \subset \omega_K} \theta_{K'}^2 \right\}^{\frac{1}{2}} + c_* (\lambda_{\mathcal{T}} + \lambda) \|u - u_{\mathcal{T}}\|_{\omega_K} + c_* |\lambda - \lambda_{\mathcal{T}}| .$$

The terms $\|u - u_{\mathcal{T}}\|$, $\|u - u_{\mathcal{T}}\|_{\omega_K}$, and $|\lambda - \lambda_{\mathcal{T}}|$ are higher order perturbations and can be estimated as in (4.35). Both constants c^* and c_* depend on the domain Ω, the relative size of the diffusion A to the reaction b, the ratios of the maximal to the minimal eigenvalues of A, and the shape parameter $C_{\mathcal{T}}$ of $\mathcal{T}$. The constant c_* in addition depends on the polynomial degree of $u_{\mathcal{T}}$.

Remark 4.46 In Section 5.3 (p. 299) we will consider problem (4.32) from a nonlinear perspective. When comparing Theorems 4.45 and 5.23 (p. 300) we observe that in the nonlinear framework we can dispense with the higher order terms of Theorem 4.45. This, on the other hand, has to be paid for by the condition that $\lambda_{\mathcal{T}}, u_{\mathcal{T}}$ has to be sufficiently close to λ, u which essentially means that $|\lambda - \lambda_{\mathcal{T}}|$ must be smaller than the distance of λ to its neighbouring eigenvalues.

Remark 4.47 The presentation of this section is inspired by [**129**]. Other approaches to a posteriori error estimates for eigenvalue problems may be found in [**24, 73, 127, 141, 151, 176, 189, 277**].

4.8 Mixed Formulation of the Poisson Equation

In this section we consider once again the model problem (1.1) (p. 4) in a bounded polyhedral two- or three-dimensional domain Ω with Lipschitz boundary Γ. But, we now impose pure homogeneous Dirichlet boundary conditions and, most important, write problem (1.1) as a first-order system by introducing ∇u as an additional unknown:

$$\begin{aligned} \operatorname{div} \sigma &= -f && \text{in } \Omega \\ \sigma &= \nabla u && \text{in } \Omega \\ u &= 0 && \text{on } \Gamma. \end{aligned} \tag{4.36}$$

Our interest in problem (4.36) is twofold:

- its finite element discretisation introduced below allows the direct approximation of ∇u without resorting to a differentiation of the finite element approximation $u_{\mathcal{T}}$ considered so far;
- its analysis prepares the a posteriori error analysis of the equations of linear elasticity considered in the next section where mixed methods are mandatory to avoid locking phenomena.

4.8.1 Variational Formulation

For the variational formulation of problem (4.36), we recall the definition of the space

$$H(\operatorname{div}; \Omega) = \left\{\sigma \in L^2(\Omega)^d : \operatorname{div} \sigma \in L^2(\Omega)\right\}$$

and of its norm

$$\|\sigma\|_{H(\operatorname{div};\Omega)} = \left\{\|\sigma\|^2 + \|\operatorname{div} \sigma\|^2\right\}^{\frac{1}{2}}$$

cf. [**90**, Section III.1] and Section 3.1.2 (p. 80). Next, we multiply the first equation of (4.36) by a function $v \in L^2(\Omega)$ and the second equation by a vector field $\tau \in H(\operatorname{div}; \Omega)$, integrate both expressions over Ω, and use integration by parts for the integral involving ∇u. We thus arrive at the problem:

find $\sigma \in H(\mathrm{div};\Omega)$ and $u \in L^2(\Omega)$ such that

$$\begin{aligned} \int_\Omega \sigma \cdot \tau + \int_\Omega u \,\mathrm{div}\,\tau &= 0 \\ \int_\Omega \mathrm{div}\,\sigma\, v &= -\int_\Omega f v \end{aligned} \tag{4.37}$$

holds for all $\tau \in H(\mathrm{div};\Omega)$ and $v \in L^2(\Omega)$.

Problems (4.36) and (4.37) are equivalent in the usual weak sense: every classical solution of (4.36) is a solution of (4.37) and every sufficiently regular solution of (4.37) is a classical solution of (4.36).

Problem (4.37) fits into the abstract framework of Section 4.1 (p. 151) with

$$\begin{aligned} X = Y &= H(\mathrm{div};\Omega) \times L^2(\Omega), \\ \|(\tau,v)\|_X = \|(\tau,v)\|_Y &= \left\{ \|\tau\|^2_{H(\mathrm{div};\Omega)} + \|v\|^2 \right\}^{\frac{1}{2}}, \\ B((\sigma,u),(\tau,v)) &= \int_\Omega \sigma\cdot\tau + \int_\Omega u\,\mathrm{div}\,\tau + \int_\Omega \mathrm{div}\,\sigma\, v, \\ \langle \ell, (\tau,v)\rangle_Y &= -\int_\Omega f v. \end{aligned}$$

The following result shows that problem (4.37) fulfils the assumptions of Proposition 4.1 (p. 152). We give an elementary proof and do not aim at obtaining the optimal value for the constant β in the inf–sup condition (4.5) (p. 152).

Proposition 4.48 *The bi-linear form B defined above satisfies the inf–sup condition*

$$\begin{aligned} \inf_{(\tau,v)\in X\setminus\{0\}} \sup_{(\rho,w)\in X\setminus\{0\}} \frac{B((\tau,v),(\rho,w))}{\|(\tau,v)\|_X \|(\rho,w)\|_X} &= \inf_{(\rho,w)\in X\setminus\{0\}} \sup_{(\tau,v)\in X\setminus\{0\}} \frac{B((\tau,v),(\rho,w))}{\|(\tau,v)\|_X \|(\rho,w)\|_X} \\ &\geq \frac{4}{\sqrt{6}(12+\mathrm{diam}(\Omega)^2)}. \end{aligned}$$

Proof The equality follows from the symmetry of B and Remark 4.4 (p. 153).

To prove the inf–sup condition, fix a pair $(\tau,v) \in X \setminus \{0\}$. From the definition of B we immediately obtain

$$\begin{aligned} B((\tau,v),(\tau,-v+\mathrm{div}\,\tau)) &= B((\tau,v),(\tau,-v)) + B((\tau,v),(0,\mathrm{div}\,\tau)) \\ &= \int_\Omega \tau\cdot\tau + \int_\Omega \mathrm{div}\,\tau\,\mathrm{div}\,\tau \\ &= \|\tau\|^2_{H(\mathrm{div};\Omega)}. \end{aligned}$$

Next, we choose a point $x^* \in \Omega$ and a number $R > 0$ such that Ω is contained in

$$B_{|\cdot|_\infty}(x^*,R) = \left\{ x \in \mathbb{R}^d : |x - x^*|_\infty < R \right\},$$

where $|\cdot|_\infty$ denotes the maximum norm on $\mathbb{R}^d$. This is possible since Ω is supposed to be bounded. We extend v by 0 outside Ω and define a vector field ρ by

$$\rho_i(x) = \frac{1}{d}\int_{x_i^*-R}^{x_i} v(x_1,\ldots,x_{i-1},t,x_{i+1},\ldots,x_d)\mathrm{d}t\ ,\ 1 \le i \le d. \tag{4.38}$$

Obviously we have

$$\operatorname{div}\rho = v.$$

The Cauchy–Schwarz inequality, on the other hand, yields for every component i of ρ

$$\begin{aligned}
|\rho_i(x)|^2 &= \frac{1}{d^2}\left|\int_{x_i^*-R}^{x_i} v(x_1,\ldots,x_{i-1},t,x_{i+1},\ldots,x_d)\mathrm{d}t\right|^2 \\
&\le \frac{1}{d^2}(x_i - x_i^* + R)\int_{x_i^*-R}^{x_i} v(x_1,\ldots,x_{i-1},t,x_{i+1},\ldots,x_d)^2\mathrm{d}t \\
&\le \frac{1}{d^2}(x_i - x_i^* + R)\int_{x_i^*-R}^{x_i^*+R} v(x_1,\ldots,x_{i-1},t,x_{i+1},\ldots,x_d)^2\mathrm{d}t
\end{aligned}$$

and therefore

$$\int_{x_i^*-R}^{x_i^*+R} |\rho_i(x)|^2\,\mathrm{d}x_i \le \frac{2R^2}{d^2}\int_{x_i^*-R}^{x_i^*+R} v(x_1,\ldots,x_{i-1},t,x_{i+1},\ldots,x_d)^2\mathrm{d}t.$$

Combining this estimate with Fubini's theorem, we arrive at

$$\begin{aligned}
\|\rho\|^2 &= \sum_{i=1}^d \int_\Omega |\rho_i(x)|^2 \\
&\le \sum_{i=1}^d \int_{B_{|\cdot|_\infty}(x^*,R)} |\rho_i(x)|^2 \\
&\le \frac{2R^2}{d}\|v\|^2 \\
&\le R^2\|v\|^2.
\end{aligned}$$

The Cauchy–Schwarz inequality and the estimate $-ab \ge -\frac{1}{2}a^2 - \frac{1}{2}b^2$ with $a = R\|\tau\|$ and $b = \|v\|$ therefore imply

$$\begin{aligned}
B((\tau,v),(\rho,0)) &= \int_\Omega \tau\cdot\rho + \int_\Omega v\operatorname{div}\rho \\
&\ge -\|\tau\|\,\|\rho\| + \|v\|^2 \\
&\ge -\|\tau\|\,R\,\|v\| + \|v\|^2 \\
&\ge -\frac{R^2}{2}\|\tau\|^2 + \frac{1}{2}\|v\|^2.
\end{aligned}$$

Hence we have for every positive parameter δ

$$
\begin{aligned}
B((\tau,v),(\tau+\delta\rho,-v+\operatorname{div}\tau)) &= B((\tau,v),(\tau,-v+\operatorname{div}\tau)) + B((\tau,v),(\delta\rho,0))\\
&\geq \|\tau\|^2_{H(\operatorname{div};\Omega)} + \frac{\delta}{2}\|v\|^2 - \frac{\delta R^2}{2}\|\tau\|^2\\
&\geq \left(1-\frac{\delta R^2}{2}\right)\|\tau\|^2_{H(\operatorname{div};\Omega)} + \frac{\delta}{2}\|v\|^2.
\end{aligned}
$$

The choice $\delta = \frac{2}{R^2+1}$ yields

$$
\begin{aligned}
B((\tau,v),(\tau+\delta\rho,-v+\operatorname{div}\tau)) &\geq \frac{1}{R^2+1}\left\{\|\tau\|^2_{H(\operatorname{div};\Omega)} + \|v\|^2\right\}\\
&= \frac{1}{R^2+1}\left\|(\tau,v)\right\|^2_X.
\end{aligned}
$$

The previous estimates and the Cauchy–Schwarz inequality for sums on the other hand imply

$$
\begin{aligned}
\left\|(\tau+\delta\rho,-v+\operatorname{div}\tau)\right\|_X &\leq \left\|(\tau,-v)\right\|_X + \left\|(0,\operatorname{div}\tau)\right\|_X + \left\|(\delta\rho,0)\right\|_X\\
&\leq \left\|(\tau,v)\right\|_X + \|\tau\|_{H(\operatorname{div};\Omega)} + \delta\sqrt{1+R^2}\,\|v\|\\
&\leq \sqrt{3}\left\{2\|\tau\|^2_{H(\operatorname{div};\Omega)} + [1+\delta^2(1+R^2)]\,\|v\|^2\right\}^{\frac{1}{2}}\\
&\leq \sqrt{3}\sqrt{2+\delta^2(1+R^2)}\left\{\|\tau\|^2_{H(\operatorname{div};\Omega)} + \|v\|^2\right\}^{\frac{1}{2}}\\
&\leq \sqrt{\frac{6(3+R^2)}{1+R^2}}\left\|(\tau,v)\right\|_X.
\end{aligned}
$$

This proves that

$$
\begin{aligned}
\sup_{(\rho,w)\in X\setminus\{0\}} \frac{B((\tau,v),(\rho,w))}{\left\|(\tau,v)\right\|_X\left\|(\rho,w)\right\|_X} &\geq \frac{B((\tau,v),(\tau+\delta\rho,-v+\operatorname{div}\tau))}{\left\|(\tau,v)\right\|_X\left\|(\tau+\delta\rho,-v+\operatorname{div}\tau)\right\|_X}\\
&\geq \sqrt{\frac{1+R^2}{6(3+R^2)}}\,\frac{1}{1+R^2}\\
&= \frac{1}{\sqrt{6(3+R^2)(1+R^2)}}\\
&\geq \frac{1}{\sqrt{6}(3+R^2)}.
\end{aligned}
$$

Since the point x^* was arbitrary and all estimates are invariant under rotations of the coordinate system, we can choose $R = \frac{1}{2}\operatorname{diam}(\Omega)$ and obtain the assertion. □

4.8.2 Mixed Finite Element Approximation

To keep the exposition as simple as possible, we only consider the simplest discretisation of problem (4.37) which is given by the lowest order *Raviart–Thomas spaces*, cf. [**90**, Sections III.3.1 and III.3.2].

We assume that $\mathcal{T}$ exclusively consists of simplices. For every element $K \in \mathcal{T}$ we set

$$\mathrm{RT}_0(K) = R_0(K)^d + R_0(K) \begin{pmatrix} x_1 \\ \vdots \\ x_d \end{pmatrix}$$

and

$$\mathrm{RT}_0(\mathcal{T}) = \left\{ \sigma_{\mathcal{T}} : \sigma_{\mathcal{T}}\big|_K \in \mathrm{RT}_0(K) \text{ for all } K \in \mathcal{T}, \int_E \mathbb{J}_E(\mathbf{n}_E \cdot \sigma_{\mathcal{T}}) = 0 \text{ for all } E \in \mathcal{E}_\Omega \right\}.$$

The degrees of freedom associated with $\mathrm{RT}_0(\mathcal{T})$ are the values of the normal components of the $\sigma_{\mathcal{T}}$ evaluated at the midpoints of edges, if $d = 2$, or the barycentres of faces, if $d = 3$. Since the normal components $\mathbf{n}_E \cdot \sigma_{\mathcal{T}}$ are constant on the edges or faces, the condition

$$\int_E \mathbb{J}_E(\mathbf{n}_E \cdot \sigma_{\mathcal{T}}) = 0 \quad \text{for all } E \in \mathcal{E}_\Omega$$

ensures that the space $\mathrm{RT}_0(\mathcal{T})$ is contained in $H(\mathrm{div};\Omega)$.

The *mixed finite element approximation* of problem (4.37) is then given by:

find $\sigma_{\mathcal{T}} \in \mathrm{RT}_0(\mathcal{T})$ and $u_{\mathcal{T}} \in S^{0,-1}(\mathcal{T})$ such that

$$\begin{aligned} \int_\Omega \sigma_{\mathcal{T}} \cdot \tau_{\mathcal{T}} + \int_\Omega u_{\mathcal{T}} \operatorname{div} \tau_{\mathcal{T}} &= 0 \\ \int_\Omega \operatorname{div} \sigma_{\mathcal{T}} v_{\mathcal{T}} &= -\int_\Omega f v_{\mathcal{T}} \end{aligned} \tag{4.39}$$

holds for all $\tau_{\mathcal{T}} \in \mathrm{RT}_0(\mathcal{T})$ and $v_{\mathcal{T}} \in S^{0,-1}(\mathcal{T})$.

Problem (4.39) fits into the framework of Section 4.1 (p. 151) with

$$\begin{aligned} X_{\mathcal{T}} = Y_{\mathcal{T}} &= \mathrm{RT}_0(\mathcal{T}) \times S^{0,-1}(\mathcal{T}), \\ B_{\mathcal{T}}((\tau_{\mathcal{T}}, v_{\mathcal{T}}), (\rho_{\mathcal{T}}, w_{\mathcal{T}})) &= B((\tau_{\mathcal{T}}, v_{\mathcal{T}}), (\rho_{\mathcal{T}}, w_{\mathcal{T}})), \\ \langle \ell_{\mathcal{T}}, (\tau_{\mathcal{T}}, v_{\mathcal{T}}) \rangle_{Y_{\mathcal{T}}} &= \langle \ell, (\tau_{\mathcal{T}}, v_{\mathcal{T}}) \rangle_Y . \end{aligned}$$

From the definition of $\mathrm{RT}_0(\mathcal{T})$ we conclude that

$$\operatorname{div}(\mathrm{RT}_0(\mathcal{T})) \subset S^{0,-1}(\mathcal{T})$$

and that the vector field ρ of equation (4.38) is in $\mathrm{RT}_0(\mathcal{T})$ if the function v is in $S^{0,-1}(\mathcal{T})$. The proof of Proposition 4.48 therefore carries over to the discrete space $X_{\mathcal{T}}$ and implies that problem (4.39) fulfils the assumptions of Proposition 4.5 (p. 154).

4.8.3 Commuting Diagram Property

As in previous sections, we denote by $P_{\mathcal{T}} : L^2(\Omega) \to S^{0,-1}(\mathcal{T})$ the L^2-projection which, for all $v \in L^2(\Omega)$ and all $w_{\mathcal{T}} \in S^{0,-1}(\Omega)$, is defined by

$$\int_\Omega P_{\mathcal{T}} v w_{\mathcal{T}} = \int_\Omega v w_{\mathcal{T}}.$$

For any $s > d$, we in addition define a quasi-interpolation operator $\mathcal{J}_{\mathcal{T}} : H(\mathrm{div};\Omega) \cap L^s(\Omega)^d \to \mathrm{RT}_0(\mathcal{T})$ [**90**, Section III.3.3] by setting for all $\sigma \in H(\mathrm{div};\Omega) \cap L^s(\Omega)^d$ and all edges or faces $E \in \mathcal{E}$

$$\int_E \mathbf{n}_E \cdot (\mathcal{J}_{\mathcal{T}}\sigma) = \int_E \mathbf{n}_E \cdot \sigma.$$

Remark 4.49 The additional regularity $\sigma \in L^s(\Omega)^d$ with $s > d$ is needed to ensure that σ has well-defined normal components on the edges or faces of the elements in the sense of traces and that $\mathcal{J}_{\mathcal{T}}$ is continuous.

The following result will be a fundamental tool in the subsequent analysis.

Proposition 4.50 *The operators $\mathcal{J}_{\mathcal{T}}$ and $P_{\mathcal{T}}$ satisfy the* commuting diagram property, *cf. Figure 4.6*

$$\int_\Omega \mathrm{div}(\mathcal{J}_{\mathcal{T}}\tau) v = \int_\Omega P_{\mathcal{T}}(\mathrm{div}\,\tau) v$$

for all $\tau \in H(\mathrm{div};\Omega)$ and all $v \in L^2(\Omega)$ or in short

$$\mathrm{div} \circ \mathcal{J}_{\mathcal{T}} = P_{\mathcal{T}} \circ \mathrm{div}.$$

Proof Since $\mathrm{div}(\mathrm{RT}_0(\mathcal{T})) \subset S^{0,-1}(\mathcal{T})$, we have for all $v \in L^2(\Omega)$ and all $\tau \in H(\mathrm{div};\Omega)$

$$\int_\Omega \mathrm{div}(\mathcal{J}_{\mathcal{T}}\tau) v = \int_\Omega \mathrm{div}(\mathcal{J}_{\mathcal{T}}\tau) P_{\mathcal{T}} v.$$

Integration by parts element-wise and the definition of $\mathcal{J}_{\mathcal{T}}$ therefore yield

$$\begin{aligned}
\int_\Omega \mathrm{div}(\mathcal{J}_{\mathcal{T}}\tau) v &= \sum_{K\in\mathcal{T}} \int_K \mathrm{div}(\mathcal{J}_{\mathcal{T}}\tau) P_{\mathcal{T}} v \\
&= \sum_{K\in\mathcal{T}} \int_{\partial K} \mathbf{n}_K \cdot (\mathcal{J}_{\mathcal{T}}\tau) P_{\mathcal{T}} v \\
&= \sum_{E\in\mathcal{E}_\Omega} \int_E \mathbf{n}_E \cdot (\mathcal{J}_{\mathcal{T}}\tau) \mathbb{J}_E(P_{\mathcal{T}} v) + \sum_{E\in\mathcal{E}_\Gamma} \int_E \mathbf{n}_E \cdot (\mathcal{J}_{\mathcal{T}}\tau) P_{\mathcal{T}} v \\
&= \sum_{E\in\mathcal{E}_\Omega} \int_E \mathbf{n}_E \cdot \tau \mathbb{J}_E(P_{\mathcal{T}} v) + \sum_{E\in\mathcal{E}_\Gamma} \int_E \mathbf{n}_E \cdot \tau P_{\mathcal{T}} v \\
&= \sum_{K\in\mathcal{T}} \int_{\partial K} \mathbf{n}_K \cdot \tau P_{\mathcal{T}} v \\
&= \sum_{K\in\mathcal{T}} \int_K \mathrm{div}\,\tau P_{\mathcal{T}} v \\
&= \int_\Omega \mathrm{div}\,\tau P_{\mathcal{T}} v.
\end{aligned}$$

□

$$\begin{array}{ccc} H(\mathrm{div};\Omega) & \xrightarrow{\mathrm{div}} & L^2(\Omega) \\ \mathcal{J}_{\mathcal{T}} \downarrow & & \downarrow P_{\mathcal{T}} \\ \mathrm{RT}_0(\mathcal{T}) & \xrightarrow{\mathrm{div}} & S^{0,-1}(\mathcal{T}) \end{array}$$

Figure 4.6 Commuting diagram property $\mathrm{div} \circ \mathcal{J}_{\mathcal{T}} = P_{\mathcal{T}} \circ \mathrm{div}$

4.8.4 *The Helmholtz Decomposition*

As usual, we define the so-called *curl operator* curl by

$$\operatorname{curl} \tau = \nabla \times \tau$$

$$= \begin{pmatrix} \dfrac{\partial \tau_3}{\partial x_2} - \dfrac{\partial \tau_2}{\partial x_3} \\ \dfrac{\partial \tau_1}{\partial x_3} - \dfrac{\partial \tau_3}{\partial x_1} \\ \dfrac{\partial \tau_2}{\partial x_1} - \dfrac{\partial \tau_1}{\partial x_2} \end{pmatrix} \qquad \text{if } \tau : \Omega \to \mathbb{R}^3,$$

$$\operatorname{curl} \tau = \frac{\partial \tau_2}{\partial x_1} - \frac{\partial \tau_1}{\partial x_2} \qquad \text{if } \tau : \Omega \to \mathbb{R}^2,$$

$$\operatorname{curl} v = \begin{pmatrix} \dfrac{\partial v}{\partial x_2} \\ -\dfrac{\partial v}{\partial x_1} \end{pmatrix} \qquad \text{if } v : \Omega \to \mathbb{R} \text{ and } d = 2.$$

To write the following results in a compact form independently of the space dimension, we set

$$n_d = \begin{cases} 1 & \text{if } d = 2, \\ 3 & \text{if } d = 3. \end{cases}$$

Since

$$\operatorname{div}(\operatorname{curl} \varphi) = 0$$

for all $\varphi \in H^1(\Omega)^{n_d}$, we have

$$\operatorname{curl} \in \mathcal{L}(H^1(\Omega)^{n_d}, H(\mathrm{div};\Omega)).$$

Moreover, the Gauss theorem yields the following integration by parts formula

$$\int_\omega \operatorname{curl} \varphi \cdot \tau = \int_\omega \varphi \cdot \operatorname{curl} \tau + \int_{\partial\omega} (\tau - (\tau \cdot \mathbf{n})\mathbf{n}) \cdot \varphi \qquad (4.40)$$

for all subsets ω of Ω with Lipschitz boundary $\partial\omega$, all $\varphi \in H^1(\Omega)^{n_d}$, and all $\tau \in H^1(\Omega)^d$. Here, $\mathbf{n}$ is a unit normal vector of $\partial\omega$ and $\tau - (\tau \cdot \mathbf{n})\mathbf{n}$ is the tangential component of τ. It is independent of the orientation of $\mathbf{n}$ and is identified with a scalar function in two dimensions.

The canonical graph space associated with the curl operator is denoted by

$$H(\text{curl}; \Omega) = \left\{\varphi \in L^2(\Omega)^{n_d} : \text{curl}\,\varphi \in L^2(\Omega)^d\right\}$$

and is equipped with the natural graph norm

$$\|\varphi\|_{H(\text{curl};\Omega)} = \left\{\|\varphi\|^2 + \|\text{curl}\,\varphi\|^2\right\}^2.$$

The following proposition gives a fundamental characterisation of solenoidal vector fields.

Proposition 4.51 *For every vector field $\tau \in H(\text{div}; \Omega)$, the following properties are equivalent.*

(1) *The vector field τ is* solenoidal, *i.e.* $\text{div}\,\tau = 0$, *and* $\int_\gamma \tau \cdot n = 0$ *for every connected component* γ *of* Γ.

(2) *There is a* stream function $\varphi \in H^1(\Omega)$, *if* $d = 2$, *or a* vector potential $\varphi \in H^1(\Omega)^d$, *if* $d = 3$, *such that* $\tau = \text{curl}\,\varphi$ *and, if* $d = 3$, $\text{div}\,\varphi = 0$.

Proof See [**145**, Theorem 3.1] for two dimensions and [**21**, Theorem 3.12], [**145**, Theorem 3.4] for three dimensions. □

Proposition 4.51 implies the so-called Helmholtz decomposition of vector fields.

Proposition 4.52 *There are two operators*

$$G \in \mathcal{L}(H(\text{div}; \Omega), H^1(\Omega)), \quad R \in \mathcal{L}(H(\text{div}; \Omega), H(\text{curl}; \Omega))$$

such that the Helmholtz decomposition

$$\tau = \nabla(G\tau) + \text{curl}(R\tau)$$

holds for all $\tau \in H(\text{div}; \Omega)$. *If in addition* Ω *is convex,* G *and* R *map* $H(\text{div}; \Omega)$ *continuously into* $H^2(\Omega)$ *and* $H^1(\Omega)^{n_d}$, *respectively.*

Proof Fix a vector field $\tau \in H(\text{div}; \Omega)$ and denote by $G\tau \in H^1(\Omega)$ the unique solution of the following Neumann problem for the Laplacian

$$\int_\Omega \nabla(G\tau) \cdot \nabla w = \int_\Omega \text{div}\,\tau w$$

$$\int_\Omega G\tau = 0$$

for all $w \in H^1(\Omega)$. Integration by parts of the right-hand side of the first equation gives

$$\int_{\Omega} \nabla(G\tau) \cdot \nabla w = -\int_{\Omega} \tau \cdot \nabla w$$

for all $w \in H_0^1(\Omega)$. Hence we have $\nabla(G\tau) = \operatorname{div}\tau$ in the weak sense and $\mathbf{n} \cdot \nabla(G\tau) = \mathbf{n} \cdot \tau$ in the sense of traces on Γ. Therefore, we may apply Proposition 4.51 to $\tau - \nabla(G\tau)$ and obtain the existence of $R\tau$.

The first continuity result follows from the definitions of the operators and of the corresponding spaces and norms.

The second, stronger statement follows for the operator G from the H^2-regularity of the Laplace operator with Neumann boundary conditions in convex domains. For the operator R it follows for general two-dimensional polygons from the identity $\|\nabla(R\tau)\| = \|\operatorname{curl}(R\tau)\|$ and for convex three-dimensional polyhedra from [**21**, Theorems 2.17 and 3.12]. □

4.8.5 A Residual A Posteriori Error Indicator

We denote by (σ, u) and $(\sigma_{\mathcal{T}}, u_{\mathcal{T}})$ the unique solutions of problems (4.37) and (4.39), respectively.

We first derive a suitable L^2-representation of the residual which is based on the following observations.

- The residual splits into two contributions corresponding to the equations $\sigma = \nabla u$ and $\operatorname{div}\sigma = -f$, respectively.
- The contribution associated with the equation $\sigma = \nabla u$ can be split further by using the Helmholtz decomposition $\tau = \nabla(G\tau) + \operatorname{curl}(R\tau)$ for the test function τ.
- The terms involving products of $\sigma_{\mathcal{T}}$ or $u_{\mathcal{T}}$ with $G\tau$ can be combined into a single term involving $\sigma_{\mathcal{T}}$ alone by using the commuting diagram property.
- The terms involving $R\tau$ can be reformulated using integration by parts element-wise.

We choose an arbitrary pair $(\tau, v) \in H(\operatorname{div}; \Omega) \times L^2(\Omega)$ and obtain

$$\begin{aligned}\langle \ell - L(\sigma_{\mathcal{T}}, u_{\mathcal{T}}), (\tau, v)\rangle_Y = &-\int_{\Omega} \sigma_{\mathcal{T}} \cdot \tau - \int_{\Omega} u_{\mathcal{T}} \operatorname{div}\tau \\ &- \int_{\Omega} (f + \operatorname{div}\sigma_{\mathcal{T}})v.\end{aligned} \tag{4.41}$$

Propositions 4.51 and 4.52 yield for the first two terms on the right-hand side of equation (4.41)

$$\begin{aligned}\int_{\Omega} \sigma_{\mathcal{T}} \cdot \tau + \int_{\Omega} u_{\mathcal{T}} \operatorname{div}\tau = &\int_{\Omega} \sigma_{\mathcal{T}} \cdot \operatorname{curl}(R\tau) + \int_{\Omega} \sigma_{\mathcal{T}} \cdot \nabla(G\tau) \\ &+ \int_{\Omega} u_{\mathcal{T}} \operatorname{div}(\nabla(G\tau)).\end{aligned}$$

From Proposition 4.50 and the first equation of (4.39) we conclude that

$$\begin{aligned}\int_\Omega u_{\mathcal{T}} \operatorname{div}(\nabla(G\tau)) &= \int_\Omega u_{\mathcal{T}} P_{\mathcal{T}}(\operatorname{div}(\nabla(G\tau))) \\ &= \int_\Omega u_{\mathcal{T}} \operatorname{div}(\mathcal{J}_{\mathcal{T}}(\nabla(G\tau))) \\ &= -\int_\Omega \sigma_{\mathcal{T}} \cdot \mathcal{J}_{\mathcal{T}}(\nabla(G\tau)).\end{aligned}$$

We insert the last two identities in equation (4.41) and obtain the intermediate result

$$\begin{aligned}\langle \ell - L(\sigma_{\mathcal{T}}, u_{\mathcal{T}}), (\tau, v)\rangle_Y &= -\int_\Omega \sigma_{\mathcal{T}} \cdot \operatorname{curl}(R\tau) \\ &\quad - \int_\Omega \sigma_{\mathcal{T}} \cdot (Id - \mathcal{J}_{\mathcal{T}})(\nabla(G\tau)) \\ &\quad - \int_\Omega (f + \operatorname{div} \sigma_{\mathcal{T}}) v.\end{aligned}$$

In a last step, we apply the integration by parts formula (4.40) element-wise to the first term on the right-hand side of this equation. This gives the desired representation of the residual

$$\begin{aligned}\langle \ell - L(\sigma_{\mathcal{T}}, u_{\mathcal{T}}), (\tau, v)\rangle_Y &= -\sum_{K \in \mathcal{T}} \int_K \operatorname{curl} \sigma_{\mathcal{T}} \cdot (R\tau) \\ &\quad - \sum_{E \in \mathcal{E}_\Omega} \int_E \mathbb{J}_E(\sigma_{\mathcal{T}} - (\sigma_{\mathcal{T}} \cdot \mathbf{n}_E)\mathbf{n}_E) \cdot (R\tau) \\ &\quad - \sum_{E \in \mathcal{E}_\Gamma} \int_E (\sigma_{\mathcal{T}} - (\sigma_{\mathcal{T}} \cdot \mathbf{n}_E)\mathbf{n}_E) \cdot (R\tau) \\ &\quad - \int_\Omega \sigma_{\mathcal{T}} \cdot (Id - \mathcal{J}_{\mathcal{T}})(\nabla(G\tau)) \\ &\quad - \int_\Omega (f + \operatorname{div} \sigma_{\mathcal{T}}) v.\end{aligned} \tag{4.42}$$

In order to profit from the Galerkin orthogonality, we now set

$$\tau_{\mathcal{T}} = \operatorname{curl}(I_{\mathcal{T}}(R\tau)),$$

where $I_{\mathcal{T}}$ is the quasi-interpolation operator of equation (1.4) (p. 7) with the modification that we take the sum with respect to *all* vertices $z \in \mathcal{N}$ and that we apply $I_{\mathcal{T}}$ componentwise to a vector field.

From the first equation of (4.37) and (4.39), respectively, we then have

$$\int_\Omega (\sigma - \sigma_{\mathcal{T}}) \cdot \tau_{\mathcal{T}} + \int_\Omega (u - u_{\mathcal{T}}) \operatorname{div} \tau_{\mathcal{T}} = 0.$$

Hence, we may replace τ by $\tau - \tau_{\mathcal{T}}$ in equation (4.41) and obtain

$$\langle \ell - L(\sigma_{\mathcal{T}}, u_{\mathcal{T}}), (\tau, v)\rangle_Y = \langle \ell - L(\sigma_{\mathcal{T}}, u_{\mathcal{T}}), (\tau - \tau_{\mathcal{T}}, v)\rangle_Y .$$

We combine this relation with equation (4.42). Taking into account that Proposition 4.51 implies $G\tau_{\mathcal{T}} = 0$, this yields

$$\begin{aligned}\langle \ell - L(\sigma_{\mathcal{T}}, u_{\mathcal{T}}), (\tau, v)\rangle_Y = &-\sum_{K\in\mathcal{T}} \int_K \operatorname{curl} \sigma_{\mathcal{T}} \cdot (Id - I_{\mathcal{T}})(R\tau) \\ &- \sum_{E\in\mathcal{E}_\Omega} \int_E \mathbb{J}_E(\sigma_{\mathcal{T}} - (\sigma_{\mathcal{T}} \cdot \mathbf{n}_E)\mathbf{n}_E) \cdot (Id - I_{\mathcal{T}})(R\tau) \\ &- \sum_{E\in\mathcal{E}_\Gamma} \int_E (\sigma_{\mathcal{T}} - (\sigma_{\mathcal{T}} \cdot \mathbf{n}_E)\mathbf{n}_E) \cdot (Id - I_{\mathcal{T}})(R\tau) \\ &- \int_\Omega \sigma_{\mathcal{T}} \cdot (Id - \mathcal{J}_{\mathcal{T}})(\nabla(G\tau)) \\ &- \int_\Omega (f + \operatorname{div} \sigma_{\mathcal{T}})v.\end{aligned} \tag{4.43}$$

Next, we estimate the terms on the right-hand side of equation (4.43). To this end and in view of Proposition 4.52 we assume that Ω is convex. Proposition 1.3 (p. 7) then yields for all elements K

$$\int_K \operatorname{curl} \sigma_{\mathcal{T}} \cdot (Id - I_{\mathcal{T}})(R\tau) \le c_{A,1}\, h_K \, \|\operatorname{curl} \sigma_{\mathcal{T}}\|_K \, \|\nabla R\tau\|_{\tilde{\omega}_K} ,$$

for all interior faces E

$$\int_E \mathbb{J}_E(\sigma_{\mathcal{T}} - (\sigma_{\mathcal{T}} \cdot \mathbf{n}_E)\mathbf{n}_E) \cdot (Id - I_{\mathcal{T}})(R\tau) \le c_{A,2}\, h_E^{\frac{1}{2}} \left\|\mathbb{J}_E(\sigma_{\mathcal{T}} - (\sigma_{\mathcal{T}} \cdot \mathbf{n}_E)\mathbf{n}_E)\right\|_E \|\nabla R\tau\|_{\tilde{\omega}_E}$$

and for all boundary faces E

$$\int_E (\sigma_{\mathcal{T}} - (\sigma_{\mathcal{T}} \cdot \mathbf{n}_E)\mathbf{n}_E) \cdot (Id - I_{\mathcal{T}})(R\tau) \le c_{A,2}\, h_E^{\frac{1}{2}} \left\|\sigma_{\mathcal{T}} - (\sigma_{\mathcal{T}} \cdot \mathbf{n}_E)\mathbf{n}_E\right\|_E \|\nabla R\tau\|_{\tilde{\omega}_E} .$$

Proposition 4.52 and [**90**, Proposition 3.6] on the other hand imply that

$$\begin{aligned}\int_K \sigma_{\mathcal{T}} \cdot (Id - \mathcal{J}_{\mathcal{T}})(\nabla(G\tau)) &\le \|\sigma_{\mathcal{T}}\|_K \left\|(Id - \mathcal{J}_{\mathcal{T}})(\nabla(G\tau))\right\|_K \\ &\le c_1 h_K \|\sigma_{\mathcal{T}}\|_K \left\|\nabla(G\tau)\right\|_{1;K} \\ &\le c_1 h_K \|\sigma_{\mathcal{T}}\|_K \|G\tau\|_{2;K}\end{aligned}$$

holds for all elements K with a constant c_1 that only depends on the shape parameter $C_{\mathcal{T}}$ of $\mathcal{T}$.

Finally, for the last term on the right-hand side of equation (4.43), we obviously have

$$\int_K (f + \operatorname{div} \sigma_{\mathcal{T}})v \le \left\|f + \operatorname{div} \sigma_{\mathcal{T}}\right\|_K \|v\|_K$$

for all elements K.

Now, we insert these estimates in equation (4.43), use the Cauchy–Schwarz inequality for sums, take into account Proposition 4.52, and recall that $(\tau, v) \in H(\text{div}; \Omega) \times L^2(\Omega)$ was chosen arbitrarily. Thus we arrive at the upper bound

$$\begin{aligned}\left\| \ell - L(\sigma_{\mathcal{T}}, u_{\mathcal{T}}) \right\|_{Y*} \le c \Bigg\{ &\sum_{K\in\mathcal{T}} h_K^2 \left\|\operatorname{curl} \sigma_{\mathcal{T}}\right\|_K^2 \\ &+ \sum_{E\in\mathcal{E}_\Omega} h_E \left\| \mathbb{J}_E(\sigma_{\mathcal{T}} - (\sigma_{\mathcal{T}} \cdot \mathbf{n}_E)\mathbf{n}_E) \right\|_E^2 \\ &+ \sum_{E\in\mathcal{E}_\Gamma} h_E \left\| \sigma_{\mathcal{T}} - (\sigma_{\mathcal{T}} \cdot \mathbf{n}_E)\mathbf{n}_E \right\|_E^2 \\ &+ \sum_{K\in\mathcal{T}} h_K^2 \left\|\sigma_{\mathcal{T}}\right\|_K^2 + \sum_{K\in\mathcal{T}} \left\| f + \operatorname{div} \sigma_{\mathcal{T}} \right\|_K^2 \Bigg\}^{\frac{1}{2}}\end{aligned}$$

with a constant c which only depends on the shape parameter $C_{\mathcal{T}}$ of $\mathcal{T}$ and the domain Ω. This establishes the first condition of Theorem 4.7 (p. 156) with $\theta_{\mathcal{T}} = 0$.

Since $B_{\mathcal{T}}$ and $\ell_{\mathcal{T}}$ are the restrictions of B and ℓ, respectively, we have Galerkin orthogonality and the second condition of Theorem 4.7 is automatically satisfied with $c_C = 0$.

To prove the third condition of Theorem 4.7, we proceed in the standard way and insert suitable test functions in equations (4.41) and (4.42).

Consider an arbitrary element K, denote by χ_K its characteristic function, and set

$$v_K = -(f + \operatorname{div} \sigma_{\mathcal{T}})\chi_K.$$

Since $v_K \in L^2(\Omega)$, we may insert $(\tau, v) = (0, v_K)$ in equation (4.42) and obtain

$$\begin{aligned}\left\| f + \operatorname{div} \sigma_{\mathcal{T}} \right\|_K^2 &= - \int_\Omega (f + \operatorname{div} \sigma_{\mathcal{T}}) v_K \\ &= \left\langle \ell - L(\sigma_{\mathcal{T}}, u_{\mathcal{T}}),\, (0, v_K) \right\rangle_Y\end{aligned}$$

and

$$\left\| (0, v_K) \right\|_Y = \left\| f + \operatorname{div} \sigma_{\mathcal{T}} \right\|_K.$$

Next, we set

$$\tau_K = -\operatorname{curl}(\psi_K \operatorname{curl} \sigma_{\mathcal{T}})$$

and insert $(\tau, v) = (\tau_K, 0)$ in equation (4.42). Since

$$G\tau_K = 0, \quad R\tau_K = -\psi_K \operatorname{curl} \sigma_{\mathcal{T}},$$

this yields

$$\begin{aligned}\int_K \psi_K \,|\operatorname{curl}\sigma_\mathcal{T}|^2 &= -\int_K \operatorname{curl}\sigma_\mathcal{T}\cdot(-\psi_K \operatorname{curl}\sigma_\mathcal{T})\\ &= -\int_K \operatorname{curl}\sigma_\mathcal{T}\cdot(R\tau_K)\\ &= \langle \ell - L(\sigma_\mathcal{T},u_\mathcal{T}),\,(\tau_K,0)\rangle_Y .\end{aligned}$$

From Proposition 1.4 (p. 9) we conclude that

$$\|\operatorname{curl}\sigma_\mathcal{T}\|_K^2 \le c_{I,1}^2 \int_K \psi_K\,|\operatorname{curl}\sigma_\mathcal{T}|^2$$

and

$$\begin{aligned}\|(\tau_K,0)\|_Y &= \|\tau_K\|_{H(\operatorname{div};\Omega)} = \|\tau_K\|_K\\ &\le c_{I,2}\, h_K^{-1}\,\|\operatorname{curl}\sigma_\mathcal{T}\|_K .\end{aligned}$$

Now, we consider an arbitrary interior face $E \in \mathcal{E}_\Omega$, set

$$\tau_E = -\operatorname{curl}(\psi_E \mathbb{J}_E(\sigma_\mathcal{T} - (\sigma_\mathcal{T}\cdot\mathbf{n}_E)\mathbf{n}_E)),$$

and insert $(\tau,v) = (\tau_E,0)$ in equation (4.42). Since

$$G\tau_E = 0, \quad R\tau_E = -\psi_E\mathbb{J}_E(\sigma_\mathcal{T} - (\sigma_\mathcal{T}\cdot\mathbf{n}_E)\mathbf{n}_E),$$

we obtain

$$\begin{aligned}&\int_E \psi_E\,\left|\mathbb{J}_E(\sigma_\mathcal{T} - (\sigma_\mathcal{T}\cdot\mathbf{n}_E)\mathbf{n}_E)\right|^2\\ &= -\int_E \mathbb{J}_E(\sigma_\mathcal{T} - (\sigma_\mathcal{T}\cdot\mathbf{n}_E)\mathbf{n}_E)\cdot(-\psi_E\mathbb{J}_E(\sigma_\mathcal{T} - (\sigma_\mathcal{T}\cdot\mathbf{n}_E)\mathbf{n}_E))\\ &= -\int_E \mathbb{J}_E(\sigma_\mathcal{T} - (\sigma_\mathcal{T}\cdot\mathbf{n}_E)\mathbf{n}_E)\cdot(R\tau_E)\\ &= \langle \ell - L(\sigma_\mathcal{T},u_\mathcal{T}),\,(\tau_E,0)\rangle_Y + \sum_{K\subset\omega_E}\int_K \operatorname{curl}\sigma_\mathcal{T}\cdot(R\tau_E)\\ &= \langle \ell - L(\sigma_\mathcal{T},u_\mathcal{T}),\,(\tau_E,0)\rangle_Y - \sum_{K\subset\omega_E}\int_K \operatorname{curl}\sigma_\mathcal{T}\cdot(-\psi_E\mathbb{J}_E(\sigma_\mathcal{T} - (\sigma_\mathcal{T}\cdot\mathbf{n}_E)\mathbf{n}_E)).\end{aligned}$$

Proposition 1.4 (p. 9) now yields

$$\left\|\mathbb{J}_E(\sigma_\mathcal{T} - (\sigma_\mathcal{T}\cdot\mathbf{n}_E)\mathbf{n}_E)\right\|_E^2 \le c_{I,3}^2 \int_E \psi_E\,\left|\mathbb{J}_E(\sigma_\mathcal{T} - (\sigma_\mathcal{T}\cdot\mathbf{n}_E)\mathbf{n}_E)\right|^2$$

and

$$\begin{aligned}\left\|(\tau_E, 0)\right\|_Y &= \|\tau_E\|_{H(\mathrm{div};\Omega)} = \|\tau_E\|_E \\ &\le c_{I,4}\, h_E^{-\frac{1}{2}} \left\|\mathbb{J}_E(\sigma_\mathcal{T} - (\sigma_\mathcal{T}\cdot \mathbf{n}_E)\mathbf{n}_E)\right\|_E\end{aligned}$$

and for every element $K \subset \omega_E$

$$\int_K \operatorname{curl}\sigma_\mathcal{T} \cdot (-\psi_E \mathbb{J}_E(\sigma_\mathcal{T} - (\sigma_\mathcal{T}\cdot \mathbf{n}_E)\mathbf{n}_E)) \le c_{I,5}\, h_E^{\frac{1}{2}} \,\|\operatorname{curl}\sigma_\mathcal{T}\|_K \left\|\mathbb{J}_E(\sigma_\mathcal{T} - (\sigma_\mathcal{T}\cdot \mathbf{n}_E)\mathbf{n}_E)\right\|_E .$$

Edges or faces on the boundary Γ are treated in the same way with

$$\tau_E = -\operatorname{curl}(\psi_E(\sigma_\mathcal{T} - (\sigma_\mathcal{T}\cdot \mathbf{n}_E)\mathbf{n}_E)).$$

Finally, we come to the $\|\sigma_\mathcal{T}\|_K$ terms. Now, we resort to the representation (4.41) of the residual. We choose an arbitrary element K, set

$$\tau_K = -\psi_K \sigma_\mathcal{T},$$

and insert $(\tau, v) = (\tau_K, 0)$ in equation (4.41). Since $\nabla u_\mathcal{T} = 0$ on K and $\tau_K = 0$ on ∂K, integration by parts yields

$$\begin{aligned}\int_K \psi_K\, |\sigma_\mathcal{T}|^2 &= -\int_K \sigma_\mathcal{T}\cdot\tau_K \\ &= -\int_K \sigma_\mathcal{T}\cdot\tau_K + \int_K \nabla u_\mathcal{T}\cdot\tau_K \\ &= -\int_K \sigma_\mathcal{T}\cdot\tau_K - \int_K u_\mathcal{T}\operatorname{div}\tau_K \\ &= \langle \ell - L(\sigma_\mathcal{T}, u_\mathcal{T}),\, (\tau_K, 0)\rangle_Y .\end{aligned}$$

From Proposition 1.4 (p. 9) we now obtain

$$\|\sigma_\mathcal{T}\|_K^2 \le c_{I,1}^2 \int_K \psi_K\, |\sigma_\mathcal{T}|^2$$

and

$$\begin{aligned}\left\|(\tau_K, 0)\right\|_Y &= \|\tau_K\|_{H(\mathrm{div};\Omega)} \\ &= \left\{\|\psi_K\sigma_\mathcal{T}\|_K^2 + \left\|\operatorname{div}(\psi_K\sigma_\mathcal{T})\right\|_K^2\right\}^{\frac{1}{2}} \\ &\le \left\{1 + c_{I,2}^2\, h_K^{-2}\right\}^{\frac{1}{2}} \|\sigma_\mathcal{T}\|_K \\ &\le \max\{1, c_{I,2}\}\, h_K^{-1}\, \|\sigma_\mathcal{T}\|_K .\end{aligned}$$

These estimates establish the third condition of Theorem 4.7 (p. 156) with

$$\widetilde{Y}_{\mathcal{T}} = \operatorname{span}\Big\{(0,(f+\operatorname{div}\sigma_{\mathcal{T}})\chi_K),(\operatorname{curl}(\psi_K\operatorname{curl}\sigma_{\mathcal{T}}),0),\\ (\psi_K\sigma_{\mathcal{T}},0),\\ (\operatorname{curl}(\psi_E\mathbb{J}_E(\sigma_{\mathcal{T}}-(\sigma_{\mathcal{T}}\cdot\mathbf{n}_E)\mathbf{n}_E)),0),\\ (\operatorname{curl}(\psi_{E'}(\sigma_{\mathcal{T}}-(\sigma_{\mathcal{T}}\cdot\mathbf{n}_{E'})\mathbf{n}_{E'})),0):\\ K\in\mathcal{T},E\in\mathcal{E}_\Omega,E'\in\mathcal{E}_\Gamma\Big\}.$$

Theorem 4.7 therefore yields the following a posteriori error estimates.

Theorem 4.53 *Denote by (σ,u) and $(\sigma_{\mathcal{T}},u_{\mathcal{T}})$ the unique solutions of problems (4.37) and (4.39), respectively. For every element $K\in\mathcal{T}$ define the* residual a posteriori error indicator $\eta_{R,K}$ *by*

$$\eta_{R,K} = \Bigg\{h_K^2\,\|\sigma_{\mathcal{T}}\|_K^2 + \|f+\operatorname{div}\sigma_{\mathcal{T}}\|_K^2 + h_K^2\,\|\operatorname{curl}\sigma_{\mathcal{T}}\|_K^2\\ +\frac{1}{2}\sum_{E\in\mathcal{E}_{K,\Omega}} h_E\,\big\|\mathbb{J}_E(\sigma_{\mathcal{T}}-(\sigma_{\mathcal{T}}\cdot\mathbf{n}_E)\mathbf{n}_E)\big\|_E^2 + \sum_{E\in\mathcal{E}_{K,\Gamma}} h_E\,\big\|\sigma_{\mathcal{T}}-(\sigma_{\mathcal{T}}\cdot\mathbf{n}_E)\mathbf{n}_E\big\|_E^2\Bigg\}^{\frac{1}{2}}.$$

Assume that Ω is convex. Then there are two constants c^* and c_*, which only depend on Ω and the shape parameter $C_{\mathcal{T}}$ of $\mathcal{T}$, such that the estimates

$$\Big\{\|\sigma-\sigma_{\mathcal{T}}\|_{H(\operatorname{div};\Omega)}^2 + \|u-u_{\mathcal{T}}\|^2\Big\}^{\frac{1}{2}} \le c^*\Big\{\sum_{K\in\mathcal{T}}\eta_{R,K}^2\Big\}^{\frac{1}{2}}$$

and

$$\eta_{R,K}\le c_*\Big\{\|\sigma-\sigma_{\mathcal{T}}\|_{H(\operatorname{div};\omega_K)}^2 + \|u-u_{\mathcal{T}}\|_{\omega_K}^2\Big\}^{\frac{1}{2}}$$

hold for all $K\in\mathcal{T}$.

Remark 4.54 The terms $\|\operatorname{curl}\sigma_{\mathcal{T}}\|_K$, $\big\|\mathbb{J}_E(\sigma_{\mathcal{T}}-(\sigma_{\mathcal{T}}\cdot\mathbf{n}_E)\mathbf{n}_E)\big\|_E$, and $\big\|\sigma_{\mathcal{T}}-(\sigma_{\mathcal{T}}\cdot\mathbf{n}_E)\mathbf{n}_E\big\|_E$ in $\eta_{R,K}$ are the residuals of $\sigma_{\mathcal{T}}$ corresponding to the equation $\operatorname{curl}\sigma=0$. Due to the condition $\sigma=\nabla u$, this equation is a redundant one for the analytical problem. For the discrete problem, however, it is an extra condition which is not incorporated in problem (4.39).

Remark 4.55 Theorem 4.53 can be extended to higher order discretisations which still satisfy the commuting diagram property of Proposition 4.50. Examples of finite element spaces falling into this category are the higher order Raviart–Thomas (RT), Brezzi–Douglas–Marini (BDM), or Brezzi–Douglas–Marini–Fortin (BDMF) spaces, cf. [**90**, Section III.3] and the next section. When using higher order elements, in $\eta_{R,K}$, one must replace the terms $\|\sigma_{\mathcal{T}}\|_K$ by $\|\sigma_{\mathcal{T}}-\nabla u_{\mathcal{T}}\|_K$.

Remark 4.56 The analysis of this section follows the lines of [**91**]. A posteriori error estimates for mixed finite element approximations of the model problem were first analysed in [**84**]. There, the Helmholtz decomposition of Proposition 4.52 was not used. The error was measured in two different norms:

- the norm $\|\sigma\|_{H(\mathrm{div};\Omega)} + \|u\|$ which we use in this section and which in a certain sense is the natural norm associated with problem (4.37) and
- a mesh-dependent norm approximating $\|\sigma\| + \|u\|_1$.

For the first choice of norms, there was a gap of h^{-1} between the upper and lower bounds for the error. This is due to the fact that functions in $H(\mathrm{div};\Omega)$ have a lesser regularity than those in $H^1(\Omega)$. For the second choice of norms, this drawback was overcome. Yet, this norm is too strong for problem (4.37). Moreover, for both norms, the analysis was based on a saturation assumption which cannot be checked in practice. These drawbacks were overcome in [**91**, **116**] by using the Helmholtz decomposition.

4.9 The Equations of Linear Elasticity

The equations of linear elasticity are given by the boundary value problem

$$\begin{aligned} \varepsilon &= Du && \text{in } \Omega \\ \sigma &= C\varepsilon && \text{in } \Omega \\ -\operatorname{div}\sigma &= f && \text{in } \Omega \\ \operatorname{as}(\sigma) &= 0 && \text{in } \Omega \\ u &= 0 && \text{on } \Gamma_D \\ \sigma\cdot\mathbf{n} &= 0 && \text{on } \Gamma_N \end{aligned} \tag{4.44}$$

where the various quantities are

- $\Omega \subset \mathbb{R}^d$, $d \in \{2,3\}$, a bounded connected polyhedral domain with Lipschitz boundary Γ consisting of two disjoint parts Γ_D and Γ_N,
- $u : \Omega \to \mathbb{R}^d$ the *displacement,*
- $Du = \frac{1}{2}(\nabla u + \nabla u^T) = \frac{1}{2}\left(\frac{\partial u_i}{\partial x_j} + \frac{\partial u_j}{\partial x_i}\right)_{1\le i,j\le d}$ the *deformation tensor* or *symmetric gradient,*
- $\varepsilon : \Omega \to \mathbb{R}^{d\times d}$ the *strain tensor,*
- $\sigma : \Omega \to \mathbb{R}^{d\times d}$ the *stress tensor,*
- C the *elasticity tensor,*
- $f : \Omega \to \mathbb{R}^d$ the given *body load,*
- $\operatorname{as}(\tau) = \frac{1}{2}(\tau - \tau^T)$ the *skew symmetric part* of a given tensor.

The most prominent example of an elasticity tensor is given by

$$C\varepsilon = \lambda\,\mathrm{tr}(\varepsilon)I + 2\mu\varepsilon \tag{4.45}$$

where $I \in \mathbb{R}^{d\times d}$ is the unit tensor, $\mathrm{tr}(\varepsilon) = \sum_i \varepsilon_{ii}$ denotes the trace of ε, and λ, $\mu > 0$ are the *Lamé parameters.* To simplify the presentation, we assume throughout this section that C is given by (4.45). We are particularly interested in estimates which are uniform with respect to the Lamé parameters.

4.9.1 Displacement Formulation

The simplest discretisation of problem (4.44) is based on its *displacement formulation*.

Inserting the first and second equation of (4.44) in its third one, we obtain the boundary value problem

$$\begin{aligned} -\operatorname{div}(CDu) &= f \quad \text{in } \Omega \\ u &= 0 \quad \text{on } \Gamma_D \\ \mathbf{n}\cdot CDu &= 0 \quad \text{on } \Gamma_N. \end{aligned} \tag{4.46}$$

Its variational formulation fits into the abstract framework of Section 4.1 (p. 151) with

$$\begin{aligned} X = Y &= H_D^1(\Omega)^d, \\ B(u,v) &= \int_\Omega Du : CDv, \\ \langle \ell\,,\, v\rangle_Y &= \int_\Omega f\cdot v. \end{aligned}$$

Here, $\sigma : \tau$ denotes the inner product of two tensors, i.e.

$$\sigma : \tau = \sum_{1\le i,j\le d} \sigma_{ij}\tau_{ij}.$$

The variational problem (4.2) (p. 152) is the Euler–Lagrange equation corresponding to the problem of minimising the *total energy*

$$J(u) = \frac{1}{2}\int_\Omega Du : CDu - \int_\Omega f\cdot u.$$

The finite element discretisation of problem (4.46) is straightforward and fits into the abstract framework of Section 4.1 with

$$\begin{aligned} X_{\mathcal{T}} = Y_{\mathcal{T}} &= S_D^{k,0}(\mathcal{T})^d, \\ B_{\mathcal{T}}(u_{\mathcal{T}}, v_{\mathcal{T}}) &= B(u_{\mathcal{T}}, v_{\mathcal{T}}), \\ \langle \ell_{\mathcal{T}}\,,\, v_{\mathcal{T}}\rangle_{Y_{\mathcal{T}}} &= \langle \ell\,,\, v_{\mathcal{T}}\rangle_Y\,. \end{aligned}$$

The methods of Sections 1.4 (p. 10) and 4.2 (p. 157) directly carry over to this situation. They yield the residual a posteriori error indicator

$$\eta_{R,K} = \left\{ h_K^2 \left\| f_{\mathcal{T}} + \operatorname{div}(CDu_{\mathcal{T}})\right\|_K^2 + \frac{1}{2}\sum_{E\in\mathcal{E}_{K,\Omega}} h_E \left\|\mathbb{J}_E(\mathbf{n}_E\cdot CDu_{\mathcal{T}})\right\|_E^2 + \sum_{E\in\mathcal{E}_{K,\Gamma_N}} h_E \left\|\mathbf{n}_E\cdot CDu_{\mathcal{T}}\right\|_E^2 \right\}^{\frac{1}{2}}$$

which gives global upper and local lower bounds for the H^1-norm of the error in the displacements.

Similarly, the methods of Sections 1.7 (p. 25) and 1.8 (p. 31) can be extended to give indicators based on the solution of auxiliary local discrete problems and hierarchical indicators. Now, of course, the auxiliary problems are elasticity problems in displacement form. An overview of a posteriori error estimates for displacement methods is given in [**262**].

Though appealing by its simplicity, the displacement formulation of problem (4.44) and the corresponding finite element discretisations suffer from serious drawbacks.

- The displacement formulation and its discretisation break down for nearly incompressible materials, i.e. $\lambda \gg 1$, which is reflected by the so-called locking phenomenon.
- The quality of the a posteriori error estimates in the sense of Remark 4.8 (p. 156) deteriorates in the incompressible limit. More precisely, the quantity

$$\|L\|_{\mathcal{L}(X,Y^*)} \left\|L^{-1}\right\|_{\mathcal{L}(Y^*,X)} (c_A + c_C)c_I$$

depends on the Lamé parameter λ and tends to infinity for large values of λ.
- Often, the displacement field is not of primary interest, but the stress tensor is of physical interest. This quantity, however, is not directly discretised in displacement methods and must a posteriori be extracted from the displacement field which often leads to unsatisfactory results.

These drawbacks can be overcome by suitable mixed formulations of problem (4.44) and appropriate mixed finite element discretisations. Correspondingly these are in the focus of the following subsections. We are primarily interested in discretisations and a posteriori error estimates which are robust in the sense that their quality does not deteriorate in the incompressible limit. This is reflected by the need for estimates which are uniform with respect to the Lamé parameters λ and μ.

4.9.2 *The Hellinger–Reissner Principle*

To simplify the notation, we introduce the spaces

$$\begin{aligned}
H &= H(\mathrm{div};\Omega)^d = \left\{\sigma \in L^2(\Omega)^{d\times d} \mid \mathrm{div}\,\sigma \in L^2(\Omega)^d\right\},\\
V &= L^2(\Omega)^d,\\
W &= \left\{\gamma \in L^2(\Omega)^{d\times d} \mid \gamma + \gamma^T = 0\right\}
\end{aligned}$$

and equip them with their natural norms

$$\begin{aligned}
\|\sigma\|_H &= \left\{\|\sigma\|^2 + \|\mathrm{div}\,\sigma\|^2\right\}^{\frac{1}{2}},\\
\|u\|_V &= \|u\|,\\
\|\gamma\|_W &= \|\gamma\|.
\end{aligned}$$

Here, the divergence of a tensor σ is taken row by row, i.e.

$$(\mathrm{div}\,\sigma)_i = \sum_{1\le j\le n} \frac{\partial \sigma_{ij}}{\partial x_j}.$$

The *Hellinger–Reissner principle* is a mixed variational formulation of problem (4.44), in which the strain ε is eliminated. It fits into the abstract framework of Section 4.1 (p. 151) with

$$X = Y = H \times V \times W,$$

$$B((\sigma, u, \gamma), (\tau, v, \eta)) = \int_\Omega C^{-1}\sigma : \tau$$
$$+ \int_\Omega \operatorname{div} \tau \cdot u + \int_\Omega \tau : \gamma$$
$$+ \int_\Omega \operatorname{div} \sigma \cdot v + \int_\Omega \sigma : \eta,$$
$$\langle \ell, (\tau, v, \eta) \rangle_Y = - \int_\Omega f \cdot v.$$

Notice that $\eta \in W$ implies $\eta^T = -\eta$ and therefore

$$\int_\Omega \sigma : \eta = \int_\Omega \operatorname{as}(\sigma) : \eta. \tag{4.47}$$

The following result is due to D. N. Arnold and R. S. Falk [**28**].

Proposition 4.57 *Set*

$$Z = \left\{ \sigma \in H : \int_\Omega \operatorname{div} \sigma \cdot v + \int_\Omega \sigma : \eta = 0 \text{ for all } v \in V, \eta \in W \right\}.$$

Then there are constants c_a and c_b which do not depend on the Lamé parameter λ such that for all $\sigma, \tau \in H$, $\rho \in Z$, $v \in V$, $\eta \in W$

$$\int_\Omega C^{-1}\sigma : \tau \le \frac{1}{\mu} \|\sigma\|_H \|\tau\|_H,$$
$$\int_\Omega C^{-1}\rho : \rho \ge c_a \|\rho\|_H^2,$$
$$\int_\Omega \operatorname{div} \sigma \cdot v + \int_\Omega \sigma : \eta \le \|\sigma\|_H \left\|(v, \eta)\right\|_{V \times W},$$
$$\sup_{\sigma \in H \setminus \{0\}} \frac{\int_\Omega \operatorname{div} \sigma \cdot v + \int_\Omega \sigma : \eta}{\|\sigma\|_H} \ge c_b \left\|(v, \eta)\right\|_{V \times W}.$$

Proposition 4.57 and the techniques used in the proof of Proposition 4.48 (p. 209) show that the bi-linear form B defined above fulfils the conditions of Proposition 4.1 (p. 152). For a detailed proof we refer to [**182**, Lemma 2.9 and Lemma 2.10].

Proposition 4.58 *Set*

$$c_D = \sqrt{2} \max \left\{ 1, \frac{1}{\mu} \right\}, \quad c_d = \frac{1}{\sqrt{5}} \min \left\{ c_a, \frac{\mu c_b^2}{\sqrt{2\mu^2 c_b^2 + 8}} \right\}.$$

Then B satisfies the estimates

$$\sup_{(\sigma,u,\gamma) \in X \setminus \{0\}} \sup_{(\tau,v,\eta) \in X \setminus \{0\}} \frac{B((\sigma, u, \gamma), (\tau, v, \eta))}{\left\|(\sigma, u, \gamma)\right\|_X \left\|(\tau, v, \eta)\right\|_X} \le c_D$$
$$\inf_{(\sigma,u,\gamma) \in X \setminus \{0\}} \sup_{(\tau,v,\eta) \in X \setminus \{0\}} \frac{B((\sigma, u, \gamma), (\tau, v, \eta))}{\left\|(\sigma, u, \gamma)\right\|_X \left\|(\tau, v, \eta)\right\|_X} \ge c_d.$$

Thanks to this stability result, all forthcoming constants are independent of the Lamé parameter λ. Hence, the corresponding estimates are robust for nearly incompressible materials, i.e. $\lambda \gg 1$.

4.9.3 *PEERS and BDMS Elements*

We consider two types of mixed finite element discretisations of problem (4.44):

- the *PEERS element* and
- the *BDMS elements.*

Both families have proved to be particularly well suited to avoid locking phenomena. For their description we recall the curl operator of Section 4.8.4 (p. 214) and the Raviart–Thomas space

$$\mathrm{RT}_0(K) = R_0(K)^d + R_0(K)\begin{pmatrix} x_1 \\ \vdots \\ x_d \end{pmatrix}$$

of Section 4.8.2 (p. 212) and set for every integer $k \in \mathbb{N}$ and every element $K \in \mathcal{T}$

$$B_k(K) = \left\{\sigma \in \mathbb{R}^{d\times d} : (\sigma_{i1}, \dots, \sigma_{id}) = \mathrm{curl}(\psi_K w_i),\ \ w_i \in R_k(K)^{2d-3}, 1 \le i \le d\right\},$$

$$\mathrm{BDM}_k(K) = R_k(K)^d.$$

Both types of discretisations fit into the abstract framework of Section 4.1 (p. 151) with

$$\begin{aligned} X_{\mathcal{T}} = Y_{\mathcal{T}} &= H_{\mathcal{T}} \times V_{\mathcal{T}} \times W_{\mathcal{T}}, \\ B_{\mathcal{T}}((\tau_{\mathcal{T}}, v_{\mathcal{T}}, \eta_{\mathcal{T}}), (\rho_{\mathcal{T}}, w_{\mathcal{T}}, \nu_{\mathcal{T}}) &= B((\tau_{\mathcal{T}}, v_{\mathcal{T}}, \eta_{\mathcal{T}}), (\rho_{\mathcal{T}}, w_{\mathcal{T}}, \nu_{\mathcal{T}})), \\ \langle \ell_{\mathcal{T}}\,,\, (\tau_{\mathcal{T}}, v_{\mathcal{T}}, \eta_{\mathcal{T}}) \rangle_{Y_{\mathcal{T}}} &= \langle \ell\,,\, (\tau_{\mathcal{T}}, v_{\mathcal{T}}, \eta_{\mathcal{T}}) \rangle_{Y}. \end{aligned}$$

They differ by the spaces $H_{\mathcal{T}}$, $V_{\mathcal{T}}$, and $W_{\mathcal{T}}$ which are given by

$$\begin{aligned} H_{\mathcal{T}} &= \{\sigma_{\mathcal{T}} \in H : \sigma_{\mathcal{T}}|_K \in \mathrm{RT}_0(K)^d \oplus \mathrm{B}_0(K), K \in \mathcal{T},\ \ \sigma_{\mathcal{T}} \cdot n = 0 \text{ on } \Gamma_N\}, \\ V_{\mathcal{T}} &= \{v_{\mathcal{T}} \in V : v_{\mathcal{T}}|_K \in R_0(K)^d, K \in \mathcal{T}\}, \\ W_{\mathcal{T}} &= \{\eta_{\mathcal{T}} \in W \cap C(\Omega)^{d\times d} : \eta_{\mathcal{T}}|_K \in R_1(K)^{d\times d}, K \in \mathcal{T}\}, \end{aligned}$$

for the *PEERS element* and

$$\begin{aligned} H_{\mathcal{T}} &= \{\sigma_{\mathcal{T}} \in H : \sigma_{\mathcal{T}}|_K \in \mathrm{BDM}_k(K)^d \oplus B_{k-1}(K), K \in \mathcal{T},\ \ \sigma_{\mathcal{T}} \cdot n = 0 \text{ on } \Gamma_N\}, \\ V_{\mathcal{T}} &= \{v_{\mathcal{T}} \in V : v_{\mathcal{T}}|_K \in R_{k-1}(K)^d, K \in \mathcal{T}\}, \\ W_{\mathcal{T}} &= \{\eta_{\mathcal{T}} \in W \cap C(\Omega)^{d\times d} : \eta_{\mathcal{T}}|_K \in R_k(K)^{d\times d}, K \in \mathcal{T}\} \end{aligned}$$

for the *BDMS elements.*

Remark 4.59 For both elements, the space $W_{\mathcal{T}}$ is isomorphic to the space $\{\varphi : \varphi|_K \in R_k(K)^{2d-3}, K \in \mathcal{T}\}$. The corresponding isomorphism is given by

$$\varphi \mapsto \begin{pmatrix} 0 & -\varphi \\ \varphi & 0 \end{pmatrix} \qquad \text{if } d = 2,$$

$$\begin{pmatrix} \varphi_1 \\ \varphi_2 \\ \varphi_3 \end{pmatrix} \mapsto \begin{pmatrix} 0 & \varphi_1 & \varphi_2 \\ -\varphi_1 & 0 & \varphi_3 \\ -\varphi_2 & -\varphi_3 & 0 \end{pmatrix} \qquad \text{if } d = 3.$$

We define a discrete analogue of the space Z by

$$Z_{\mathcal{T}} = \left\{ \sigma_{\mathcal{T}} \in H_{\mathcal{T}} : \int_\Omega \operatorname{div} \sigma_{\mathcal{T}} \cdot v_{\mathcal{T}} + \int_\Omega \sigma_{\mathcal{T}} : \eta_{\mathcal{T}} = 0 \right.$$
$$\left. \text{for all } v_{\mathcal{T}} \in V_{\mathcal{T}}, \eta \in W_{\mathcal{T}} \right\}.$$

Note that $Z_{\mathcal{T}}$ is *not* a subspace of Z.

For the PEERS element the following analogue of Proposition 4.57 is established in [26]:

$$\int_\Omega C^{-1} \rho_{\mathcal{T}} : \rho_{\mathcal{T}} \geq c_a^{PEERS} \|\rho_{\mathcal{T}}\|_H^2,$$

$$\sup_{\sigma_{\mathcal{T}} \in H \setminus \{0\}} \frac{\int_\Omega \operatorname{div} \sigma_{\mathcal{T}} \cdot v_{\mathcal{T}} + \int_\Omega \sigma_{\mathcal{T}} : \eta_{\mathcal{T}}}{\|\sigma_{\mathcal{T}}\|_H} \geq c_b^{PEERS} \left\| (v_{\mathcal{T}}, \eta_{\mathcal{T}}) \right\|_{V \times W}$$

for all $\rho_{\mathcal{T}} \in Z_{\mathcal{T}}, v_{\mathcal{T}} \in V_{\mathcal{T}}, \eta_{\mathcal{T}} \in W_{\mathcal{T}}$.

A similar result is established for the BDMS elements with corresponding constants c_a^{BDMS} and c_b^{BDMS} in [183].

These estimates and the techniques used in the proof of Proposition 4.48 (p. 209) prove that the discrete problem fulfils the assumptions of Proposition 4.5 (p. 154) and admits a unique solution.

4.9.4 *Residual A Posteriori Error Estimates*

In what follows we always denote by $(\sigma, u, \gamma) \in X$ the unique solution of the mixed variational formulation of problem (4.44) based on the Hellinger–Reissner principle and by $(\sigma_{\mathcal{T}}, u_{\mathcal{T}}, \gamma_{\mathcal{T}}) \in X_{\mathcal{T}}$ its finite element approximation using the PEERS or BDMS elements. We want to derive a posteriori error estimates which are robust in the sense that the multiplicative constants in the estimates are uniformly bounded with respect to the Lamé parameters λ and μ. This in particular yields estimates which do not deteriorate in the incompressible limit $\lambda \to \infty$. We first consider the case $\mu = 1$ and then treat the general case by a scaling argument.

We first derive a suitable representation of the residual by adapting the arguments of Section 4.8.5 (p. 216). To this end, we choose an arbitrary element $(\tau, v, \eta) \in Y = X$ and keep it fixed in what follows. From the definition of the bi-linear form B and of the linear functional ℓ and relation (4.47) we then immediately obtain the following representation of the residual

$$
\begin{aligned}
\langle \ell - L(\sigma_{\mathcal{T}}, u_{\mathcal{T}}, \gamma_{\mathcal{T}}), (\tau, v, \eta) \rangle_Y = &- \int_\Omega (C^{-1}\sigma_{\mathcal{T}} + \gamma_{\mathcal{T}}) : \tau \\
&- \int_\Omega \operatorname{div} \tau \cdot u_{\mathcal{T}} \\
&- \int_\Omega (f + \operatorname{div} \sigma_{\mathcal{T}}) \cdot v \\
&- \int_\Omega \operatorname{as}(\sigma_{\mathcal{T}}) : \eta.
\end{aligned}
\tag{4.48}
$$

Now, we apply the Helmholtz decomposition of Proposition 4.52 (p. 215) to the rows of the tensor τ and obtain

$$
\tau = \nabla(G\tau) + \operatorname{curl}(R\tau)
$$

with

$$
G\tau \in H^1(\Omega)^d, \quad R\tau \in H(\operatorname{curl}; \Omega)^d.
$$

This gives for the first two terms on the right-hand side of equation (4.48)

$$
\begin{aligned}
\int_\Omega (C^{-1}\sigma_{\mathcal{T}} + \gamma_{\mathcal{T}}) : \tau + \int_\Omega \operatorname{div} \tau \cdot u_{\mathcal{T}} = &\int_\Omega (C^{-1}\sigma_{\mathcal{T}} + \gamma_{\mathcal{T}}) : \operatorname{curl}(R\tau) \\
&+ \int_\Omega (C^{-1}\sigma_{\mathcal{T}} + \gamma_{\mathcal{T}}) : \nabla(G\tau) \\
&+ \int_\Omega \operatorname{div}(\nabla(G\tau)) \cdot u_{\mathcal{T}}.
\end{aligned}
$$

From [**183**] we know that the spaces $H_{\mathcal{T}}$ and $V_{\mathcal{T}}$ satisfy a commuting diagram property similar to Proposition 4.50 (p. 213). More precisely, there is an operator $\mathcal{J}_{\mathcal{T}} : H \to H_{\mathcal{T}}$ such that the identities

$$
\begin{aligned}
\int_K \operatorname{div} \rho \cdot v_{\mathcal{T}} &= \int_K \operatorname{div}(\mathcal{J}_{\mathcal{T}} \rho) \cdot v_{\mathcal{T}} \\
\int_E \mathbf{n}_E \cdot \rho \cdot v_{\mathcal{T}} &= \int_E \mathbf{n}_E \cdot (\mathcal{J}_{\mathcal{T}} \rho) \cdot v_{\mathcal{T}}
\end{aligned}
$$

hold for all $\rho \in H$, all $v_{\mathcal{T}} \in V_{\mathcal{T}}$, all elements $K \in \mathcal{T}$, and all edges or faces $E \in \mathcal{E}$.

These identities and integration by parts yield for all $\rho \in H$, all $v_{\mathcal{T}} \in V_{\mathcal{T}}$, and all elements $K \in \mathcal{T}$

$$
\begin{aligned}
\int_K \nabla v_{\mathcal{T}} : (Id - \mathcal{J}_{\mathcal{T}})\rho &= - \int_K \operatorname{div}((Id - \mathcal{J}_{\mathcal{T}})\rho) \cdot v_{\mathcal{T}} + \sum_{E \in \mathcal{E}_K} \int_E \mathbf{n}_K \cdot (Id - \mathcal{J}_{\mathcal{T}})\rho \cdot v_{\mathcal{T}} \\
&= 0.
\end{aligned}
$$

Here, as usual, $\mathbf{n}_K$ denotes the unit exterior normal of K.

Inserting $(\mathcal{J}_{\mathcal{T}}(\nabla(G\tau)), 0, 0)$ as a test function in the discrete problem, we conclude that

$$\int_{\Omega} \operatorname{div}(\mathcal{J}_{\mathcal{T}}(\nabla(G\tau))) \cdot u_{\mathcal{T}} = -\int_{\Omega} (C^{-1}\sigma_{\mathcal{T}} + \gamma_{\mathcal{T}}) : \mathcal{J}_{\mathcal{T}}(\nabla(G\tau)).$$

Combining these results, we obtain for the first two terms on the right-hand side of equation (4.48)

$$\begin{aligned}
\int_{\Omega} (C^{-1}\sigma_{\mathcal{T}} + \gamma_{\mathcal{T}}) : \tau + \int_{\Omega} \operatorname{div}\tau \cdot u_{\mathcal{T}} &= \int_{\Omega} (C^{-1}\sigma_{\mathcal{T}} + \gamma_{\mathcal{T}}) : \operatorname{curl}(R\tau) \\
&\quad + \int_{\Omega} (C^{-1}\sigma_{\mathcal{T}} + \gamma_{\mathcal{T}}) : \nabla(G\tau) + \int_{\Omega} \operatorname{div}(\nabla(G\tau)) \cdot u_{\mathcal{T}} \\
&= \int_{\Omega} (C^{-1}\sigma_{\mathcal{T}} + \gamma_{\mathcal{T}}) : \operatorname{curl}(R\tau) + \int_{\Omega} (C^{-1}\sigma_{\mathcal{T}} + \gamma_{\mathcal{T}}) : \nabla(G\tau) \\
&\quad + \int_{\Omega} \operatorname{div}(\mathcal{J}_{\mathcal{T}}(\nabla(G\tau)) \cdot u_{\mathcal{T}} \\
&= \int_{\Omega} (C^{-1}\sigma_{\mathcal{T}} + \gamma_{\mathcal{T}}) : \operatorname{curl}(R\tau) \\
&\quad + \int_{\Omega} (C^{-1}\sigma_{\mathcal{T}} + \gamma_{\mathcal{T}}) : (Id - \mathcal{J}_{\mathcal{T}})(\nabla(G\tau)) \\
&= \int_{\Omega} (C^{-1}\sigma_{\mathcal{T}} + \gamma_{\mathcal{T}}) : \operatorname{curl}(R\tau) \\
&\quad + \int_{\Omega} (C^{-1}\sigma_{\mathcal{T}} + \gamma_{\mathcal{T}} - \nabla u_{\mathcal{T}}) : (Id - \mathcal{J}_{\mathcal{T}})(\nabla(G\tau)).
\end{aligned}$$

We insert this result in equation (4.48) and perform integration by parts element-wise for the terms involving $\operatorname{curl}(R\tau)$. Denoting by γ_E the trace operator in the tangential direction to an edge or face E, i.e.

$$\gamma_E(\rho) = \rho - (\rho \cdot \mathbf{n}_E)\mathbf{n}_E,$$

we thus arrive at the following representation of the residual which is similar to equation (4.42) (p. 217)

$$\begin{aligned}
\langle \ell - L(\sigma_{\mathcal{T}}, u_{\mathcal{T}}, \gamma_{\mathcal{T}}), (\tau, v, \eta) \rangle_Y &= -\int_{\Omega} (C^{-1}\sigma_{\mathcal{T}} + \gamma_{\mathcal{T}} - \nabla u_{\mathcal{T}}) : (Id - \mathcal{J}_{\mathcal{T}})(\nabla(G\tau)) \\
&\quad - \int_{\Omega} (f + \operatorname{div}\sigma_{\mathcal{T}}) \cdot v - \int_{\Omega} \operatorname{as}(\sigma_{\mathcal{T}}) : \eta \\
&\quad - \sum_{K\in\mathcal{T}} \int_K \operatorname{curl}(C^{-1}\sigma_{\mathcal{T}} + \gamma_{\mathcal{T}}) \cdot (R\tau) \\
&\quad - \sum_{E\in\mathcal{E}_\Omega} \int_E \mathbb{J}_E(\gamma_E(C^{-1}\sigma_{\mathcal{T}} + \gamma_{\mathcal{T}})) \cdot (R\tau) \\
&\quad - \sum_{E\in\mathcal{E}_\Gamma} \int_E \gamma_E(C^{-1}\sigma_{\mathcal{T}} + \gamma_{\mathcal{T}}) \cdot (R\tau).
\end{aligned} \tag{4.49}$$

To obtain an upper bound for the dual norm of the residual, we estimate the individual terms in equation (4.49) in exactly the same way as in the previous section. To handle the terms involving $\mathcal{J}_{\mathcal{T}}$ we must assume that Ω is *convex*. From Proposition 4.52 (p. 215) and [**183**], cf. also [**90**, Proposition 3.6], we then obtain

$$\begin{aligned}\left\|(Id - \mathcal{J}_{\mathcal{T}})(\nabla(G\tau))\right\|_K &\leq c_1 h_K \left\|\nabla(G\tau)\right\|_{1;K} \\ &\leq c_1 h_K \left\|G\tau\right\|_{2;K}\end{aligned}$$

with a constant c_1 which only depends on the shape parameter $C_{\mathcal{T}}$ of $\mathcal{T}$. We thus arrive at the upper bound

$$\begin{aligned}\left\|\ell - L(\sigma_{\mathcal{T}}, u_{\mathcal{T}})\right\|_{Y^*} \leq c \Bigg\{ &\sum_{K\in\mathcal{T}} \Big[h_K^2 \left\|C^{-1}\sigma_{\mathcal{T}} + \gamma_{\mathcal{T}} - \nabla u_{\mathcal{T}}\right\|_K^2 + \left\|f + \operatorname{div}\sigma_{\mathcal{T}}\right\|_K^2 \\ &+ \left\|\operatorname{as}(\sigma_{\mathcal{T}})\right\|_K^2 + h_K^2 \left\|\operatorname{curl}(C^{-1}\sigma_{\mathcal{T}} + \gamma_{\mathcal{T}})\right\|_K^2 \Big] \\ &+ \sum_{E\in\mathcal{E}_\Omega} h_E \left\|\mathbb{J}_E(\gamma_E(C^{-1}\sigma_{\mathcal{T}} + \gamma_{\mathcal{T}}))\right\|_E^2 + \sum_{E\in\mathcal{E}_\Gamma} h_E \left\|\gamma_E(C^{-1}\sigma_{\mathcal{T}} + \gamma_{\mathcal{T}})\right\|_E^2 \Bigg\}^{\frac{1}{2}}\end{aligned}$$

with a constant c which only depends on the shape parameter $C_{\mathcal{T}}$ of $\mathcal{T}$ and the domain Ω.

For the error indicator we will as usual replace f by some approximation $f_{\mathcal{T}}$. The above estimate therefore establishes the first condition of Theorem 4.7 (p. 156) with

$$\theta_{\mathcal{T}} = \left\|f - f_{\mathcal{T}}\right\|$$

and a constant c_A which only depends on the shape parameter $C_{\mathcal{T}}$ of $\mathcal{T}$.

Since $B_{\mathcal{T}}$ and $\ell_{\mathcal{T}}$ are the restrictions of B and ℓ, respectively, we have Galerkin orthogonality and the second condition of Theorem 4.7 is automatically satisfied with $c_C = 0$.

The third condition of Theorem 4.7 is established in exactly the same way as in the previous section. The only new term is the one involving $\operatorname{as}(\sigma_{\mathcal{T}})$. This is handled similarly to the term involving $f_{\mathcal{T}} + \operatorname{div}\sigma_{\mathcal{T}}$ using the local test function

$$\eta_K = -\operatorname{as}(\sigma_{\mathcal{T}})\chi_K$$

where again χ_K denotes the characteristic function of the element K. The space $\widetilde{Y}_{\mathcal{T}}$ is now given by

$$\begin{aligned}\widetilde{Y}_{\mathcal{T}} = \operatorname{span} \Big\{ &(0, (f + \operatorname{div}\sigma_{\mathcal{T}})\chi_K, 0), (0, 0, \operatorname{as}(\sigma_{\mathcal{T}})\chi_K), \\ &(\operatorname{curl}(\psi_K \operatorname{curl}(C^{-1}\sigma_{\mathcal{T}} + \gamma_{\mathcal{T}})), 0, 0), \\ &(\psi_K(C^{-1}\sigma_{\mathcal{T}} + \gamma_{\mathcal{T}} - \nabla u_{\mathcal{T}}), 0, 0), \\ &(\operatorname{curl}(\psi_E \mathbb{J}_E(\gamma_E(C^{-1}\sigma_{\mathcal{T}} + \gamma_{\mathcal{T}}))), 0, 0), \\ &(\operatorname{curl}(\psi_{E'}\gamma_{E'}(C^{-1}\sigma_{\mathcal{T}} + \gamma_{\mathcal{T}})), 0, 0) : \\ &K \in \mathcal{T}, E \in \mathcal{E}_\Omega, E' \in \mathcal{E}_\Gamma \Big\}.\end{aligned}$$

These estimates and Theorem 4.7 yield upper and lower bounds for the mixed finite element discretisation of problem (4.44) with the Lamé parameter $\mu = 1$ and constants which are independent of the Lamé parameter λ.

To handle the case of arbitrary μ, we use a simple scaling argument. Denote by (σ, u, γ) the weak solution of problem (4.44) with arbitrary Lamé parameters λ and μ and by $(\sigma_{\mathcal{T}}, u_{\mathcal{T}}, \gamma_{\mathcal{T}})$ its finite element approximation with the PEERS or BDMS elements. Set

$$\begin{aligned} \overline{\sigma} &= \frac{1}{\mu}\sigma, & \overline{\sigma}_{\mathcal{T}} &= \frac{1}{\mu}\sigma_{\mathcal{T}}, \\ \overline{u} &= u, & \overline{u}_{\mathcal{T}} &= u_{\mathcal{T}}, \\ \overline{\gamma} &= \gamma, & \overline{\gamma}_{\mathcal{T}} &= \gamma_{\mathcal{T}}, \\ \overline{f} &= \frac{1}{\mu}f, & \overline{f}_{\mathcal{T}} &= \frac{1}{\mu}f_{\mathcal{T}}, \\ \overline{C} &= \frac{1}{\mu}C. \end{aligned}$$

Then $(\overline{\sigma}, \overline{u}, \overline{\gamma})$ and $(\overline{\sigma}_{\mathcal{T}}, \overline{u}_{\mathcal{T}}, \overline{\gamma}_{\mathcal{T}})$ are the weak solutions of problem (4.44) corresponding to the Lamé parameters $\overline{\lambda} = \frac{\lambda}{\mu}$ and $\overline{\mu} = 1$ and of the corresponding mixed finite element approximation, respectively. Since

$$\begin{aligned} \|\overline{\sigma} - \overline{\sigma}_{\mathcal{T}}\| &= \frac{1}{\mu}\|\sigma - \sigma_{\mathcal{T}}\| & \overline{f}_{\mathcal{T}} + \operatorname{div}\overline{\sigma}_{\mathcal{T}} &= \frac{1}{\mu}(f_{\mathcal{T}} + \operatorname{div}\sigma_{\mathcal{T}}) \\ \operatorname{as}(\overline{\sigma}_{\mathcal{T}}) &= \frac{1}{\mu}\operatorname{as}(\sigma_{\mathcal{T}}) & \overline{C}^{-1}\overline{\sigma}_{\mathcal{T}} &= C^{-1}\sigma_{\mathcal{T}} \end{aligned}$$

the previous results for the case $\mu = 1$ imply the following robust a posteriori error estimates.

Theorem 4.60 *Denote by $(\sigma, u, \gamma) \in X$ the unique solution of the mixed variational formulation of problem (4.44) based on the Hellinger–Reissner principle and by $(\sigma_{\mathcal{T}}, u_{\mathcal{T}}, \gamma_{\mathcal{T}}) \in X_{\mathcal{T}}$ its finite element approximation using the PEERS or BDMS elements. For every element $K \in \mathcal{T}$ define the* residual a posteriori error indicator $\eta_{R,K}$ *by*

$$\begin{aligned} \eta_{R,K} = \Bigg\{ & h_K^2 \left\|C^{-1}\sigma_{\mathcal{T}} + \gamma_{\mathcal{T}} - \nabla u_{\mathcal{T}}\right\|_K^2 + \frac{1}{\mu^2}\left\|f + \operatorname{div}\sigma_{\mathcal{T}}\right\|_K^2 + \frac{1}{\mu^2}\left\|\operatorname{as}(\sigma_{\mathcal{T}})\right\|_K^2 \\ & + h_K^2 \left\|\operatorname{curl}(C^{-1}\sigma_{\mathcal{T}} + \gamma_{\mathcal{T}})\right\|_K^2 + \sum_{E \in \mathcal{E}_{K,\Omega}} h_E \left\|\mathbb{J}_E(\gamma_E(C^{-1}\sigma_{\mathcal{T}} + \gamma_{\mathcal{T}}))\right\|_E^2 \\ & + \sum_{E \in \mathcal{E}_{K,\Gamma}} h_E \left\|\gamma_E(C^{-1}\sigma_{\mathcal{T}} + \gamma_{\mathcal{T}})\right\|_E^2 \Bigg\}^{\frac{1}{2}}. \end{aligned}$$

Assume that Ω is convex. Then there are two constants c^* and c_*, which only depend on Ω and the shape parameter $C_{\mathcal{T}}$ of $\mathcal{T}$ but which are independent of the Lamé parameters λ and μ, such that the estimates

$$\left\{\frac{1}{\mu^2}\|\sigma - \sigma_{\mathcal{T}}\|^2_{H(\operatorname{div};\Omega)} + \|u - u_{\mathcal{T}}\|^2 + \|\gamma - \gamma_{\mathcal{T}}\|^2\right\}^{\frac{1}{2}} \le c^* \left\{\sum_{K \in \mathcal{T}} \left[\eta_{R,K}^2 + \|f - f_{\mathcal{T}}\|_K^2\right]\right\}^{\frac{1}{2}}$$

and

$$\eta_{R,K} \le c_* \left\{ \frac{1}{\mu^2} \|\sigma - \sigma_{\mathcal{T}}\|^2_{H(\operatorname{div};\omega_K)} + \|u - u_{\mathcal{T}}\|^2_{\omega_K} + \|\gamma - \gamma_{\mathcal{T}}\|^2_{\omega_K} + \left\|f - f_{\mathcal{T}}\right\|^2_K \right\}^{\frac{1}{2}}$$

hold for all $K \in \mathcal{T}$. Here, $f_{\mathcal{T}}$ is any piecewise polynomial approximation of f.

4.9.5 Local Neumann Problems

We want to treat local auxiliary problems which are based on a single element $K \in \mathcal{T}$. Furthermore we want to impose pure Neumann boundary conditions. Since the displacement of a linear elasticity problem with pure Neumann boundary conditions is unique only up to *rigid body motions*, we must factor out the rigid body motions $\mathcal{R}_K$ of the element K. These are given by

$$\mathcal{R}_K = \begin{cases} \left\{v = (a,b) + c(-x_2, x_1) : a,b,c \in \mathbb{R}\right\} & \text{if } d = 2, \\ \{v = a + b \times x : a,b \in \mathbb{R}^3\} & \text{if } d = 3. \end{cases}$$

We set

$$\begin{aligned} H_K &= \mathrm{BDM}_m(K)^d \oplus B_{m-1}(K), \\ V_K &= R_{m-1}(K)^d / \mathcal{R}_K, \\ W_K &= \left\{\eta_K \in R_m(K)^{d\times d} : \eta_K + \eta_K^T = 0\right\}, \\ X_K &= H_K \times V_K \times W_K, \end{aligned}$$

where the polynomial degree m will be determined later, cf. Theorem 4.63. Furthermore we denote by B_K the restriction of the bi-linear form B to $X_K \times X_K$, i.e.

$$\begin{aligned} B_K((\sigma,u,\gamma),(\tau,v,\eta)) = &\int_K C^{-1}\sigma : \tau \\ &+ \int_K \operatorname{div}\tau \cdot u + \int_K \tau : \gamma \\ &+ \int_K \operatorname{div}\sigma \cdot v + \int_K \sigma : \eta. \end{aligned}$$

Lemma 4.61 *The bi-linear form B_K satisfies the estimates*

$$\sup_{(\sigma,u,\gamma)\in X_K\setminus\{0\}} \sup_{(\tau,v,\eta)\in X_K\setminus\{0\}} \frac{B_K((\sigma,u,\gamma),(\tau,v,\eta))}{\left\|(\sigma,u,\gamma)\right\|_X \left\|(\tau,v,\eta)\right\|_X} \le \bar{c}_D$$

$$\inf_{(\sigma,u,\gamma)\in X_K\setminus\{0\}} \sup_{(\tau,v,\eta)\in X_K\setminus\{0\}} \frac{B_K((\sigma,u,\gamma),(\tau,v,\eta))}{\left\|(\sigma,u,\gamma)\right\|_X \left\|(\tau,v,\eta)\right\|_X} \ge \bar{c}_d$$

with constants $\bar{c}_D$ and $\bar{c}_d$ which only depend on the shape parameter $C_{\mathcal{T}}$ of $\mathcal{T}$ and the Lamé parameter μ.

Proof Lemma 4.61 is proved by transforming to the reference element, invoking the stability of the linear elasticity problem there, and transforming back to the element K, cf. [**182**] for details. □

Due to Lemma 4.61 there is a unique element $(\sigma_K, u_K, \gamma_K) \in X_K$ such that

$$B_K((\sigma_K, u_K, \gamma_K), (\tau_K, v_K, \eta_K)) = \langle \ell - L(\sigma_{\mathcal{T}}, u_{\mathcal{T}}, \gamma_{\mathcal{T}}), (\tau_K, v_K, \eta_K) \rangle_Y \tag{4.50}$$

holds for all $(\tau_K, v_K, \eta_K) \in X_K$. We then define the error indicator $\eta_{N,K}$ by

$$\eta_{N,K} = \left\{ \frac{1}{\mu^2} \|\sigma_K\|^2_{H(\mathrm{div};K)} + \|v_K\|^2_K + \|\gamma_K\|^2_K \right\}^{\frac{1}{2}}. \tag{4.51}$$

Remark 4.62 Problem (4.50) is a discrete linear elasticity problem with pure Neumann boundary conditions on the single element K. In order to implement the error indicator $\eta_{N,K}$ one has to construct a basis for the space V_K. This can be done by taking the standard basis of $R_m(K)^d$ and dropping those degrees of freedom that belong to the rigid body motions. Afterwards one has to compute the stiffness matrix for each element K and solve the associated local auxiliary problem.

Theorem 4.63 *In addition to the conditions of Theorem 4.60 assume that the discretisation of problem (4.44) is based on the BDMS element with polynomial degree $k \geq 2$ and that the polynomial degree m associated with the spaces X_K is at least $k + 2d$. Then the error indicator $\eta_{N,K}$ of equation (4.51) yields the global upper bound*

$$\left\{ \frac{1}{\mu^2} \|\sigma - \sigma_{\mathcal{T}}\|^2_{H(\mathrm{div};\Omega)} + \|u - u_{\mathcal{T}}\|^2 + \|\gamma - \gamma_{\mathcal{T}}\|^2 \right\}^{\frac{1}{2}} \leq \widehat{c}^* \left\{ \sum_{K\in\mathcal{T}} \left[\eta^2_{N,K} + \|f - f_{\mathcal{T}}\|^2_K \right] \right\}^{\frac{1}{2}}$$

and the local lower bound

$$\eta_{N,K} \leq \widehat{c}_* \left\{ \frac{1}{\mu^2} \|\sigma - \sigma_{\mathcal{T}}\|^2_{H(\mathrm{div};K)} + \|u - u_{\mathcal{T}}\|^2_K + \|\gamma - \gamma_{\mathcal{T}}\|^2_K \right\}^{\frac{1}{2}}.$$

Here, $f_{\mathcal{T}}$ is any piecewise polynomial approximation of f. The constants $\widehat{c}^*$ and $\widehat{c}_*$ only depend on the shape parameter $C_{\mathcal{T}}$ of $\mathcal{T}$ but are independent of the Lamé parameters λ and μ.

Proof The lower bound for the error follows from the identity

$$\langle \ell - L(\sigma_{\mathcal{T}}, u_{\mathcal{T}}, \gamma_{\mathcal{T}}), (\tau_K, v_K, \eta_K) \rangle_Y = B((\sigma - \sigma_{\mathcal{T}}, u - u_{\mathcal{T}}, \gamma - \gamma_{\mathcal{T}}), (\tau_K, v_K, \eta_K)),$$

Lemma 4.61, and the definition of $\eta_{N,K}$.

The assumption $m \geq k + 2d$ implies that

$$\widetilde{Y}_{\mathcal{T}} \subset \bigoplus_{K\in\mathcal{T}} X_K.$$

Hence, we have

$$\left\{ \sum_{K\in\mathcal{T}} \eta^2_{R,K} \right\}^{\frac{1}{2}} \leq c_I \left\| \ell - L(\sigma_{\mathcal{T}}, u_{\mathcal{T}}, \gamma_{\mathcal{T}}) \right\|_{\widetilde{Y}^*_{\mathcal{T}}} \leq \bar{c}_D c_I \left\{ \sum_{K\in\mathcal{T}} \eta^2_{N,K} \right\}^{\frac{1}{2}}.$$

Together with Theorem 4.60 this implies the upper bound for the error.
The stability of the estimates with respect to the Lamé parameters λ and μ is obtained as for $\eta_{R,K}$ by first considering the case $\mu = 1$ and then applying the scaling argument. □

4.9.6 *Local Dirichlet Problems*

Now we want to construct an error indicator which is similar to $\eta_{D,K}$ defined in equation (1.41) (p. 28) and which is based on the solution of discrete linear elasticity problems with Dirichlet boundary conditions on the patches ω_K. To this end, we associate with every element K the spaces

$$\begin{aligned}
\widetilde{H}_K &= \left\{\sigma_K \in H(\mathrm{div};\omega_K)^d : \sigma_{\mathcal{T}}|_{K'} \in \mathrm{BDM}_m(K')^d \oplus B_{m-1}(K'),\ K' \in \mathcal{T}, \mathcal{E}_{K'} \cap \mathcal{E}_K \neq \emptyset\right\},\\
\widetilde{V}_K &= \left\{v_{\mathcal{T}} \in L^2(\omega_K)^d : v_{\mathcal{T}}|_{K'} \in R_{m-1}(K')^d,\ K' \in \mathcal{T}, \mathcal{E}_{K'} \cap \mathcal{E}_K \neq \emptyset\right\},\\
\widetilde{W}_K &= \left\{\eta_{\mathcal{T}} \in L^2(\omega_K)^{d\times d} \cap C(\omega_K)^{d\times d} : \eta_{\mathcal{T}} + \eta_{\mathcal{T}}^T = 0,\ \eta_{\mathcal{T}}|_{K'} \in R_m(K')^{d\times d},\right.\\
&\qquad\qquad \left. K' \in \mathcal{T}, \mathcal{E}_{K'} \cap \mathcal{E}_K \neq \emptyset\right\},\\
\widetilde{X}_K &= \widetilde{H}_K \times \widetilde{V}_K \times \widetilde{W}_K,
\end{aligned}$$

where the polynomial degree will be determined later, cf. Theorem 4.65, and denote by B_{ω_K} the restriction of B to ω_K, i.e.

$$\begin{aligned}
B_{\omega_K}((\sigma,u,\gamma),(\tau,v,\eta)) &= \int_{\omega_K} C^{-1}\sigma : \tau\\
&\quad + \int_{\omega_K} \mathrm{div}\,\tau \cdot u + \int_{\omega_K} \tau : \gamma\\
&\quad + \int_{\omega_K} \mathrm{div}\,\sigma \cdot v + \int_{\omega_K} \sigma : \eta.
\end{aligned}$$

Lemma 4.64 *The bi-linear form B_{ω_K} satisfies the estimates*

$$\begin{aligned}
\sup_{(\sigma,u,\gamma)\in\widetilde{X}_K\setminus\{0\}} \sup_{(\tau,v,\eta)\in\widetilde{X}_K\setminus\{0\}} \frac{B_{\omega_K}((\sigma,u,\gamma),(\tau,v,\eta))}{\|(\sigma,u,\gamma)\|_X \|(\tau,v,\eta)\|_X} &\le \widetilde{c}_D\\
\inf_{(\sigma,u,\gamma)\in\widetilde{X}_K\setminus\{0\}} \sup_{(\tau,v,\eta)\in\widetilde{X}_K\setminus\{0\}} \frac{B_{\omega_K}((\sigma,u,\gamma),(\tau,v,\eta))}{\|(\sigma,u,\gamma)\|_X \|(\tau,v,\eta)\|_X} &\ge \widetilde{c}_d
\end{aligned}$$

with constants $\widetilde{c}_D$ and $\widetilde{c}_d$ which only depend on the shape parameter $C_{\mathcal{T}}$ of $\mathcal{T}$ and the Lamé parameter μ.

Proof Lemma 4.64 is proved by applying the affine transformation which maps K onto the reference element $\widehat{K}$ to the complete patch ω_K, invoking the stability of the linear elasticity problem on the resulting patch $\widehat{\omega}_{\widehat{K}}$ and transforming back to the original patch ω_K, cf. Figure 4.7. Here, we must take into account that every patch ω_K gives rise to a different patch $\widehat{\omega}_{\widehat{K}}$ and thus different constants $\widetilde{c}_d$ and $\widetilde{c}_D$. Due to the shape regularity of $\mathcal{T}$, however, the set of all possible patches $\widehat{\omega}_{\widehat{K}}$ is compact. Since $\widetilde{c}_d$ and $\widetilde{c}_D$ depend continuously on $\widehat{\omega}_{\widehat{K}}$ they are both uniformly bounded, the bounds being independent of the Lamé parameter λ. □

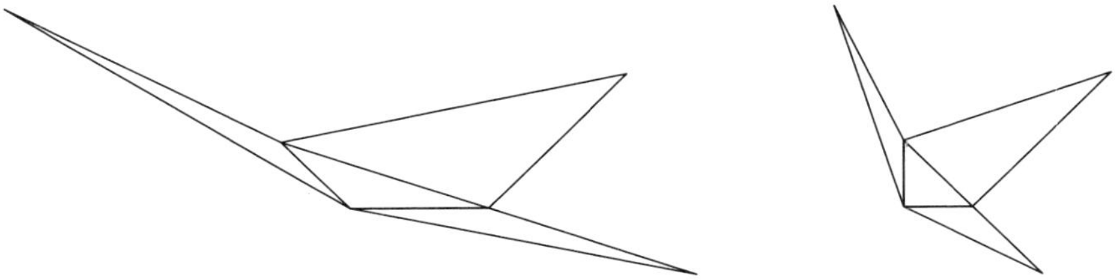

Figure 4.7 Transformation of ω_K (left) to $\widehat{\omega}_{\widehat{K}}$ (right)

Due to Lemma 4.64 there is a unique element $(\widetilde{\sigma}_K, \widetilde{u}_K, \widetilde{\gamma}_K) \in \widetilde{X}_K$ such that

$$B_{\omega_K}((\widetilde{\sigma}_K, \widetilde{u}_K, \widetilde{\gamma}_K), (\tau_K, v_K, \eta_K)) = \langle \ell - L(\sigma_{\mathcal{T}}, u_{\mathcal{T}}, \gamma_{\mathcal{T}}), (\tau_K, v_K, \eta_K)\rangle_Y$$

holds for all $(\tau_K, v_K, \eta_K) \in \widetilde{X}_K$. We then define the error indicator $\eta_{D,K}$ by

$$\eta_{D,K} = \left\{ \frac{1}{\mu^2} \|\widetilde{\sigma}_K\|^2_{H(\mathrm{div};\omega_K)} + \|\widetilde{v}_K\|^2_{\omega_K} + \|\widetilde{\gamma}_K\|^2_{\omega_K} \right\}^{\frac{1}{2}}. \tag{4.52}$$

With the same arguments as in the previous subsection, one can prove that $\eta_{D,K}$ provides a reliable and efficient a posteriori error indicator which is robust with respect to nearly incompressible materials.

Theorem 4.65 *In addition to the conditions of Theorem 4.60 assume that the discretisation of problem (4.44) is based on the BDMS element with polynomial degree $k \geq 2$ and that the polynomial degree m associated with the spaces $\widetilde{X}_K$ is at least $k + 2d$. Then the error indicator $\eta_{D,K}$ of equation (4.52) yields the global upper bound*

$$\left\{ \frac{1}{\mu^2} \|\sigma - \sigma_{\mathcal{T}}\|^2_{H(\mathrm{div};\Omega)} + \|u - u_{\mathcal{T}}\|^2 + \|\gamma - \gamma_{\mathcal{T}}\|^2 \right\}^{\frac{1}{2}} \leq \widetilde{c}^{\,*} \left\{ \sum_{K\in\mathcal{T}} \left[\eta^2_{D,K} + \|f - f_{\mathcal{T}}\|^2_K \right] \right\}^{\frac{1}{2}}$$

and the local lower bound

$$\eta_{D,K} \leq \widetilde{c}_* \left\{ \frac{1}{\mu^2} \|\sigma - \sigma_{\mathcal{T}}\|^2_{H(\mathrm{div};\omega_K)} + \|u - u_{\mathcal{T}}\|^2_{\omega_K} + \|\gamma - \gamma_{\mathcal{T}}\|^2_{\omega_K} \right\}^{\frac{1}{2}}.$$

Here, $f_{\mathcal{T}}$ is any piecewise polynomial approximation of f. The constants $\widetilde{c}^{\,*}$ and $\widetilde{c}_*$ only depend on the shape parameter $C_{\mathcal{T}}$ of $\mathcal{T}$ but are independent of the Lamé parameters λ and μ.

Remark 4.66 The analysis of this section is due to [**182**] and was first published in [**183**]. It builds upon earlier results in [**18**] and [**91**].

4.10 The Stokes Equations

The *Stokes equations*

$$\begin{aligned} -\Delta \mathbf{u} + \nabla p &= \mathbf{f} \quad \text{in } \Omega \\ \operatorname{div} \mathbf{u} &= 0 \quad \text{in } \Omega \\ \mathbf{u} &= 0 \quad \text{on } \Gamma \end{aligned} \tag{4.53}$$

are a very simple model for the viscous flow of a fluid with velocity field $\mathbf{u}$ and pressure p under the influence of an exterior force f. They are the Euler–Lagrange equations of the constrained minimisation problem:

minimise

$$J(\mathbf{u}) = \frac{1}{2}\int_\Omega \nabla \mathbf{u} : \nabla \mathbf{u} - \int_\Omega \mathbf{f} \cdot \mathbf{u}$$

subject to the constraint

$$\operatorname{div} \mathbf{u} = 0.$$

Since the space

$$V = \left\{ \mathbf{u} \in H_0^1(\Omega)^d : \operatorname{div} \mathbf{u} = 0 \right\}$$

of solenoidal velocity fields is a closed subspace of $H_0^1(\Omega)^d$, it is very tempting to base the discretisation of problem (4.53) on the variational problem:

find $\mathbf{u} \in V$ such that

$$\int_\Omega \nabla \mathbf{u} : \nabla \mathbf{v} = \int_\Omega \mathbf{f} \cdot \mathbf{v}$$

holds for all $\mathbf{v} \in V$.

Yet, it is well known that this naive approach leads to a dead-end road:

- there are no low order conforming finite element subspaces of V;
- the above variational problem gives no information on the pressure which is often the physically most important quantity.

Instead, one considers a saddle-point formulation of the Stokes equations with the pressure as the Lagrange multiplier corresponding to the divergence constraint and corresponding mixed finite element approximations which simultaneously yield information on the velocity and the pressure.

4.10.1 Saddle-Point Formulation

Since equations (4.53) at best determine the pressure up to an additive constant, we introduce the space

$$L_0^2(\Omega) = \left\{ p \in L^2(\Omega) : \int_\Omega p = 0 \right\} \tag{4.54}$$

which is a closed subspace of $L^2(\Omega)$. With this space the saddle-point formulation of the Stokes equations fits into the abstract framework of Section 4.1 (p. 151) with

$$X = Y = H_0^1(\Omega)^d \times L_0^2(\Omega),$$

$$\left\| (\mathbf{u}, p) \right\|_X = \left\{ \|\nabla \mathbf{u}\|^2 + \|p\|^2 \right\}^{\frac{1}{2}},$$

$$B((\mathbf{u}, p), (\mathbf{v}, q)) = \int_\Omega \nabla \mathbf{u} : \nabla \mathbf{v} - \int_\Omega p \operatorname{div} \mathbf{v} + \int_\Omega q \operatorname{div} \mathbf{u},$$

$$\langle \ell, (\mathbf{v}, q) \rangle_Y = \int_\Omega \mathbf{f} \cdot \mathbf{v}.$$

The well-posedness of the saddle-point problem crucially hinges on the following well-known result [**145**, Corollary I.2.4 and inequality (I.5.14)].

Proposition 4.67 *The divergence operator* div *is an isomorphism of*

$$V^\perp = \left\{ \mathbf{v} \in H_0^1(\Omega)^d : \int_\Omega \nabla \mathbf{v} : \nabla \mathbf{w} = 0 \textit{ for all } \mathbf{w} \in V \right\}$$

onto $L_0^2(\Omega)$ *and*

$$\beta = \inf_{q \in L_0^2(\Omega) \setminus \{0\}} \sup_{\mathbf{v} \in H_0^1(\Omega)^d \setminus \{0\}} \frac{\int_\Omega q \operatorname{div} \mathbf{v}}{\|\nabla \mathbf{v}\| \, \|q\|} > 0.$$

Remark 4.68 Usually, Proposition 4.67 is proved by functional analytical arguments based on results by J. Nečas [**201**] and G. de Rham [**120**]. These arguments give no information on the actual size of the quantity β which is the inverse of the norm of the maximal right-inverse of the divergence operator. A different, less well-known proof is due to M. E. Bogovskii [**72**]. It is based on the explicit construction of the maximal right-inverse of the divergence operator using a Green's kernel. This proof provides an explicit estimate of β: for bounded domains β is proportional to the inverse of the diameter of Ω.

The following proposition shows that the bi-linear form B fulfils the conditions of Proposition 4.1 (p. 152).

Proposition 4.69 *The bi-linear form B given above fulfils the estimates*

$$\begin{aligned} \inf_{(\mathbf{w},r) \in X \setminus \{0\}} \sup_{(\mathbf{v},q) \in X \setminus \{0\}} \frac{B((\mathbf{v}, q), (\mathbf{w}, r))}{\left\| (\mathbf{v}, q) \right\|_X \left\| (\mathbf{w}, r) \right\|_X} &= \inf_{(\mathbf{v},q) \in X \setminus \{0\}} \sup_{(\mathbf{w},r) \in X \setminus \{0\}} \frac{B((\mathbf{v}, q), (\mathbf{w}, r))}{\left\| (\mathbf{v}, q) \right\|_X \left\| (\mathbf{w}, r) \right\|_X} \\ &\geq \frac{\beta^2}{(1 + \beta)^2}, \end{aligned}$$

where β is the constant in Proposition 4.67.

Proof The first equation follows from the identity

$$B((\mathbf{v},q),(\mathbf{w},r)) = B((\mathbf{w},r),(\mathbf{v},-q))$$

for all $(\mathbf{v},q),(\mathbf{w},r) \in X$ and Remark 4.4 (p. 153).

To prove the inequality, we choose an arbitrary element $(\mathbf{v},q) \in X \setminus \{0\}$ and keep it fixed in what follows. The definition of B immediately implies that

$$B((\mathbf{v},q),(\mathbf{v},q)) = \|\nabla \mathbf{v}\|^2 .$$

Due to Proposition 4.67, there is a velocity field $\mathbf{w}_q$ with

$$\left\|\nabla \mathbf{w}_q\right\| = 1 \quad \text{and} \quad \int_\Omega q \operatorname{div} \mathbf{w}_q \geq \beta \left\|q\right\| .$$

We therefore obtain for every $\delta > 0$

$$\begin{aligned}
B((\mathbf{v},q),(\mathbf{v}-\delta \left\|q\right\| \mathbf{w}_q,q)) &= B((\mathbf{v},q),(\mathbf{v},q)) - \delta \left\|q\right\| B((\mathbf{v},q),(\mathbf{w}_q,0)) \\
&= \|\nabla \mathbf{v}\|^2 - \delta \left\|q\right\| \int_\Omega \nabla \mathbf{v} : \nabla \mathbf{w}_q + \delta \left\|q\right\| \int_\Omega q \operatorname{div} \mathbf{w}_q \\
&\geq \|\nabla \mathbf{v}\|^2 - \delta \|\nabla \mathbf{v}\| \left\|q\right\| + \delta\beta \left\|q\right\|^2 \\
&\geq \left(1 - \frac{\delta}{2\beta}\right) \|\nabla \mathbf{v}\|^2 + \frac{1}{2}\delta\beta \left\|q\right\|^2 .
\end{aligned}$$

The choice $\delta = \frac{2\beta}{1+\beta^2}$ yields

$$B((\mathbf{v},q),(\mathbf{v}-\delta \left\|q\right\| \mathbf{w}_q,q)) \geq \frac{\beta^2}{1+\beta^2} \left\|(\mathbf{v},q)\right\|_X^2 .$$

On the other hand we have

$$\begin{aligned}
\left\|(\mathbf{v}-\delta \left\|q\right\| \mathbf{w}_q,q)\right\|_X &\leq \left\|(\mathbf{v},q)\right\|_X + \left\|(\delta \left\|q\right\| \mathbf{w}_q,0)\right\|_X \\
&= \left\|(\mathbf{v},q)\right\|_X + \delta \left\|q\right\| \left\|\nabla \mathbf{w}_q\right\| \\
&= \left\|(\mathbf{v},q)\right\|_X + \delta \left\|q\right\| \\
&\leq (1+\delta) \left\|(\mathbf{v},q)\right\|_X \\
&= \frac{1+\beta^2+2\beta}{1+\beta^2} \left\|(\mathbf{v},q)\right\|_X .
\end{aligned}$$

Combining these estimates we arrive at

$$\begin{aligned}
\sup_{(\mathbf{w},r)\in X\setminus\{0\}} \frac{B((\mathbf{v},q),(\mathbf{w},r))}{\left\|(\mathbf{v},q)\right\|_X \left\|(\mathbf{w},r)\right\|_X} &\geq \frac{B((\mathbf{v},q),(\mathbf{v}-\delta \left\|q\right\| \mathbf{w}_q,q))}{\left\|(\mathbf{v},q)\right\|_X \left\|(\mathbf{v}-\delta \left\|q\right\| \mathbf{w}_q,q)\right\|_X} \\
&\geq \frac{\beta^2}{(1+\beta)^2}.
\end{aligned}$$

Since $(\mathbf{v},q) \in X \setminus \{0\}$ was arbitrary, this proves Proposition 4.69. □

4.10.2 *Finite Element Approximation*

Conforming finite element approximations of problem (4.53) can be grouped into two categories:

- stable mixed methods and
- stabilised schemes.

Both categories can be formulated in a common form which fits into the abstract framework of Section 4.1 (p. 151). To do this, we assume that Ω is a bounded connected polyhedral domain with Lipschitz boundary Γ and denote by $V_{\mathcal{T}} \subset H_0^1(\Omega)^d$ and $P_{\mathcal{T}} \subset L_0^2(\Omega)$ two finite element spaces associated with an admissible, affine-equivalent, shape-regular partition $\mathcal{T}$ of Ω which approximate the velocity field and the pressure, respectively. Examples for such spaces will be given below. Within the framework of Section 4.1, the finite element approximation of the Stokes equations then corresponds to the choice

$$
\begin{aligned}
X_{\mathcal{T}} &= Y_{\mathcal{T}} = V_{\mathcal{T}} \times P_{\mathcal{T}}, \\
B_{\mathcal{T}}((\mathbf{v}_{\mathcal{T}}, q_{\mathcal{T}}), (\mathbf{w}_{\mathcal{T}}, r_{\mathcal{T}})) &= \int_\Omega \nabla \mathbf{v}_{\mathcal{T}} : \nabla \mathbf{w}_{\mathcal{T}} - \int_\Omega q_{\mathcal{T}} \operatorname{div} \mathbf{w}_{\mathcal{T}} + \int_\Omega r_{\mathcal{T}} \operatorname{div} \mathbf{v}_{\mathcal{T}} \\
&\quad + \sum_{K \in \mathcal{T}} \vartheta_K h_K^2 \int_K [-\Delta \mathbf{v}_{\mathcal{T}} + \nabla q_{\mathcal{T}}] \cdot \nabla r_{\mathcal{T}} \\
&\quad + \sum_{E \in \mathcal{E}_\Omega} \vartheta_E h_E \int_E \mathbb{J}_E(q_{\mathcal{T}}) \mathbb{J}_E(r_{\mathcal{T}}) \\
&\quad + \sum_{K \in \mathcal{T}} \widetilde{\vartheta}_K \int_K \operatorname{div} \mathbf{v}_{\mathcal{T}} \operatorname{div} \mathbf{w}_{\mathcal{T}}, \\
\langle \ell_{\mathcal{T}}, (\mathbf{v}_{\mathcal{T}}, q_{\mathcal{T}}) \rangle_{Y_{\mathcal{T}}} &= \int_\Omega \mathbf{f} \cdot \mathbf{v}_{\mathcal{T}} + \sum_{K \in \mathcal{T}} \vartheta_K h_K^2 \int_K \mathbf{f} \cdot \nabla r_{\mathcal{T}}.
\end{aligned}
$$

Here, ϑ_K, ϑ_E, and $\widetilde{\vartheta}_K$ are non-negative stabilisation parameters.

Stable mixed methods correspond to the choice

$$
\vartheta_K = \vartheta_E = \widetilde{\vartheta}_K = 0
$$

for all elements K and all edges or faces E. They have to satisfy the following *stability condition*:

> there is a positive constant $\widetilde{\beta}$ such that
>
> $$\inf_{q_{\mathcal{T}} \in P_{\mathcal{T}} \setminus \{0\}} \sup_{\mathbf{v}_{\mathcal{T}} \in V_{\mathcal{T}} \setminus \{0\}} \frac{\int_\Omega q_{\mathcal{T}} \operatorname{div} \mathbf{v}_{\mathcal{T}}}{\|\nabla \mathbf{v}_{\mathcal{T}}\| \, \|q_{\mathcal{T}}\|} \geq \widetilde{\beta}$$
>
> holds uniformly with respect to all meshes which are obtained by successive uniform or adaptive refinement of any given initial mesh.

Prominent examples falling into this category are:

- the *mini element*, cf. [**27**] and [**145**, Section II.4.1]

$$V_{\mathcal{T}} = \left[S_0^{1,0}(\mathcal{T}) \oplus \operatorname{span}\{\psi_K : K \in \mathcal{T}\}\right]^d,$$
$$P_{\mathcal{T}} = S^{1,0}(\mathcal{T}) \cap L_0^2(\Omega);$$

- the classical *Hood–Taylor element*, cf. [**152**] and [**249**]

$$V_{\mathcal{T}} = S_0^{2,0}(\mathcal{T})^d,$$
$$P_{\mathcal{T}} = S^{1,0}(\mathcal{T}) \cap L_0^2(\Omega);$$

- the *modified Hood–Taylor element*, cf. [**58**] and [**249**]

$$V_{\mathcal{T}} = S_0^{2,0}(\mathcal{T}/2)^d,$$
$$P_{\mathcal{T}} = S^{1,0}(\mathcal{T}) \cap L_0^2(\Omega),$$

  where $\mathcal{T}/2$ denotes the partition which is obtained from $\mathcal{T}$ by one step of uniform refinement;
- the *higher order Hood–Taylor elements* [**89**]

$$V_{\mathcal{T}} = S_0^{k,0}(\mathcal{T})^d,$$
$$P_{\mathcal{T}} = S^{k-1,0}(\mathcal{T}) \cap L_0^2(\Omega),$$

  where $k \geq 3$;
- the *Bernardi–Raugel element*, cf. [**65**, Example 3.1], [**68**, Section II] and [**145**, Sections II.2.1, II.2.3, II.3.2]

$$V_{\mathcal{T}} = \left[S_0^{1,0}(\mathcal{T}) \oplus \operatorname{span}\{\psi_E \mathbf{n}_E : E \in \mathcal{E}_\Omega\}\right]^d,$$
$$P_{\mathcal{T}} = S^{0,-1}(\mathcal{T}) \cap L_0^2(\Omega);$$

- the spaces [**236**]

$$V_{\mathcal{T}} = S_0^{k,0}(\mathcal{T})^d,$$
$$P_{\mathcal{T}} = S^{k-1,-1}(\mathcal{T}) \cap L_0^2(\Omega),$$

  where $k \geq 2$ and where the mesh $\mathcal{T}$ has to satisfy a certain non-degeneracy condition.

Stabilised methods [**71, 125, 154**] correspond to the choices

$$\begin{aligned} &\vartheta_K > 0, \\ &\vartheta_E > 0, \quad \text{if the pressure approximation is discontinuous,} \\ &\vartheta_E = 0, \quad \text{if the pressure approximation is continuous,} \\ &\widetilde{\vartheta}_K \geq 0. \end{aligned}$$

The most prominent examples are:

- the so-called *equal order interpolation*

$$V_{\mathcal{T}} = S_0^{k,0}(\mathcal{T})^d,$$
$$P_{\mathcal{T}} = S^{k,0}(\mathcal{T}) \cap L_0^2(\Omega),$$

  where $k \geq 1$;
- the spaces

$$V_{\mathcal{T}} = S_0^{k,0}(\mathcal{T})^d,$$
$$P_{\mathcal{T}} = S^{k-1,-1}(\mathcal{T}) \cap L_0^2(\Omega),$$

  where $k \geq 1$.

When using stable mixed methods or stabilised methods with an appropriate choice of the stabilisation parameters, the resulting discrete problem admits a unique solution. As in the proof of Proposition 4.69, one can then prove that the bi-linear form $B_{\mathcal{T}}$ fulfils the conditions of Proposition 4.5 (p. 154) [**257**].

4.10.3 Residual A Posteriori Error Estimates

In what follows we denote by $(\mathbf{u}, p) \in X$ the unique solution of the saddle-point formulation of the Stokes equations and assume that

- the finite element approximation of the Stokes equations admits a unique solution $(\mathbf{u}_{\mathcal{T}}, p_{\mathcal{T}}) \in X_{\mathcal{T}}$ and
- the velocity space $V_{\mathcal{T}}$ contains the space $S_0^{1,0}(\mathcal{T})^d$ of continuous piecewise (multi-) linear vector fields.

These conditions are fulfilled for all approximations used in practice.

We choose an arbitrary element $(\mathbf{v}, q) \in X$ and keep it fixed in what follows. Integration by parts element-wise then yields the following L^2-representation of the residual

$$\begin{aligned}\langle \ell - L(\mathbf{u}_{\mathcal{T}}, p_{\mathcal{T}}), (\mathbf{v}, q)\rangle_Y &= \sum_{K \in \mathcal{T}} \int_K (\mathbf{f} + \Delta \mathbf{u}_{\mathcal{T}} - \nabla p_{\mathcal{T}}) \cdot \mathbf{v} \\ &\quad - \sum_{E \in \mathcal{E}_\Omega} \int_E \mathbb{J}_E(\mathbf{n}_E \cdot (\nabla \mathbf{u}_{\mathcal{T}} - p_{\mathcal{T}} I)) \cdot \mathbf{v} \\ &\quad - \sum_{K \in \mathcal{T}} \int_K q \operatorname{div} \mathbf{u}_{\mathcal{T}},\end{aligned}$$

where $I \in \mathbb{R}^{d \times d}$ denotes the identity matrix.

Next, we define the restriction operator $Q_{\mathcal{T}} : X \to X_{\mathcal{T}}$ by

$$Q_{\mathcal{T}}(\mathbf{v}, q) = (I_{\mathcal{T}} \mathbf{v}, 0),$$

where $I_{\mathcal{T}} : L^1(\Omega)^d \to S_0^{1,0}(\mathcal{T})^d$ denotes the quasi-interpolation operator defined in Section 3.5.1 (p. 108) applied to the components of the velocity field. Proposition 3.33 (p. 109) and the Cauchy–Schwarz inequality then imply

$$
\begin{aligned}
\langle \ell - L(\mathbf{u}_\mathcal{T}, p_\mathcal{T}), (Id - Q_\mathcal{T})(\mathbf{v}, q)\rangle_Y &\le \sum_{K\in\mathcal{T}} \left\|\mathbf{f} + \Delta\mathbf{u}_\mathcal{T} - \nabla p_\mathcal{T}\right\|_K \|\mathbf{v} - I_\mathcal{T}\mathbf{v}\|_K \\
&\quad + \sum_{E\in\mathcal{E}_\Omega} \left\|\mathbb{J}_E(\mathbf{n}_E \cdot (\nabla\mathbf{u}_\mathcal{T} - p_\mathcal{T} I))\right\|_E \|\mathbf{v} - I_\mathcal{T}\mathbf{v}\|_E \\
&\quad + \sum_{K\in\mathcal{T}} \|\operatorname{div}\mathbf{u}_\mathcal{T}\|_K \left\|q\right\|_K \\
&\le \sum_{K\in\mathcal{T}} \left\|\mathbf{f} + \Delta\mathbf{u}_\mathcal{T} - \nabla p_\mathcal{T}\right\|_K C_{A,2,2}(K) h_K \|\nabla\mathbf{v}\|_{\tilde{\omega}_K} \\
&\quad + \sum_{E\in\mathcal{E}_\Omega} \left\|\mathbb{J}_E(\mathbf{n}_E \cdot (\nabla\mathbf{u}_\mathcal{T} - p_\mathcal{T} I))\right\|_E C_{A,4,2}(E) h_E^{\frac{1}{2}} \|\nabla\mathbf{v}\|_{\tilde{\omega}_E} \\
&\quad + \sum_{K\in\mathcal{T}} \|\operatorname{div}\mathbf{u}_\mathcal{T}\|_K \left\|q\right\|_K \\
&\le c \max\left\{C_{A,2,2}(K), C_{A,4,2}(E)\right\} \left\|(\mathbf{v}, q)\right\|_X \\
&\quad \cdot \left\{\sum_{K\in\mathcal{T}} h_K^2 \left\|\mathbf{f} + \Delta\mathbf{u}_\mathcal{T} - \nabla p_\mathcal{T}\right\|_K^2 + \sum_{K\in\mathcal{T}} \|\operatorname{div}\mathbf{u}_\mathcal{T}\|_K^2 \right. \\
&\qquad \left. + \sum_{E\in\mathcal{E}_\Omega} h_E \left\|\mathbb{J}_E(\mathbf{n}_E \cdot (\nabla\mathbf{u}_\mathcal{T} - p_\mathcal{T} I))\right\|_E^2 \right\}^{\frac{1}{2}},
\end{aligned}
$$

where the constant c only depends on the shape parameter $C_\mathcal{T}$ of $\mathcal{T}$ and takes into account that every element is counted several times.

Since $(\mathbf{v}, q) \in X$ was arbitrary, this estimate establishes the first condition of Theorem 4.7 (p. 156) with

$$
\begin{aligned}
\eta_\mathcal{T} &= \left\{\sum_{K\in\mathcal{T}} h_K^2 \left\|\mathbf{f}_\mathcal{T} + \Delta\mathbf{u}_\mathcal{T} - \nabla p_\mathcal{T}\right\|_K^2 + \sum_{K\in\mathcal{T}} \|\operatorname{div}\mathbf{u}_\mathcal{T}\|_K^2 + \sum_{E\in\mathcal{E}_\Omega} h_E \left\|\mathbb{J}_E(\mathbf{n}_E \cdot (\nabla\mathbf{u}_\mathcal{T} - p_\mathcal{T} I))\right\|_E^2 \right\}^{\frac{1}{2}}, \\
\theta_\mathcal{T} &= \left\{\sum_{K\in\mathcal{T}} h_K^2 \|\mathbf{f} - \mathbf{f}_\mathcal{T}\|_K^2 \right\}^{\frac{1}{2}}
\end{aligned}
$$

and a constant c_A which only depends on the shape parameter $C_\mathcal{T}$ of $\mathcal{T}$. Here, as usual, $\mathbf{f}_\mathcal{T}$ denotes any piecewise polynomial approximation of $\mathbf{f}$.

To prove the second condition of Theorem 4.7, we observe that the definitions of $Q_\mathcal{T}$, ℓ, B, and $B_\mathcal{T}$ and the identity

$$
\begin{aligned}
\langle \ell_\mathcal{T} - L_\mathcal{T}(\mathbf{u}_\mathcal{T}, p_\mathcal{T}), Q_\mathcal{T}(\mathbf{v}, q)\rangle_{Y_\mathcal{T}} &= \langle \ell_\mathcal{T}, Q_\mathcal{T}(\mathbf{v}, q)\rangle_{Y_\mathcal{T}} - B_\mathcal{T}((\mathbf{u}_\mathcal{T}, p_\mathcal{T}), Q_\mathcal{T}(\mathbf{v}, q)) \\
&= 0
\end{aligned}
$$

imply that

$$\begin{aligned}\langle \ell - L(\mathbf{u}_\mathcal{T}, p_\mathcal{T}), Q_\mathcal{T}(\mathbf{v}, q)\rangle_Y &= \langle \ell - \ell_\mathcal{T}, Q_\mathcal{T}(\mathbf{v}, q)\rangle_Y + \langle L_\mathcal{T}(\mathbf{u}_\mathcal{T}, p_\mathcal{T}) - L(\mathbf{u}_\mathcal{T}, p_\mathcal{T}), Q_\mathcal{T}(\mathbf{v}, q)\rangle_Y \\ &= \sum_{K\in\mathcal{T}} \widetilde{\vartheta}_K \int_K \operatorname{div} \mathbf{u}_\mathcal{T} \operatorname{div}(I_\mathcal{T}\mathbf{v}).\end{aligned}$$

From Proposition 3.33 (p. 109) we conclude that

$$\begin{aligned}\left\|\operatorname{div}(I_\mathcal{T}\mathbf{v})\right\|_K &\leq \sqrt{d}\left\|\nabla(I_\mathcal{T}\mathbf{v})\right\|_K \\ &\leq \sqrt{d}(1 + C_{A,3,2}(K))\, \|\nabla\mathbf{v}\|_{\widetilde{\omega}_K}\end{aligned}$$

holds for all elements K where the constant $C_{A,3,2}(K)$ only depends on the shape-parameter $C_\mathcal{T}$ of $\mathcal{T}$. We therefore obtain

$$\begin{aligned}\langle \ell - L(\mathbf{u}_\mathcal{T}, p_\mathcal{T}), Q_\mathcal{T}(\mathbf{v}, q)\rangle_Y &\leq \sum_{K\in\mathcal{T}} \widetilde{\vartheta}_K \|\operatorname{div}\mathbf{u}_\mathcal{T}\|_K \left\|\operatorname{div}(I_\mathcal{T}\mathbf{v})\right\|_K \\ &\leq \max_{K\in\mathcal{T}} \widetilde{\vartheta}_K(1 + C_{A,3,2}(K)) \sum_{K\in\mathcal{T}} \|\operatorname{div}\mathbf{u}_\mathcal{T}\|_K \|\nabla\mathbf{v}\|_{\widetilde{\omega}_K} \\ &\leq c \max_{K\in\mathcal{T}} \widetilde{\vartheta}_K\, \eta_\mathcal{T}\, \|\nabla\mathbf{v}\|\end{aligned}$$

with a constant c which only depends on the shape parameter $C_\mathcal{T}$ of $\mathcal{T}$. This establishes the second condition of Theorem 4.7 (p. 156) with $c_C = c \max_{K\in\mathcal{T}} \widetilde{\vartheta}_K$. Notice that in many applications $\widetilde{\vartheta}_K = 0$ holds for all elements K.

The third condition of Theorem 4.7 is established in the standard way by successively inserting the test functions

$$\begin{aligned}\psi_K(\mathbf{f}_\mathcal{T} + \Delta\mathbf{u}_\mathcal{T} - \nabla p_\mathcal{T}) &\quad \text{for } K \in \mathcal{T}, \\ -\psi_E \mathbb{J}_E(\mathbf{n}_E \cdot (\nabla\mathbf{u}_\mathcal{T} - p_\mathcal{T} I)) &\quad \text{for } E \in \mathcal{E}_\Omega, \\ -\psi_K \operatorname{div}\mathbf{u}_\mathcal{T} &\quad \text{for } K \in \mathcal{T}\end{aligned} \tag{4.55}$$

in the L^2-representation of the residual and using Proposition 3.37 (p. 112). Here, ψ_K, ψ_E are the local cut-off functions of Section 3.2.3 (p. 83). Note that the polynomial degree of these test functions and correspondingly the value of the parameter k in Proposition 3.37 depend on the polynomial degrees of $\mathbf{u}_\mathcal{T}$ and $p_\mathcal{T}$.

Theorem 4.7 (p. 156) therefore yields the following a posteriori error estimates.

Theorem 4.70 *Denote by $(\mathbf{u}, p)$ the unique solution of the saddle-point formulation of the Stokes equations and by $(\mathbf{u}_\mathcal{T}, p_\mathcal{T}) \in X_\mathcal{T}$ its finite element approximation. For every element $K \in \mathcal{T}$ define the* residual a posteriori error indicator *$\eta_{R,K}$ by*

$$\eta_{R,K} = \left\{ h_K^2 \left\|\mathbf{f}_\mathcal{T} + \Delta\mathbf{u}_\mathcal{T} - \nabla p_\mathcal{T}\right\|_K^2 + \|\operatorname{div}\mathbf{u}_\mathcal{T}\|_K^2 + \frac{1}{2} \sum_{E\in\mathcal{E}_{K,\Omega}} h_E \left\|\mathbb{J}_E(\mathbf{n}_E \cdot (\nabla\mathbf{u}_\mathcal{T} - p_\mathcal{T} I))\right\|_E^2 \right\}^{\frac{1}{2}},$$

where $\mathbf{f}_{\mathcal{T}}$ is any piecewise polynomial approximation of $\mathbf{f}$*. There are two constants c^* and c_* such that the estimates*

$$\left\{ \left\| \nabla(\mathbf{u} - \mathbf{u}_{\mathcal{T}}) \right\|^2 + \left\| p - p_{\mathcal{T}} \right\|^2 \right\}^{\frac{1}{2}} \leq c^* \left\{ \sum_{K \in \mathcal{T}} \eta_{R,K}^2 + \sum_{K \in \mathcal{T}} h_K^2 \left\| \mathbf{f} - \mathbf{f}_{\mathcal{T}} \right\|_K^2 \right\}^{\frac{1}{2}}$$

and

$$\eta_{R,K} \leq c_* \left\{ \left\| \nabla(\mathbf{u} - \mathbf{u}_{\mathcal{T}}) \right\|_{\omega_K}^2 + \left\| p - p_{\mathcal{T}} \right\|_{\omega_K}^2 + \sum_{K' \subset \omega_K} h_{K'}^2 \left\| \mathbf{f} - \mathbf{f}_{\mathcal{T}} \right\|_{K'}^2 \right\}^{\frac{1}{2}}$$

hold for all $K \in \mathcal{T}$. Both constants c^* and c_* depend on the shape parameter $C_{\mathcal{T}}$. The constant c_* in addition depends on the polynomial degrees of $\mathbf{u}_{\mathcal{T}}$, $p_{\mathcal{T}}$, and $\mathbf{f}_{\mathcal{T}}$; the constant c^* in addition depends on the stabilisation parameters of the finite element approximation through $\max_{K \in \mathcal{T}} \widetilde{\vartheta}_K$.

4.10.4 Auxiliary Local Problems

As in the preceding sections, one can devise a posteriori error indicators for the Stokes problem which are based on the solution of auxiliary local discrete Stokes equations. Of course one must guarantee the unique solvability of these problems. This can most easily be achieved by choosing the polynomial degree of the local velocity space large enough such that it contains the gradients of the local pressure approximation. If in addition the polynomial degree of the local velocity and pressure approximation is chosen large enough such that the local spaces contain the test functions of equation (4.55), one can compare the new error indicator with the residual indicator defined above.

To make things precise, we choose three integers n_p, $n_{\mathbf{u},E}$, and $n_{\mathbf{u},K}$ such that the following conditions are satisfied for all elements K, all edges or faces E, and all $q_K \in R_{n_p}(K)$:

$$\begin{aligned}
(\operatorname{div} \mathbf{u}_{\mathcal{T}})|_K &\in R_{n_p}(K), \\
\mathbb{J}_E(\mathbf{n}_E \cdot (\nabla \mathbf{u}_{\mathcal{T}} - p_{\mathcal{T}} I)) &\in R_{n_{\mathbf{u},E}}(E), \\
(\mathbf{f}_{\mathcal{T}} + \Delta \mathbf{u}_{\mathcal{T}} - \nabla p_{\mathcal{T}})|_K &\in R_{n_{\mathbf{u},K}}(K), \\
\nabla(\psi_K q_K) &\in R_{n_{\mathbf{u},K}}(K).
\end{aligned}$$

If, e.g. $\mathcal{T}$ consists of d-simplices and we have

$$\mathbf{f}_{\mathcal{T}} \in S^{0,-1}(\mathcal{T})^d, \quad V_{\mathcal{T}} \subset S^{k,0}(\mathcal{T})^d, \quad P_{\mathcal{T}} \subset S^{k-1,0}(\mathcal{T})$$

for some integer $k \geq 1$, the choices

$$n_p = k - 1, \quad n_{\mathbf{u},E} = k - 1, \quad n_{\mathbf{u},K} = d + k - 1$$

do the job.

For every element K, we now define the spaces V_K, P_K, and X_K by

$$V_K = \operatorname{span}\left\{\psi_E \mathbf{v}_E, \psi_K \mathbf{v}_K : \mathbf{v}_E \in R_{n_{\mathbf{u},E}}(E), \mathbf{v}_K \in R_{n_{\mathbf{u},K}}(K),\ E \in \mathcal{E}_K \cap \mathcal{E}_\Omega\right\},$$

$$P_K = \operatorname{span}\left\{\psi_K q_K : q_K \in R_{n_p}(K)\right\},$$

$$X_K = V_K \times P_K$$

and denote by B_K the restriction of the bi-linear form B to K, i.e.

$$B_K((\mathbf{v},q),(\mathbf{w},r)) = \int_K \nabla \mathbf{v} : \nabla \mathbf{w} - \int_K q \operatorname{div} \mathbf{w} + \int_K r \operatorname{div} \mathbf{v}.$$

Lemma 4.71 *The bi-linear form B_K fulfils the estimates*

$$\inf_{(\mathbf{w},r)\in X_K\setminus\{0\}} \sup_{(\mathbf{v},q)\in X_K\setminus\{0\}} \frac{B_K((\mathbf{v},q),(\mathbf{w},r))}{\|(\mathbf{v},q)\|_X \|(\mathbf{w},r)\|_X} = \inf_{(\mathbf{v},q)\in X_K\setminus\{0\}} \sup_{(\mathbf{w},r)\in X_K\setminus\{0\}} \frac{B_K((\mathbf{v},q),(\mathbf{w},r))}{\|(\mathbf{v},q)\|_X \|(\mathbf{w},r)\|_X}$$
$$\geq \frac{\beta_K^2}{(1+\beta_K)^2},$$

where $\beta_K > 0$ only depends on the shape parameter $C_\mathcal{T}$ of $\mathcal{T}$.

Proof Choose an arbitrary pressure $q \in P_K$ and set

$$\mathbf{v}_q = -\psi_K(\nabla q).$$

From Proposition 3.37 (p. 112) we then obtain

$$\begin{aligned}\int_K q \operatorname{div} \mathbf{v}_q &= -\int_K \mathbf{v}_q \cdot \nabla q \\ &= \int_K \psi_K |\nabla q|^2 \\ &\geq C_{I,1,n_p+d-1}(K)^{-2} \|\nabla q\|_K^2\end{aligned}$$

and

$$\|\nabla \mathbf{v}_q\|_K \leq C_{I,2,n_p+d-1}(K) h_K^{-1} \|\nabla q\|_K.$$

Since q is of the form $\psi_K r$ with $r \in R_{n_p}(K)$, a scaling argument and Friedrichs' inequality, cf. [**3**], [**109**, inequality (2.2)], and Proposition 3.28 (p. 105), on the reference element imply

$$\|\nabla q\|_K \geq c_{n_p} h_K^{-1} \|q\|_K.$$

Since $\mathbf{v}_q \in V_K$, this proves that

$$\sup_{\mathbf{w}\in V_K\setminus\{0\}} \frac{\int_K q \operatorname{div}\mathbf{w}}{\|\nabla\mathbf{w}\|_K \|q\|_K} \geq \frac{\int_K q \operatorname{div}\mathbf{v}_q}{\|\nabla\mathbf{v}_q\|_K \|q\|_K}$$
$$\geq \frac{c_{n_p}}{C_{I,1,n_p+d-1}(K)^2\, C_{I,2,n_p+d-1}(K)}.$$

Now, the proof proceeds in exactly the same way as the proof of Proposition 4.69 with β replaced by

$$\beta_K = \frac{c_{n_p}}{C_{I,1,n_p+d-1}(K)^2\, C_{I,2,n_p+d-1}(K)}.$$

□

Due to Lemma 4.71, there is a unique element $(\mathbf{u}_K, p_K) \in X_K$ such that

$$B_K((\mathbf{u}_K, p_K), (\mathbf{v}, q)) = \int_K (\mathbf{f}_\mathcal{T} + \Delta\mathbf{u}_\mathcal{T} - \nabla p_\mathcal{T})\cdot\mathbf{v} - \frac{1}{2}\sum_{E\in\mathcal{E}_{K,\Omega}} \int_E \mathbb{J}_E(\mathbf{n}_E\cdot(\nabla\mathbf{u}_\mathcal{T} - p_\mathcal{T} I))\cdot\mathbf{v}$$
$$- \int_K \operatorname{div}\mathbf{u}_\mathcal{T}\, q$$

holds for all $(\mathbf{v}, q) \in X_K$. Now we define

$$\eta_{N,K} = \left\{\|\nabla\mathbf{u}_K\|_K^2 + \|p_K\|_K^2\right\}^{\frac{1}{2}}. \tag{4.56}$$

Since, by construction, the test functions of equation (4.55) are contained in the spaces V_K and P_K, respectively, we can compare the indicators $\eta_{R,K}$ and $\eta_{N,K}$ as in Section 1.7.3 (p. 29). The continuity of the bi-linear form B_K, Lemma 4.71, and the proof of Theorem 4.70 then yield the following result.

Theorem 4.72 *Denote by $(\mathbf{u}, p)$ the unique solution of the saddle point formulation of the Stokes equations and by $(\mathbf{u}_\mathcal{T}, p_\mathcal{T}) \in X_\mathcal{T}$ its finite element approximation. Then the estimates*

$$\eta_{N,K} \leq c_1 \eta_{R,K},$$

$$\eta_{R,K} \leq c_2 \left\{\sum_{K'\subset\omega_K} \eta_{N,K'}^2\right\}^{\frac{1}{2}},$$

$$\eta_{N,K} \leq c_3 \left\{\|\nabla(\mathbf{u}-\mathbf{u}_\mathcal{T})\|_{\omega_K}^2 + \|p - p_\mathcal{T}\|_{\omega_K}^2 + \sum_{K'\subset\omega_K} h_{K'}^2 \|\mathbf{f}-\mathbf{f}_\mathcal{T}\|_{K'}^2\right\}^{\frac{1}{2}}$$

and

$$\left\{\|\nabla(\mathbf{u}-\mathbf{u}_\mathcal{T})\|^2 + \|p - p_\mathcal{T}\|^2\right\}^{\frac{1}{2}} \leq c_4 \left\{\sum_{K\in\mathcal{T}} \eta_{N,K}^2 + \sum_{K\in\mathcal{T}} h_K^2 \|\mathbf{f}-\mathbf{f}_\mathcal{T}\|_K^2\right\}^{\frac{1}{2}}$$

hold for all $K \in \mathcal{T}$. Here, $\mathbf{f}_{\mathcal{T}}$ and $\eta_{R,K}$ are as in Theorem 4.70 (p. 244), and $\eta_{N,K}$ is defined by equation (4.56). The constants $c_1, \ldots, c_4$ only depend on the shape parameter $C_{\mathcal{T}}$ of $\mathcal{T}$ and the polynomial degree of $\mathbf{u}_{\mathcal{T}}$ and $p_{\mathcal{T}}$; the constant c_4 in addition depends on the stabilisation parameters of the finite element approximation through $\max_{K \in \mathcal{T}} \widetilde{\vartheta}_K$.

Error indicators based on the solution of auxiliary local discrete Stokes problems with Dirichlet boundary conditions on a patch of elements can be devised similarly. Since, following the lines indicated above, the arguments are straightforward modifications of those formerly used for the model problem or for reaction–diffusion equations, we leave the details to the reader.

Remark 4.73 The results of Section 5.4 (p. 301) for the stationary Navier–Stokes equations also extend to the simpler Stokes equations. In particular the results of Section 5.4.5 (p. 307) yield L^2-error estimates for the velocity and H^{-1}-error estimates for the pressure. The element and face residuals in $\eta_{R,K}$ must then be multiplied by additional factors h_K and h_E, respectively, and the solutions of the auxiliary local discrete Stokes problems must be evaluated with respect to a local L^2-norm for the auxiliary velocity and a scaled local L^2-norm mimicking a local H^{-1}-norm for the auxiliary pressure.

Remark 4.74 A posteriori error estimates for conforming mixed finite element approximations of the Stokes problem were first given in [**251**]. Non-conforming discretisations are analysed in [**253**].

4.11 The Bi-harmonic Equation

As an example for higher order elliptic equations we consider the bi-harmonic equation in a bounded connected *two-dimensional* polygonal domain with Lipschitz boundary

$$\begin{aligned} \Delta^2 u &= f && \text{in } \Omega \\ u &= 0 && \text{on } \Gamma \\ \frac{\partial u}{\partial n} &= 0 && \text{on } \Gamma. \end{aligned} \tag{4.57}$$

This models the vertical displacement u of the mid-surface of a thin clamped plate under the influence of a vertical load f. The variational formulation of the bi-harmonic equation based on an energy principle is straightforward. Its conforming finite element discretisation, however, needs C^1-elements which are rather expensive due to the required high polynomial degree [**108**, Section 6.1], [**109**, Sections 44–47]. To overcome this difficulty one often tries to relax the C^1-constraint by either considering a mixed variational formulation [**90**, Section IV.4.1], [**108**, Section 7.1] or non-conforming approximations [**108**, Section 6.2], [**109**, Section 49] each having its own benefits and drawbacks. In what follows we will discuss all three approaches and present corresponding residual a posteriori error estimates.

4.11.1 Conforming Discretisations

The variational formulation of problem (4.57) based on an energy principle and its conforming finite element discretisation fit into the abstract framework of Section 4.1 (p. 151) with

$$
\begin{aligned}
X = Y &= H_0^2(\Omega), \\
\|\cdot\|_X = \|\cdot\|_Y &= \|\cdot\|_2, \\
B(u, v) &= \int_\Omega \Delta u \Delta v, \\
\langle \ell, v \rangle &= \int_\Omega fv, \\
X_{\mathcal{T}} &= Y_{\mathcal{T}} \subset H_0^2(\Omega), \\
B_{\mathcal{T}}(u_{\mathcal{T}}, v_{\mathcal{T}}) &= B(u_{\mathcal{T}}, v_{\mathcal{T}}), \\
\langle \ell_{\mathcal{T}}, v_{\mathcal{T}} \rangle &= \langle \ell, v_{\mathcal{T}} \rangle .
\end{aligned}
$$

We do not specify the finite element space $X_{\mathcal{T}}$ in detail. We only assume that it satisfies the interpolation error estimates of Proposition 3.35 (p. 112) and that the finite element functions are contained in $H^4(K)$ element-wise. Both assumptions are satisfied for the examples in [**109**, Sections 45, 47]. The second assumption is violated for the composite elements in [**109**, Section 46]. Then the L^2-representation (4.58) of the residual also incorporates edge residuals on the interfaces of the sub-elements. These can be controlled in the same way as the edge residuals on the element boundaries.

Obviously the consistency error of the discretisation vanishes and we have Galerkin orthogonality. To derive an L^2-representation of the residual we must perform integration by parts element-wise twice. This yields for all $v \in H_0^2(\Omega)$

$$
\langle R, v \rangle = \int_\Omega rv + \int_\Sigma j_1 \mathbf{n}_E \cdot \nabla v + \int_\Sigma j_2 v \tag{4.58}
$$

with

$$
\begin{aligned}
r|_K &= f - \Delta^2 u_{\mathcal{T}} && \text{on } K \in \mathcal{T}, \\
j_1|_E &= \begin{cases} -\mathbb{J}_E(\Delta u_{\mathcal{T}}) & \text{if } E \in \mathcal{E}_\Omega, \\ 0 & \text{if } E \in \mathcal{E}_\Gamma, \end{cases} \\
j_2|_E &= \begin{cases} \mathbb{J}_E(\mathbf{n}_E \cdot \nabla \Delta u_{\mathcal{T}}) & \text{if } E \in \mathcal{E}_\Omega, \\ 0 & \text{if } E \in \mathcal{E}_\Gamma. \end{cases}
\end{aligned}
$$

The L^2-representation and Proposition 3.35 (p. 112) establish condition (4.9) (p. 156) with $Q_{\mathcal{T}} = 0$ and $c_C = 0$. To verify condition (4.10) (p. 156) we must still specify the space $\widetilde{Y}_{\mathcal{T}}$. It must now be contained in $H_0^2(\Omega)$, i.e. its elements must be continuously differentiable globally, and it must balance three different kinds of contributions: the element residuals $f_{\mathcal{T}} - \Delta^2 u_{\mathcal{T}}$, the edge residuals $\mathbb{J}_E(\Delta u_{\mathcal{T}})$, and the edge residuals $\mathbb{J}_E(\mathbf{n}_E \cdot \nabla \Delta u_{\mathcal{T}})$. This is achieved with the help of the smooth cut-off functions $\psi_{K,1}$, $\psi_{E,1}$, and $\Psi_{E,1}$ of Section 3.2.5 (p. 85) and Proposition 3.51 (p. 129). More precisely, the element residuals are controlled by inserting in (4.58) functions of the form $\psi_{K,1}(f_{\mathcal{T}} - \Delta^2 u_{\mathcal{T}})$. These do not interact with the edge residuals, since $\psi_{K,1}$ vanishes on ∂K together with its first-order derivatives. The edge residuals $\mathbb{J}_E(\Delta u_{\mathcal{T}})$ are controlled by inserting in (4.58) functions of the form $\Psi_{E,1}\mathbb{J}_E(\Delta u_{\mathcal{T}})$. These interact with the element residuals on ω_E but not with the edge residual $\mathbb{J}_E(\mathbf{n}_E \cdot \nabla \Delta u_{\mathcal{T}})$, since $\Psi_{E,1}$ vanishes on E. Finally, the edge residuals $\mathbb{J}_E(\mathbf{n}_E \cdot \nabla \Delta u_{\mathcal{T}})$ are controlled by inserting in (4.58) functions of the form $\psi_{E,1}\mathbb{J}_E(\mathbf{n}_E \cdot \nabla \Delta u_{\mathcal{T}})$. These interact with both the element residuals on ω_E and the edge residual $\mathbb{J}_E(\Delta u_{\mathcal{T}})$.

Theorem 4.7 therefore yields the following a posteriori error estimates.

Theorem 4.75 *Denote by u the weak solution of the bi-harmonic equation (4.57) and by $u_{\mathcal{T}} \in X_{\mathcal{T}}$ its conforming finite element approximation. Define the* residual a posteriori error indicator $\eta_{R,K}$ *by*

$$\eta_{R,K} = \left\{ h_K^4 \left\| f_{\mathcal{T}} - \Delta^2 u_{\mathcal{T}} \right\|_K^2 + \sum_{E \in \mathcal{E}_{K,\Omega}} h_E \left\| \mathbb{J}_E(\Delta u_{\mathcal{T}}) \right\|_E^2 + \sum_{E \in \mathcal{E}_{K,\Omega}} h_E^3 \left\| \mathbb{J}_E(\mathbf{n}_E \cdot \nabla \Delta u_{\mathcal{T}}) \right\|_E^2 \right\}^{\frac{1}{2}},$$

where $f_{\mathcal{T}}$ is any piecewise polynomial approximation of f. Then the error is bounded from above by

$$\|u - u_{\mathcal{T}}\|_2 \le c^* \left\{ \sum_{K \in \mathcal{T}} \eta_{R,K}^2 + \sum_{K \in \mathcal{T}} h_K^4 \left\| f - f_{\mathcal{T}} \right\|_K^2 \right\}^{\frac{1}{2}}$$

and, for every element K separately, from below by

$$\eta_{R,K} \le c_* \left\{ \|u - u_{\mathcal{T}}\|_{2;\omega_K}^2 + \sum_{K' \subset \omega_K} h_K^4 \left\| f - f_{\mathcal{T}} \right\|_{K'}^2 \right\}^{\frac{1}{2}}.$$

Both constants c^* and c_* depend on the shape parameter $C_{\mathcal{T}}$ of $\mathcal{T}$. The constant c_* in addition depends on the polynomial degree of $u_{\mathcal{T}}$.

4.11.2 *Mixed Discretisations*

The mixed formulation of the bi-harmonic equation (4.57) and associated mixed finite element discretisations correspond to problems (4.2) (p. 152) and (4.6) (p. 153) with

$$\begin{aligned}
X = Y &= H^1(\Omega) \times H_0^1(\Omega), \\
\left\|(\varphi, u)\right\|_X &= \left\|(\varphi, u)\right\|_Y = \left\{ \|\varphi\|_1^2 + \|u\|_1^2 \right\}^{\frac{1}{2}}, \\
B((\varphi, u), (\psi, v)) &= \int_\Omega \varphi\psi + \int_\Omega \nabla\psi \cdot \nabla u + \int_\Omega \nabla\varphi \cdot \nabla v, \\
\langle \ell, (\psi, v) \rangle &= -\int_\Omega fv, \\
X_{\mathcal{T}} &= Y_{\mathcal{T}} = V_{\mathcal{T}} \times W_{\mathcal{T}}, \\
B_{\mathcal{T}}((\varphi_{\mathcal{T}}, u_{\mathcal{T}}), (\psi_{\mathcal{T}}, v_{\mathcal{T}})) &= B((\varphi_{\mathcal{T}}, u_{\mathcal{T}}), (\psi_{\mathcal{T}}, v_{\mathcal{T}})), \\
\langle \ell_{\mathcal{T}}, (\psi_{\mathcal{T}}, v_{\mathcal{T}}) \rangle &= \langle \ell, (\psi_{\mathcal{T}}, v_{\mathcal{T}}) \rangle.
\end{aligned}$$

We do not specify the finite element spaces $V_{\mathcal{T}}$ and $W_{\mathcal{T}}$ in detail. We only assume that

$$\begin{aligned}
W_{\mathcal{T}} &\subset V_{\mathcal{T}}, \\
S_0^{2,0}(\mathcal{T}) &\subset W_{\mathcal{T}} \subset H_0^1(\Omega), \\
S^{2,0}(\mathcal{T}) &\subset V_{\mathcal{T}} \subset H^1(\Omega).
\end{aligned} \tag{4.59}$$

The bi-linear form B is continuous on $X \times Y$, but *it does not satisfy the inf–sup condition* (4.5) (p. 152). This is due to the fact that the bi-linear form $\varphi, \psi \mapsto \int_\Omega \varphi\psi$ is not coercive on the space of harmonic functions which is the kernel of the bi-linear form $\varphi, u \mapsto \int_\Omega \nabla\varphi \cdot \nabla u$. Thus the mixed formulation of the bi-harmonic problem does not fit into the abstract framework of [**29**, **88**] for saddle-point problems. This deficit is reflected in sub-optimal a priori error estimates and, as we will see below, non-standard a posteriori error estimates. Nevertheless, the mixed formulation of the bi-harmonic equation always admits at most one solution. This easily follows from the relations

$$B((\varphi, u), (\varphi, -u)) = \|\varphi\|^2,$$
$$B((\varphi, u), (u, 0)) = \|\nabla u\|^2 + \int_\Omega \varphi u$$

which hold for all $(\varphi, u) \in X$. If, on the other hand, Ω is convex, the mixed problem also admits at least one solution. To prove this, simply observe that the unique weak solution $u \in H_0^2(\Omega)$ of the bi-harmonic problem in its standard variational formulation is contained in $H^3(\Omega)$ if Ω is convex and therefore yields the solution $(\Delta u, u)$ of the mixed problem.

The mixed finite element discretisation violates the inf–sup condition of Proposition 4.5 (p. 154) but it is well-posed in the sense that it always admits a unique solution. The latter follows from the above identities for the bi-linear form B profiting from the property $W_\mathcal{T} \subset V_\mathcal{T}$.

In the sequel we will therefore always assume that Ω is *convex* and that $(\varphi, u) = (\Delta u, u)$ and $(\varphi_\mathcal{T}, u_\mathcal{T})$ are the unique solutions of the mixed formulation of the bi-harmonic equations and its mixed finite element discretisation, respectively. We associate with $(\varphi_\mathcal{T}, u_\mathcal{T})$ *two* residuals R_1 and R_2 by setting for every $\psi \in H^1(\Omega)$ and every $v \in H_0^1(\Omega)$

$$\langle R_1\, , \, \psi\rangle = \int_\Omega \varphi_\mathcal{T}\psi + \int_\Omega \nabla u_\mathcal{T} \cdot \nabla\psi,$$
$$\langle R_2\, , \, v\rangle = \int_\Omega \nabla\varphi_\mathcal{T} \cdot \nabla v + \int_\Omega fv.$$

Due to the assumptions $V_\mathcal{T} \subset H^1(\Omega)$ and $W_\mathcal{T} \subset H_0^1(\Omega)$, the residuals satisfy the Galerkin orthogonality

$$\langle R_1\, , \, \psi_\mathcal{T}\rangle = 0, \quad \langle R_2\, , \, v_\mathcal{T}\rangle = 0$$

for all $\psi_\mathcal{T} \in V_\mathcal{T}$ and all $v_\mathcal{T} \in W_\mathcal{T}$. Moreover they are related to the error by

$$B((\varphi_\mathcal{T} - \varphi, u_\mathcal{T} - u), (\psi, v)) = \langle R_1\, , \, \psi\rangle + \langle R_2\, , \, v\rangle \tag{4.60}$$

for all $(\psi, v) \in X$. Integration by parts element-wise finally yields the L^2-representations

$$\begin{aligned} \langle R_1\, , \, \psi\rangle &= \sum_{K\in\mathcal{T}} \int_K (\varphi_\mathcal{T} - \Delta u_\mathcal{T})\, \psi + \sum_{E\in\mathcal{E}} \int_E \mathbb{J}_E(\mathbf{n}_E \cdot \nabla u_\mathcal{T})\psi, \\ \langle R_2\, , \, v\rangle &= \sum_{K\in\mathcal{T}} \int_K (f - \Delta\varphi_\mathcal{T})\, v + \sum_{E\in\mathcal{E}_\Omega} \int_E \mathbb{J}_E(\mathbf{n}_E \cdot \nabla\varphi_\mathcal{T})v. \end{aligned} \tag{4.61}$$

Thus, the a posteriori error analyses were completely straightforward, if B were to satisfy the inf–sup condition (4.5) (p. 152). Since this is not the case, we have to invest in more work.

We start by deriving lower bounds for the error in order to get an idea of suitable error indicators and norms for the error. In order to simplify the notation, we denote in what follows by $c, c_1, c_2, \ldots,$ various constants which only depend on the shape parameter $C_{\mathcal{T}}$ of $\mathcal{T}$ and the polynomial degree of $\varphi_{\mathcal{T}}$ and $u_{\mathcal{T}}$.

Inspired by our previous results, we choose an arbitrary element K and insert the functions $\psi = \psi_K(\varphi_{\mathcal{T}} - \Delta u_{\mathcal{T}})$, $v = 0$ in (4.60) and (4.61). This yields

$$\|\varphi_{\mathcal{T}} - \Delta u_{\mathcal{T}}\|_K^2 \leq c_1 h_K^{-1} \left\|\nabla(u - u_{\mathcal{T}})\right\|_K \|\varphi_{\mathcal{T}} - \Delta u_{\mathcal{T}}\|_K + c_2 \|\varphi - \varphi_{\mathcal{T}}\|_K \|\varphi_{\mathcal{T}} - \Delta u_{\mathcal{T}}\|_K$$

and thus

$$h_K \|\varphi_{\mathcal{T}} - \Delta u_{\mathcal{T}}\|_K \leq c\left\{\left\|\nabla(u - u_{\mathcal{T}})\right\|_K + h_K \|\varphi - \varphi_{\mathcal{T}}\|_K\right\}.$$

Choosing an arbitrary edge $E \in \mathcal{E}$ and inserting $\psi = \psi_E \mathbb{J}_E(\mathbf{n}_E \cdot \nabla u_{\mathcal{T}})$, $v = 0$ in (4.60) and (4.61), we similarly obtain

$$h_E^{\frac{1}{2}} \left\|\mathbb{J}_E(\mathbf{n}_E \cdot \nabla\varphi_{\mathcal{T}})\right\|_E \leq c\left\{\left\|\nabla(u - u_{\mathcal{T}})\right\|_{\omega_E} + h_E \|\varphi - \varphi_{\mathcal{T}}\|_{\omega_E}\right\}.$$

This shows that the mesh-dependent norm $\left\{\|\nabla v\|^2 + \sum_K h_K^2 \|\psi\|_K^2\right\}^{\frac{1}{2}}$ may be a candidate for measuring the error. This choice, however, causes some problems when deriving bounds for R_2, since the latter is linked to $\nabla(\varphi - \varphi_{\mathcal{T}})$. An obvious remedy is to perform integration by parts for the corresponding term. This, however, requires test functions in $H^2(\Omega)$. We therefore choose an arbitrary element K and insert the functions $\psi = 0$, $v = v_K = \psi_{K,1}(f_{\mathcal{T}} - \Delta u_{\mathcal{T}})$ in (4.60) and (4.61), where $f_{\mathcal{T}}$ is a piecewise polynomial approximation of f and $\psi_{K,1}$ is the smooth cut-off function of Section 3.2.5 (p. 85). Taking into account Proposition 3.51 (p. 129) this yields

$$\begin{aligned}
c_1^{-1} \left\|f_{\mathcal{T}} - \Delta u_{\mathcal{T}}\right\|_K^2 &\leq \langle R_2, v_K\rangle + \int_K (f - f_{\mathcal{T}})v_K \\
&= \int_K \nabla(\varphi_{\mathcal{T}} - \varphi) \cdot \nabla v_K + \int_K (f - f_{\mathcal{T}})v_K \\
&= \int_K (\varphi - \varphi_{\mathcal{T}})\Delta v_K + \int_K (f - f_{\mathcal{T}})v_K \\
&\leq c_2 h_K^{-2} \|\varphi - \varphi_{\mathcal{T}}\|_K \left\|f_{\mathcal{T}} - \Delta u_{\mathcal{T}}\right\|_K + \left\|f - f_{\mathcal{T}}\right\|_K \left\|f_{\mathcal{T}} - \Delta u_{\mathcal{T}}\right\|_K
\end{aligned}$$

and thus

$$h_K^3 \left\|f_{\mathcal{T}} - \Delta u_{\mathcal{T}}\right\|_K \leq c\left\{h_K \|\varphi - \varphi_{\mathcal{T}}\|_K + h_K^3 \left\|f - f_{\mathcal{T}}\right\|_K\right\}.$$

Choosing an arbitrary edge $E \in \mathcal{E}_\Omega$ and inserting the functions $\psi = 0$, $v = \psi_{E,1}\mathbb{J}_E(\mathbf{n}_E \cdot \nabla\varphi_{\mathcal{T}})$ in (4.60) and (4.61), we similarly obtain

$$h_E^{\frac{5}{2}} \left\|\mathbb{J}_E(\mathbf{n}_E \cdot \nabla\varphi_{\mathcal{T}})\right\|_E \leq c\left\{h_E \|\varphi - \varphi_{\mathcal{T}}\|_{\omega_E} + h_E^3 \left\|f - f_{\mathcal{T}}\right\|_{\omega_E}\right\}.$$

Combining all results, we thus get the following lower bound for the error

$$\left\{h_K^2 \|\varphi_{\mathcal{T}} - \Delta u_{\mathcal{T}}\|_K^2 + \sum_{E\in\mathcal{E}_K} h_E \left\|\mathbb{J}_E(\mathbf{n}_E \cdot \nabla u_{\mathcal{T}})\right\|_E^2 \right.$$

$$\left. +h_K^6 \left\|f_{\mathcal{T}} - \Delta\varphi_{\mathcal{T}}\right\|_K^2 + \sum_{E\in\mathcal{E}_{K,\Omega}} h_E^5 \left\|\mathbb{J}_E(\mathbf{n}_E \cdot \nabla\varphi_{\mathcal{T}})\right\|_E^2 \right\}^{\frac{1}{2}}$$

$$\leq c\left\{\left\|\nabla(u - u_{\mathcal{T}})\right\|_{\omega_K}^2 + h_K^2 \|\varphi - \varphi_{\mathcal{T}}\|_{\omega_K}^2 + h_K^6 \left\|f - f_{\mathcal{T}}\right\|_{\omega_K}^2\right\}^{\frac{1}{2}}.$$

To derive a similar upper bound for the error, we set for brevity

$$\eta_{\mathcal{T},1} = \left\{\sum_{K\in\mathcal{T}} h_K^2 \|\varphi_{\mathcal{T}} - \Delta u_{\mathcal{T}}\|_K^2 + \sum_{E\in\mathcal{E}} h_E \left\|\mathbb{J}_E(\mathbf{n}_E \cdot \nabla u_{\mathcal{T}})\right\|_E^2\right\}^{\frac{1}{2}},$$

$$\eta_{\mathcal{T},2} = \left\{\sum_{K\in\mathcal{T}} h_K^6 \left\|f_{\mathcal{T}} - \Delta\varphi_{\mathcal{T}}\right\|_K^2 + \sum_{E\in\mathcal{E}_\Omega} h_E^5 \left\|\mathbb{J}_E(\mathbf{n}_E \cdot \nabla\varphi_{\mathcal{T}})\right\|_E^2\right\}^{\frac{1}{2}},$$

$$\theta_{\mathcal{T}} = \left\{\sum_{K\in\mathcal{T}} h_K^6 \left\|f - f_{\mathcal{T}}\right\|_{\omega_K}^2\right\}^{\frac{1}{2}}.$$

The estimation of $\left\|\nabla(u - u_{\mathcal{T}})\right\|$ is based on a duality argument. Consider an arbitrary function $g \in H^{-1}(\Omega)$ and denote by u_g the unique weak solution of the bi-harmonic equation (4.57) in its standard variational formulation with right-hand side g and set $\varphi_g = \Delta u_g$. Since Ω is convex, we have

$$\left\|\varphi_g\right\|_1 + \left\|u_g\right\|_3 \leq c_\Omega \left\|g\right\|_{-1} \tag{4.62}$$

and

$$B((\varphi_g, u_g), (\psi, v)) = \langle g\,,\, v\rangle$$

for all $(\psi, v) \in H^1(\Omega) \times H_0^1(\Omega)$. Inserting $\psi = \varphi - \varphi_{\mathcal{T}}$ and $v = u - u_{\mathcal{T}}$ in this equation, taking into account relation (4.60), the Galerkin orthogonality of the residuals and their L^2-representations as well as assumption (4.59), and using standard interpolation error estimates for the space $S_0^{2,0}(\mathcal{T})$ [**109**, Theorem 16.1], we obtain

$$\begin{aligned}
\langle g\,,\, u - u_{\mathcal{T}}\rangle &= B((\varphi_g, u_g), (\varphi - \varphi_{\mathcal{T}}, u - u_{\mathcal{T}})) \\
&= B((\varphi - \varphi_{\mathcal{T}}, u - u_{\mathcal{T}}), (\varphi_g, u_g)) \\
&= \langle R_1\,,\, \varphi_g\rangle + \langle R_2\,,\, u_g\rangle \\
&\leq c\left\{\eta_{\mathcal{T},1} \left\|\varphi_g\right\|_1 + \eta_{\mathcal{T},2} \left\|u_g\right\|_3 + \theta_{\mathcal{T}} \left\|u_g\right\|_3\right\}.
\end{aligned}$$

Since $g \in H^{-1}(\Omega)$ was arbitrary, this proves

$$\left\|\nabla(u - u_{\mathcal{T}})\right\| \leq c\left\{\eta_{\mathcal{T},1} + \eta_{\mathcal{T},2} + \theta_{\mathcal{T}}\right\}. \tag{4.63}$$

In order to derive an upper bound for $\left\{\sum_K h_K^2 \|\varphi - \varphi_{\mathcal{T}}\|_K^2\right\}^{\frac{1}{2}}$, we would like to insert the function $\sum_K h_K^2(\varphi - \varphi_{\mathcal{T}})$ as a test function in equation (4.60). This is not possible, since this function is not contained in $H^1(\Omega)$. We therefore replace the piecewise constant function h_K by a suitable smooth mesh-function. More precisely, we assume that the partition $\mathcal{T}$ satisfies the following *mesh-smoothness condition*:

> there is a function $h \in W^{1,\infty}(\Omega)$ such that the mesh-size h_K can locally be bounded from above and from below by h, i.e. the parameter
>
> $$\widetilde{c}_{\mathcal{T}} = \max_{K\in\mathcal{T}} \max_{x\in K} \max\left\{\frac{h(x)}{h_K}, \frac{h_K}{h(x)}\right\}$$
>
> is of moderate size.

With this assumption we may now insert $\psi = h^2(\varphi - \varphi_{\mathcal{T}})$ and $v = -h^2(u - u_{\mathcal{T}})$ in equation (4.60) and thus obtain

$$\begin{aligned}\langle R_1, h^2(\varphi - \varphi_{\mathcal{T}})\rangle - \langle R_2, h^2(u - u_{\mathcal{T}})\rangle &= B((\varphi - \varphi_{\mathcal{T}}, u - u_{\mathcal{T}}), (h^2(\varphi - \varphi_{\mathcal{T}}), -h^2(u - u_{\mathcal{T}})))\\ &= \left\|h(\varphi - \varphi_{\mathcal{T}})\right\|^2 + \int_\Omega \nabla\left(h^2(\varphi - \varphi_{\mathcal{T}})\right)\cdot\nabla(u - u_{\mathcal{T}})\\ &\quad - \int_\Omega \nabla(\varphi - \varphi_{\mathcal{T}})\cdot\nabla\left(h^2(u - u_{\mathcal{T}})\right)\\ &= \left\|h(\varphi - \varphi_{\mathcal{T}})\right\|^2 + 2\int_\Omega h(\varphi - \varphi_{\mathcal{T}})\nabla h\cdot\nabla(u - u_{\mathcal{T}})\\ &\quad - 2\int_\Omega h(u - u_{\mathcal{T}})\nabla(\varphi - \varphi_{\mathcal{T}})\cdot\nabla h.\end{aligned}$$

Regrouping terms this proves

$$\begin{aligned}\left\|h(\varphi - \varphi_{\mathcal{T}})\right\|^2 &= \langle R_1, h^2(\varphi - \varphi_{\mathcal{T}})\rangle - \langle R_2, h^2(u - u_{\mathcal{T}})\rangle\\ &\quad - 2\int_\Omega h(\varphi - \varphi_{\mathcal{T}})\nabla h\cdot\nabla(u - u_{\mathcal{T}})\\ &\quad + 2\int_\Omega h(u - u_{\mathcal{T}})\nabla(\varphi - \varphi_{\mathcal{T}})\cdot\nabla h.\end{aligned} \tag{4.64}$$

Next we bound the four terms on the right-hand side of this estimate.

Since R_1 admits an L^2-representation and satisfies the Galerkin orthogonality, Theorem 3.57 (p. 135) yields

$$\langle R_1, h^2(\varphi - \varphi_{\mathcal{T}})\rangle \leq c\eta_{\mathcal{T},1}\left\|\nabla(h^2(\varphi - \varphi_{\mathcal{T}}))\right\|.$$

The product rule for differentiation and the triangle inequality imply that

$$\left\|\nabla(h^2(\varphi - \varphi_{\mathcal{T}}))\right\| \leq 2\left\|\nabla h\right\|_\infty \left\|h(\varphi - \varphi_{\mathcal{T}})\right\| + \left\|h^2\nabla(\varphi - \varphi_{\mathcal{T}})\right\|.$$

To bound the second term on the right-hand side of this estimate, we split $\varphi - \varphi_{\mathcal{T}}$ in $\varphi - I_{\mathcal{T}}\varphi$ and $I_{\mathcal{T}}\varphi - \varphi_{\mathcal{T}}$ where $I_{\mathcal{T}}$ is the quasi-interpolation operator of equation (3.22) (p. 108). Using the error estimates of Proposition 3.33 (p. 109), a local inverse estimate for $I_{\mathcal{T}}\varphi - \varphi_{\mathcal{T}}$, the mesh-smoothness condition, and the a priori bound $\|\varphi\|_1 \le c_\Omega \|f\|_{-1}$ we then obtain

$$\begin{aligned}\left\|h^2\nabla(\varphi - \varphi_{\mathcal{T}})\right\| &\le \left\|h^2\nabla(\varphi - I_{\mathcal{T}}\varphi)\right\| + \left\|h^2\nabla(I_{\mathcal{T}}\varphi - \varphi_{\mathcal{T}})\right\| \\ &\le \left\|h^2\nabla(\varphi - I_{\mathcal{T}}\varphi)\right\| + c_1\left\|h(I_{\mathcal{T}}\varphi - \varphi_{\mathcal{T}})\right\| \\ &\le \left\|h^2\nabla(\varphi - I_{\mathcal{T}}\varphi)\right\| + c_1\left\|h(\varphi - I_{\mathcal{T}}\varphi)\right\| \\ &\quad + c_1\left\|h(\varphi - \varphi_{\mathcal{T}})\right\| \\ &\le c_2 \max_{K\in\mathcal{T}} h_K^2 \|f\|_{-1} + c_1\left\|h(\varphi - \varphi_{\mathcal{T}})\right\|.\end{aligned}$$

Since R_2 admits an L^2-representation and satisfies the Galerkin orthogonality, the mesh-smoothness condition and the arguments used in Section 1.4.4 (p. 11) to prove estimate (1.17) (p. 12) yield

$$\begin{aligned}\left\langle R_2, h^2(u - u_{\mathcal{T}})\right\rangle &\le c_1 \sum_{K\in\mathcal{T}} h_K^3 \|f - \Delta\varphi_{\mathcal{T}}\|_K h_K^{-2} \left\|\nabla\left(h^2(u - u_{\mathcal{T}})\right)\right\|_{\widetilde{\omega}_K} \\ &\quad + c_2 \sum_{E\in\mathcal{T}_\Omega} h_E^{\frac{5}{2}} \left\|\mathbb{J}_E(\mathbf{n}_E\cdot\nabla\varphi_{\mathcal{T}})\right\|_E h_K^{-2} \left\|\nabla\left(h^2(u - u_{\mathcal{T}})\right)\right\|_{\widetilde{\omega}_E} \\ &\le c_3\{\eta_{\mathcal{T},2} + \theta_{\mathcal{T}}\} \left\|h^{-2}\nabla\left(h^2(u - u_{\mathcal{T}})\right)\right\|.\end{aligned}$$

The product rule for differentiation and the triangle inequality now imply that

$$\left\|h^{-2}\nabla\left(h^2(u - u_{\mathcal{T}})\right)\right\| \le \left\|\nabla(u - u_{\mathcal{T}})\right\| + 2\|\nabla h\|_\infty \left\|h^{-1}(u - u_{\mathcal{T}})\right\|.$$

Estimate (4.63) provides an upper bound for the first term on the right-hand side of this inequality. To estimate the second one, we split $u - u_{\mathcal{T}}$ in $u - i_{\mathcal{T}}u$ and $i_{\mathcal{T}}u - u_{\mathcal{T}}$ where $i_{\mathcal{T}}$ is the standard nodal interpolation operator on $S_0^{2,0}(\mathcal{T})$. Using standard interpolation error estimates [**109**, Theorem 16.1], a local inverse estimate for $i_{\mathcal{T}}u - u_{\mathcal{T}}$, the mesh-smoothness condition, and the a priori bound $\|u\|_3 \le c_\Omega \|f\|_{-1}$ we thus obtain

$$\begin{aligned}\left\|h^{-1}(u - u_{\mathcal{T}})\right\| &\le \left\|h^{-1}(u - i_{\mathcal{T}}u)\right\| + \left\|h^{-1}(i_{\mathcal{T}}u - u_{\mathcal{T}})\right\| \\ &\le \left\|h^{-1}(u - i_{\mathcal{T}}u)\right\| + c_1\left\|\nabla(i_{\mathcal{T}}u - u_{\mathcal{T}})\right\| \\ &\le \left\|h^{-1}(u - i_{\mathcal{T}}u)\right\| + c_1\left\|\nabla(u - i_{\mathcal{T}}u)\right\| + c_1\left\|\nabla(u - u_{\mathcal{T}})\right\| \\ &\le c_2 \max_{K\in\mathcal{T}} h_K^2 \|f\|_{-1} + c_1\left\|\nabla(u - u_{\mathcal{T}})\right\|.\end{aligned}$$

For the third and fourth term on the right-hand side of estimate (4.64) finally, the Cauchy–Schwarz inequality yields the bounds

$$\int_\Omega h(\varphi - \varphi_{\mathcal{T}})\nabla h \cdot \nabla(u - u_{\mathcal{T}}) \le \|\nabla h\|_\infty \left\|h(\varphi - \varphi_{\mathcal{T}})\right\| \left\|\nabla(u - u_{\mathcal{T}})\right\|$$

$$\int_\Omega h(u - u_{\mathcal{T}})\nabla(\varphi - \varphi_{\mathcal{T}}) \cdot \nabla h \le \|\nabla h\|_\infty \left\|h^2\nabla(\varphi - \varphi_{\mathcal{T}})\right\| \cdot \left\|h^{-1}(u - u_{\mathcal{T}})\right\|$$

which only involve terms that are already estimated.

Since the mesh-smoothness condition implies that $\|h(\varphi - \varphi_{\mathcal{T}})\|$ is an upper bound for the mesh-dependent norm $\{\sum_K h_K^2 \|\varphi - \varphi_{\mathcal{T}}\|_K^2\}^{\frac{1}{2}}$, the above estimates prove that

$$\left\{\sum_K h_K^2 \|\varphi - \varphi_{\mathcal{T}}\|_K^2\right\}^{\frac{1}{2}} \le c\left\{\eta_{\mathcal{T},1} + \eta_{\mathcal{T},2} + \theta_{\mathcal{T}} + \max_{K\in\mathcal{T}} h_K^2 \|f\|_{-1}\right\}.$$

The results of this subsection can be summarised as follows.

Theorem 4.76 *Assume that Ω is convex, that $\mathcal{T}$ satisfies the mesh-smoothness condition, and that the finite element spaces $V_{\mathcal{T}}$ and $W_{\mathcal{T}}$ have the property (4.59). Denote by (φ, u) and by $(\varphi_{\mathcal{T}}, u_{\mathcal{T}}) \in V_{\mathcal{T}} \times W_{\mathcal{T}}$ the unique solutions of the mixed formulation of the bi-harmonic equation (4.57) and its mixed finite element discretisation, respectively. Define the* residual a posteriori error indicator *$\eta_{R,K}$ by*

$$\eta_{R,K} = \left\{h_K^2 \|\varphi_{\mathcal{T}} - \Delta u_{\mathcal{T}}\|_K^2 + \sum_{E\in\mathcal{E}_K} h_E \|\mathbb{J}_E(\mathbf{n}_E \cdot \nabla u_{\mathcal{T}})\|_E^2 \right.$$
$$\left. + h_K^6 \|f_{\mathcal{T}} - \Delta\varphi_{\mathcal{T}}\|_K^2 + \sum_{E\in\mathcal{E}_{K,\Omega}} h_E^5 \|\mathbb{J}_E(\mathbf{n}_E \cdot \nabla\varphi_{\mathcal{T}})\|_E^2\right\}^{\frac{1}{2}},$$

where $f_{\mathcal{T}}$ is any piecewise polynomial approximation of f. Then the error is bounded from above by

$$\left\{\sum_K h_K^2 \|\varphi - \varphi_{\mathcal{T}}\|_K^2\right\}^{\frac{1}{2}} + \|\nabla(u - u_{\mathcal{T}})\|$$
$$\le c^*\left\{\sum_{K\in\mathcal{T}} \eta_{R,K}^2 + \sum_{K\in\mathcal{T}} h_K^6 \|f - f_{\mathcal{T}}\|_K^2\right\}^{\frac{1}{2}} + c^* \max_{K\in\mathcal{T}} h_K^2 \|f\|_{-1}$$

and, on every element K separately, from below by

$$\eta_{R,K} \le c_*\left\{\|\nabla(u - u_{\mathcal{T}})\|_{\omega_K}^2 + h_K^2 \|\varphi - \varphi_{\mathcal{T}}\|_{\omega_K}^2 + h_K^6 \|f - f_{\mathcal{T}}\|_{\omega_K}^2\right\}^{\frac{1}{2}}.$$

The constant c^* depends on the constant c_Ω in the a priori bound (4.62), the shape parameter $C_{\mathcal{T}}$ of $\mathcal{T}$, and the parameters $\|\nabla h\|_\infty$ and $\tilde{c}_{\mathcal{T}}$ in the mesh-smoothness condition. The constant c_* depends on the shape parameter $C_{\mathcal{T}}$ and the polynomial degree of $\varphi_{\mathcal{T}}$ and $u_{\mathcal{T}}$.

Remark 4.77 The above result is not standard in the sense that it requires the mesh-smoothness condition, involves the a priori bound (4.62), and contains the *global mesh-size* $\max_K h_K$. These restrictions reflect the missing inf–sup stability of the mixed variational problem. The arguments of this subsection are inspired by [**104**]. There, however, the analysis is restricted to *uniform* meshes and to the choice $W_{\mathcal{T}} = S_0^{2,0}(\mathcal{T})$. Thus, Theorem 4.76 is slightly stronger than the results of [**104**] although the mesh-smoothness condition imposes some restrictions on the adaptive process.

4.11.3 Non-conforming Discretisations

We only consider a representative example. The variational formulation is given by problem (4.2) (p. 152) with

$$X = Y = H_0^2(\Omega),$$
$$\|u\|_X = \|u\|_Y = \left\|D^2 u\right\|,$$
$$B(u,v) = \int_\Omega D^2 u : D^2 v,$$
$$\langle \ell\,,\, v\rangle = \int_\Omega fv.$$

Here, $\Omega \subset \mathbb{R}^2$ is a bounded connected polygonal domain, $D^2 u = \left(\frac{\partial^2 u}{\partial x_i \partial x_j}\right)_{ij}$ denotes the Hessian matrix of u, and $A : B = \sum_{ij} A_{ij}B_{ij}$ is the inner product of matrices. Obviously, the bi-linear form B is continuous and coercive. Hence, the variational problem admits a unique solution u.

The basic idea of the discretisation is to impose only C^0-continuity on the finite element functions and to compensate for the missing C^1-continuity by suitable averages and jumps on the element boundaries. To this end we define for every interior face E and every piecewise continuous function φ its *average* on E by

$$\mathbb{A}_E(\varphi)(x) = \frac{1}{2}\left(\lim_{t\to 0+} v(x - t\mathbf{n}_E) + \lim_{t\to 0+} v(x + t\mathbf{n}_E)\right). \tag{4.65}$$

For faces on the boundary, the second term on the right-hand side of (4.65) is suppressed. Note that

$$\mathbb{J}_E(\varphi\psi) = \mathbb{J}_E(\varphi)\mathbb{A}_E(\psi) + \mathbb{A}_E(\varphi)\mathbb{J}_E(\psi)$$

holds for all piecewise continuous functions φ and ψ.

With these notations the discretisation is given by problem (4.6) (p. 153) with

$$X_{\mathcal{T}} = Y_{\mathcal{T}} = S_0^{2,0}(\mathcal{T}),$$
$$B_{\mathcal{T}}(u_{\mathcal{T}}, v_{\mathcal{T}}) = \sum_{K\in\mathcal{T}} \int_K D^2 u_{\mathcal{T}} : D^2 v_{\mathcal{T}} + \sum_{E\in\mathcal{E}} \int_E \mathbb{A}_E(\mathbf{n}_E \cdot D^2 u_{\mathcal{T}} \mathbf{n}_E)\mathbb{J}_E(\mathbf{n}_E \cdot \nabla v_{\mathcal{T}})$$
$$+ \sum_{E\in\mathcal{E}} \int_E \mathbb{J}_E(\mathbf{n}_E \cdot \nabla u_{\mathcal{T}})\mathbb{A}_E(\mathbf{n}_E \cdot D^2 v_{\mathcal{T}} \mathbf{n}_E)$$
$$+ \sigma \sum_{E\in\mathcal{E}} h_E^{-1} \int_E \mathbb{J}_E(\mathbf{n}_E \cdot \nabla u_{\mathcal{T}})\mathbb{J}_E(\mathbf{n}_E \cdot \nabla v_{\mathcal{T}}),$$
$$\langle \ell_{\mathcal{T}}\,,\, v_{\mathcal{T}}\rangle = \int_\Omega f v_{\mathcal{T}}.$$

Here, $\mathcal{T}$ is an admissible, shape-regular *triangulation* of Ω and σ denotes a positive stabilisation or penalty parameter. The space $X_{\mathcal{T}}$ is equipped with the mesh-dependent semi-norm

$$|v_{\mathcal{T}}|_{\mathcal{T}} = \left\{\sum_{K\in\mathcal{T}} \left\|D^2 v_{\mathcal{T}}\right\|_K^2\right\}^{\frac{1}{2}}$$

and the mesh-dependent norm

$$|||v_{\mathcal{T}}|||_{\mathcal{T}} = \left\{ |v_{\mathcal{T}}|_{\mathcal{T}}^2 + \sigma \sum_{E \in \mathcal{E}} h_E^{-1} \left\| \mathbb{J}_E(\mathbf{n}_E \cdot \nabla v_{\mathcal{T}}) \right\|_E^2 \right\}^{\frac{1}{2}} .$$

With this choice of norms the bi-linear form $B_{\mathcal{T}}$ is uniformly continuous and coercive provided σ is sufficiently large [**87**, **133**]. Hence, the discrete problem admits a unique solution $u_{\mathcal{T}}$. Integration by parts element-wise shows that the discrete problem is also consistent in the sense that the weak solution u of the bi-harmonic problem (4.57) satisfies for all $v_{\mathcal{T}} \in X_{\mathcal{T}}$

$$B_{\mathcal{T}}(u, v_{\mathcal{T}}) = \langle \ell_{\mathcal{T}}, v_{\mathcal{T}} \rangle .$$

Since the solution u of the variational problem is in $H_0^2(\Omega)$, we have

$$|||u - u_{\mathcal{T}}|||_{\mathcal{T}} = \left\{ |u - u_{\mathcal{T}}|_{\mathcal{T}}^2 + \sigma \sum_{E \in \mathcal{E}} h_E^{-1} \left\| \mathbb{J}_E(\mathbf{n}_E \cdot \nabla u_{\mathcal{T}}) \right\|_E^2 \right\}^{\frac{1}{2}} .$$

The second term on the right-hand side is a potential ingredient of an error indicator.

To control the first term, we compare the error $u - u_{\mathcal{T}}$ with the error $u - E_{\mathcal{T}} u_{\mathcal{T}}$ of a suitable lifting of $u_{\mathcal{T}}$ to a *conforming* finite element subspace of $H_0^2(\Omega)$. More precisely, denote by $V_{\mathcal{T}} \subset H_0^2(\Omega)$ the Hsieh–Clough–Tougher space associated with $\mathcal{T}$ [**109**, Section 46]. Its degrees of freedom are the function values and the first-order derivatives at the vertices and the normal derivatives at the midpoints of edges. The lifting $E_{\mathcal{T}}$ is defined in a similar way as the averaged gradient $Gu_{\mathcal{T}}$ in Section 1.9 (p. 36): the degrees of freedom of $E_{\mathcal{T}} u_{\mathcal{T}} \in V_{\mathcal{T}}$ are the averages of the corresponding values of $u_{\mathcal{T}}$ [**85**, equation (2.8)], i.e.

$$\begin{aligned} E_{\mathcal{T}} u_{\mathcal{T}}(z) &= u_{\mathcal{T}}(z), \\ \nabla(E_{\mathcal{T}} u_{\mathcal{T}})(z) &= \frac{1}{\#\{K \subset \omega_z\}} \sum_{K \subset \omega_z} \nabla u_{\mathcal{T}}|_K(z), \\ \mathbf{n}_E \cdot \nabla(E_{\mathcal{T}} u_{\mathcal{T}})(z_E) &= \mathbb{A}_E(\mathbf{n}_E \cdot u_{\mathcal{T}})(z_E) \end{aligned}$$

for all vertices z and all midpoints of edges z_E.

Since $E_{\mathcal{T}} u_{\mathcal{T}}$ is in $H_0^2(\Omega)$, the triangle inequality implies

$$|u - u_{\mathcal{T}}|_{\mathcal{T}} \le \left\| D^2(u - E_{\mathcal{T}} u_{\mathcal{T}}) \right\| + |E_{\mathcal{T}} u_{\mathcal{T}} - u_{\mathcal{T}}|_{\mathcal{T}} .$$

To bound $|E_{\mathcal{T}} u_{\mathcal{T}} - u_{\mathcal{T}}|_{\mathcal{T}}$ we first use a standard local inverse estimate and obtain

$$|u_{\mathcal{T}} - E_{\mathcal{T}} u_{\mathcal{T}}|_{\mathcal{T}} \le c \left\{ \sum_{K \in \mathcal{T}} h_K^{-4} \left\| u_{\mathcal{T}} - E_{\mathcal{T}} u_{\mathcal{T}} \right\|_K^2 \right\}^{\frac{1}{2}} .$$

Then a scaling argument yields for all elements K

$$h_K^{-4} \left\| u_{\mathcal{T}} - E_{\mathcal{T}} u_{\mathcal{T}} \right\|_K^2 \le c \sum_{z \in \mathcal{N}_K} \left| \nabla(u_{\mathcal{T}} - E_{\mathcal{T}} u_{\mathcal{T}})(z) \right|^2 + c \left\{ \sum_{E \in \mathcal{E}_K} h_E^{-1} \left\| \mathbb{J}_E(\mathbf{n}_E \cdot \nabla u_{\mathcal{T}}) \right\|_E^2 \right\}^{\frac{1}{2}} .$$

The methods used in Section 1.9 (p. 36) to control the averaged gradient finally show that the first term on the right-hand side can be bounded by the second one [**85**, Section 2].

Thus we are left with the task of bounding $\|D^2(u - E_{\mathcal{T}} u_{\mathcal{T}})\|$. Since $u - E_{\mathcal{T}} u_{\mathcal{T}}$ is contained in $H_0^2(\Omega)$, we have

$$\|D^2(u - E_{\mathcal{T}} u_{\mathcal{T}})\| = \sup_{v \in H_0^2(\Omega)\setminus\{0\}} \frac{B(u - E_{\mathcal{T}} u_{\mathcal{T}}, v)}{\|D^2 v\|}.$$

We therefore consider an arbitrary element $v \in H_0^2(\Omega)$ and denote by $i_{\mathcal{T}} v \in S_0^{2,0}(\mathcal{T})$ its standard nodal interpolation. The definition of the variational and discrete problems and of the bi-linear form $B_{\mathcal{T}}$ then imply

$$\begin{aligned} B(u - E_{\mathcal{T}} u_{\mathcal{T}}, v) = &\int_\Omega f(v - i_{\mathcal{T}} v) - \sum_{K \in \mathcal{T}} \int_K D^2(E_{\mathcal{T}} u_{\mathcal{T}} - u_{\mathcal{T}}) : D^2 v - \sum_{K \in \mathcal{T}} \int_K D^2 u_{\mathcal{T}} : D^2(v - i_{\mathcal{T}} v) \\ &+ \sum_{E \in \mathcal{E}} \int_E \mathbb{A}_E(\mathbf{n}_E \cdot D^2 u_{\mathcal{T}} \mathbf{n}_E) \mathbb{J}_E(\mathbf{n}_E \cdot \nabla i_{\mathcal{T}} v) \\ &+ \sum_{E \in \mathcal{E}} \int_E \mathbb{J}_E(\mathbf{n}_E \cdot \nabla u_{\mathcal{T}}) \mathbb{A}_E(\mathbf{n}_E \cdot D^2 i_{\mathcal{T}} v \mathbf{n}_E) \\ &+ \sigma \sum_{E \in \mathcal{E}} h_E^{-1} \int_E \mathbb{J}_E(\mathbf{n}_E \cdot \nabla u_{\mathcal{T}}) \mathbb{J}_E(\mathbf{n}_E \cdot \nabla i_{\mathcal{T}} v). \end{aligned}$$

The first two terms on the right-hand side can be bounded with the help of Proposition 3.35 (p. 112) and the above estimate of $|E_{\mathcal{T}} u_{\mathcal{T}} - u_{\mathcal{T}}|_{\mathcal{T}}$ by

$$c \left\{ \sum_{K \in \mathcal{T}} h_K^4 \|f\|_K^2 \right\}^{\frac{1}{2}} \|D^2 v\|$$

and

$$c \left\{ \sum_{E \in \mathcal{E}} h_E^{-1} \|\mathbb{J}_E(\mathbf{n}_E \cdot \nabla v_{\mathcal{T}})\|_E^2 \right\}^{\frac{1}{2}} \|D^2 v\|.$$

Using integration by parts element-wise and taking into account that the third-order derivatives of $u_{\mathcal{T}}$ vanish, the last four terms on the right-hand side of the previous estimate can be written in the form

$$\begin{aligned} &\sum_{E \in \mathcal{E}} \int_E \mathbb{A}_E(\mathbf{n}_E \cdot D^2(i_{\mathcal{T}} v) \mathbf{n}_E) \mathbb{J}_E(\mathbf{n}_E \cdot \nabla u_{\mathcal{T}}) + \sigma \sum_{E \in \mathcal{E}} h_E^{-1} \int_E \mathbb{J}_E(\mathbf{n}_E \cdot \nabla u_{\mathcal{T}}) \mathbb{J}_E(\mathbf{n}_E \cdot \nabla i_{\mathcal{T}} v) \\ &\quad + \sum_{E \in \mathcal{E}_\Omega} \int_E \mathbb{A}_E(\mathbf{n}_E \cdot \nabla(v - i_{\mathcal{T}} v)) \mathbb{J}_E(\mathbf{n}_E \cdot D^2 u_{\mathcal{T}} \mathbf{n}_E). \end{aligned}$$

Proposition 3.35 (p. 112) and the trace inequalities of Proposition 3.5 (p. 90) and Remark 3.6 (p. 90) finally enable us to bound this expression by

$$\left\{\sigma^2 \sum_{E\in\mathcal{E}} h_E^{-1} \left\|\mathbb{J}_E(\mathbf{n}_E \cdot \nabla u_{\mathcal{T}})\right\|_E^2 + \sum_{E\in\mathcal{E}_\Omega} h_E \left\|\mathbb{J}_E(\mathbf{n}_E \cdot D^2 u_{\mathcal{T}} \mathbf{n}_E)\right\|_E^2\right\}^{\frac{1}{2}} \left\|D^2 v\right\|.$$

Collecting all estimates, we obtain the upper bound

$$\begin{aligned}|||u - u_{\mathcal{T}}|||_{\mathcal{T}} \le c^* \Bigg\{ &\sum_{K\in\mathcal{T}} h_K^4 \left\|f\right\|_K^2 + \sigma^2 \sum_{E\in\mathcal{E}} h_E^{-1} \left\|\mathbb{J}_E(\mathbf{n}_E \cdot \nabla u_{\mathcal{T}})\right\|_E^2 \\ &+ \sum_{E\in\mathcal{E}_\Omega} h_E \left\|\mathbb{J}_E(\mathbf{n}_E \cdot D^2 u_{\mathcal{T}} \mathbf{n}_E)\right\|_E^2 \Bigg\}^{\frac{1}{2}}.\end{aligned}$$

A similar lower bound is proved in the standard way by successively inserting the functions $v = \psi_{K,1} f_{\mathcal{T}}$, $v = \Psi_{E,1}\mathbb{J}_E(\mathbf{n}_E \cdot \nabla u_{\mathcal{T}})$, and $v = \psi_{E,1}\mathbb{J}_E(\mathbf{n}_E \cdot D^2 u_{\mathcal{T}} \mathbf{n}_E)$ in the above representation of $B(u - E_{\mathcal{T}} u_{\mathcal{T}}, v)$ [**85**, Section 4], where $\psi_{K,1}$, $\Psi_{E,1}$, and $\psi_{E,1}$ are the smooth cut-off functions of Section 3.2.5 (p. 85) and Proposition 3.51 (p. 129).

These results can be summarised as follows.

Theorem 4.78 *Denote by u the unique weak solution of the bi-harmonic equation* (4.57) *and by* $u_{\mathcal{T}} \in S_0^{2,0}(\mathcal{T})$ *its non-conforming approximation. Define the* residual a posteriori error indicator *by*

$$\eta_{R,K} = \left\{ h_K^4 \left\|f_{\mathcal{T}}\right\|_K^2 + \sigma^2 \sum_{E\in\mathcal{E}_K} h_E^{-1} \left\|\mathbb{J}_E(\mathbf{n}_E \cdot \nabla u_{\mathcal{T}})\right\|_E^2 + \sum_{E\in\mathcal{E}_{K,\Omega}} h_E \left\|\mathbb{J}_E(\mathbf{n}_E \cdot D^2 u_{\mathcal{T}} \mathbf{n}_E)\right\|_E^2 \right\}^{\frac{1}{2}},$$

where $f_{\mathcal{T}}$ *is any piecewise polynomial approximation of* f. *Then the error is bounded from above by*

$$|||u - u_{\mathcal{T}}|||_{\mathcal{T}} \le c^* \left\{ \sum_{K\in\mathcal{T}} \eta_{R,K}^2 + \sum_{K\in\mathcal{T}} h_K^4 \left\|f - f_{\mathcal{T}}\right\|_K^2 \right\}^{\frac{1}{2}}$$

and, for every element K separately, from below by

$$\eta_{R,K} \le c_* \left\{ \left\|D^2(u - u_{\mathcal{T}})\right\|_K^2 + h_K^4 \left\|f - f_{\mathcal{T}}\right\|_K^2 \right\}^{\frac{1}{2}}.$$

Both constants c^* *and* c_* *depend on the shape parameter* $C_{\mathcal{T}}$ *of* $\mathcal{T}$*; the constant* c_* *in addition depends on the stability parameter* σ.

Remark 4.79 Subsection 4.11.1 is a slightly improved variant of [**258**, Section 3.7]. The approach of Subsection 4.11.2 is inspired by [**104**]. The analysis of Subsection 4.11.3 is based upon [**85**]. Further a posteriori error estimates for various discretisations of the bi-harmonic equation and its relatives may be found in [**55, 56, 57, 178, 203, 223, 278**].

4.12 Non-Conforming Discretisations

To simplify the presentation and to better exploit the fundamental differences to conforming discretisations, we first consider the two-dimensional Poisson equation (1.1) (p. 4) with homogeneous Dirichlet boundary conditions as a model problem. In Remark 4.81 below we will then discuss the necessary modifications for general diffusion equations, mixed boundary conditions, and three space dimensions.

In what follows $X_{\mathcal{T}}$ denotes a finite element space of piecewise polynomials associated with an admissible, affine-equivalent, shape-regular partition $\mathcal{T}$ of a polygonal domain Ω in $\mathbb{R}^2$. $X_{\mathcal{T}}$ is arbitrary up to the following two conditions.

- It is rich enough in the sense that

$$S_0^{1,0}(\mathcal{T}) \subset X_{\mathcal{T}}. \tag{4.66}$$

- Its functions are weakly continuous and satisfy the homogeneous boundary condition weakly in the sense that

$$\begin{aligned} \int_E \mathbb{J}_E(u) &= 0 \quad \text{for all } E \in \mathcal{E}_\Omega, \\ \int_E u &= 0 \quad \text{for all } E \in \mathcal{E}_\Gamma. \end{aligned} \tag{4.67}$$

There are many examples of spaces satisfying these conditions [99, 100]. The simplest and probably most prominent one is the Crouzeix–Raviart element. It consists of piecewise affine functions on triangles which are continuous at the midpoints of interior edges and which vanish at the midpoints of boundary edges. Since the midpoint rule for integration is exact for first order polynomials, conditions (4.66) and (4.67) are satisfied.

The discrete problem corresponds to problem (4.6) (p. 153) with

$$\begin{aligned} Y_{\mathcal{T}} &= X_{\mathcal{T}}, \\ \|v\|_{X_{\mathcal{T}}} &= \left\{ \sum_{K\in\mathcal{T}} \|\nabla v\|_K^2 \right\}^{\frac{1}{2}}, \\ B_{\mathcal{T}}(u_{\mathcal{T}}, v_{\mathcal{T}}) &= \sum_{K\in\mathcal{T}} \int_K \nabla u_{\mathcal{T}} \cdot \nabla v_{\mathcal{T}}, \\ \langle \ell_{\mathcal{T}}, v_{\mathcal{T}} \rangle &= \sum_{K\in\mathcal{T}} \int_K f v_{\mathcal{T}}. \end{aligned}$$

Condition (4.67) implies that $\|\cdot\|_{X_{\mathcal{T}}}$ is a norm on $X_{\mathcal{T}}$. Since $B_{\mathcal{T}}$ is continuous and coercive with respect to $\|\cdot\|_{X_{\mathcal{T}}}$, the discrete problem admits a unique solution $u_{\mathcal{T}}$. Notice that $\|\cdot\|_{X_{\mathcal{T}}}$, $B_{\mathcal{T}}$, and $\ell_{\mathcal{T}}$ coincide on $H_0^1(\Omega)$ with $\|\nabla\cdot\|$, B, and ℓ, respectively.

The main tool for the estimation of $\|u - u_{\mathcal{T}}\|_{X_{\mathcal{T}}}$ is a Helmholtz-type decomposition of the element-wise gradient of $u_{\mathcal{T}}$. Unfortunately, we cannot apply Proposition 4.52 (p. 215) directly to the element-wise gradient of $u_{\mathcal{T}}$ since in general this function is not contained in $H(\text{div}; \Omega)$. Its Ritz projection onto $H_0^1(\Omega)$, however, allows us to use Proposition 4.51 (p. 215). More precisely, denote by $\nabla_{\mathcal{T}} u_{\mathcal{T}}$ the element-wise gradient of $u_{\mathcal{T}}$ and by $v \in H_0^1(\Omega)$ its Ritz projection. The latter is the unique solution of

$$\int_\Omega \nabla v \cdot \nabla w = \int_\Omega \nabla_{\mathcal{T}} u_{\mathcal{T}} \cdot \nabla w \tag{4.68}$$

for all $w \in H_0^1(\Omega)$. Proposition 4.51 (p. 215) can then be applied to $\nabla v - \nabla_{\mathcal{T}} u_{\mathcal{T}}$ and yields a stream function $\varphi \in H^1(\Omega)$ with

$$\nabla_{\mathcal{T}} u_{\mathcal{T}} = \nabla v - \operatorname{curl} \varphi \tag{4.69}$$

and

$$\|u - u_{\mathcal{T}}\|_{X_{\mathcal{T}}}^2 = \left\|\nabla(u - v)\right\|^2 + \|\operatorname{curl} \varphi\|^2 .$$

The decomposition of $\nabla_{\mathcal{T}} u_{\mathcal{T}}$ implies that

$$\begin{aligned} B_{\mathcal{T}}(u - u_{\mathcal{T}}, u - u_{\mathcal{T}}) &= B_{\mathcal{T}}(u - u_{\mathcal{T}}, u - v) \\ &\quad - \int_\Omega \nabla_{\mathcal{T}}(u - u_{\mathcal{T}}) \cdot \operatorname{curl} \varphi. \end{aligned} \tag{4.70}$$

The second term on the right-hand side describes the non-conformity of the discretisation. For conforming discretisations considered so far this term vanishes.

We first consider the first term on the right-hand side of (4.70). For every $w \in H_0^1(\Omega)$ we have

$$B_{\mathcal{T}}(u - u_{\mathcal{T}}, w) = \int_\Omega fw - B_{\mathcal{T}}(u_{\mathcal{T}}, w). \tag{4.71}$$

The right-hand side of (4.71) defines a residual. Thanks to condition (4.66) it satisfies the Galerkin orthogonality (3.30) (p. 132). Integration by parts element-wise shows that it also satisfies the L^2-representation (3.29) (p. 132) with

$$r|_K = f + \Delta u_{\mathcal{T}} \quad \text{on every } K \in \mathcal{T}$$

$$j|_E = \begin{cases} -\mathbb{J}_E(\mathbf{n}_E \cdot \nabla u_{\mathcal{T}}) & \text{if } E \in \mathcal{E}_\Omega, \\ 0 & \text{if } E \in \mathcal{E}_\Gamma. \end{cases}$$

Therefore, the results of Section 3.8 (p. 132) immediately yield upper and lower bounds for the dual norm of this residual. Since $u - v$ is contained in $H^1(\Omega)$, the upper bound for the dual norm of the residual provides an upper bound for the first term on the right-hand side of equation (4.70). Thanks to equation (4.71), the lower bound for the dual norm of the residual in turn yields a lower bound for $\|u - u_{\mathcal{T}}\|_{X_{\mathcal{T}}}$.

To bound the second term on the right-hand side of equation (4.70), we first observe that

$$\int_\Omega \nabla_{\mathcal{T}} u \cdot \operatorname{curl} \psi = \int_\Omega \nabla u \cdot \operatorname{curl} \psi = 0$$

holds for all $\psi \in H^1(\Omega)$. Next we claim that

$$\int_\Omega \nabla_{\mathcal{T}} u_{\mathcal{T}} \cdot \operatorname{curl} \psi_{\mathcal{T}} = 0$$

holds for all $\psi_{\mathcal{T}} \in S^{1,0}(\mathcal{T})$. To prove this, we perform integration by parts element-wise, use the identity div $\circ$ curl $= 0$, take into account that $\mathbf{t}_E \cdot \nabla \psi_{\mathcal{T}}$, the tangential derivative of $\psi_{\mathcal{T}}$, is constant on the edges, and profit from condition (4.67). We thus obtain

$$\begin{aligned}\int_\Omega \nabla_{\mathcal{T}} u_{\mathcal{T}} \cdot \operatorname{curl} \psi_{\mathcal{T}} &= \sum_{E \in \mathcal{E}_\Omega} \int_E \mathbb{J}_E(u_{\mathcal{T}})\mathbf{t}_E \cdot \nabla \psi + \sum_{E \in \mathcal{E}_\Gamma} \int_E u_{\mathcal{T}} \mathbf{t}_E \cdot \nabla \psi \\ &= 0.\end{aligned}$$

Combining both properties we conclude that

$$-\int_\Omega \nabla_{\mathcal{T}}(u - u_{\mathcal{T}}) \cdot \operatorname{curl} \psi = \int_\Omega \nabla_{\mathcal{T}} u_{\mathcal{T}} \cdot \operatorname{curl}(\psi - I_{\mathcal{T}} \psi)$$

holds for all $\psi \in H^1(\Omega)$. Here, $I_{\mathcal{T}}$ is the quasi-interpolation operator of (3.22) (p. 108) with the modification that the first part of (3.21) (p. 108) is used for all vertices, i.e. vertices on the boundary are treated in the same way as interior vertices. To estimate the right-hand side of this identity, we once again perform integration by parts element-wise, but now profit from the relation curl $\circ \nabla = 0$. This yields

$$\int_\Omega \nabla_{\mathcal{T}} u_{\mathcal{T}} \cdot \operatorname{curl}(\psi - I_{\mathcal{T}} \psi) = \sum_{E \in \mathcal{E}_\Omega} \int_E \mathbb{J}_E(\mathbf{t}_E \cdot \nabla u_{\mathcal{T}})(\psi - I_{\mathcal{T}} \psi) + \sum_{E \in \mathcal{E}_\Gamma} \int_E (\mathbf{t}_E \cdot \nabla u_{\mathcal{T}})(\psi - I_{\mathcal{T}} \psi).$$

The right-hand side can be estimated in the standard way using the Cauchy–Schwarz inequality, Proposition 3.33 (p. 109), and the identity $\|\nabla \psi\| = \|\operatorname{curl} \psi\|$ which holds in two dimensions. This gives an upper bound for the second term on the right-hand side of (4.70).

Adding the upper bounds for the two terms on the right-hand side of (4.70) yields an upper bound for $\|u - u_{\mathcal{T}}\|_{X_{\mathcal{T}}}$ in terms of the element residuals $f + \Delta u_{\mathcal{T}}$, the edge residuals $\mathbb{J}_E(\mathbf{n}_E \cdot \nabla u_{\mathcal{T}})$, and, additionally to conforming methods, the edge residuals $\mathbb{J}_E(\mathbf{t}_E \cdot \nabla u_{\mathcal{T}})$ and $\mathbf{t}_E \cdot \nabla u_{\mathcal{T}}$ on interior and boundary edges, respectively. To obtain a similar lower bound for $\|u - u_{\mathcal{T}}\|_{X_{\mathcal{T}}}$ we still have to control the last two terms involving the tangential derivatives. This is easily achieved by successively inserting the functions $\psi = \psi_E \mathbb{J}_E(\mathbf{t}_E \cdot \nabla u_{\mathcal{T}})$ for $E \in \mathcal{E}_\Omega$ and $\psi = \psi_E \mathbf{t}_E \cdot \nabla u_{\mathcal{T}}$ for $E \in \mathcal{E}_\Gamma$ in the identity

$$-\int_\Omega \nabla_{\mathcal{T}}(u - u_{\mathcal{T}}) \cdot \operatorname{curl} \psi = \sum_{E \in \mathcal{E}_\Omega} \int_E \mathbb{J}_E(\mathbf{t}_E \cdot \nabla u_{\mathcal{T}})\psi + \sum_{E \in \mathcal{E}_\Gamma} \int_E (\mathbf{t}_E \cdot \nabla u_{\mathcal{T}})\psi.$$

The above results can be summarised as follows.

Theorem 4.80 *Denote by u the unique weak solution of the Poisson equation with homogeneous Dirichlet boundary conditions and by $u_{\mathcal{T}} \in X_{\mathcal{T}}$ its non-conforming finite element approximation. Assume that the space $X_{\mathcal{T}}$ satisfies conditions* (4.66) *and* (4.67). *Define the* residual a posteriori error indicator *$\eta_{R,K}$ by*

$$\begin{aligned}\eta_{R,K} = \Bigg\{ & h_K^2 \left\|f_{\mathcal{T}} + \Delta u_{\mathcal{T}}\right\|_K^2 + \sum_{E \in \mathcal{E}_\Omega} h_E \left\|\mathbb{J}_E(\mathbf{n}_E \cdot \nabla u_{\mathcal{T}})\right\|_E^2 \\ & + \sum_{E \in \mathcal{E}_\Omega} h_E \left\|\mathbb{J}_E(\mathbf{t}_E \cdot \nabla u_{\mathcal{T}})\right\|_E^2 + \sum_{E \in \mathcal{E}_\Gamma} h_E \left\|\mathbf{t}_E \cdot \nabla u_{\mathcal{T}}\right\|_E^2 \Bigg\}^{\frac{1}{2}},\end{aligned}$$

where $f_{\mathcal{T}}$ is any piecewise polynomial approximation of f. Then the error can be bounded from above by

$$\|u - u_{\mathcal{T}}\|_{X_{\mathcal{T}}} \le c^* \left\{ \sum_{K \in \mathcal{T}} \eta_{R,K}^2 + \sum_{K \in \mathcal{T}} h_K^2 \|f - f_{\mathcal{T}}\|_K^2 \right\}^{\frac{1}{2}}$$

and, on every element K separately, from below by

$$\eta_{R,K} \le c_* \left\{ \sum_{K' \subset \omega_K} \|\nabla(u - u_{\mathcal{T}})\|_{K'}^2 + \sum_{K' \subset \omega_K} h_{K'}^2 \|f - f_{\mathcal{T}}\|_{K'}^2 \right\}^{\frac{1}{2}}.$$

Both constants c^* and c_* depend on the shape parameter $C_{\mathcal{T}}$ of $\mathcal{T}$. The constant c_* in addition depends on the polynomial degree of $u_{\mathcal{T}}$.

Remark 4.81 At the expense of slightly more technical arguments, Theorem 4.80 can be extended to general diffusion equations, mixed boundary conditions, and three space dimensions [**94**].

When replacing the differential operator Δ by a general diffusion operator $\operatorname{div}(A\nabla)$, the Ritz projection must by computed with respect to the associated bi-linear form, i.e. equations (4.68) and (4.69) must be replaced by $\int_\Omega \nabla v \cdot A\nabla w = \int_\Omega \nabla_{\mathcal{T}} u_{\mathcal{T}} \cdot A\nabla w$ and $A\nabla_{\mathcal{T}} u_{\mathcal{T}} = A\nabla v - \operatorname{curl} \varphi$, respectively. Notice that the contributions of the tangential derivatives to the error indicator do not change when passing to general diffusion.
In the presence of Neumann boundary conditions, the contributions of the φ and ψ terms on the Neumann boundary require special attention [**94**, Lemmas 3.3, 3.4].

In three space dimensions finally, the decomposition (4.69) involves the vector potential. Correspondingly the spaces $H^1(\Omega)$ and $S^{1,0}(\mathcal{T})$ must be replaced by $H^1(\Omega)^3$ and $S^{1,0}(\mathcal{T})^3$, respectively. The operator $I_{\mathcal{T}}$ then of course is applied to the components of the corresponding vector fields.

Remark 4.82 The presentation of this section is inspired by [**94**]. A general framework for the a posteriori error analysis of non-conforming discretisations is given in [**99, 100**]. Further a posteriori error estimates for non-conforming methods involving other differential equations and other types of indicators may be found in [**5, 6, 14, 80, 92, 93, 96, 116, 132, 146, 161, 174, 176, 177, 179, 204, 205, 206, 234**].

4.13 Convergence of the Adaptive Process II

Following [**200**] we extend in this section the results of Section 1.14 (p. 58) and prove the convergence of a generic adaptive algorithm for general linear elliptic equations. Moreover we briefly address the question of the optimal complexity of the adaptive process. The present analysis will be extended to time-dependent problems in Section 6.8 (p. 362).

4.13.1 The Variational Problem

We consider the abstract variational problem (4.2) (p. 152) assuming that the conditions of Proposition 4.1 (p. 152) are satisfied. Moreover we assume that the spaces X, Y, their norms, and the bi-linear form B satisfy the following additional conditions which will be discussed in Subsection 4.13.3 below.

- X and Y are Hilbert spaces and are contained in $L^2(\Omega)^n$ for a suitable bounded connected set $\Omega \subset \mathbb{R}^d$ with Lipschitz boundary and suitable integers $d \geq 2$ and $n \geq 1$.
- Given any open connected subset ω of Ω with Lipschitz boundary, the norms $\|\cdot\|_X$ and $\|\cdot\|_Y$ admit natural restrictions to the spaces of restrictions to ω of functions in X and Y, respectively which are denoted by $\|\cdot\|_{X(\omega)}$ and $\|\cdot\|_{Y(\omega)}$.
- The norms $\|\cdot\|_X$ and $\|\cdot\|_Y$ are *sub-additive*, i.e.

$$\|\cdot\|_{X(\omega_1)}^2 + \|\cdot\|_{X(\omega_2)}^2 \leq \|\cdot\|_X^2, \quad \|\cdot\|_{Y(\omega_1)}^2 + \|\cdot\|_{Y(\omega_2)}^2 \leq \|\cdot\|_Y^2$$

holds for any pair of disjoint open connected subsets with Lipschitz boundaries and $\overline{\Omega} = \overline{\omega}_1 \cup \overline{\omega}_2$.
- The norm $\|\cdot\|_X$ is *absolutely continuous* with respect to the Lebesgue measure μ_d, i.e. $\mu_d(\omega) \to 0$ implies $\|\varphi\|_X \to 0$ for every $\varphi \in X$.
- The bi-linear form B is *locally continuous*, i.e. there is a constant C_B such that

$$\left|B(\varphi, \psi)\right| \leq C_B \|\varphi\|_{X(\omega)} \|\psi\|_Y$$

holds for every $\psi \in Y$ with $\operatorname{supp}\psi \subset \omega$ and every open connected subset ω of Ω with Lipschitz boundary.

Notice that these assumptions are satisfied for the examples considered in the previous sections.

4.13.2 The Adaptive Algorithm and its Components

In order to specify the required properties of the adaptive algorithm and its components, we write Algorithm 1.1 (p. 2) in a functional form. Notice that we drop the stopping criterion. Algorithm 4.83 produces an infinite sequence $(\varphi_k)_{k\in\mathbb{N}}$ of approximate solutions. Theorem 4.84 (p. 269) shows that, under the conditions given below, this sequence converges to the solution of the variational problem. Consequently, under these conditions, Algorithm 1.1 will terminate after a finite number of steps depending on the tolerance ε.

Algorithm 4.83 (Adaptive algorithm). *Given an initial partition $\mathcal{T}_0$, set $k = 0$.*

(1) $\varphi_k = SOLVE(X_{\mathcal{T}_k}, Y_{\mathcal{T}_k})$

(2) $(\eta_{k,K})_{K\in\mathcal{T}_k} = ESTIMATE(\varphi_k, \mathcal{T}_k)$, $\eta_k = \left\{\sum_{K\in\mathcal{T}_k} \eta_{k,K}^2\right\}^{\frac{1}{2}}$

(3) $\mathcal{M}_k = MARK((\eta_{k,K})_{K\in\mathcal{T}_k}, \mathcal{T}_k)$

(4) $\mathcal{T}_{k+1} = REFINE(\mathcal{T}_k, \mathcal{M}_k)$, *increase k by 1 and return to step* (1).

Next we will specify the required properties of the components of Algorithm 4.83. These assumptions will be discussed in Subsection 4.13.3 below.

4.13.2.1 Framework

We call a partition $\mathcal{T}'$ a *refinement* of another partition $\mathcal{T}$ if $\mathcal{T}'$ is obtained from $\mathcal{T}$ by a finite number of steps of *REFINE* and denote by

$$\mathcal{P} = \{\mathcal{T} : \mathcal{T} \text{ is a refinement of } \mathcal{T}_0\}$$

the collection of all possible partitions produced by Algorithm 4.83.

We assume that refinement relies on a *quasi-regular subdivision of elements*. More precisely, there exist two constants q_* and q^* in $(0,1)$ such that, for all $\mathcal{T} \in \mathcal{P}$, every $K \in \mathcal{T}$ can be subdivided into $n_K \geq 2$ sub-elements $K'_1, \ldots, K'_{n_K}$ such that

$$K = \bigcup_{1 \leq i \leq n_K} K'_i, \quad \mu_d(K) = \sum_{1 \leq i \leq n_K} \mu_d(K'_i)$$

and, for all $1 \leq i \leq n_K$,

$$q_* \mu_d(K) \leq \mu_d(K'_i) \leq q^* \mu_d(K).$$

These subdivisions generate a *forest* $\mathbb{F}$ of infinite trees. Each node corresponds to an element, its direct successors to its sub-elements, and the roots to the elements of the initial partition $\mathcal{T}_0$. A subforest $\widehat{\mathbb{F}} \subset \mathbb{F}$ is called finite if it has a finite number of nodes. Any finite tree may have interior nodes which have successors and leaf nodes which don't have successors.

A subdivision $\mathcal{S}$ of Ω is called subordinate to $\mathcal{T}_0$ if every element of $\mathcal{S}$ is contained in an element of $\mathcal{T}_0$. With every such subdivision we may associate a finite subforest $\mathbb{F}(\mathcal{S})$ whose leaf nodes are the elements of the subdivision. A partition $\mathcal{T}'$ then is a refinement of $\mathcal{T}$ if and only if $\mathbb{F}(\mathcal{T}) \subset \mathbb{F}(\mathcal{T}')$.

For any two elements K and K' in a tree $\mathbb{T} \subset \mathbb{F}$, we denote by $\operatorname{dist}(K, K')$ the distance of the nodes K, K' in $\mathbb{T}$. Given an integer $n \geq 1$ and a subset $\widehat{\mathcal{S}}$ of a subdivision $\mathcal{S}$ of Ω subordinate to $\mathcal{T}_0$, we denote be $\mathbb{F}^n(\mathcal{S}, \widehat{\mathcal{S}})$ the subforest of $\mathbb{F}$ which consists of $\mathbb{F}(\mathcal{S})$ and all successors of elements in $\widehat{\mathcal{S}}$ with distance at most n.

We assume that $\mathcal{P}$ is a subclass of the subdivisions of Ω subordinate to $\mathcal{T}_0$ and that it is *locally quasi-uniform* in the sense that

$$\sup_{\mathcal{T} \in \mathcal{P}} \max_{K \in \mathcal{T}} \# \mathcal{U}_{\mathcal{T},K} \leq c_1, \quad \sup_{\mathcal{T} \in \mathcal{P}} \max_{K \in \mathcal{T}} \max_{K' \in \mathcal{U}_{\mathcal{T},K}} \frac{\mu_d(K)}{\mu_d(K')} \leq c_2 .$$

Here,

$$\mathcal{U}_{\mathcal{T},K} = \{K' \in \mathcal{T} : \mathcal{N}_{K'} \cap \mathcal{N}_K \neq \emptyset\}$$

denotes the collection of all neighbours of K in $\mathcal{T}$. The partitions in $\mathcal{P}$ may have additional properties such as admissibility or affine-equivalence.

In summary $\mathbb{F}$ contains *all possible* subdivisions subordinate to $\mathcal{T}_0$ while $\bigcup_k \mathbb{F}(\mathcal{T}_k)$ consists of all subdivisions associated with a concrete realisation of Algorithm 4.83 and may be a proper subforest of $\mathbb{F}$.

4.13.2.2 *SOLVE*

We assume that, for all $\mathcal{T} \in \mathcal{P}$, the discretisations are conforming in the sense that $X_{\mathcal{T}} \subset X$ and $Y_{\mathcal{T}} \subset Y$, that the discrete bi-linear forms $B_{\mathcal{T}}$ and the linear functionals $\ell_{\mathcal{T}}$ are the restrictions of B and ℓ, and that B is *uniformly stable* on the spaces $X_{\mathcal{T}}$, $Y_{\mathcal{T}}$ in the sense that the inf–sup condition of Proposition 4.5 (p. 154) holds with a fixed constant $\beta_* > 0$ independently of $\mathcal{T}$. Moreover, we suppose that the discretisations are *nested* in the sense that $X_{\mathcal{T}} \subset X_{\mathcal{T}'}$ and $Y_{\mathcal{T}} \subset Y_{\mathcal{T}'}$ whenever $\mathcal{T}'$ is a refinement of $\mathcal{T}$. Finally, we require that φ_k, the output of *SOLVE*, is the exact solution of the discrete problem (4.6) (p. 153). Consequently, we always have Galerkin orthogonality. Given any $\mathcal{T} \in \mathcal{P}$, we denote by $R_{\mathcal{T}}$ the *residual* corresponding to $\varphi_{\mathcal{T}}$, i.e.

$$\langle R_{\mathcal{T}}\,,\,\psi\rangle = \langle \ell\,,\,\psi\rangle - B(\varphi_{\mathcal{T}}, \psi) \tag{4.72}$$

for all $\psi \in Y$. For simplicity we write R_k instead of $R_{\mathcal{T}_k}$.

4.13.2.3 *ESTIMATE*

We assume that *ESTIMATE* always yields a global upper bound for the error and a local lower bound up to data oscillation. More precisely, for all partitions $\mathcal{T} \subset \mathcal{P}$, the output $(\eta_{\mathcal{T},K})_{K\in\mathcal{T}}$ of *ESTIMATE*$(\varphi_{\mathcal{T}}, \mathcal{T})$ satisfies

$$\|\varphi - \varphi_{\mathcal{T}}\|_X \le c^* \left\{ \sum_{K\in\mathcal{T}} \eta_{\mathcal{T},K}^2 \right\}^{\frac{1}{2}}$$

and there is a fixed integer $n \ge 1$ such that

$$\eta_{\mathcal{T},K} \le c_* \left\{ \sup_{\psi \in Y_{\mathcal{T}'}(\widetilde{\omega}_{\mathcal{T},K})\setminus\{0\}} \frac{\langle R_{\mathcal{T}}\,,\,\psi\rangle}{\|\psi\|_Y} + \operatorname{osc}_{\mathcal{T}}(K) \right\}$$

holds for all $K \in \mathcal{T}$ and all $\mathcal{T}' \in \mathcal{P}$ with $\mathbb{F}(\mathcal{T}') \supset \mathbb{F}^n(\mathcal{T}, \mathcal{U}_{\mathcal{T},K})$. Here, $\widetilde{\omega}_{\mathcal{T},K}$ is the union of all elements in $\mathcal{U}_{\mathcal{T},K}$ and $Y_{\mathcal{T}'}(\widetilde{\omega}_{\mathcal{T},K})$ denotes the subspace of all functions in $Y_{\mathcal{T}'}$ having their support contained in this set. We assume that the oscillation term $\operatorname{osc}_{\mathcal{T}}(K)$ satisfies an estimate of the form

$$\operatorname{osc}_{\mathcal{T}}(K) \le m(\mu_d(K)) \left\{ \|\varphi_{\mathcal{T}}\|_{X(\widetilde{\omega}_{\mathcal{T},K})} + \|\delta\|_{Z(\widetilde{\omega}_{\mathcal{T},K})} \right\}.$$

Here, $m : \mathbb{R}_+ \to \mathbb{R}_+$ is a continuous non-decreasing function, δ solely depends on the data of the variational problem, and Z is another Hilbert space with the same properties as the spaces X and Y.

Notice that, due to the local continuity of B, the above lower bound for the local dual norm of the residual implies the local lower bound

$$\eta_{\mathcal{T},K} \le c_* \left\{ C_B \|\varphi - \varphi_{\mathcal{T}}\|_{X(\widetilde{\omega}_{\mathcal{T},K})} + \operatorname{osc}_{\mathcal{T}}(K) \right\} \tag{4.73}$$

for the error.

4.13.2.4 *MARK*

We assume that, for every $\mathcal{T} \in \mathcal{P}$, the output $\mathcal{M}$ of *MARK*$((\eta_{\mathcal{T},K})_{K\in\mathcal{T}}, \mathcal{T})$ has the property

$$\eta_{\mathcal{T},K} \le g\left(\left\{ \sum_{K'\in\mathcal{M}} \eta_{\mathcal{T},K'}^2 \right\}^{\frac{1}{2}} \right)$$

for all $K \in \mathcal{T} \setminus \mathcal{M}$ with a fixed continuous function $g : \mathbb{R}_+ \to \mathbb{R}_+$ satisfying $g(0) = 0$.

4.13.2.5 *REFINE*

We suppose that, for every $\mathcal{T} \in \mathcal{P}$, the output $\mathcal{T}'$ of *REFINE*$(\mathcal{T}, \mathcal{M})$ satisfies the minimal requirement

$$\mathbb{F}(\mathcal{T}') \supset \mathbb{F}^1(\mathcal{T}, \mathcal{M}).$$

That is, every marked element of the input partition is subdivided at least once in the output partition. Additional elements in $\mathcal{T} \setminus \mathcal{M}$ may be subdivided in order to ensure the local quasi-uniformity or admissibility of the refined partition.

4.13.3 *Discussion of the Assumptions*

The assumptions of Subsection 4.13.1 are satisfied for all variational problems considered in this chapter.

The assumption of a quasi-regular subdivision stated in Subsection 4.13.2.1 is satisfied with $q_* = \frac{1}{4}$, $q^* = \frac{1}{2}$ for the regular-irregular refinement of Sections 2.1.2 (p. 65) and 2.1.3 (p. 66) and with $q_* = q^* = \frac{1}{2}$ for the marked edge bisection of Section 2.1.4 (p. 68). The assumption of local quasi-uniformity of Subsection 4.13.2.1 is equivalent to the condition (3.3) (p. 82) of shape regularity.

The assumptions of Subsection 4.13.2.2 concerning the component *SOLVE* are satisfied for all *conforming* discretisations of this chapter as long as they do not contain any stabilisation terms. This excludes the SUPG stabilisation of Section 4.4.2 (p. 166) and the stabilised Stokes elements of Section 4.10.2 (p. 240). Since in both cases the stabilisation terms can be controlled by the error indicators, this gap can be bridged by a suitably adapted, slightly more complicated reformulation of the assumptions of Subsection 4.13.2.2.

The error indicators of this chapter all satisfy the assumptions of Subsection 4.13.2.3 concerning the component *REFINE*. The upper bound immediately follows from the results of this chapter by either basing the error indicators on the exact data or by combining the given indicators and the oscillation terms. The lower bound follows from the observation that the local cut-off functions ψ_K and ψ_E can be replaced by suitable nodal shape functions associated with a sufficiently refined partition. This replacement is responsible for the condition $\mathbb{F}(\mathcal{T}') \supset \mathbb{F}^n(\mathcal{T}, \mathcal{U}_{\mathcal{T},K})$. Recall that the convergence result of Section 1.14 (p. 58) is also based on this modification. Since $\mu_d(K) \leq h_K^{\frac{1}{d}}$ holds for all elements K, the assumption concerning the data oscillation is always satisfied with $m(x) = x^{\frac{1}{d}}$ and $Z = L^2(\Omega)$ or $Z = H^1(\Omega)$. The latter form of Z and the term $\|\varphi\|_X$ on the right-hand side of the upper bound for the data oscillation are needed when approximating coefficients of the differential operator. In [**200**] P. Morin, K. G. Siebert, and A. Veeser prove that the assumptions of Subsection 4.13.2.3 are satisfied for all error indicators of Chapter 1 applied to the model problem (1.1) (p. 4) with Dirichlet boundary conditions.

The maximum strategy for marking, Algorithm 2.1 (p. 64), and the equilibration strategy, Algorithm 2.2 (p. 65), satisfy the assumptions of Subsection 4.13.2.4 with $g(x) = \theta x$ and $g(x) = x\sqrt{\frac{1}{\theta} - 1}$, respectively.

The conditions of Subsection 4.13.2.5 finally only require that every marked element is refined at least once. This is satisfied by the refinement techniques of Section 2.1 (p. 64).

4.13.4 *A Convergence Proof*

In this subsection we prove the following general convergence result. The proof will be discussed in Subsection 4.13.5 below.

Theorem 4.84 *Suppose that the conditions of Proposition 4.1 (p. 152) are satisfied and denote by φ the unique solution of the variational problem (4.2) (p. 152). Moreover assume that the assumptions of Subsections 4.13.2.1–4.13.2.5 are fulfilled. Then the approximate solutions and error indicators produced by Algorithm 4.83 converge, i.e.*

$$\|\varphi - \varphi_k\|_X \to 0 \quad \textit{and} \quad \eta_k \to 0 \quad \textit{as} \quad k \to \infty.$$

Algorithm 4.83 produces a sequence $(\mathcal{T}_k)_{k\in\mathbb{N}}$ of partitions. This sequence is accompanied with the sequence $(\mathbb{F}_k)_{k\in\mathbb{N}}$ of the associated finite forests, the sequence $(\varphi_k)_{k\in\mathbb{N}}$ of the associated discrete solutions, and the sequence $(h_k)_{k\in\mathbb{N}} \subset L^\infty(\Omega)$ of associated mesh-functions which are defined by

$$h_k(x) = \mu_d(\omega_{k,x})^{\frac{1}{d}} \quad \text{with} \quad \omega_{k,x} = \bigcup \{K \in \mathcal{T}_k : x \in K\}.$$

We will first prove that these three sequences are convergent and will exploit some properties of their limits. Based on these results we will then establish the convergence of the error indicators. In a final step we will use the upper bound of Subsection 4.13.2.3 to prove that the limit of the discrete solutions actually is the solution of the variational problem.

For simplicity we replace subscripts $\mathcal{T}_k$ by k in what follows and write $a \lesssim b$ if $a \leq Cb$ with a constant C which only depends on the constants β_*, C_B, q_*, q^*, c_1, c_2, c_*, and c^*, and the data δ introduced above.

4.13.4.1 Convergence of the forests

We may consider the forest

$$\mathbb{F}_\infty = \bigcup_{k\in\mathbb{N}} \mathbb{F}_k$$

as the limit of the sequence $(\mathbb{F}_k)_{k\in\mathbb{N}}$. This subforest of the forest $\mathbb{F}$ contains all element subdivisions occurring in the course of Algorithm 4.83. Due to the adaptive marking, $\mathbb{F}_\infty$ in general is a proper subforest of $\mathbb{F}$ and may have leaf nodes. The set $\mathcal{T}^+$ of all leaf nodes can be expressed in terms of the sequence $(\mathcal{T}_k)_{k\in\mathbb{N}}$ of partitions by

$$\mathcal{T}^+ = \bigcup_{k\in\mathbb{N}} \bigcap_{\ell\geq k} \mathcal{T}_\ell.$$

The following lemma shows that neighbours of leaf nodes eventually are leaf nodes too.

Lemma 4.85 *Let $K \in \mathbb{F}_\infty$ be a leaf node, i.e. $K \in \mathcal{T}^+$. Then there exists an iteration number $m_K \in \mathbb{N}$ such that for all $m \geq m_K$ the neighbourhoods of K in $\mathcal{T}_m$ coincide and consist of leaf nodes, i.e.*

$$\mathcal{U}_{m,K} = \mathcal{U}_{m_K,K} \subset \mathcal{T}^+$$

for all $m \geq m_K$.

Proof Consider an arbitrary leaf node $K \in \mathcal{T}^+$ and denote by k the smallest index such that $K \in \mathcal{T}_k$.

For every index $m \geq k$ and every neighbour K' of K in $\mathcal{U}_{m,K}$ there is an element $K_r \in \mathcal{U}_{k,K}$ such that K_r is the root of a subtree $\mathbb{T}_r \subset \mathbb{F}_\infty$ with $K' \in \mathbb{T}_r$. Due to the quasi-regularity and local quasi-uniformity of the subdivision we have

$$\begin{aligned}\mu_d(K) &\leq c_2\, \mu_d(K') \\ &\leq c_2\, (q^*)^{\operatorname{dist}(K',K_r)}\, \mu_d(K_r) \\ &\leq c_2^2\, (q^*)^{\operatorname{dist}(K',K_r)}\, \mu_d(K).\end{aligned}$$

Hence there is a fixed constant M such that $\operatorname{dist}(K', K_r) \leq M$. The set $\bigcup_{m\geq k} \mathcal{U}_{m,K}$ of all possible neighbours of K is thus contained in $\mathbb{F}^M(K, \mathcal{U}_{k,K})$. Therefore, if we choose the index m_K large enough such that the finite forest $\mathbb{F}^M(K, \mathcal{U}_{k,K}) \cap \mathbb{F}_\infty$ is contained in $\mathbb{F}_{m_K}$, all possible neighbours of K are contained in $\mathbb{F}_{m_K}$. Moreover, every neighbour K' of K in $\mathcal{U}_{m_K,K}$ must be a leaf node. To check this, observe that, due to the quasi-regularity of the subdivision, every subdivision of such a neighbour K' would create at least one new neighbour which could not be contained in $\mathbb{F}_{m_K}$. □

4.13.4.2 Convergence of the discrete solutions

Next, we prove that the discrete solutions φ_k converge to some limit function φ_∞. Later we will show that this limit function really is the solution of the variational problem. To this end we denote by

$$X_\infty = \overline{\bigcup_{k\in\mathbb{N}} X_k} \quad \text{and} \quad Y_\infty = \overline{\bigcup_{k\in\mathbb{N}} Y_k}$$

the limit spaces of the discrete spaces $(X_k)_{k\in\mathbb{N}}$ and $(Y_k)_{k\in\mathbb{N}}$. Here, the completion is taken with respect to the norm of X and Y, respectively. Since the spaces X_k and Y_k are nested, X_∞ and Y_∞ are Hilbert spaces.

In combination with Proposition 4.1 (p. 152) the following lemma implies the well-posedness of the variational problem corresponding to B, ℓ, X_∞, and Y_∞.

Lemma 4.86 *Assume that the conditions of Subsections 4.13.1 and 4.13.2.2 are satisfied. Then there is a positive constant β_∞ such that*

$$\inf_{\sigma\in X_\infty\setminus\{0\}} \sup_{\tau\in Y_\infty\setminus\{0\}} \frac{B(\sigma,\tau)}{\|\sigma\|_X\, \|\tau\|_Y} \geq \beta_\infty$$

$$\inf_{\tau\in Y_\infty\setminus\{0\}} \sup_{\sigma\in X_\infty\setminus\{0\}} \frac{B(\sigma,\tau)}{\|\sigma\|_X\, \|\tau\|_Y} \geq \beta_\infty.$$

Proof Consider an arbitrary $\sigma \in X_\infty$ with $\|\sigma\|_X = 1$. By definition $\bigcup_k X_k$ is dense in X_∞. Hence there is an index k and an element $\sigma_k \in X_k$ such that

$$\|\sigma - \sigma_k\|_X \leq \frac{\beta_*}{2(\beta_* + C_B)}$$

where β_* and C_B are the constants from the uniform stability 4.13.2.2 and local continuity 4.13.1 of B. Due to the uniform stability of B there is a $\tau_k \in Y_k$ with

$$\|\tau_k\|_Y = 1 \quad \text{and} \quad B(\sigma_k, \tau_k) \geq \beta_*\, \|\sigma_k\|_X\,.$$

Hence we have

$$
\begin{aligned}
\sup_{\tau \in Y_\infty \setminus \{0\}} \frac{B(\sigma,\tau)}{\|\tau\|_Y} &\geq B(\sigma,\tau_k) \\
&= B(\sigma_k,\tau_k) + B(\sigma - \sigma_k,\tau_k) \\
&\geq \beta_* \|\sigma_k\|_X - C_B \|\sigma - \sigma_k\|_X \\
&\geq \beta_* - (\beta_* + C_B) \|\sigma - \sigma_k\|_X \\
&\geq \frac{1}{2}\beta_*.
\end{aligned}
$$

This proves the first inf–sup condition of the lemma with $\beta_\infty \geq \frac{1}{2}\beta_*$.

Proposition 4.5 (p. 154) implies that, for every k, the linear operator associated via (4.1) (p. 152) with the restriction of B to X_k, Y_k is an isomorphism of X_k onto Y_k^* and that the norm of its inverse is at most β_*^{-1}. Hence, the adjoint of this operator is an isomorphism of Y_k onto X_k^* and the inverse of this adjoint also has norm at most β_*^{-1}. Thus, for every k and every $\tau_k \in Y_k \setminus \{0\}$, there is a $\sigma_k \in X_k$ with

$$
\|\sigma_k\|_X = 1 \quad \text{and} \quad B(\sigma_k,\tau_k) \geq \beta_* \|\tau_k\|_Y .
$$

Reversing the roles of X_∞ and Y_∞, this property and the arguments of the first part of this proof establish the second inf–sup condition of the lemma. □

The next lemma establishes the convergence of the discrete solutions φ_k and characterises the limit.

Lemma 4.87 *Assume that the conditions of Subsections 4.13.1 and 4.13.2.2 are satisfied. Then the sequence $(\varphi_k)_{k\in\mathbb{N}}$ of discrete solutions generated by Algorithm 4.83 converges in the norm of X to the solution $\varphi_\infty \in X_\infty$ of the variational problem*

$$
B(\varphi_\infty,\psi) = \langle \ell\,,\,\psi\rangle \tag{4.74}
$$

for all $\psi \in Y_\infty$.

Proof Lemma 4.86 and Proposition 4.1 (p. 152) imply that the above variational problem admits a unique solution $\varphi_\infty \in X_\infty$. Moreover, for every k, φ_k is the Galerkin approximation of φ_∞ in the subspace X_k of X_∞. Hence, we have for every $k \in \mathbb{N}$ and every $\sigma_k \in X_k$

$$
\|\varphi_\infty - \varphi_k\|_X \leq \|\varphi_\infty - \sigma_k\|_X + \|\sigma_k - \varphi_k\|_X
$$

and

$$
\begin{aligned}
\beta_\infty \|\sigma_k - \varphi_k\|_X &\leq \sup_{\tau_k \in Y_k \setminus \{0\}} \frac{B(\sigma_k - \varphi_k,\tau_k)}{\|\tau_k\|_Y} \\
&= \sup_{\tau_k \in Y_k \setminus \{0\}} \frac{B(\sigma_k - \varphi_\infty,\tau_k)}{\|\tau_k\|_Y} \\
&\leq C_B \|\varphi_\infty - \sigma_k\|_X .
\end{aligned}
$$

For every $k \in \mathbb{N}$ this implies

$$\|\varphi_\infty - \varphi_k\|_X \leq \left(1 + \frac{C_B}{\beta_\infty}\right) \inf_{\sigma_k \in X_k} \|\varphi_\infty - \sigma_k\|_X .$$

Since by construction $\bigcup_k X_k$ is dense in X_∞, the right-hand side of this estimate converges to 0 for $k \to \infty$. □

Thanks to Lemma 4.87, we must still prove that φ_∞ is the solution of the variational problem associated with B, ℓ, X, and Y, i.e. $\varphi_\infty = \varphi$. This would be obvious if we could show that $X_\infty = X$ and Y_∞ is dense in Y. Unfortunately, the spaces X_∞ and Y_∞ depend in a complicated way on the data of the variational problem and the adaptive algorithm 4.83. In general they are proper subspaces of X and Y, respectively. Hence, we still have to do some work to complete the proof of Theorem 4.84.

4.13.4.3 Convergence of the mesh-functions

Recall that in the classical, non-adaptive setting one has $\|h_k\|_\infty \to 0$. In the present adaptive setting this, however, is not true in general and we must prove the L^∞-convergence of the mesh-functions to a suitable limit function. To this end observe that, for every k, the skeleton Σ_k of $\mathcal{T}_k$ has d-dimensional Lebesgue measure 0 and that $\# \{K \in \mathcal{T}_k : x \in K\} = 1$ for all $x \in \Omega \setminus \Sigma_k$. Thus the limiting skeleton

$$\Sigma_\infty = \bigcup_{k \in \mathbb{N}} \Sigma_k$$

has d-dimensional Lebesgue measure 0, too and the sequence $(h_k(x))_{k\in\mathbb{N}}$ is monotonically decreasing and bounded from below by 0 for every $x \in \Omega \setminus \Sigma_\infty$. Hence,

$$h_\infty(x) = \lim_{k\to\infty} h_k(x)$$

is well defined on $\Omega \setminus \Sigma_\infty$ and defines a function in $L^\infty(\Omega)$. The following lemma shows that this point-wise convergence almost everywhere actually holds in $L^\infty(\Omega)$.

Lemma 4.88 *The sequence $(h_k)_{k\in\mathbb{N}}$ of mesh-functions converges to h_∞ uniformly on $\Omega \setminus \Sigma_\infty$, i.e.*

$$\lim_{k\to\infty} \|h_\infty - h_k\|_\infty = 0.$$

Proof Consider an arbitrary $\varepsilon > 0$ and denote by $m = m_\varepsilon$ the smallest integer such that $M^{\frac{1}{d}} (q^*)^{\frac{m}{d}} \leq \varepsilon$, where M is the maximal measure of elements in $\mathcal{T}_0$. Obviously, $\mathbb{F}_\infty \cap \mathbb{F}^m(\mathcal{T}_0, \mathcal{T}_0)$ is a subdivision of Ω with finite depth, i.e. every node in $\mathbb{F}_\infty \cap \mathbb{F}^m(\mathcal{T}_0, \mathcal{T}_0)$ is created by at most m subdivisions of an element in $\mathcal{T}_0$. Hence, there is an index $k = k_\varepsilon$ such that $\mathbb{F}_\infty \cap \mathbb{F}^m(\mathcal{T}_0, \mathcal{T}_0) \subset \mathbb{F}_k$. We claim that

$$\|h_\infty - h_n\|_\infty \leq \|h_\infty - h_k\|_\infty \leq \varepsilon$$

holds for all $n \geq k$. To prove this we first observe that

$$0 \leq h_n - h_\infty \leq h_k - h_\infty$$

holds on $\Omega \setminus \Sigma_\infty$ for all $n \geq k$. Next, consider an arbitrary leaf node $K \in \mathcal{T}_k$ of $\mathbb{F}_k$ and denote by $K_0 \in \mathcal{T}_0$ the ancestor of K, i.e. the element that generates K by subdivisions. To estimate $(h_k - h_\infty)|_K$, we consider two cases.

Case 1: $\operatorname{dist}(K, K_0) < m$. This implies that K is a leaf node of $\mathbb{F}_\infty$ and thus $h_k|_K = h_\infty|_K$.

Case 2: $\operatorname{dist}(K, K_0) \geq m$. Hence, K is obtained by at least m subdivisions of K_0. The monotonicity of the mesh-function, the quasi-regularity of the subdivisions, and the definition of m imply that

$$
\begin{aligned}
0 \leq (h_k - h_\infty)|_K \leq h_k|_K &= \mu_d(K)^{\frac{1}{d}} \\
&\leq (q^*)^{\frac{m}{d}} \mu_d(K_0)^{\frac{1}{d}} \leq (q^*)^{\frac{m}{d}} M^{\frac{1}{d}} \leq \varepsilon.
\end{aligned}
$$

This completes the proof. □

4.13.4.4 A splitting of the error indicators

In order to prove that the sequence $(\eta_k)_{k\in\mathbb{N}}$ of the error indicators converges to zero, we split every η_k into the form

$$
\eta_k = \left\{ \left(\eta_k^0\right)^2 + (\eta_k^+)^2 + (\eta_k^*)^2 \right\}^{\frac{1}{2}}
$$

with

$$
\begin{aligned}
\left(\eta_k^0\right)^2 &= \sum_{K\in\mathcal{T}_k^0} \eta_{k,K}^2, \quad \mathcal{T}_k^0 = \left\{K \in \mathcal{T}_k : \mathbb{F}^n(\mathcal{T}_k, \mathcal{U}_{k,K}) \subset \mathbb{F}_\infty\right\}, \\
(\eta_k^+)^2 &= \sum_{K\in\mathcal{T}_k^+} \eta_{k,K}^2, \quad \mathcal{T}_k^+ = \{K \in \mathcal{T}_k : \mathcal{U}_{k,K} \subset \mathcal{T}^+\}, \\
(\eta_k^*)^2 &= \sum_{K\in\mathcal{T}_k^*} \eta_{k,K}^2, \quad \mathcal{T}_k^* = \mathcal{T}_k \setminus \left(\mathcal{T}_k^0 \cup \mathcal{T}_k^+\right).
\end{aligned}
$$

Here, n is the parameter of Assumption 4.13.2.3 concerning the local lower bound. Thus we can apply the lower bound of Subsection 4.13.2.3 with a successive finite element space to η_k^0. The elements in $\mathcal{T}_k^+$ are no longer marked. Thus, we will have to exploit properties of the marking strategy for the indicators η_k^+. The elements in $\mathcal{T}_k^*$ and the corresponding indicators are between the preceding two extreme cases.

4.13.4.5 Convergence of $(\eta_k^0)_{k\in\mathbb{N}}$

In order to prove the convergence of $(\eta_k^0)_{k\in\mathbb{N}}$, we exploit the properties 4.13.1 of the variational problem, the local quasi-uniformity 4.13.2.1 of the partitions, the assumptions 4.13.2.3 concerning the estimation, and the auxiliary results in Lemmas 4.87 and 4.88. The marking strategy does not play any role.

We first prove an auxiliary result which is a more or less direct consequence of Lemma 4.88.

Lemma 4.89 *The restrictions of the mesh-functions to the sets*

$$\Omega_k^0 = \bigcup \left\{ \widetilde{\omega}_{k,K} : K \in \mathcal{T}_k^0 \right\}$$

of all neighbours of all elements in $\mathcal{T}_k^0$ converge to 0, i.e.

$$\lim_{k\to\infty} \|h_k\|_{\infty;\Omega_k^0} = 0.$$

Proof Consider an arbitrary element K in $\mathcal{T}_k^0$. By definition all its neighbours are subdivided at least n times. Hence, the monotonicity of the sequence of mesh-functions and the quasi-regularity 4.13.2.1 of the subdivision imply in $\widetilde{\omega}_{k,K}$

$$h_\infty \leq (q^*)^n h_k \quad \text{and} \quad h_k \leq \alpha\,(h_k - h_\infty) \quad \text{with} \quad \alpha = \frac{1}{1-(q^*)^n}.$$

Since $K \in \mathcal{T}_k^0$ is arbitrary, this yields

$$\|h_k\|_{\infty;\Omega_k^0} \leq \alpha\, \|h_k - h_\infty\|_{\infty;\Omega_k^0} \leq \alpha\, \|h_k - h_\infty\|_\infty \to 0$$

for $k \to \infty$ due to Lemma 4.88. □

Proposition 4.90 *The sequence $(\eta_k^0)_{k\in\mathbb{N}}$ converges to 0.*

Proof Consider an arbitrary element $K \in \mathcal{T}_k^0$ and choose an index $m > k$ such that $\mathbb{F}_m \supset \mathbb{F}^n(\mathcal{T}_k, \mathcal{U}_{k,K})$. The definition of $\mathcal{T}_k^0$ then implies that the error indicator $\eta_{k,K}$ satisfies the local lower bound 4.13.2.3 with $\mathcal{T}' = \mathcal{T}_m$. Hence, there is $\psi \in Y_m$ with $\operatorname{supp}\psi \subset \widetilde{\omega}_{k,K}$ and

$$\eta_{k,K} \leq c_* \left\{ \langle R_k\,,\, \psi\rangle + \operatorname{osc}_{\mathcal{T}}(K) \right\}.$$

The definitions (4.72) of R_k and (4.74) of φ_∞ imply that

$$\langle R_k\,,\, \psi\rangle = B(\varphi_\infty - \varphi_k, \psi).$$

The local continuity of B and Assumption 4.13.2.3 for the oscillation therefore yield

$$\eta_{k,K} \leq c_*\, C_B\, \|\varphi_\infty - \varphi_k\|_{X(\widetilde{\omega}_{k,K})} + c_*\, m(\mu_d(K)) \left\{ \|\varphi_k\|_{X(\widetilde{\omega}_{k,K})} + \|\delta\|_{Z(\widetilde{\omega}_{k,K})} \right\}.$$

When squaring this estimate and taking the sum with respect to all elements in $\mathcal{T}_k^0$, we obtain an upper bound for η_k^0. In doing so, the contribution to the right-hand side of every element in Ω_k^0 is counted several times. Due to the local quasi-uniformity of the partitions, however, this multiplicity is at most $c_1^2 + 1$. Hence, we obtain

$$\eta_k^0 \lesssim \|\varphi_\infty - \varphi_k\|_X + \left\| m\left(h_k^d\right)\right\|_{\infty;\Omega_k^0} \left\{ \|\varphi_k\|_X + \|\delta\|_Z \right\}.$$

Lemma 4.87 implies that the first term on the right-hand side converges to 0. To prove the same for the second term, we invoke the monotonicity and continuity of the function m and Lemma 4.88. This gives

$$\left\| m\left(h_k^d\right)\right\|_{\infty;\Omega_k^0} \le m\left(\|h_k\|^d_{\infty;\Omega_k^0}\right) \to 0$$

for $k \to \infty$ and thus completes the proof. □

4.13.4.6 Convergence of $(\eta_k^*)_{k\in\mathbb{N}}$

Next, we prove the convergence of the sequence $(\eta_k^*)_{k\in\mathbb{N}}$ associated with the partitions $\mathcal{T}_k^*$. As in the previous subsection, the proof exploits the properties 4.13.1 of the variational problem, the local quasi-uniformity 4.13.2.1 of the partitions, the assumptions 4.13.2.3 concerning the estimation, and the auxiliary results in Lemmas 4.87 and 4.88. Contrary to the previous subsection, however, we now compare the discrete solutions φ_k with the solution φ of the variational problem and not with the solution φ_∞ of problem 4.74. The marking strategy again does not play any role.

We again start with an auxiliary result.

Lemma 4.91 *The d-dimensional Lebesgue measure of the set*

$$\Omega_k^* = \bigcup\left\{\widetilde{\omega}_{k,K} : K \in \mathcal{T}_k^*\right\}$$

of all neighbours of all elements in $\mathcal{T}_k^*$ converges to 0, i.e.

$$\lim_{k\to\infty} \mu_d\left(\Omega_k^*\right) = 0.$$

Proof Consider an arbitrary element $K \in \mathcal{T}_k^*$. We have to distinguish two cases with respect to the location of the neighbours $\mathcal{U}_{k,K}$ of K within the forest $\mathbb{F}_\infty$.

Case 1: $\mathcal{U}_{k,K} \cap \mathcal{T}^+ = \emptyset$. Then all neighbours of K are interior nodes of $\mathbb{F}_\infty$. Since $K \notin \mathcal{T}_k^0$ and thus $\mathbb{F}^n(\mathcal{T}_k, \mathcal{U}_{k,K}) \setminus \mathbb{F}_\infty \neq \emptyset$, there is an element $K' \in \mathcal{U}_{k,K}$ and a leaf node $S_K \subset K'$ of $\mathbb{F}_\infty$ with $0 < \operatorname{dist}(S_K, K') \le n$. Due to the quasi-regularity and local quasi-uniformity of the subdivisions we have

$$\mu_d(\widetilde{\omega}_{k,K}) \le c_1 c_2 \mu_d(K') \le c_1 c_2 q_*^{-n} \mu_d(S_K).$$

Hence, S_K has the following three properties

$$S_K \in \mathcal{T}^+ \setminus \mathcal{T}_k, \quad S_K \subset \widetilde{\omega}_{k,K}, \quad \mu_d(\widetilde{\omega}_{k,K}) \lesssim \mu_d(S_K). \tag{4.75}$$

Case 2: $\mathcal{U}_{k,K} \cap \mathcal{T}^+ \neq \emptyset$. Now at least one element in $\mathcal{U}_{k,K}$ is a leaf node of $\mathbb{F}_\infty$. Since $K \in \mathcal{T}_k^*$, not all elements in $\mathcal{U}_{k,K}$ are leaf nodes of $\mathbb{F}_\infty$. Hence, we can choose two elements K' and K'' in the patch $\widetilde{\omega}_{k,K}$ such that

$$K' \in \mathcal{T}^+, \quad K'' \notin \mathcal{T}^+, \quad K' \cap K'' \neq \emptyset.$$

According to Lemma 4.85 there exists a successor S_K of K'' such that $S_K \in \mathcal{T}^+$ and $S_K \cap K' \neq \emptyset$. The quasi-regularity and local quasi-uniformity of the subdivisions again imply that

$$\mu_d(\widetilde{\omega}_{k,K}) \lesssim \mu_d(S_K).$$

Thus, also in this case, S_K verifies (4.75).

Using (4.75) in both cases and the fact that the second part of the local quasi-uniformity of the partitions entails for every $\widetilde{K} \in \mathcal{T}^+ \setminus \mathcal{T}_k$

$$\#\left\{K \in \mathcal{T}_k^* : S_K = \widetilde{K}\right\} \lesssim 1,$$

we obtain

$$\mu_d(\Omega_k^*) \leq \sum_{K \in \mathcal{T}_k^*} \mu_d(\widetilde{\omega}_{k,K}) \lesssim \sum_{K \in \mathcal{T}_k^*} \mu_d(S_K) \lesssim \sum_{\widetilde{K} \in \mathcal{T}^+ \setminus \mathcal{T}_k} \mu_d(\widetilde{K}).$$

The last term in this estimate is the tail of the series $\sum_{\widetilde{K} \in \mathcal{T}^+} \mu_d(\widetilde{K})$. The latter is convergent since it has positive terms and all partial sums are bounded by $\mu_d(\Omega)$. This proves the lemma. □

Proposition 4.92 *The sequence* $(\eta_k^*)_{k \in \mathbb{N}}$ *converges to* 0.

Proof The local lower bound (4.73) for the error and the estimate $\mu_d(K) \leq \mu_d(\Omega)$ imply for all $K \in \mathcal{T}_k^*$ the estimate

$$\begin{aligned}
\eta_{k,K} &\lesssim \|\varphi - \varphi_k\|_{X(\widetilde{\omega}_{k,K})} + m(\mu_d(K)) \left\{\|\varphi_k\|_{X(\widetilde{\omega}_{k,K})} + \|\delta\|_{Z(\widetilde{\omega}_{k,K})}\right\} \\
&\lesssim \|\varphi_\infty - \varphi_k\|_{X(\widetilde{\omega}_{k,K})} + \|\varphi\|_{X(\widetilde{\omega}_{k,K})} + \|\varphi_\infty\|_{X(\widetilde{\omega}_{k,K})} \\
&\quad + m(\mu_d(\Omega)) \left\{\|\varphi_\infty\|_{X(\widetilde{\omega}_{k,K})} + \|\delta\|_{Z(\widetilde{\omega}_{k,K})}\right\}.
\end{aligned}$$

Squaring this estimate, taking the sum with respect to all elements in $\mathcal{T}_k^*$ and taking into account multiple counting of elements as in the proof of Proposition 4.90, we obtain

$$\eta_k^* \lesssim \left\{\|\varphi_\infty - \varphi_k\|_X + \|\varphi\|_{X(\Omega_k^*)} + \|\varphi_\infty\|_{X(\Omega_k^*)} + \|\delta\|_{Z(\Omega_k^*)}\right\}.$$

The first term on the right-hand side of this estimate converges to 0 due to Lemma 4.87. Lemma 4.91 implies that the remaining three terms converge to 0, too. □

4.13.4.7 Convergence of $(\eta_k^+)_{k \in \mathbb{N}}$

We now turn to the convergence of the sequence $(\eta_k^+)_{k \in \mathbb{N}}$. Here, in addition to the already exploited properties, Assumption 4.13.2.4 for the marking strategy is crucial.

Proposition 4.93 *The sequence* $(\eta_k^+)_{k \in \mathbb{N}}$ *converges to* 0.

Proof The proof proceeds in five steps.

In the *first step* we prove the element-wise convergence to 0 of $\eta_{k,K}$ for all elements in $\mathcal{T}^+$. To prove this, observe that elements in $\mathcal{T}^+$ are not subdivided and that Assumption 4.13.2.4 requires that marked elements are subdivided. Hence, we have $\mathcal{M}_k \subset \mathcal{T}_k \setminus \mathcal{T}_k^+ = \mathcal{T}_k^0 \cup \mathcal{T}_k^*$. Propositions 4.90 and 4.92 therefore imply that

$$
\begin{aligned}
0 \le \liminf_{k\to\infty} \left\{ \sum_{K\in\mathcal{M}_k} \eta_{k,K}^2 \right\}^{\frac{1}{2}} &\le \limsup_{k\to\infty} \left\{ \sum_{K\in\mathcal{M}_k} \eta_{k,K}^2 \right\}^{\frac{1}{2}} \\
&\le \limsup_{k\to\infty} \left\{ \left(\eta_k^0\right)^2 + (\eta_k^*)^2 \right\}^{\frac{1}{2}} \\
&= \lim_{k\to\infty} \left\{ \left(\eta_k^0\right)^2 + (\eta_k^*)^2 \right\}^{\frac{1}{2}} = 0.
\end{aligned}
$$

The assumption 4.13.2.4 on the marking strategy then yields the postulated element-wise convergence.

The remaining four steps of the proof aim at deducing the convergence of $(\eta_k^+)_{k\in\mathbb{N}}$ from this element-wise convergence by rewriting η_k^+ in an integral form and using a suitable variant of the dominated convergence theorem.

In the *second step* we prepare the integral reformulation of η_k^+. To this end we observe the following properties of any given element $K \in \mathcal{T}_k^+$. First, in view of the definition of $\mathcal{T}_k^+$, we have $\widetilde{\omega}_{k,K} = \widetilde{\omega}_{m,K}$ for every $m \ge k$. Next, exploiting the local lower bound 4.13.2.3 for the true error and the bound 4.13.2.3 for the oscillation term as in the proof of Proposition 4.92, we obtain

$$
\begin{aligned}
\eta_{k,K}^2 &\lesssim \left\{ \|\varphi_k - \varphi_\infty\|_{X(\widetilde{\omega}_{k,K})}^2 + \|\varphi_\infty\|_{X(\widetilde{\omega}_{k,K})}^2 + \|\varphi\|_{X(\widetilde{\omega}_{k,K})}^2 + \|\delta\|_{Z(\widetilde{\omega}_{k,K})}^2 \right\}^{\frac{1}{2}} \\
&= \|\varphi_k - \varphi_\infty\|_{X(\widetilde{\omega}_{k,K})}^2 + C_K^2,
\end{aligned}
$$

where the quantity C_K is independent of k. Note that, as a consequence of Lemma 4.87, the right-hand side tends to C_K^2 for $k \to \infty$. Moreover, arguing once again as in the proof of Proposition 4.90, we obtain for the sum of the C_K^2 the bound

$$
\begin{aligned}
\sum_{K\in\mathcal{T}_k^+} C_K^2 &= \sum_{K\in\mathcal{T}_k^+} \left(\|\varphi_\infty\|_{X(\widetilde{\omega}_{k,K})}^2 + \|\varphi\|_{X(\widetilde{\omega}_{k,K})}^2 + \|\delta\|_{Z(\widetilde{\omega}_{k,K})}^2 \right) \\
&\lesssim \left(\|\varphi_\infty\|_X^2 + \|\varphi\|_X^2 + \|\delta\|_Z^2 \right) \lesssim 1.
\end{aligned}
$$

In the *third step* we now give the integral reformulation of η_k^+. To this end set

$$
\Omega^+ = \bigcup_{K\in\mathcal{T}^+} K.
$$

Lemma 4.85 implies that $\mathcal{T}^+ = \bigcup_{k\in\mathbb{N}} \mathcal{T}_k^+$ and that the sets $\mathcal{T}_k^+$ are nested. Given $x \in \Omega^+$, denote by $m = m_x$ the smallest iteration number such that there is a $K \in \mathcal{T}_m^+$ with $x \in K$. Set

$$
\varepsilon_k(x) = M_k(x) = 0
$$

for $k < m$ and

$$
\begin{aligned}
\varepsilon_k(x) &= \frac{1}{\mu_d(K)} \eta_{k,K}^2, \\
M_k(x) &= \frac{1}{\mu_d(K)} \left(\|\varphi_k - \varphi_\infty\|_{X(\widetilde{\omega}_{k,K})}^2 + C_K^2 \right)
\end{aligned}
$$

for $k \geq m$. For every k we then have

$$(\eta_k^+)^2 = \int_{\Omega^+} \varepsilon_k$$

and the first step implies the point-wise convergence of ε_k in Ω^+ to 0. Moreover, we have

$$\sum_{K \in \mathcal{T}_k^+} \|\varphi_k - \varphi_\infty\|_{X(\widetilde{\omega}_{k,K})}^2 \lesssim \|\varphi_k - \varphi_\infty\|_X^2$$

and the second step ensures that each M_k is a summable majorant for ε_k.

In the *fourth step*, we want to prove that the majorants $(M_k)_{k\in\mathbb{N}}$ converge in $L^1(\Omega^+)$ to the function M which, for all $x \in K$ and all $K \in \mathcal{T}^+$, is defined by

$$M = \frac{1}{\mu_d(K)} C_K^2 .$$

The monotone convergence theorem [**285**, Appendix (19b), p. 1016] and the definition of M_k imply

$$\|M - M_k\|_{1;\Omega^+} = \sum_{K \in \mathcal{T}_k^+} \|M - M_k\|_{1;K} + \sum_{K \in \mathcal{T}^+ \setminus \mathcal{T}_k^+} \|M\|_{1;K} .$$

Due to Lemma 4.87, the first term satisfies

$$\sum_{K \in \mathcal{T}_k^+} \|M - M_k\|_{1;K} \lesssim \sum_{K \in \mathcal{T}_k^+} \|\varphi_k - \varphi_\infty\|_{X(\widetilde{\omega}_{k,K})}^2 \lesssim \|\varphi_k - \varphi_\infty\|_X^2 \to 0$$

for $k \to \infty$. The second term is the tail of the series

$$\sum_{K \in \mathcal{T}^+} \|M\|_{1;K} = \sum_{K \in \mathcal{T}^+} C_K^2$$

which is finite due to the second step. This implies the postulated convergence of $(M_k)_{K\in\mathbb{N}}$.

The properties of $(\varepsilon_k)_{K\in\mathbb{N}}$, $(M_k)_{K\in\mathbb{N}}$, and M derived in the third and fourth step allow us to apply in the final *fifth step* the generalised majorised convergence theorem [**285**, Appendix (19a), p. 1015]. This yields

$$\lim_{k\to\infty} (\eta_k^+)^2 = \lim_{k\to\infty} \int_{\Omega^+} \varepsilon_k = 0$$

and thus completes the proof. □

4.13.4.8 Final step of the proof of Theorem 4.84

Propositions 4.90, 4.92, and 4.93 imply that

$$\lim_{k\to\infty} \eta_k = \lim_{k\to\infty} \left\{ (\eta_k^0)^2 + (\eta_k^+)^2 + (\eta_k^*)^2 \right\}^{\frac{1}{2}} = 0.$$

Thus, the upper bound of Subsection 4.13.2.3 for the error implies that the discrete solutions φ_k converge to the solution φ of the variational problem.

4.13.5 *Discussion of the Convergence Proof*

The upper bound 4.13.2.3 for the error is only needed in the final step of the convergence proof to deduce the convergence of the discrete solutions from the convergence of the error indicators. The latter solely follows from the local lower bound 4.13.2.3 and the properties of the marking strategy, of the oscillation term, and of the variational problem and its discretisation. The size of the constants c_* and c^* in the lower and upper bounds is irrelevant as long as both are independent of the particular partition. In particular the robustness of the error indicator is never required. This is due to the fact that Theorem 4.84 only gives a qualitative result and not a quantitative one; it gives no information whatsoever on the convergence speed.

The above proof aims at controlling the error $\|\varphi_k - \varphi_{k+n}\|_X$ for a fixed, sufficiently large n contrary to older results which try to control the error $\|\varphi_k - \varphi_{k+1}\|_X$ corresponding to two successive partitions. This is a basic reason for the generality and flexibility of Theorem 4.84. On the other hand, it is responsible for the lack of information on the convergence speed.

Lemma 4.89 has the flavour of an a priori result in that it proves the convergence to 0 of the mesh-function at least on a suitable subset. This result, however, is only needed to control the data oscillation. This gives rise to the question of whether the oscillation could be avoided at all. The answer is clearly negative. Even a completely exact error indicator will always give rise to some data oscillation since by definition it only has access to the data of a finite-dimensional problem which can never completely reflect all information of an infinite-dimensional problem.

Assumption 4.13.2.4 on the marking strategy is crucial for the convergence proof. In a certain sense it is also necessary for the convergence of the adaptive process. More precisely, P. Morin, K. G. Siebert, and A. Veeser prove in [**200**] that the property

$$\lim_{k\to\infty} \left\{ \sum_{K'\in\mathcal{M}} \eta^2_{\mathcal{T},K'} \right\}^{\frac{1}{2}} = 0 \quad \Longrightarrow \quad \lim_{k\to\infty} \eta_{k,K} = 0 \text{ for all } K \in \mathcal{T}^+$$

is a necessary and sufficient condition for the convergence of the adaptive process. Using this result, they can extend the above convergence proof to other marking strategies which either also depend on the result of previous refinement steps or which terminate the marking process once a given tolerance is attained.

4.13.6 *Optimal Complexity*

One of the main objectives of the adaptive process is to attain a discrete solution with a prescribed tolerance using a minimal amount of work. Theorem 4.84 shows that the adaptive process will reach any prescribed positive tolerance within a finite number of iterations. Since it is only a qualitative result and gives no information on the convergence speed, however, it is dubious whether the required work is minimal.

To clarify this point we must first properly define the notions "minimal work" and "optimal complexity". This requires the notions of best n-term approximation and of approximation classes.

Given any initial partition $\mathcal{T}_0$ and the solution $\varphi \in X$ of the variational problem (4.2) (p. 152), the error of its *best n-term approximation* is defined by

$$\sigma_n(\varphi) = \inf_{\mathcal{T} \in \mathbb{F}^n(\mathcal{T}_0, \mathcal{T}_0)} \inf_{\psi_\mathcal{T} \in X_\mathcal{T}} \left(\|\varphi - \psi_\mathcal{T}\|_X + \left\{ \sum_{K \in \mathcal{T}} \mathrm{osc}_\mathcal{T}(K)^2 \right\}^{\frac{1}{2}} \right).$$

Notice that the first infimum is taken with respect to all partitions which can be obtained from the initial partition $\mathcal{T}_0$ by at most n steps of local or global refinement.

Given any positive real number s, the function φ belongs to the *approximation class* $\mathcal{A}^s$ if and only if

$$\limsup_{n \to \infty} n^s \sigma_n(\varphi) < \infty.$$

The adaptive algorithm then has *optimal complexity* if and only if it produces a sequence $(\varphi_k)_{k \in \mathbb{N}}$ of discrete solutions with

$$\|\varphi - \varphi_k\|_X \le C\,(\#\mathcal{T}_k - \#\mathcal{T}_0)^{-s}$$

for all $\varphi \in \mathcal{A}^s$.

Notice that every function in $H^{1+r}(\Omega)$ belongs to $\mathcal{A}^{\frac{r}{d}}$ [**102**, Remark 6.2] and every function in $W^{2,1}(\Omega)$ belongs to $\mathcal{A}^{\frac{1}{2}}$ [**102**, Remark 6.3]. Further notice that the solution of a linear elliptic equation of second order with sufficiently smooth coefficients and right-hand side $f \in L^2(\Omega)$ on a bounded domain with Lipschitz boundary is contained in $W^{2,1}(\Omega)$ if $d = 2$ [**147**, Theorem 5.2.2] and in $H^{1+r}(\Omega)$ with $r > 0$ depending on Ω if $d \ge 3$ [**156**, Theorem 3].

The definition of optimal complexity clearly shows that any result which aims at establishing the optimal complexity of the adaptive algorithm must provide information on the convergence speed of the algorithm and must simultaneously control the size of the created partitions. Theorem 4.84 evidently does not fulfil this requirement. This is the price we have to pay for its generality and flexibility. J. M. Cascon, C. Kreuzer, R. H. Nochetto, and K. G. Siebert [**102**] prove a convergence result of the required kind for variational problems with symmetric, coercive bi-linear form B. This excludes the mixed methods of Sections 4.8 (p. 208), 4.9 (p. 223), and 4.10 (p. 237). Currently it is an open question if this restriction is a technical or a structural one.

Remark 4.94 The first analysis of an adaptive algorithm for one-dimensional symmetric coercive Sturm–Liouville problems is probably due to I. Babuška and M. Vogelius [**34**]. The first multi-dimensional convergence result is due to W. Dörfler [**126**]. This seminal work gave rise to a series of articles generalising its results in various directions [**81, 82, 97, 98, 103, 192, 193, 198, 199, 200, 239, 246**]. The complexity analysis of the adaptive algorithm is strongly influenced by the seminal article [**70**] of P. Binev, W. Dahmen, and R. DeVore. Optimal complexity results for the adaptive algorithm are derived in [**102, 121, 162, 192, 193, 195, 196, 197, 198, 241**].

5

Nonlinear Elliptic Equations

In this chapter we extend the results of Chapter 4 to nonlinear stationary problems. We again start with an abstract result which establishes the basic equivalence of error and residual and which provides a framework for a posteriori error estimation using the results of Chapter 3, in particular Section 3.8. The main tool now is the implicit function theorem along with modifications for bifurcation and turning points. We apply the abstract results to quasilinear elliptic equations and the stationary incompressible Navier–Stokes equations. The nonlinear approach also enables us to improve the results of Section 4.7 for linear elliptic eigenvalue problems.

5.1 Abstract Nonlinear Problems

In this section we extend the abstract results of Section 4.1 (p. 151) to nonlinear problems and thus lay the ground for the a posteriori error estimates of this chapter.

5.1.1 *The Variational Problem and its Discretisation*

As in the previous chapter, we denote by X and Y two Banach spaces with norms $\|\cdot\|_X$ and $\|\cdot\|_Y$, respectively. Given a continuous mapping $F : X \to Y^*$, we now consider problems of the form:

find $\varphi \in X$ such that

$$F(\varphi) = 0, \tag{5.1}$$

or equivalently:

find $\varphi \in X$ such that

$$\langle F(\varphi), \psi \rangle_Y = 0. \tag{5.2}$$

holds for all $\psi \in Y$.

The abstract linear problem (4.3) (p. 152) fits into this framework with

$$F(\varphi) = L\varphi - \ell$$

or equivalently

$$\langle F(\varphi), \psi \rangle_Y = B(\varphi, \psi) - \langle \ell, \psi \rangle_Y .$$

To further elucidate the relationship of problems (5.1) and (5.2) on one hand and problems (4.2) (p. 152) and (4.3) on the other hand, note that the mapping

$$\varphi, \psi \mapsto \langle F(\varphi), \psi \rangle_Y$$

is linear in its second argument and that on the other hand the mapping

$$\varphi, \psi \mapsto B(\varphi, \psi) - \langle \ell, \psi \rangle_Y$$

is linear in its second argument and affine in its first argument.

For the discretisation of problem (5.1) we choose, similarly to Section 4.1.3 (p. 153), finite-dimensional subspaces $X_{\mathcal{T}}$ of X and $Y_{\mathcal{T}}$ of Y and a continuous mapping $F_{\mathcal{T}} : X_{\mathcal{T}} \to Y_{\mathcal{T}}^*$. The discrete analogue of problem (5.1) is then:

find $\varphi_{\mathcal{T}} \in X_{\mathcal{T}}$ such that

$$F_{\mathcal{T}}(\varphi_{\mathcal{T}}) = 0, \tag{5.3}$$

or equivalently:

find $\varphi_{\mathcal{T}} \in X_{\mathcal{T}}$ such that

$$\langle F_{\mathcal{T}}(\varphi_{\mathcal{T}}), \psi_{\mathcal{T}} \rangle_{Y_{\mathcal{T}}} = 0. \tag{5.4}$$

holds for all $\psi_{\mathcal{T}} \in Y_{\mathcal{T}}$.

Again, the abstract discrete linear problem (4.6) (p. 153) corresponds to problems (5.3) and (5.4) with

$$F_{\mathcal{T}}(\varphi_{\mathcal{T}}) = L_{\mathcal{T}}\varphi_{\mathcal{T}} - \ell_{\mathcal{T}}$$

or equivalently

$$\langle F_{\mathcal{T}}(\varphi_{\mathcal{T}}), \psi_{\mathcal{T}} \rangle_{Y_{\mathcal{T}}} = B_{\mathcal{T}}(\varphi_{\mathcal{T}}, \psi_{\mathcal{T}}) - \langle \ell_{\mathcal{T}}, \psi_{\mathcal{T}} \rangle_{Y_{\mathcal{T}}} .$$

5.1.2 *Regular Solutions*

Given an element φ_0 of X and a positive number R, we denote by

$$B_X(\varphi_0, R) = \{\varphi \in X : \|\varphi - \varphi_0\|_X < R\}$$

the open ball with centre φ_0 and radius R.

A solution φ_0 of problem (5.1) is called *regular* if and only if F is differentiable in φ_0 and $DF(\varphi_0)$ is an isomorphism of X onto Y^*.

For elements close to a regular solution the following result yields error estimates in terms of the residual.

Proposition 5.1 *Assume that φ_0 is a regular solution of problem (5.1) and that DF is locally Lipschitz continuous at φ_0, i.e. there is a number $R_0 > 0$ such that*

$$\gamma = \sup_{\varphi \in B_X(\varphi_0, R_0)} \frac{\|DF(\varphi) - DF(\varphi_0)\|_{\mathcal{L}(X,Y^*)}}{\|\varphi - \varphi_0\|_X} < \infty.$$

Set

$$R = \min\left\{R_0, \gamma^{-1}\left\|DF(\varphi_0)^{-1}\right\|^{-1}_{\mathcal{L}(Y^*,X)}, 2\gamma^{-1}\left\|DF(\varphi_0)\right\|_{\mathcal{L}(X,Y^*)}\right\}.$$

Then, the following error estimates hold for all $\varphi \in B_X(\varphi_0, R)$:

$$\begin{aligned}\frac{1}{2}\left\|DF(\varphi_0)\right\|^{-1}_{\mathcal{L}(X,Y^*)}\left\|F(\varphi)\right\|_{Y^*} &\leq \|\varphi - \varphi_0\|_X \\ &\leq 2\left\|DF(\varphi_0)^{-1}\right\|_{\mathcal{L}(Y^*,X)}\left\|F(\varphi)\right\|_{Y^*}.\end{aligned} \tag{5.5}$$

Proof For every $\varphi \in B_X(\varphi_0, R)$, we have

$$\begin{aligned}\varphi - \varphi_0 &= DF(\varphi_0)^{-1}\left\{DF(\varphi_0)(\varphi - \varphi_0) + F(\varphi) - F(\varphi) + F(\varphi_0)\right\} \\ &= DF(\varphi_0)^{-1}\left\{F(\varphi) + \int_0^1 \left[DF(\varphi_0) - DF(\varphi_0 + t(\varphi - \varphi_0))\right](\varphi - \varphi_0)\mathrm{d}t\right\}\end{aligned}$$

and thus

$$\begin{aligned}\|\varphi - \varphi_0\|_X &\leq \left\|DF(\varphi_0)^{-1}\right\|_{\mathcal{L}(Y^*,X)}\left\{\left\|F(\varphi)\right\|_{Y^*}\right. \\ &\qquad \left. + \int_0^1 \left\|DF(\varphi_0) - DF(\varphi_0 + t(\varphi - \varphi_0))\right\|_{\mathcal{L}(X,Y^*)}\|\varphi - \varphi_0\|_X\,\mathrm{d}t\right\} \\ &\leq \left\|DF(\varphi_0)^{-1}\right\|_{\mathcal{L}(Y^*,X)}\left\{\left\|F(\varphi)\right\|_{Y^*} + \frac{1}{2}\gamma\|\varphi - \varphi_0\|_X^2\right\} \\ &\leq \left\|DF(\varphi_0)^{-1}\right\|_{\mathcal{L}(Y^*,X)}\left\|F(\varphi)\right\|_{Y^*} + \frac{1}{2}\|\varphi - \varphi_0\|_X.\end{aligned}$$

This yields the second inequality in (5.5).

On the other hand, we have

$$F(\varphi) = DF(\varphi_0)(\varphi - \varphi_0) + \int_0^1 \left[DF(\varphi_0 + t(\varphi - \varphi_0)) - DF(\varphi_0)\right](\varphi - \varphi_0)\mathrm{d}t$$

and thus

$$\begin{aligned}\left\|F(\varphi)\right\|_{Y^*} &\le \left\|DF(\varphi_0)\right\|_{\mathcal{L}(X,Y^*)} \|\varphi - \varphi_0\|_X \\ &\quad + \int_0^1 \left\|DF(\varphi_0 + t(\varphi - \varphi_0)) - DF(\varphi_0)\right\|_{\mathcal{L}(X,Y^*)} \|\varphi - \varphi_0\|_X \, dt \\ &\le \left\|DF(\varphi_0)\right\|_{\mathcal{L}(X,Y^*)} \|\varphi - \varphi_0\|_X + \frac{1}{2}\gamma \|\varphi - \varphi_0\|_X^2 \\ &\le 2\left\|DF(\varphi_0)\right\|_{\mathcal{L}(X,Y^*)} \|\varphi - \varphi_0\|_X .\end{aligned}$$

This proves the first inequality in (5.5). □

Remark 5.2 Assume that Y is reflexive. Since the mapping

$$\varphi, \psi \mapsto \langle DF(\varphi_0)\varphi \, , \, \psi\rangle_Y$$

is bi-linear, Proposition 4.1 (p. 152) and Remark 4.2 (p. 153) yield

$$\left\|DF(\varphi_0)\right\|_{\mathcal{L}(X,Y^*)} = \sup_{\varphi \in X\setminus\{0\}} \sup_{\psi \in Y\setminus\{0\}} \frac{\langle DF(\varphi_0)\varphi \, , \, \psi\rangle_Y}{\|\varphi\|_X \|\psi\|_Y}$$

and

$$\left\|DF(\varphi_0)^{-1}\right\|_{\mathcal{L}(Y^*,X)}^{-1} = \inf_{\varphi \in X\setminus\{0\}} \sup_{\psi \in Y\setminus\{0\}} \frac{\langle DF(\varphi_0)\varphi \, , \, \psi\rangle_Y}{\|\varphi\|_X \|\psi\|_Y}.$$

Remark 5.3 The conditions on F can be weakened. Assume, e.g. that F is continuous, that φ_0 solves problem (5.1), and that there are a number $R > 0$ and two monotonically increasing homeomorphisms ϱ, σ of $[0, \infty)$ onto itself such that

$$\varrho(\|\varphi - \varphi_0\|_X) \le \left\|F(\varphi)\right\|_{Y^*} \le \sigma(\|\varphi - \varphi_0\|_X) \tag{5.6}$$

holds for all $\varphi \in B_X(\varphi_0, R)$. Then we have for all $\varphi \in B_X(\varphi_0, R)$

$$\sigma^{-1}(\left\|F(\varphi)\right\|_{Y^*}) \le \|\varphi - \varphi_0\|_X \le \varrho^{-1}(\left\|F(\varphi)\right\|_{Y^*}).$$

The first inequality in (5.6) is satisfied if, e.g. F is strongly monotone in a neighbourhood of φ_0. The second inequality in (5.6) holds if, e.g. F is Hölder-continuous at φ_0.

Remark 5.4 When applying Proposition 5.1 to elliptic partial differential equations, one obtains bounds for the error measured in a $W^{1,p}$-norm. For problems in continuum mechanics, e.g. this corresponds to estimating the strain energy. Sometimes, however, one is more interested in an error control with respect to an L^p-norm. For problems in fluid mechanics, e.g. this corresponds to estimating the kinetic energy. In order to cover this case by the present abstract framework, we must enlarge the space X and restrict the space Y. We therefore consider three additional Banach spaces X_-, X_+, and Y_+ such that $X_+ \subset X \subset X_-$ and $Y_+ \subset Y$ with continuous and dense injections. Here, the +/− sign indicates a space with a stronger/weaker norm. We assume that X_- is reflexive. In addition to the assumptions of Proposition 5.1, we now suppose that φ_0 is in

X_+, that $DF(\varphi_0)^*$, the adjoint of $DF(\varphi_0)$, is an isomorphism of Y_+ onto X_-^*, and that there are two numbers $R_0 > 0$ and $\beta > 0$ such that

$$\left\| [DF(\varphi_0) - DF(\varphi_0 + t\varphi)]\varphi \right\|_{Y_+^*} \leq \beta t \|\varphi\|_{X_+} \|\varphi\|_{X_-}$$

for all $t \in [0, 1]$ and all $\varphi \in B_{X_+}(0, R_0)$. Set

$$R = \min \left\{ R_0, \beta^{-1} \left\| DF(\varphi_0)^{*-1} \right\|_{\mathcal{L}(X_-^*, Y_+)}^{-1}, 2\beta^{-1} \left\| DF(\varphi_0)^* \right\|_{\mathcal{L}(Y_+, X_-^*)} \right\}.$$

Then, a straightforward modification of the proof of Proposition 5.1 [259, Proposition 2.1] yields, for all $\varphi \in B_{X_+}(\varphi_0, R)$, the estimates

$$\begin{aligned} \frac{1}{2} \left\| DF(\varphi_0)^* \right\|_{\mathcal{L}(Y_+, X_-^*)}^{-1} \left\| F(\varphi_0) \right\|_{Y^*} &\leq \|\varphi - \varphi_0\|_{X_-} \\ &\leq 2 \left\| DF(\varphi_0)^{*-1} \right\|_{\mathcal{L}(X_-^*, Y_+)} \left\| F(\varphi) \right\|_{Y_+^*}. \end{aligned}$$

The conditions of Proposition 5.1 imply that $DF(\varphi_0)^*$ is an isomorphism of Y^{**} onto X^*. The condition that $DF(\varphi_0)^*$ is an isomorphism of Y_+ onto X_-^* is more restrictive. For partial differential equations, it is equivalent to an additional regularity assumption.

For linear problems when DF is constant, one may extend F by continuity to a continuously differentiable map of X_- to Y_+^*. Then, the space X_+ is not needed. For nonlinear problems, however, this extension is often impossible or the derivative of the extension is no longer Lipschitz-continuous. This is the place where the space X_+ comes into play.

5.1.3 *Branches of Solutions*

In this subsection we briefly outline how to extend the above results to branches of solutions of equation (5.1) including simple limit and bifurcation points. To this end, we assume that $X = \mathbb{R}^m \times V$, $m \geq 1$, and that $\varphi_0 = (\lambda_0, v_0)$ is a solution of equation (5.1).

We first consider the case that φ_0 is a *regular point*, i.e. $D_V F(\varphi_0)$, the derivative of F with respect to V, is an isomorphism of V onto Y^*. The implicit function theorem then implies that there are neighbourhoods I of λ_0 in $\mathbb{R}^m$ and U of v_0 in V and a continuous map $\lambda \mapsto v_\lambda$ from I into U such that $v_{\lambda_0} = v_0$ and every $\varphi_\lambda = (\lambda, v_\lambda)$ is a solution of equation (5.1) with $D_V F(\varphi_\lambda)$ being an isomorphism of V onto Y^*. Assume that $D_V F$ is locally Lipschitz-continuous, i.e. there is an $R_0^* > 0$ such that

$$\gamma^* = \sup_{\lambda \in I} \sup_{v \in B_V(v_\lambda, R_0^*)} \frac{\left\| D_V F(\lambda, v) - D_V F(\lambda, v_\lambda) \right\|_{\mathcal{L}(V, Y^*)}}{\|v - v_\lambda\|_V} < \infty$$

and set

$$R^* = \min \left\{ R_0^*, \gamma^{*^{-1}} \sup_{\lambda \in I} \left\| D_V F(\varphi_\lambda)^{-1} \right\|_{\mathcal{L}(Y^*, V)}^{-1}, \ 2\gamma^{*^{-1}} \sup_{\lambda \in I} \left\| D_V F(\varphi_\lambda) \right\|_{\mathcal{L}(V, Y^*)} \right\}.$$

With the same arguments as in the proof of Proposition 5.1, we then obtain for all $\lambda \in I$ and all $v \in B_V(v_\lambda, R^*)$ the estimates

$$\begin{aligned}\frac{1}{2}\left\|D_V F(\varphi_\lambda)\right\|^{-1}_{\mathcal{L}(V,Y^*)}\left\|F(\lambda,v)\right\|_{Y^*} &\leq \|v - v_\lambda\|_V \\ &\leq 2\left\|D_V F(u_\lambda)^{-1}\right\|_{\mathcal{L}(Y^*,V)}\left\|F(\lambda,v)\right\|_{Y^*}.\end{aligned} \tag{5.7}$$

As described in [**114**], the case that φ_0 is not a regular point, but a simple limit or bifurcation point, may be reduced to the case of a regular point by suitably blowing-up the spaces X and Y and modifying the function F. For completeness, we briefly describe this procedure.

Consider first the case that φ_0 is a *simple limit point*, i.e. $DF(\varphi_0)$ is a Fredholm operator of X onto Y^* with index m and

$$\mathrm{range}(DF(\varphi_0)) = Y^*$$

but $D_V F(\varphi_0)$ is not an isomorphism of V onto Y^*.

Choose a linear operator $L \in \mathcal{L}(X, \mathbb{R}^m)$ with

$$\ker(L) \cap \ker(DF(\varphi_0)) = \{0\}$$

and define $\Phi \in C^1(\mathbb{R}^m \times X, \mathbb{R}^m \times Y^*)$ by

$$\Phi(t, \varphi) = (L(\varphi - \varphi_0) - t, F(\varphi)).$$

Then, $(0, \varphi_0)$ is a regular point of Φ with respect to the parameter t and we are back to the situation described in the first part of this subsection. Since L is linear, conditions concerning the Lipschitz continuity of $D\Phi$ reduce to those on DF. In this case estimate (5.7) takes the form

$$\begin{aligned}\underline{c}\left\{\left\|L(\varphi - \varphi_0) - t\right\|_{\mathbb{R}^m} + \left\|F(\varphi)\right\|_{Y^*}\right\} &\leq \|\lambda - \lambda_t\|_{\mathbb{R}^m} + \|v - v_t\|_V \\ &\leq \bar{c}\left\{\left\|L(\varphi - \varphi_0) - t\right\|_{\mathbb{R}^m} + \left\|F(\varphi)\right\|_{Y^*}\right\}\end{aligned}$$

for all t in a suitable neighbourhood of 0 and all $\varphi = (\lambda, v)$ in a suitable neighbourhood of $\varphi_t = (\lambda_t, v_t)$. Here, $t \mapsto \varphi_t$ is a regular branch of solutions of $\Phi(t, \varphi) = 0$. Note that $L\varphi_0$ is often known explicitly and that the estimation of $\left\|L(\varphi - \varphi_0) - t\right\|_{\mathbb{R}^m}$ is straightforward since it is a low-dimensional maximisation problem. The term $\left\|F(\varphi)\right\|_{Y^*}$, on the other hand, may be estimated by the methods of the next subsection as in the case of regular solutions.

Next, we consider the case of a *simple bifurcation from the trivial branch*, i.e. we assume that $\varphi_0 = (\lambda_0, 0)$ and that $D_V F(\varphi_0)$ is a Fredholm operator with index 0 and

$$\dim \ker(D_V F(\varphi_0)) = 1.$$

Choose a $w_0 \in \ker(D_V F(\varphi_0)) \setminus \{0\}$ and a linear functional $\ell \in V^*$ with

$$\langle \ell, w_0 \rangle_V = 1.$$

Define the function $\Phi \in C(\mathbb{R} \times X, \mathbb{R} \times Y^*)$ by

$$\Phi(t, \varphi) = \begin{cases} (\langle \ell, v\rangle_V - 1, \frac{1}{t}F(\lambda, tv)) & \text{if } t \neq 0, \varphi = (\lambda, v) \in X, \\ (\langle \ell, v\rangle_V - 1, D_V F(\lambda, 0)v) & \text{if } t = 0, \varphi = (\lambda, v) \in X. \end{cases}$$

Conditions concerning the Lipschitz continuity of $D\Phi$ now reduce to those on D^2F. Obviously, we have $\Phi(0,\widetilde{\varphi}_0) = 0$ where $\widetilde{\varphi}_0 = (\lambda_0, w_0)$.

If F is of class C^2 in a neighbourhood of φ_0 and

$$D^2_{\lambda v}F(\varphi_0)w_0 \notin \operatorname{range} D_V F(\varphi_0),$$

we conclude that $\widetilde{\varphi}_0$ is a regular point and we are once more back to the situation described in the first part of this subsection. Estimate (5.7) now takes the form

$$\begin{aligned}\underline{c}\left\{|\langle \ell, w\rangle_V - 1| + \left\|D_V F(\lambda, 0)w\right\|_{Y^*}\right\} &\leq \|\lambda - \lambda_0\|_{\mathbb{R}^m} + \|w - w_0\|_V \\ &\leq \bar{c}\left\{|\langle \ell, w\rangle_V - 1| + \left\|D_V F(\lambda, 0)w\right\|_{Y^*}\right\}\end{aligned}$$

for all (λ, w) in a suitable neighbourhood of $\widetilde{\varphi}_0$ and

$$\begin{aligned}\underline{c}\left\{|\langle \ell, w\rangle_V - 1| + \left\|\frac{1}{t}F(\lambda, tw)\right\|_{Y^*}\right\} &\leq \|\lambda - \lambda_t\|_{\mathbb{R}^m} + \|w - w_t\|_V \\ &\leq \bar{c}\left\{|\langle \ell, w\rangle_V - 1| + \left\|\frac{1}{t}F(\lambda, tw)\right\|_{Y^*}\right\}\end{aligned}$$

for all $t \neq 0$ in a neighbourhood of 0 and all (λ, w) in a suitable neighbourhood of $\widetilde{\varphi}_t = (\lambda_t, w_t)$. Here, $t \mapsto \widetilde{\varphi}_t$ is a regular branch of solutions of $\Phi(t,\varphi) = 0$. Note that the constants in these estimates depend on the second derivatives of F.

Finally, we consider the case of a *simple bifurcation point*, i.e. $D_V F(\varphi_0)$ is a Fredholm operator of index 0 and

$$q = \dim(\ker DF(\varphi_0)) - m \geq 1.$$

Choose a basis $\varphi_1^*, \ldots, \varphi_q^*$ of $Y^* \setminus \operatorname{range}(DF(\varphi_0))$, set

$$\widehat{X} = \mathbb{R}^q \times X, \quad \widehat{\varphi}_0 = (0, \varphi_0),$$

and define the function $\widehat{F} \in C^1(\widehat{X}, Y^*)$ by

$$\widehat{F}(\widehat{\varphi}) = F(\varphi) - \sum_{i=1}^{q} f_i \varphi_i^*$$

for all $\widehat{\varphi} = (f, \varphi) \in \widehat{X}$.

Obviously, we have

$$\widehat{F}(\widehat{\varphi}_0) = 0.$$

Moreover, $D\widehat{F}(\widehat{\varphi}_0)$ is a Fredholm operator with index $m + q$ and

$$\operatorname{range}(D\widehat{F}(\widehat{\varphi}_0)) = Y^*.$$

Replacing X, φ_0, and F by $\widehat{X}$, $\widehat{\varphi}_0$, and $\widehat{F}$, respectively, we are thus back to the situation considered in the second part of this subsection.

5.1.4 A Posteriori Error Estimation

Proposition 5.1 implies that, for all φ sufficiently close to a solution φ_0 of problem (5.1), the error $\|\varphi - \varphi_0\|_X$ is equivalent to the residual $\|F(\varphi)\|_{Y^*}$. The residual, however, can be estimated as in the linear case: a suitable restriction operator $Q_{\mathcal{T}} : Y \to Y_{\mathcal{T}}$ yields the upper bound

$$\|F(\varphi)\|_{Y^*} \leq \|(Id_Y - Q_{\mathcal{T}})^* F(\varphi)\|_{Y^*} + \|Q_{\mathcal{T}}^* F(\varphi)\|_{Y^*}$$

and a suitable finite-dimensional subspace $\widetilde{Y}_{\mathcal{T}}$ of Y yields the lower bound

$$\|F(\varphi)\|_{\widetilde{Y}_{\mathcal{T}}^*} \leq \|F(\varphi)\|_{Y^*}$$

Combining this observation with Proposition 5.1, we therefore immediately obtain the following nonlinear counterpart of Theorem 4.7 (p. 156).

Theorem 5.5 *Assume that:*

- φ_0 *is a solution of problem* (5.1)*;*
- φ_0 *and* F *satisfy the conditions of Proposition 5.1;*
- $\varphi_{\mathcal{T}0}$ *is a solution of the discrete problem* (5.3)*;*
- $\varphi_{\mathcal{T}0}$ *is in* $B_X(\varphi_0, R)$ *where* R *is as in Proposition 5.1;*
- $Q_{\mathcal{T}}$ *is a continuous restriction operator of* Y *in* $Y_{\mathcal{T}}$*;*
- $\widetilde{Y}_{\mathcal{T}}$ *is a finite-dimensional subspace of* Y *with* $Y_{\mathcal{T}} \subset \widetilde{Y}_{\mathcal{T}} \subset Y$*;*
- $\eta_{\mathcal{T}}$ *is an error indicator which only depends on the discrete solution* $\varphi_{\mathcal{T}0}$ *and the given data of the variational problem* (5.1)*;*
- $\theta_{\mathcal{T}}$ *is a data error which only depends on the data of the variational problem;*
- *the estimates*

$$\begin{aligned} \|(Id_Y - Q_{\mathcal{T}})^* F(\varphi_{\mathcal{T}0})\|_{Y^*} &\leq c_A(\eta_{\mathcal{T}} + \theta_{\mathcal{T}}) \\ \|Q_{\mathcal{T}}^* F(\varphi_{\mathcal{T}0})\|_{Y^*} &\leq c_C(\eta_{\mathcal{T}} + \theta_{\mathcal{T}}) \\ \eta_{\mathcal{T}} &\leq c_I(\|F(\varphi_{\mathcal{T}0})\|_{\widetilde{Y}_{\mathcal{T}}^*} + \theta_{\mathcal{T}}) \end{aligned} \tag{5.8}$$

are fulfilled.

Then, the error $\varphi_0 - \varphi_{\mathcal{T}0}$ *can be estimated from above by*

$$\|\varphi_0 - \varphi_{\mathcal{T}0}\|_X \leq 2\,\|DF(\varphi_0)^{-1}\|_{\mathcal{L}(Y^*,X)}\,(c_A + c_C)(\eta_{\mathcal{T}} + \theta_{\mathcal{T}})$$

and from below by

$$\eta_{\mathcal{T}} \leq c_I\left(2\,\|DF(\varphi_0)\|_{\mathcal{L}(X,Y^*)}\,\|\varphi_0 - \varphi_{\mathcal{T}0}\|_X + \theta_{\mathcal{T}}\right).$$

Remark 5.6 The quantity

$$\|DF(\varphi_0)\|_{\mathcal{L}(X,Y^*)}\,\|DF(\varphi_0)^{-1}\|_{\mathcal{L}(Y^*,X)}\,(c_A + c_C)c_I$$

is a measure for the quality of the error indicator $\eta_{\mathcal{T}}$. It is the nonlinear counterpart of the quantity

$$\|L\|_{\mathcal{L}(X,Y^*)} \left\|L^{-1}\right\|_{\mathcal{L}(Y^*,X)} (c_A + c_C)c_I$$

in Remark 4.8 (p. 156). It should be uniformly bounded with respect to families of discretisations of the same variational problem and with respect to parameters such as, e.g. the size of reaction or diffusion terms or mesh-Péclet-numbers, which are inherent in the differential equation.

For linear problems, we have devised in several places error indicators that are based on the solution of discrete auxiliary problems. When extending this device to nonlinear problems, it is crucial that the auxiliary problems can be chosen to be suitable linearisations of the original nonlinear problem. The following result gives an abstract framework for this approach.

Theorem 5.7 *In addition to the assumptions of Theorem 5.5 assume that:*

- $\widehat{X}_{\mathcal{T}}$ *is a finite-dimensional subspace of* X;
- $\widehat{Y}_{\mathcal{T}}$ *is a finite-dimensional subspace of* Y *with* $\widetilde{Y}_{\mathcal{T}} \subset \widehat{Y}_{\mathcal{T}}$;
- *there is a constant* $\widehat{c}$ *such that*

$$\left\|F(\varphi_{\mathcal{T}0})\right\|_{\widehat{Y}^*_{\mathcal{T}}} \leq \widehat{c}\left\|F(\varphi_{\mathcal{T}0})\right\|_{\widetilde{Y}^*_{\mathcal{T}}};$$

- $\widehat{L}$ *is an isomorphism of* $\widehat{X}_{\mathcal{T}}$ *onto* $\widehat{Y}^*_{\mathcal{T}}$.

Denote by $\widehat{\varphi}_{\mathcal{T}} \in \widehat{X}_{\mathcal{T}}$ *the unique solution of*

$$\widehat{L}\widehat{\varphi}_{\mathcal{T}} = F(\varphi_{\mathcal{T}0})$$

and set

$$\widehat{\eta}_{\mathcal{T}} = \|\widehat{\varphi}_{\mathcal{T}}\|_{\widehat{X}^*_{\mathcal{T}}}.$$

Then the indicators $\eta_{\mathcal{T}}$ *and* $\widehat{\eta}_{\mathcal{T}}$ *are related by*

$$\widehat{\eta}_{\mathcal{T}} \leq \left\|\widehat{L}^{-1}\right\|_{\mathcal{L}(\widehat{Y}^*_{\mathcal{T}},\widehat{X}_{\mathcal{T}})} (c_A + c_C)(\eta_{\mathcal{T}} + \theta_{\mathcal{T}}),$$
$$\eta_{\mathcal{T}} \leq c_I \left(\left\|\widehat{L}\right\|_{\mathcal{L}(\widehat{X}_{\mathcal{T}},\widehat{Y}^*_{\mathcal{T}})} \widehat{\eta}_{\mathcal{T}} + \theta_{\mathcal{T}}\right).$$

Moreover, the error is bounded from above by

$$\|\varphi_0 - \varphi_{\mathcal{T}0}\|_X \leq 2\left\|DF(\varphi_0)^{-1}\right\|_{\mathcal{L}(Y^*,X)} (c_A + c_C)\cdot\left\{c_I\left\|\widehat{L}\right\|_{\mathcal{L}(\widehat{X}_{\mathcal{T}},\widehat{Y}^*_{\mathcal{T}})} \widehat{\eta}_{\mathcal{T}} + (1 + c_I)\theta_{\mathcal{T}}\right\}$$

and from below by

$$\widehat{\eta}_{\mathcal{T}} \leq \left\|\widehat{L}^{-1}\right\|_{\mathcal{L}(\widehat{Y}^*_{\mathcal{T}},\widehat{X}_{\mathcal{T}})} (c_A + c_C)\cdot\left\{2\left\|DF(\varphi_0)\right\|_{\mathcal{L}(X,Y^*)} c_I\|\varphi_0 - \varphi_{\mathcal{T}0}\|_X + (1 + c_I)\theta_{\mathcal{T}}\right\}.$$

Proof Since $\widetilde{Y}_{\mathcal{T}} \subset \widehat{Y}_{\mathcal{T}} \subset Y$, we have

$$\left\|F(\varphi_{\mathcal{T}0})\right\|_{\widetilde{Y}^*_{\mathcal{T}}} \leq \left\|F(\varphi_{\mathcal{T}0})\right\|_{\widehat{Y}^*_{\mathcal{T}}} \leq \left\|F(\varphi_{\mathcal{T}0})\right\|_{Y^*}.$$

Assumption (5.8) of Theorem 5.5 on the other hand implies

$$\begin{aligned}\left\|F(\varphi_{\mathcal{T}0})\right\|_{Y^*} &\le \left\|(Id_Y - Q_{\mathcal{T}})^* F(\varphi_{\mathcal{T}0})\right\|_{Y^*} + \left\|Q_{\mathcal{T}}^* F(\varphi_{\mathcal{T}0})\right\|_{Y^*} \\ &\le (c_A + c_C)(\eta_{\mathcal{T}} + \theta_{\mathcal{T}}) \\ &\le c_I(c_A + c_C)\left\|F(\varphi_{\mathcal{T}0})\right\|_{\widetilde{Y}_{\mathcal{T}}^*} + (1 + c_I)(c_A + c_C)\theta_{\mathcal{T}}.\end{aligned}$$

Since $\widehat{L}$ is an isomorphism of $\widehat{X}_{\mathcal{T}}$ onto $\widehat{Y}_{\mathcal{T}}^*$, we have

$$\left\|\widehat{L}\right\|_{\mathcal{L}(\widehat{X}_{\mathcal{T}}, \widehat{Y}_{\mathcal{T}}^*)}^{-1} \left\|F(\varphi_{\mathcal{T}0})\right\|_{\widehat{Y}_{\mathcal{T}}^*} \le \widehat{\eta}_{\mathcal{T}} \le \left\|\widehat{L}^{-1}\right\|_{\mathcal{L}(\widehat{Y}_{\mathcal{T}}^*, \widehat{X}_{\mathcal{T}})} \left\|F(\varphi_{\mathcal{T}0})\right\|_{\widehat{Y}_{\mathcal{T}}^*}.$$

Combining this with Assumption (5.8) of Theorem 5.5 proves the first two estimates of Theorem 5.7.

The error estimates of Theorem 5.7 follow from the previous estimates and Proposition 5.1. □

Remark 5.8 Theorems 5.5 and 5.7 can easily be modified such that they cover the situation described in Remark 5.4 and yield upper and lower bounds on $\left\|F(\varphi_{\mathcal{T}})\right\|_{Y_+^*}$ [**259**, Proposition 2.7]. To this end, one has to proceed as follows:

- replace Y by Y_+;
- equip the spaces $\widetilde{Y}_{\mathcal{T}}$ and $\widehat{Y}_{\mathcal{T}}$ with the norm of Y_+;
- equip the space $\widehat{X}_{\mathcal{T}}$ with the norm of X_-.

The condition $\widetilde{Y}_{\mathcal{T}} \subset \widehat{Y}_{\mathcal{T}} \subset Y_+$ is stronger than the assumption $\widetilde{Y}_{\mathcal{T}} \subset \widehat{Y}_{\mathcal{T}} \subset Y$ needed in Theorems 5.5 and 5.7. When applied to partial differential equations of second order it implies that the functions in $\widetilde{Y}_{\mathcal{T}}$ and $\widehat{Y}_{\mathcal{T}}$ must at least be continuously differentiable across inter-element boundaries. The space $Y_{\mathcal{T}}$, however, must only be contained in Y. Furthermore, Remark 5.4 requires that $X_{\mathcal{T}} \subset X_+$. For practical applications, however, this is not a restriction since usually $X_+ \subset W^{1,p}(\Omega)$ with $p \le \infty$ and $X_{\mathcal{T}} \subset W^{1,\infty}(\Omega)$.

Remark 5.9 The abstract framework of this section is due to [**254**, **259**]. Independently, a similar approach was developed in [**220**].

5.2 Quasilinear Equations of Second Order

In this section we consider boundary value problems of the form

$$\begin{aligned} -\operatorname{div} \mathbf{a}(x, u, \nabla u) &= b(x, u, \nabla u) && \text{in } \Omega \\ u &= 0 && \text{on } \Gamma \end{aligned} \tag{5.9}$$

where $b \in C(\Omega \times \mathbb{R} \times \mathbb{R}^d, \mathbb{R})$ and $\mathbf{a} \in C^1(\Omega \times \mathbb{R} \times \mathbb{R}^d, \mathbb{R}^d)$ are such that the matrix

$$A(x, y, z) = \frac{1}{2}\left(\frac{\partial a_i(x, y, z)}{\partial z_j} + \frac{\partial a_j(x, y, z)}{\partial z_i}\right)_{1 \le i,j \le d}$$

is positive definite for all $x \in \Omega, y \in \mathbb{R}, z \in \mathbb{R}^d$.

5.2.1 Variational Formulation and Discretisation

Under suitable growth conditions on $\mathbf{a}$, b, and their derivatives there are real numbers $1 < r,\ p' < \infty$ such that the weak formulation of problem (5.9) fits into the framework of Section 5.1 (p. 281) with

$$\begin{aligned} X &= W_0^{1,r}(\Omega), \\ Y &= W_0^{1,p'}(\Omega), \\ \langle F(\varphi), \psi\rangle_Y &= \int_\Omega \mathbf{a}(x,\varphi,\nabla\varphi)\cdot\nabla\psi - \int_\Omega b(x,\varphi,\nabla\varphi)\psi. \end{aligned}$$

Note that $DF(\varphi)$ is an isomorphism of X onto $Y^* = W^{-1,p}(\Omega)$ if and only if the linear boundary value problem

$$\begin{aligned} -\operatorname{div}(A(x,\varphi,\nabla\varphi)\nabla v) - \operatorname{div}(\partial_y \mathbf{a}(x,\varphi,\nabla\varphi)v) \quad & \\ -\nabla_z b(x,\varphi,\nabla\varphi)\cdot\nabla v - \partial_y b(x,\varphi,\nabla\varphi)v = f \quad & \text{in } \Omega \\ v = 0 \quad & \text{on } \Gamma \end{aligned}$$

admits for every $f \in Y^*$ a unique weak solution $v \in X$ which depends continuously on f.

We do not specify the discretisation of problem (5.9) in detail. We only assume that $X_\mathcal{T} \subset X \cap W^{1,\infty}(\Omega)$ and $Y_\mathcal{T} \subset Y \cap W^{1,\infty}(\Omega)$ are finite element spaces corresponding to a partition $\mathcal{T}$ which satisfies the assumptions of Section 3.2 (p. 81), that $Y_\mathcal{T}$ contains the space $S_0^{1,0}(\mathcal{T})$, and that $F_\mathcal{T}$ is the restriction of F, i.e.

$$\langle F_\mathcal{T}(\varphi_\mathcal{T}), \psi_\mathcal{T}\rangle_{Y_\mathcal{T}} = \langle F(\varphi_\mathcal{T}), \psi_\mathcal{T}\rangle_Y \tag{5.10}$$

holds for all $\varphi_\mathcal{T} \in X_\mathcal{T}$ and $\psi_\mathcal{T} \in Y_\mathcal{T}$. The last assumption can considerably be weakened but simplifies the subsequent analysis.

5.2.2 Examples

We now give some examples of problems which fit into the present abstract framework.

Example 5.10 The *equations of prescribed mean curvature*

$$\begin{aligned} \mathbf{a}(x,u,\nabla u) &= \left[1 + |\nabla u|^2\right]^{-\frac{1}{2}} \nabla u, \\ b(x,u,\nabla u) &= f(x) \in L^2(\Omega), \\ r &= p = 2. \end{aligned}$$

Example 5.11 The α*-Laplacian*

$$\begin{aligned} \mathbf{a}(x,u,\nabla u) &= |\nabla u|^{\alpha-2}\nabla u, \quad \text{with } \alpha > 1, \\ b(x,u,\nabla u) &= f(x) \in L^p(\Omega), \\ r &= p' = \alpha, \\ p &= \frac{\alpha}{\alpha-1}. \end{aligned}$$

This example fits into the framework of Proposition 5.1 (p. 283) if $\alpha \geq 2$. In the case $1 < \alpha < 2$, the corresponding function F is no longer differentiable. Yet, it still fits into the framework of Remark 5.3 (p. 284) with [**109**, Section 5.3]

$$\varrho(t) = \underline{c}\,\{\|u_0\|_X + t\}^{\alpha-2}\, t, \quad \sigma(t) = \bar{c}t^{\alpha-1}.$$

Example 5.12 The *subsonic flow of an irrotational, ideal, compressible gas*

$$\begin{aligned} \mathbf{a}(x,u,\nabla u) &= \left[1 - \frac{\gamma-1}{2}|\nabla u|^2\right]^{\frac{1}{\gamma-1}} \nabla u, \quad \text{with } \gamma > 1, \\ b(x,u,\nabla u) &= f(x) \in L^p(\Omega), \\ r = p' &= \frac{2\gamma}{\gamma-1}, \\ p &= \frac{2\gamma}{\gamma+1}. \end{aligned}$$

Example 5.13 The *stationary heat equation with convection and nonlinear diffusion coefficient*

$$\begin{aligned} \mathbf{a}(x,u,\nabla u) &= k(u)\nabla u, \\ b(x,u,\nabla u) &= f - \mathbf{c}\cdot\nabla u, \\ f &\in L^\infty(\Omega), \\ \mathbf{c} &\in C(\Omega,\mathbb{R}^d), \\ k &\in C^2(\mathbb{R}), \\ k(s) &\geq \alpha > 0, \\ \left|k^{(\ell)}(s)\right| &\leq \gamma, \quad \text{for all } s \in \mathbb{R}, \ell = 0,1,2, \\ r &= p \in (2,4). \end{aligned}$$

Example 5.14 *Bratu's equation*

$$\begin{aligned} \mathbf{a}(x,u,\nabla u) &= \nabla u, \\ b(x,u,\nabla u) &= \lambda e^u, \quad \text{with } \lambda > 0, \\ r &= p > d. \end{aligned}$$

In this example there is a critical parameter $\lambda^* > 0$ such that problem (5.9) admits two weak solutions if $0 < \lambda < \lambda^*$, exactly one weak solution if $\lambda = \lambda^*$, and no solution if $\lambda > \lambda^*$. The solution corresponding to $\lambda = \lambda^*$ is a turning point and fits into the framework of Section 5.1.3 (p. 285) [**114**].

Example 5.15 A *nonlinear eigenvalue problem*

$$\begin{aligned} \mathbf{a}(x,u,\nabla u) &= \nabla u, \\ b(x,u,\nabla u) &= \lambda u - u^\beta, \quad \text{with } \beta \geq d, \\ r &= p \geq d. \end{aligned}$$

This example always admits the trivial solution. If λ is a simple eigenvalue of the Laplacian, there is a simple bifurcation fitting into the framework Section 5.1.3 (p. 285) [**114**].

5.2.3 Residual A Posteriori Error Estimates

In what follows we denote by $u \in X$ a weak solution of problem (5.9) and by $u_{\mathcal{T}} \in X_{\mathcal{T}}$ a solution of the corresponding discrete problem, where $F_{\mathcal{T}}$ is given by equation (5.10).

Thanks to equation (5.10), we have for all $v_{\mathcal{T}} \in Y_{\mathcal{T}}$

$$\langle F(u_{\mathcal{T}}), v_{\mathcal{T}} \rangle_Y = \langle F_{\mathcal{T}}(u_{\mathcal{T}}), v_{\mathcal{T}} \rangle_{Y_{\mathcal{T}}} = 0.$$

Hence, the residual

$$R = F(u_{\mathcal{T}})$$

satisfies the condition (3.30) (p. 132) of Galerkin orthogonality. Correspondingly, we do not need the restriction operator $Q_{\mathcal{T}}$ and the second inequality of condition (5.8) of Theorem 5.5 (p. 288) is automatically fulfilled with $c_C = 0$.

Integration by parts element-wise shows that R admits the L^p-representation (3.29) (p. 132) with

$$\begin{aligned} r|_K &= -\operatorname{div} \mathbf{a}(\cdot, u_{\mathcal{T}}, \nabla u_{\mathcal{T}}) - b(\cdot, u_{\mathcal{T}}, \nabla u_{\mathcal{T}}), \\ j|_E &= \mathbb{J}_E(\mathbf{n}_E \cdot \mathbf{a}(\cdot, u_{\mathcal{T}}, \nabla u_{\mathcal{T}})) \end{aligned}$$

for all elements $K \in \mathcal{T}$ and all faces $E \in \mathcal{E}$.

As usual, we must approximate the functions r and j by piecewise polynomials. To this end, we introduce piecewise polynomial approximations $\mathbf{a}_{\mathcal{T}}(\cdot, u_{\mathcal{T}}, \nabla u_{\mathcal{T}})$ and $b_{\mathcal{T}}(\cdot, u_{\mathcal{T}}, \nabla u_{\mathcal{T}})$ of $\mathbf{a}(\cdot, u_{\mathcal{T}}, \nabla u_{\mathcal{T}})$ and $b(\cdot, u_{\mathcal{T}}, \nabla u_{\mathcal{T}})$, respectively, and set

$$\begin{aligned} r_{\mathcal{T}}|_K &= -\operatorname{div} \mathbf{a}_{\mathcal{T}}(\cdot, u_{\mathcal{T}}, \nabla u_{\mathcal{T}}) - b_{\mathcal{T}}(\cdot, u_{\mathcal{T}}, \nabla u_{\mathcal{T}}), \\ j_{\mathcal{E}}|_E &= \mathbb{J}_E(\mathbf{n}_E \cdot \mathbf{a}_{\mathcal{T}}(\cdot, u_{\mathcal{T}}, \nabla u_{\mathcal{T}})) \end{aligned}$$

for all elements $K \in \mathcal{T}$ and all faces $E \in \mathcal{E}$.

Theorem 3.57 (p. 135) then proves the first inequality of condition (5.8) of Theorem 5.5 (p. 288) with

$$\eta_{\mathcal{T}} = \left\{ \sum_{K \in \mathcal{T}} h_K^p \|r_{\mathcal{T}}\|_{p;K}^p + \sum_{E \in \mathcal{E}_\Omega} h_E \|j_{\mathcal{E}}\|_{p;E}^p \right\}^{\frac{1}{p}},$$

$$\theta_{\mathcal{T}} = \left\{ \sum_{K \in \mathcal{T}} h_K^p \|r_{\mathcal{T}} - r\|_{p;K}^p + \sum_{E \in \mathcal{E}_\Omega} h_E \|j_{\mathcal{E}} - j\|_{p;E}^p \right\}^{\frac{1}{p}}$$

and a constant c_A which only depends on the shape parameter $C_{\mathcal{T}}$ of $\mathcal{T}$ and the Lebesgue exponent p.

Denote by m and n the polynomial degrees of $\mathbf{a}_{\mathcal{T}}(\cdot, u_{\mathcal{T}}, \nabla u_{\mathcal{T}})$ and $b_{\mathcal{T}}(\cdot, u_{\mathcal{T}}, \nabla u_{\mathcal{T}})$, respectively and set

$$\widetilde{Y}_{\mathcal{T}} = \operatorname{span}\left\{ \psi_K v, \psi_E \sigma : v \in R_{\max\{m,n\}}(K), \sigma \in R_m(E),\ K \in \mathcal{T}, E \in \mathcal{E}_\Omega \right\}.$$

Theorem 3.59 (p. 137) then proves the third inequality of condition (5.8) of Theorem 5.5 (p. 288).

Theorem 5.5 therefore yields the following a posteriori error estimates.

Theorem 5.16 *Denote by u a weak solution of problem (5.9) and by $u_{\mathcal{T}}$ a solution of the discrete problem $F_{\mathcal{T}}(u_{\mathcal{T}}) = 0$. Assume that:*

- *u and F satisfy the conditions of Proposition 5.1 (p. 283);*
- *$u_{\mathcal{T}}$ is in $B_X(u, R)$ with R as in Proposition 5.1;*
- *$F_{\mathcal{T}}$ is given by equation (5.10).*

Denote by $\mathbf{a}_{\mathcal{T}}(\cdot, u_{\mathcal{T}}, \nabla u_{\mathcal{T}})$ and $b_{\mathcal{T}}(\cdot, u_{\mathcal{T}}, \nabla u_{\mathcal{T}})$ piecewise polynomial approximations of $\mathbf{a}(\cdot, u_{\mathcal{T}}, \nabla u_{\mathcal{T}})$ and $b(\cdot, u_{\mathcal{T}}, \nabla u_{\mathcal{T}})$, respectively, and define the residual error indicator $\eta_{R,K}$ *and the data error θ_K by*

$$\eta_{R,K} = \left\{ h_K^p \left\| \operatorname{div} \mathbf{a}_{\mathcal{T}}(\cdot, u_{\mathcal{T}}, \nabla u_{\mathcal{T}}) + b_{\mathcal{T}}(\cdot, u_{\mathcal{T}}, \nabla u_{\mathcal{T}}) \right\|_{p;K}^p \right.$$
$$\left. + \frac{1}{2} \sum_{E \in \mathcal{E}_{K,\Omega}} h_E \left\| \mathbb{J}_E(\mathbf{n}_E \cdot \mathbf{a}_{\mathcal{T}}(\cdot, u_{\mathcal{T}}, \nabla u_{\mathcal{T}})) \right\|_{p;E}^p \right\}^{\frac{1}{p}}$$

and

$$\theta_K = \left\{ h_K^p \left\| (\operatorname{div} \mathbf{a} - \operatorname{div} \mathbf{a}_{\mathcal{T}} + b - b_{\mathcal{T}})(\cdot, u_{\mathcal{T}}, \nabla u_{\mathcal{T}}) \right\|_{p;K}^p \right.$$
$$\left. + \sum_{E \in \mathcal{E}_{K,\Omega}} h_E \left\| \mathbb{J}_E(\mathbf{n}_E \cdot (\mathbf{a} - \mathbf{a}_{\mathcal{T}})(\cdot, u_{\mathcal{T}}, \nabla u_{\mathcal{T}})) \right\|_{p;E}^p \right\}^{\frac{1}{p}}.$$

Then there are two constants c^* and c_*, which only depend on the shape parameter $C_{\mathcal{T}}$ of $\mathcal{T}$, the Lebesgue exponent p, and the polynomial degrees of $u_{\mathcal{T}}$, $\mathbf{a}_{\mathcal{T}}$, and $b_{\mathcal{T}}$, such that the error is bounded from above by

$$\|u - u_{\mathcal{T}}\|_{1,r} \le c^* \left\{ \sum_{K \in \mathcal{T}} \left(\eta_{R,K}^p + \theta_K^p \right) \right\}^{\frac{1}{p}}$$

and, for every element $K \in \mathcal{T}$, from below by

$$\eta_{R,K} \le c_* \|u - u_{\mathcal{T}}\|_{1,r;\omega_K} + \left\{ \sum_{K' \subset \omega_K} \theta_{K'}^p \right\}^{\frac{1}{p}}.$$

Remark 5.17 For Examples 5.10–5.15, the data error θ_K can easily be estimated explicitly when using a piecewise linear approximation, i.e. $X_{\mathcal{T}} = S_0^{1,0}(\mathcal{T})$ and $\mathcal{T}$ exclusively consists of triangles or tetrahedrons. Then $\nabla u_{\mathcal{T}}$ is piecewise constant and we obtain

$$\theta_K = h_K \|f - f_{\mathcal{T}}\|_{p;K}$$

in Examples 5.10, 5.11, and 5.12. Here, as usual, $f_{\mathcal{T}}$ is any piecewise polynomial approximation of f such as its L^2-projection on $S^{0,-1}(\mathcal{T})$. Similarly we obtain

$$\theta_K \le c h_K^{2(1-\frac{d}{p})} \left\| \nabla u_{\mathcal{T}} \right\|_{p;K}$$

for Example 5.13 and

$$\theta_K \le c h_K^2 \left\| \nabla u_{\mathcal{T}} \right\|_{p;K} \exp(\left\| u_{\mathcal{T}} \right\|_{\infty;K})$$

for Example 5.14. Finally, in Example 5.15, we have

$$\theta_K = 0$$

if β is an integer.

Remark 5.18 Theorem 5.16 can easily be extended to the case of mixed Dirichlet and Neumann boundary conditions. One only has to add to $\eta_{R,K}$ the term $h_E \left\| \mathbf{n}_E \cdot \mathbf{a}_{\mathcal{T}}(\cdot, u_{\mathcal{T}}, \nabla u_{\mathcal{T}}) \right\|_{p;E}^p$ and to θ_K the term $h_E \left\| \mathbf{n}_E \cdot (\mathbf{a}(\cdot, u_{\mathcal{T}}, \nabla u_{\mathcal{T}}) - \mathbf{a}_{\mathcal{T}}(\cdot, u_{\mathcal{T}}, \nabla u_{\mathcal{T}})) \right\|_{p;E}^p$ both corresponding to the faces on the Neumann boundary.

The first estimate of Theorem 5.16 also holds if $\eta_{R,K}$ is defined using the original functions $\mathbf{a}$ and b. The data error then of course disappears. The proof of the lower bound, however, necessarily requires an approximation of $\mathbf{a}$ and b.

As mentioned before, Theorem 5.16 can be applied to Example 5.11 only in the case $\alpha \ge 2$. Observing that for $1 < \alpha < 2$ the strong monotonicity of F implies the unique solvability of the corresponding weak problem, we obtain from Remark 5.3 and the above estimates the following result which complements the results of [**48**].

Theorem 5.19 *Assume that* $1 < \alpha < 2$ *and denote by* $u \in W_0^{1,\alpha}(\Omega)$ *the unique solution of*

$$\int_\Omega |\nabla u|^{\alpha-2} \nabla u \nabla v = \int_\Omega f v$$

for all $v \in W_0^{1,\alpha}(\Omega)$ *and by* $u_{\mathcal{T}} \in X_{\mathcal{T}}$ *a solution of a discretisation of the above problem with* $X_{\mathcal{T}} \subset W_0^{1,\infty}(\Omega)$. *Then the following a posteriori error estimates hold:*

$$\left\| u - u_{\mathcal{T}} \right\|_{1,\alpha} \le c_1 \left\{ \sum_{K \in \mathcal{T}} \eta_{R,K}^\alpha \right\}^{\frac{1}{\alpha}} + c_2 \left\{ \sum_{K \in \mathcal{T}} \theta_K^\alpha \right\}^{\frac{1}{\alpha}}$$

and

$$\left\{ \sum_{K \in \mathcal{T}} \eta_{R,K}^\alpha \right\}^{\frac{1}{\alpha}} \le c_5 \left\| u - u_{\mathcal{T}} \right\|_{1,\alpha}^{\frac{1}{\alpha-1}} + c_6 \left\{ \sum_{K \in \mathcal{T}} \theta_K^\alpha \right\}^{\frac{1}{\alpha(\alpha-1)}}.$$

Here, $\eta_{R,K}$ *and* θ_K *are as in Theorem 5.16. Moreover, the data error satisfies*

$$\theta_K = h_K \left\| f - f_{\mathcal{T}} \right\|_{\alpha;K}$$

if $u_{\mathcal{T}}$ *is piecewise linear.*

As mentioned before, Example 5.15 exhibits a simple bifurcation from the trivial branch at the simple eigenvalues of the Laplacian. Choosing

$$\langle \ell, v \rangle_V = \int_\Omega v$$

in Section 5.2.3 (p. 293) and combining the results of Section 5.1.3 with those of this section, we obtain the following a posteriori error estimates.

Theorem 5.20 *Denote by $\lambda^* \in \mathbb{R}$ and $u^* \in W_0^{1,p}(\Omega)$ with $p > d$ a weak solution of the nonlinear eigenvalue problem of Example 5.15 with $\int_\Omega u^* = 1$. Denote by $\lambda_\mathcal{T} \in \mathbb{R}$ and $u_\mathcal{T} \in X_\mathcal{T}$ a solution of*

$$\int_\Omega \nabla u_\mathcal{T} \cdot \nabla v_\mathcal{T} - \lambda_\mathcal{T} \int_\Omega u_\mathcal{T} v_\mathcal{T} + \int_\Omega u_\mathcal{T}^\beta v_\mathcal{T} = 0$$

for all $v_\mathcal{T} \in X_\mathcal{T}$. Assume that $X_\mathcal{T} \subset W_0^{1,\infty}(\Omega)$ is a finite element space corresponding to $\mathcal{T}$ consisting of piecewise polynomials and that $\beta \in \mathbb{N}$ with $\beta \geq d$. Define the error indicator $\eta_{R,K}$ by

$$\eta_{R,K} = \left\{ h_K^p \left\| \Delta u_\mathcal{T} + \lambda_\mathcal{T} u_\mathcal{T} - u_\mathcal{T}^\beta \right\|_{p;K}^p + \frac{1}{2} \sum_{E \in \mathcal{E}_{K,\Omega}} h_E \left\| \mathbb{J}_E(\mathbf{n}_E \cdot \nabla u_\mathcal{T}) \right\|_{p;E}^p \right\}^{\frac{1}{p}}.$$

Then, if $\lambda_\mathcal{T}$ and $u_\mathcal{T}$ are sufficiently close to λ^* and u^*, the following a posteriori error estimates hold:

$$|\lambda_\mathcal{T} - \lambda^*| + \|u_\mathcal{T} - u^*\|_{1,p} \leq c_1 \left\{ \left| \int_\Omega u_\mathcal{T} - 1 \right| + \left\{ \sum_{K \in \mathcal{T}} \eta_{R,K}^p \right\}^{\frac{1}{p}} \right\}$$

and

$$\left| \int_\Omega u_\mathcal{T} - 1 \right| + \left\{ \sum_{K \in \mathcal{T}} \eta_{R,K}^p \right\}^{\frac{1}{p}} \leq c_2 \left\{ |\lambda_\mathcal{T} - \lambda^*| + \|u_\mathcal{T} - u^*\|_{1,p} \right\}.$$

When comparing Theorems 5.16 and 5.20, we observe that the latter only yields global lower bounds on the error. This is due to the global nature of the functional ℓ defined above.

5.2.4 Error Estimates Based on the Solution of Auxiliary Local Problems

In what follows we give a simple example of an a posteriori error indicator which is based on the solution of auxiliary local problems and which generalises the indicator $\eta_{D,z}$ of Section 1.7.1 (p. 26). For simplicity we assume that $p = p' = r = 2$.

We choose an arbitrary vertex z_0 in $\mathcal{N}_\Omega$ and keep it fixed in what follows. Denote by $\omega_0 = \omega_{z_0}$ and $\sigma_0 = \sigma_{z_0}$ the union of all elements and faces, respectively, which have z_0 as a vertex and set

$$\begin{aligned}\widehat{X}_\mathcal{T} &= \widehat{Y}_\mathcal{T} = \widetilde{Y}_\mathcal{T}|_{\omega_0} \\ &= \operatorname{span}\left\{ \psi_K v, \psi_E \sigma : v \in R_{\max\{m,n\}}(K), \sigma \in R_m(E),\ K \in \mathcal{T}, z_0 \in \mathcal{N}_K, E \in \mathcal{E}_\Omega, z_0 \in \mathcal{N}_E \right\}\end{aligned}$$

and define the operator $\widehat{L} \in \mathcal{L}(\widehat{X}_\mathcal{T}, \widehat{Y}_\mathcal{T}^*)$ by

$$\left\langle \widehat{L}u, v \right\rangle_{\widehat{Y}_\mathcal{T}} = \int_{\omega_0} \nabla v \cdot A_0 \nabla u$$

for all $u \in \widehat{X}_\mathcal{T}$ and $v \in \widehat{Y}_\mathcal{T}$ where

$$A_0 = A\left(z_0, u_\mathcal{T}(z_0), \overline{(\nabla u_\mathcal{T})}_{z_0}\right).$$

Note that the operator $\widehat{L}$ is obtained by first linearising around $u_{\mathcal{T}}$ the differential operator associated with problem (5.9), then freezing the coefficients of the resulting linear operator at z_0, and finally retaining only the principal part of the linear constant-coefficient operator. Since $\nabla u_{\mathcal{T}}$ may be discontinuous, its value at z_0 is approximated by the average $\overline{(\nabla u_{\mathcal{T}})_{z_0}}$. Other constructions are of course possible, too.

Since the matrix $A(x, y, z)$ is symmetric and positive definite for all $x \in \Omega, y \in \mathbb{R}, z \in \mathbb{R}^d$ and since the functions in $\widehat{X}_{\mathcal{T}} = \widehat{Y}_{\mathcal{T}}$ vanish on $\partial\omega_0$, we immediately obtain from Friedrichs' inequality, cf. [3], [**109**, inequality (2.2)], inequality (1.66) (p. 47), and Section 3.4.7 (p. 105), that $\widehat{L}$ is an isomorphism of $\widehat{X}_{\mathcal{T}}$ onto $\widehat{Y}_{\mathcal{T}}^*$.

For brevity, we define the approximation $\widetilde{F}_{\mathcal{T}}(u_{\mathcal{T}})$ of $F(u_{\mathcal{T}})$ by

$$\langle \widetilde{F}_{\mathcal{T}}(u_{\mathcal{T}}), v \rangle_Y = \int_{\omega_0} \mathbf{a}_{\mathcal{T}}(\cdot, u_{\mathcal{T}}, \nabla u_{\mathcal{T}}) \cdot \nabla v - \int_{\omega_0} b_{\mathcal{T}}(\cdot, u_{\mathcal{T}}, \nabla u_{\mathcal{T}}) v$$

for all $v \in \widehat{Y}_{\mathcal{T}}$. Then, we denote by $u_0 \in \widehat{X}_{\mathcal{T}}$ the unique solution of

$$\langle \widehat{L} u_0 , v \rangle_{\widehat{Y}_{\mathcal{T}}} = \langle \widetilde{F}_{\mathcal{T}}(u_{\mathcal{T}}), v \rangle_Y \tag{5.11}$$

for all $v \in \widehat{Y}_{\mathcal{T}}$ and set

$$\eta_{D,z_0} = \|\nabla u_0\|_{\omega_0} . \tag{5.12}$$

Integration by parts element-wise yields for all $v \in \widehat{Y}_{\mathcal{T}}$

$$\begin{aligned} \langle \widetilde{F}_{\mathcal{T}}(u_{\mathcal{T}}), v \rangle_Y &= \sum_{K \subset \omega_0} \int_K (-\operatorname{div} \mathbf{a}_{\mathcal{T}}(\cdot, u_{\mathcal{T}}, \nabla u_{\mathcal{T}}) - b_{\mathcal{T}}(\cdot, u_{\mathcal{T}}, \nabla u_{\mathcal{T}}))\, v \\ &+ \sum_{E \subset \sigma_0} \int_E \mathbb{J}_E(\mathbf{n}_E \cdot \mathbf{a}_{\mathcal{T}}(\cdot, u_{\mathcal{T}}, \nabla u_{\mathcal{T}}))v. \end{aligned}$$

The proof of Theorem 5.16 therefore implies

$$\underline{c}\, \|\widetilde{F}_{\mathcal{T}}(u_{\mathcal{T}})\|_{\widehat{Y}_{\mathcal{T}}^*} \leq \left\{ \sum_{K \subset \omega_{z_0}} \eta_{R,K}^2 \right\}^{\frac{1}{2}} \leq \bar{c}\, \|\widetilde{F}_{\mathcal{T}}(u_{\mathcal{T}})\|_{\widehat{Y}_{\mathcal{T}}^*}$$

with constants $\underline{c}$ and $\bar{c}$ which only depend on the shape parameter $C_{\mathcal{T}}$ of $\mathcal{T}$ and the polynomial degrees of $u_{\mathcal{T}}$, $\mathbf{a}_{\mathcal{T}}$, and $b_{\mathcal{T}}$. Together with Theorem 5.7 (p. 289) this yields the following result.

Theorem 5.21 *Denote by z_0 an arbitrary vertex in $\mathcal{N}_\Omega$. Then there are two constants c_1, c_2, which only depend on the polynomial degree of the space $X_{\mathcal{T}}$ and on the shape parameter $C_{\mathcal{T}}$ of $\mathcal{T}$, such that the following inequalities hold*

$$c_1 \left\{ \sum_{K \subset \omega_0} \eta_{R,K}^2 \right\}^{\frac{1}{2}} \leq \eta_{D,z_0} \leq c_2 \left\{ \sum_{K \subset \omega_0} \eta_{R,K}^2 \right\}^{\frac{1}{2}} .$$

Here, $\eta_{R,K}$ and η_{D,z_0} are given by Theorem 5.16 and equation (5.12), respectively.

5.2.5 L^r-Error Estimates

In order to obtain bounds for the error measured in the L^r-norm, we apply Remarks 5.4 (p. 284) and 5.8 (p. 290) and the results of Section 3.2.5 (p. 85). More precisely, we set

$$\begin{aligned} X_- &= L^r(\Omega), & \|\cdot\|_{X_-} &= \|\cdot\|_r\,, \\ X_+ &= W^{1,s}(\Omega)\cap X, & \|\cdot\|_{X_+} &= \|\cdot\|_{1,s}\,, \\ Y_+ &= W^{2,p'}(\Omega)\cap Y, & \|\cdot\|_{Y_+} &= \|\cdot\|_{2,p'} \end{aligned}$$

with $r \le s \le \infty$. The choice of s depends on the particular example.

The condition that $DF(u)^*$ is an isomorphism of Y_+ onto X_-^* is equivalent to the assumption that the adjoint linearised problem

$$\begin{aligned} -\operatorname{div}(A(x,u,\nabla u)\nabla v) + \partial_y \mathbf{a}(x,u,\nabla u)\cdot\nabla v & \\ + \operatorname{div}(\nabla_z b(x,u,\nabla u)v) - \partial_y b(x,u,\nabla u)v &= f \quad \text{in } \Omega \\ v &= 0 \quad \text{on } \Gamma \end{aligned}$$

admits for every $f \in L^{r'}(\Omega)$ a unique weak solution $v \in Y_+$ which depends continuously on f. Here, as usual, $r' = \frac{r}{r-1}$ denotes the dual exponent of r. This assumption is satisfied, if the coefficients $\mathbf{a}$ and b are sufficiently smooth and Ω is convex.

The spaces $\widetilde{Y}_\mathcal{T}$, $\widehat{Y}_\mathcal{T}$, and $\widehat{X}_\mathcal{T}$ are defined as before; but now the functions ψ_K and ψ_E are replaced by their C^1-counterparts $\psi_{K,1}$ and $\psi_{E,1}$, respectively. The arguments for checking condition (5.8) of Theorem 5.5 (p. 288) carry over immediately. Since Y_+ and $\widetilde{Y}_\mathcal{T}$ are equipped with the $W^{2,p'}(\Omega)$ norm, we may now apply Propositions 3.35 (p. 112) and 3.51 (p. 129). We therefore always gain an additional power of h.

In order to check that the operator $\widehat{L}$ is an isomorphism of $\widehat{X}_\mathcal{T}$ onto $\widehat{Y}_\mathcal{T}^*$ with this choice of norms too, we proceed as follows. Since A_0 is symmetric positive definite, the Cauchy–Schwarz inequality and Proposition 3.51 (p. 129) imply for all $u \in \widehat{X}_\mathcal{T}$ and all $v \in \widehat{Y}_\mathcal{T}$

$$\begin{aligned} \left\langle \widehat{L}u\,,\, v\right\rangle_{\widehat{Y}_\mathcal{T}} &= \int_{\omega_0} \nabla v\cdot A_0\nabla u \\ &\le \lambda_{\max}(A_0)\,\|\nabla u\|_{\omega_0}\,\|\nabla v\|_{\omega_0} \\ &\le \lambda_{\max}(A_0)c_1 h_K^{-2}\,\|u\|_{\omega_0}\,\|v\|_{\omega_0} \\ &\le \lambda_{\max}(A_0)c_2\,\|u\|_{\omega_0}\,\|v\|_{2;\omega_0} \end{aligned}$$

and

$$\begin{aligned} \left\langle \widehat{L}u\,,\, u\right\rangle_{\widehat{Y}_\mathcal{T}} &= \int_{\omega_0} \nabla u\cdot A_0\nabla u \\ &\ge \lambda_{\min}(A_0)\,\|\nabla u\|_{\omega_0}^2 \\ &\ge \lambda_{\min}(A_0)c_3 h_K^{-2}\,\|u\|_{\omega_0}^2 \\ &\ge \lambda_{\min}(A_0)c_4\,\|u\|_{\omega_0}\,\|u\|_{2;\omega_0}\,, \end{aligned}$$

where $\lambda_{\max}(A_0)$ and $\lambda_{\min}(A_0)$ denote the maximal and minimal eigenvalue of A_0 and where the constants $c_1, \ldots, c_4$ depend on the constants in Proposition 3.51 (p. 129).

Collecting all these results, we obtain the following error estimates

$$\|u - u_{\mathcal{T}}\|_r \leq c^{\sharp} \left\{ \sum_{K \in \mathcal{T}} \left(h_K^p \eta_{R,K}^p + h_K^p \theta_K^p \right) \right\}^{\frac{1}{p}}$$

$$h_K \eta_{R,K} \leq c_{\sharp} \left\{ \|u - u_{\mathcal{T}}\|_{r;\omega_K} + \left[\sum_{K' \subset \omega_K} h_{K'}^p \theta_{K'}^p \right]^{\frac{1}{p}} \right\}$$

and

$$\widetilde{c}_1 \left\{ \sum_{K \subset \omega_0} h_K^2 \eta_{R,K}^2 \right\}^{\frac{1}{2}} \leq \widetilde{\eta}_{D,z_0} \leq \widetilde{c}_2 \left\{ \sum_{K \subset \omega_0} h_K^2 \eta_{R,K}^2 \right\}^{\frac{1}{2}}.$$

Here, $\eta_{R,K}$ and θ_K are as in Theorem 5.16 and $\widetilde{\eta}_{D,z_0}$ is given by

$$\widetilde{\eta}_{D,z_0} = \|u_0\|_{\omega_0}$$

where $u_0 \in \widehat{X}_{\mathcal{T}}$ is the solution of the local auxiliary problem (5.11).

Remark 5.22 The present analysis follows the lines of [**254**, **259**]. Examples 5.11 and 5.13 were originally analysed in [**48**] and [**220**], respectively. There, however, only upper bounds on the error are established. Further a posteriori error estimates for the α-Laplacian can be found in [**113**, **179**, **180**, **181**]. A more detailed analysis of the equations of prescribed mean curvature and of Bratu's problem can be found in [**111**] and [**136**], respectively.

5.3 Eigenvalue Problems Revisited

We now analyse the eigenvalue problem (4.32) (p. 205) within the nonlinear framework of this chapter. When considering λ as a parameter, problem (4.32) can be treated as a bifurcation problem similar to example 5.15 (p. 292). Here, we adopt a different strategy and define

$$X = Y = \mathbb{R} \times H_0^1(\Omega),$$

$$\|\cdot\|_X = \|\cdot\|_Y = \left\{ |\cdot|^2 + \|\nabla \cdot\|^2 \right\}^{\frac{1}{2}},$$

$$\langle F((\lambda, u)), (\mu, v) \rangle_Y = \int_\Omega \{ \nabla v \cdot A \nabla u + buv - \lambda uv \} + \mu \left\{ \int_\Omega u^2 - 1 \right\}.$$

Then, $(\lambda, u) \in X$ with $\int_\Omega u^2 = 1$, is a solution of problem (4.33) (p. 205) if and only if it is a solution of problem (5.1) (p. 281). Moreover, one easily checks that (λ, u) is a regular solution in the sense of Proposition 5.1 (p. 283) if and only if λ is a simple eigenvalue of the differential operator associated with problem (4.32).

As in the previous section and Section 4.7 (p. 205), we do not specify the discretisation of problem (4.32) in detail. We only assume that

$$X_{\mathcal{T}} = Y_{\mathcal{T}} = \mathbb{R} \times V_{\mathcal{T}} \subset Y,$$
$$\langle F_{\mathcal{T}}((\lambda_{\mathcal{T}}, u_{\mathcal{T}})), (\mu_{\mathcal{T}}, v_{\mathcal{T}})\rangle_{Y_{\mathcal{T}}} = \langle F((\lambda_{\mathcal{T}}, u_{\mathcal{T}})), (\mu_{\mathcal{T}}, v_{\mathcal{T}})\rangle_{Y}$$

for all $(\lambda_{\mathcal{T}}, u_{\mathcal{T}}) \in X_{\mathcal{T}}$, $(\mu_{\mathcal{T}}, v_{\mathcal{T}}) \in Y_{\mathcal{T}}$, where $V_{\mathcal{T}}$ is a finite element space corresponding to $\mathcal{T}$ which contains the space $S_0^{1,0}(\mathcal{T})$. Obviously, the consistency error of the above discretisation vanishes. Moreover, $(\lambda_{\mathcal{T}}, u_{\mathcal{T}}) \in X_{\mathcal{T}}$ is a solution of problem (5.3) (p. 282) if and only if it solves the discrete eigenvalue problem (4.33) (p. 205).

Equation (4.33) and integration by parts element-wise yield for all $(\mu, v) \in Y$

$$\langle F((\lambda_{\mathcal{T}}, u_{\mathcal{T}})), (\mu, v)\rangle_{Y} = \sum_{K\in\mathcal{T}} \int_K \{-\nabla \cdot (A\nabla u_{\mathcal{T}}) + bu_{\mathcal{T}} - \lambda_{\mathcal{T}} u_{\mathcal{T}}\} v + \sum_{E\in\mathcal{E}_\Omega} \int_E \mathbb{J}_E(\mathbf{n}_E \cdot (A\nabla u_{\mathcal{T}}))v.$$

As usual we denote by $A_{\mathcal{T}}$ and $b_{\mathcal{T}}$ piecewise polynomial approximations of A and b, respectively, and define the restriction operator $Q_{\mathcal{T}}$ by

$$Q_{\mathcal{T}}(\mu, v) = (0, I_{\mathcal{T}} v),$$

where $I_{\mathcal{T}}$ is the quasi-interpolation operator of Section 3.5.1 (p. 108).

The above L^2-representation of the residual and Theorem 3.57 (p. 135) then establish the first and second inequalities of Assumption (5.8) of Theorem 5.5 (p. 288) with

$$\eta_{\mathcal{T}} = \left\{\sum_{K\in\mathcal{T}} \eta_{R,K}^2\right\}^{\frac{1}{2}}, \quad \theta_{\mathcal{T}} = \left\{\sum_{K\in\mathcal{T}} \theta_K^2\right\}^{\frac{1}{2}},$$

$c_C = 0$, and a constant c_A which only depends on the shape parameter $C_{\mathcal{T}}$ of $\mathcal{T}$, where $\eta_{R,K}$ and θ_K are as in Theorem 4.45 (p. 207).

With the same arguments as in the previous sections, Theorem 3.59 (p. 137) implies the third inequality of Assumption (5.8) of Theorem 5.5 (p. 288) with

$$\widetilde{Y}_{\mathcal{T}} = \mathbb{R} \times \operatorname{span}\{\psi_K v, \psi_E \sigma : v \in R_k(K), \sigma \in R_k(E), K \in \mathcal{T}, E \in \mathcal{E}_\Omega\}$$

and a constant c_I which only depends on the shape parameter $C_{\mathcal{T}}$ of $\mathcal{T}$ and the polynomial degree k which in turn depends on the polynomial degrees of $u_{\mathcal{T}}, A_{\mathcal{T}}$, and $b_{\mathcal{T}}$.

Theorem 5.5 (p. 288) therefore yields the following a posteriori error estimate for problem (4.32) (p. 205).

Theorem 5.23 *Assume that λ is a simple eigenvalue of the differential operator associated with problem (4.32) (p. 205) and that u is a corresponding eigenfunction with $\int_\Omega u^2 = 1$. Suppose that $(\lambda_{\mathcal{T}}, u_{\mathcal{T}}) \in X_{\mathcal{T}}$ is a solution of problem (4.33) (p. 205) which is sufficiently close to (λ, u) in the sense of Proposition 5.1 (p. 283). Then the following a posteriori error estimates hold:*

$$|\lambda - \lambda_{\mathcal{T}}| + \|\nabla(u - u_{\mathcal{T}})\| \le c_1 \left\{\sum_{K\in\mathcal{T}} \eta_{R,K}^2\right\}^{\frac{1}{2}} + c_2 \left\{\sum_{K\in\mathcal{T}} \theta_K^2\right\}^{\frac{1}{2}}$$

and

$$\left\{\sum_{K\in\mathcal{T}} \eta_{R,K}^2\right\}^{\frac{1}{2}} \le c_3 \left\{|\lambda - \lambda_{\mathcal{T}}| + \|\nabla(u - u_{\mathcal{T}})\|\right\} + c_4 \left\{\sum_{K\in\mathcal{T}} \theta_K^2\right\}^{\frac{1}{2}},$$

where the constants $c_1, \ldots, c_4$ only depend on the polynomial degrees of $u_{\mathcal{T}}$, $A_{\mathcal{T}}$, and $b_{\mathcal{T}}$ and on the shape parameter $C_{\mathcal{T}}$ of $\mathcal{T}$ and $\eta_{R,K}$ and θ_K are as in Theorem 4.45 (p. 207).

Remark 5.24 The condition that $(\lambda_{\mathcal{T}}, u_{\mathcal{T}})$ has to be sufficiently close to (λ, u) essentially means that $|\lambda - \lambda_{\mathcal{T}}|$ has to be smaller than the distance of λ to its neighbouring eigenvalues. In contrast to Theorems 5.16 (p. 294) and 4.45 (p. 207), we obtain in Theorem 5.23 only a global lower bound on the error. This is due to the global nature of the constraint $\int_\Omega u^2 = 1$ inherent in the definition of F. When comparing Theorems 5.23 and 4.45 (p. 207) we observe that we could now abolish the higher order terms of Theorem 4.45. This, however, has to be paid for by the condition that $(\lambda_{\mathcal{T}}, u_{\mathcal{T}})$ has to be sufficiently close to (λ, u).

Remark 5.25 Other approaches to a posteriori error estimates for eigenvalue problems may be found in [**24, 73, 127, 129, 141, 151, 176, 189, 277**].

5.4 The Stationary Navier–Stokes Equations

The *stationary, incompressible Navier–Stokes equations* are given by

$$\begin{aligned} -\nu \Delta \mathbf{u} + (\mathbf{u} \cdot \nabla)\mathbf{u} + \nabla p &= \mathbf{f} \quad \text{in } \Omega \\ \nabla \cdot \mathbf{u} &= 0 \quad \text{in } \Omega \\ \mathbf{u} &= 0 \quad \text{on } \Gamma \end{aligned} \tag{5.13}$$

where $\nu > 0$ is the constant *viscosity* of the fluid and $\mathbf{u}$ and p denote its *velocity* and *pressure*, respectively.

5.4.1 Variational Formulation

Problem (5.13) fits into the abstract framework of Section 5.1 (p. 281) with

$$\begin{aligned} X &= Y = H_0^1(\Omega)^d \times L_0^2(\Omega), \\ \|\cdot\|_X &= \|\cdot\|_Y = \left\{ \|\nabla \cdot\|^2 + \|\cdot\|^2 \right\}^{\frac{1}{2}}, \\ \langle F((\mathbf{u}, p)), (\mathbf{v}, q) \rangle_Y &= \nu \int_\Omega \nabla \mathbf{u} : \nabla \mathbf{v} + \int_\Omega (\mathbf{u} \cdot \nabla)\mathbf{u} \cdot \mathbf{v} - \int_\Omega p \nabla \cdot \mathbf{v} + \int_\Omega q \nabla \cdot \mathbf{u} - \int_\Omega \mathbf{f} \cdot \mathbf{v}, \end{aligned}$$

where $L_0^2(\Omega)$ denotes the space of all L^2-functions with mean value zero, cf. equation (4.54) (p. 238).

Problem (5.13) and its variational formulation (5.1) have the following properties [**145, 243**].

- The variational problem (5.1) admits at least one solution.
- Every solution of the variational problem (5.1) satisfies the a priori bound $\|\nabla \mathbf{u}\| \le \frac{1}{\nu} \|\mathbf{f}\|$.
- The variational problem (5.1) admits a unique solution provided $\frac{1}{\nu} \|\mathbf{f}\| < c(\Omega)$, where the constant $c(\Omega)$ only depends on the diameter of Ω and the space dimension d.
- Every solution of the variational problem (5.1) has the same regularity properties as the solution of the Stokes equations (4.53) (p. 237).
- The mapping which associates with ν a solution of the variational problem (5.1) is differentiable. Its derivative with respect to ν is a continuous linear operator which is invertible with a

continuous inverse for all but finitely many values of ν, i.e. there are only finitely many turning or bifurcation points and for all but finitely many values of ν every weak solution of problem (5.13) is regular in the sense of Proposition 5.1 (p. 283).

5.4.2 *Finite Element Approximation*

As in Section 4.10 (p. 237) we denote by $V_{\mathcal{T}} \subset H_0^1(\Omega)^d$ and $P_{\mathcal{T}} \subset L_0^2(\Omega)$ two finite element spaces associated with an admissible, affine-equivalent, shape-regular partition $\mathcal{T}$ of Ω which approximate the velocity field and the pressure, respectively. Examples of these spaces are given in Section 4.10.2 (p. 240). The finite element approximation of problem (5.13) then fits into the abstract framework of Section 5.1 (p. 281) with

$$
\begin{aligned}
&X_{\mathcal{T}} = Y_{\mathcal{T}} = V_{\mathcal{T}} \times P_{\mathcal{T}},\\
&\langle F_{\mathcal{T}}((\mathbf{u}_{\mathcal{T}}, p_{\mathcal{T}})), (\mathbf{v}_{\mathcal{T}}, q_{\mathcal{T}})\rangle_{Y_{\mathcal{T}}} = \big\langle F((\mathbf{u}_{\mathcal{T}}, p_{\mathcal{T}})), (\mathbf{v}_{\mathcal{T}}, q_{\mathcal{T}})\big\rangle_Y\\
&\quad + \sum_{K\in\mathcal{T}} \vartheta_K h_K^2 \int_K \big\{-\nu\Delta\mathbf{u}_{\mathcal{T}} + (\mathbf{u}_{\mathcal{T}}\cdot\nabla)\mathbf{u}_{\mathcal{T}} + \nabla p_{\mathcal{T}} - \mathbf{f}\big\} \cdot \big\{(\mathbf{u}_{\mathcal{T}}\cdot\nabla)\mathbf{v}_{\mathcal{T}} + \nabla q_{\mathcal{T}}\big\}\\
&\quad + \sum_{E\in\mathcal{E}_\Omega} \vartheta_E h_E \int_E \mathbb{J}_E(p_{\mathcal{T}})\mathbb{J}_E(q_{\mathcal{T}}) + \sum_{K\in\mathcal{T}} \widetilde{\vartheta}_K \int_K \operatorname{div}\mathbf{u}_{\mathcal{T}} \operatorname{div}\mathbf{v}_{\mathcal{T}}.
\end{aligned}
$$

Here, as in Section 4.10.2 (p. 240), ϑ_K, ϑ_E, and $\widetilde{\vartheta}_K$ are non-negative stabilisation parameters. If these are strictly positive, the above discretisation is capable of stabilising both the influence of the convection term and of the divergence constraint without any conditions on the spaces $V_{\mathcal{T}}$, $P_{\mathcal{T}}$, or the mesh-Péclet-number $h_K\nu^{-1}$ [**76**, **77**, **140**, **184**, **185**, **190**, **244**]. The case $\vartheta_K = \vartheta_E = \widetilde{\vartheta}_K = 0$ corresponds to the standard mixed finite element discretisation of problem (5.13). The spaces $V_{\mathcal{T}}$ and $P_{\mathcal{T}}$ then have to satisfy the Babuška–Brezzi condition

$$
\inf_{p_{\mathcal{T}}\in P_{\mathcal{T}}\setminus\{0\}} \sup_{\mathbf{u}_{\mathcal{T}}\in V_{\mathcal{T}}\setminus\{0\}} \frac{\int_\Omega p_{\mathcal{T}} \operatorname{div}\mathbf{u}_{\mathcal{T}}}{\|p_{\mathcal{T}}\| \, \|\nabla\mathbf{u}_{\mathcal{T}}\|} \geq \beta > 0
$$

uniformly with respect to all families of partitions $\mathcal{T}$ obtained by uniform or adaptive refinement. Moreover, the mesh-Péclet-number $h_K\nu^{-1}$ must then be sufficiently small in order to balance the influence of the convection term.

5.4.3 *Residual A Posteriori Error Estimates*

In what follows we denote by $(\mathbf{u}, p) \in X$ and $(\mathbf{u}_{\mathcal{T}}, p_{\mathcal{T}}) \in X_{\mathcal{T}}$ a weak solution of problem (5.13) and a corresponding finite element approximation and assume that the space $V_{\mathcal{T}}$ contains the space $S_0^{1,0}(\mathcal{T})^d$ of continuous piecewise linear vector fields.

As in Section 4.10.3 (p. 242), we can then define the restriction operator $Q_{\mathcal{T}} : X \to X_{\mathcal{T}}$ by

$$
Q_{\mathcal{T}}(\mathbf{v}, q) = (I_{\mathcal{T}}\mathbf{v}, 0),
$$

where $I_{\mathcal{T}} : L^1(\Omega)^d \to S_0^{1,0}(\mathcal{T})^d$ denotes the quasi-interpolation operator defined in Section 3.5.1 (p. 108) applied to the components of the velocity field.

We choose an arbitrary element $(\mathbf{v}, q) \in X$ and keep it fixed in what follows. Integration by parts element-wise then yields the following L^2-representation of the residual

$$\langle F(\mathbf{u}_{\mathcal{T}}, p_{\mathcal{T}}), (\mathbf{v}, q)\rangle_Y = \sum_{K\in\mathcal{T}} \int_K (-\nu\Delta\mathbf{u}_{\mathcal{T}} + (\mathbf{u}_{\mathcal{T}}\cdot\nabla)\mathbf{u}_{\mathcal{T}} + \nabla p_{\mathcal{T}} - \mathbf{f})\cdot\mathbf{v}$$
$$+ \sum_{E\in\mathcal{E}_\Omega} \int_E \mathbb{J}_E(\mathbf{n}_E\cdot(\nabla\mathbf{u}_{\mathcal{T}} - p_{\mathcal{T}}I))\cdot\mathbf{v} + \sum_{K\in\mathcal{T}} \int_K q \operatorname{div}\mathbf{u}_{\mathcal{T}},$$

where $I \in \mathbb{R}^{d\times d}$ denotes the identity matrix.

Proposition 3.33 (p. 109) and the Cauchy–Schwarz inequality imply in the standard way

$$\langle F(\mathbf{u}_{\mathcal{T}}, p_{\mathcal{T}}), (Id - Q_{\mathcal{T}})(\mathbf{v}, q)\rangle_Y$$
$$\leq c\max\left\{C_{A,2,2}(K), C_{A,4,2}(E)\right\} \|(\mathbf{v}, q)\|_X$$
$$\cdot\left\{\sum_{K\in\mathcal{T}} h_K^2 \|\mathbf{f} + \nu\Delta\mathbf{u}_{\mathcal{T}} - (\mathbf{u}_{\mathcal{T}}\cdot\nabla)\mathbf{u}_{\mathcal{T}} - \nabla p_{\mathcal{T}}\|_K^2\right.$$
$$\left.+ \sum_{E\in\mathcal{E}_\Omega} h_E \|\mathbb{J}_E(\mathbf{n}_E\cdot(\nabla\mathbf{u}_{\mathcal{T}} - p_{\mathcal{T}}I))\|_E^2 + \sum_{K\in\mathcal{T}} \|\operatorname{div}\mathbf{u}_{\mathcal{T}}\|_K^2\right\}^{\frac{1}{2}},$$

where the constant c only depends on the shape parameter $C_{\mathcal{T}}$ of $\mathcal{T}$ and takes into account that every element is counted several times.

Since $(\mathbf{v}, q) \in X$ was arbitrary, this estimate establishes the first inequality of condition (5.8) of Theorem 5.5 (p. 288) with

$$\eta_{\mathcal{T}} = \left\{\sum_{K\in\mathcal{T}} h_K^2 \|\mathbf{f}_{\mathcal{T}} + \nu\Delta\mathbf{u}_{\mathcal{T}} - (\mathbf{u}_{\mathcal{T}}\cdot\nabla)\mathbf{u}_{\mathcal{T}} - \nabla p_{\mathcal{T}}\|_K^2 + \sum_{E\in\mathcal{E}_\Omega} h_E \|\mathbb{J}_E(\mathbf{n}_E\cdot(\nabla\mathbf{u}_{\mathcal{T}} - p_{\mathcal{T}}I))\|_E^2\right.$$
$$\left.+ \sum_{K\in\mathcal{T}} \|\operatorname{div}\mathbf{u}_{\mathcal{T}}\|_K^2\right\}^{\frac{1}{2}},$$

and

$$\theta_{\mathcal{T}} = \left\{\sum_{K\in\mathcal{T}} h_K^2 \|\mathbf{f} - \mathbf{f}_{\mathcal{T}}\|_K^2\right\}^{\frac{1}{2}}$$

and a constant c_A which only depends on the shape parameter $C_{\mathcal{T}}$ of $\mathcal{T}$. Here, as usual, $\mathbf{f}_{\mathcal{T}}$ denotes any piecewise polynomial approximation of $\mathbf{f}$.

To prove the second inequality of condition (5.8) of Theorem 5.5 (p. 288), we observe that the definitions of $Q_{\mathcal{T}}$, F, and $F_{\mathcal{T}}$ and the identity

$$\langle F_{\mathcal{T}}((\mathbf{u}_{\mathcal{T}}, p_{\mathcal{T}})), Q_{\mathcal{T}}(\mathbf{v}, q)\rangle_{Y_{\mathcal{T}}} = 0$$

imply that

$$\begin{aligned}\langle F((\mathbf{u}_\mathcal{T},p_\mathcal{T})),\, Q_\mathcal{T}(\mathbf{v},q)\rangle_Y &= \langle F((\mathbf{u}_\mathcal{T},p_\mathcal{T})),\, Q_\mathcal{T}(\mathbf{v},q)\rangle_Y - \langle F_\mathcal{T}((\mathbf{u}_\mathcal{T},p_\mathcal{T})),\, Q_\mathcal{T}(\mathbf{v},q)\rangle_{Y_\mathcal{T}} \\ &= -\sum_{K\in\mathcal{T}} \vartheta_K h_K^2 \int_K \{-\nu\Delta\mathbf{u}_\mathcal{T}+(\mathbf{u}_\mathcal{T}\cdot\nabla)\mathbf{u}_\mathcal{T}+\nabla p_\mathcal{T}-\mathbf{f}\}\cdot\{(\mathbf{u}_\mathcal{T}\cdot\nabla)(I_\mathcal{T}\mathbf{v})\} \\ &\quad -\sum_{K\in\mathcal{T}} \widetilde{\vartheta}_K \int_K \operatorname{div}\mathbf{u}_\mathcal{T}\,\operatorname{div}(I_\mathcal{T}\mathbf{v}).\end{aligned}$$

This identity and standard inverse estimates establish the second inequality of condition (5.8) of Theorem 5.5 with

$$c_C = c\max_{K\in\mathcal{T}}\max\left\{\vartheta_K, \widetilde{\vartheta}_K\right\}\{1+\|\nabla\mathbf{u}_\mathcal{T}\|\}$$

and a constant c which only depends on the polynomial degree of $\mathbf{u}_\mathcal{T}$ and the shape parameter $C_\mathcal{T}$ of $\mathcal{T}$.

The third inequality of condition (5.8) of Theorem 5.5 is proved in the standard way using the L^2-representation of the residual and the space

$$\widetilde{Y}_\mathcal{T} = \operatorname{span}\left\{(\psi_K\mathbf{v},0),(\psi_E\mathbf{w},0),(0,\psi_K q) : K\in\mathcal{T}, E\in\mathcal{E}_\Omega,\ \mathbf{v}\in R_k(K)^d, \mathbf{w}\in R_\ell(E)^d,\ q\in R_m(K)\right\},$$

where the polynomial degrees k, ℓ, and m must be chosen such that

$$\begin{aligned}(\mathbf{f}_\mathcal{T}+\nu\Delta\mathbf{u}_\mathcal{T}-(\mathbf{u}_\mathcal{T}\cdot\nabla)\mathbf{u}_\mathcal{T}-\nabla p_\mathcal{T})|_K &\in R_k(K)^d, \\ (\mathbb{J}_E(\mathbf{n}_E\cdot(\nabla\mathbf{u}_\mathcal{T}-p_\mathcal{T}I)))|_E &\in R_\ell(E)^d, \\ (\operatorname{div}\mathbf{u}_\mathcal{T})|_K &\in R_m(K).\end{aligned}$$

Theorem 5.5 therefore yields the following a posteriori error estimates.

Theorem 5.26 *Denote by $(\mathbf{u},p)$ a weak solution of problem (5.13) which is regular in the sense of Proposition 5.1 (p. 283) and by $(\mathbf{u}_\mathcal{T},p_\mathcal{T})\in X_\mathcal{T}$ a solution of the discrete problem $F_\mathcal{T}((\mathbf{u}_\mathcal{T},p_\mathcal{T}))=0$ which is sufficiently close to $(\mathbf{u},p)$ in the sense of Proposition 5.1. Define the residual a posteriori error indicator $\eta_{R,K}$ by*

$$\eta_{R,K} = \left\{h_K^2\left\|\mathbf{f}_\mathcal{T}+\nu\Delta\mathbf{u}_\mathcal{T}-(\mathbf{u}_\mathcal{T}\cdot\nabla)\mathbf{u}_\mathcal{T}-\nabla p_\mathcal{T}\right\|_K^2+\|\operatorname{div}\mathbf{u}_\mathcal{T}\|_K^2 \right.$$
$$\left. +\frac{1}{2}\sum_{E\in\mathcal{E}_{K,\Omega}} h_E\left\|\mathbb{J}_E(\mathbf{n}_E\cdot(\nabla\mathbf{u}_\mathcal{T}-p_\mathcal{T}I))\right\|_E^2\right\}^{\frac{1}{2}},$$

where $\mathbf{f}_{\mathcal{T}}$ is any piecewise polynomial approximation of $\mathbf{f}$. *Then the following a posteriori error estimates hold:*

$$\left\{ \left\| \nabla(\mathbf{u} - \mathbf{u}_{\mathcal{T}}) \right\|^2 + \left\| p - p_{\mathcal{T}} \right\|^2 \right\}^{\frac{1}{2}}$$

$$\leq c_1 \left[1 + \max_{K \in \mathcal{T}} \max \left\{ \vartheta_K, \widetilde{\vartheta}_K \right\} \{ 1 + \| \nabla \mathbf{u}_{\mathcal{T}} \| \} \right] \left\{ \sum_{K \in \mathcal{T}} \eta_{R,K}^2 \right\}^{\frac{1}{2}}$$

$$+ c_2 \left\{ \sum_{K \in \mathcal{T}} h_K^2 \left\| \mathbf{f} - \mathbf{f}_{\mathcal{T}} \right\|_K^2 \right\}^{\frac{1}{2}},$$

and

$$\eta_{R,K} \leq c_3 \left\{ \left\| \nabla(\mathbf{u} - \mathbf{u}_{\mathcal{T}}) \right\|_{\omega_K}^2 + \left\| p - p_{\mathcal{T}} \right\|_{\omega_K}^2 \right\}^{\frac{1}{2}} + c_4 \left\{ \sum_{K' \subset \omega_K} h_{K'}^2 \left\| \mathbf{f} - \mathbf{f}_{\mathcal{T}} \right\|_{K'}^2 \right\}^{\frac{1}{2}}.$$

The constants $c_1, \ldots, c_4$ only depend on the polynomial degrees of the spaces $V_{\mathcal{T}}$ and $P_{\mathcal{T}}$ and on the shape parameter $C_{\mathcal{T}}$ of $\mathcal{T}$.

Remark 5.27 Theorem 5.26 can be extended to the case of the *slip boundary condition*

$$\mathbf{u} \cdot \mathbf{n} = \underline{\mathbf{T}}(\nu \mathbf{u}, p) - [\mathbf{n} \cdot \underline{\mathbf{T}}(\nu \mathbf{u}, p) \cdot \mathbf{n}]\mathbf{n} = 0$$

on some part Γ_N of Γ. Here

$$\underline{\mathbf{T}}(\mathbf{u}, p) = \left(\frac{1}{2} \left(\frac{\partial u_j}{\partial x_i} + \frac{\partial u_i}{\partial x_j} \right) - p\delta_{ij} \right)_{1 \leq i,j \leq d}$$

denotes the stress tensor. In the definition of $\eta_{R,K}$, one then has to replace the term $\nu \nabla \mathbf{u}_{\mathcal{T}} - p_{\mathcal{T}} I$ by $\underline{\mathbf{T}}(\nu \mathbf{u}_{\mathcal{T}}, p_{\mathcal{T}})$. Moreover, for every face E contained in $\Gamma_N \cap \mathcal{E}_K$, one has to add the term

$$h_E \left\| \underline{\mathbf{T}}(\nu \mathbf{u}_{\mathcal{T}}, p_{\mathcal{T}}) - [\mathbf{n} \cdot \underline{\mathbf{T}}(\nu \mathbf{u}_{\mathcal{T}}, p_{\mathcal{T}}) \cdot \mathbf{n}]\mathbf{n} \right\|_E^2$$

to $\eta_{R,K}^2$. Of course, the discretisation then also has to take account of the different boundary condition [**250**, **252**].

Remark 5.28 The previous results can also be extended to non-Newtonian fluids. Combining the arguments used to establish Theorems 5.16 (p. 294), 5.19 (p. 295), and 5.26, one can prove that the error indicator of [**48**] also yields local lower bounds similar to the second estimate of Theorem 5.26.

5.4.4 Auxiliary Local Problems

We proceed as in Section 4.10.4 (p. 245) and solve local discrete Stokes problems. But now, of course, the right-hand sides are the residuals of the nonlinear problem.

More precisely, we choose as in Section 4.10.4 three integers n_p, $n_{\mathbf{u},E}$, and $n_{\mathbf{u},K}$ such that the following conditions are satisfied for all elements K, all faces E, and all pressures $q_K \in R_{n_p}(K)$

$$\begin{aligned}(\operatorname{div}\mathbf{u}_{\mathcal{T}})|_K &\in R_{n_p}(K),\\ \mathbb{J}_E(\mathbf{n}_E\cdot(\nabla\mathbf{u}_{\mathcal{T}}-p_{\mathcal{T}}I)) &\in R_{n_{\mathbf{u},E}}(E),\\ (\mathbf{f}_{\mathcal{T}}+\nu\Delta\mathbf{u}_{\mathcal{T}}-(\mathbf{u}_{\mathcal{T}}\cdot\nabla)\mathbf{u}_{\mathcal{T}}-\nabla p_{\mathcal{T}})|_K &\in R_{n_{\mathbf{u},K}}(K),\\ \nabla(\psi_K q_K) &\in R_{n_{\mathbf{u},K}}(K).\end{aligned}$$

For every element K we define the spaces V_K, P_K, and X_K as in Section 4.10.4 by

$$\begin{aligned}V_K &= \operatorname{span}\{\psi_E\mathbf{v}_E,\psi_K\mathbf{v}_K:\mathbf{v}_E\in R_{n_{\mathbf{u},E}}(E),\mathbf{v}_K\in R_{n_{\mathbf{u},K}}(K),E\in\mathcal{E}_K\cap\mathcal{E}_\Omega\},\\ P_K &= \operatorname{span}\{\psi_K q_K : q_K\in R_{n_p}(K)\},\\ X_K &= V_K\times P_K\end{aligned}$$

and denote by $B_K : X_K\times X_K\to\mathbb{R}$ the bi-linear form

$$B_K((\mathbf{v},q),(\mathbf{w},r)) = \int_K \nabla\mathbf{v}:\nabla\mathbf{w} - \int_K q\operatorname{div}\mathbf{w} + \int_K r\operatorname{div}\mathbf{v}.$$

Lemma 4.71 (p. 246) implies that the corresponding linear operator $L_K : X_K\to X_K^*$ is an isomorphism. Therefore, there is a unique element $(\mathbf{u}_K,p_K)\in X_K$ such that

$$\begin{aligned}B_K((\mathbf{u}_K,p_K),(\mathbf{v},q)) = &\int_K(\mathbf{f}_{\mathcal{T}}+\nu\Delta\mathbf{u}_{\mathcal{T}}-(\mathbf{u}_{\mathcal{T}}\cdot\nabla)\mathbf{u}_{\mathcal{T}}-\nabla p_{\mathcal{T}})\cdot\mathbf{v}\\ &-\frac{1}{2}\sum_{E\in\mathcal{E}_{K,\Omega}}\int_E\mathbb{J}_E(\mathbf{n}_E\cdot(\nabla\mathbf{u}_{\mathcal{T}}-p_{\mathcal{T}}I))\cdot\mathbf{v}-\int_K\operatorname{div}\mathbf{u}_{\mathcal{T}}q\end{aligned}$$

holds for all $(\mathbf{v},q)\in X_K$. Now, we define

$$\eta_{N,K} = \left\{\|\nabla\mathbf{u}_K\|_K^2+\|p_K\|_K^2\right\}^{\frac{1}{2}}. \tag{5.14}$$

The choice of n_p, $n_{\mathbf{u},E}$, and $n_{\mathbf{u},K}$ ensures that the functions of $\widetilde{Y}_{\mathcal{T}}$ corresponding to K and to the faces in $\mathcal{E}_{K,\Omega}$ are contained in X_K. Therefore, we can compare the indicators $\eta_{R,K}$ and $\eta_{N,K}$ as in Sections 1.7.3 (p. 29) and 4.10.4 (p. 245). The continuity of the bi-linear form B_K, Lemma 4.71, and the proof of Theorem 5.26 then yield the following a posteriori error estimates.

Theorem 5.29 *Denote by $(\mathbf{u},p)$ a weak solution of problem (5.13) and by $(\mathbf{u}_{\mathcal{T}},p_{\mathcal{T}})\in X_{\mathcal{T}}$ a solution of the corresponding discrete problem which is sufficiently close to $(\mathbf{u},p)$ in the sense of Proposition 5.1. Then the estimates*

$$\eta_{N,K} \leq c_1 \eta_{R,K},$$

$$\eta_{R,K} \leq c_2 \left\{ \sum_{K' \subset \omega_K} \eta_{N,K'}^2 \right\}^{\frac{1}{2}},$$

$$\eta_{N,K} \leq c_3 \left\{ \left\| \nabla(\mathbf{u} - \mathbf{u}_{\mathcal{T}}) \right\|_{\omega_K}^2 + \left\| p - p_{\mathcal{T}} \right\|_{\omega_K}^2 + \sum_{K' \subset \omega_K} h_{K'}^2 \left\| \mathbf{f} - \mathbf{f}_{\mathcal{T}} \right\|_{K'}^2 \right\}^{\frac{1}{2}}$$

hold for all elements $K \in \mathcal{T}$. Moreover, the error is bounded from above by

$$\left\{ \left\| \nabla(\mathbf{u} - \mathbf{u}_{\mathcal{T}}) \right\|^2 + \left\| p - p_{\mathcal{T}} \right\|^2 \right\}^{\frac{1}{2}} \leq c_4 \left\{ \sum_{K \in \mathcal{T}} \eta_{N,K}^2 + \sum_{K \in \mathcal{T}} h_K^2 \left\| \mathbf{f} - \mathbf{f}_{\mathcal{T}} \right\|_K^2 \right\}^{\frac{1}{2}}.$$

Here, $\mathbf{f}_{\mathcal{T}}$ and $\eta_{R,K}$ are as in Theorem 5.26, and $\eta_{N,K}$ is defined by equation (5.14). The constants $c_1, \ldots, c_4$ only depend on the shape parameter $C_{\mathcal{T}}$ of $\mathcal{T}$ and the polynomial degree of $\mathbf{u}_{\mathcal{T}}$ and $p_{\mathcal{T}}$; the constant c_4 in addition depends on the stabilisation parameters of the finite element approximation through $\max_{K \in \mathcal{T}} \max \left\{ \vartheta_K, \tilde{\vartheta}_K \right\}$.

Error indicators based on the solution of auxiliary local discrete Stokes problems with Dirichlet boundary conditions on a patch of elements can be devised similarly. Since, following the lines indicated above, the arguments are straightforward modifications of those formerly used for the model problem or for reaction–diffusion equations, we leave the details to the reader.

5.4.5 Error Estimates with Respect to Weaker Norms

Remarks 5.4 (p. 284) and 5.8 (p. 290) yield estimates for the error measured in the L^2-norm for the velocity and in the H^{-1}-norm for the pressure. More precisely, we set

$$X_- = L^2(\Omega)^d \times [H^1(\Omega) \cap L_0^2(\Omega)]^*,$$

$$X_+ = [W^{1,3}(\Omega)^d \cap H_0^1(\Omega)^d] \times L_0^2(\Omega),$$

$$Y_+ = [H^2(\Omega)^d \cap H_0^1(\Omega)^d] \times [H^1(\Omega) \cap L_0^2(\Omega)].$$

The choice of X_+ and the Sobolev embedding theorem ensure the Lipschitz continuity of $DF((\mathbf{u}, p))$. The condition that $DF((\mathbf{u}, p))^*$ is an isomorphism of X_- onto Y_+^* is equivalent to the assumption that the linearised adjoint Navier–Stokes equations

$$\begin{aligned} -\nu\Delta\mathbf{v} - (\mathbf{u} \cdot \nabla)\mathbf{v} + (\mathbf{v} \cdot \nabla)\mathbf{u} - \nabla q &= \mathbf{w} && \text{in } \Omega \\ -\nabla \cdot \mathbf{v} &= r && \text{in } \Omega \\ \mathbf{v} &= 0 && \text{on } \Gamma \end{aligned}$$

admit for each right-hand side $(\mathbf{w}, r) \in L^2(\Omega)^d \times [H^1(\Omega) \cap L_0^2(\Omega)]$ a unique weak solution $(\mathbf{v}, q) \in Y_+$ which depends continuously on $(\mathbf{w}, r)$. This assumption in particular is fulfilled if Ω is convex and $\nu^{-2} \|\mathbf{f}\| < 1$.

The spaces $\widetilde{Y}_{\mathcal{T}}$ and X_K are defined as before; but now, the functions ψ_K and ψ_E are replaced by their C^1-counterparts $\psi_{K,1}$ and $\psi_{E,1}$, respectively, defined in Section 3.2.5 (p. 85). The arguments used to check condition (5.8) (p. 288) of Theorem 5.5 (p. 288) with the modified norms carry over immediately. Yet now, we always gain an additional power of h. Similarly, the proof of Lemma 4.71 (p. 246) immediately carries over to the modified norms and shows that L_K is an isomorphism of $\widehat{X}_K$ onto $\widehat{X}_K^*$ also with the present choice of norms. Assuming that the consistency error $\left\|F((\mathbf{u}_{\mathcal{T}}, p_{\mathcal{T}})) - F_{\mathcal{T}}((\mathbf{u}_{\mathcal{T}}, p_{\mathcal{T}}))\right\|_{Y_{\mathcal{T}}^*}$ vanishes, i.e. $\vartheta_K = \widetilde{\vartheta}_K = \vartheta_E = 0$ for all K and E, we then obtain the following error estimates

$$c_1 h_K \eta_{R,K} \le \widetilde{\eta}_{N,K} \le c_2 h_K \eta_{R,K},$$

$$h_K \eta_{R,K} \le c_3 \left\{ \|\mathbf{u} - \mathbf{u}_{\mathcal{T}}\|_{\omega_K}^2 + \left\|p - p_{\mathcal{T}}\right\|_{-1;\omega_K}^2 \right\}^{\frac{1}{2}} + c_4 \left\{ \sum_{K' \subset \omega_K} h_{K'}^4 \, \|\mathbf{f} - \mathbf{f}_{\mathcal{T}}\|_{K'}^2 \right\}^{\frac{1}{2}}$$

and

$$\|\mathbf{u} - \mathbf{u}_{\mathcal{T}}\| + \left\|p - p_{\mathcal{T}}\right\|_{-1} \le c_5 \left\{ \sum_{K \in \mathcal{T}} h_K^2 \eta_{R,K}^2 \right\}^{\frac{1}{2}} + c_6 \left\{ \sum_{K \in \mathcal{T}} h_K^4 \, \|\mathbf{f} - \mathbf{f}_{\mathcal{T}}\|_K^2 \right\}^{\frac{1}{2}}.$$

Here, $\eta_{R,K}$ is as in Theorem 5.26 and $\widetilde{\eta}_{N,K}$ is given by

$$\widetilde{\eta}_{N,K} = \left\{ \|\mathbf{u}_K\|_K^2 + h_K^2 \left\|p_K\right\|_K^2 \right\}^{\frac{1}{2}}$$

with $(\mathbf{u}_K, p_K)$ denoting the solution of the auxiliary local problem of the preceding subsection. Note that, in order to obtain an expression for $\widetilde{\eta}_{N,K}$ which is easy to compute, we have used the fact that

$$\left\{ \|\mathbf{v}_K\|_K^2 + \left\|q_K\right\|_{-1;K}^2 \right\}^{\frac{1}{2}} \quad \text{and} \quad \left\{ \|\mathbf{v}_K\|_K^2 + h_K^2 \left\|q_K\right\|_K^2 \right\}^{\frac{1}{2}}$$

define equivalent norms on X_K.

Remark 5.30 The present analysis follows the lines of [**254**]. Residual a posteriori error estimates for the Stokes and Navier–Stokes equations have been considered in [**1, 43, 44, 64, 251, 253**]. Estimates based on the solution of auxiliary local problems are considered in [**43, 44, 64, 279**]. Further results may be found in [**25, 52, 155, 157, 159**].

6

Parabolic Equations

In this chapter we derive a posteriori error estimates for space–time finite element and certain finite volume discretisations of linear and nonlinear parabolic equations of second order. The starting point of our analysis again is the equivalence of error and residual. For linear problems, this equivalence follows from standard parabolic energy estimates; for nonlinear problems, we resort to the abstract results of Section 5.1. The residuals consist of two contributions which can be associated with the temporal and the spatial discretisation, respectively, and which accordingly are labelled temporal and spatial residual. These contributions can be estimated separately thanks to the results of Section 3.7. The estimation of the temporal residual is often straightforward, while bounds for the spatial residual are derived from the results of Chapter 3, in particular Section 3.8. In Section 6.8, finally, we prove the convergence of an adaptive algorithm which simultaneously adjusts the time-steps and the spatial meshes using the error indicators of this chapter.

6.1 The Heat Equation

As a simple model problem, we consider the *heat equation*

$$\begin{aligned} \partial_t u - \Delta u &= f && \text{in } \Omega \times (0, T] \\ u &= 0 && \text{on } \Gamma \times (0, T] \\ u(\cdot, 0) &= u_0 && \text{in } \Omega \end{aligned} \tag{6.1}$$

in a bounded space–time cylinder with a polygonal cross-section $\Omega \subset \mathbb{R}^d$, $d \geq 2$, having a Lipschitz boundary Γ. The final time T is arbitrary, but kept fixed in what follows. For simplicity, the right-hand side f is assumed to be measurable and square integrable on $\Omega \times (0, T]$ and to be continuous with respect to time; the initial datum u_0 is assumed to be measurable and square integrable on Ω.

6.1.1 Variational Formulation

Throughout this section, we equip the space $H_0^1(\Omega)$ with the norm $\|\nabla \cdot\|$ and denote by $H^{-1}(\Omega)$ the corresponding dual space. In particular we have

$$\|\ell\|_{-1} = \sup_{v \in H_0^1(\Omega) \setminus \{0\}} \frac{\langle \ell, v \rangle}{\|\nabla v\|} \tag{6.2}$$

for all $\ell \in H^{-1}(\Omega)$. With the notation of Section 3.1.3 (p. 81), the variational formulation of problem (6.1) then takes the form:

find $u \in W^2(0,T;H_0^1(\Omega),H^{-1}(\Omega))$ such that $u(\cdot,0) = u_0$ in $H^{-1}(\Omega)$ and for almost every $t \in (0,T)$ and all $v \in H_0^1(\Omega)$

$$\int_\Omega \partial_t uv + \int_\Omega \nabla u \cdot \nabla v = \int_\Omega fv. \tag{6.3}$$

The assumptions on f and u_0 imply that problem (6.3) admits a unique solution, cf. [**20**, Sections II.4 and V.2] or [**119**, Theorems XVIII.3.1 and XVIII.3.2].

6.1.2 Finite Element Discretisation

For the finite element discretisation of problem (6.1), we choose a partition $\mathcal{I}$ of $[0,T]$, corresponding partitions $\mathcal{T}_n$ of Ω and associated finite element spaces X_n, which satisfy the conditions of Section 3.2.6 (p. 86), and a parameter $\theta \in [\frac{1}{2},1]$, and denote by π_n the L^2-projection onto X_n. With the abbreviation

$$f^n = f(\cdot,t_n), \quad f^{n\theta} = \theta f^n + (1-\theta)f^{n-1}$$

the finite element discretisation of problem (6.1) is then given by:

find $u_{\mathcal{T}_n}^n \in X_n, 0 \le n \le N_{\mathcal{I}}$, such that

$$u_{\mathcal{T}_0}^0 = \pi_0 u_0 \tag{6.4}$$

and, for $n = 1,\ldots,N_{\mathcal{I}}$ and $u^{n\theta} = \theta u_{\mathcal{T}_n}^n + (1-\theta)u_{\mathcal{T}_{n-1}}^{n-1}$,

$$\int_\Omega \frac{1}{\tau_n}\left(u_{\mathcal{T}_n}^n - u_{\mathcal{T}_{n-1}}^{n-1}\right) v_{\mathcal{T}_n} + \int_\Omega \nabla u^{n\theta} \cdot \nabla v_{\mathcal{T}_n} = \int_\Omega f^{n\theta} v_{\mathcal{T}_n} \tag{6.5}$$

for all $v_{\mathcal{T}_n} \in X_n$.

This is the popular *A*-stable *θ-scheme* which in particular yields the *Crank–Nicolson scheme* if $\theta = \frac{1}{2}$ and the *implicit Euler scheme* if $\theta = 1$.

The Lax–Milgram lemma [**109**, Theorem 1.1] implies that the discrete problem admits a unique solution $(u_{\mathcal{T}_n}^n)_{0\le n\le N_{\mathcal{I}}}$. With this sequence we associate the function $u_{\mathcal{I}}$ which is *piecewise affine* on the time intervals $[t_{n-1},t_n]$, $1 \le n \le N_{\mathcal{I}}$, and which equals $u_{\mathcal{T}_n}^n$ at time t_n, $0 \le n \le N_{\mathcal{I}}$, i.e.

$$u_{\mathcal{I}}(\cdot,t) = \frac{t_n - t}{\tau_n} u_{\mathcal{T}_{n-1}}^{n-1} + \frac{t - t_{n-1}}{\tau_n} u_{\mathcal{T}_n}^n \quad \text{on } [t_{n-1},t_n]. \tag{6.6}$$

Note that

$$\partial_t u_{\mathcal{I}} = \frac{1}{\tau_n}\left(u_{\mathcal{T}_n}^n - u_{\mathcal{T}_{n-1}}^{n-1}\right) \quad \text{on } (t_{n-1},t_n]. \tag{6.7}$$

Similarly we denote by $f_{\mathcal{I}}$ the function which is *piecewise constant* on the time intervals and which, on each interval $(t_{n-1}, t_n]$, is equal to the L^2-projection of $\theta f^n + (1-\theta)f^{n-1}$ onto the finite element space X_n, i.e.

$$f_{\mathcal{I}}(\cdot, t) = \pi_n f^{n\theta} \quad \text{on } (t_{n-1}, t_n].$$

6.1.3 The Equivalence of Error and Residual

With the function $u_{\mathcal{I}}$ we associate the residual $R(u_{\mathcal{I}}) \in L^2(0, T; H^{-1}(\Omega))$ by setting for all $v \in H_0^1(\Omega)$

$$\langle R(u_{\mathcal{I}}), v\rangle = \int_\Omega fv - \int_\Omega \partial_t u_{\mathcal{I}} v - \int_\Omega \nabla u_{\mathcal{I}} \cdot \nabla v.$$

The following proposition shows that this residual and the error $u - u_{\mathcal{I}}$ are equivalent. Its proof is based on a standard parabolic energy estimate.

Proposition 6.1 *The error can be bounded from below by*

$$\|R(u_{\mathcal{I}})\|_{L^2(0,T;H^{-1}(\Omega))} \le \Big\{\|u - u_{\mathcal{I}}\|^2_{L^\infty(0,T;L^2(\Omega))} + \|u - u_{\mathcal{I}}\|^2_{L^2(0,T;H_0^1(\Omega))} + \|\partial_t(u - u_{\mathcal{I}})\|^2_{L^2(0,T;H^{-1}(\Omega))}\Big\}^{\frac{1}{2}}$$

and, for every $n \in \{1, \dots, N_{\mathcal{I}}\}$, from above by

$$\Big\{\|u - u_{\mathcal{I}}\|^2_{L^\infty(0,t_n;L^2(\Omega))} + \|u - u_{\mathcal{I}}\|^2_{L^2(0,t_n;H_0^1(\Omega))} + \|\partial_t(u - u_{\mathcal{I}})\|^2_{L^2(0,t_n;H^{-1}(\Omega))}\Big\}^{\frac{1}{2}}$$
$$\le \Big\{4\|u_0 - \pi_0 u_0\|^2 + 6\|R(u_{\mathcal{I}})\|^2_{L^2(0,t_n;H^{-1}(\Omega))}\Big\}^{\frac{1}{2}}.$$

Proof Equation (6.3) and the definition of R imply for all $v \in H_0^1(\Omega)$

$$\int_\Omega \partial_t(u - u_{\mathcal{I}})v + \int_\Omega \nabla(u - u_{\mathcal{I}}) \cdot \nabla v = \langle R(u_{\mathcal{I}}), v\rangle. \tag{6.8}$$

This immediately yields the first estimate of the proposition.

To prove the second estimate, we choose an integer $n \in \{1, \dots, N_{\mathcal{I}}\}$ and a time $t \in [0, t_n]$ and insert $v = (u - u_{\mathcal{I}})(\cdot, t)$ in equation (6.8). This gives

$$\begin{aligned}
\frac{1}{2}\frac{\mathrm{d}}{\mathrm{d}t}&\|(u - u_{\mathcal{I}})(\cdot,t)\|^2 + \|\nabla(u - u_{\mathcal{I}})(\cdot,t)\|^2 \\
&= \int_\Omega \partial_t(u - u_{\mathcal{I}})(\cdot,t)(u - u_{\mathcal{I}})(\cdot,t) + \int_\Omega \nabla(u - u_{\mathcal{I}})(\cdot,t) \cdot \nabla(u - u_{\mathcal{I}})(\cdot,t) \\
&= \langle R(u_{\mathcal{I}})(\cdot,t), (u - u_{\mathcal{I}})(\cdot,t)\rangle \\
&\le \|R(u_{\mathcal{I}})(\cdot,t)\|_{-1}\|\nabla(u - u_{\mathcal{I}})(\cdot,t)\| \\
&\le \frac{1}{2}\|R(u_{\mathcal{I}})(\cdot,t)\|^2_{-1} + \frac{1}{2}\|\nabla(u - u_{\mathcal{I}})(\cdot,t)\|^2
\end{aligned}$$

and thus

$$\frac{\mathrm{d}}{\mathrm{d}t}\left\|(u-u_{\mathcal{I}})(\cdot,t)\right\|^2+\left\|\nabla(u-u_{\mathcal{I}})(\cdot,t)\right\|^2\le\left\|R(u_{\mathcal{I}})(\cdot,t)\right\|_{-1}^2.$$

Integrating this estimate from 0 to t implies

$$\begin{aligned}\left\|(u-u_{\mathcal{I}})(\cdot,t)\right\|^2-\|u_0-\pi_0u_0\|^2+\int_0^t\left\|\nabla(u-u_{\mathcal{I}})(\cdot,s)\right\|^2\mathrm{d}s\\ \le\left\|R(u_{\mathcal{I}})\right\|_{L^2(0,t;H^{-1}(\Omega))}^2\\ \le\left\|R(u_{\mathcal{I}})\right\|_{L^2(0,t_n;H^{-1}(\Omega))}^2.\end{aligned}$$

Since $t\in(0,t_n]$ is arbitrary, this yields

$$\|u-u_{\mathcal{I}}\|_{L^\infty(0,t_n;L^2(\Omega))}^2\le\|u_0-\pi_0u_0\|^2+\left\|R(u_{\mathcal{I}})\right\|_{L^2(0,t_n;H^{-1}(\Omega))}^2 \tag{6.9}$$

and

$$\|u-u_{\mathcal{I}}\|_{L^2(0,t_n;H_0^1(\Omega))}^2\le\|u_0-\pi_0u_0\|^2+\left\|R(u_{\mathcal{I}})\right\|_{L^2(0,t_n;H^{-1}(\Omega))}^2. \tag{6.10}$$

Equation (6.8), on the other hand, implies

$$\left\|\partial_t(u-u_{\mathcal{I}})\right\|_{-1}\le\left\|R(u_{\mathcal{I}})\right\|_{-1}+\left\|\nabla(u-u_{\mathcal{I}})\right\|.$$

Taking the square of this inequality, integrating from 0 to t_n, and inserting estimate (6.10), we arrive at the bound

$$\begin{aligned}\left\|\partial_t(u-u_{\mathcal{I}})\right\|_{L^2(0,t_n;H^{-1}(\Omega))}^2&\le2\left\|R(u_{\mathcal{I}})\right\|_{L^2(0,t_n;H^{-1}(\Omega))}^2+2\|u-u_{\mathcal{I}}\|_{L^2(0,t_n;H_0^1(\Omega))}^2\\&\le2\|u_0-\pi_0u_0\|^2+4\left\|R(u_{\mathcal{I}})\right\|_{L^2(0,t_n;H^{-1}(\Omega))}^2.\end{aligned} \tag{6.11}$$

Combining estimates (6.9), (6.10), and (6.11) proves the second estimate of the proposition. □

Remark 6.2 Proposition 6.1 shows that the a posteriori error estimation should be based on the norm

$$\left\{\|v\|_{L^\infty(0,T;L^2(\Omega))}^2+\|v\|_{L^2(0,T;H_0^1(\Omega))}^2+\|\partial_tv\|_{L^2(0,T;H^{-1}(\Omega))}^2\right\}^{\frac{1}{2}}.$$

Since the space $W^2(0,T;H_0^1(\Omega),H^{-1}(\Omega))$ is continuously embedded into $L^\infty(0,T;L^2(\Omega))$, cf. [**119**, Chapter XVIII, Section 1, Theorem 1] and [**230**, Theorem 7.2], this norm is equivalent to the standard norm

$$\left\{\|v\|_{L^2(0,T;H_0^1(\Omega))}^2+\|\partial_tv\|_{L^2(0,T;H^{-1}(\Omega))}^2\right\}^{\frac{1}{2}}$$

of $W^2(0,T;H_0^1(\Omega),H^{-1}(\Omega))$.

6.1.4 Decomposition of the Residual

The subsequent analysis relies on an appropriate decomposition of the residual $R(u_{\mathcal{I}})$. To this end, we define a temporal residual $R_\tau(u_{\mathcal{I}}) \in L^2(0,T;H^{-1}(\Omega))$ and a spatial residual $R_h(u_{\mathcal{I}}) \in L^2(0,T;H^{-1}(\Omega))$ by setting on $(t_{n-1},t_n]$ for all $n \in \{1,\dots,N_{\mathcal{I}}\}$ and all $v \in H_0^1(\Omega)$

$$\langle R_\tau(u_{\mathcal{I}}), v\rangle = \int_\Omega \nabla\left[\theta u_{\mathcal{T}_n}^n + (1-\theta)u_{\mathcal{T}_{n-1}}^{n-1} - u_{\mathcal{I}}\right]\cdot\nabla v$$

and

$$\langle R_h(u_{\mathcal{I}}), v\rangle = \int_\Omega f_{\mathcal{I}} v - \int_\Omega \frac{1}{\tau_n}\left(u_{\mathcal{T}_n}^n - u_{\mathcal{T}_{n-1}}^{n-1}\right)v - \int_\Omega \nabla\left(\theta u_{\mathcal{T}_n}^n + (1-\theta)u_{\mathcal{T}_{n-1}}^{n-1}\right)\cdot\nabla v. \qquad (6.12)$$

Since $\partial_t u_{\mathcal{I}}$ is piecewise constant and equals $\frac{1}{\tau_n}(u_{\mathcal{T}_n}^n - u_{\mathcal{T}_{n-1}}^{n-1})$ on $(t_{n-1},t_n]$, we obtain the decomposition

$$R(u_{\mathcal{I}}) = f - f_{\mathcal{I}} + R_\tau(u_{\mathcal{I}}) + R_h(u_{\mathcal{I}}).$$

Since $f - f_{\mathcal{I}}$ describes oscillations of the known data, the task of deriving upper and lower bounds for the $L^2(t_{n-1},t_n;H^{-1}(\Omega))$-norms of $R(u_{\mathcal{I}})$ reduces to the estimation of the corresponding norms of $R_\tau(u_{\mathcal{I}}) + R_h(u_{\mathcal{I}})$. The following lemma shows that this can be achieved by estimating the contributions of $R_\tau(u_{\mathcal{I}})$ and $R_h(u_{\mathcal{I}})$ separately.

Lemma 6.3 *For every $n \in \{1,\dots,N_{\mathcal{I}}\}$ we have*

$$\begin{aligned}
&\sqrt{\frac{5}{14}}\left(1-\frac{\sqrt{3}}{2}\right)\left\{\left\|R_\tau(u_{\mathcal{I}})\right\|_{L^2(t_{n-1},t_n;H^{-1}(\Omega))}^2 + \left\|R_h(u_{\mathcal{I}})\right\|_{L^2(t_{n-1},t_n;H^{-1}(\Omega))}^2\right\}^{\frac{1}{2}} \\
&\le \left\|R_\tau(u_{\mathcal{I}}) + R_h(u_{\mathcal{I}})\right\|_{L^2(t_{n-1},t_n;H^{-1}(\Omega))} \\
&\le \left\|R_\tau(u_{\mathcal{I}})\right\|_{L^2(t_{n-1},t_n;H^{-1}(\Omega))} + \left\|R_h(u_{\mathcal{I}})\right\|_{L^2(t_{n-1},t_n;H^{-1}(\Omega))}.
\end{aligned}$$

Proof We define $r^n \in H^{-1}(\Omega)$ by setting for every $v \in H_0^1(\Omega)$

$$\langle r^n, v\rangle = \int_\Omega \nabla\left(u_{\mathcal{T}_n}^n - u_{\mathcal{T}_{n-1}}^{n-1}\right)\cdot\nabla v. \qquad (6.13)$$

For every $n \in \{1,\dots,N_{\mathcal{I}}\}$ we then have on $(t_{n-1},t_n]$

$$\langle R_\tau(u_{\mathcal{I}}), v\rangle = \left(\theta - \frac{t-t_{n-1}}{\tau_n}\right)\langle r^n, v\rangle. \qquad (6.14)$$

The assertion therefore follows from Lemma 3.53 (p. 130) and Remark 3.54 (p. 132) with $p = 2$, $Y = H_0^1(\Omega)$, $\|\cdot\|_Y = \|\nabla\cdot\|$, $\varphi = R_h(u_{\mathcal{I}})$, and $\psi = r^n$. □

6.1.5 Estimation of the Temporal Residual

The definition of the dual norm, equations (6.13) and (6.14), the identity

$$\int_{t_{n-1}}^{t_n} \left(\theta - \frac{t - t_{n-1}}{\tau_n}\right)^2 dt = \frac{\tau_n}{3}\left(\theta^3 + (1-\theta)^3\right),$$

and the estimate $\frac{1}{4} \leq \theta^3 + (1-\theta)^3 \leq 1$ for all $\theta \in [0,1]$ yield the following upper and lower bounds for the temporal residual.

Lemma 6.4 *For every $n \in \{1,\ldots,N_{\mathcal{I}}\}$, the temporal residual can be bounded from above and from below by*

$$\sqrt{\frac{\tau_n}{12}}\left\|\nabla\left(u^n_{\mathcal{T}_n} - u^{n-1}_{\mathcal{T}_{n-1}}\right)\right\| \leq \left\|R_\tau(u_{\mathcal{I}})\right\|_{L^2(t_{n-1},t_n;H^{-1}(\Omega))} \leq \sqrt{\frac{\tau_n}{3}}\left\|\nabla\left(u^n_{\mathcal{T}_n} - u^{n-1}_{\mathcal{T}_{n-1}}\right)\right\|.$$

6.1.6 Estimation of the Spatial Residual

Thanks to the identity

$$\left\|R_h(u_{\mathcal{I}})\right\|_{L^2(t_{n-1},t_n;H^{-1}(\Omega))} = \sqrt{\tau_n}\left\|R_h(u_{\mathcal{I}})\right\|_{-1} \tag{6.15}$$

for all $n \in \{1,\ldots,N_{\mathcal{I}}\}$, the estimation of the spatial residual reduces to the evaluation of $\left\|R_h(u_{\mathcal{I}})\right\|_{-1}$ on the time intervals. Since $R_h(u_{\mathcal{I}})$ corresponds to the stationary problems (6.5), its estimation is standard and follows along the lines of Section 3.8 (p. 132). The only minor technical difficulty arises from the simultaneous presence of finite element functions on different spatial meshes. This is the place where we need the transition condition of Section 3.2.6 (p. 86).

Lemma 6.5 *For every $n \in \{1,\ldots,N_{\mathcal{I}}\}$ define the* spatial error indicators *$\eta^n_{\mathcal{T}_n}$ by*

$$\eta^n_{\mathcal{T}_n} = \left\{\sum_{K\in\widetilde{\mathcal{T}}_n} h_K^2 \left\|f_{\mathcal{I}} - \frac{1}{\tau_n}\left(u^n_{\mathcal{T}_n} - u^{n-1}_{\mathcal{T}_{n-1}}\right) + \theta\Delta u^n_{\mathcal{T}_n} + (1-\theta)\Delta u^{n-1}_{\mathcal{T}_{n-1}}\right\|_K^2 \right.$$
$$\left. + \sum_{E\in\widetilde{\mathcal{E}}_{n,\Omega}} h_E \left\|\mathbb{J}_E\left(\theta \mathbf{n}_E\cdot\nabla u^n_{\mathcal{T}_n} + (1-\theta)\mathbf{n}_E\cdot\nabla u^{n-1}_{\mathcal{T}_{n-1}}\right)\right\|_E^2\right\}^{\frac{1}{2}},$$

where $\widetilde{\mathcal{E}}_{n,\Omega}$ denotes the set of all interior faces corresponding to $\widetilde{\mathcal{T}}_n$. Then the estimates

$$c_\dagger \eta^n_{\mathcal{T}_n} \leq \left\|R_h(u_{\mathcal{I}})\right\|_{-1} \leq c^\dagger \eta^n_{\mathcal{T}_n}.$$

hold on each interval $(t_{n-1}, t_n]$. The constants $c^\dagger$ and $c_\dagger$ depend on the shape parameters $C_{\mathcal{T}_n}$. The constant $c^\dagger$ in addition depends on the constant in the transition condition of Section 3.2.6 (p. 86) and the constant $c_\dagger$ also depends on the maximum of the polynomial degrees of the finite element functions.

Proof Choose an integer $n \in \{1, \dots, N_{\mathcal{I}}\}$ and keep it fixed in what follows. Integration by parts element-wise on the elements of $\widetilde{\mathcal{T}}_n$ yields the L^2-representation (3.29) (p. 132) with

$$r|_K = f_{\mathcal{I}} - \frac{1}{\tau_n}\left(u^n_{\mathcal{T}_n} - u^{n-1}_{\mathcal{T}_{n-1}}\right) + \theta \Delta u^n_{\mathcal{T}_n} + (1-\theta)\Delta u^{n-1}_{\mathcal{T}_{n-1}}$$
$$j|_E = -\mathbb{J}_E\left(\theta \mathbf{n}_E \cdot \nabla u^n_{\mathcal{T}_n} + (1-\theta)\mathbf{n}_E \cdot \nabla u^{n-1}_{\mathcal{T}_{n-1}}\right)$$

for all $K \in \widetilde{\mathcal{T}}_n$ and all $E \in \widetilde{\mathcal{E}}_{n,\Omega}$. The definition of $f_{\mathcal{I}}$ and equations (6.5) and (6.12) imply the Galerkin orthogonality

$$\left\langle R_h(u_{\mathcal{I}}),\, v_{\mathcal{T}_n}\right\rangle = 0$$

for all $v_{\mathcal{T}_n} \in X_n$. Due to the degree condition, X_n contains the space $S_0^{1,0}(\mathcal{T}_n)$. Hence, the Galerkin orthogonality (3.30) (p. 132) holds for $\mathcal{T} = \mathcal{T}_n$. Now, we may repeat the arguments of Section 3.8.1 (p. 133) with $\mathcal{N} = \mathcal{N}_n$, the set of vertices corresponding to $\mathcal{T}_n$. We only have to take into account that the sets ω_z and σ_z split into several elements and faces in $\widetilde{\mathcal{T}}_n$ and $\widetilde{\mathcal{E}}_{n,\Omega}$, respectively, and that, due to the transition condition, the diameter of any set ω_z can be bounded by the diameter of every element of $\widetilde{\mathcal{T}}_n$ contained in ω_z. This proves the upper bound. The lower bound is proved in exactly the same way as Theorem 3.59 (p. 137). □

6.1.7 A Residual A Posteriori Error Estimator

Proposition 6.1, Lemmas 6.3, 6.4, and 6.5, and equation (6.15) yield the following a posteriori error estimates.

Theorem 6.6 *For every $n \in \{1, \dots, N_{\mathcal{I}}\}$ define the* error indicators η^n *by*

$$\eta^n = \left\{ \sum_{K \in \widetilde{\mathcal{T}}_n} \tau_n h_K^2 \left\| f_{\mathcal{I}} - \frac{1}{\tau_n}\left(u^n_{\mathcal{T}_n} - u^{n-1}_{\mathcal{T}_{n-1}}\right) + \theta\Delta u^n_{\mathcal{T}_n} + (1-\theta)\Delta u^{n-1}_{\mathcal{T}_{n-1}} \right\|_K^2 \right.$$
$$\left. + \sum_{E \in \widetilde{\mathcal{E}}_{n,\Omega}} \tau_n h_E \left\| \mathbb{J}_E\left(\theta \mathbf{n}_E \cdot \nabla u^n_{\mathcal{T}_n} + (1-\theta)\mathbf{n}_E \cdot \nabla u^{n-1}_{\mathcal{T}_{n-1}}\right)\right\|_E^2 + \sum_{K \in \widetilde{\mathcal{T}}_n} \tau_n \left\| \nabla\left(u^n_{\mathcal{T}_n} - u^{n-1}_{\mathcal{T}_{n-1}}\right)\right\|_K^2 \right\}^{\frac{1}{2}}.$$

Then, for all $n \in \{1, \dots, N_{\mathcal{I}}\}$, the following a posteriori error estimates hold for the weak solution u of problem (6.1) and the function $u_{\mathcal{I}}$ associated with the solutions of problems (6.4), (6.5):

$$\left\{ \|u - u_{\mathcal{I}}\|^2_{L^\infty(0,t_n;L^2(\Omega))} + \|u - u_{\mathcal{I}}\|^2_{L^2(0,t_n;H_0^1(\Omega))} + \left\|\partial_t(u - u_{\mathcal{I}})\right\|^2_{L^2(0,t_n;H^{-1}(\Omega))} \right\}^{\frac{1}{2}}$$
$$\le c^* \left\{ \sum_{m=1}^{n} (\eta^m)^2 + \|f - f_{\mathcal{I}}\|^2_{L^2(0,t_n;H^{-1}(\Omega))} + \|u_0 - \pi_0 u_0\|^2 \right\}^{\frac{1}{2}}$$

and

$$\eta^n \le c_* \left\{ \|u - u_{\mathcal{I}}\|^2_{L^\infty(t_{n-1},t_n;L^2(\Omega))} + \|u - u_{\mathcal{I}}\|^2_{L^2(t_{n-1},t_n;H_0^1(\Omega))} + \left\|\partial_t(u - u_{\mathcal{I}})\right\|^2_{L^2(t_{n-1},t_n;H^{-1}(\Omega))} \right.$$
$$\left. + \|f - f_{\mathcal{I}}\|^2_{L^2(t_{n-1},t_n;H^{-1}(\Omega))} \right\}^{\frac{1}{2}}.$$

The constants c^* and c_* depend on the shape parameters of the corresponding meshes. The constant c^* in addition depends on the constant in the transition condition of Section 3.2.6 (p. 86) and the constant c_* also depends on the maximum of the polynomial degrees of the finite element functions. All constants are independent of the final time T.

Remark 6.7 The third term in η^n can be interpreted as a measure for the error of the time discretisation. Correspondingly, it can be used for controlling the step size in time. The first two terms, on the other hand, can be viewed as a measure for the error of the spatial discretisation and can be used to adapt the mesh-size in space, cf. Section 6.8.2 (p. 363).

Remark 6.8 The present error indicator equals that of [**59**]. Theorem 6.6, however, is stronger than the estimates of [**59**], since it bounds the space–time error which is smaller than the sum of the time error and of the space error estimated in [**59**]. Moreover, we use a different method of proof, in particular when establishing the lower bound.

Remark 6.9 The present indicator is also similar to that of [**218**]. Yet our results are stronger than those of [**218**] since we do not need a condition of the form 'spatial mesh-size sufficiently small', do not need a CFL condition linking the mesh-sizes in space and time, and consider a larger class of discretisations. In [**218**], the transition condition of Section 3.2.6 (p. 86) is replaced by the assumption that $\mathcal{T}_n$ is always a refinement of $\mathcal{T}_{n-1}$. In view of problems with moving fronts, this condition, however, is not realistic.

Remark 6.10 The first term in η^n is the element residual of the discrete solution on a space–time cylinder $Q = K \times (t_{n-1}, t_n]$. The second and third terms are the face residuals on the lateral surface of Q and on its bottom, respectively. Using an inverse estimate and assuming that the mesh-sizes of $\mathcal{T}_n$ and of $\mathcal{T}_{n-1}$ are comparable, the third term can be bounded by $\sum_{K\in\mathcal{T}_n} \tau_n h_K^{-2} \left\| u_{\mathcal{T}_n}^n - u_{\mathcal{T}_{n-1}}^{n-1} \right\|_K^2$. With this modification η^n is identical to the indicator of [**260**].

The lower error bound of Theorem 6.6 is local in time, but global in space. The corresponding local bounds of [**260**], in contrast, are local in space and time.

In [**260**] the ratio between the constants in the upper and lower bounds is proportional to $1 + \tau_n^{-1} h_n^2 + \tau_n h_n^{-2}$ with h_n denoting the maximal mesh-size in $\mathcal{T}_n$. Moreover, in two dimensions, there is an additional loss of a factor $|\ln h_n|$ due to the non-local nature of the H^{-1}-norm. Both drawbacks are overcome by the present analysis.

The time discretisation in [**260**] is not a standard one. It is based on the very weak formulation of the differential equation and can be interpreted as a family of implicit Runge–Kutta schemes. For the lowest order it amounts to the Crank–Nicolson scheme. Here, we use the popular A-stable θ-schemes which in particular yield the Crank–Nicolson scheme, if $\theta = \frac{1}{2}$, and the implicit Euler scheme, if $\theta = 1$.

The present analysis is more restricted than that of [**260**] which covers general quasilinear parabolic equations. This is due to the fact that the present analysis is based on a simple parabolic energy estimate, whereas [**260**] uses a more general variational approach based on the concept of very weak solutions. This restriction is compensated by a simpler analysis and stronger results; it will partially be overcome in Section 6.6 (p. 347).

Remark 6.11 For the Crank–Nicolson scheme, i.e. $\theta = \frac{1}{2}$, the present analysis is not optimal since it does not reflect the second-order accuracy with respect to time of this discretisation. This drawback can be overcome by a more refined analysis based on a suitable higher order interpolation of the discrete solution $u_{\mathcal{T}_n}^n$ labelled *elliptic reconstruction* [**17, 173, 188**].

Remark 6.12 The transition condition may be dropped at the expense of an additional consistency error with respect to time. This term is explicitly controlled and incorporated in the a posteriori error estimates in [**69**].

Remark 6.13 Theorem 6.6 was first proved in [**267**]. The present analysis is inspired by [**63**].

6.2 Time-Dependent Convection–Diffusion Equations

We consider non-stationary convection–diffusion equations

$$\begin{aligned} \partial_t u - \varepsilon \Delta u + \mathbf{a} \cdot \nabla u + bu &= f && \text{in } \Omega \times (0,T] \\ u &= 0 && \text{on } \Gamma_D \times (0,T] \\ \varepsilon \frac{\partial u}{\partial n} &= g && \text{on } \Gamma_N \times (0,T] \\ u(\cdot, 0) &= u_0 && \text{in } \Omega \end{aligned} \tag{6.16}$$

in a bounded space–time cylinder with a polygonal cross-section $\Omega \subset \mathbb{R}^d$, $d \geq 2$, having a Lipschitz boundary Γ consisting of two disjoint parts Γ_D and Γ_N. The final time T is arbitrary, but kept fixed in what follows. We assume that the data satisfy the following conditions.

- $f \in C(0,T; L^2(\Omega))$, $g \in C(0,T; L^2(\Gamma_N))$,
  $\mathbf{a} \in C(0,T; W^{1,\infty}(\Omega)^d)$, $b \in C(0,T; L^\infty(\Omega))$.
- $0 < \varepsilon \ll 1$.
- There are two constants $\beta \geq 0$ and $c_b \geq 0$, which do not depend on ε, such that $-\frac{1}{2} \operatorname{div} \mathbf{a} + b \geq \beta$ in $\Omega \times (0,T]$ and $\|b\|_\infty \leq c_b \beta$ in $(0,T]$.
- The Dirichlet boundary Γ_D has positive $(d-1)$-dimensional Hausdorff measure and includes the inflow boundary

$$\bigcup_{0<t\leq T} \left\{ x \in \Gamma : \mathbf{a}(x,t) \cdot \mathbf{n}(x) < 0 \right\}.$$

The third assumption allows us to simultaneously handle the case of a non-vanishing zero-order reaction term and that of absent reaction; the latter one corresponding to $\beta = 0$. In the case $\beta = 0$ we set $c_b = 0$. The second assumption of course means that we are interested in the convection-dominated regime. The first assumption can be replaced by weaker conditions concerning the temporal smoothness. Its present form, however, simplifies the analysis.

6.2.1 Variational Formulation

As in Section 1.2 (p. 4) we set $H^1_D(\Omega) = \left\{ v \in H^1(\Omega) : v = 0 \text{ on } \Gamma_D \right\}$. But, now, we equip $H^1_D(\Omega)$ with the norm, cf. Section 4.4.1 (p. 164)

$$|||v||| = \left\{ \varepsilon \|\nabla v\|^2 + \beta \|v\|^2 \right\}^{\frac{1}{2}}. \tag{6.17}$$

Due to the third and fourth assumption, this is the natural energy norm of problem (6.16). As usual, the dual space of $H^1_D(\Omega)$ is denoted by $H^1_D(\Omega)^*$ and equipped with the norm

$$|||\varphi|||_* = \sup_{v\in H^1_D(\Omega)\setminus\{0\}} \frac{\langle \varphi\, , \, v\rangle}{|||v|||}. \tag{6.18}$$

The space of Γ_N-traces of H^1-functions is denoted by $H^{\frac{1}{2}}(\Gamma_N)$ and is equipped with the trace norm induced by the energy norm, i.e.

$$\|\varphi\|_{\frac{1}{2};\Gamma_N} = \inf\left\{|||v||| \, : v \in H^1_D(\Omega)\, , \; v = \varphi \text{ on } \Gamma_N\right\}.$$

The dual space of $H^{\frac{1}{2}}(\Gamma_N)$ is denoted by $H^{-\frac{1}{2}}(\Gamma_N)$ and is equipped with the corresponding dual norm. Thus the norms of $H^{\frac{1}{2}}(\Gamma_N)$ and $H^{-\frac{1}{2}}(\Gamma_N)$ depend on the energy norm and consequently on the parameters ε and β.

With these notations, the variational formulation of problem (6.16) takes the form:

find $u \in W^2(0, T; H^1_D(\Omega), H^1_D(\Omega)^*)$ such that $u(\cdot, 0) = u_0$ in $H^1_D(\Omega)^*$ and for almost every $t \in (0, T)$ and all $v \in H^1_D(\Omega)$

$$\int_\Omega \partial_t uv + \varepsilon \int_\Omega \nabla u \cdot \nabla v + \int_\Omega \mathbf{a} \cdot \nabla uv + \int_\Omega buv = \int_\Omega fv + \int_{\Gamma_N} gv. \tag{6.19}$$

The above assumptions imply that problem (6.19) admits a unique solution, cf. [**20**, Sections II.4 and V.2] or [**119**, Theorems XVIII.3.1 and XVIII.3.2].

For later use we note that integration by parts, the third and fourth assumption, and the definition of the dual norm imply for all $v, w \in H^1_D(\Omega)$

$$\varepsilon \int_\Omega \nabla v \cdot \nabla v + \int_\Omega \mathbf{a} \cdot \nabla vv + \int_\Omega bvv \geq |||v|||^2 \tag{6.20}$$

and

$$\varepsilon \int_\Omega \nabla v \cdot \nabla w + \int_\Omega bvw \leq \max\{c_b, 1\}\; |||v|||\; |||w||| \, . \tag{6.21}$$

6.2.2 *Finite Element Discretisation*

For the finite element discretisation of problem (6.16), we choose as in Section 6.1 (p. 309) a partition $\mathcal{I}$ of $[0, T]$, corresponding partitions $\mathcal{T}_n$ of Ω and associated finite element spaces X_n, which satisfy the conditions of Section 3.2.6 (p. 86), and a parameter $\theta \in [\frac{1}{2}, 1]$ and denote by π_n the L^2-projection onto X_n. With the abbreviation

$$\begin{aligned} f^{n\theta} &= \theta f(\cdot, t_n) + (1-\theta) f(\cdot, t_{n-1}), & g^{n\theta} &= \theta g(\cdot, t_n) + (1-\theta) g(\cdot, t_{n-1}), \\ \mathbf{a}^{n\theta} &= \theta \mathbf{a}(\cdot, t_n) + (1-\theta)\mathbf{a}(\cdot, t_{n-1}), & b^{n\theta} &= \theta b(\cdot, t_n) + (1-\theta) b(\cdot, t_{n-1}) \end{aligned}$$

the finite element discretisation of problem (6.16) is given by:

find $u^n_{\mathcal{T}_n} \in X_n, 0 \le n \le N_{\mathcal{I}}$, such that

$$u^0_{\mathcal{T}_0} = \pi_0 u_0 \tag{6.22}$$

and, for $n = 1, \ldots, N_{\mathcal{I}}$ and $u^{n\theta} = \theta u^n_{\mathcal{T}_n} + (1-\theta)u^{n-1}_{\mathcal{T}_{n-1}}$,

$$\begin{aligned}
&\int_\Omega \frac{1}{\tau_n}\left(u^n_{\mathcal{T}_n} - u^{n-1}_{\mathcal{T}_{n-1}}\right) v_{\mathcal{T}_n} + \varepsilon \int_\Omega \nabla u^{n\theta} \cdot \nabla v_{\mathcal{T}_n} + \int_\Omega \mathbf{a}^{n\theta} \cdot \nabla u^{n\theta} v_{\mathcal{T}_n} + \int_\Omega b^{n\theta} u^{n\theta} v_{\mathcal{T}_n} \\
&\quad + \sum_{K \in \widetilde{\mathcal{T}}_n} \vartheta_K \int_K \left(\frac{1}{\tau_n}\left(u^n_{\mathcal{T}_n} - u^{n-1}_{\mathcal{T}_{n-1}}\right) - \varepsilon \Delta u^{n\theta} + \mathbf{a}^{n\theta} \cdot \nabla u^{n\theta} + b^{n\theta} u^{n\theta} \right) \mathbf{a}^{n\theta} \cdot \nabla v_{\mathcal{T}_n} \\
&= \int_\Omega f^{n\theta} v_{\mathcal{T}_n} + \int_{\Gamma_N} g^{n\theta} v_{\mathcal{T}_n} + \sum_{K \in \widetilde{\mathcal{T}}_n} \vartheta_K \int_K f^{n\theta} \mathbf{a}^{n\theta} \cdot \nabla v_{\mathcal{T}_n}.
\end{aligned} \tag{6.23}$$

for all $v_{\mathcal{T}_n} \in X_n$.

The ϑ_K are non-negative stabilisation parameters. The choice $\vartheta_K = 0$ for all K yields the standard Galerkin discretisation; the choice $\vartheta_K > 0$ for all K corresponds to the SUPG discretisations, cf. [**139, 153, 154**] and Section 4.4.2 (p. 166). In what follows we will always assume that

$$\vartheta_K \left\| \mathbf{a}^{n\theta} \right\|_{\infty;K} \le h_K \tag{6.24}$$

holds for all $K \in \widetilde{\mathcal{T}}_n$ and all $n \in \{1, \ldots, N_{\mathcal{I}}\}$. This condition is satisfied for all choices of ϑ_K used in practice.

The third and fourth assumption on the data, condition (6.24), and standard arguments for SUPG discretisations [**139, 153, 154**] imply that problems (6.22) and (6.23) admit a unique solution $(u^n_{\mathcal{T}_n})_{0 \le n \le N_{\mathcal{I}}}$. With this sequence we associate as in Section 6.1 (p. 309) the function $u_{\mathcal{I}}$ which is *piecewise affine* on the time intervals $[t_{n-1}, t_n]$, $1 \le n \le N_{\mathcal{I}}$, and which equals $u^n_{\mathcal{T}_n}$ at time t_n, $0 \le n \le N_{\mathcal{I}}$, cf. equations (6.6) and (6.7) (p. 310).

6.2.3 *The Equivalence of Error and Residual*

With the function $u_{\mathcal{I}}$ we associate the residual $R(u_{\mathcal{I}}) \in L^2(0, T; H^1_D(\Omega)^*)$ by setting for all $v \in H^1_D(\Omega)$

$$\langle R(u_{\mathcal{I}}), v\rangle = \int_\Omega fv + \int_{\Gamma_N} gv - \int_\Omega \partial_t u_{\mathcal{I}} v - \varepsilon \int_\Omega \nabla u_{\mathcal{I}} \cdot \nabla v - \int_\Omega \mathbf{a} \cdot \nabla u_{\mathcal{I}} v - \int_\Omega b u_{\mathcal{I}} v.$$

The following proposition shows that this residual and the error $u - u_{\mathcal{I}}$ are equivalent. Its proof is based on a standard parabolic energy estimate and is similar to the proof of Proposition 6.1 (p. 311). Recall that $H^1_D(\Omega)$ and its dual space $H^1_D(\Omega)^*$ are equipped with the energy norm $|||\cdot|||$ and the dual norm $|||\cdot|||_*$ respectively.

Proposition 6.14 *The error can be bounded from below by*

$$\left\| R(u_{\mathcal{I}}) \right\|_{L^2(0,T;H_D^1(\Omega)^*)} \leq \sqrt{2} \max\{1, c_b\} \Big\{ \|u - u_{\mathcal{I}}\|^2_{L^\infty(0,T;L^2(\Omega))} + \|u - u_{\mathcal{I}}\|^2_{L^2(0,T;H_D^1(\Omega))} + \left\| \partial_t(u - u_{\mathcal{I}}) + \mathbf{a} \cdot \nabla(u - u_{\mathcal{I}}) \right\|^2_{L^2(0,T;H_D^1(\Omega)^*)} \Big\}^{\frac{1}{2}}$$

and, for every $n \in \{1, \ldots, N_{\mathcal{I}}\}$, from above by

$$\Big\{ \|u - u_{\mathcal{I}}\|^2_{L^\infty(0,t_n;L^2(\Omega))} + \|u - u_{\mathcal{I}}\|^2_{L^2(0,t_n;H_D^1(\Omega))} + \left\| \partial_t(u - u_{\mathcal{I}}) + \mathbf{a} \cdot \nabla(u - u_{\mathcal{I}}) \right\|^2_{L^2(0,t_n;H_D^1(\Omega)^*)} \Big\}^{\frac{1}{2}}$$
$$\leq \Big\{ 2(1 + \max\{1, c_b\}^2) \|u_0 - \pi_0 u_0\|^2 + 2(2 + \max\{1, c_b\}^2) \left\| R(u_{\mathcal{I}}) \right\|^2_{L^2(0,t_n;H_D^1(\Omega)^*)} \Big\}^{\frac{1}{2}}.$$

Here c_b is the constant of the third assumption on the data.

Proof Equation (6.19) and the definition of R imply for all $v \in H_D^1(\Omega)$

$$\langle R(u_{\mathcal{I}}), v \rangle = \int_\Omega \partial_t(u - u_{\mathcal{I}})v + \varepsilon \int_\Omega \nabla(u - u_{\mathcal{I}}) \cdot \nabla v + \int_\Omega \mathbf{a} \cdot \nabla(u - u_{\mathcal{I}})v + \int_\Omega b(u - u_{\mathcal{I}})v. \quad (6.25)$$

This identity, the definitions of the norms $|||\cdot|||$ and $|||\cdot|||_*$, and inequality (6.21) yield the first estimate of the proposition.

To prove the second estimate, we proceed as in the proof of Proposition 6.1 (p. 311). We choose an integer $n \in \{1, \ldots, N_{\mathcal{I}}\}$ and a time $t \in [0, t_n]$, insert $v = (u - u_{\mathcal{I}})(\cdot, t)$ in equation (6.25), and integrate the resulting estimate from 0 to t. Taking into account inequality (6.20), this yields

$$\|u - u_{\mathcal{I}}\|^2_{L^\infty(0,t_n;L^2(\Omega))} \leq \|u_0 - \pi_0 u_0\|^2 + \left\| R(u_{\mathcal{I}}) \right\|^2_{L^2(0,t_n;H_D^1(\Omega; *)} \quad (6.26)$$

and

$$\|u - u_{\mathcal{I}}\|^2_{L^2(0,t_n;H_D^1(\Omega))} \leq \|u_0 - \pi_0 u_0\|^2 + \left\| R(u_{\mathcal{I}}) \right\|^2_{L^2(0,t_n;H_D^1(\Omega)^*)}. \quad (6.27)$$

Equation (6.25) and inequality (6.21), on the other hand, imply

$$\left|\left|\left| \partial_t(u - u_{\mathcal{I}}) + \mathbf{a} \cdot \nabla(u - u_{\mathcal{I}}) \right|\right|\right|_* \leq \left|\left|\left| R(u_{\mathcal{I}}) \right|\right|\right|_* + \max\{1, c_b\} \, |||u - u_{\mathcal{I}}||| \, .$$

Taking the square of this inequality, integrating from 0 to t_n, and inserting estimate (6.27), we arrive at

$$\left\| \partial_t(u - u_{\mathcal{I}}) + \mathbf{a} \cdot \nabla(u - u_{\mathcal{I}}) \right\|^2_{L^2(0,t_n;H_D^1(\Omega)^*)}$$
$$\leq 2 \max\{1, c_b\}^2 \|u_0 - \pi_0 u_0\|^2 + 2(1 + \max\{1, c_b\}^2) \left\| R(u_{\mathcal{I}}) \right\|^2_{L^2(0,t_n;H_D^1(\Omega)^*)}. \quad (6.28)$$

Combining estimates (6.26), (6.27), and (6.28) proves the second estimate of the proposition. □

Remark 6.15 Proposition 6.14 shows that the a posteriori error estimation should be based on the norm

$$\left\{\|v\|^2_{L^\infty(0,T;L^2(\Omega))} + \|v\|^2_{L^2(0,T;H^1_D(\Omega))} + \|\partial_t v + \mathbf{a}\cdot\nabla v\|^2_{L^2(0,T;H^1_D(\Omega)^*)}\right\}^{\frac{1}{2}}$$

and that we can strive for a robust error indicator. The continuous embedding of $W^2(0,T;H^1_D(\Omega),H^1_D(\Omega)^*)$ into $L^\infty(0,T;L^2(\Omega))$, cf. [**119**, Chapter XVIII, Section 1, Theorem 1] and [**230**, Theorem 7.2], and the assumptions on the data imply that this norm is equivalent to the standard norm

$$\left\{\|v\|^2_{L^2(0,T;H^1_D(\Omega))} + \|\partial_t v\|^2_{L^2(0,T;H^1_D(\Omega)^*)}\right\}^{\frac{1}{2}}$$

of $W^2(0,T;H^1_D(\Omega),H^1_D(\Omega)^*)$. Yet, the corresponding estimates are not uniform with respect to ε. Hence, error estimates based on the standard norm cannot be robust.

6.2.4 *Decomposition of the Residual*

As for the heat equation, the subsequent analysis relies on an appropriate decomposition of the residual $R(u_{\mathcal{I}})$. To this end we define a temporal residual $R_\tau(u_{\mathcal{I}}) \in L^2(0,T;H^1_D(\Omega)^*)$, a spatial residual $R_h(u_{\mathcal{I}}) \in L^2(0,T;H^1_D(\Omega)^*)$, and a temporal data residual $R_D(u_{\mathcal{I}}) \in L^2(0,T;H^1_D(\Omega)^*)$ by setting on $(t_{n-1},t_n]$ for all $n\in\{1,\dots,N_{\mathcal{I}}\}$ and all $v\in H^1_D(\Omega)$

$$\begin{aligned}
\langle R_\tau(u_{\mathcal{I}}), v\rangle &= \varepsilon\int_\Omega \nabla\left[\theta u^n_{\mathcal{T}_n} + (1-\theta)u^{n-1}_{\mathcal{T}_{n-1}} - u_{\mathcal{I}}\right]\cdot\nabla v + \int_\Omega \mathbf{a}^{n\theta}\cdot\nabla\left[\theta u^n_{\mathcal{T}_n} + (1-\theta)u^{n-1}_{\mathcal{T}_{n-1}} - u_{\mathcal{I}}\right]v \\
&\quad + \int_\Omega b^{n\theta}\left[\theta u^n_{\mathcal{T}_n} + (1-\theta)u^{n-1}_{\mathcal{T}_{n-1}} - u_{\mathcal{I}}\right]v, \\
\langle R_h(u_{\mathcal{I}}), v\rangle &= \int_\Omega f^{n\theta}v + \int_{\Gamma_N} g^{n\theta}v - \int_\Omega \frac{1}{\tau_n}\left(u^n_{\mathcal{T}_n} - u^{n-1}_{\mathcal{T}_{n-1}}\right)v - \varepsilon\int_\Omega\nabla\left[\theta u^n_{\mathcal{T}_n} + (1-\theta)u^{n-1}_{\mathcal{T}_{n-1}}\right]\cdot\nabla v \\
&\quad - \int_\Omega \mathbf{a}^{n\theta}\cdot\nabla\left[\theta u^n_{\mathcal{T}_n} + (1-\theta)u^{n-1}_{\mathcal{T}_{n-1}}\right]v - \int_\Omega b^{n\theta}\left[\theta u^n_{\mathcal{T}_n} + (1-\theta)u^{n-1}_{\mathcal{T}_{n-1}}\right]v, \\
\langle R_D(u_{\mathcal{I}}), v\rangle &= \int_\Omega (f-f^{n\theta})v + \int_{\Gamma_N}(g-g^{n\theta})v + \int_\Omega(\mathbf{a}^{n\theta}-\mathbf{a})\cdot\nabla u_{\mathcal{I}}v + \int_\Omega (b^{n\theta}-b)u_{\mathcal{I}}v.
\end{aligned}$$

Since $\partial_t u_{\mathcal{I}}$ is piecewise constant and equals $\frac{1}{\tau_n}(u^n_{\mathcal{T}_n} - u^{n-1}_{\mathcal{T}_{n-1}})$ on $(t_{n-1},t_n]$, we obtain the decomposition

$$R(u_{\mathcal{I}}) = R_D(u_{\mathcal{I}}) + R_\tau(u_{\mathcal{I}}) + R_h(u_{\mathcal{I}}). \tag{6.29}$$

Since $R_D(u_{\mathcal{I}})$ describes temporal oscillations of the known data, the task of deriving upper and lower bounds for the $L^2(t_{n-1},t_n;H^1_D(\Omega)^*)$-norms of $R(u_{\mathcal{I}})$ reduces to the estimation of the corresponding norms of $R_\tau(u_{\mathcal{I}}) + R_h(u_{\mathcal{I}})$. The following lemma shows that this can be achieved by estimating the contributions of $R_\tau(u_{\mathcal{I}})$ and $R_h(u_{\mathcal{I}})$ separately.

Lemma 6.16 *For every $n \in \{1, \dots, N_{\mathcal{I}}\}$ we have*

$$\sqrt{\frac{5}{14}}\left(1-\frac{\sqrt{3}}{2}\right)\left\{\left\|R_\tau(u_{\mathcal{I}})\right\|^2_{L^2(t_{n-1},t_n;H^1_D(\Omega)^*)} + \left\|R_h(u_{\mathcal{I}})\right\|^2_{L^2(t_{n-1},t_n;H^1_D(\Omega)^*)}\right\}^{\frac{1}{2}}$$
$$\le \left\|R_\tau(u_{\mathcal{I}}) + R_h(u_{\mathcal{I}})\right\|_{L^2(t_{n-1},t_n;H^1_D(\Omega)^*)}$$
$$\le \left\|R_\tau(u_{\mathcal{I}})\right\|_{L^2(t_{n-1},t_n;H^1_D(\Omega)^*)} + \left\|R_h(u_{\mathcal{I}})\right\|_{L^2(t_{n-1},t_n;H^1_D(\Omega)^*)}.$$

Proof For every $n \in \{1, \dots, N_{\mathcal{I}}\}$ we now set for all $v \in H^1_D(\Omega)$

$$\langle r^n, v\rangle = \varepsilon \int_\Omega \nabla\left(u^n_{\mathcal{T}_n} - u^{n-1}_{\mathcal{T}_{n-1}}\right)\cdot \nabla v + \int_\Omega \mathbf{a}^{n\theta}\cdot\nabla\left(u^n_{\mathcal{T}_n} - u^{n-1}_{\mathcal{T}_{n-1}}\right) v + \int_\Omega b^{n\theta}\left(u^n_{\mathcal{T}_n} - u^{n-1}_{\mathcal{T}_{n-1}}\right) v. \tag{6.30}$$

The assertion then follows from Lemma 3.53 (p. 130) and Remark 3.54 (p. 132) with $p = 2$, $Y = H^1_D(\Omega)$, $\|\cdot\|_Y = |||\cdot|||$, $\varphi = R_h(u_{\mathcal{I}})$, and $\psi = r^n$. □

6.2.5 *Estimation of the Temporal Residual*

Thanks to Lemma 6.4 (p. 314), bounding the $L^2(t_{n-1}, t_n; H^1_D(\Omega)^*)$-norms of the temporal residual $R_\tau(u_{\mathcal{I}})$ reduces to the estimation of the norms $|||r^n|||_*$ of the residuals r^n defined in equation (6.30). Here, we may benefit from Proposition 4.17 (p. 165). Lemma 6.4 (p. 314), equation (6.30), and Proposition 4.17 with $\mathbf{a}$ and b replaced by $\mathbf{a}^{n\theta}$ and $b^{n\theta}$, respectively, prove the following estimates for the temporal residual.

Lemma 6.17 *For every $n \in \{1, \dots, N_{\mathcal{I}}\}$, the temporal residual can be bounded from above and from below by*

$$\frac{\sqrt{\tau_n}}{\sqrt{12}(2+\max\{c_b, 1\})}\left\{\left|\left|\left|u^n_{\mathcal{T}_n} - u^{n-1}_{\mathcal{T}_{n-1}}\right|\right|\right| + \left|\left|\left|\mathbf{a}^{n\theta}\cdot\nabla\left(u^n_{\mathcal{T}_n} - u^{n-1}_{\mathcal{T}_{n-1}}\right)\right|\right|\right|_*\right\}$$
$$\le \left\|R_\tau(u_{\mathcal{I}})\right\|_{L^2(t_{n-1},t_n;H^1_D(\Omega)^*)}$$
$$\le \frac{\sqrt{\tau_n}}{\sqrt{3}\max\{c_b, 1\}}\left\{\left|\left|\left|u^n_{\mathcal{T}_n} - u^{n-1}_{\mathcal{T}_{n-1}}\right|\right|\right| + \left|\left|\left|\mathbf{a}^{n\theta}\cdot\nabla\left(u^n_{\mathcal{T}_n} - u^{n-1}_{\mathcal{T}_{n-1}}\right)\right|\right|\right|_*\right\}.$$

In contrast to $\left|\left|\left|u^n_{\mathcal{T}_n} - u^{n-1}_{\mathcal{T}_{n-1}}\right|\right|\right|$ the term $\left|\left|\left|\mathbf{a}^{n\theta}\cdot\nabla(u^n_{\mathcal{T}_n} - u^{n-1}_{\mathcal{T}_{n-1}})\right|\right|\right|_*$ is not suited as an error indicator since it involves a dual norm. Standard approaches bound this term by inverse estimates, if need be, combined with integration by parts, cf. estimate (6.34) (p. 329). This, however, leads to estimates which incorporate a factor $\varepsilon^{-\frac{1}{2}}$ and which are not robust. The idea which leads to computable robust indicators is as follows.

> Due to the definition of the dual norm, the quantities $\left|\left|\left|\mathbf{a}^{n\theta}\cdot\nabla(u^n_{\mathcal{T}_n} - u^{n-1}_{\mathcal{T}_{n-1}})\right|\right|\right|_*$ equal the energy norm of the weak solutions of suitable stationary reaction–diffusion equations. These solutions are approximated by suitable finite element functions. The error of the approximations is estimated by robust error indicators for reaction–diffusion equations.

Lemma 6.18 *For every $n \in \{1,\ldots,N_{\mathcal{I}}\}$ denote by $\widetilde{X}_n = S_D^{1,0}(\widetilde{\mathcal{T}}_n)$ the space of continuous piecewise (multi-) linear functions corresponding to $\widetilde{\mathcal{T}}_n$ and vanishing on Γ_D and by $\widetilde{u}^n_{\mathcal{T}_n} \in \widetilde{X}_n$ the unique solution of the discrete reaction–diffusion problem*

$$\varepsilon \int_\Omega \nabla \widetilde{u}^n_{\mathcal{T}_n} \cdot \nabla v_{\mathcal{T}_n} + \beta \int_\Omega \widetilde{u}^n_{\mathcal{T}_n} v_{\mathcal{T}_n} = \int_\Omega \mathbf{a}^{n\theta} \cdot \nabla \left(u^n_{\mathcal{T}_n} - u^{n-1}_{\mathcal{T}_{n-1}} \right) v_{\mathcal{T}_n} \tag{6.31}$$

for all $v_{\mathcal{T}_n} \in \widetilde{X}_n$. Define the error indicator $\widetilde{\eta}^n_{\mathcal{T}_n}$ by

$$\widetilde{\eta}^n_{\mathcal{T}_n} = \left\{ \sum_{K \in \widetilde{\mathcal{T}}_n} \hbar_K^2 \left\| \mathbf{a}^{n\theta} \cdot \nabla \left(u^n_{\mathcal{T}_n} - u^{n-1}_{\mathcal{T}_{n-1}} \right) + \varepsilon \Delta \widetilde{u}^n_{\mathcal{T}_n} - \beta \widetilde{u}^n_{\mathcal{T}_n} \right\|_K^2 + \sum_{E \in \widetilde{\mathcal{E}}_{n,\Omega} \cup \widetilde{\mathcal{E}}_{n,\Gamma_N}} \varepsilon^{-\frac{1}{2}} \hbar_E \left\| \mathbb{J}_E \left(\mathbf{n}_E \cdot \nabla \widetilde{u}^n_{\mathcal{T}_n} \right) \right\|_E^2 \right\}^{\frac{1}{2}}, \tag{6.32}$$

where $\hbar_S$ is defined as in (3.4) (p. 82). Then there are two constants $\widetilde{c}_\dagger$ and $\widetilde{c}^\dagger$ which only depend on the shape parameter of $\widetilde{\mathcal{T}}_n$ such that the following estimates are valid

$$\widetilde{c}_\dagger \left\{ \left|\left|\left| \widetilde{u}^n_{\mathcal{T}_n} \right|\right|\right| + \widetilde{\eta}^n_{\mathcal{T}_n} \right\} \le \left|\left|\left| \mathbf{a}^{n\theta} \cdot \nabla \left(u^n_{\mathcal{T}_n} - u^{n-1}_{\mathcal{T}_{n-1}} \right) \right|\right|\right|_* \le \widetilde{c}^\dagger \left\{ \left|\left|\left| \widetilde{u}^n_{\mathcal{T}_n} \right|\right|\right| + \widetilde{\eta}^n_{\mathcal{T}_n} \right\}.$$

Proof We choose $n \in \{1,\ldots,N_{\mathcal{I}}\}$ and keep it fixed in what follows. Denote by $U^n \in H^1_D(\Omega)$ the unique solution of the stationary reaction–diffusion equation

$$\varepsilon \int_\Omega \nabla U^n \cdot \nabla v + \beta \int_\Omega U^n v = \int_\Omega \mathbf{a}^{n\theta} \cdot \nabla \left(u^n_{\mathcal{T}_n} - u^{n-1}_{\mathcal{T}_{n-1}} \right) v$$

for all $v \in H^1_D(\Omega)$. The definitions (6.17) and (6.18) of the energy norm $|||\cdot|||$ and of the dual norm $|||\cdot|||_*$, respectively, imply that

$$|||U^n||| = \left|\left|\left| \mathbf{a}^{n\theta} \cdot \nabla \left(u^n_{\mathcal{T}_n} - u^{n-1}_{\mathcal{T}_{n-1}} \right) \right|\right|\right|_*.$$

Inserting $v_{\mathcal{T}_n} = \widetilde{u}^n_{\mathcal{T}_n}$ as a test function in the discrete problem (6.31), we obtain

$$\left|\left|\left| \widetilde{u}^n_{\mathcal{T}_n} \right|\right|\right| \le \left|\left|\left| \mathbf{a}^{n\theta} \cdot \nabla \left(u^n_{\mathcal{T}_n} - u^{n-1}_{\mathcal{T}_{n-1}} \right) \right|\right|\right|_*.$$

The triangle inequality therefore yields

$$\begin{aligned} \frac{1}{3} \left\{ \left|\left|\left| \widetilde{u}^n_{\mathcal{T}_n} \right|\right|\right| + \left|\left|\left| U^n - \widetilde{u}^n_{\mathcal{T}_n} \right|\right|\right| \right\} &\le \left|\left|\left| \mathbf{a}^{n\theta} \cdot \nabla \left(u^n_{\mathcal{T}_n} - u^{n-1}_{\mathcal{T}_{n-1}} \right) \right|\right|\right|_* \\ &\le \left\{ \left|\left|\left| \widetilde{u}^n_{\mathcal{T}_n} \right|\right|\right| + \left|\left|\left| U^n - \widetilde{u}^n_{\mathcal{T}_n} \right|\right|\right| \right\}. \end{aligned}$$

Since $\mathbf{a}^{n\theta} \cdot \nabla (u^n_{\mathcal{T}_n} - u^{n-1}_{\mathcal{T}_{n-1}})$ is piecewise a polynomial, we know from Theorem 4.10 (p. 160) that $\widetilde{\eta}^n_{\mathcal{T}_n}$ yields upper and lower bounds for $\left|\left|\left| U^n - \widetilde{u}^n_{\mathcal{T}_n} \right|\right|\right|$ with multiplicative constants that only depend on the shape parameter of $\widetilde{\mathcal{T}}_n$. This proves Lemma 6.18. □

6.2.6 Estimation of the Spatial Residual

The estimation of the spatial residual follows the arguments of Section 6.1.6 (p. 314). We only have to take into account that now $H_0^1(\Omega)$ and $\|\nabla\cdot\|$ are replaced by $H_D^1(\Omega)$ and $|||\cdot|||$. This results in different weighting factors for the element and face residuals. They take into account that $|||\cdot|||$ is the energy norm corresponding to the reaction–diffusion operator $u \mapsto -\varepsilon\Delta u + \beta u$. Therefore, we can benefit from the results of Section 4.3 (p. 159).

For brevity, we denote by $f_{\mathcal{I}}$, $g_{\mathcal{I}}$, $\mathbf{a}_{\mathcal{I}}$, and $b_{\mathcal{I}}$ functions which are *piecewise constant* on the time intervals and which, on each interval $(t_{n-1}, t_n]$, equal the L^2-projection of $f^{n\theta}$, $g^{n\theta}$, $\mathbf{a}^{n\theta}$, and $b^{n\theta}$, respectively, onto the space of piecewise constant functions or vector fields corresponding to $\mathcal{T}_n$. Recall that $\widetilde{\mathcal{E}}_n$ denotes the set of faces corresponding to the partition $\widetilde{\mathcal{T}}_n$.

With this notation, we define

element residuals R_K, $K \in \widetilde{\mathcal{T}}_n$, $1 \le n \le N_{\mathcal{I}}$, by

$$\begin{aligned} R_K &= f_{\mathcal{I}} - \frac{1}{\tau_n}\left(u_{\mathcal{T}_n}^n - u_{\mathcal{T}_{n-1}}^{n-1}\right) + \varepsilon\Delta\left(\theta u_{\mathcal{T}_n}^n + (1-\theta)u_{\mathcal{T}_{n-1}}^{n-1}\right) \\ &\quad - \mathbf{a}_{\mathcal{I}}\cdot\nabla\left(\theta u_{\mathcal{T}_n}^n + (1-\theta)u_{\mathcal{T}_{n-1}}^{n-1}\right) - b_{\mathcal{I}}\left(\theta u_{\mathcal{T}_n}^n + (1-\theta)u_{\mathcal{T}_{n-1}}^{n-1}\right), \end{aligned}$$

face residuals R_E, $E \in \widetilde{\mathcal{E}}_n$, $1 \le n \le N_{\mathcal{I}}$, by

$$R_E = \begin{cases} -\mathbb{J}_E\left(\varepsilon\mathbf{n}_E\cdot\nabla(\theta u_{\mathcal{T}_n}^n + (1-\theta)u_{\mathcal{T}_{n-1}}^{n-1}\right) & \text{if } E \in \widetilde{\mathcal{E}}_{n,\Omega}, \\ g_{\mathcal{I}} - \varepsilon\mathbf{n}_E\cdot\nabla\left(\theta u_{\mathcal{T}_n}^n + (1-\theta)u_{\mathcal{T}_{n-1}}^{n-1}\right) & \text{if } E \in \widetilde{\mathcal{E}}_{n,\Gamma_N}, \\ 0 & \text{if } E \in \widetilde{\mathcal{E}}_{n,\Gamma_D}, \end{cases}$$

element-wise data errors D_K, $K \in \widetilde{\mathcal{T}}_n$, $1 \le n \le N_{\mathcal{I}}$, by

$$D_K = \left\{f^{n\theta} - f_{\mathcal{I}} + (\mathbf{a}_{\mathcal{I}} - \mathbf{a}^{n\theta})\nabla\cdot\left(\theta u_{\mathcal{T}_n}^n + (1-\theta)u_{\mathcal{T}_{n-1}}^{n-1}\right) + (b_{\mathcal{I}} - b^{n\theta})\left(\theta u_{\mathcal{T}_n}^n + (1-\theta)u_{\mathcal{T}_{n-1}}^{n-1}\right)\right\},$$

and *face-wise data errors* D_E, $E \in \widetilde{\mathcal{E}}_{n,\Gamma_N}$, $1 \le n \le N_{\mathcal{I}}$, by

$$D_E = g^{n\theta} - g_{\mathcal{I}}.$$

Here, of course, $(u_{\mathcal{T}_n}^n)_{0\le n\le N}$ denotes the solution of problems (6.22) and (6.23).

Lemma 6.19 *For every $n \in \{1,\ldots,N_{\mathcal{I}}\}$, define a spatial error indicator $\eta_{\mathcal{T}_n}^n$ by*

$$\eta_{\mathcal{T}_n}^n = \left\{\sum_{K\in\widetilde{\mathcal{T}}_n} \hbar_K^2 \|R_K\|_K^2 + \sum_{E\in\widetilde{\mathcal{E}}_n} \varepsilon^{-\frac{1}{2}}\hbar_E \|R_E\|_E^2\right\}^{\frac{1}{2}}$$

and a spatial data error indicator by

$$\Theta_{\mathcal{T}_n}^n = \left\{\sum_{K\in\widetilde{\mathcal{T}}_n} \hbar_K^2 \|D_K\|_K^2 + \sum_{E\in\widetilde{\mathcal{E}}_{n,\Gamma_N}} \varepsilon^{-\frac{1}{2}}\hbar_E \|D_E\|_E^2\right\}^{\frac{1}{2}},$$

where $\hbar_K$ and $\hbar_E$ are given by (3.4) (p. 82) with $\omega = K$ and $\omega = E$, respectively. Then, there are constants $c^\dagger$ and $c_\dagger$ such that the estimates

$$c_\dagger \eta^n_{\mathcal{T}_n} \le \left|\left|\left| R_h(u_\mathcal{I}) \right|\right|\right|_* + \Theta^n_{\mathcal{T}_n}$$
$$\left|\left|\left| R_h(u_\mathcal{I}) \right|\right|\right|_* \le c^\dagger \left\{ \eta^n_{\mathcal{T}_n} + \Theta^n_{\mathcal{T}_n} \right\}.$$

hold on each interval $(t_{n-1}, t_n]$. The constants $c^\dagger$ and $c_\dagger$ depend on the shape parameters $C_{\widetilde{\mathcal{T}}_n}$. The constant $c^\dagger$ in addition depends on the constant in the transition condition of Section 3.2.6 (p. 86) and the constant $c_\dagger$ also depends on the maximum of the polynomial degrees of the finite element functions.

Proof The proof of Lemma 6.19 is similar to that of Lemma 6.5 (p. 314) and follows from Theorems 3.57 (p. 135) and 3.59 (p. 137). □

6.2.7 *A Robust A posteriori Error Indicator*

Proposition 6.14 and Lemmas 6.16, 6.17, 6.18, and 6.19, and equation (6.15) (p. 314) yield the following robust a posteriori error estimates.

Theorem 6.20 *The error between the solution u of problem (6.21) and the solution $u_\mathcal{I}$ of problems (6.22), (6.23) is bounded from above by*

$$\begin{aligned}
&\left\{ \|u - u_\mathcal{I}\|^2_{L^\infty(0,T;L^2(\Omega))} + \|u - u_\mathcal{I}\|^2_{L^2(0,T;H^1_D(\Omega))} + \left\| \partial_t(u - u_\mathcal{I}) + \mathbf{a}\cdot\nabla(u - u_\mathcal{I}) \right\|^2_{L^2(0,T;H^1_D(\Omega)^*)} \right\}^{\frac{1}{2}} \\
&\le \widetilde{c}^* \left\{ \|u_0 - \pi_0 u_0\|^2 + \sum_{n=1}^{N_\mathcal{I}} \tau_n \left[\left(\eta^n_{\mathcal{T}_n}\right)^2 + \left|\left|\left| u^n_{\mathcal{T}_n} - u^{n-1}_{\mathcal{T}_{n-1}} \right|\right|\right|^2 + \left(\widetilde{\eta}^n_{\mathcal{T}_n}\right)^2 + \left|\left|\left| \widetilde{u}^n_{\mathcal{T}_n} \right|\right|\right|^2 \right] \right. \\
&\quad + \sum_{n=1}^{N_\mathcal{I}} \tau_n \left(\Theta^n_{\mathcal{T}_n}\right)^2 + \left\| f - f^{n\theta} - (\mathbf{a} - \mathbf{a}^{n\theta})\cdot\nabla u_\mathcal{I} - (b - b^{n\theta}) u_\mathcal{I} \right\|^2_{L^2(0,T;H^1_D(\Omega)^*)} \\
&\quad \left. + \left\| g - g_\mathcal{I} \right\|^2_{L^2(0,T;H^{-\frac{1}{2}}(\Gamma_N))} \right\}^{\frac{1}{2}}
\end{aligned}$$

and on each interval $(t_{n-1}, t_n]$, $1 \le n \le N_\mathcal{I}$, from below by

$$\begin{aligned}
&\tau_n^{\frac{1}{2}} \left\{ \left(\eta^n_{\mathcal{T}_n}\right)^2 + \left|\left|\left| u^n_{\mathcal{T}_n} - u^{n-1}_{\mathcal{T}_{n-1}} \right|\right|\right|^2 + \left(\widetilde{\eta}^n_{\mathcal{T}_n}\right)^2 + \left|\left|\left| \widetilde{u}^n_{\mathcal{T}_n} \right|\right|\right|^2 \right\}^{\frac{1}{2}} \\
&\le \widetilde{c}_* \left\{ \|u - u_\mathcal{I}\|^2_{L^\infty(t_{n-1},t_n;L^2(\Omega))} + \|u - u_\mathcal{I}\|^2_{L^2(t_{n-1},t_n;H^1_D(\Omega))} \right. \\
&\quad + \left\| \partial_t(u - u_\mathcal{I}) + \mathbf{a}\cdot\nabla(u - u_\mathcal{I}) \right\|^2_{L^2(t_{n-1},t_n;H^1_D(\Omega)^*)} + \tau_n \left(\Theta^n_{\mathcal{T}_n}\right)^2 \\
&\quad \left. + \left\| f - f^{n\theta} - (\mathbf{a} - \mathbf{a}^{n\theta})\cdot\nabla u_\mathcal{I} - (b - b^{n\theta}) u_\mathcal{I} \right\|^2_{L^2(t_{n-1},t_n;H^1_D(\Omega)^*)} + \left\| g - g_\mathcal{I} \right\|^2_{L^2(t_{n-1},t_n;H^{-\frac{1}{2}}(\Gamma_N))} \right\}^{\frac{1}{2}}.
\end{aligned}$$

The functions $\widetilde{u}^n_{\mathcal{T}_n}$ and the indicators $\widetilde{\eta}^n_{\mathcal{T}_n}$ are defined in Lemma 6.18, and the quantities $\eta^n_{\mathcal{T}_n}$ and $\Theta^n_{\mathcal{T}_n}$ are as in Lemma 6.19. The constants $\widetilde{c}^*$ and $\widetilde{c}_*$ depend on the shape parameters $C_{\widetilde{\mathcal{T}}_n}$. The constant $\widetilde{c}^*$ in addition depends on the constant in the transition condition of Section 3.2.6 (p. 86) and the *constant $\tilde{c}_*$ also depends on the maximum of the polynomial degrees of the finite element functions. All constants are independent of the final time T, the viscosity ε, and the parameter β.*

Remark 6.21 Theorem 6.20 shows that the quantity

$$\tau_n^{\frac{1}{2}} \left\{ \left(\eta_{\mathcal{T}_n}^n\right)^2 + \left|\left|\left| u_{\mathcal{T}_n}^n - u_{\mathcal{T}_{n-1}}^{n-1} \right|\right|\right|^2 + \left(\tilde{\eta}_{\mathcal{T}_n}^n\right)^2 + \left|\left|\left| \tilde{u}_{\mathcal{T}_n}^n \right|\right|\right|^2 \right\}^{\frac{1}{2}}$$

is a robust error indicator in the sense of Section 4.3 (p. 159). The remaining terms on the right-hand side of the upper bound for the error and the second and third term on the right-hand side of the lower bound for the error are data errors. They can be bounded a priori by computable norms involving the data f, g, $\mathbf{a}$, and b. The term $\tau_n^{\frac{1}{2}} \eta_{\mathcal{T}_n}^n$ can be interpreted as a spatial error indicator. The terms

$$\tau_n^{\frac{1}{2}} \left\{ \left|\left|\left| u_{\mathcal{T}_n}^n - u_{\mathcal{T}_{n-1}}^{n-1} \right|\right|\right|^2 + \left(\tilde{\eta}_{\mathcal{T}_n}^n\right)^2 + \left|\left|\left| \tilde{u}_{\mathcal{T}_n}^n \right|\right|\right|^2 \right\}^{\frac{1}{2}}$$

on the other hand can be viewed as temporal error indicators.

Remark 6.22 The presentation of this section follows [**63**, **270**].

6.3 Linear Parabolic Equations of Second Order

In this section we extend the results of the previous section to general linear parabolic equations of second order:

$$\begin{aligned} \partial_t u - \operatorname{div}(A\nabla u) + \mathbf{a} \cdot \nabla u + bu &= f && \text{in } \Omega \times (0, T] \\ u &= 0 && \text{on } \Gamma_D \times (0, T] \\ \mathbf{n} \cdot A\nabla u &= g && \text{on } \Gamma_N \times (0, T] \\ u(\cdot, 0) &= u_0 && \text{in } \Omega. \end{aligned} \tag{6.33}$$

Here, as in Section 6.2, $\Omega \subset \mathbb{R}^d$, $d \geq 2$, is a bounded polygonal cross-section with a Lipschitz boundary Γ consisting of two disjoint parts Γ_D and Γ_N. The final time T is arbitrary, but kept fixed in what follows.

We assume that the data satisfy the following conditions.

- $f \in C(0, T; L^2(\Omega)), g \in C(0, T; L^2(\Gamma_N))$,
  $A \in C(0, T; L^\infty(\Omega)^{d\times d}), \mathbf{a} \in C(0, T; W^{1,\infty}(\Omega)^d)$,
  $b \in C(0, T; L^\infty(\Omega))$.
- The diffusion coefficient A is symmetric, uniformly positive definite, and uniformly isotropic, i.e.

$$\varepsilon = \inf_{0<t\leq T, x\in\Omega} \min_{z\in\mathbb{R}^d\setminus\{0\}} \frac{z^T A(x,t) z}{z^T z} > 0$$

and the constant

$$\kappa = \varepsilon^{-1} \sup_{0<t\leq T, x\in\Omega} \max_{z\in\mathbb{R}^d\setminus\{0\}} \frac{z^T A(x,t) z}{z^T z}$$

is of moderate size.

- There is a constant $\beta \geq 0$ such that $b - \frac{1}{2}\operatorname{div}\mathbf{a} \geq \beta$ on $\Omega \times (0, T]$. Moreover, there is a constant $c_b \geq 0$ of moderate size such that $\|b\|_{L^\infty((0,T)\times\Omega)} \leq c_b\beta$.
- The Dirichlet boundary Γ_D has positive $(d-1)$-dimensional Hausdorff measure and includes the inflow boundary

$$\bigcup_{0<t\leq T} \left\{x \in \Gamma : \mathbf{a}(x,t) \cdot \mathbf{n}(x) < 0\right\}.$$

These assumptions guarantee that problem (6.33) is a well-posed parabolic problem. The parameter κ is introduced in order to stress that the ratio of the constants in the error estimates depends on the condition number of the diffusion matrix. If this condition number is exceedingly large, length scales such as element diameters must be measured in a diffusion-dependent metric in order to recover robust estimates as described in Section 4.6 (p. 191) for elliptic equations. The assumptions cover a wide range of different regimes:

- *dominant diffusion*: $\|\mathbf{a}\|_{L^\infty(0,T;W^{1,\infty}(\Omega))} \leq c_c\varepsilon$ and $\beta \leq c_b'\varepsilon$ with constants of moderate size;
- *dominant reaction*: $\|\mathbf{a}\|_{L^\infty(0,T;W^{1,\infty}(\Omega))} \leq c_c\varepsilon$ and $\beta \gg \varepsilon$ with a constant c_c of moderate size;
- *dominant convection*: $\|\mathbf{a}\|_{L^\infty(0,T;W^{1,\infty}(\Omega))} \gg \varepsilon^{\frac{1}{2}} \max\{\varepsilon, \beta\}^{\frac{1}{2}}$.

The variational formulation of problem (6.33) is given by equation (6.19) (p. 318) with

$$\varepsilon \int_\Omega \nabla u \cdot \nabla v \quad \text{replaced by} \quad \int_\Omega \nabla u \cdot A \cdot \nabla v.$$

Again the previous assumptions imply that the variational problem admits a unique solution, cf. [**20**, Sections II.4 and V.2] or [**119**, Theorems XVIII.3.1 and XVIII.3.2].

In the finite element discretisation (6.23) (p. 319) the terms

$$\varepsilon \int_\Omega \nabla \left(\theta u^n_{\mathcal{T}_n} + (1-\theta) u^{n-1}_{\mathcal{T}_{n-1}}\right) \cdot \nabla v_{\mathcal{T}_n}, \quad \varepsilon\Delta\left(\theta u^n_{\mathcal{T}_n} + (1-\theta)u^{n-1}_{\mathcal{T}_{n-1}}\right)$$

must be replaced by

$$\int_\Omega \nabla\left(\theta u^n_{\mathcal{T}_n} + (1-\theta)u^{n-1}_{\mathcal{T}_{n-1}}\right) \cdot A^{n\theta} \cdot \nabla v_{\mathcal{T}_n},$$
$$\operatorname{div}\left(A^{n\theta}\nabla\left(\theta u^n_{\mathcal{T}_n} + (1-\theta)u^{n-1}_{\mathcal{T}_{n-1}}\right)\right),$$

respectively, where $A^{n\theta} = \theta A(\cdot, t_n) + (1-\theta)A(\cdot, t_{n-1})$.

In the definition of the residual $R(u_\mathcal{T})$, the term $\varepsilon \int_\Omega \nabla u_\mathcal{T} \cdot \nabla v$ must be replaced by $\int_\Omega \nabla u_\mathcal{T} \cdot A \cdot \nabla v$.

Proposition 6.14 (p. 320) carries over immediately, but the constants now depend on the parameters c_b and κ. The splitting (6.29) (p. 321) of the residual and Lemma 6.16 (p. 322) remain valid. Due to the variable diffusion, the data residual must be augmented by the term $\int_\Omega (A^{n\theta} - A)\nabla u_{\mathcal{I}} \cdot \nabla v$.

The estimation of the temporal residual, Lemma 6.17 (p. 322), carries over immediately with the modification that the constants also depend on the parameter κ.

In the definition of the element residuals R_K and the face residuals R_E the terms

$$\Delta\left(\theta u^n_{\mathcal{T}_n} + (1-\theta)u^{n-1}_{\mathcal{T}_{n-1}}\right), \quad \varepsilon \mathbf{n}_E \cdot \nabla\left(\theta u^n_{\mathcal{T}_n} + (1-\theta)u^{n-1}_{\mathcal{T}_{n-1}}\right)$$

must be replaced by

$$\operatorname{div}\left(A\nabla\left(\theta u^n_{\mathcal{T}_n} + (1-\theta)u^{n-1}_{\mathcal{T}_{n-1}}\right)\right), \quad \mathbf{n}_E \cdot A\nabla\left(\theta u^n_{\mathcal{T}_n} + (1-\theta)u^{n-1}_{\mathcal{T}_{n-1}}\right).$$

The element-wise data errors D_K and the face-wise data errors D_E must be augmented by the terms $\operatorname{div}((A^{n\theta}_{\mathcal{T}_n} - A^{n\theta})\nabla(\theta u^n_{\mathcal{T}_n} + (1-\theta)u^{n-1}_{\mathcal{T}_{n-1}}))$ and $\mathbf{n}_E \cdot ((A^{n\theta}_{\mathcal{T}_n} - A^{n\theta})\nabla(\theta u^n_{\mathcal{T}_n} - (1-\theta)u^{n-1}_{\mathcal{T}_{n-1}}))$ for faces on Γ_N and $\mathbb{J}_E\left(\mathbf{n}_E \cdot ((A^{n\theta}_{\mathcal{T}_n} - A^{n\theta})\nabla(\theta u^n_{\mathcal{T}_n} + (1-\theta)u^{n-1}_{\mathcal{T}_{n-1}}))\right)$ for interior faces where $A^{n\theta}_{\mathcal{T}_n}$ is the L^2-projection of $A^{n\theta}$ onto $S^{0,-1}(\mathcal{T}_n)^{d\times d}$. Moreover, the spatial data error indicator $\Theta^n_{\mathcal{T}_n}$ must be augmented by

$$\sum_{E\in\widetilde{\mathcal{E}}_{n,\Omega}} \varepsilon^{-\frac{1}{2}} \hbar_E \|D_E\|^2_E.$$

With these modifications, Lemma 6.19 (p. 324) remains valid. The constants $c^\dagger$ and $c_\dagger$ now depend on the parameter κ too.

For the derivation of the final, computable a posteriori error indicator which does not incorporate the dual norms of the convection terms, we must now distinguish two cases:

- *large convection*: $\|\mathbf{a}\|_{L^\infty(0,T;L^\infty(\Omega))} \gg \varepsilon^{\frac{1}{2}} \max\{\varepsilon, \beta\}^{\frac{1}{2}}$;
- *small convection*: $\|\mathbf{a}\|_{L^\infty(0,T;L^\infty(\Omega))} \le c_c \varepsilon^{\frac{1}{2}} \max\{\varepsilon, \beta\}^{\frac{1}{2}}$ with a constant c_c of moderate size.

In the first case, we proceed exactly as in Section 6.2.7 (p. 325) and solve an auxiliary reaction–diffusion equation. Theorem 6.20 (p. 325) immediately carries over with the only modification that the constants $\widetilde{c}^*$ and $\widetilde{c}_*$ also depend on the parameter κ.

In the second case, we use an inverse estimate to bound the critical term $\left|\left|\left|\mathbf{a}^{n\theta} \cdot \nabla(u^n_{\mathcal{T}_n} - u^{n-1}_{\mathcal{T}_{n-1}})\right|\right|\right|_*$ by $\left|\left|\left|u^n_{\mathcal{T}_n} - u^{n-1}_{\mathcal{T}_{n-1}}\right|\right|\right|$ times a constant of moderate size. More precisely, denote by c_F the constant in Friedrichs' inequality, cf. [3], [**109**, inequality (2.2)], inequality (1.66) (p. 47), and Section 3.4.7 (p. 105)

$$\|v\| \le c_F \|\nabla v\|$$

for all $v \in H^1_D(\Omega)$. This estimate and the definition (6.17) of the energy norm yield for all v, $w \in H^1_D(\Omega)$ the estimate

$$\begin{aligned}
&\int_\Omega \mathbf{a}^{n\theta} \cdot \nabla v w \\
&\leq \left\|\mathbf{a}^{n\theta}\right\|_\infty \|\nabla v\| \, \|w\| \\
&\leq \|\mathbf{a}\|_{L^\infty(0,T;L^\infty(\Omega))} \, \varepsilon^{-\frac{1}{2}} \, |||v||| \min\left\{\beta^{-\frac{1}{2}}, c_F \varepsilon^{-\frac{1}{2}}\right\} |||w||| \\
&\leq \max\{1, c_F\} \|\mathbf{a}\|_{L^\infty(0,T;L^\infty(\Omega))} \, \varepsilon^{-\frac{1}{2}} \min\left\{\beta^{-\frac{1}{2}}, \varepsilon^{-\frac{1}{2}}\right\} |||v||| \, |||w||| \\
&= \max\{1, c_F\} \|\mathbf{a}\|_{L^\infty(0,T;L^\infty(\Omega))} \, \varepsilon^{-\frac{1}{2}} \max\{\beta, \varepsilon\}^{-\frac{1}{2}} |||v||| \, |||w||| \\
&\leq \max\{1, c_F\} \, c_c \, |||v||| \, |||w||| \, .
\end{aligned}$$

Recalling the definition (6.18) (p. 318) of the dual norm, this implies that

$$\left|\left|\left|\mathbf{a} \cdot \nabla \left(u^n_{\mathcal{T}_n} - u^{n-1}_{\mathcal{T}_{n-1}}\right)\right|\right|\right|_* \leq \max\{1, c_F\} \, c_c \left|\left|\left|u^n_{\mathcal{T}_n} - u^{n-1}_{\mathcal{T}_{n-1}}\right|\right|\right| . \tag{6.34}$$

Therefore, in the case of small convection, we may bound the convection term in the second estimate of Lemma 6.17 using estimate (6.34) and drop it in the first estimate of Lemma 6.17.

With these modifications, Theorem 6.20 (p. 325) remains valid. The constants $\widetilde{c}^*$ and $\widetilde{c}_*$ now also depend on the parameter κ. Moreover, the terms $\widetilde{\eta}^n_{\mathcal{T}_n}$ and $\left|\left|\left|\widetilde{u}^n_{\mathcal{T}_n}\right|\right|\right|$ may be omitted in the case of a dominant diffusion or reaction.

Remark 6.23 A more detailed analysis may be found in [**268**].

6.4 The Method of Characteristics

We again consider general linear parabolic equations of second order (6.33) (p. 326). In order to simplify the presentation and to avoid unnecessary technical difficulties, we now assume that we have pure Dirichlet boundary conditions, i.e. $\Gamma_D = \Gamma$, and impose the more restrictive condition

$$\begin{aligned}
\operatorname{div} \mathbf{a} &= 0 \quad \text{in } \Omega \times (0, T], \\
\mathbf{a} &= 0 \quad \text{on } \Gamma \times (0, T]
\end{aligned} \tag{6.35}$$

on the convection. These conditions in particular imply that the fourth assumption of Section 6.3 is automatically satisfied.

As in Section 6.1 (p. 309), the dual space of $H^1_0(\Omega)$ is denoted by $H^{-1}(\Omega)$. But, now, these spaces are equipped with the energy norm $|||\cdot|||$ and the dual norm $|||\cdot|||_*$, respectively, defined in (6.17) (p. 317) and (6.18) (p. 318).

6.4.1 Finite Element Discretisation

The key idea of the method of characteristics relies on a different formulation of problem (6.33). Since the function $\mathbf{a}$ is Lipschitz continuous with respect to the spatial variable and vanishes on the boundary Γ, for every $(x^*, t^*) \in \Omega \times (0, T]$, standard global existence results for the flows of ordinary differential equations [**19**, Theorem II.7.6] imply that the *characteristic equation*

$$\begin{aligned} \frac{\mathrm{d}}{\mathrm{d}t} x(t; x^*, t^*) &= \mathbf{a}(x(t; x^*, t^*), t), \quad t \in (0, t^*), \\ x(t^*; x^*, t^*) &= x^* \end{aligned} \tag{6.36}$$

has a unique solution $x(\cdot; x^*, t^*)$ which exists for all $t \in [0, t^*]$ and stays within $\Omega \cup \Gamma$. Hence, we may set $U(x^*, t) = u(x(t; x^*, t^*), t)$. The total derivative $\frac{\mathrm{d}}{\mathrm{d}t} U$ satisfies

$$\frac{\mathrm{d}}{\mathrm{d}t} U = \partial_t u + \mathbf{a} \cdot \nabla u.$$

Therefore, the first equation of problem (6.33) can equivalently be written as

$$\frac{\mathrm{d}}{\mathrm{d}t} U - \operatorname{div}(A \nabla u) + bu = f \quad \text{in } \Omega \times (0, T). \tag{6.37}$$

The discretisation by the method of characteristics relies on a separate treatment of equations (6.36) and (6.37).

As in the previous sections, we choose a partition $\mathcal{I}$ of $[0, T]$, corresponding partitions $\mathcal{T}_n$ of Ω, and associated finite element spaces X_n which satisfy the conditions of Section 3.2.6 (p. 86). In addition, we need the following condition which is not too restrictive:

- for every n there is a set $\mathcal{V}_n$ of *nodes* such that every function in X_n is uniquely defined by its values in the nodes (*Lagrange condition*).

Every set $\mathcal{V}_n$ splits into the subsets $\mathcal{V}_{n,\Omega}$ and $\mathcal{V}_{n,\Gamma}$ of all nodes inside Ω and on Γ, respectively. For every intermediate time t_n, $1 \leq n \leq N_{\mathcal{I}}$, and every node $z \in \mathcal{V}_{n,\Omega}$ we compute an approximation x_z^{n-1} to $x(t_{n-1}; z, t_n)$ by applying an arbitrary but fixed ODE solver such as the explicit Euler scheme to the characteristic equation (6.36) with $(x^*, t^*) = (z, t_n)$, cf. Figure 6.1. We assume that

- the time-step τ_n and the ODE solver are chosen such that x_z^{n-1} lies within $\Omega \cup \Gamma$ for every $n \in \{1, \dots, N_{\mathcal{I}}\}$ and every $z \in \mathcal{V}_{n,\Omega}$.

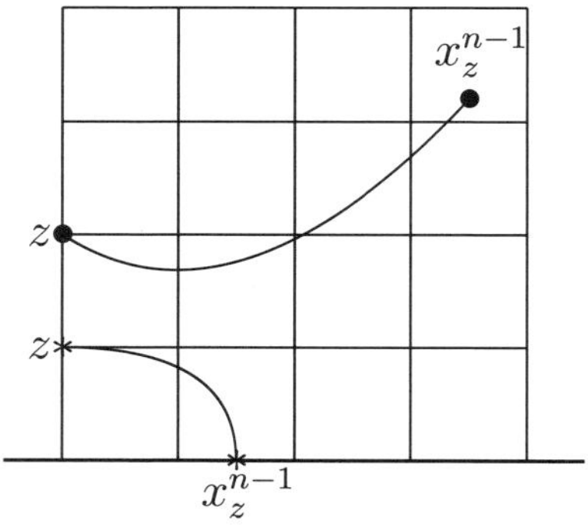

Figure 6.1 Computation of x_z^{n-1} in the method of characteristics

The assumptions on the convection $\mathbf{a}$ imply that this condition is satisfied for a single explicit Euler step if $\tau_n < 1/\left\|\mathbf{a}(\cdot,t_n)\right\|_{1,\infty}$ [232, Proposition 1]. Note that $x_z^{n-1} = z - \tau_n \mathbf{a}(z,t_n)$ in this case. We stress that our results hold for *every* ODE solver which satisfies the previous condition on the x_z^{n-1}.

As in the previous sections, for every $n \in \{0,\ldots,N_{\mathcal{I}}\}$, we denote by π_n the L^2-projection onto X_n and set $A^n = A(\cdot,t_n)$, $\mathbf{a}^n = \mathbf{a}(\cdot,t_n)$, $b^n = b(\cdot,t_n)$, $f^n = f(\cdot,t_n)$. With this notation the discretisation of problem (6.35) is given by:

set

$$u_{\mathcal{T}_0}^0 = \pi_0 u_0 \tag{6.38}$$

and, for $n = 1,\ldots,N_{\mathcal{I}}$, successively compute $\widetilde{u}_{\mathcal{T}_n}^{n-1} \in X_n$ such that

$$\widetilde{u}_{\mathcal{T}_n}^{n-1}(z) = \begin{cases} u_{\mathcal{T}_{n-1}}^{n-1}\left(x_z^{n-1}\right) & \text{if } z \in \mathcal{V}_{n,\Omega}, \\ 0 & \text{if } z \in \mathcal{V}_{n,\Gamma}, \end{cases} \tag{6.39}$$

and find $u_{\mathcal{T}_n}^n \in X_n$ such that

$$\int_\Omega \frac{1}{\tau_n}\left(u_{\mathcal{T}_n}^n - \widetilde{u}_{\mathcal{T}_n}^{n-1}\right) v_{\mathcal{T}_n} + \int_\Omega \nabla u_{\mathcal{T}_n}^n \cdot A^n \cdot \nabla v_{\mathcal{T}_n} + \int_\Omega b^n u_{\mathcal{T}_n}^n v_{\mathcal{T}_n} = \int_\Omega f^n v_{\mathcal{T}_n} \tag{6.40}$$

holds for all $v_{\mathcal{T}_n} \in X_n$.

The Lagrange condition and the property $x_z^{n-1} \in \Omega \cup \Gamma$ for all $n \in \{1,\ldots,N_{\mathcal{I}}\}$ and all $z \in \mathcal{V}_{n,\Omega}$ imply that the $\widetilde{u}_{\mathcal{T}_n}^{n-1}$ are well-defined. Similarly, the assumptions on A and b imply that, for every n, problem (6.40) admits a unique solution $u_{\mathcal{T}_n}^n$. Hence the above discretisation yields a unique sequence $(u_{\mathcal{T}_n}^n)_{0 \le n \le N_{\mathcal{I}}}$. With this we associate as in the previous sections the function $u_{\mathcal{I}}$ which is continuous and piecewise affine with respect to time and which equals $u_{\mathcal{T}_n}^n$ at time t_n, $0 \le n \le N_{\mathcal{I}}$, cf. equations (6.6) and (6.7) (p. 310).

6.4.2 *Properties of the Error and the Residual*

Similarly to the previous sections, we associate with $u_{\mathcal{I}}$ the residual $R(u_{\mathcal{I}}) \in L^2(0,T;H^{-1}(\Omega))$ by setting for every $v \in H_0^1(\Omega)$

$$\left\langle R(u_{\mathcal{I}}), v\right\rangle = \int_\Omega fv - \int_\Omega \partial_t u_{\mathcal{I}} v - \int_\Omega \nabla u_{\mathcal{I}} \cdot A \cdot \nabla v - \int_\Omega \mathbf{a} \cdot \nabla u_{\mathcal{I}} v - \int_\Omega b u_{\mathcal{I}} v.$$

The equivalence of the error $u - u_{\mathcal{I}}$ and of the residual $R(u_{\mathcal{I}})$, Proposition 6.14 (p. 320), remains valid since its proof only depends on properties of the differential equation.

As in the previous sections, we may write the residual $R(u_{\mathcal{I}})$ as the sum of three contributions: a term $R_D(u_{\mathcal{I}})$ representing oscillations of the data and the coefficients with respect to time, a temporal residual $R_\tau(u_{\mathcal{I}})$, and a spatial residual $R_h(u_{\mathcal{I}})$. For every $v \in H_0^1(\Omega)$ and every interval $(t_{n-1},t_n]$ these are defined by

$$\left\langle R_D(u_{\mathcal{I}}), v\right\rangle = \int_\Omega (f - f^n)v + \int_\Omega \nabla u_{\mathcal{I}} \cdot (A^n - A) \cdot \nabla v + \int_\Omega (\mathbf{a}^n - \mathbf{a}) \cdot \nabla u_{\mathcal{I}} v + \int_\Omega (b^n - b) u_{\mathcal{I}} v,$$

$$\langle R_\tau(u_\mathcal{I}), v\rangle = \int_\Omega \nabla\left(u^n_{\mathcal{T}_n} - u_\mathcal{I}\right)\cdot A^n \cdot \nabla v + \int_\Omega \mathbf{a}^n \cdot \nabla\left(u^n_{\mathcal{T}_n} - u_\mathcal{I}\right) v + \int_\Omega b^n \left(u^n_{\mathcal{T}_n} - u_\mathcal{I}\right) v,$$

$$\langle R_h(u_\mathcal{I}), v\rangle = \int_\Omega f^n v - \int_\Omega \partial_t u_\mathcal{I} v - \int_\Omega \nabla u^n_{\mathcal{T}_n}\cdot A^n \cdot \nabla v - \int_\Omega \mathbf{a}^n \cdot \nabla u^n_{\mathcal{T}_n} v - \int_\Omega b^n u^n_{\mathcal{T}_n} v. \tag{6.41}$$

Since $R_D(u_\mathcal{I})$ only involves the known discrete solution $u_\mathcal{I}$ and temporal oscillations of the data and coefficients of (6.33), the task of deriving upper and lower bounds for the $L^2(t_{n-1}, t_n; H^{-1}(\Omega))$-norms of $R(u_\mathcal{I})$ again reduces to the estimation of the corresponding norms of $R_\tau(u_\mathcal{I}) - R_h(u_\mathcal{I})$. As in the previous sections, this can be done separately. Lemma 6.16 (p. 322) remains valid since it only uses the property that the spatial and the temporal residual are piecewise constant and piecewise affine with respect to time. Note that the present situation corresponds to the choice $\theta = 1$ in Lemma 3.53 (p. 130).

The temporal residual $R_\tau(u_\mathcal{I})$ can be estimated as before. Lemma 6.17 (p. 322) and estimate (6.34) (p. 329) remain valid since they only use properties of the stationary convection–diffusion equation associated with problem (6.33).

As in the previous sections, the estimation of the spatial residual requires, on each time interval $(t_{n-1}, t_n]$ separately, a reliable, efficient, and robust error indicator for the restriction of $R_h(u_\mathcal{I})$ to $(t_{n-1}, t_n]$. Using integration by parts element-wise we easily conclude from the definition (6.41) of $R_h(u_\mathcal{I})$ that it admits an L^2-representation consisting of element residuals and inter-element jumps. Thus it falls into the class of functionals considered in Section 3.8 (p. 132). A comparison of equations (6.40) and (6.41), however, reveals that $R_h(u_\mathcal{I})$ does not satisfy the Galerkin orthogonality, i.e. its kernel does not contain the space $S_0^{1,0}(\mathcal{T}_n)$. Instead, for every $n \in \{1, \ldots, N_\mathcal{I}\}$ and every $v_{\mathcal{T}_n} \in X_n$, we have

$$\langle R_h(u_\mathcal{I}), v_{\mathcal{T}_n}\rangle = \int_\Omega \left(\frac{1}{\tau_n}\left(u^{n-1}_{\mathcal{T}_n} - \widetilde{u}^{n-1}_{\mathcal{T}_n}\right) - \mathbf{a}^n \cdot \nabla u^n_{\mathcal{T}_n}\right) v_{\mathcal{T}_n}. \tag{6.42}$$

This term describes the difference between the method of characteristics and the backward Euler scheme and must be taken into account by the error indicator. The following lemma shows a possible way to achieve this. The underlying idea is the same as for Lemma 6.18 (p. 323).

For brevity we denote by $f_\mathcal{I}$, $b_\mathcal{I}$, $\mathbf{a}_\mathcal{I}$, and $A_\mathcal{I}$ the piecewise constant functions with respect to time which, for each n, equal on $[t_{n-1}, t_n]$ the L^2-projection of f^n, b^n, $\mathbf{a}^n$, and A^n, respectively, on the space of piecewise constant functions with respect to space corresponding to the partition $\mathcal{T}_n$. Set $\widehat{X}_n = S_0^{1,0}(\mathcal{T}_n)$ and denote by $\widehat{u}^n_{\mathcal{T}_n} \in \widehat{X}_n$ the unique solution of the discrete diffusion–reaction problem

$$\varepsilon \int_\Omega \nabla \widehat{u}^n_{\mathcal{T}_n} \cdot \nabla v_{\mathcal{T}_n} + \beta \int_\Omega \widehat{u}^n_{\mathcal{T}_n} v_{\mathcal{T}_n} = \int_\Omega \left[\frac{1}{\tau_n}\left(u^{n-1}_{\mathcal{T}_n} - \widetilde{u}^{n-1}_{\mathcal{T}_n}\right) - \mathbf{a}^n \cdot \nabla u^n_{\mathcal{T}_n}\right] v_{\mathcal{T}_n} \tag{6.43}$$

for all $v_{\mathcal{T}_n} \in \widehat{X}_n$. Furthermore, define the error indicator $\widehat{\eta}^n_{\mathcal{T}_n}$ by

$$\widehat{\eta}^n_{\mathcal{T}_n} = \left\{\sum_{K\in\mathcal{T}_n} \hbar_K^2 \left\| f_\mathcal{I} - \frac{1}{\tau_n}\left(u^n_{\mathcal{T}_n} - u^{n-1}_{\mathcal{T}_{n-1}}\right) + \operatorname{div}\left(A_\mathcal{I}\nabla u^n_{\mathcal{T}_n}\right) - \mathbf{a}_\mathcal{I}\cdot\nabla u^n_{\mathcal{T}_n} - b_\mathcal{I} u^n_{\mathcal{T}_n} + \varepsilon\Delta\widehat{u}^n_{\mathcal{T}_n} - \beta\widehat{u}^n_{\mathcal{T}_n}\right\|_K^2\right.$$

$$\left. + \sum_{E\in\mathcal{E}_n} \varepsilon^{-\frac{1}{2}} \hbar_E \left\| \mathbb{J}_E\left(\varepsilon^{-1}\mathbf{n}_E\cdot\left(A_\mathcal{I}\nabla u^n_{\mathcal{T}_n}\right) + \mathbf{n}_E\cdot\nabla\widehat{u}^n_{\mathcal{T}_n}\right)\right\|_E^2\right\}^{\frac{1}{2}} \tag{6.44}$$

and the data error $\widehat{\Theta}^n_{\mathcal{T}_n}$ by

$$\widehat{\Theta}^n_{\mathcal{T}_n} = \Bigg\{ \sum_{K\in\mathcal{T}_n} \hbar_K^2 \| f^n - f_{\mathcal{I}} + \operatorname{div}\left((A^n - A_{\mathcal{I}})\nabla u^n_{\mathcal{T}_n}\right) - (\mathbf{a}^n - \mathbf{a}_{\mathcal{I}})\cdot\nabla u^n_{\mathcal{T}_n} - (b^n - b_{\mathcal{I}})u^n_{\mathcal{T}_n} \|_K^2 \\ + \sum_{E\in\mathcal{E}_n} \varepsilon^{-\frac{1}{2}} \hbar_E \left\| \mathbb{J}_E\left(\varepsilon^{-1}\mathbf{n}_E\cdot\left((A^n - A_{\mathcal{I}})\nabla u^n_{\mathcal{T}_n}\right)\right)\right\|_E^2 \Bigg\}^{\frac{1}{2}}, \tag{6.45}$$

where $\hbar_S$, $S \in \mathcal{T}_n \cup \mathcal{E}_n$, is as in (3.4) (p. 82).

Lemma 6.24 *There are two constants $c_{\dagger}$ and $c^{\dagger}$ which only depend on the shape parameters $C_{\mathcal{T}_n}$ such that the following estimates are valid for every $n \in \{1, \ldots, N_{\mathcal{I}}\}$*

$$|||\widehat{u}^n_{\mathcal{T}_n}||| + c_{\dagger}\left\{\widehat{\eta}^n_{\mathcal{T}_n} - \widehat{\Theta}^n_{\mathcal{T}_n}\right\} \le |||R_h(u_{\mathcal{I}})|||_* \le |||\widehat{u}^n_{\mathcal{T}_n}||| + c^{\dagger}\left\{\widehat{\eta}^n_{\mathcal{T}_n} + \widehat{\Theta}^n_{\mathcal{T}_n}\right\}.$$

Proof Due to the degree condition the space $\widehat{X}_n$ is contained in X_n. Hence, we conclude from (6.42) that

$$\varepsilon \int_\Omega \nabla \widehat{u}^n_{\mathcal{T}_n}\cdot\nabla v_{\mathcal{T}_n} + \beta \int_\Omega \widehat{u}^n_{\mathcal{T}_n} v_{\mathcal{T}_n} = \left\langle R_h(u_{\mathcal{I}}),\, v_{\mathcal{T}_n}\right\rangle$$

holds for all $v_{\mathcal{T}_n} \in \widehat{X}_n$. This in particular implies

$$|||\widehat{u}^n_{\mathcal{T}_n}||| \le |||R_h(u_{\mathcal{I}})|||_*.$$

Denote by $U^n \in H_0^1(\Omega)$ the unique solution of the weak diffusion–reaction equation

$$\varepsilon \int_\Omega \nabla U^n \cdot \nabla v + \beta \int_\Omega U^n v = \left\langle R_h(u_{\mathcal{I}}),\, v\right\rangle$$

for all $v \in H_0^1(\Omega)$. Then we have

$$|||U^n||| = |||R_h(u_{\mathcal{I}})|||_*$$

and therefore

$$\frac{1}{3}\left\{|||\widehat{u}^n_{\mathcal{T}_n}||| + |||U^n - \widehat{u}^n_{\mathcal{T}_n}|||\right\} \le |||R_h(u_{\mathcal{I}})|||_* \le \left\{|||\widehat{u}^n_{\mathcal{T}_n}||| + |||U^n - \widehat{u}^n_{\mathcal{T}_n}|||\right\}.$$

Integration by parts element-wise yields the following L^2-representation for the residual of $\widehat{u}^n_{\mathcal{T}_n}$

$$\begin{aligned}\left\langle R_h(u_{\mathcal{I}}),\, \varphi\right\rangle &- \varepsilon \int_\Omega \nabla \widehat{u}^n_{\mathcal{T}_n}\cdot\nabla v - \beta\int_\Omega \widehat{u}^n_{\mathcal{T}_n} v \\ &= \sum_{K\in\mathcal{T}_n}\int_K \Bigg(f^n - \frac{1}{\tau_n}\left(u^n_{\mathcal{T}_n} - u^{n-1}_{\mathcal{T}_{n-1}}\right) + \operatorname{div}\left(A^n\nabla u^n_{\mathcal{T}_n}\right) \\ &\qquad - \mathbf{a}^n\cdot\nabla u^n_{\mathcal{T}_n} - b^n u^n_{\mathcal{T}_n} + \varepsilon\Delta\widehat{u}^n_{\mathcal{T}_n} - \beta\widehat{u}^n_{\mathcal{T}_n}\Bigg) v \\ &\quad - \sum_{E\in\mathcal{E}_n}\int_E \mathbb{J}_E\left(\mathbf{n}_E\cdot\left(A^n\nabla u^n_{\mathcal{T}_n}\right) + \varepsilon\mathbf{n}_E\cdot\nabla\widehat{u}^n_{\mathcal{T}_n}\right) v.\end{aligned}$$

Theorem 4.10 (p. 160) therefore yields

$$c_{\dagger}\left\{\widehat{\eta}^{n}_{\mathcal{T}_n}-\widehat{\Theta}^{n}_{\mathcal{T}_n}\right\}\le\left|\!\left|\!\left|U^n-\widehat{u}^{n}_{\mathcal{T}_n}\right|\!\right|\!\right|\le c^{\dagger}\left\{\widehat{\eta}^{n}_{\mathcal{T}_n}+\widehat{\Theta}^{n}_{\mathcal{T}_n}\right\}$$

and thus proves the lemma. □

6.4.3 A Posteriori Error Estimates

Proposition 6.14 and Lemmas 6.16, 6.17, and 6.24 yield the following a posteriori error estimates.

Theorem 6.25 *Denote by u the solution of the variational problem associated with (6.33) (p. 326) and by $u_{\mathcal{I}}$ its approximation associated with the sequence $(u^n_{\mathcal{T}_n})_{0\le n\le N_{\mathcal{I}}}$ defined by the discrete problems (6.38), (6.39), (6.40). Then the error $u-u_{\mathcal{I}}$ can be bounded from above by*

$$\begin{aligned}
&\left\{\|u-u_{\mathcal{I}}\|^2_{L^\infty(0,T;L^2(\Omega))}+\|u-u_{\mathcal{I}}\|^2_{L^2(0,T;H^1_0(\Omega))}+\left\|\partial_t(u-u_{\mathcal{I}})+\mathbf{a}\cdot\nabla(u-u_{\mathcal{I}})\right\|^2_{L^2(0,T;H^{-1}(\Omega))}\right\}^{\frac12}\\
&\le c_{\sharp}\left\{\left\|u_0-u^0_{\mathcal{T}_0}\right\|^2+\sum_{n=1}^{N_{\mathcal{I}}}\tau_n\left[\left|\!\left|\!\left|u^n_{\mathcal{T}_n}-u^{n-1}_{\mathcal{T}_{n-1}}\right|\!\right|\!\right|^2+\left|\!\left|\!\left|\widetilde{u}^n_{\mathcal{T}_n}\right|\!\right|\!\right|^2+(\widetilde{\eta}^n_{\mathcal{T}_n})^2\right]\right.\\
&\quad+\sum_{n=1}^{N_{\mathcal{I}}}\tau_n\left[\left|\!\left|\!\left|\widehat{u}^n_{\mathcal{T}_n}\right|\!\right|\!\right|^2+(\widehat{\eta}^n_{\mathcal{T}_n})^2\right]\\
&\quad+\sum_{n=1}^{N_{\mathcal{I}}}\|f-f^n+\operatorname{div}((A-A^n)\nabla u_{\mathcal{I}}-(\mathbf{a}-\mathbf{a}^n)\cdot\nabla u_{\mathcal{I}}-(b-b^n)u_{\mathcal{I}}\|^2_{L^2(t_{n-1},t_n;H^{-1}(\Omega))}\\
&\quad\left.+\sum_{n=1}^{N_{\mathcal{I}}}\tau_n(\widehat{\Theta}^n_{\mathcal{T}_n})^2\right\}^{\frac12}
\end{aligned}$$

and, for every $n\in\{1,\ldots,N_{\mathcal{I}}\}$, from below by

$$\begin{aligned}
&\tau_n^{\frac12}\left\{\left|\!\left|\!\left|u^n_{\mathcal{T}_n}-u^{n-1}_{\mathcal{T}_{n-1}}\right|\!\right|\!\right|+\left|\!\left|\!\left|\widetilde{u}^n_{\mathcal{T}_n}\right|\!\right|\!\right|+\widetilde{\eta}^n_{\mathcal{T}_n}+\left|\!\left|\!\left|\widehat{u}^n_{\mathcal{T}_n}\right|\!\right|\!\right|+\widehat{\eta}^n_{\mathcal{T}_n}\right\}\\
&\le c^{\sharp}\left\{\|u-u_{\mathcal{I}}\|^2_{L^\infty(t_{n-1},t_n;L^2(\Omega))}+\|u-u_{\mathcal{I}}\|^2_{L^2(t_{n-1},t_n;H^1_0(\Omega))}\right.\\
&\quad+\left\|\partial_t(u-u_{\mathcal{I}})+\mathbf{a}\cdot\nabla(u-u_{\mathcal{I}})\right\|^2_{L^2(t_{n-1},t_n;H^{-1}(\Omega))}\\
&\quad+\|f-f^n+\operatorname{div}((A-A^n)\nabla u_{\mathcal{I}}-(\mathbf{a}-\mathbf{a}^n)\cdot\nabla u_{\mathcal{I}}-(b-b^n)u_{\mathcal{I}}\|^2_{L^2(t_{n-1},t_n;H^{-1}(\Omega))}\\
&\quad\left.+\tau_n(\widehat{\Theta}^n_{\mathcal{T}_n})^2\right\}^{\frac12}.
\end{aligned}$$

The functions $\widetilde{u}^n_{\mathcal{T}_n}$ and $\widehat{u}^n_{\mathcal{T}_n}$ and the quantities $\widetilde{\eta}^n_{\mathcal{T}_n}$, $\widehat{\eta}^n_{\mathcal{T}_n}$ and $\widehat{\Theta}^n_{\mathcal{T}_n}$ are defined in equations (6.31) (p. 323), (6.43), (6.32) (p. 323), (6.44), (6.45), respectively. The constants $c_{\sharp}$ and $c^{\sharp}$ depend on the shape parameters $C_{\mathcal{T}_n}$ but are independent of any relative size of the diffusion with respect to the convection or reaction. If the diffusion or reaction are dominant, the quantities $\left|\!\left|\!\left|\widetilde{u}^n_{\mathcal{T}_n}\right|\!\right|\!\right|$ and $\widetilde{\eta}^n_{\mathcal{T}_n}$ can be omitted.

Remark 6.26 The a posteriori error estimate of Theorem 6.25 is more complex than that of Theorem 6.20 (p. 325) in that it requires the solution of the auxiliary discrete diffusion–reaction problems (6.43). This is compensated by the simpler structure of the discrete problems (6.40) when compared with the θ-scheme (6.23) (p. 319).

Remark 6.27 The method of characteristics was initiated in [219] as an alternative to standard space–time finite elements with or without up-winding or SUPG stabilisation. Its main advantages are that it implicitly includes an up-winding and thus stabilising effect and that it requires only the solution of symmetric coercive problems. On the other hand, it needs an efficient way for back-tracking the characteristics and re-interpolating finite element functions. Depending on the data structure of the implementation, this can be costly or not. The initial method was of order one but it has been rapidly extended to higher order in [75]. We refer to [61, 62, 75, 232, 245] for a more detailed description of the method including a priori error estimates.

Remark 6.28 The presentation of this section follows [63] which gives the first a posteriori error analysis of the method of characteristics.

6.5 The Time-Dependent Stokes Equations

We consider the time-dependent Stokes equations

$$\begin{aligned} \partial_t \mathbf{u} - \nu\Delta\mathbf{u} + \nabla p &= \mathbf{f} && \text{in } \Omega \times (0,T) \\ \operatorname{div}\mathbf{u} &= 0 && \text{in } \Omega \times (0,T) \\ \mathbf{u} &= 0 && \text{on } \Gamma \times (0,T) \\ \mathbf{u}(\cdot,0) &= \mathbf{u}_0 && \text{in } \Omega \end{aligned} \tag{6.46}$$

in a bounded space–time cylinder with a polygonal cross-section $\Omega \subset \mathbb{R}^d$, $d \geq 2$, having a Lipschitz boundary Γ. The final time T is arbitrary, but kept fixed in what follows. The unknowns are the velocity $\mathbf{u}$ and the pressure p; the data are the distribution $\mathbf{f}$ which represents a density of body forces and the initial velocity $\mathbf{u}_0$, while the viscosity ν is a positive constant.

Problem (6.46) could be formulated as a heat equation in the space $V = \{\mathbf{u} \in H_0^1(\Omega)^d : \operatorname{div}\mathbf{u} = 0\}$ of solenoidal velocity fields. Since this space has the same analytical properties as $H_0^1(\Omega)^d$, the results of Section 6.1 (p. 309) would carry over immediately. This approach, however, is not feasible for the a posteriori error analysis of problem (6.46) since virtually all finite element discretisations used in practice are non-conforming in the sense that the discrete velocities are either discontinuous and thus not contained in $H_0^1(\Omega)^d$ or are not solenoidal and thus not contained in V, cf. Section 4.10 (p. 237). Therefore, we must develop particular techniques for problem (6.46). As we will see in Subsection 6.5.3 below, the crucial point here is the equivalence of error and residual.

6.5.1 Variational Formulation

Recalling the space $L_0^2(\Omega)$ of square integrable functions with vanishing mean value, cf. (4.54) (p. 238), problem (6.46) admits the following variational formulation:

find $\mathbf{u} \in W^2(0,T; H_0^1(\Omega)^d, H^{-1}(\Omega)^d), p \in L^2(0,T; L_0^2(\Omega))$ such that

$$\mathbf{u}(\cdot,0) = \mathbf{u}_0 \tag{6.47}$$

in $H^{-1}(\Omega)^d$ and

$$\begin{aligned}\int_\Omega \partial_t \mathbf{u}\cdot\mathbf{v} + \nu\int_\Omega \nabla\mathbf{u} : \nabla\mathbf{v} - \int_\Omega p\operatorname{div}\mathbf{v} &= \int_\Omega \mathbf{f}\cdot\mathbf{v} \\ \int_\Omega q\operatorname{div}\mathbf{u} &= 0\end{aligned} \tag{6.48}$$

for almost all $t \in (0,T)$ and all $\mathbf{v} \in H_0^1(\Omega)^d, q \in L_0^2(\Omega)$.

Here and in the sequel, $H_0^1(\Omega)^d$ and its dual space $H^{-1}(\Omega)^d$ are equipped with the norm $\|\nabla\cdot\|$ and the corresponding dual norm (6.2) (p. 309).

The following existence, uniqueness, and stability result can be derived from [**144**, Chapter V, Section 1.3] and [**243**, Chapter III, Theorem 1.1]. Note that the stability estimate of Proposition 6.29 is proved in the same way as the corresponding inequality in Proposition 6.1 (p. 311) using the fact that the velocity is solenoidal.

Proposition 6.29 *For every pair of data $\mathbf{f} \in L^2(0,T;H^{-1}(\Omega)^d)$ and $\mathbf{u}_0 \in L^2(\Omega)^d$ such that $\mathbf{u}_0$ is solenoidal in Ω, problem (6.47), (6.48) has a unique solution $(\mathbf{u},p)$ with $\partial_t\mathbf{u} + \nabla p \in L^2(0,T;H^{-1}(\Omega)^d)$. Moreover, this solution satisfies the stability estimate*

$$\begin{aligned}&\left\{\left\|\partial_t\mathbf{u} + \nabla p\right\|^2_{L^2(0,T;H^{-1}(\Omega)^d)} + \|\mathbf{u}\|^2_{L^\infty(0,T;L^2(\Omega)^d)} + \nu\,\|\mathbf{u}\|^2_{L^2(0,T;H_0^1(\Omega)^d)}\right\}^{\frac{1}{2}} \\ &\le \left\{\left(4 + \frac{2}{\nu}\right)\|\mathbf{f}\|^2_{L^2(0,T;H^{-1}(\Omega)^d)} + (4\nu+2)\,\|\mathbf{u}_0\|^2\right\}^{\frac{1}{2}}.\end{aligned}$$

Remark 6.30 Propositions 6.29 and 6.32 below show that the a posteriori error estimation can be based on the norm

$$\left\{\left\|\partial_t\mathbf{u} + \nabla p\right\|^2_{L^2(0,T;H^{-1}(\Omega)^d)} + \|\mathbf{u}\|^2_{L^\infty(0,T;L^2(\Omega)^d)} + \nu\,\|\mathbf{u}\|^2_{L^2(0,T;H_0^1(\Omega)^d)}\right\}^{\frac{1}{2}}.$$

Yet, this norm is not equivalent the standard norm

$$\left\{\|\partial_t\mathbf{u}\|^2_{L^2(0,T;H^{-1}(\Omega)^d)} + \|\mathbf{u}\|^2_{L^2(0,T;H_0^1(\Omega)^d)} + \|p\|^2_{L^2(0,T;L^2(\Omega))}\right\}^{\frac{1}{2}}$$

since it does not provide separate bounds for the pressure and the time-derivative of the velocity field. Nonetheless, a posteriori error estimates with respect to the standard norm are possible. Yet, they do not fit into our abstract framework and yield weaker results, cf. Remark 6.38 below.

6.5.2 *Finite Element Discretisation*

We choose a partition $\mathcal{I}$ of $[0,T]$, corresponding affine-equivalent, admissible, and shape-regular partitions $\mathcal{T}_n$ of Ω, and associated finite element spaces V_n and P_n for the approximation of the velocity and pressure. The partitions $\mathcal{T}_n$ and the spaces V_n, P_n must now satisfy the following modifications of the transition and degree conditions of Section 3.2.6 (p. 86):

- For every $n \in \{1,\ldots,N_{\mathcal{I}}\}$ there are *two* affine-equivalent, admissible, shape-regular partitions $\widetilde{\mathcal{T}}_n$ and $\widehat{\mathcal{T}}_n$ such that $\widetilde{\mathcal{T}}_n$ is a common refinement of $\mathcal{T}_n$ and $\mathcal{T}_{n-1}$, both $\mathcal{T}_n$ and $\mathcal{T}_{n-1}$ are common refinements of $\widehat{\mathcal{T}}_n$, and

$$\sup_{1\le n\le N_{\mathcal{I}}} \max\left\{ \sup_{K\in\widetilde{\mathcal{T}}_n} \sup_{\substack{K'\in\mathcal{T}_n\\ K\subset K'}} \frac{h_{K'}}{h_K}, \sup_{K\in\mathcal{T}_n\cup\mathcal{T}_{n-1}} \sup_{\substack{K'\in\widetilde{\mathcal{T}}_n\\ K\subset K'}} \frac{h_{K'}}{h_K} \right\} < \infty$$

holds uniformly with respect to all partitions $\mathcal{I}$ which are obtained by adaptive or uniform refinement of any initial partition of $[0, T]$ (*modified transition condition*).

- The spaces V_n and P_n consist of functions which are piecewise polynomials, the degrees being bounded uniformly with respect to all partitions $\mathcal{T}_n$ and $\mathcal{I}$. Every V_n contains the space $S_0^{1,0}(\mathcal{T}_n)^d$ and every P_n either contains the space $S^{1,0}(\mathcal{T}_n) \cap L_0^2(\Omega)$, if it consists of continuous functions, or the space $S^{0,-1}(\mathcal{T}_n) \cap L_0^2(\Omega)$, if it consists of discontinuous functions (*modified degree condition*).

The modified transition condition is stronger than the transition condition of Section 3.2.6. When passing from $\mathcal{T}_{n-1}$ to $\mathcal{T}_n$ it not only restricts a possible coarsening but also a possible refinement. For practical use, however, it is not too restrictive. It is satisfied when constructing $\mathcal{T}_n$ from $\mathcal{T}_{n-1}$ using Algorithms 2.1 (p. 64) or 2.2 (p. 65) for refinement and 2.3 (p. 69) for coarsening.

Keeping in mind the mixed finite element discretisation of the stationary Stokes equations of Section 4.10 (p. 237), the modified degree condition is an obvious adjustment of the degree condition to the present situation. Similarly to the stationary case, the spaces V_n and P_n will have to fulfil an additional stability condition given below depending on the choice of the parameters ϑ_K, ϑ_E, and $\widetilde{\vartheta}_K$ introduced below.

We choose a parameter $\theta \in [\frac{1}{2}, 1]$, denote by π_n the L^2-projection onto V_n, and set for abbreviation $\mathbf{f}^n = \mathbf{f}(\cdot, t_n)$, $\mathbf{f}^{n\theta} = \theta\mathbf{f}^n + (1-\theta)\mathbf{f}^{n-1}$. The finite element discretisation of problem (6.46) is then given by:

set

$$\mathbf{u}_{\mathcal{T}_0}^0 = \pi_0 \mathbf{u}_0 \tag{6.49}$$

and, for $n = 1, \ldots, N_{\mathcal{I}}$ and $\mathbf{u}^{n\theta} = \theta\mathbf{u}_{\mathcal{T}_n}^n + (1-\theta)\mathbf{u}_{\mathcal{T}_{n-1}}^{n-1}$, find $\mathbf{u}_{\mathcal{T}_n}^n \in V_n$ and $p_{\mathcal{T}_n}^n \in P_n$ such that

$$\begin{aligned}
&\int_\Omega \frac{1}{\tau_n}\left(\mathbf{u}_{\mathcal{T}_n}^n - \mathbf{u}_{\mathcal{T}_{n-1}}^{n-1}\right)\cdot\mathbf{v}_{\mathcal{T}_n} + \nu\int_\Omega \nabla\mathbf{u}^{n\theta} : \nabla\mathbf{v}_{\mathcal{T}_n} - \int_\Omega p_{\mathcal{T}_n}^n \operatorname{div}\mathbf{v}_{\mathcal{T}_n} + \int_\Omega q_{\mathcal{T}_n}\operatorname{div}\mathbf{u}_{\mathcal{T}_n}^n \\
&+ \sum_{K\in\mathcal{T}_n} \vartheta_K h_K^2 \int_K \left[\frac{\mathbf{u}_{\mathcal{T}_n}^n - \mathbf{u}_{\mathcal{T}_{n-1}}^{n-1}}{\tau_n} - \nu\Delta\mathbf{u}^{n\theta} + \nabla p_{\mathcal{T}_n}^n\right]\cdot\nabla q_{\mathcal{T}_n} + \sum_{E\in\mathcal{E}_{n,\Omega}} \vartheta_E h_E \int_E \mathbb{J}_E\left(p_{\mathcal{T}_n}^n\right)\mathbb{J}_E\left(q_{\mathcal{T}_n}\right) \\
&+ \sum_{K\in\mathcal{T}_n} \widetilde{\vartheta}_K \int_K \operatorname{div}\mathbf{u}_{\mathcal{T}_n}^n \operatorname{div}\mathbf{v}_{\mathcal{T}_n} \\
&\quad = \int_\Omega \mathbf{f}^{n\theta}\cdot\mathbf{u}_{\mathcal{T}_n} + \sum_{K\in\mathcal{T}_n}\vartheta_K h_K^2 \int_K \mathbf{f}^{n\theta}\cdot\nabla q_{\mathcal{T}_n}
\end{aligned} \tag{6.50}$$

for all $\mathbf{v}_{\mathcal{T}_n} \in V_n$, $q_{\mathcal{T}_n} \in P_n$.

The parameters $\theta = \frac{1}{2}$ and $\theta = 1$ correspond to the the *Crank–Nicolson scheme* and the *implicit Euler scheme*, respectively.

As for the stationary Stokes equations, ϑ_K, ϑ_E, and $\widetilde{\vartheta}_K$ are non-negative stabilisation parameters. *Stabilised methods* correspond to the choices $\vartheta_K > 0$, $\widetilde{\vartheta}_K \ge 0$, and $\vartheta_E > 0$, if the pressure approximation is discontinuous, or $\vartheta_E = 0$, if the pressure approximation is continuous. *Stable mixed methods* correspond to the choice $\vartheta_K = \vartheta_E = \widetilde{\vartheta}_K = 0$ for all elements K and all faces E. They have to satisfy the following *stability condition*:

> there is a positive constant $\widetilde{\beta}$ such that
>
> $$\inf_{q_{\mathcal{T}_n} \in P_n \setminus \{0\}} \sup_{\mathbf{v}_{\mathcal{T}_n} \in V_n \setminus \{0\}} \frac{\int_\Omega q_{\mathcal{T}_n} \operatorname{div} \mathbf{v}_{\mathcal{T}_n}}{\left\| \nabla \mathbf{v}_{\mathcal{T}_n} \right\| \left\| q_{\mathcal{T}_n} \right\|} \geq \widetilde{\beta}$$
>
> holds for all $n \in \{1, \ldots, N_{\mathcal{I}}\}$ uniformly with respect to all partitions $\mathcal{I}$ which are obtained by adaptive or uniform refinement of any initial partition of $[0, T]$.

Examples of appropriate spaces V_n, P_n are given in Section 4.10.2 (p. 240), both for stabilised and stable mixed methods.

For both types of methods, standard results [**145**, **154**] imply that problems (6.49) and (6.50) admit unique solutions and yield two sequences $(\mathbf{u}^n_{\mathcal{T}_n})_{0 \leq n \leq N_{\mathcal{I}}}$ and $(p^n_{\mathcal{T}_n})_{1 \leq n \leq N_{\mathcal{I}}}$. With these we associate two functions $\mathbf{u}_{\mathcal{I}}$ and $p_{\mathcal{I}}$. The function $\mathbf{u}_{\mathcal{I}}$ is continuous and piecewise affine with respect to time and equals $\mathbf{u}^n_{\mathcal{T}_n}$ at time t_n, $0 \leq n \leq N_{\mathcal{I}}$, cf. equations (6.6) and (6.7) (p. 310); the function $p_{\mathcal{I}}$ is piecewise constant with respect to time and equals $p^n_{\mathcal{T}_n}$ on $(t_{n-1}, t_n]$, $1 \leq n \leq N_{\mathcal{I}}$.

6.5.3 The Equivalence of Error and Residual

We associate two residuals $R_{\mathrm{m}}(\mathbf{u}_{\mathcal{I}}, p_{\mathcal{I}}) \in L^2(0, T; H^{-1}(\Omega)^d)$ and $R_{\mathrm{c}}(\mathbf{u}_{\mathcal{I}}, p_{\mathcal{I}}) \in L^2(0, T; L^2(\Omega))$ with the functions $\mathbf{u}_{\mathcal{I}}$ and $p_{\mathcal{I}}$. These residuals correspond to the momentum (first) and continuity (second) equation in (6.46) and are defined for almost all $t \in (0, T)$ and all $\mathbf{v} \in H^1_0(\Omega)^d$, $q \in L^2_0(\Omega)$ by

$$\begin{aligned}
\left\langle R_{\mathrm{m}}(\mathbf{u}_{\mathcal{I}}, p_{\mathcal{I}}), \mathbf{v} \right\rangle &= \int_\Omega \mathbf{f} \cdot \mathbf{v} - \int_\Omega \partial_t \mathbf{u}_{\mathcal{I}} \cdot \mathbf{v} - \nu \int_\Omega \nabla \mathbf{u}_{\mathcal{I}} : \nabla \mathbf{v} + \int_\Omega p_{\mathcal{I}} \operatorname{div} \mathbf{v} \\
\left\langle R_{\mathrm{c}}(\mathbf{u}_{\mathcal{I}}, p_{\mathcal{I}}), q \right\rangle &= -\int_\Omega q \operatorname{div} \mathbf{u}_{\mathcal{I}}.
\end{aligned}$$

If the discretisation (6.50) were fully conservative, i.e. $\operatorname{div} \mathbf{u}^n_{\mathcal{T}_n} = 0$ for all n, the stability estimate of Proposition 6.29 would immediately yield the equivalence of

$$\left\{ \left\| \partial_t(\mathbf{u} - \mathbf{u}_{\mathcal{I}}) + \nabla(p - p_{\mathcal{I}}) \right\|^2_{L^2(0,T;H^{-1}(\Omega)^d)} + \left\| \mathbf{u} - \mathbf{u}_{\mathcal{I}} \right\|^2_{L^\infty(0,T;L^2(\Omega)^d)} + \nu \left\| \mathbf{u} - \mathbf{u}_{\mathcal{I}} \right\|^2_{L^2(0,T;H^1_0(\Omega)^d)} \right\}^{\frac{1}{2}}$$

and

$$\left\| R_{\mathrm{m}}(\mathbf{u}_{\mathcal{I}}, p_{\mathcal{I}}) \right\|_{L^2(0,T;H^{-1}(\Omega)^d)}.$$

Unfortunately, virtually all discretisations used in practice are not fully conservative and lead to a non-vanishing residual $R_{\mathrm{c}}(\mathbf{u}_{\mathcal{I}}, p_{\mathcal{I}})$. Therefore, we must establish a stability estimate for problem (6.46) with a non-homogeneous continuity equation. To achieve this, we need the *Stokes projection*

Π which maps $H_0^1(\Omega)^d$ onto the orthogonal complement $V^\perp$ of the space V. For every $\mathbf{v} \in H_0^1(\Omega)^d$ its projection $\Pi\mathbf{v}$ is the unique solution of the Stokes problem

$$\begin{aligned}\int_\Omega \nabla\Pi\mathbf{v} : \nabla\mathbf{w} - \int_\Omega q \operatorname{div}\mathbf{w} &= 0 \\ \int_\Omega r \operatorname{div}\Pi\mathbf{v} &= \int_\Omega r \operatorname{div}\mathbf{v}\end{aligned} \tag{6.51}$$

for all $\mathbf{w} \in H_0^1(\Omega)^d$, $r \in L_0^2(\Omega)$.

The following lemma collects some properties of Π. It uses the notion of a *re-entrant corner* and of a *discretely solenoidal vector field*. A corner $x^* \in \Gamma$ of the polygonal domain Ω is called re-entrant, if for every $\varepsilon > 0$ the set $\{x \in \Omega : |x - x^*| < \varepsilon\}$ is not convex. A vector field $\mathbf{v}$ is called discretely solenoidal if there is an admissible, affine-equivalent, shape-regular partition $\mathcal{T}$ such that

$$\int_\Omega q_\mathcal{T} \operatorname{div}\mathbf{v} = 0$$

holds for all $q_\mathcal{T} \in S^{1,0}(\mathcal{T}) \cap L_0^2(\Omega)$ or all $q_\mathcal{T} \in S^{0,-1}(\mathcal{T}) \cap L_0^2(\Omega)$.

Lemma 6.31 *For every $\mathbf{v} \in H_0^1(\Omega)^d$ we have*

$$\|\nabla\Pi\mathbf{v}\| \le \frac{1}{\beta}\|\operatorname{div}\mathbf{v}\|,$$

where β is the inf–sup constant in Proposition 4.67 (p. 238). If $\mathbf{v}$ is discretely solenoidal we in addition have

$$\|\Pi\mathbf{v}\| \le c_\Pi \left\{\sum_{K\in\mathcal{T}} h_K^{2\alpha_K}\|\operatorname{div}\mathbf{v}\|_K^2\right\}^{\frac{1}{2}}.$$

Here, the exponent α_K equals 1, if $K \cap \mathcal{U} = \emptyset$, and $\frac{1}{2}$, if $K \cap \mathcal{U} \ne \emptyset$, where $\mathcal{U}$ is a fixed neighbourhood of the re-entrant corners of Ω. The constant c_Π only depends on the shape parameter $C_\mathcal{T}$ and the neighbourhood $\mathcal{U}$.

Proof Inserting $\Pi\mathbf{v}$ as test function $\mathbf{w}$ in (6.51) yields

$$\|\nabla\Pi\mathbf{v}\|^2 = \int_\Omega q \operatorname{div}\Pi\mathbf{v} = \int_\Omega q \operatorname{div}\mathbf{v} \le \|\operatorname{div}\mathbf{v}\|\,\|q\|.$$

The inf–sup condition of Proposition 4.67 (p. 238) on the other hand implies

$$\beta\,\|q\| \le \sup_{\mathbf{w}\in H_0^1(\Omega)^d} \frac{\int_\Omega q \operatorname{div}\mathbf{w}}{\|\nabla\mathbf{w}\|} = \sup_{\mathbf{w}\in H_0^1(\Omega)^d} \frac{\int_\Omega \nabla\Pi\mathbf{v} : \nabla\mathbf{w}}{\|\nabla\mathbf{w}\|} \le \|\nabla\Pi\mathbf{v}\|.$$

This proves the first part of the lemma.

To prove the second part, we rely on a duality argument. For every $\mathbf{v} \in H_0^1(\Omega)^d$ there is a unique pair $\mathbf{z} \in H_0^1(\Omega)^d$, $s \in L_0^2(\Omega)$ which solves the Stokes problem

$$\begin{aligned} \int_\Omega \nabla \mathbf{z} : \nabla \mathbf{w} - \int_\Omega s \operatorname{div} \mathbf{w} &= \int_\Omega \Pi \mathbf{v} \cdot \mathbf{w} \\ \int_\Omega r \operatorname{div} \mathbf{z} &= 0 \end{aligned} \tag{6.52}$$

for all $\mathbf{w} \in H_0^1(\Omega)^d$, $r \in L_0^2(\Omega)$. We insert $\Pi\mathbf{v}$ as test function $\mathbf{w}$ in (6.52) and take into account (6.51). This yields

$$\begin{aligned} \|\Pi \mathbf{v}\|^2 &= \int_\Omega \nabla \mathbf{z} : \nabla \Pi \mathbf{v} - \int_\Omega s \operatorname{div} \Pi \mathbf{v} \\ &= \int_\Omega q \operatorname{div} \mathbf{z} - \int_\Omega s \operatorname{div} \mathbf{v} \\ &= - \int_\Omega s \operatorname{div} \mathbf{v}. \end{aligned}$$

Next, we choose an arbitrary function $s_{\mathcal{T}}$ in $S^{1,0}(\mathcal{T}) \cap L_0^2(\Omega)$ or in $S^{0,-1}(\mathcal{T}) \cap L_0^2(\Omega)$ depending on the properties of $\mathbf{v}$. Using the Cauchy–Schwarz inequalities for integrals and sums we thus obtain

$$\begin{aligned} \|\Pi \mathbf{v}\|^2 &= - \int_\Omega (s - s_{\mathcal{T}}) \operatorname{div} \mathbf{v} \\ &\le \left\{ \sum_{K \in \mathcal{T}} h_K^{-2\alpha_K} \|s - s_{\mathcal{T}}\|_K^2 \right\}^{\frac{1}{2}} \left\{ \sum_{K \in \mathcal{T}} h_K^{2\alpha_K} \|\operatorname{div} \mathbf{v}\|_K^2 \right\}^{\frac{1}{2}}. \end{aligned}$$

Regularity theorems for differential equations in polygonal domains with corners [**112**, **117**, **147**] imply that $s \in H^1(\Omega \setminus \mathcal{U})$ and $s \in H^{\frac{1}{2}}(\mathcal{U})$ and that the corresponding norms are bounded by $\|\Pi\mathbf{v}\|$. Standard approximation results for finite element spaces [**90**, **109**] therefore prove that $s_{\mathcal{T}}$ can be chosen such that

$$\left\{ \sum_{K \in \mathcal{T}} h_K^{-2\alpha_K} \|s - s_{\mathcal{T}}\|_K^2 \right\}^{\frac{1}{2}} \le c \, \|\Pi \mathbf{v}\| .$$

Combining these estimates proves the second part of the lemma. □

With the help of Lemma 6.31 we can now prove the following result relating the error and the residual:

Proposition 6.32 *The error can be bounded from below by*

$$\begin{aligned} &\left\| R_{\mathrm{m}}(\mathbf{u}_{\mathcal{I}}, p_{\mathcal{I}}) \right\|_{L^2(0,T;H^{-1}(\Omega)^d)} + \left\| R_{\mathrm{c}}(\mathbf{u}_{\mathcal{I}}, p_{\mathcal{I}}) \right\|_{L^2(0,T;L^2(\Omega))} \\ &\quad \le \Big\{ 2 \left\| \partial_t(\mathbf{u} - \mathbf{u}_{\mathcal{I}}) + \nabla(p - p_{\mathcal{I}}) \right\|^2_{L^2(0,T;H^{-1}(\Omega)^d)} \\ &\qquad + \|\mathbf{u} - \mathbf{u}_{\mathcal{I}}\|^2_{L^\infty(0,T;L^2(\Omega)^d)} + (2\nu^2 + d) \, \|\mathbf{u} - \mathbf{u}_{\mathcal{I}}\|^2_{L^2(0,T;H_0^1(\Omega)^d)} \Big\}^{\frac{1}{2}} \end{aligned}$$

and from above by

$$\left\{ \left\| \partial_t(\mathbf{u} - \mathbf{u}_{\mathcal{I}}) + \nabla(p - p_{\mathcal{I}}) \right\|^2_{L^2(0,T;H^{-1}(\Omega)^d)} + \|\mathbf{u} - \mathbf{u}_{\mathcal{I}}\|^2_{L^\infty(0,T;L^2(\Omega)^d)} + \nu \|\mathbf{u} - \mathbf{u}_{\mathcal{I}}\|^2_{L^2(0,T;H^1_0(\Omega)^d)} \right\}^{\frac{1}{2}}$$

$$\leq \left\{ \left(\frac{12}{\nu} + 14 \right) \left\| R_{\mathrm{m}}(\mathbf{u}_{\mathcal{I}}, p_{\mathcal{I}}) \right\|^2_{L^2(0,T;H^{-1}(\Omega)^d)} \right.$$

$$+ \frac{12\nu(\nu+1)}{\beta^2} \left\| R_{\mathrm{c}}(\mathbf{u}_{\mathcal{I}}, p_{\mathcal{I}}) \right\|^2_{L^2(0,T;L^2(\Omega))} + 4(\nu+1) \left\| \mathbf{u}_0 - \mathbf{u}^0_{\mathcal{T}_0} \right\|^2$$

$$+ 64(\nu+1)c_\Pi^2 \max_{0 \leq n \leq N_{\mathcal{I}}} \sum_{K \in \mathcal{T}_n} h_K^{2\alpha_K} \left\| \operatorname{div} \mathbf{u}^n_{\mathcal{T}_n} \right\|^2_K$$

$$\left. +16(\nu+1)c_\Pi^2 \left(\sum_{n=1}^{N_{\mathcal{I}}} \left[\sum_{K \in \mathcal{T}_n} h_K^{2\alpha_K} \left\| \operatorname{div} \left(\mathbf{u}^n_{\mathcal{T}_n} - \mathbf{u}^{n-1}_{\mathcal{T}_{n-1}} \right) \right\|^2_K \right]^{\frac{1}{2}} \right)^2 \right\}^{\frac{1}{2}},$$

where c_Π is the constant of Lemma 6.31.

Proof For almost all $t \in (0, T)$ and all $\mathbf{v} \in H^1_0(\Omega)^d$, $q \in L^2_0(\Omega)$ we have

$$\begin{aligned} \langle R_{\mathrm{m}}(\mathbf{u}_{\mathcal{I}}, p_{\mathcal{I}}), \mathbf{v} \rangle &= \langle \partial_t(\mathbf{u} - \mathbf{u}_{\mathcal{I}}), \mathbf{v} \rangle + \nu \int_\Omega \nabla(\mathbf{u} - \mathbf{u}_{\mathcal{I}}) : \nabla \mathbf{v} - \int_\Omega (p - p_{\mathcal{I}}) \operatorname{div} \mathbf{v} \\ &= \langle \partial_t(\mathbf{u} - \mathbf{u}_{\mathcal{I}}) + \nabla(p - p_{\mathcal{I}}), \mathbf{v} \rangle + \nu \int_\Omega \nabla(\mathbf{u} - \mathbf{u}_{\mathcal{I}}) : \nabla \mathbf{v}, \\ \langle R_{\mathrm{c}}(\mathbf{u}_{\mathcal{I}}, p_{\mathcal{I}}), q \rangle &= \int_\Omega q \operatorname{div}(\mathbf{u} - \mathbf{u}_{\mathcal{I}}). \end{aligned}$$

These identities combined with the definition (6.2) (p. 309) of $\|\cdot\|_{-1}$, the Cauchy–Schwarz inequality in $\mathbb{R}^2$, and the inequality $\|\operatorname{div} \mathbf{v}\| \leq \sqrt{d}\, \|\nabla \mathbf{v}\|$ imply the lower bound for the error.

For the proof of the upper bound, we first observe that

$$\begin{aligned} \Pi \mathbf{u}_{\mathcal{I}} &= \frac{t - t_{n-1}}{\tau_n} \Pi \mathbf{u}^n_{\mathcal{T}_n} + \frac{t_n - t}{\tau_n} \Pi \mathbf{u}^{n-1}_{\mathcal{T}_{n-1}} \\ \partial_t \Pi \mathbf{u}_{\mathcal{I}} &= \frac{1}{\tau_n} \left(\Pi \mathbf{u}^n_{\mathcal{T}_n} - \Pi \mathbf{u}^{n-1}_{\mathcal{T}_{n-1}} \right) = \Pi(\partial_t \mathbf{u}_{\mathcal{I}}) \end{aligned} \tag{6.53}$$

holds on all intervals $(t_{n-1}, t_n]$. Next we insert $\mathbf{u} - (\mathbf{u}_{\mathcal{I}} - \Pi \mathbf{u}_{\mathcal{I}})$ as test function $\mathbf{v}$ in the above identity for $R_{\mathrm{m}}(\mathbf{u}_{\mathcal{I}}, p_{\mathcal{I}})$. Since $\mathbf{u} - (\mathbf{u}_{\mathcal{I}} - \Pi \mathbf{u}_{\mathcal{I}})$ is solenoidal, this gives

$$\begin{aligned} &\langle R_{\mathrm{m}}(\mathbf{u}_{\mathcal{I}}, p_{\mathcal{I}}), \mathbf{u} - (\mathbf{u}_{\mathcal{I}} - \Pi \mathbf{u}_{\mathcal{I}}) \rangle \\ &\quad = \langle \partial_t(\mathbf{u} - \mathbf{u}_{\mathcal{I}}), \mathbf{u} - (\mathbf{u}_{\mathcal{I}} - \Pi \mathbf{u}_{\mathcal{I}}) \rangle + \nu \int_\Omega \nabla(\mathbf{u} - \mathbf{u}_{\mathcal{I}}) : \nabla(\mathbf{u} - (\mathbf{u}_{\mathcal{I}} - \Pi \mathbf{u}_{\mathcal{I}})) \\ &\quad = \frac{1}{2} \frac{\mathrm{d}}{dt} \|\mathbf{u} - \mathbf{u}_{\mathcal{I}}\|^2 + \langle \partial_t(\mathbf{u} - \mathbf{u}_{\mathcal{I}}), \Pi \mathbf{u}_{\mathcal{I}} \rangle + \nu \left\| \nabla(\mathbf{u} - \mathbf{u}_{\mathcal{I}}) \right\|^2 + \nu \int_\Omega \nabla(\mathbf{u} - \mathbf{u}_{\mathcal{I}}) : \nabla \Pi \mathbf{u}_{\mathcal{I}}. \end{aligned}$$

Regrouping terms this is equivalent to

$$\frac{1}{2}\frac{\mathrm{d}}{dt}\|\mathbf{u}-\mathbf{u}_{\mathcal{I}}\|^2+\nu\left\|\nabla(\mathbf{u}-\mathbf{u}_{\mathcal{I}})\right\|^2$$
$$=\left\langle R_{\mathrm{m}}(\mathbf{u}_{\mathcal{I}},p_{\mathcal{I}}),\,\mathbf{u}-\mathbf{u}_{\mathcal{I}}\right\rangle+\left\langle R_{\mathrm{m}}(\mathbf{u}_{\mathcal{I}},p_{\mathcal{I}}),\,\Pi\mathbf{u}_{\mathcal{I}}\right\rangle-\left\langle\partial_t(\mathbf{u}-\mathbf{u}_{\mathcal{I}}),\,\Pi\mathbf{u}_{\mathcal{I}}\right\rangle-\nu\int_\Omega\nabla(\mathbf{u}-\mathbf{u}_{\mathcal{I}}):\nabla\Pi\mathbf{u}_{\mathcal{I}}.$$

Recalling the definition of the dual norm and Lemma 6.31 and using the inequality $ab\le\frac{\varepsilon}{2}a^2+\frac{1}{2\varepsilon}b^2$ with appropriate values of ε, a, and b several times, this implies that

$$\begin{aligned}
\frac{1}{2}\frac{\mathrm{d}}{dt}&\|\mathbf{u}-\mathbf{u}_{\mathcal{I}}\|^2+\nu\left\|\nabla(\mathbf{u}-\mathbf{u}_{\mathcal{I}})\right\|^2\\
&\le\left\|R_{\mathrm{m}}(\mathbf{u}_{\mathcal{I}},p_{\mathcal{I}})\right\|_{-1}\left\|\nabla(\mathbf{u}-\mathbf{u}_{\mathcal{I}})\right\|+\left\|R_{\mathrm{m}}(\mathbf{u}_{\mathcal{I}},p_{\mathcal{I}})\right\|_{-1}\|\nabla\Pi\mathbf{u}_{\mathcal{I}}\|\\
&\quad-\left\langle\partial_t(\mathbf{u}-\mathbf{u}_{\mathcal{I}}),\,\Pi\mathbf{u}_{\mathcal{I}}\right\rangle+\nu\left\|\nabla(\mathbf{u}-\mathbf{u}_{\mathcal{I}})\right\|\|\nabla\Pi\mathbf{u}_{\mathcal{I}}\|\\
&\le\left\|R_{\mathrm{m}}(\mathbf{u}_{\mathcal{I}},p_{\mathcal{I}})\right\|_{-1}\left\|\nabla(\mathbf{u}-\mathbf{u}_{\mathcal{I}})\right\|+\left\|R_{\mathrm{m}}(\mathbf{u}_{\mathcal{I}},p_{\mathcal{I}})\right\|_{-1}\frac{1}{\beta}\|\operatorname{div}\mathbf{u}_{\mathcal{I}}\|\\
&\quad-\left\langle\partial_t(\mathbf{u}-\mathbf{u}_{\mathcal{I}}),\,\Pi\mathbf{u}_{\mathcal{I}}\right\rangle+\nu\left\|\nabla(\mathbf{u}-\mathbf{u}_{\mathcal{I}})\right\|\frac{1}{\beta}\|\operatorname{div}\mathbf{u}_{\mathcal{I}}\|\\
&\le\frac{\nu}{2}\left\|\nabla(\mathbf{u}-\mathbf{u}_{\mathcal{I}})\right\|^2+\frac{3}{2\nu}\left\|R_{\mathrm{m}}(\mathbf{u}_{\mathcal{I}},p_{\mathcal{I}})\right\|_{-1}^2+\frac{3\nu}{2\beta^2}\|\operatorname{div}\mathbf{u}_{\mathcal{I}}\|^2-\left\langle\partial_t(\mathbf{u}-\mathbf{u}_{\mathcal{I}}),\,\Pi\mathbf{u}_{\mathcal{I}}\right\rangle.
\end{aligned}$$

Since $\operatorname{div}\mathbf{u}_{\mathcal{I}}=-R_{\mathrm{c}}(\mathbf{u}_{\mathcal{I}},p_{\mathcal{I}})$, this is equivalent to

$$\frac{\mathrm{d}}{\mathrm{d}t}\|\mathbf{u}-\mathbf{u}_{\mathcal{I}}\|^2+\nu\left\|\nabla(\mathbf{u}-\mathbf{u}_{\mathcal{I}})\right\|^2\le\frac{3}{\nu}\left\|R_{\mathrm{m}}(\mathbf{u}_{\mathcal{I}},p_{\mathcal{I}})\right\|_{-1}^2+\frac{3\nu}{\beta^2}\left\|R_{\mathrm{c}}(\mathbf{u}_{\mathcal{I}},p_{\mathcal{I}})\right\|^2-2\left\langle\partial_t(\mathbf{u}-\mathbf{u}_{\mathcal{I}}),\,\Pi\mathbf{u}_{\mathcal{I}}\right\rangle.$$

Next we choose an arbitrary $t\in(0,T)$, integrate this inequality from 0 to t, and use integration by parts for the term $\left\langle\partial_t(\mathbf{u}-\mathbf{u}_{\mathcal{I}}),\,\Pi\mathbf{u}_{\mathcal{I}}\right\rangle$. We thus obtain

$$\begin{aligned}
&\left\|(\mathbf{u}-\mathbf{u}_{\mathcal{I}})(\cdot,t)\right\|^2-\left\|\mathbf{u}_0-\mathbf{u}_{\mathcal{T}_0}^0\right\|^2+\nu\int_0^t\left\|\nabla(\mathbf{u}-\mathbf{u}_{\mathcal{I}})(\cdot,s)\right\|^2\mathrm{d}s\\
&\le\frac{3}{\nu}\int_0^t\left\|R_{\mathrm{m}}(\mathbf{u}_{\mathcal{I}},p_{\mathcal{I}})(\cdot,s)\right\|_{-1}^2\mathrm{d}s+\frac{3\nu}{\beta^2}\int_0^t\left\|R_{\mathrm{c}}(\mathbf{u}_{\mathcal{I}},p_{\mathcal{I}})(\cdot,s)\right\|^2\mathrm{d}s\\
&\quad+2\int_0^t\left\langle(\mathbf{u}-\mathbf{u}_{\mathcal{I}})(\cdot,s),\,\partial_t\Pi\mathbf{u}_{\mathcal{I}}(\cdot,s)\right\rangle\mathrm{d}s-2\left\langle(\mathbf{u}-\mathbf{u}_{\mathcal{I}})(\cdot,t),\,\Pi\mathbf{u}_{\mathcal{I}}(\cdot,t)\right\rangle+2\left\langle(\mathbf{u}_0-\mathbf{u}_{\mathcal{T}_0}^0),\,\Pi\mathbf{u}_{\mathcal{T}_0}^0\right\rangle\\
&\le\frac{3}{\nu}\left\|R_{\mathrm{m}}(\mathbf{u}_{\mathcal{I}},p_{\mathcal{I}})\right\|_{L^2(0,T;H^{-1}(\Omega)^d)}^2+\frac{3\nu}{\beta^2}\left\|R_{\mathrm{c}}(\mathbf{u}_{\mathcal{I}},p_{\mathcal{I}})\right\|_{L^2(0,T;L^2(\Omega))}^2\\
&\quad+2\,\|\mathbf{u}-\mathbf{u}_{\mathcal{I}}\|_{L^\infty(0,T;L^2(\Omega)^d)}\|\partial_t\Pi\mathbf{u}_{\mathcal{I}}\|_{L^1(0,T;L^2(\Omega)^d)}\\
&\quad+4\,\|\mathbf{u}-\mathbf{u}_{\mathcal{I}}\|_{L^\infty(0,T;L^2(\Omega)^d)}\|\Pi\mathbf{u}_{\mathcal{I}}\|_{L^\infty(0,T;L^2(\Omega)^d)}\\
&\le\frac{3}{\nu}\left\|R_{\mathrm{m}}(\mathbf{u}_{\mathcal{I}},p_{\mathcal{I}})\right\|_{L^2(0,T;H^{-1}(\Omega)^d)}^2+\frac{3\nu}{\beta^2}\left\|R_{\mathrm{c}}(\mathbf{u}_{\mathcal{I}},p_{\mathcal{I}})\right\|_{L^2(0,T;L^2(\Omega))}^2\\
&\quad+4\,\|\partial_t\Pi\mathbf{u}_{\mathcal{I}}\|_{L^1(0,T;L^2(\Omega)^d)}^2+16\,\|\Pi\mathbf{u}_{\mathcal{I}}\|_{L^\infty(0,T;L^2(\Omega)^d)}^2+\frac{1}{2}\,\|\mathbf{u}-\mathbf{u}_{\mathcal{I}}\|_{L^\infty(0,T;L^2(\Omega)^d)}^2.
\end{aligned}$$

Since $t \in (0, T)$ is arbitrary, this proves

$$\begin{aligned}
&\|\mathbf{u} - \mathbf{u}_{\mathcal{I}}\|^2_{L^\infty(0,T;L^2(\Omega)^d)} \\
&\quad \le \frac{6}{\nu} \left\|R_{\mathrm{m}}(\mathbf{u}_{\mathcal{I}}, p_{\mathcal{I}})\right\|^2_{L^2(0,T;H^{-1}(\Omega)^d)} + \frac{6\nu}{\beta^2} \left\|R_{\mathrm{c}}(\mathbf{u}_{\mathcal{I}}, p_{\mathcal{I}})\right\|^2_{L^2(0,T;L^2(\Omega))} \\
&\qquad + 8 \|\partial_t \Pi \mathbf{u}_{\mathcal{I}}\|^2_{L^1(0,T;L^2(\Omega)^d)} + 32 \|\Pi \mathbf{u}_{\mathcal{I}}\|^2_{L^\infty(0,T;L^2(\Omega)^d)} + 2 \left\|\mathbf{u}_0 - \mathbf{u}^0_{\mathcal{T}_0}\right\|^2
\end{aligned}$$

and

$$\begin{aligned}
&\nu \|\mathbf{u} - \mathbf{u}_{\mathcal{I}}\|^2_{L^2(0,T;H^1_0(\Omega)^d)} \\
&\quad \le \frac{6}{\nu} \left\|R_{\mathrm{m}}(\mathbf{u}_{\mathcal{I}}, p_{\mathcal{I}})\right\|^2_{L^2(0,T;H^{-1}(\Omega)^d)} + \frac{6\nu}{\beta^2} \left\|R_{\mathrm{c}}(\mathbf{u}_{\mathcal{I}}, p_{\mathcal{I}})\right\|^2_{L^2(0,T;L^2(\Omega))} \\
&\qquad + 8 \|\partial_t \Pi \mathbf{u}_{\mathcal{I}}\|^2_{L^1(0,T;L^2(\Omega)^d)} + 32 \|\Pi \mathbf{u}_{\mathcal{I}}\|^2_{L^\infty(0,T;L^2(\Omega)^d)} + 2 \left\|\mathbf{u}_0 - \mathbf{u}^0_{\mathcal{T}_0}\right\|^2 .
\end{aligned}$$

The representation of $R_{\mathrm{m}}(\mathbf{u}_{\mathcal{I}}, p_{\mathcal{I}})$ on the other hand implies that

$$\left\|\partial_t(\mathbf{u} - \mathbf{u}_{\mathcal{I}}) + \nabla(p - p_{\mathcal{I}})\right\|_{-1} \le \left\|R_{\mathrm{m}}(\mathbf{u}_{\mathcal{I}}, p_{\mathcal{I}})\right\|_{-1} + \nu \left\|\nabla(\mathbf{u} - \mathbf{u}_{\mathcal{I}})\right\| .$$

Squaring this inequality, integrating the result from 0 to T, and taking into account the estimate for $\nu \|\mathbf{u} - \mathbf{u}_{\mathcal{I}}\|^2_{L^2(0,T;H^1_0(\Omega)^d)}$, we obtain

$$\begin{aligned}
&\left\|\partial_t(\mathbf{u} - \mathbf{u}_{\mathcal{I}}) + \nabla(p - p_{\mathcal{I}})\right\|^2_{L^2(0,T;H^{-1}(\Omega)^d)} \\
&\quad \le 2 \left\|R_{\mathrm{m}}(\mathbf{u}_{\mathcal{I}}, p_{\mathcal{I}})\right\|^2_{L^2(0,T;H^{-1}(\Omega)^d)} + 2\nu^2 \|\mathbf{u} - \mathbf{u}_{\mathcal{I}}\|^2_{L^2(0,T;H^1_0(\Omega)^d)} \\
&\quad \le 14 \left\|R_{\mathrm{m}}(\mathbf{u}_{\mathcal{I}}, p_{\mathcal{I}})\right\|^2_{L^2(0,T;H^{-1}(\Omega)^d)} + \frac{12\nu^2}{\beta^2} \left\|R_{\mathrm{c}}(\mathbf{u}_{\mathcal{I}}, p_{\mathcal{I}})\right\|^2_{L^2(0,T;L^2(\Omega))} \\
&\qquad + 16\nu \|\partial_t \Pi \mathbf{u}_{\mathcal{I}}\|^2_{L^1(0,T;L^2(\Omega)^d)} + 64\nu \|\Pi \mathbf{u}_{\mathcal{I}}\|^2_{L^\infty(0,T;L^2(\Omega)^d)} + 4\nu \left\|\mathbf{u}_0 - \mathbf{u}^0_{\mathcal{T}_0}\right\|^2 .
\end{aligned}$$

Combining these estimates, taking into account (6.53), and using Lemma 6.31 establishes the upper bound for the error. □

6.5.4 *Decomposition of the Residuals*

We define temporal residuals

$$R_{\mathrm{m},\tau}(\mathbf{u}_{\mathcal{I}}, p_{\mathcal{I}}) \in L^2(0, T; H^{-1}(\Omega)^d), \quad R_{\mathrm{c},\tau}(\mathbf{u}_{\mathcal{I}}, p_{\mathcal{I}}) \in L^2(0, T; L^2(\Omega))$$

and spatial residuals

$$R_{\mathrm{m},h}(\mathbf{u}_{\mathcal{I}}, p_{\mathcal{I}}) \in L^2(0, T; H^{-1}(\Omega)^d), \quad R_{\mathrm{c},h}(\mathbf{u}_{\mathcal{I}}, p_{\mathcal{I}}) \in L^2(0, T; L^2(\Omega))$$

by setting on $(t_{n-1}, t_n]$ for all $n \in \{1, \dots, N_{\mathcal{I}}\}$ and all $\mathbf{v} \in H^1_0(\Omega)^d$, $q \in L^2_0(\Omega)$

$$\left\langle R_{\mathrm{m},\tau}(\mathbf{u}_{\mathcal{I}}, p_{\mathcal{I}}), \mathbf{v}\right\rangle = \nu \int_\Omega \nabla\left[\mathbf{u}^{n\theta} - \mathbf{u}_{\mathcal{I}}\right] : \nabla \mathbf{v},$$

$$\langle R_{c,\tau}(\mathbf{u}_{\mathcal{I}}, p_{\mathcal{I}}), q\rangle = \int_{\Omega} q \operatorname{div}\left[\mathbf{u}^n_{\mathcal{T}_n} - \mathbf{u}_{\mathcal{I}}\right],$$

$$\langle R_{m,h}(\mathbf{u}_{\mathcal{I}}, p_{\mathcal{I}}), \mathbf{v}\rangle = \int_{\Omega} \mathbf{f}^{n\theta} \cdot \mathbf{v} - \int_{\Omega} \frac{1}{\tau_n}\left(\mathbf{u}^n_{\mathcal{T}_n} - \mathbf{u}^{n-1}_{\mathcal{T}_{n-1}}\right) \cdot \mathbf{v} - \nu \int_{\Omega} \nabla \mathbf{u}^{n\theta} : \nabla \mathbf{v} + \int_{\Omega} p^n_{\mathcal{T}_n} \operatorname{div} \mathbf{v},$$

$$\langle R_{c,h}(\mathbf{u}_{\mathcal{I}}, p_{\mathcal{I}}), q\rangle = -\int_{\Omega} q \operatorname{div} \mathbf{u}^n_{\mathcal{T}_n}.$$

We then have

$$R_m(\mathbf{u}_{\mathcal{I}}, p_{\mathcal{I}}) = \mathbf{f} - \mathbf{f}^{n\theta} + R_{m,\tau}(\mathbf{u}_{\mathcal{I}}, p_{\mathcal{I}}) + R_{m,h}(\mathbf{u}_{\mathcal{I}}, p_{\mathcal{I}}),$$
$$R_c(\mathbf{u}_{\mathcal{I}}, p_{\mathcal{I}}) = R_{c,\tau}(\mathbf{u}_{\mathcal{I}}, p_{\mathcal{I}}) + R_{c,h}(\mathbf{u}_{\mathcal{I}}, p_{\mathcal{I}}).$$

Since $\mathbf{f} - \mathbf{f}^{n\theta}$ describes temporal oscillations of the known data, we have to find computable upper and lower bounds for appropriate norms of $R_{m,\tau}(\mathbf{u}_{\mathcal{I}}, p_{\mathcal{I}}) + R_{m,h}(\mathbf{u}_{\mathcal{I}}, p_{\mathcal{I}})$ and $R_{c,\tau}(\mathbf{u}_{\mathcal{I}}, p_{\mathcal{I}}) + R_{c,h}(\mathbf{u}_{\mathcal{I}}, p_{\mathcal{I}})$. Again, the following lemma shows that this can be achieved by estimating the norms of all contributions separately.

Lemma 6.33 *For every $n \in \{1, \ldots, N_{\mathcal{I}}\}$ we have*

$$\sqrt{\frac{5}{14}}\left(1 - \frac{\sqrt{3}}{2}\right)\left\{\left\|R_{m,\tau}(\mathbf{u}_{\mathcal{I}}, p_{\mathcal{I}})\right\|^2_{L^2(t_{n-1},t_n;H^{-1}(\Omega)^d)} + \left\|R_{m,h}(\mathbf{u}_{\mathcal{I}}, p_{\mathcal{I}})\right\|^2_{L^2(t_{n-1},t_n;H^{-1}(\Omega)^d)}\right\}^{\frac{1}{2}}$$
$$\leq \left\|R_{m,\tau}(\mathbf{u}_{\mathcal{I}}, p_{\mathcal{I}}) + R_{m,h}(\mathbf{u}_{\mathcal{I}}, p_{\mathcal{I}})\right\|_{L^2(t_{n-1},t_n;H^{-1}(\Omega)^d)}$$
$$\leq \left\|R_{m,\tau}(\mathbf{u}_{\mathcal{I}}, p_{\mathcal{I}})\right\|_{L^2(t_{n-1},t_n;H^{-1}(\Omega)^d)} + \left\|R_{m,h}(\mathbf{u}_{\mathcal{I}}, p_{\mathcal{I}})\right\|_{L^2(t_{n-1},t_n;H^{-1}(\Omega)^d)}$$

and

$$\sqrt{\frac{5}{14}}\left(1 - \frac{\sqrt{3}}{2}\right)\left\{\left\|R_{c,\tau}(\mathbf{u}_{\mathcal{I}}, p_{\mathcal{I}})\right\|^2_{L^2(t_{n-1},t_n;L^2(\Omega))} + \left\|R_{c,h}(\mathbf{u}_{\mathcal{I}}, p_{\mathcal{I}})\right\|^2_{L^2(t_{n-1},t_n;L^2(\Omega))}\right\}^{\frac{1}{2}}$$
$$\leq \left\|R_{c,\tau}(\mathbf{u}_{\mathcal{I}}, p_{\mathcal{I}}) + R_{c,h}(\mathbf{u}_{\mathcal{I}}, p_{\mathcal{I}})\right\|_{L^2(t_{n-1},t_n;L^2(\Omega))}$$
$$\leq \left\|R_{c,\tau}(\mathbf{u}_{\mathcal{I}}, p_{\mathcal{I}})\right\|_{L^2(t_{n-1},t_n;L^2(\Omega))} + \left\|R_{c,h}(\mathbf{u}_{\mathcal{I}}, p_{\mathcal{I}})\right\|_{L^2(t_{n-1},t_n;L^2(\Omega))}.$$

Proof The assertion follows from Lemma 3.53 (p. 130) and Remark 3.54 (p. 132) with $p = 2$ and $Y = H^1_0(\Omega)^d$, $\|\cdot\|_Y = \|\nabla \cdot\|$ for the residuals associated with the momentum equation and $Y = L^2_0(\Omega)$, $\|\cdot\|_Y = \|\cdot\|$ for the residuals associated with the continuity equation. □

6.5.5 *Estimation of the Temporal Residuals*

Obviously, we have for all $n \in \{1, \ldots, N_{\mathcal{I}}\}$ and all $\mathbf{v} \in H^1_0(\Omega)^d$, $q \in L^2_0(\Omega)$

$$\langle R_{m,\tau}(\mathbf{u}_{\mathcal{I}}, p_{\mathcal{I}}), \mathbf{v}\rangle = \nu\left(\theta - \frac{t - t_{n-1}}{\tau_n}\right)\int_{\Omega} \nabla\left[\mathbf{u}^n_{\mathcal{T}_n} - \mathbf{u}^{n-1}_{\mathcal{T}_{n-1}}\right] : \nabla \mathbf{v}$$

$$\langle R_{c,\tau}(\mathbf{u}_{\mathcal{I}}, p_{\mathcal{I}}), q\rangle = \frac{t_n - t}{\tau_n}\int_{\Omega} q \operatorname{div}\left[\mathbf{u}^n_{\mathcal{T}_n} - \mathbf{u}^{n-1}_{\mathcal{T}_{n-1}}\right]$$

on $(t_{n-1}, t_n]$. The definition (6.2) (p. 309) of the dual norm and a straightforward calculation therefore prove the following estimates.

Lemma 6.34 *For every $n \in \{1, \dots, N_{\mathcal{I}}\}$, the norms of the temporal residuals satisfy*

$$\nu\sqrt{\frac{\tau_n}{12}}\left\|\nabla(\mathbf{u}^n_{\mathcal{T}_n} - \mathbf{u}^{n-1}_{\mathcal{T}_{n-1}})\right\| \le \left\|R_{\mathrm{m},\tau}(\mathbf{u}_{\mathcal{I}}, p_{\mathcal{I}})\right\|_{L^2(t_{n-1},t_n;H^{-1}(\Omega)^d)} \le \nu\sqrt{\frac{\tau_n}{3}}\left\|\nabla(\mathbf{u}^n_{\mathcal{T}_n} - \mathbf{u}^{n-1}_{\mathcal{T}_{n-1}})\right\|$$

and

$$\left\|R_{\mathrm{c},\tau}(\mathbf{u}_{\mathcal{I}}, p_{\mathcal{I}})\right\|_{L^2(t_{n-1},t_n;L^2(\Omega))} = \sqrt{\frac{\tau_n}{3}}\left\|\operatorname{div}(\mathbf{u}^n_{\mathcal{T}_n} - \mathbf{u}^{n-1}_{\mathcal{T}_{n-1}})\right\|.$$

6.5.6 Estimation of the Spatial Residuals

The spatial residuals $R_{\mathrm{m},h}(\mathbf{u}_{\mathcal{I}}, p_{\mathcal{I}})$, $R_{\mathrm{c},h}(\mathbf{u}_{\mathcal{I}}, p_{\mathcal{I}})$ are associated with the discrete stationary Stokes equations (6.50) and its analytical counterpart. This problem differs from the Stokes equations considered in Section 4.10 (p. 237) by the additional zero-order term in the momentum equation. The methods of Section 4.10 can easily be adapted to this modification and yield the following estimates.

Lemma 6.35 *For every $n \in \{1, \dots, N_{\mathcal{I}}\}$ define a spatial error indicator by*

$$\eta^n = \left\{\sum_{K\in\mathcal{T}_n} h_K^2 \left\|\mathbf{f}^{n\theta}_{\mathcal{T}_n} - \frac{1}{\tau_n}\left(\mathbf{u}^n_{\mathcal{T}_n} - \mathbf{u}^{n-1}_{\mathcal{T}_{n-1}}\right) + \nu\Delta\mathbf{u}^{n\theta} - \nabla p^n_{\mathcal{T}_n}\right\|_K^2 + \sum_{E\in\mathcal{E}_{n,\Omega}} h_E \left\|\mathbb{J}_E\left(\mathbf{n}_E\cdot(\nu\nabla\mathbf{u}^{n\theta} - p^n_{\mathcal{T}_n}I)\right)\right\|_E^2 + \sum_{K\in\mathcal{T}_n}\left\|\operatorname{div}\mathbf{u}^n_{\mathcal{T}_n}\right\|_K^2\right\}^{\frac{1}{2}} \tag{6.54}$$

and a spatial data error indicator by

$$\Theta^n = \left\{\sum_{K\in\mathcal{T}} h_K^2 \left\|\mathbf{f} - \mathbf{f}^{n\theta}_{\mathcal{T}_n}\right\|_K^2\right\}^{\frac{1}{2}}, \tag{6.55}$$

where $\mathbf{f}^{n\theta}_{\mathcal{T}_n}$ is any piecewise polynomial approximation of $\mathbf{f}^{n\theta}$. There are two constants c^* and c_* such that the norms of the spatial residuals can be bounded from above by

$$\left\|R_{\mathrm{m},h}(\mathbf{u}_{\mathcal{I}}, p_{\mathcal{I}})\right\|_{-1} + \left\|R_{\mathrm{c},h}(\mathbf{u}_{\mathcal{I}}, p_{\mathcal{I}})\right\| \le c^*\{\eta^n + \Theta^n\}$$

and from below by

$$\eta^n \le c_*\left\{\left\|R_{\mathrm{m},h}(\mathbf{u}_{\mathcal{I}}, p_{\mathcal{I}})\right\|_{-1} + \left\|R_{\mathrm{c},h}(\mathbf{u}_{\mathcal{I}}, p_{\mathcal{I}})\right\| + \Theta^n\right\}.$$

Both constants c^* and c_* depend on the shape parameters $C_{\mathcal{T}_n}$ and the viscosity ν. The constant c_* in addition depends on the polynomial degrees of the functions $\mathbf{u}^n_{\mathcal{T}_n}$, $p^n_{\mathcal{T}_n}$, and $\mathbf{f}^{n\theta}_{\mathcal{T}_n}$; the constant c^* in addition depends on the stabilisation parameters of the finite element approximation through $\max_{1\le n\le N_{\mathcal{I}}}\max_{K\in\mathcal{T}_n}\widetilde{\vartheta}_K$.

6.5.7 A Posteriori Error Estimates

Proposition 6.32 and Lemmas 6.33, 6.34, and 6.35 yield the following a posteriori error estimates.

Theorem 6.36 *Denote by* $\mathbf{u}$, p *the solutions of problems* (6.47) *and* (6.48) *and by* $\mathbf{u}_{\mathcal{I}}$, $p_{\mathcal{I}}$ *their finite element approximations associated with the solutions of problems* (6.49) *and* (6.50). *Then the error can be bounded from above by*

$$\left\{\left\|\partial_t(\mathbf{u}-\mathbf{u}_{\mathcal{I}})+\nabla(p-p_{\mathcal{I}})\right\|^2_{L^2(0,T;H^{-1}(\Omega)^d)}+\|\mathbf{u}-\mathbf{u}_{\mathcal{I}}\|^2_{L^\infty(0,T;L^2(\Omega)^d)}+\nu\,\|\mathbf{u}-\mathbf{u}_{\mathcal{I}}\|^2_{L^2(0,T;H^1_0(\Omega)^d)}\right\}^{\frac{1}{2}}$$

$$\leq c^\sharp\left\{\sum_{n=1}^{N_{\mathcal{I}}}\tau_n\left[\left\|\nabla\left(\mathbf{u}^n_{\mathcal{T}_n}-\mathbf{u}^{n-1}_{\mathcal{T}_{n-1}}\right)\right\|^2+\left\|\operatorname{div}\left(\mathbf{u}^n_{\mathcal{T}_n}-\mathbf{u}^{n-1}_{\mathcal{T}_{n-1}}\right)\right\|^2+(\eta^n)^2\right]+\left\|\mathbf{u}_0-\mathbf{u}^0_{\mathcal{T}_0}\right\|^2\right.$$

$$+\sum_{n=1}^{N_{\mathcal{I}}}\tau_n\,(\Theta^n)^2+\max_{0\leq n\leq N_{\mathcal{I}}}\sum_{K\in\mathcal{T}_n}h_K^{2\alpha_K}\left\|\operatorname{div}\mathbf{u}^n_{\mathcal{T}_n}\right\|^2_K$$

$$\left.+\left(\sum_{n=1}^{N_{\mathcal{I}}}\left[\sum_{K\in\mathcal{T}_n}h_K^{2\alpha_K}\left\|\operatorname{div}\left(\mathbf{u}^n_{\mathcal{T}_n}-\mathbf{u}^{n-1}_{\mathcal{T}_{n-1}}\right)\right\|^2_K\right]^{\frac{1}{2}}\right)^2\right\}^{\frac{1}{2}}$$

and from below by

$$\left\{\sum_{n=1}^{N_{\mathcal{I}}}\tau_n\left[\left\|\nabla\left(\mathbf{u}^n_{\mathcal{T}_n}-\mathbf{u}^{n-1}_{\mathcal{T}_{n-1}}\right)\right\|^2+\left\|\operatorname{div}(\mathbf{u}^n_{\mathcal{T}_n}-\mathbf{u}^{n-1}_{\mathcal{T}_{n-1}})\right\|^2+(\eta^n)^2\right]\right\}^{\frac{1}{2}}$$

$$\leq c_\sharp\left\{\left\|\partial_t(\mathbf{u}-\mathbf{u}_{\mathcal{I}})+\nabla(p-p_{\mathcal{I}})\right\|^2_{L^2(0,T;H^{-1}(\Omega)^d)}\right.$$

$$\left.+\|\mathbf{u}-\mathbf{u}_{\mathcal{I}}\|^2_{L^\infty(0,T;L^2(\Omega)^d)}+\nu\,\|\mathbf{u}-\mathbf{u}_{\mathcal{I}}\|^2_{L^2(0,T;H^1_0(\Omega)^d)}+\left\|\mathbf{u}_0-\mathbf{u}^0_{\mathcal{T}_0}\right\|^2+\sum_{n=1}^{N_{\mathcal{I}}}\tau_n\,(\Theta^n)^2\right\}^{\frac{1}{2}}.$$

The indicators η^n, Θ^n *are defined in equations* (6.54), (6.55); *the quantities* α_k *are as in Lemma 6.31. The constants* $c^\sharp$ *and* $c_\sharp$ *depend on the viscosity* ν, *the inf–sup constant* β *of Proposition 4.67 (p. 238), the constant* c_Π *of Lemma 6.31, the shape parameters* $C_{\mathcal{T}_n}$, *the constant in the modified transition condition, and the polynomial degrees of the spaces* V_n *and* P_n.

Remark 6.37 The terms

$$\sqrt{\tau_n}\left[\left\|\nabla\left(\mathbf{u}^n_{\mathcal{T}_n}-\mathbf{u}^{n-1}_{\mathcal{T}_{n-1}}\right)\right\|^2+\left\|\operatorname{div}\left(\mathbf{u}^n_{\mathcal{T}_n}-\mathbf{u}^{n-1}_{\mathcal{T}_{n-1}}\right)\right\|^2\right]^{\frac{1}{2}}$$

control the temporal error and can be used to adapt the time-steps, cf. Section 6.8.3 (p. 364). Of course, $\left\|\operatorname{div}(\mathbf{u}^n_{\mathcal{T}_n}-\mathbf{u}^{n-1}_{\mathcal{T}_{n-1}})\right\|$ can be bounded by $\sqrt{d}\left\|\nabla(\mathbf{u}^n_{\mathcal{T}_n}-\mathbf{u}^{n-1}_{\mathcal{T}_{n-1}})\right\|$.

The terms $\sqrt{\tau_n}\eta^n$ control the spatial errors and can be used to adapt the spatial meshes, cf. Section 6.8.3.

If $h_K^{2\alpha_K} \le \tau_n$ for all elements $K \in \mathcal{T}_n$ and all n, the term

$$\max_{0 \le n \le N_{\mathcal{I}}} \sum_{K \in \mathcal{T}_n} h_K^{2\alpha_K} \left\| \operatorname{div} \mathbf{u}_{\mathcal{T}_n}^n \right\|_K^2$$

can be bounded by $\sum_n \tau_n (\eta^n)^2$. Since α_K is at least $\frac{1}{2}$ for all K and since usually τ_n is of the order of the spatial mesh-size, this condition is fulfilled even in the vicinity of re-entrant corners of Ω.

If, on the other hand, $h_K^{2\alpha_K} \le \tau_n^2$ for all elements $K \in \mathcal{T}_n$ and all n, the term

$$\left(\sum_{n=1}^{N_{\mathcal{I}}} \left[\sum_{K \in \mathcal{T}_n} h_K^{2\alpha_K} \left\| \operatorname{div} \left(\mathbf{u}_{\mathcal{T}_n}^n - \mathbf{u}_{\mathcal{T}_{n-1}}^{n-1} \right) \right\|_K^2 \right]^{\frac{1}{2}} \right)^2$$

can be bounded by

$$T \sum_{n=1}^{N_{\mathcal{I}}} \tau_n \left\| \operatorname{div} \left(\mathbf{u}_{\mathcal{T}_n}^n - \mathbf{u}_{\mathcal{T}_{n-1}}^{n-1} \right) \right\|^2 .$$

This condition is usually fulfilled only away from re-entrant corners and in particular is satisfied for convex domains.

Remark 6.38 The error estimates of Theorem 6.36 are weaker than those of [**67**, Theorem 5.1] in that they do not provide separate bounds for the error $p - p_{\mathcal{I}}$ of the pressure approximation. They are stronger in that they apply to a larger class of discretisations, require weaker assumptions on the spatial and temporal partitions, and yield direct control on the error of full space–time discretisations. In contrast, [**67**, Theorem 5.1] only gives control on the accumulated error of semi-discretisations in time and space which is usually much larger than the error of a full space–time discretisation.

Remark 6.39 The presentation of this section is based on [**273**].

6.6 Nonlinear Parabolic Equations of Second Order

In this section we consider quasilinear parabolic equations

$$\begin{aligned} \partial_t u - \operatorname{div} \mathbf{a}(x, u, \nabla u) + b(x, u, \nabla u) &= 0 && \text{in } \Omega \times (0, T] \\ u &= 0 && \text{on } \Gamma \times (0, T] \\ u(\cdot, 0) &= u_0 && \text{in } \Omega \end{aligned} \tag{6.56}$$

in a bounded space–time cylinder with a polygonal cross-section $\Omega \subset \mathbb{R}^d$, $d \ge 2$, having a Lipschitz boundary Γ. The final time T is arbitrary, but kept fixed in what follows. The coefficients $\mathbf{a} : \Omega \times \mathbb{R} \times \mathbb{R}^d \to \mathbb{R}^d$ and $b : \Omega \times \mathbb{R} \times \mathbb{R}^d \to \mathbb{R}$ must be continuously differentiable with Lipschitz-continuous derivatives. They have to satisfy suitable growth conditions so that problem (6.56) admits an appropriate variational formulation, cf. Subsections 6.6.1 and 6.6.3 for details. Some sample problems satisfying these conditions are given in Subsection 6.6.3. Some results of Subsections 6.6.7 and 6.6.8 in addition require that the space dimension d equals 2 and that the cross-section Ω is convex.

6.6.1 Variational Formulation

In what follows r, p, ρ, π are Lebesgue exponents in $(1, \infty)$ which must satisfy the condition

$$\frac{1}{\pi} - \frac{1}{\rho} > -\frac{2}{d}. \tag{6.57}$$

The following variational formulation of problem (6.56) is inspired by [**20**] and [**119**, Section XVIII.1.4].

A function $u \in W^r(0, T; W_0^{1,\rho}(\Omega), W^{-1,\pi}(\Omega))$ is called a *weak solution* of problem (6.56), if

$$u(\cdot, 0) = u_0 \tag{6.58}$$

in $W^{-1,\pi}(\Omega)$ and for all $v \in L^{p'}(0, T; W_0^{1,\pi'}(\Omega))$

$$0 = \int_0^T \langle \partial_t u(\cdot, t), v(\cdot, t) \rangle \, dt + \int_0^T \int_\Omega \mathbf{a}(x, u, \nabla u) \cdot \nabla v + b(x, u, \nabla u) v \, dxdt. \tag{6.59}$$

Condition (6.57) ensures the continuous embedding of $W_0^{1,\rho}(\Omega)$ in $W^{-1,\pi}(\Omega)$. Since $r > 1$, this guarantees that for every function $w \in W^r(0, T; W_0^{1,\rho}(\Omega), W^{-1,\pi}(\Omega))$ its trace $w(\cdot, 0)$ is defined as an element of $W^{-1,\pi}(\Omega)$ [**119**, Chapter XVIII, Section 1, Proposition 9]. Note that in two dimensions, $d = 2$, condition (6.57) is fulfilled for all $\rho, \pi \in (1, \infty)$.

6.6.2 Finite Element Discretisation

We again choose a partition $\mathcal{I}$ of $[0, T]$, corresponding partitions $\mathcal{T}_n$ of Ω and associated finite element spaces X_n, which satisfy the conditions of Section 3.2.6 (p. 86), and a parameter $\theta \in [\frac{1}{2}, 1]$, and denote by π_n the L^2-projection onto X_n. Then, the finite element discretisation of problem (6.56) is given by:

find $u_{\mathcal{T}_n}^n \in X_n, 0 \le n \le N_{\mathcal{I}}$, such that

$$u_{\mathcal{T}_0}^0 = \pi_0 u_0 \tag{6.60}$$

and, for $n = 1, \ldots, N_{\mathcal{I}}$ and $u^{n\theta} = \theta u_{\mathcal{T}_n}^n + (1 - \theta) u_{\mathcal{T}_{n-1}}^{n-1}$,

$$0 = \int_\Omega \frac{1}{\tau_n} \left(u_{\mathcal{T}_n}^n - u_{\mathcal{T}_{n-1}}^{n-1} \right) v_{\mathcal{T}_n} dx + \int_\Omega \mathbf{a}\left(x, u^{n\theta}, \nabla u^{n\theta}\right) \cdot \nabla v_{\mathcal{T}_n} + b\left(x, u^{n\theta}, \nabla u^{n\theta}\right) v_{\mathcal{T}_n} \, dx \tag{6.61}$$

for all $v_{\mathcal{T}_n} \in X_n$.

The choice $\theta = \frac{1}{2}$ corresponds to the popular *midpoint rule*.

With every solution $(u_{\mathcal{T}_n}^n)_{0 \le n \le N_{\mathcal{I}}}$ of problems (6.60) and (6.61) we associate two functions $u_{\mathcal{I}}$ and $\widetilde{u}_{\mathcal{I}}$. The function $u_{\mathcal{I}}$ is *piecewise affine* with respect to time and equals $u_{\mathcal{T}_n}^n$ at time t_n, $0 \le n \le N_{\mathcal{I}}$, cf. equations (6.6) and (6.7) (p. 310); the function $\widetilde{u}_{\mathcal{I}}$ is *piecewise constant* with respect to time and equals $u^{n\theta}$ on $(t_{n-1}, t_n]$, $1 \le n \le N_{\mathcal{I}}$.

6.6.3 The Equivalence of Error and Residual

As for nonlinear elliptic equations, the a posteriori error analysis of problem (6.56) is based on the abstract results of Section 5.1 (p. 281), in particular Proposition 5.1 (p. 283).

For brevity, we define a function

$$G : L^r(0, T; W_0^{1,\rho}(\Omega)) \to L^p(0, T; W^{-1,\pi}(\Omega))$$

by setting for all $v \in W_0^{1,\pi'}(\Omega)$ and almost all $t \in (0, T)$

$$\langle G(u), v \rangle = \int_\Omega \mathbf{a}(x, u, \nabla u)\nabla v + b(x, u, \nabla u)v \, dx. \tag{6.62}$$

Then the variational formulation (6.58), (6.59) of problem (6.56) fits into the abstract framework of Section 5.1 with

$$\begin{aligned}
X &= W^r(0, T; W_0^{1,\rho}(\Omega), W^{-1,\pi}(\Omega)), \\
Y &= W_0^{1,\pi'}(\Omega) \times L^{p'}(0, T; W_0^{1,\pi'}(\Omega)), \\
\langle F(u), (v_1, v_2) \rangle &= \begin{pmatrix} \langle u(\cdot, 0) - u_0 , v_1 \rangle \\ \int_0^T \langle \partial_t u, v_2 \rangle + \langle G(u), v_2 \rangle \, dt \end{pmatrix}.
\end{aligned} \tag{6.63}$$

Proposition 5.1 (p. 283) therefore yields the following estimates.

Proposition 6.40 *Denote by u a solution of problems (6.58) and (6.59) and by $u_\mathcal{I}$ an approximation of u associated with solutions $u_{\mathcal{T}_n}^n$ of problems (6.60) and (6.61). Assume that u and the function F of equation (6.63) satisfy the conditions of Proposition 5.1 (p. 283) and that*

$$\| u - u_\mathcal{I} \|_{W^r(0,T;W_0^{1,\rho}(\Omega),W^{-1,\pi}(\Omega))} \le R.$$

Then, there are two constants c_* and c^* such that

$$\begin{aligned}
&c_* \left\{ \left\| u_0 - u_{\mathcal{T}_0}^0 \right\|_{-1,\pi} + \left\| \partial_t u_\mathcal{I} + G(u_\mathcal{I}) \right\|_{L^p(0,T;W^{-1,\pi}(\Omega))} \right\} \\
&\le \| u - u_\mathcal{I} \|_{W^r(0,T;W_0^{1,\rho}(\Omega),W^{-1,\pi}(\Omega))} \\
&\le c^* \left\{ \left\| u_0 - u_{\mathcal{T}_0}^0 \right\|_{-1,\pi} + \left\| \partial_t u_\mathcal{I} + G(u_\mathcal{I}) \right\|_{L^p(0,T;W^{-1,\pi}(\Omega))} \right\}.
\end{aligned}$$

Remark 6.41 Assume that DG is locally Lipschitz continuous at u, i.e. there are two constants $\gamma > 0$ and $R_0 > 0$ such that

$$\begin{aligned}
&\left\| DG(v) - DG(w) \right\|_{\mathcal{L}(L^r(0,T;W_0^{1,\rho}(\Omega)),L^p(0,T;W^{-1,\pi}(\Omega)))} \\
&\le \gamma \, \| v - w \|_{L^r(0,T;W_0^{1,\rho}(\Omega))}
\end{aligned}$$

holds for all $v, w \in L^r(0, T; W_0^{1,\rho}(\Omega))$ with

$$\| u - v \|_{L^r(0,T;W_0^{1,\rho}(\Omega))} \le R_0 \quad \text{and} \quad \| u - w \|_{L^r(0,T;W_0^{1,\rho}(\Omega))} \le R_0.$$

Then the function F of equation (6.63) satisfies the Lipschitz condition of Proposition 5.1 with the same constants γ and R_0.

Next, we give some examples of problems which fit into the present abstract framework.

Example 6.42 The *heat equation with convection and nonlinear diffusion coefficient*, cf. Example 5.13 (p. 292):

$$
\begin{aligned}
\mathbf{a}(x,u,\nabla u) &= k(u)\nabla u,\\
b(x,u,\nabla u) &= \mathbf{c}\cdot\nabla u - f,\\
f &\in L^\infty(\Omega),\\
\mathbf{c} &\in C(\Omega,\mathbb{R}^d),\\
k &\in C^2(\mathbb{R}),\\
k(s) &\geq \alpha > 0,\\
\left|k^{(\ell)}(s)\right| &\leq \gamma \quad \text{for all } s\in\mathbb{R}, \ell\in\{0,1,2\},\\
\rho = \pi &\in (2,4),\\
p > 2, &\quad r \geq 2p.
\end{aligned}
$$

Example 6.43 The *non-stationary equation of prescribed mean curvature*, cf. Example 5.10 (p. 291):

$$
\begin{aligned}
\mathbf{a}(x,u,\nabla u) &= \left[1+|\nabla u|^2\right]^{-\frac{1}{2}}\nabla u,\\
b(x,u,\nabla u) &= -f \in L^2(\Omega),\\
\rho > 2, &\quad \pi = \frac{\rho}{2},\\
r \geq 2\rho, &\quad p = \frac{r}{2}.
\end{aligned}
$$

Example 6.44 The *non-stationary α-Laplacian*, cf. Example 5.11 (p. 291):

$$
\begin{aligned}
\mathbf{a}(x,u,\nabla u) &= |\nabla u|^{\alpha-2}\nabla u \quad \text{with } \alpha\geq 2,\\
b(x,u,\nabla u) &= -f \in L^{\alpha'}(\Omega),\\
\rho = \alpha, &\quad \pi = \alpha',\\
r > 6, &\quad p = \frac{r}{3}.
\end{aligned}
$$

Example 6.45 The *non-stationary subsonic flow of an irrotational ideal compressible gas*, cf. Example 5.12 (p. 292):

$$
\begin{aligned}
\mathbf{a}(x,u,\nabla u) &= \left[1-\frac{\gamma-1}{2}|\nabla u|^2\right]^{\frac{1}{\gamma-1}}\nabla u \quad \text{with } \gamma > 1,\\
b(x,u,\nabla u) &= -f \in L^\pi(\Omega),\\
\rho &= \frac{2\gamma}{\gamma-1}, \pi = \frac{2\gamma}{\gamma+1},\\
r > 6, &\quad p = \frac{r}{3}.
\end{aligned}
$$

6.6.4 Decomposition of the Residual

Proposition 6.40 states that the error $u - u_{\mathcal{I}}$ and the residual $R(u_{\mathcal{I}}) = \partial_t u_{\mathcal{I}} + G(u_{\mathcal{I}})$ are equivalent. As for linear problems, we decompose $R(u_{\mathcal{I}})$ into a spatial residual $R_h(u_{\mathcal{I}})$ and a temporal residual $R_\tau(u_{\mathcal{I}})$ which are defined by setting for all $v \in W_0^{1,\pi'}(\Omega)$ and almost all $t \in (0,T)$

$$\langle R_h(u_{\mathcal{I}}), v\rangle = \langle \partial_t u_{\mathcal{I}} + G(\widetilde{u}_{\mathcal{I}}), v\rangle$$
$$\langle R_\tau(u_{\mathcal{I}}), v\rangle = \langle G(u_{\mathcal{I}}) - G(\widetilde{u}_{\mathcal{I}}), v\rangle .$$

Contrary to linear problems, we now further decompose the temporal residual by setting for all $v \in W_0^{1,\pi'}(\Omega)$ and almost all $t \in (0,T)$

$$\left\langle R_\tau^{(1)}(u_{\mathcal{I}}), v\right\rangle = \langle DG(\widetilde{u}_{\mathcal{I}})(u_{\mathcal{I}} - \widetilde{u}_{\mathcal{I}}), v\rangle$$
$$\left\langle R_\tau^{(2)}(u_{\mathcal{I}}), v\right\rangle = \langle G(u_{\mathcal{I}}) - G(\widetilde{u}_{\mathcal{I}}) - DG(\widetilde{u}_{\mathcal{I}})(u_{\mathcal{I}} - \widetilde{u}_{\mathcal{I}}), v\rangle .$$

The term $R_\tau^{(1)}(u_{\mathcal{I}})$ can be interpreted as a linearisation of $R_\tau(u_{\mathcal{I}})$; the term $R_\tau^{(2)}(u_{\mathcal{I}})$ vanishes for linear problems when G is affine. Obviously we have

$$\partial_t u_{\mathcal{I}} + G(u_{\mathcal{I}}) = R_h(u_{\mathcal{I}}) + R_\tau^{(1)}(u_{\mathcal{I}}) + R_\tau^{(2)}(u_{\mathcal{I}}).$$

The following lemma shows that, up to the term $R_\tau^{(2)}(u_{\mathcal{I}})$, the $L^p(0,T;W^{-1,\pi}(\Omega))$-norm of $R(u_{\mathcal{I}})$ is equivalent to the sum of the corresponding norms of $R_h(u_{\mathcal{I}})$ and $R_\tau^{(1)}(u_{\mathcal{I}})$. The results of the next subsection show that the contribution of $R_\tau^{(2)}(u_{\mathcal{I}})$ is of higher order when compared to $R_\tau^{(1)}(u_{\mathcal{I}})$, cf. Remark 6.55 too.

Lemma 6.46 *For every Lebesgue exponent $p \in (1,\infty)$ and every parameter $\theta \in [\frac{1}{2}, 1]$ we have*

$$\begin{aligned}
\beta_{p,\theta} &\left\{ \|R_h(u_{\mathcal{I}})\|^p_{L^p(0,T;W^{-1,\pi}(\Omega))} + \left\|R_\tau^{(1)}(u_{\mathcal{I}})\right\|^p_{L^p(0,T;W^{-1,\pi}(\Omega))} \right\}^{\frac{1}{p}} - \left\|R_\tau^{(2)}(u_{\mathcal{I}})\right\|_{L^p(0,T;W^{-1,\pi}(\Omega))} \\
&\le \|\partial_t u_{\mathcal{I}} + G(u_{\mathcal{I}})\|_{L^p(0,T;W^{-1,\pi}(\Omega))} \\
&\le \|R_h(u_{\mathcal{I}})\|_{L^p(0,T;W^{-1,\pi}(\Omega))} + \left\|R_\tau^{(1)}(u_{\mathcal{I}})\right\|_{L^p(0,T;W^{-1,\pi}(\Omega))} + \left\|R_\tau^{(2)}(u_{\mathcal{I}})\right\|_{L^p(0,T;W^{-1,\pi}(\Omega))},
\end{aligned}$$

where $\beta_{p,\theta}$ is the constant of Lemma 3.53 (p. 130).

Proof The triangle inequality implies the upper bound of the lemma and

$$\|\partial_t u_{\mathcal{I}} + G(u_{\mathcal{I}})\|_{L^p(0,T;W^{-1,\pi}(\Omega))} \ge \left\|R_h(u_{\mathcal{I}}) + R_\tau^{(1)}(u_{\mathcal{I}})\right\|_{L^p(0,T;W^{-1,\pi}(\Omega))} - \left\|R_\tau^{(2)}(u_{\mathcal{I}})\right\|_{L^p(0,T;W^{-1,\pi}(\Omega))}.$$

Hence, we still have to verify that

$$\begin{aligned}
&\left\|R_h(u_{\mathcal{I}}) + R_\tau^{(1)}(u_{\mathcal{I}})\right\|_{L^p(0,T;W^{-1,\pi}(\Omega))} \\
&\quad\ge \beta_{p,\theta} \left\{ \|R_h(u_{\mathcal{I}})\|^p_{L^p(0,T;W^{-1,\pi}(\Omega))} + \left\|R_\tau^{(1)}(u_{\mathcal{I}})\right\|^p_{L^p(0,T;W^{-1,\pi}(\Omega))} \right\}^{\frac{1}{p}} .
\end{aligned} \tag{6.64}$$

For every $n \in \{1, \ldots, N_{\mathcal{I}}\}$ we define $r^n \in W^{-1,\pi}(\Omega)$ by setting for all $v \in W_0^{1,\pi'}(\Omega)$

$$\langle r^n \,,\, v\rangle = \left\langle DG(\widetilde{u}_{\mathcal{I}})\left(u^n_{\mathcal{T}_n} - u^{n-1}_{\mathcal{T}_{n-1}}\right) \,,\, v\right\rangle .$$

The definitions of r^n, $u_{\mathcal{I}}$, $\widetilde{u}_{\mathcal{I}}$, and $u^{n\theta}$ imply that

$$R^{(1)}_\tau(u_{\mathcal{I}}) = \left(\frac{t - t_{n-1}}{\tau_n} - \theta\right) r^n \quad \text{on } (t_{n-1}, t_n) \tag{6.65}$$

holds for every $n \in \{1, \ldots, N_{\mathcal{I}}\}$. Inequality (6.64) therefore follows from Lemma 3.53 (p. 130) and Remark 3.54 (p. 132) applied to the intervals (t_{n-1}, t_n) with $Y = W_0^{1,\pi'}(\Omega)$, $\varphi = R_h(u_{\mathcal{I}})$ and $\psi = r^n$. □

For later use we note that

$$\begin{aligned}\langle r^n \,,\, v\rangle = \int_\Omega &\nabla v \cdot \mathbf{a}_p\left(x, u^{n\theta}, \nabla u^{n\theta}\right) \cdot \nabla\left(u^n_{\mathcal{T}_n} - u^{n-1}_{\mathcal{T}_{n-1}}\right) + \nabla v \cdot \mathbf{a}_u\left(x, u^{n\theta}, \nabla u^{n\theta}\right)\left(u^n_{\mathcal{T}_n} - u^{n-1}_{\mathcal{T}_{n-1}}\right)\\ &+ v b_p\left(x, u^{n\theta}, \nabla u^{n\theta}\right) \cdot \nabla\left(u^n_{\mathcal{T}_n} - u^{n-1}_{\mathcal{T}_{n-1}}\right) + v b_u\left(x, u^{n\theta}, \nabla u^{n\theta}\right)\left(u^n_{\mathcal{T}_n} - u^{n-1}_{\mathcal{T}_{n-1}}\right)\, dx.\end{aligned} \tag{6.66}$$

Here, as usual, $\mathbf{a}_u$, b_u denote the partial derivatives of $\mathbf{a}$, b with respect to the second argument and $\mathbf{a}_p$, b_p those with respect to the third argument.

6.6.5 *Estimation of the Temporal Residual*

Since

$$\frac{1}{2^p(p+1)} \le \frac{1}{p+1}\left(\theta^{p+1} + (1-\theta)^{p+1}\right) \le \frac{1}{p+1}$$

holds for all $\theta \in [\frac{1}{2}, 1]$, equation (6.65) and a straightforward calculation immediately yield the following estimates.

Lemma 6.47 *The contribution $R^{(1)}_\tau(u_{\mathcal{I}})$ to the temporal residual $R_\tau(u_{\mathcal{I}})$ is bounded from above and from below by*

$$\frac{1}{2(p+1)^{\frac{1}{p}}}\left\{\sum_{n=1}^{N_{\mathcal{I}}} \tau_n \|r^n\|^p_{-1,\pi}\right\}^{\frac{1}{p}} \le \left\|R^{(1)}_\tau(u_{\mathcal{I}})\right\|_{L^p(0,T;W^{-1,\pi}(\Omega))} \le \frac{1}{(p+1)^{\frac{1}{p}}}\left\{\sum_{n=1}^{N_{\mathcal{I}}} \tau_n \|r^n\|^p_{-1,\pi}\right\}^{\frac{1}{p}}.$$

Here, the residual r^n is given by equation (6.66).

The next lemma shows that the contribution of $R^{(2)}_\tau(u_{\mathcal{I}})$ to the temporal residual $R_\tau(u_{\mathcal{I}})$ is of higher order when compared with $R^{(1)}_\tau(u_{\mathcal{I}})$, cf. Remark 6.55 too.

Lemma 6.48 *Assume that the function G defined in equation (6.62) satisfies the Lipschitz condition of Remark 6.41 and that*

$$\|u - u_{\mathcal{I}}\|_{L^r(0,T;W_0^{1,\rho}(\Omega))} \le R_0, \quad \|u - \widetilde{u}_{\mathcal{I}}\|_{L^r(0,T;W_0^{1,\rho}(\Omega))} \le R_0.$$

Then the contribution of $R_\tau^{(2)}(u_{\mathcal{I}})$ to the temporal residual $R_\tau(u_{\mathcal{I}})$ can be bounded from above by

$$\left\| R_\tau^{(2)}(u_{\mathcal{I}}) \right\|_{L^p(0,T;W^{-1,\pi}(\Omega))} \leq \frac{\gamma}{2(r+1)^{\frac{2}{r}}} \left\{ \sum_{n=1}^{N_{\mathcal{I}}} \tau_n \left\| u_{\mathcal{T}_n}^n - u_{\mathcal{T}_{n-1}}^{n-1} \right\|_{1,\rho}^r \right\}^{\frac{2}{r}}.$$

Proof From the definition of $R_\tau^{(2)}(u_{\mathcal{I}})$ we conclude that

$$R_\tau^{(2)}(u_{\mathcal{I}}) = \int_0^1 \left[DG(\widetilde{u}_{\mathcal{I}} + s(u_{\mathcal{I}} - \widetilde{u}_{\mathcal{I}})) - DG(\widetilde{u}_{\mathcal{I}}) \right] (u_{\mathcal{I}} - \widetilde{u}_{\mathcal{I}}) \, \mathrm{d}s.$$

The Lipschitz condition of Remark 6.41 therefore implies

$$\left\| R_\tau^{(2)}(u_{\mathcal{I}}) \right\|_{L^p(0,T;W^{-1,\pi}(\Omega))} \leq \frac{\gamma}{2} \left\| u_{\mathcal{I}} - \widetilde{u}_{\mathcal{I}} \right\|_{L^r(0,T;W_0^{1,\rho}(\Omega))}^2.$$

The definitions of $u_{\mathcal{I}}$, $\widetilde{u}_{\mathcal{I}}$, and $u^{n\theta}$, on the other hand, yield

$$u_{\mathcal{I}} - \widetilde{u}_{\mathcal{I}} = \left(\frac{t - t_{n-1}}{\tau_n} - \theta \right) \left(u_{\mathcal{T}_n}^n - u_{\mathcal{T}_{n-1}}^{n-1} \right)$$

on every interval (t_{n-1}, t_n), $1 \leq n \leq N_{\mathcal{I}}$. A straightforward calculation gives

$$\left\| u_{\mathcal{I}} - \widetilde{u}_{\mathcal{I}} \right\|_{L^r(0,T;W_0^{1,\rho}(\Omega))} \leq \frac{1}{(r+1)^{\frac{1}{r}}} \left\{ \sum_{n=1}^{N_{\mathcal{I}}} \tau_n \left\| u_{\mathcal{T}_n}^n - u_{\mathcal{T}_{n-1}}^{n-1} \right\|_{1,\rho}^r \right\}^{\frac{1}{r}}. \qquad \square$$

6.6.6 *Estimation of the Spatial Residual*

The estimation of the spatial residual follows along the lines of Sections 6.1.6 (p. 314) and 6.2.6 (p. 324) and is based on Theorems 3.57 (p. 135) and 3.59 (p. 137).

Since $\partial_t u_{\mathcal{I}} = \frac{1}{\tau_n}(u_{\mathcal{T}_n}^n - u_{\mathcal{T}_{n-1}}^{n-1})$ and $\widetilde{u}_{\mathcal{I}} = u^{n\theta}$ on each interval $(t_{n-1}, t_n]$, problem (6.61) is equivalent to

$$\left\langle R_h(u_{\mathcal{I}}), v_{\mathcal{T}_n} \right\rangle = 0$$

for all $v_{\mathcal{T}_n} \in X_n$ and all $n \in \{1, \ldots, N_{\mathcal{I}}\}$. Thanks to the degree condition, the spatial residual therefore satisfies the Galerkin orthogonality (3.30) (p. 132).

The L^p-representation (3.29) (p. 132) of the spatial residual is obtained in the standard way using integration by parts element-wise on the elements of $\widetilde{\mathcal{T}}_n$.

For the approximation of the residuals r and j, we choose an integer ℓ and denote for every $n \in \{1, \ldots, N_{\mathcal{I}}\}$ by $\mathbf{a}_{\mathcal{T}_n}(x, u^{n\theta}, \nabla u^{n\theta})$ and $b_{\mathcal{T}_n}(x, u^{n\theta}, \nabla u^{n\theta})$ the L^2-projections of the vector field $\mathbf{a}(x, u^{n\theta}, \nabla u^{n\theta})$ and of the function $b(x, u^{n\theta}, \nabla u^{n\theta})$ onto discontinuous vector fields and functions, respectively, which are piecewise polynomials of degree ℓ on the elements of $\widetilde{\mathcal{T}}_n$. With this notation, we define

element residuals R_K, $K \in \widetilde{\mathcal{T}}_n$, $1 \leq n \leq N_{\mathcal{I}}$, by

$$R_K = \frac{1}{\tau_n} \left(u_{\mathcal{T}_n}^n - u_{\mathcal{T}_{n-1}}^{n-1} \right) - \operatorname{div} \mathbf{a}_{\mathcal{T}_n} \left(x, u^{n\theta}, \nabla u^{n\theta} \right) + b_{\mathcal{T}_n} \left(x, u^{n\theta}, \nabla u^{n\theta} \right),$$

face residuals $R_E, E \in \widetilde{\mathcal{E}}_n, 1 \leq n \leq N_{\mathcal{I}}$, by

$$R_E = \begin{cases} \mathbb{J}_E\left(\mathbf{n}_E \cdot \mathbf{a}_{\mathcal{T}_n}(x, u^{n\theta}, \nabla u^{n\theta})\right) & \text{if } E \in \widetilde{\mathcal{E}}_{n,\Omega}, \\ 0 & \text{if } E \in \widetilde{\mathcal{E}}_{n,\Gamma}, \end{cases}$$

element-wise data errors $D_K, K \in \widetilde{\mathcal{T}}_n, 1 \leq n \leq N_{\mathcal{I}}$, by

$$D_K = \mathbf{a}\left(x, u^{n\theta}, \nabla u^{n\theta}\right) - \mathbf{a}_{\mathcal{T}_n}\left(x, u^{n\theta}, \nabla u^{n\theta}\right) + b\left(x, u^{n\theta}, \nabla u^{n\theta}\right) - b_{\mathcal{T}_n}\left(x, u^{n\theta}, \nabla u^{n\theta}\right),$$

and *face-wise data errors* $D_E, E \in \widetilde{\mathcal{E}}_n, 1 \leq n \leq N_{\mathcal{I}}$, by

$$D_E = \begin{cases} \mathbb{J}_E\left(\mathbf{n}_E \cdot (\mathbf{a}(x, u^{n\theta}, \nabla u^{n\theta}) - \mathbf{a}_{\mathcal{T}_n}(x, u^{n\theta}, \nabla u^{n\theta}))\right) & \text{if } E \in \widetilde{\mathcal{E}}_{n,\Omega}, \\ 0 & \text{if } E \in \widetilde{\mathcal{E}}_{n,\Gamma}. \end{cases}$$

The choice of the parameter ℓ is influenced by the polynomial degree of the finite element spaces X_n and by the smoothness of $\mathbf{a}$ and b. The simplest choice of course is $\ell = 0$.

With these notations, Theorems 3.57 (p. 135) and 3.59 (p. 137) and the arguments used in the proof of Lemmas 6.5 (p. 314) and 6.19 (p. 324) give the following estimates.

Lemma 6.49 *For every* $n \in \{1, \ldots, N_{\mathcal{I}}\}$*, define a spatial error indicator* $\eta^n_{\mathcal{T}_n}$ *by*

$$\eta^n_{\mathcal{T}_n} = \left\{ \sum_{K \in \widetilde{\mathcal{T}}_n} h_K^\pi \|R_K\|^\pi_{\pi;K} + \sum_{E \in \widetilde{\mathcal{E}}_n} h_E \|R_E\|^\pi_{\pi;E} \right\}^{\frac{1}{\pi}} \tag{6.67}$$

and a spatial data error indicator $\Theta^n_{\mathcal{T}_n}$ *by*

$$\Theta^n_{\mathcal{T}_n} = \left\{ \sum_{K \in \widetilde{\mathcal{T}}_n} h_K^\pi \|D_K\|^\pi_{\pi;K} + \sum_{E \in \widetilde{\mathcal{E}}_n} h_E \|D_E\|^\pi_{\pi;E} \right\}^{\frac{1}{\pi}}. \tag{6.68}$$

Then, for every $n \in \{1, \ldots, N_{\mathcal{I}}\}$*, the spatial residual* $R_h(u_{\mathcal{I}})$ *can be bounded from above and from below by*

$$c_\dagger \eta^n_{\mathcal{T}_n} - \Theta^n_{\mathcal{T}_n} \leq \left\| R_h(u_{\mathcal{I}}) \right\|_{-1,\pi} \leq c^\dagger \left\{ \eta^n_{\mathcal{T}_n} + \Theta^n_{\mathcal{T}_n} \right\}.$$

The constants $c_\dagger$ *and* $c^\dagger$ *depend on the shape parameters* $C_{\widetilde{\mathcal{T}}_n}$*. The constant* $c^\dagger$ *in addition depends on the constant in the transition condition of Section 3.2.6 (p. 86); the constant* $c_\dagger$ *also depends on the polynomial degree* ℓ*.*

6.6.7 $W^{1,q}$-Stability Results for the Laplacian

Proposition 6.40, Lemmas 6.46, 6.47, 6.48, and 6.49, and the identity

$$\left\| R_h(u_{\mathcal{I}}) \right\|_{L^p(0,T;W^{-1,\pi}(\Omega))} = \left\{ \sum_{n=1}^{N_{\mathcal{I}}} \tau_n \left\| R_h(u_{\mathcal{I}}) \right\|^p_{-1,\pi} \right\}^{\frac{1}{p}} \tag{6.69}$$

yield reliable and efficient a posteriori error estimates for the error $u - u_{\mathcal{I}}$. These, however, are not suited for practical computations since they involve the $W^{-1,\pi}$-norms of the residuals r^n defined in equation (6.66). Using inverse estimates these can be replaced by L^π-norms scaled with negative powers of the local mesh-size in space. But, the efficiency of the error estimates is thus lost. In order to obtain error estimates which are efficient, reliable, and computable at the same time, we proceed as in Lemmas 6.18 (p. 323) and 6.24 (p. 333) and solve auxiliary discrete Poisson equations, cf. Lemma 6.53 below. In order to prove that the $W^{1,\pi'}$-norm of the solutions of the auxiliary problems is equivalent to the $W^{-1,\pi}$-norms of the r^n, we need the $W^{1,\pi'}$-stability of the Laplacian both in its analytical and discrete form. In the case $\pi = 2$, this is obvious. In the general case $\pi \neq 2$, however, this is not evident. In order to handle this case, we have to assume that Ω is a convex polygon in two space dimensions.

We start with the analytical case. The following result is well-known for domains with smooth C^1-boundary [**240**]. For polygonal domains, however, we are not aware of a proof.

Proposition 6.50 *For every convex bounded polygonal domain $\Omega \subset \mathbb{R}^2$ and every $q \in [1, \infty]$, there is a constant $\alpha_q > 0$ such that*

$$\inf_{v \in W_0^{1,q}(\Omega)} \sup_{w \in W_0^{1,q'}(\Omega)} \frac{\int_\Omega \nabla v \cdot \nabla w}{\|\nabla v\|_q \|\nabla w\|_{q'}} \geq \alpha_q. \tag{6.70}$$

The constant α_q only depends on q and on the maximum interior angle at the vertices of Ω.

Proof Inequality (6.70) is proved in [**240**] for domains $\Omega \subset \mathbb{R}^d$, $d \geq 2$, with smooth C^1-boundary. The proof is based on the following three auxiliary results:

- inequality (6.70) holds for $\mathbb{R}^d$;
- inequality (6.70) holds for the half-space $H_+ = \{x \in \mathbb{R}^d : x_1 > 0\}$;
- inequality (6.70) holds for domains $H_\omega = \{x \in \mathbb{R}^d : x_1 > \omega(x_2, \ldots, x_d)\}$ with functions $\omega \in C^1(\mathbb{R}^{d-1})$ satisfying $\omega(0) = 0$ and $\|\nabla \omega\|_{\infty;\mathbb{R}^{n-1}} \ll 1$.

The third result is the only point where the smoothness of the boundary comes into play. The smoothness condition on ω can be relaxed to the condition that ω should be Lipschitz continuous and that its Lipschitz constant is sufficiently small. This, however, does not help us since it would require that the interior angles at the vertices of Ω should be sufficiently close to π. Instead, we must prove that inequality (6.70) holds for domains $H_c = \{x \in \mathbb{R}^2 : |x_2| \leq cx_1\}$ with $c > 0$.

To verify this, choose a parameter $c > 0$ and keep it fixed. Then introduce polar coordinates to transform H_c to the strip $\{(r, \varphi) : r > 0, |\varphi| \leq \alpha\}$ where $\alpha = \arctan c$. Next, apply the scaling $r \mapsto \frac{2}{\pi} r$, $\varphi \mapsto \frac{2\alpha}{\pi}\varphi$ to transform to the strip $\{(s, \psi) : s > 0, |\psi| \leq \frac{\pi}{2}\}$. Then transform back to Cartesian coordinates. The combination of these transformations maps H_c to the half-space H_+. Now, we already know that inequality (6.70) holds for H_+. Hence it also holds in polar coordinates. Since the left-hand side of (6.70) is invariant under scalings, inequality (6.70) holds in polar coordinates on the strip $\{(r, \varphi) : r > 0, |\varphi| \leq \alpha\}$ and thus on H_c.

Once we know that we may replace H_ω by H_c, the remaining part of the proof of Proposition 6.50 proceeds as in [**240**]. □

Remark 6.51 When Ω is not convex but has a re-entrant corner with angle $\alpha > \pi$, Proposition 6.50 can at best hold for Lebesgue exponents $q \in [1, \frac{2\alpha}{\alpha - \pi})$. This is due to the fact that the singular solution $r^{\frac{\pi}{\alpha}} \sin(\frac{\pi}{\alpha}\varphi)$ of the Laplacian is in $W^{1,q}(\Omega)$ only for this realm of Lebesgue exponents. In the proof of Proposition 6.50 the convexity is reflected by the fact that the transformation from polar to Cartesian coordinates is globally invertible in the vicinity of convex corners. For non-convex corners it is only locally invertible.

Now, we come to the discrete case.

Proposition 6.52 *Consider a convex, bounded, polygonal domain $\Omega \subset \mathbb{R}^2$ and an arbitrary affine-equivalent, admissible, and shape-regular partition $\mathcal{T}$ of Ω. Then, for every $q \in [1, \infty]$, there is a constant $\beta_q > 0$ such that*

$$\inf_{v_{\mathcal{T}} \in S_0^{1,0}(\mathcal{T})} \sup_{w_{\mathcal{T}} \in S_0^{1,0}(\mathcal{T})} \frac{\int_\Omega \nabla v_{\mathcal{T}} \cdot \nabla w_{\mathcal{T}}}{\|\nabla v_{\mathcal{T}}\|_q \|\nabla w_{\mathcal{T}}\|_{q'}} \geq \beta_q. \tag{6.71}$$

The constant β_q only depends on q, on the maximum interior angle at the vertices of Ω, and on the shape parameter $C_{\mathcal{T}}$.

Proof We denote by $R_{\mathcal{T}} : W_0^{1,1}(\Omega) \to S_0^{1,0}(\mathcal{T})$ the Ritz projection which, for all $v \in W_0^{1,1}(\Omega)$ and all $w_{\mathcal{T}} \in S_0^{1,0}(\mathcal{T})$, is defined by

$$\int_\Omega \nabla(R_{\mathcal{T}} v) \cdot \nabla w_{\mathcal{T}} = \int_\Omega \nabla v \cdot \nabla w_{\mathcal{T}}.$$

Consider first the case $q \in [1, 2]$. Then we have $q' \geq 2$. From [**86**, Chapter 7] and [**225**] we know that $R_{\mathcal{T}}$ is stable in the $W^{1,q'}$-norm, i.e. there is a constant $c_{q'} > 0$ such that for all $w \in W_0^{1,q'}(\Omega)$

$$\left\|\nabla(R_{\mathcal{T}} w)\right\|_{q'} \leq c_{q'} \|\nabla w\|_{q'}.$$

The constant $c_{q'}$ only depends on q', on the maximum interior angle at a vertex of Ω, and on the shape parameter $C_{\mathcal{T}}$.

Consider an arbitrary function $v_{\mathcal{T}} \in S_0^{1,0}(\mathcal{T})$. From Proposition 6.50 we know that there is a function $w_\delta \in W_0^{1,q'}(\Omega)$ with

$$\|\nabla w_\delta\|_{q'} = 1 \quad \text{and} \quad \int_\Omega \nabla v_{\mathcal{T}} \cdot \nabla w_\delta = \alpha_q \|\nabla v_{\mathcal{T}}\|_q.$$

Together with the stability of the Ritz projection this implies

$$\begin{aligned} \sup_{w_{\mathcal{T}} \in S_0^{1,0}(\mathcal{T})} \frac{\int_\Omega \nabla v_{\mathcal{T}} \cdot \nabla w_{\mathcal{T}}}{\|\nabla v_{\mathcal{T}}\|_q \|\nabla w_{\mathcal{T}}\|_{q'}} &\geq \frac{\int_\Omega \nabla v_{\mathcal{T}} \cdot \nabla(R_{\mathcal{T}} w_\delta)}{\|\nabla v_{\mathcal{T}}\|_q \left\|\nabla(R_{\mathcal{T}} w_\delta)\right\|_{q'}} \\ &\geq \frac{1}{c_{q'}} \frac{\int_\Omega \nabla v_{\mathcal{T}} \cdot \nabla(R_{\mathcal{T}} w_\delta)}{\|\nabla v_{\mathcal{T}}\|_q \|\nabla w_\delta\|_{q'}} \\ &= \frac{1}{c_{q'}} \frac{\int_\Omega \nabla v_{\mathcal{T}} \cdot \nabla w_\delta}{\|\nabla v_{\mathcal{T}}\|_q \|\nabla w_\delta\|_{q'}} \\ &\geq \frac{\alpha_q}{c_{q'}}. \end{aligned}$$

Since $v_{\mathcal{T}}$ is arbitrary, this proves inequality (6.71) with $\beta_q = \frac{\alpha_q}{c_{q'}}$. One easily checks that inequality (6.71) implies the stability of $R_{\mathcal{T}}$ in the $W^{1,q}$-norm with $c_q = \frac{1}{\beta_q} = \frac{c_{q'}}{\alpha_q}$.

Now consider the case $q > 2$. This implies $1 < q' < 2$. Since we have already established the stability of $R_{\mathcal{T}}$ in the $W^{1,q'}$-norm with $c_{q'} = \frac{1}{\beta_{q'}} = \frac{c_q}{\alpha_{q'}}$, we can proceed as in the case $q \geq 2$ and obtain inequality (6.71) with $\beta_q = \frac{\alpha_q}{c_{q'}} = \frac{\alpha_q \alpha_{q'}}{c_q}$. □

6.6.8 A Posteriori Error Estimates

As in Subsection 6.6.6 we choose an integer ℓ and denote for every $n \in \{1, \ldots, N_{\mathcal{I}}\}$ by

$$\begin{aligned} &\mathbf{a}_{p;\mathcal{I}}(x, u^{n\theta}, \nabla u^{n\theta}), \quad \mathbf{a}_{u;\mathcal{I}}(x, u^{n\theta}, \nabla u^{n\theta}), \\ &b_{p;\mathcal{I}}(x, u^{n\theta}, \nabla u^{n\theta}), \quad b_{u;\mathcal{I}}(x, u^{n\theta}, \nabla u^{n\theta}) \end{aligned}$$

the L^2-projections of

$$\begin{aligned} &\mathbf{a}_p(x, u^{n\theta}, \nabla u^{n\theta}), \quad \mathbf{a}_u(x, u^{n\theta}, \nabla u^{n\theta}), \\ &b_p(x, u^{n\theta}, \nabla u^{n\theta}), \quad b_u(x, u^{n\theta}, \nabla u^{n\theta}) \end{aligned}$$

onto discontinuous tensors, vector fields, and functions, respectively, which are piecewise polynomials of degree ℓ on the elements of $\widetilde{\mathcal{T}}_n$.

For brevity, we set for every element $K \in \widetilde{\mathcal{T}}_n$, every face $E \in \widetilde{\mathcal{E}}_{n,\Omega}$, and every $n \in \{1, \ldots, N_{\mathcal{I}}\}$

$$\begin{aligned} \widetilde{R}_K = &- \operatorname{div}\left[\mathbf{a}_{p;\mathcal{I}}\left(x, u^{n\theta}, \nabla u^{n\theta}\right) \cdot \nabla\left(u^n_{\mathcal{T}_n} - u^{n-1}_{\mathcal{T}_{n-1}}\right)\right] \\ &- \operatorname{div}\left[\mathbf{a}_{u;\mathcal{I}}\left(x, u^{n\theta}, \nabla u^{n\theta}\right)\left(u^n_{\mathcal{T}_n} - u^{n-1}_{\mathcal{T}_{n-1}}\right)\right] \\ &+ b_{p;\mathcal{I}}\left(x, u^{n\theta}, \nabla u^{n\theta}\right) \cdot \nabla\left(u^n_{\mathcal{T}_n} - u^{n-1}_{\mathcal{T}_{n-1}}\right) \\ &+ b_{u;\mathcal{I}}\left(x, u^{n\theta}, \nabla u^{n\theta}\right)\left(u^n_{\mathcal{T}_n} - u^{n-1}_{\mathcal{T}_{n-1}}\right), \\ \widetilde{R}_E = &\ \mathbf{a}_{p;\mathcal{I}}\left(x, u^{n\theta}, \nabla u^{n\theta}\right) \cdot \nabla\left(u^n_{\mathcal{T}_n} - u^{n-1}_{\mathcal{T}_{n-1}}\right) \\ &+ \mathbf{a}_{u;\mathcal{I}}\left(x, u^{n\theta}, \nabla u^{n\theta}\right)\left(u^n_{\mathcal{T}_n} - u^{n-1}_{\mathcal{T}_{n-1}}\right), \\ \widetilde{D}_K = &\ \operatorname{div}\left[\left(\mathbf{a}_p\left(x, u^{n\theta}, \nabla u^{n\theta}\right) - \mathbf{a}_{p;\mathcal{I}}\left(x, u^{n\theta}, \nabla u^{n\theta}\right)\right) \cdot \nabla\left(u^n_{\mathcal{T}_n} - u^{n-1}_{\mathcal{T}_{n-1}}\right)\right] \\ &- \operatorname{div}\left[\left(\mathbf{a}_u\left(x, u^{n\theta}, \nabla u^{n\theta}\right) - \mathbf{a}_{u;\mathcal{I}}\left(x, u^{n\theta}, \nabla u^{n\theta}\right)\right)\left(u^n_{\mathcal{T}_n} - u^{n-1}_{\mathcal{T}_{n-1}}\right)\right] \\ &+ \left(b_p\left(x, u^{n\theta}, \nabla u^{n\theta}\right) - b_{p;\mathcal{I}}\left(x, u^{n\theta}, \nabla u^{n\theta}\right)\right) \cdot \nabla\left(u^n_{\mathcal{T}_n} - u^{n-1}_{\mathcal{T}_{n-1}}\right) \\ &+ \left(b_u\left(x, u^{n\theta}, \nabla u^{n\theta}\right) - b_{u;\mathcal{I}}\left(x, u^{n\theta}, \nabla u^{n\theta}\right)\right)\left(u^n_{\mathcal{T}_n} - u^{n-1}_{\mathcal{T}_{n-1}}\right), \\ \widetilde{D}_E = &\left(\mathbf{a}_p\left(x, u^{n\theta}, \nabla u^{n\theta}\right) - \mathbf{a}_{p;\mathcal{I}}\left(x, u^{n\theta}, \nabla u^{n\theta}\right)\right) \cdot \nabla\left(u^n_{\mathcal{T}_n} - u^{n-1}_{\mathcal{T}_{n-1}}\right) \\ &+ \left(\mathbf{a}_u\left(x, u^{n\theta}, \nabla u^{n\theta}\right) - \mathbf{a}_{u;\mathcal{I}}\left(x, u^{n\theta}, \nabla u^{n\theta}\right)\right)\left(u^n_{\mathcal{T}_n} - u^{n-1}_{\mathcal{T}_{n-1}}\right). \end{aligned}$$

Lemma 6.53 *Assume that either the Lebesgue exponent π equals 2 or that Ω is a convex polygon in $\mathbb{R}^2$. For every $n \in \{1, \ldots, N_{\mathcal{I}}\}$, denote by $\widetilde{u}^n_{\mathcal{T}_n} \in S_0^{1,0}(\widetilde{\mathcal{T}}_n)$ the unique solution of the discrete Poisson equation*

$$\int_\Omega \nabla \widetilde{u}^n_{\mathcal{T}_n} \cdot \nabla v_{\mathcal{T}_n} = \left\langle r^n , v_{\mathcal{T}_n}\right\rangle \tag{6.72}$$

for all $v_{\mathcal{T}_n} \in S_0^{1,0}(\widetilde{\mathcal{T}}_n)$, where the residuals r^n are given by equation (6.66). Define the error indicator $\widetilde{\eta}^n_{\mathcal{T}_n}$ by

$$\widetilde{\eta}^n_{\mathcal{T}_n} = \left\{ \sum_{K\in\widetilde{\mathcal{T}}_n} h_K^\pi \left\|\widetilde{R}_K + \Delta\widetilde{u}^n_{\mathcal{T}_n}\right\|^\pi_{\pi;K} + \sum_{E\in\widetilde{\mathcal{E}}_{n,\Omega}} h_E \left\|\mathbb{J}_E\left(\mathbf{n}_E \cdot (\nabla \widetilde{u}^n_{\mathcal{T}_n} - \widetilde{R}_E)\right)\right\|^\pi_{\pi;E} \right\}^{\frac{1}{\pi}} \tag{6.73}$$

and the data error $\widetilde{\Theta}^n_{\mathcal{T}_n}$ by

$$\widetilde{\Theta}^n_{\mathcal{T}_n} = \left\{ \sum_{K\in\widetilde{\mathcal{T}}_n} h_K^\pi \left\|\widetilde{D}_K\right\|^\pi_{\pi;K} + \sum_{E\in\widetilde{\mathcal{E}}_{n,\Omega}} h_E \left\|\widetilde{D}_E\right\|^\pi_{\pi;E} \right\}^{\frac{1}{\pi}} . \tag{6.74}$$

Then, there are two constants $\widetilde{c}_\dagger$ and $\widetilde{c}^\dagger$, which only depend on the polynomial degree ℓ and on the shape parameters $C_{\widetilde{\mathcal{T}}_n}$, such that

$$\begin{aligned} \|r^n\|_{-1,\pi} &\le \widetilde{c}^\dagger \left\{ \widetilde{\eta}^n_{\mathcal{T}_n} + \left\|\nabla \widetilde{u}^n_{\mathcal{T}_n}\right\|_\pi + \widetilde{\Theta}^n_{\mathcal{T}_n} \right\}, \\ \widetilde{\eta}^n_{\mathcal{T}_n} + \left\|\nabla \widetilde{u}^n_{\mathcal{T}_n}\right\|_\pi &\le \widetilde{c}_\dagger \left\{ \|r^n\|_{-1,\pi} + \widetilde{\Theta}^n_{\mathcal{T}_n} \right\}. \end{aligned}$$

Proof If the Lebesgue exponent π equals 2, the Lemma follows with the arguments used in the proof of Lemma 6.18 (p. 323). Therefore, we only have to consider the case $\pi \neq 2$. We choose an integer $n \in \{1, \ldots, N_{\mathcal{I}}\}$ and keep it fixed in what follows. Proposition 6.50 implies that the Poisson equation

$$\int_\Omega \nabla U^n \cdot \nabla v = \langle r^n , v\rangle$$

for all $v \in W_0^{1,\pi'}(\Omega)$ admits a unique solution $U^n \in W_0^{1,\pi}(\Omega)$ and that

$$\|\nabla U^n\|_\pi \le \frac{1}{\alpha_\pi} \|r^n\|_{-1,\pi} .$$

The definition of the negative Sobolev norms on the other hand yields

$$\|r^n\|_{-1,\pi} \le \|\nabla U^n\|_\pi .$$

Proposition 6.52 similarly implies that

$$\left\|\nabla \widetilde{u}^n_{\mathcal{T}_n}\right\|_\pi \le \frac{1}{\beta_\pi} \|r^n\|_{-1,\pi} .$$

The triangle inequality therefore yields

$$\begin{aligned}\frac{1}{3}\min\{\alpha_\pi,\beta_\pi\}\left\{\left\|\nabla\widetilde{u}_{\mathcal{T}_n}^n\right\|_\pi+\left\|\nabla(U^n-\widetilde{u}_{\mathcal{T}_n}^n)\right\|_\pi\right\}&\le\|r^n\|_{-1,\pi}\\&\le\left\|\nabla\widetilde{u}_{\mathcal{T}_n}^n\right\|_\pi+\left\|\nabla(U^n-\widetilde{u}_{\mathcal{T}_n}^n)\right\|_\pi.\end{aligned}\tag{6.75}$$

With the arguments of Sections 4.2 (p. 157) and 5.2 (p. 290), we conclude that

$$c\widetilde{\eta}_{\mathcal{T}_n}^n-\widetilde{\Theta}_{\mathcal{T}_n}^n\le\left\|\nabla(U^n-\widetilde{u}_{\mathcal{T}_n}^n)\right\|_\pi\le C\left\{\widetilde{\eta}_{\mathcal{T}_n}^n+\widetilde{\Theta}_{\mathcal{T}_n}^n\right\}\tag{6.76}$$

with constants c and C which only depend on the polynomial degree ℓ and the shape parameter of $\widetilde{\mathcal{T}}_n$. Combining estimates (6.75) and (6.76), we arrive at the postulated estimate of $\|r^n\|_{-1,\pi}$. □

A standard duality argument for the L^2-projection onto finite element spaces yields

$$\|u_0-\pi_0u_0\|_{-1,\pi}\le c\left\{\sum_{K\in\mathcal{T}_0}h_K^\pi\|u_0-\pi_0u_0\|_{\pi;K}^\pi\right\}^{\frac{1}{\pi}}.$$

Combining this estimate, Proposition 6.40, Lemmas 6.46, 6.47, 6.48, 6.49, and 6.53, and equation (6.69) yields the following a posteriori error estimates.

Theorem 6.54 *Assume that either the Lebesgue exponent π equals 2 or that Ω is a convex polygon in $\mathbb{R}^2$ and that the functions F, G, u, $u_{\mathcal{I}}$, and $\widetilde{u}_{\mathcal{I}}$ fulfil the conditions of Proposition 6.40, Remark 6.41, and Lemma 6.48. Then, the error between the weak solution u of problem (6.56) and its approximation $u_{\mathcal{I}}$ associated with solutions of problems (6.60) and (6.61) is bounded from above by*

$$\begin{aligned}&\|u-u_{\mathcal{I}}\|_{W^r(0,T;W_0^{1,\rho}(\Omega),W^{-1,\pi}(\Omega))}\\&\le\widetilde{c}^\sharp\left\{\left(\sum_{n=1}^{N_{\mathcal{I}}}\tau_n\left[(\eta_{\mathcal{T}_n}^n)^p+(\widetilde{\eta}_{\mathcal{T}_n}^n)^p+\left\|\nabla\widetilde{u}_{\mathcal{T}_n}^n\right\|_\pi^p\right]\right)^{\frac{1}{p}}+\left(\sum_{n=1}^{N_{\mathcal{I}}}\tau_n\left[(\Theta_{\mathcal{T}_n}^n)^p+(\widetilde{\Theta}_{\mathcal{T}_n}^n)^p\right]\right)^{\frac{1}{p}}\right.\\&\qquad\left.+\left(\sum_{K\in\mathcal{T}_0}h_K^\pi\|u_0-\pi_0u_0\|_{\pi;K}^\pi\right)^{\frac{1}{\pi}}+\left(\sum_{n=1}^{N_{\mathcal{I}}}\tau_n\left\|u_{\mathcal{T}_n}^n-u_{\mathcal{T}_{n-1}}^{n-1}\right\|_{1,\rho}^r\right)^{\frac{2}{r}}\right\}\end{aligned}$$

and from below by

$$\begin{aligned}&\left\{\sum_{n=1}^{N_{\mathcal{I}}}\tau_n\left[(\eta_{\mathcal{T}_n}^n)^p+(\widetilde{\eta}_{\mathcal{T}_n}^n)^p+\left\|\nabla\widetilde{u}_{\mathcal{T}_n}^n\right\|_\pi^p\right]\right\}^{\frac{1}{p}}\\&\le\widetilde{c}_\sharp\left\{\|u-u_{\mathcal{I}}\|_{W^r(0,T;W_0^{1,\rho}(\Omega))}+\left(\sum_{n=1}^{N_{\mathcal{I}}}\tau_n\left[(\Theta_{\mathcal{T}_n}^n)^p+(\widetilde{\Theta}_{\mathcal{T}_n}^n)^p\right]\right)^{\frac{1}{p}}\right.\\&\qquad\left.+\left(\sum_{n+1}^{N_{\mathcal{I}}}\tau_n\left\|u_{\mathcal{T}_n}^n-u_{\mathcal{T}_{n-1}}^{n-1}\right\|_{1,\rho}^r\right)^{\frac{2}{r}}\right\}.\end{aligned}$$

The quantities $\eta_{\mathcal{T}_n}^n$, $\Theta_{\mathcal{T}_n}^n$, $\widetilde{\eta}_{\mathcal{T}_n}^n$, $\widetilde{\Theta}_{\mathcal{T}_n}^n$, *and* $\widetilde{u}_{\mathcal{T}_n}^n$ *are defined in equations (6.67), (6.68), (6.73), (6.74), and (6.72), respectively. The constants* $c^\sharp$ *and* $c_\sharp$ *only depend on the polynomial degree of the finite element spaces, the shape parameters* $C_{\widetilde{\mathcal{T}}_n}$, *and the constant in the transition condition of Section 3.2.6 (p. 86).*

Remark 6.55 The left-hand side of the lower bound for the error in Theorem 6.54 is our error indicator. Its first term controls the error of the space discretisation and can be used for adapting the spatial mesh. Its second and third term control the error of the time discretisation and can be used to adapt the temporal mesh, cf. Section 6.8.3 (p. 364). The last terms on the right-hand sides of the bounds for the error in Theorem 6.54 are not present for linear differential equations. They control the linearisation error that is implicit in the discretisation. Since they are computable, they can be used to control this linearisation error. These contributions are of higher order in the sense that, up to a different L^p-norm, they are similar to the square of the error indicator. The second and third term on the right-hand side of the upper bound for the error and the second term on the right-hand side of the lower bound for the error of Theorem 6.54 are data errors. In contrast to linear problems, they not only involve given data but the discrete solution as well.

Remark 6.56 The lower bound for the error of Theorem 6.54 is based on the lower bound of Proposition 6.40. This in turn follows from the lower bound in the abstract error estimate of Proposition 5.1 (p. 283). The latter only involves the Fréchet derivative $DF(u)$. Applied to differential equations this corresponds to a linearised differential operator. Since such operators have a local effect, the lower bounds can be localised. Therefore, as for the linear differential equations of the previous sections, the lower bound for the error of Theorem 6.54 has an analogue that is local with respect to time.

Remark 6.57 The present results should be compared with the results of [**260**].

Here, we consider standard time discretisations which in particular cover the implicit Euler scheme and the midpoint rule. The discrete problems in [**260**] are based on non-standard time discretisations which could be interpreted as implicit Runge–Kutta schemes and which cover the Crank–Nicolson scheme as method of lowest order.

Here, the ratio of upper and lower bounds for the error is independent of any mesh-size in space and time and of any relation between both parameters. In [**260**], the ratio of the upper and lower bounds for the error is proportional to $1 + h^2\tau^{-1} + h^{-2}\tau$, where h and τ denote the local mesh-sizes in space and time, respectively.

The present analysis and that in [**260**] both depart from the abstract framework of Section 4.1 (p. 151) for an abstract nonlinear equation $F(u) = 0$ with a continuously differentiable mapping $F : X \to Y^*$. Here, X carries a stronger topology than Y; in [**260**] these roles are reversed.

In [**260**] the Sobolev norms of negative order are estimated using inverse estimates. This results in the discrepancy between the constants in the upper and the lower bounds for the error. On the other hand, the solution of additional discrete problems is avoided. Thus the additional discrete problems are the price we have to pay for a truly reliable and efficient error indicator.

Remark 6.58 The present analysis follows [**269**].

6.7 Finite Volume Methods

Finite volume methods are a different popular approach for solving parabolic problems, in particular those with large convection. We refer to [**135, 175**] for an overview. For this type of discretisation, the theory of a posteriori error estimation and adaptivity is much less developed than for finite element methods. Some results may be found in [**4**, **60**, **166**, **167**, **209**, **210**, **211**, **212**, **274**]. Yet, there is an important particular case where finite volume methods can easily profit from finite element techniques. This is the case of so-called *dual finite volume meshes*. To describe the underlying idea we restrict ourselves to the two-dimensional case $d = 2$.

6.7.1 Construction of Dual Finite Volume Meshes

For constructing the finite volume mesh $\mathcal{T}$, we start from a standard finite element partition $\widetilde{\mathcal{T}}$ which satisfies the conditions of Section 3.2.1 (p. 81). Then we subdivide each element $\widetilde{K} \in \widetilde{\mathcal{T}}$ into smaller elements by either

- drawing the perpendicular bisectors at the midpoints of edges of $\widetilde{K}$, cf. Figure 6.2, or by
- connecting the barycentre of $\widetilde{K}$ with its midpoints of edges, cf. Figure 6.3.

Then the elements in $\mathcal{T}$ consist of the unions of all small elements that share a common vertex in the partition $\widetilde{\mathcal{T}}$.

Thus the elements in $\mathcal{T}$ can be associated with the vertices $\widetilde{\mathcal{N}}$ corresponding to $\widetilde{\mathcal{T}}$. Moreover, we may associate with each edge in $\mathcal{E}$, the set of edges corresponding to $\mathcal{T}$, exactly two vertices in $\widetilde{\mathcal{N}}$ such that the line connecting these vertices intersects the given edge.

The first construction has the advantage that this intersection is orthogonal. Yet this construction also has some disadvantages which are not present with the second construction.

- The perpendicular bisectors of a triangle may intersect in a point outside the triangle. The intersection point is within the triangle only if its largest angle is at most a right angle.
- The perpendicular bisectors of a quadrilateral may not intersect at all. They intersect in a common point inside the quadrilateral only if it is a rectangle.
- The first construction has no three-dimensional analogue.

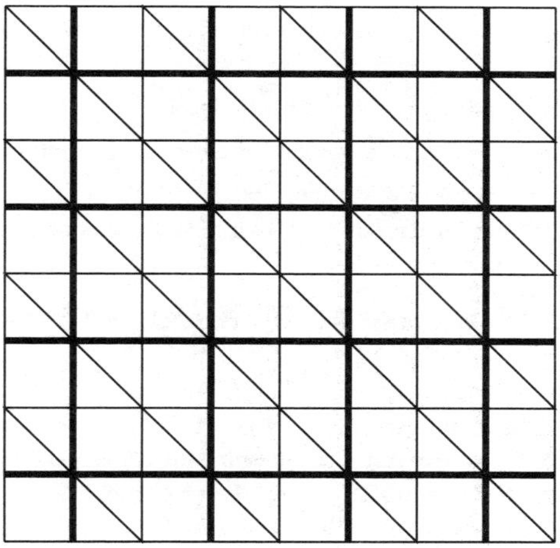

Figure 6.2 Dual mesh (thick lines) via perpendicular bisectors of primal mesh (thin lines)

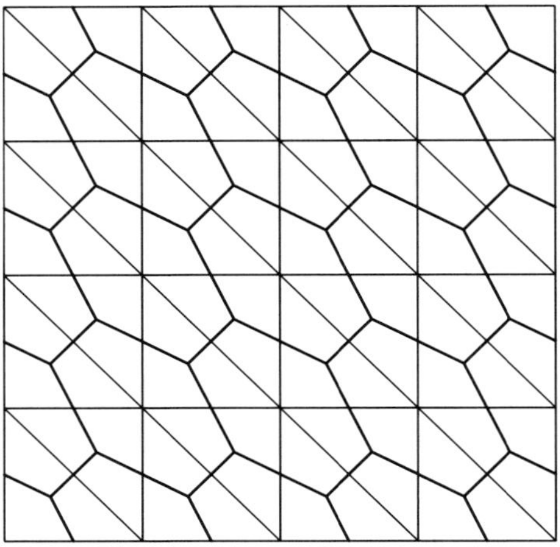

Figure 6.3 Dual mesh (thick lines) via barycentres of primal mesh (thin lines)

6.7.2 Relation to Finite Element Methods

The fact that the elements of a dual mesh can be associated with the vertices of a finite element partition gives a link between finite volume and finite element methods.

Consider a function φ that is piecewise constant on the dual mesh $\mathcal{T}$, i.e. $\varphi \in S^{0,-1}(\mathcal{T})$. With φ we associate a continuous piecewise linear function $\Phi \in S^{1,0}(\widetilde{\mathcal{T}})$ corresponding to the finite element partition $\widetilde{\mathcal{T}}$ such that $\Phi(x_K) = \varphi_K$ for the vertex $x_K \in \widetilde{\mathcal{N}}$ corresponding to $K \in \mathcal{T}$.

This link considerably simplifies the analysis of finite volume methods and suggests a very simple and natural approach to a posteriori error estimation and mesh adaptivity for finite volume methods.

- Given the solution φ of the finite volume scheme, compute the corresponding finite element function Φ.
- Compute a standard a posteriori error indicator for Φ.
- Given the error indicator, apply a standard mesh-refinement strategy to the finite element mesh $\widetilde{\mathcal{T}}$ and thus construct a new, locally refined partition $\widehat{\mathcal{T}}$.
- Use $\widehat{\mathcal{T}}$ to construct a new dual mesh $\mathcal{T}'$. This is the refinement of $\mathcal{T}$.

6.8 Convergence of the Adaptive Process III

While there are several quite satisfactory results concerning the convergence of the adaptive process for stationary linear elliptic problems, the situation is quite different for non-stationary problems. In fact we are currently aware of only two results establishing the convergence of an adaptive space–time discretisation. The first result is due to Z. Chen and J. Feng [**106**]; the second very recent one was obtained by C. Kreuzer, C. A. Möller, A. Schmidt, and K. G. Siebert [**163**].

The result of Z. Chen and J. Feng has two drawbacks. First, it does not guarantee that the final time T is reached; the adaptive process may generate a sequence of discrete times which converges to some limit less than T. Second, the result depends on the ratio of the local spatial mesh-size to the local time-step size which is highly undesirable.

These drawbacks are overcome by C. Kreuzer, C. A. Möller, A. Schmidt, and K. G. Siebert. In what follows we will report their results. To simplify the presentation, we restrict ourselves to the heat equation of Section 6.1 (p. 309); the results also hold for time-dependent diffusion–reaction equations [**163**].

The following two observations will be crucial:

- an a priori bound for the energy of the discrete solution which only depends on the data of the differential equation guarantees that the final time is actually reached within a finite number of time-steps, cf. Proposition 6.65 below;
- the coarsening of spatial meshes which is mandatory while advancing in time leads to an increase in the energy of the discrete solution which must effectively be controlled, cf. Proposition 6.66 below.

6.8.1 The Variational Problem

We consider the variational formulation (6.3) (p. 310) of the heat equation (6.1) (p. 309) and its discretisation (6.4) (p. 310) and (6.5) (p. 310) with the following minor modifications.

- The temporal discretisation is based on the *implicit Euler scheme*, i.e. $\theta = 1$.
- The transition condition of Section 3.2.6 (p. 86) concerning the partitions $\mathcal{T}_n$ is dropped.
- We now set

$$f^n = \frac{1}{\tau_n} \int_{t_{n-1}}^{t_n} f(\cdot, t)\, \mathrm{d}t.$$

Notice that the transition condition may be dropped, since, for $\theta = 1$, the term $\int_\Omega \nabla u^{n\theta} \cdot \nabla v_{\mathcal{T}_n}$ on the left-hand side of (6.5) (p. 310) only incorporates functions corresponding to the same partition $\mathcal{T}_n$.

6.8.2 Error Indicators

We consider the residual a posteriori error indicator of Theorem 6.6 (p. 315) with the following minor modifications.

- The parameter θ is frozen to the value 1.
- The auxiliary partitions $\widetilde{\mathcal{T}}_n$ are replaced by the actual partitions $\mathcal{T}_n$.
- The quantities entering in the constant c^* of the upper bound are explicitly traced and incorporated in the error indicator.
- The error indicator is split into four contributions which measure the error in the approximation of the initial data, the data oscillation, the temporal error, and the spatial error, respectively.

Denoting by $C_{F,2}(\Omega)$ and $C_{\mathcal{T}_n}$ Friedrichs' constant of Ω and the shape parameter of $\mathcal{T}_n$, respectively, we thus set

$$\eta_0 = \sqrt{3}\,\|u_0 - \pi_0 u_0\|,$$

$$\eta_{n,f} = \sqrt{15\,C_{F,2}(\Omega)}\left\{\frac{1}{\tau_n}\int_{t_{n-1}}^{t_n}\|f - f^n\|^2\,dt\right\}^{\frac{1}{2}},$$

$$\eta_{n,\tau} = \sqrt{5}\left\|\nabla\left(u_{\mathcal{T}_n}^n - u_{\mathcal{T}_{n-1}}^{n-1}\right)\right\|,$$

$$\eta_{n,h} = \sqrt{15\,C_{\mathcal{T}_n}}\left\{\sum_{K\in\mathcal{T}_n} h_K^2\left\|f^n - \frac{1}{\tau_n}\left(u_{\mathcal{T}_n}^n - u_{\mathcal{T}_{n-1}}^{n-1}\right) + \Delta u_{\mathcal{T}_n}^n\right\|_K^2 + \sum_{E\in\mathcal{E}_n} h_E\left\|\mathbb{J}_E\left(\mathbf{n}_E\cdot\nabla u_{\mathcal{T}_n}^n\right)\right\|_E^2\right\}^{\frac{1}{2}}$$

and

$$\eta = \left\{\eta_0^2 + \sum_{n=1}^{N_{\mathcal{I}}}\tau_n\left(\eta_{n,f}^2 + \eta_{n,\tau}^2 + \eta_{n,h}^2\right)\right\}^{\frac{1}{2}}.$$

The arguments of Section 6.1 (p. 309) then yield the upper bound

$$\left\{\|u - u_{\mathcal{I}}\|_{L^\infty(0,T;L^2(\Omega))}^2 + \|u - u_{\mathcal{I}}\|_{L^2(0,T;H_0^1(\Omega))}^2 + \left\|\partial_t(u - u_{\mathcal{I}})\right\|_{L^2(0,T;H^{-1}(\Omega))}^2\right\}^{\frac{1}{2}} \le \eta \tag{6.77}$$

for the error of the solution u of the heat equation and its approximation $u_{\mathcal{I}}$.

Notice that $\eta_{n,\tau}$ implicitly also measures the consistency error which arises from a potential coarsening of $\mathcal{T}_{n-1}$. To understand this, denote by $R_n : H_0^1(\Omega) \to X_n$ the Ritz projection onto X_n. If $\mathcal{T}_n$ is obtained from $\mathcal{T}_{n-1}$ by local refinement exclusively, we have $X_{n-1} \subset X_n$ and consequently $R_n u_{\mathcal{T}_{n-1}}^{n-1} = u_{\mathcal{T}_{n-1}}^{n-1}$ and

$$\eta_{n,\tau} = \sqrt{5}\left\|\nabla\left(u_{\mathcal{T}_n}^n - R_n u_{\mathcal{T}_{n-1}}^{n-1}\right)\right\|.$$

If, on the other hand, $\mathcal{T}_{n-1}$ is partially coarsened when passing to $\mathcal{T}_n$, we have $X_{n-1} \not\subset X_n$ and

$$\eta_{n,\tau} = \sqrt{5}\left\{\left\|\nabla\left(u_{\mathcal{T}_n}^n - R_n u_{\mathcal{T}_{n-1}}^{n-1}\right)\right\|^2 + \left\|\nabla\left(u_{\mathcal{T}_{n-1}}^{n-1} - R_n u_{\mathcal{T}_{n-1}}^{n-1}\right)\right\|^2\right\}^{\frac{1}{2}}.$$

As we will see in Proposition 6.66 below, a partial coarsening of the partition $\mathcal{T}_{n-1}$ may lead to an increasing energy of the discrete solution. To control this effect, we introduce the additional error indicator

$$\eta_{n,*} = \left\|\nabla\left(\pi_n u_{\mathcal{T}_{n-1}}^{n-1}\right)\right\|^2 - \left\|\nabla\left(u_{\mathcal{T}_{n-1}}^{n-1}\right)\right\|^2 - \frac{1}{\tau_n}\left\|u_{\mathcal{T}_n}^n - \pi_n u_{\mathcal{T}_{n-1}}^{n-1}\right\|^2$$

where π_n denotes the L^2-projection onto X_n. Proposition 6.66 in particular implies that the adaptive process has to ensure $\eta_{n,*} \le 0$ for all n. Notice that the third term in $\eta_{n,*}$ may compensate for a positive contribution of the first two terms and that $\eta_{n,*} \le 0$ when $\mathcal{T}_n$ is obtained from $\mathcal{T}_{n-1}$ by refinement exclusively.

6.8.3 The Adaptive Algorithm and its Components

The adaptive Algorithm 6.61 consists of several components which we will introduce next. The first four components are standard ingredients of any adaptive process and can be described very briefly.

The last two components, on the other hand, are particular to the space–time adaptivity and need a more detailed description.

Given the initial value u_0 of the heat equation, an initial partition $\widehat{\mathcal{T}_0}$ and a given tolerance ε_0, the component ADAPT_INIT($u_0, \widehat{\mathcal{T}_0}, \varepsilon_0$) constructs the first partition $\mathcal{T}_0$ and the corresponding discrete initial value $\pi_0 u_0$ such that $\eta_0 \le \varepsilon_0$. This is a standard process similar to the approximation of the right-hand side in Section 1.14.2 (p. 59).

Given $\mathcal{T}_n$, τ_n, f^n, and $u^{n-1}_{\mathcal{T}_{n-1}}$, the component SOLVE($u^{n-1}_{\mathcal{T}_{n-1}}, f^n, \tau_n, \mathcal{T}_n$) solves the discrete problem (6.5) (p. 310) on X_n. We assume that this is done exactly and that all integrals are evaluated exactly so that we have Galerkin orthogonality.

Given $\mathcal{T}_{n-1}$ and $u^{n-1}_{\mathcal{T}_{n-1}}$, the component COARSEN($u^{n-1}_{\mathcal{T}_{n-1}}, \mathcal{T}_{n-1}$) produces a partial coarsening of $\mathcal{T}_{n-1}$. It is invoked at the beginning of the n-th time-step. Afterwards only refinement is performed during the time-step. This component may not be based on an error indicator at all. In particular, its output may be independent of $u^{n-1}_{\mathcal{T}_{n-1}}$ and may equal $\mathcal{T}_{n-1}$. For efficiency reasons, however, it is desirable that COARSEN removes as many degrees of freedom as possible and at the same time keeps the difference of $u^{n-1}_{\mathcal{T}_{n-1}}$ to a suitable interpolation in the finite element space associated with the output of COARSEN at a moderate size [**163**, Section 5.2].

The component MARK_REFINE is used in two different ways depending on its input arguments.

Given a partition $\mathcal{T}$ and an error indicator η which is the Euclidean sum of contributions η_K from the elements $K \in \mathcal{T}$, MARK_REFINE($\eta, \mathcal{T}$) produces a new admissible partition such that at least one element in the subset $\operatorname{argmax}_{K\in\mathcal{T}} \eta_K$ of $\mathcal{T}$ is refined. This version of MARK_REFINE will be used in connection with the partitions $\mathcal{T}_n$ and the associated indicators $\eta_{n,h}$. All marking and refinement strategies of Chapter 2 are suited for this component.

Given a partition $\mathcal{T}$, an associated error indicator η as before, and a second partition $\mathcal{T}'$, MARK_REFINE($\eta, \mathcal{T}, \mathcal{T}'$) has the same effect as MARK_REFINE($\eta, \mathcal{T}$) with the additional condition that at least one element of $\mathcal{T} \setminus \mathcal{T}'$ which has been *coarsened* is refined. Here, we call an element $K \in \mathcal{T} \setminus \mathcal{T}'$ coarsened with respect to $\mathcal{T}'$ if $K = \bigcup_{K'\in\mathcal{T}'} K' \cap K$ with $h_{K'} < h_K$ for every involved K'. This version of MARK_REFINE will be applied to the indicators $\left\| \nabla \left(u^{n-1}_{\mathcal{T}_{n-1}} - \pi_n u^{n-1}_{\mathcal{T}_{n-1}} \right) \right\|$ and $\eta_{n,*}$. The marking and refinement strategies of Chapter 2 can easily be adapted to fulfil the additional requirements for this version of MARK_REFINE.

Given a tolerance ε_f, the right-hand side f of the heat equation, the intermediate time t_{n-1}, and a candidate τ_n for the size of the next time-step, the component CONSISTENCY($\varepsilon_f, f, t_{n-1}, \tau_n$) determines the size τ_n of the next time-step such that $\eta_{n,f} \le \varepsilon_f$. It first tries a larger time-step and then successively reduces its size until the prescribed tolerance is met. Lemma 6.64 below shows that this goal is achieved in a finite number of iterations and that the produced step sizes τ_n remain above a positive threshold which only depends on the given right-hand side f and the tolerance ε_f. The following algorithm gives a possible realisation of CONSISTENCY.

Algorithm 6.59 (CONSISTENCY). *Given: a tolerance ε_f, the right-hand side f of the heat equation, the final time T, an intermediate time t_{n-1}, a candidate τ_n for the size of the next time-step, and parameters $\sigma \in (0,1)$, $\delta \in (0,1)$, and $\vartheta > 1$.*
Sought: the size τ_n of the next time-step such that $\eta_{n,f} \le \varepsilon_f$.

(1) *Compute f^n and $\eta_{n,f}$.*

(2) *If $\eta_{n,f} < \sigma\varepsilon_f$, set $\tau_n = \min\{T - t_{n-1}, \vartheta\tau_n\}$.*

(3) *Compute f^n and $\eta_{n,f}$.*

(4) *If $\eta_{n,f} \le \varepsilon_f$,* stop. *Otherwise set $\tau_n = \delta\tau_n$ and return to step (3).*

The final component is ST_ADAPTATION. Given the previous partition $\mathcal{T}_{n-1}$, the discrete solution $u^{n-1}_{\mathcal{T}_{n-1}}$ at the previous time t_{n-1}, a candidate τ_n for the size of the next time-step, a candidate $\mathcal{T}_n$ for the next partition, the right-hand side f of the heat equation, and a tolerance $\varepsilon_{h,\tau}$, it tries to determine the size τ_n of the next time-step and the next partition $\mathcal{T}_n$ such that $\eta^2_{n,\tau} + \eta^2_{n,h} \le \varepsilon^2_{h,\tau}$. This component is an iteration which requires right at the start the solution of the discrete problem (6.5) (p. 310) with the actual partition and the actual time-step. It is a typical adaptive iteration for an adaptive time-stepping scheme subject to the following modifications. First, spatial adaptation is restricted to refinement exclusively. Second, the size of the time-step is reduced but never below a given tolerance $\tau_* > 0$. The latter is determined in step (3) of the adaptive algorithm 6.61 which calls the component ST_ADAPTATION. Notice, however, that ST_ADAPTATION may be entered with a step-size τ_n less than τ_*. The following algorithm gives a possible realisation of ST_ADAPTATION.

Algorithm 6.60 (ST_ADAPTATION). *Given: a partition $\mathcal{T}_{n-1}$ and an intermediate time t_{n-1}, the discrete solution $u^{n-1}_{\mathcal{T}_{n-1}}$ corresponding to $\mathcal{T}_{n-1}$ and t_{n-1}, a candidate τ_n for the size of the next time-step, a candidate $\mathcal{T}_n$ for the next partition, the right-hand side f of the heat equation, a tolerance $\varepsilon_{h,\tau}$, and a parameter $\delta \in (0,1)$.*
Sought: the next partition $\mathcal{T}_n$, the size τ_n of the next time-step, and the discrete solution $u^n_{\mathcal{T}_n}$ corresponding to $\mathcal{T}_n$ and τ_n.

(1) *Compute f^n, solve the discrete problem associated with $\mathcal{T}_n$ and τ_n, i.e. $u^n_{\mathcal{T}_n}$ =* SOLVE *$(u^{n-1}_{\mathcal{T}_{n-1}}, f^n, \tau_n, \mathcal{T}_n)$, and determine $\eta_{n,\tau}$ and $\eta_{n,h}$.*
(2) *If $\eta^2_{n,\tau} + \eta^2_{n,h} \le \varepsilon^2_{h,\tau}$,* stop; *otherwise continue with step (3).*
(3) *If $\eta_{n,h} > \eta_{n,\tau}$, refine $\mathcal{T}_n$ =* MARK_REFINE*$(\eta_{n,h}, \mathcal{T}_n)$ and return to step (1). Otherwise compute $\eta_{n,1} = \sqrt{5}\left\|\nabla\left(u^n_{\mathcal{T}_n} - \pi_n u^{n-1}_{\mathcal{T}_{n-1}}\right)\right\|$ and $\eta_{n,2} = \sqrt{5}\left\|\nabla\left(u^{n-1}_{\mathcal{T}_{n-1}} - \pi_n u^{n-1}_{\mathcal{T}_{n-1}}\right)\right\|$.*
(4) *If $\tau_n > \tau_*$ continue with step (5), otherwise with step (6).*
(5) *If $\eta_{n,1} > \eta_{n,2}$, set $\tau_n = \max\{\delta\tau_n, \tau_*\}$. Otherwise refine $\mathcal{T}_n$ =* MARK_REFINE*$(\eta_{n,h}, \mathcal{T}_n, \mathcal{T}_{n-1})$. Return to step (1)*
(6) *If $\eta^2_{n,h} + 2\eta^2_{n,2} \le \varepsilon^2_{h,\tau}$,* stop. *Otherwise continue with step (7).*
(7) *If $\eta_{n,h} > \sqrt{2}\eta_{n,2}$, refine $\mathcal{T}_n$ =* MARK_REFINE*$(\eta_{n,h}, \mathcal{T}_n)$. Otherwise refine $\mathcal{T}_n$ =* MARK_REFINE*$(\eta_{n,h}, \mathcal{T}_n, \mathcal{T}_{n-1})$. Return to step (1).*

Algorithm 6.60 can only be abandoned at two instances. The first one is in step (2). In this case the discrete solution $u^n_{\mathcal{T}_n}$ satisfies the condition $\eta^2_{n,\tau} + \eta^2_{n,h} \le \varepsilon^2_{h,\tau}$. This is the standard case. The second instance is in step (6). In this non-standard case, the step-size τ_n cannot be reduced further and Algorithm 6.60 only controls the spatial indicator $\eta_{n,h}$ and the indicator $\eta_{n,2}$. The uncontrolled indicator $\eta_{n,1}$ is implicitly handled in this case by calling the adaptive algorithm 6.61 in combination with Proposition 6.66 below. Notice that step (7) only takes effect if Algorithm 6.60 is called with a time-step less than τ_* and that the master Algorithm 6.61 is responsible for this.

With these preparations we can now present the adaptive algorithm.

Algorithm 6.61 (Adaptive space–time algorithm). *Given: a tolerance ε, the right-hand side f of the heat equation, the final time T, an initial time-step τ_0, and an initial partition $\widehat{\mathcal{T}_0}$.*
Sought: a discrete solution $u_{\mathcal{I}}$ such that $\eta \le \varepsilon$.

(1) *Split the tolerance into positive contributions ε_0, ε_f, $\varepsilon_{h,\tau}$, and ε_* such that $\varepsilon_0^2 + T\varepsilon_f^2 + T\varepsilon^2_{h,\tau} + \varepsilon_*^2 = \varepsilon^2$. Set $n = 0$.*

(2) *Determine the first partition* $\mathcal{T}_0$ *and the initial value* $u^0_{\mathcal{T}_0}$, *i.e.* $(u^0_{\mathcal{T}_0}, \mathcal{T}_0) = \mathtt{ADAPT\_INIT}(u_0, \widehat{\mathcal{T}}_0, \varepsilon_0)$.

(3) *Set* $\tau_* = \dfrac{\varepsilon_*^2}{2\sqrt{5}\left(\|f\|^2_{\Omega\times(0,T)} + \left\|\nabla u^0_{\mathcal{T}_0}\right\|^2\right)}$.

(4) *Increment n by* 1 *and set* $\tau_n = \min\{\tau_{n-1}, T - t_{n-1}\}$.

(5) *Compute* $(f^n, \tau_n) = \mathtt{CONSISTENCY}(\varepsilon_f, f, t_{n-1}, \tau_n)$.

(6) *Determine a first candidate for* $\mathcal{T}_n$ *by coarsening* $\mathcal{T}_{n-1}$, *i.e.* $\mathcal{T}_n = \mathtt{COARSEN}(u^{n-1}_{\mathcal{T}_{n-1}}, \mathcal{T}_{n-1})$.

(7) *Determine the size of the next time-step* τ_n, *the next partition* $\mathcal{T}_n$, *and the next discrete solution* $u^n_{\mathcal{T}_n}$, *i.e.* $(u^n_{\mathcal{T}_n}, \tau_n, \mathcal{T}_n) = \mathtt{ST\_ADAPTATION}(u^{n-1}_{\mathcal{T}_{n-1}}, \tau_n, f, \mathcal{T}_n, \mathcal{T}_{n-1}, \varepsilon_{h,\tau})$ *and compute* $\eta_{n,*}$.

(8) *If* $\eta_{n,*} \le 0$, *continue with step* (9). *Otherwise adapt* $\mathcal{T}_n$, *i.e.* $\mathcal{T}_n = \mathtt{MARK\_REFINE}(\eta_{n,*}, \mathcal{T}_n, \mathcal{T}_{n-1})$ *and return to step* (7).

(9) *If* $t_{n-1} + \tau_n = T$, stop. *Otherwise return to step* (4).

Notice that step (7) of Algorithm 6.61 may be iterated several times. The iteration is stopped once a potential increase in energy indicated by a positive value of $\eta_{n,*}$ is avoided by a sufficient refinement of $\mathcal{T}_n$ and reduction of τ_n.

6.8.4 Convergence of the Adaptive Process

In this section we prove the convergence of the adaptive Algorithm 6.61.

Theorem 6.62 *Assume that the right-hand side* f *of the heat equation is contained in* $H^1(0, T; L^2(\Omega))$ *and that the initial value* u_0 *is contained in* $L^2(\Omega)$. *Then Algorithm 6.61 produces for every initial partition* $\widehat{\mathcal{T}}_0$, *every initial time-step* τ_0, *and every positive tolerance* ε *a finite sequence* $0 = t_0 < t_1 < \cdots < t_N = T$ *of intermediate times, a sequence* $\mathcal{T}_0, \ldots, \mathcal{T}_N$ *of spatial meshes, and an associated discrete solution* $u_{\mathcal{I}}$ *of problems* (6.4) (*p.* 310) *and* (6.5) (*p.* 310) *such that the error in the solution u of the variational formulation* (6.1) (*p.* 309) *of the heat equation satisfies*

$$\left\{\|u - u_{\mathcal{I}}\|^2_{L^\infty(0,T;L^2(\Omega))} + \|u - u_{\mathcal{I}}\|^2_{L^2(0,T;H^1_0(\Omega))} + \left\|\partial_t(u - u_{\mathcal{I}})\right\|^2_{L^2(0,T;H^{-1}(\Omega))}\right\}^{\frac{1}{2}} \le \varepsilon.$$

The proof of Theorem 6.62 consists of several steps which are the subject of the following subsections.

The main difficulty of the proof may be described as follows. Due to estimate (6.77), Algorithm 6.61 strives at ensuring $\eta \le \varepsilon$. This is an L^2-type condition which requires knowledge of the indicators $\eta_{n,f}$, $\eta_{n,\tau}$, and $\eta_{n,h}$ at *all* intermediate times t_n. At the n-th time-step, however, we only have access to current values of these indicators and not its future ones. Correspondingly, the actual error control is of an L^∞-type. This difference leads to severe problems which, among others, are responsible for the sub-optimality of the results of Z. Chen and J. Feng [**106**]. To overcome this drawback, C. Kreuzer, C. A. Möller, A. Schmidt, and K. G. Siebert [**163**] explicitly control the effect of replacing the L^2-criterion by an L^∞-one. This must happen at two instances. The first one is related to the right-hand side f and can be handled quite easily in Subsection 6.8.4.1 assuming the extra regularity $f \in H^1(0, T; L^2(\Omega))$. The second one is related to the possible increase in energy due to a potential coarsening of the partitions. This point is more delicate and is the focus of Subsection 6.8.4.2.

6.8.4.1 Control of the consistency error

The following Poincaré-type estimate allows us to control the indicator $\eta_{n,f}$.

Lemma 6.63 *Assume that* $f \in H^1(0,T;L^2(\Omega))$. *Then*

$$\int_t^{t+\tau} \left\| f(s,\cdot) - \frac{1}{\tau}\int_t^{t+\tau} f(\sigma,\cdot)d\sigma \right\|^2 ds \le \tau^2 \int_t^{t+\tau} \left\| \partial_t f(s,\cdot)\right\|^2 ds$$

holds for all $t \in (0,T)$ *and all* $\tau \in (0, T-t)$. *In particular, for all* n, *we have* $\eta_{n,f} \le \varepsilon_f$ *if* $\tau_n \le \frac{\varepsilon_f^2}{15\, C_{F,2}(\Omega) \|\partial_t f\|^2_{\Omega\times(0,T)}}$.

Proof For every $t \in (0,T)$, $\tau \in (0,T-t)$, $s \in (t, t+\tau)$, and $x \in \Omega$ the fundamental theorem of calculus implies

$$f(s,x) - \frac{1}{\tau}\int_t^{t+\tau} f(\sigma,x)d\sigma = \frac{1}{\tau}\int_t^{t+\tau}\int_{\min\{s,\sigma\}}^{\max\{s,\sigma\}} \partial_t f(\rho,x)d\rho ds.$$

Taking the square of this equality, integrating with respect to x, and using the Cauchy–Schwarz inequality gives

$$\left\| f(s,\cdot) - \frac{1}{\tau}\int_t^{t+\tau} f(\sigma,\cdot)d\sigma \right\|^2 \le \tau \int_t^{t+\tau} \left\|\partial_t f(\rho,\cdot)\right\|^2 d\rho.$$

Integration with respect to t establishes the first estimate of the lemma. The second one immediately follows from the definition of $\eta_{n,f}$ and the first estimate. □

Lemma 6.63 allows us to prove the termination of Algorithm 6.59.

Lemma 6.64 *Assume that* $f \in H^1(0,T;L^2(\Omega))$ *and set*

$$\tau_f = \frac{\delta \varepsilon_f^2}{15\, C_{F,2}(\Omega) \left\|\partial_t f\right\|^2_{\Omega\times(0,T)}}. \tag{6.78}$$

Then for every $t_{n-1} \in (0,T)$ *and every candidate* $\widehat{\tau}_n \in (0, T - t_{n-1})$, *Algorithm 6.59 terminates with a time-step* τ_n *satisfying*

$$\min\{\tau_f, \widehat{\tau}_n\} \le \tau_n \le T - t_{n-1} \quad \textit{and} \quad \eta_{n,f} \le \varepsilon_f.$$

Proof Algorithm 6.59 tries to increase the time-step once. If it stops immediately in step (4), the assertion of the lemma obviously holds. Thus we only have to discuss its behaviour when it loops and decreases the time-step at least once. Lemma 6.63 and the definition of τ_f imply that $\eta_{n,f} \le \varepsilon_f$ if $\tau_n \le \frac{\tau_f}{\delta}$. In every loop the time-step is reduced by the factor δ and we enter the loop with $\frac{\tau_f}{\delta} \le \tau_n \le T - t_{n-1}$. Consequently this loop is iterated only a finite number of times and the smallest step-size produced is bounded from below by τ_f. This proves the assertion. □

6.8.4.2 Control of the discrete energy

We first prove a uniform a priori bound for the energy of the discrete solutions $u^n_{\mathcal{T}_n}$.

Proposition 6.65 *Consider an arbitrary $N \in \mathbb{N} \cup \{\infty\}$ and arbitrary intermediate times $0 = t_0 < t_1 < \cdots < t_N \leq T$ with associated step-sizes $\tau_1, \ldots, \tau_N$ and denote by $u_{\mathcal{T}_n}^n$ the solution of problem (6.5) (p. 310) with $\theta = 1$. Then for every $m \in \{1, \ldots, N\}$ there holds*

$$\sum_{n=1}^{m} \left(\frac{1}{\tau_n} \left\| u_{\mathcal{T}_n}^n - \pi_n u_{\mathcal{T}_{n-1}}^{n-1} \right\|^2 + \left\| \nabla \left(u_{\mathcal{T}_n}^n - \pi_n u_{\mathcal{T}_{n-1}}^{n-1} \right) \right\|^2 \right.$$
$$\left. + \left\| \nabla u_{\mathcal{T}_n}^n \right\|^2 - \left\| \nabla \left(\pi_n u_{\mathcal{T}_{n-1}}^{n-1} \right) \right\|^2 \right) \leq \|f\|_{\Omega \times (0,t_m)}^2 .$$

Proof We insert $v_{\mathcal{T}_n} = u_{\mathcal{T}_n}^n - \pi_n u_{\mathcal{T}_{n-1}}^{n-1}$ as test function in equation (6.5) (p. 310). Since $\theta = 1$, this yields

$$\frac{1}{\tau_n} \left\| u_{\mathcal{T}_n}^n - \pi_n u_{\mathcal{T}_{n-1}}^{n-1} \right\|^2 + \left\| \nabla \left(u_{\mathcal{T}_n}^n - \pi_n u_{\mathcal{T}_{n-1}}^{n-1} \right) \right\|^2$$
$$= \int_\Omega f^n \left(u_{\mathcal{T}_n}^n - \pi_n u_{\mathcal{T}_{n-1}}^{n-1} \right) - \int_\Omega \nabla \pi_n u_{\mathcal{T}_{n-1}}^{n-1} \cdot \nabla \left(u_{\mathcal{T}_n}^n - \pi_n u_{\mathcal{T}_{n-1}}^{n-1} \right)$$

and

$$\frac{1}{\tau_n} \left\| u_{\mathcal{T}_n}^n - \pi_n u_{\mathcal{T}_{n-1}}^{n-1} \right\|^2 + \left\| \nabla u_{\mathcal{T}_n}^n \right\|^2 = \int_\Omega f^n \left(u_{\mathcal{T}_n}^n - \pi_n u_{\mathcal{T}_{n-1}}^{n-1} \right) + \int_\Omega \nabla u_{\mathcal{T}_n}^n \cdot \nabla \left(\pi_n u_{\mathcal{T}_{n-1}}^{n-1} \right).$$

Adding both identities gives

$$\frac{2}{\tau_n} \left\| u_{\mathcal{T}_n}^n - \pi_n u_{\mathcal{T}_{n-1}}^{n-1} \right\|^2 + \left\| \nabla \left(u_{\mathcal{T}_n}^n - \pi_n u_{\mathcal{T}_{n-1}}^{n-1} \right) \right\|^2 + \left\| \nabla u_{\mathcal{T}_n}^n \right\|^2$$
$$= 2 \int_\Omega f^n \left(u_{\mathcal{T}_n}^n - \pi_n u_{\mathcal{T}_{n-1}}^{n-1} \right) + \left\| \nabla \left(\pi_n u_{\mathcal{T}_{n-1}}^{n-1} \right) \right\|^2 .$$

The Cauchy–Schwarz and Young's inequality therefore imply that

$$\frac{1}{\tau_n} \left\| u_{\mathcal{T}_n}^n - \pi_n u_{\mathcal{T}_{n-1}}^{n-1} \right\|^2 + \left\| \nabla \left(u_{\mathcal{T}_n}^n - \pi_n u_{\mathcal{T}_{n-1}}^{n-1} \right) \right\|^2 + \left\| \nabla u_{\mathcal{T}_n}^n \right\|^2$$
$$\leq \tau_n \left\| f^n \right\|^2 + \left\| \nabla \left(\pi_n u_{\mathcal{T}_{n-1}}^{n-1} \right) \right\|^2 .$$

Taking the sum of this inequality with respect to n and observing that

$$\sum_{n=1}^{m} \tau_n \left\| f^n \right\|^2 \leq \|f\|_{\Omega \times (0,t_m)}^2$$

proves the proposition. □

Proposition 6.65 allows us to control the Euclidean sum of the indicators $\eta_{n,2}$ in Algorithm 6.60.

Proposition 6.66 *In addition to the assumptions of Proposition 6.65 suppose that $\eta_{n,*} \leq 0$ holds for all n. Then*

$$\sum_{n=1}^{m} \left\| \nabla \left(u_{\mathcal{T}_n}^n - \pi_n u_{\mathcal{T}_{n-1}}^{n-1} \right) \right\|^2 \leq \|f\|_{\Omega \times (0,t_m)}^2 + \left\| \nabla u_{\mathcal{T}_0}^0 \right\|^2 - \left\| \nabla u_{\mathcal{T}_m}^m \right\|^2$$

holds for all m. In particular, the left hand-side of the above inequality is bounded irrespective of the number N of time-steps and the sequence of intermediate times.

Proof Due to the definition of $\eta_{n,*}$ we have

$$\left\| \nabla u_{\mathcal{T}_{n-1}}^{n-1} \right\|^2 - \left\| \nabla \left(\pi_n u_{\mathcal{T}_{n-1}}^{n-1} \right) \right\|^2 + \frac{1}{\tau_n} \left\| u_{\mathcal{T}_n}^n - \pi_n u_{\mathcal{T}_{n-1}}^{n-1} \right\|^2 \geq 0$$

for all n. Taking the sum of these inequalities gives

$$0 \leq \sum_{n=1}^{m} \left(\left\| \nabla u_{\mathcal{T}_{n-1}}^{n-1} \right\|^2 - \left\| \nabla \left(\pi_n u_{\mathcal{T}_{n-1}}^{n-1} \right) \right\|^2 + \frac{1}{\tau_n} \left\| u_{\mathcal{T}_n}^n - \pi_n u_{\mathcal{T}_{n-1}}^{n-1} \right\|^2 \right)$$

which is equivalent to

$$\begin{aligned} &\left\| \nabla u_{\mathcal{T}_m}^m \right\|^2 - \left\| \nabla u_{\mathcal{T}_0}^0 \right\|^2 \\ &\leq \sum_{n=1}^{m} \left(\left\| \nabla u_{\mathcal{T}_n}^n \right\|^2 - \left\| \nabla \left(\pi_n u_{\mathcal{T}_{n-1}}^{n-1} \right) \right\|^2 + \frac{1}{\tau_n} \left\| u_{\mathcal{T}_n}^n - \pi_n u_{\mathcal{T}_{n-1}}^{n-1} \right\|^2 \right). \end{aligned}$$

Adding $\sum_{n=1}^{m} \left\| \nabla \left(u_{\mathcal{T}_n}^n - \pi_n u_{\mathcal{T}_{n-1}}^{n-1} \right) \right\|^2$ to both sides of this estimate and taking into account Proposition 6.65 proves the claimed bound for the discrete energies. □

6.8.4.3 Termination of MARK_REFINE

Theorem 4.84 (p. 269) implies that, for every $n, f^n, \tau_n, \mathcal{T}_{n-1}, u_{\mathcal{T}_{n-1}}^{n-1}$, and initial candidate $\mathcal{T}_n$, the component MARK_REFINE in steps (3), (5), and (7) of Algorithm 6.60 and step (8) of Algorithm 6.61 terminates after a finite number of iterations with a partition $\mathcal{T}_n$ and a discrete solution $u_{\mathcal{T}_n}^n$ satisfying $\eta_{n,h} \leq \frac{1}{\sqrt{2}} \varepsilon_{h,\tau}$.

6.8.4.4 Termination of ST_ADAPTATION

We claim that, for every n, f^n, τ_n, $\mathcal{T}_{n-1}$, $u_{\mathcal{T}_{n-1}}^{n-1}$, and initial candidates $\widehat{\tau}_n$ and $\mathcal{T}_n$, the component ST_ADAPTATION in step (7) of Algorithm 6.61 terminates after a finite number of iterations with a time-step τ_n, a partition $\mathcal{T}_n$, and a discrete solution $u_{\mathcal{T}_n}^n$ which either satisfy

$$\min\{\widehat{\tau}_n, \tau_*\} \leq \tau_n \leq \widehat{\tau}_n \quad \text{and} \quad \eta_{n,h}^2 + \eta_{n,\tau}^2 \leq \varepsilon_{h,\tau}^2 \tag{6.79}$$

or

$$\tau_n \leq \tau_* \quad \text{and} \quad \eta_{n,h}^2 + 2\eta_{n,2}^2 \leq \varepsilon_{h,\tau}^2. \tag{6.80}$$

To prove this, we recall that Algorithm 6.60 can only be abandoned in steps (2) and (6). The corresponding stopping criterion then implies the first assertion in case of step (2) and the second one in case of step (6). As long as Algorithm 6.60 does not branch to its step (7), the properties of MARK_REFINE imply that both situations are reached after a finite number of iterations since in each iteration either the size of the time-step is reduced by a fixed factor or the partition $\mathcal{T}_n$ is refined. Thus the only possibility for an infinite cycling of Algorithm 6.60 could be its step (7). Since Algorithm 6.60 is invoked with an initial candidate $\mathcal{T}_n$ which is obtained by a coarsening of $\mathcal{T}_{n-1}$, a finite number of executions of the statement $\mathcal{T}_n$ = MARK_REFINE($\eta_{n,h}$, $\mathcal{T}_n$, $\mathcal{T}_{n-1}$) in step (7) produces a partition $\mathcal{T}_n$ which is a refinement of $\mathcal{T}_{n-1}$. Then we have $X_{n-1} \subset X_n$ and consequently $\eta_{n,2} \leq 0$. Once this situation is reached, we always refine the partition $\mathcal{T}_n$ keeping the time-step τ_n fixed. The properties of MARK_REFINE therefore again imply that after a finite number of iterations $\eta_{n,h}$ is less than $\varepsilon_{h,\tau}$ so that Algorithm 6.60 is abandoned in step (6).

6.8.4.5 Proof of Theorem 6.62

We claim that Algorithm 6.61 produces a finite number N of intermediate times $0 = t_0 < t_1 < \cdots < t_N = T$ and that $\tau_n \geq \min\{\tau_f, \tau_*, \tau_0\}$ for $n \in \{1, \ldots, N-1\}$. Recall that τ_0 is an input parameter of Algorithm 6.61, that τ_* is determined in step (3) of this algorithm, and that τ_f is given by (6.78). Our claim follows from the previous results if we can ascertain that a positive value of $\eta_{n,*}$ in step (8) of Algorithm 6.61 neither leads to an infinite cycling nor spoils the lower bound for the size of the time-step. To prove this, we observe that τ_n is not changed in step (8) and that a finite number of executions of the statement $\mathcal{T}_n$ = MARK_REFINE($\eta_{n,h}$, $\mathcal{T}_n$, $\mathcal{T}_{n-1}$) leads to $X_{n-1} \subset X_n$ which entails $\eta_{n,*} \leq 0$.

Thus we are left with establishing the error estimate. Due to the upper bound (6.77), the latter is a consequence of $\eta \leq \varepsilon$. To prove this inequality, we observe that $\eta_0 \leq \varepsilon_0$ is guaranteed by the properties of the module ADAPT_INIT. The properties of the module CONSISTENCY, on the other hand, ensure that $\eta_{n,f} \leq \varepsilon_f$ for all n which entails

$$\sum_{n=1}^{N} \tau_n \eta_{n,f}^2 \leq T\varepsilon_f^2.$$

When finalising a time-step, either inequality (6.79) or (6.80) is valid. In the second case, we split $\eta_{n,\tau}$ into its contributions $\eta_{n,1}$ and $\eta_{n,2}$ and obtain from (6.80)

$$\tau_n \eta_{n,h}^2 + \tau_n \eta_{n,\tau}^2 \leq \tau_n \eta_{n,h}^2 + 2\tau_n \eta_{n,1}^2 + 2\tau_n \eta_{n,2}^2 \leq \tau_n \varepsilon_{h,\tau}^2 + 2\tau_* \eta_{n,1}^2.$$

Multiplying the second inequality of (6.79) by τ_n and then adding $2\tau_* \eta_{n,1}^2$ to the right-hand side, we see that the above estimate holds for both cases. Summing over all time-steps we therefore obtain

$$\sum_{n=1}^{N} \left(\tau_n \eta_{n,h}^2 + \tau_n \eta_{n,\tau}^2\right) \leq \sum_{n=1}^{N} \left(\tau_n \varepsilon_{h,\tau}^2 + 2\tau_* \eta_{n,1}^2\right).$$

The first term on the right-hand side is bounded by $T\varepsilon_{h,\tau}^2$. To bound the second term, we observe that $\eta_{n,*} \leq 0$ holds at the end of every time-step. Proposition 6.66 and the definitions of τ_* and $\eta_{n,1}$, therefore imply that the second term is bounded by ε_*^2. This proves that

$$\sum_{n=1}^{N} \left(\tau_n \eta_{n,h}^2 + \tau_n \eta_{n,\tau}^2\right) \leq T\varepsilon_{h,\tau}^2 + \varepsilon_*^2.$$

Collecting all bounds and taking into account the splitting of the tolerance ε proves the desired result $\eta \leq \varepsilon$ and completes the proof of Theorem 6.62.

REFERENCES

[1] E. M. Abdalass, *Resolution performante du problème de Stokes par mini-éléments, maillages auto-adaptifs et méthodes multigrilles – applications*, Ph.D. thesis, Ecole Centrale de Lyon, 1987.

[2] G. Acosta and R. G. Durán, *An optimal Poincaré inequality in L^1 for convex domains*, Proc. Amer. Math. Soc. **132** (2004), no. 1, 195–202.

[3] R. A. Adams, *Sobolev Spaces*, Academic Press, New York, 1975.

[4] M. Afif, A. Bergam, Z. Mghazli, and R. Verfürth, *A posteriori estimators for the finite volume discretization of an elliptic problem*, Numer. Algorithms **34** (2003), no. 2–4, 127–136, International Conference on Numerical Algorithms, Vol. II (Marrakesh, 2001).

[5] A. Agouzal, *A posteriori error estimators for nonconforming approximation*, Int. J. Numer. Anal. Model. **5** (2008), no. 1, 77–85.

[6] M. Ainsworth, *Robust a posteriori error estimation for nonconforming finite element approximation*, SIAM J. Numer. Anal. **42** (2005), no. 6, 2320–2341.

[7] M. Ainsworth, *A posteriori error estimation for discontinuous Galerkin finite element approximation*, SIAM J. Numer. Anal. **45** (2007), no. 4, 1777–1798.

[8] M. Ainsworth and I. Babuška, *Reliable and robust a posteriori error estimating for singularly perturbed reaction–diffusion problems*, SIAM J. Numer. Anal. **36** (1999), no. 2, 331–353.

[9] M. Ainsworth, L. Demkowicz, and C. Kim, *Analysis of the equilibrated residual method for a posteriori error estimation on meshes with hanging nodes*, Comput. Methods Appl. Mech. Engrg. **196** (2007), 3493–3507.

[10] M. Ainsworth and J. T. Oden, *A posteriori error estimators for second order elliptic systems. I. Theoretical foundations and a posteriori error analysis*, Comput. Math. Appl. **25** (1993), no. 2, 101–113.

[11] M. Ainsworth and J. T. Oden, *A posteriori error estimators for second order elliptic systems. II. An optimal order process for calculating self-equilibrating fluxes*, Comput. Math. Appl. **26** (1993), no. 9, 75–87.

[12] M. Ainsworth and J. T. Oden, *A unified approach to a posteriori error estimation using element residual methods*, Numer. Math. **65** (1993), 23–50.

[13] M. Ainsworth and J. T. Oden, *A Posteriori Error Estimation in Finite Element Analysis*, Pure and Applied Mathematics (New York), Wiley-Interscience, New York, 2000.

[14] M. Ainsworth and R. Rankin, *Robust a posteriori error estimation for the nonconforming Fortin–Soulie finite element approximation*, Math. Comp. 77 (2008), no. 264, 1917–1939.

[15] M. Ainsworth and T. Vejchodský, *Fully computable robust a posteriori error bounds for singularly perturbed reaction–diffusion problems*, Numer. Math. **119** (2011), no. 2, 219–243.

[16] M. Ainsworth, J. Z. Zhu, A. W. Craig, and O. C. Zienkiewicz, *Analysis of the Zienkiewicz–Zhu a posteriori error estimator in the finite element method*, Internat. J. Numer. Methods Engrg. **28** (1989), no. 9, 2161–2174.

[17] G. Akrivis, C. Makridakis, and R. H. Nochetto, *A posteriori error estimates for the Crank–Nicolson method for parabolic equations*, Math. Comp. **75** (2006), no. 254, 511–531.

[18] A. Alonso, *Error estimators for a mixed method*, Numer. Math. **74** (1996), 385–395.

[19] H. Amann, *Ordinary Differential Equations*, de Gruyter Studies in Mathematics, vol. 13, Walter de Gruyter, Berlin, 1990.

[20] H. Amann, *Linear and Quasilinear Parabolic Problems. Volume I: Abstract Linear Theory*, Birkhäuser, Basel, 1995.

[21] C. Amrouche, C. Bernardi, M. Dauge, and V. Girault, *Vector potentials in three-dimensional nonsmooth domains*, Math. Meth. Appl. Sci. **21** (1998), 823–864.

[22] T. Apel, *Anisotropic interpolation error estimates for isoparametric quadrilateral finite elements*, Computing **60** (1998), no. 2, 157–174.

[23] T. Apel, *Interpolation of non-smooth functions on anisotropic finite element meshes*, M2AN Math. Model. Numer. Anal. **33** (1999), no. 6, 1149–1185.

[24] M. G. Armentano and C. Padra, *A posteriori error estimates for the Steklov eigenvalue problem*, Appl. Numer. Math. **58** (2008), no. 5, 593–601.

[25] D. Arnica and C. Padra, *A posteriori error estimators for the steady incompressible Navier–Stokes equations*, Numer. Methods Partial Differential Equations **13** (1997), no. 5, 561–574.

[26] D. N. Arnold, F. Brezzi, and J. Douglas Jr., *PEERS: A new mixed finite element for plane elasticity*, Japan J. Appl. Math. **1** (1984), no. 2, 347–367.

[27] D. N. Arnold, F. Brezzi, and M. Fortin, *A stable finite element for the Stokes equations*, Calcolo **21** (1984), 337–344.

[28] D. N. Arnold and R. S. Falk, *Well-posedness of the fundamental boundary value problems for constrained anisotropic elastic materials*, Arch. Ration. Mech. Anal. **98** (1987), 143–190.

[29] I. Babuška, *The finite element method with Lagrange multipliers*, Numer. Math. **20** (1973), 179–192.

[30] I. Babuška, R. G. Durán, and R. Rodriguez, *Analysis of the efficiency of an a posteriori error estimator for linear triangular elements*, SIAM J. Numer. Anal. **29** (1992), 947–964.

[31] I. Babuška and W. C. Rheinboldt, *Error estimates for adaptive finite element computations*, SIAM J. Numer. Anal. **15** (1978), 736–754.

[32] I. Babuška and W. C. Rheinboldt, *A posteriori error estimates for the finite element method*, Internat. J. Numer. Meth. Engrg. **12** (1978), 1597–1615.

[33] I. Babuška and R. Rodriguez, *The problem of the selection of an a posteriori error indicator based on smoothing techniques*, Internat. J. Numer. Math. Engrg. **36** (1993), 539–567.

[34] I. Babuška and M. Vogelius, *Feedback and adaptive finite element solution of one-dimensional boundary value problems*, Numer. Math. **44** (1984), 75–102.

[35] G. Bangerth and R. Rannacher, *Adaptive Finite Element Methods for Differential Equations*, Birkhäuser, Basel, 2003.

[36] W. Bangerth, R. Hartmann, and G. Kanschat, *deal.II differential equations analysis library, technical reference*, `www.dealii.org`.

[37] W. Bangerth, R. Hartmann, and G. Kanschat, *deal.II — a general-purpose object-oriented finite element library*, ACM Trans. Math. Softw. **33** (2007), no. 4, Art. 24, 27.

[38] R. E. Bank, *PLTMG: A software package for solving elliptic partial differential equations. User's Guide 6.0*, SIAM, Philadelphia, 1990.

[39] R. E. Bank, *PLTMG: A software package for solving elliptic partial differential equations. User's Guide 10.0*, Department of Mathematics, University of California at San Diego, 2007.

[40] R. E. Bank and K. Smith, *A posteriori error estimates based on hierarchical bases*, SIAM J. Numer. Anal. **30** (1993), 921–935.

[41] R. E. Bank and K. Smith, *Mesh smoothing using a posteriori error estimates*, SIAM J. Numer. Anal. **34** (1997), 979–997.

[42] R. E. Bank and A. Weiser, *Some a posteriori error estimators for elliptic partial differential equations*, Math. Comput. **44** (1985), 283–301.

[43] R. E. Bank and B. D. Welfert, *A posteriori error estimates for the Stokes equations: a comparison*, Comput. Methods Appl. Mech. Engrg. **82** (1990), 323–340.

[44] R. E. Bank and B. D. Welfert, *A posteriori error estimates for the Stokes problem*, SIAM J. Numer. Anal. **28** (1991), 591–623.

[45] R. E. Bank and J. Xu, *Asymptotically exact a posteriori error estimators. I. Grids with superconvergence*, SIAM J. Numer. Anal. **41** (2003), no. 6, 2294–2312.

[46] R. E. Bank and J. Xu, *Asymptotically exact a posteriori error estimators. II. General unstructured grids*, SIAM J. Numer. Anal. **41** (2003), no. 6, 2313–2332.

[47] E. Bänsch, *Local mesh refinement in 2 and 3 dimensions*, IMPACT Comput. Sci. Engrg. **3** (1991), 181–191.

[48] J. Baranger and H. El Amri, *Estimateur a posteriori d'erreur pour le calcul adaptatif d'écoulements quasi-Newtoniens*, RAIRO M2AN **25** (1991), 31–48.

[49] S. Bartels and C. Carstensen, *Each averaging technique yields reliable a posteriori error control in FEM on unstructured grids. II. Higher order FEM*, Math. Comp. **71** (2002), no. 239, 971–994.

[50] M. Bebendorf, *A note on the Poincaré inequality for convex domains*, Z. Anal. Anwendungen **22** (2003), no. 4, 751–756.

[51] C. Becker, S. Buijssen, and S. Turek, *FEAST: development of HPC technologies for FEM applications*, High Performance Computing in Science and Engineering '07, Springer, Berlin, 2008, pp. 503–516.

[52] R. Becker, *An optimal-control approach to a posteriori error estimation for finite element discretizations of the Navier–Stokes equations*, East–West J. Numer. Math. **8** (2000), no. 4, 257–274.

[53] R. Becker and R. Rannacher, *Weighted a posteriori error control in FE methods*, ENUMATH 97 (Singapore) (H. G. Bock *et al.*, eds.), ENUMATH-95, World Scientific, Singapore, 1998, pp. 612–637.

[54] R. Becker and R. Rannacher, *An optimal control approach to error estimation and mesh adaptation in finite element methods*, Acta Numerica (A. Iserles, ed.), vol. 2000, Cambridge University Press, 2001, pp. 1–101.

[55] L. Beirão da Veiga, J. Niiranen, and R. Stenberg, *A family of C^0 finite elements for Kirchhoff plates. I. Error analysis*, SIAM J. Numer. Anal. **45** (2007), no. 5, 2047–2071.

[56] L. Beirão da Veiga, J. Niiranen, and R. Stenberg, *A posteriori error estimates for the Morley plate bending element*, Numer. Math. **106** (2007), no. 2, 165–179.

[57] L. Beirão da Veiga, J. Niiranen, and R. Stenberg, *A posteriori error analysis for the Morley plate element with general boundary conditions*, Internat. J. Numer. Methods Engrg. **83** (2010), no. 1, 1–26.

[58] M. Bercovier and O. Pironneau, *Error estimates for finite element method solution of the Stokes problem in the primitive variables*, Numer. Math. **33** (1979), 211–224.

[59] A. Bergam, C. Bernardi, and Z. Mghazli, *A posteriori analysis of the finite element discretization of some parabolic equations*, Math. Comp. **74** (2005), no. 251, 1117–1138.

[60] A. Bergam, Z. Mghazli, and R. Verfürth, *Estimations a posteriori d'un schéma de volumes finis pour un problème non linéaire*, Numer. Math. **95** (2003), no. 4, 599–624.

[61] A. Bermùdez, M. R. Nogueiras, and C. Vàzquez, *Numerical analysis of convection–diffusion–reaction problems with higher order characteristics/finite elements. Part I. Time discretization*, SIAM J. Numer. Anal. **44** (2006), 1829–1853.

[62] A. Bermùdez, M. R. Nogueiras, and C. Vàzquez, *Numerical analysis of convection–diffusion–reaction problems with higher order characteristics/finite elements. Part II. Fully discretized schemes and quadrature formulas*, SIAM J. Numer. Anal. **44** (2006), 1854–1876.

[63] C. Bernardi, F. Hecht, and R. Verfürth, *A posteriori error analysis of the method of characteristics*, Math. Models Methods Appl. Sci. **21** (2011), no. 6, 1355–1376.

[64] C. Bernardi, B. Métivet, and R. Verfürth, *Analyse numérique d'indicateurs d'erreur*, Maillage et Adaptation (Paris) (P. L. George, ed.), Hermès, Paris 2002, pp. 251–278.

[65] C. Bernardi and G. Raugel, *A conforming finite element method for the time-dependent Navier–Stokes equations*, SIAM J. Numer. Anal. **22** (1985), no. 3, 455–473.

[66] C. Bernardi and R. Verfürth, *Adaptive finite element methods for elliptic equations with non-smooth coefficients*, Numer. Math. **85** (2000), no. 4, 579–608.

[67] C. Bernardi and R. Verfürth, *A posteriori error analysis of the fully discretized time-dependent Stokes equations*, M2AN Math. Model. Numer. Anal. **38** (2004), no. 3, 437–455.

[68] G. Bernardi and C. Raugel, *Analysis of some finite elements for the Stokes problem*, Math. Comput. **44** (1985), 71–79.

[69] S. Berrone, *Skipping transition conditions in a posteriori error estimates for finite element discretizations of parabolic equations*, M2AN Math. Model. Numer. Anal. **44** (2010), no. 3, 455–484.

[70] P. Binev, W. Dahmen, and R. DeVore, *Adaptive finite element methods with convergence rates*, Numer. Math. **97** (2004), 219–268.

[71] P. B. Bochev, C. R. Dohrmann, and M. D. Gunzburger, *Stabilization of low-order mixed finite elements for the Stokes equations*, SIAM J. Numer. Anal. **44** (2006), no. 1, 82–101.

[72] M. E. Bogovskii, *Solution of the first boundary value problem for the equation of continuity of an incompressible medium*, Soviet Math. Dokl. **20** (1979), 1094–1098.

[73] N. V. Bonnaillie, *A posteriori error estimator for the eigenvalue problem associated to the Schrödinger operator with magnetic field*, Numer. Math. **99** (2004), no. 2, 325–348.

[74] F. A. Bornemann, B. Erdmann, and R. Kornhuber, *A posteriori error estimates for elliptic problems in two and three space dimensions*, SIAM J. Numer. Anal. **33** (1996), no. 3, 1188–1204.

[75] K. Boukir, Y. Maday, and B. Métivet, *A high order characteristics method for the incompressible Navier–Stokes equations*, Comput. Methods Appl. Mech. Engrg. **116** (1994), no. 1–4, 211–218.

[76] M. Braack, E. Burman, V. John, and G. Lube, *Stabilized finite element methods for the generalized Oseen problem*, Comput. Methods Appl. Mech. Engrg. **196** (2007), no. 4–6, 853–866.

[77] M. Braack and G. Lube, *Finite elements with local projection stabilization for incompressible flow problems*, J. Comput. Math. **27** (2009), no. 2–3, 116–147.

[78] D. Braess, *The contraction number of a multigrid method for solving the Poisson equation*, Numer. Math. **37** (1981), no. 3, 387–404.

[79] D. Braess, *Finite Elements: Theory, Fast Solvers and Applications in Solid Mechanics*, 3rd ed., Cambridge University Press, 2007.

[80] D. Braess, *An a posteriori error estimate and a comparison theorem for the nonconforming P_1 element*, Calcolo **46** (2009), no. 2, 149–155.

[81] D. Braess, C. Carstensen, and R. H. W. Hoppe, *Convergence analysis of a conforming adaptive finite element method for an obstacle problem*, Numer. Math. **107** (2007), no. 3, 455–471.

[82] D. Braess, C. Carstensen, and R. H. W. Hoppe, *Error reduction in adaptive finite element approximations of elliptic obstacle problems*, J. Comput. Math. **27** (2009), no. 2–3, 148–169.

[83] D. Braess and J. Schöberl, *Equilibrated residual error estimator for edge elements*, Math. Comput. **77** (2008), no. 262, 651–672.

[84] D. Braess and R. Verfürth, *A posteriori error estimators for the Raviart-Thomas element*, SIAM J. Numer. Anal. **33** (1996), no. 6, 2431–2444.

[85] S. C. Brenner, T. Gudi, and L.-Y. Sung, *An a posteriori error estimator for a quadratic C^0-interior penalty method for the biharmonic problem*, IMA J. Numer. Anal. **30** (2010), no. 3, 777–798.

[86] S. C. Brenner and L. R. Scott, *The Mathematical Theory of Finite Element Methods*, Springer, Berlin, 1994.

[87] S. C. Brenner and L.-Y. Sung, *C^0 interior penalty methods for fourth order elliptic boundary value problems on polygonal domains,* J. Sci. Comput. **22/23** (2005), 83–118.
[88] F. Brezzi, *On the existence, uniqueness, and approximation of saddle-point problems arising from Lagrangian multipliers,* RAIRO Anal Numér. **8** (1974), 129–151.
[89] F. Brezzi and R. S. Falk, *Stability of a higher order Hood–Taylor element,* SIAM J. Numer. Anal. **28** (1991), 581–590.
[90] F. Brezzi and M. Fortin, *Mixed and Hybrid Finite Element Methods,* Springer Series in Computational Mathematics, vol. 15, Springer, Berlin, 1991.
[91] C. Carstensen, *A posteriori error estimate for the mixed finite element method,* Math. Comp. **66** (1997), no. 218, 465–476.
[92] C. Carstensen, *Residual-based a posteriori error estimate for a nonconforming Reissner–Mindlin plate finite element,* SIAM J. Numer. Anal. **39** (2002), no. 6, 2034–2044.
[93] C. Carstensen and S. Bartels, *Each averaging technique yields reliable a posteriori error control in FEM on unstructured grids. I. Low order conforming, nonconforming, and mixed FEM,* Math. Comp. **71** (2002), no. 239, 945–969.
[94] C. Carstensen, S. Bartels, and S. Jansche, *A posteriori error estimates for nonconforming finite element methods,* Numer. Math. **92** (2002), no. 2, 233–256.
[95] C. Carstensen and S. A. Funken, *Fully reliable localized error control in the FEM,* SIAM J. Sci. Comput. **21** (1999/2000), no. 4, 1465–1484.
[96] C. Carstensen and S. A. Funken, *Averaging technique for a posteriori error control in elasticity. III. Locking-free nonconforming FEM,* Comput. Methods Appl. Mech. Engrg. **191** (2001), no. 8–10, 861–877.
[97] C. Carstensen and R. H. W. Hoppe, *Convergence analysis of an adaptive nonconforming finite element method,* Numer. Math. **103** (2006), no. 2, 251–266.
[98] C. Carstensen and R. H. W. Hoppe, *Error reduction and convergence for an adaptive mixed finite element method,* Math. Comp. **75** (2006), no. 255, 1033–1042.
[99] C. Carstensen and J. Hu, *A unifying theory of a posteriori error control for nonconforming finite element methods,* Numer. Math. **107** (2007), no. 3, 473–502.
[100] C. Carstensen, J. Hu, and A. Orlando, *Framework for the a posteriori error analysis of nonconforming finite elements,* SIAM J. Numer. Anal. **45** (2007), no. 1, 68–82.
[101] C. Carstensen and R. Verfürth, *Edge residuals dominate a posteriori error estimates for low order finite element methods,* SIAM J. Numer. Anal. **36** (1999), no. 5, 1571–1587.
[102] J. M. Cascon, C. Kreuzer, R. H. Nochetto, and K. G. Siebert, *Quasi-optimal convergence rate for an adaptive finite element method,* SIAM J. Numer. Anal. **46** (2008), no. 5, 2524–2550.
[103] J. M. Cascon, R. H. Nochetto, and K. G. Siebert, *Design and convergence of AFEM in $H(\text{div})$,* Math. Models Methods Appl. Sci. **17** (2007), no. 11, 1849–1881.
[104] A. Charbonneau, K. Dossou, and R. Pierre, *A residual-based a posteriori error estimator for the Ciarlet–Raviart formulation of the first biharmonic problem,* Numer. Methods Partial Differential Equations **13** (1997), no. 1, 93–111.
[105] I. Cheddadi, R. Fučík, M. I. Prieto, and M. Vohralík, *Guaranteed and robust a posteriori error estimates for singularly perturbed reaction–diffusion problems,* M2AN Math. Model. Numer. Anal. **43** (2009), no. 5, 867–888.
[106] Z. Chen and J. Feng, *An adaptive finite element algorithm with reliable and efficient error control for linear parabolic problems,* Math. Comp. **73** (2004), no. 247, 1167–1193.
[107] S.-K. Chua and R. L. Wheeden, *Estimates for best constants in weighted Poincaré inequalities for convex domains,* Proc. London Math. Soc. **93** (2006), no. 3, 197–226.
[108] P. G. Ciarlet, *The Finite Element Method for Elliptic Problems,* North Holland, Amsterdam, 1978.

[109] P. G. Ciarlet, *Basic Error Estimates for Elliptic Problems*, Handbook of Numerical Analysis (P. G. Ciarlet and J. L. Lions, eds.), vol. 2, North-Holland, Amsterdam, 1991, pp. 17–351.

[110] P. Clément, *Approximation by finite element functions using local regularization*, RAIRO Anal. Numér. **2** (1975), 77–84.

[111] K. A. Cliffe, E. J. C. Hall, P. Houston, E. T. Phipps, and A. G. Salinger, *Adaptivity and a posteriori error control for bifurcation problems I: the Bratu problem*, Commun. Comput. Phys. **8** (2010), no. 4, 845–865.

[112] M. Costabel and M. Dauge, *Construction of corner singularities for Agmon–Douglis–Nirenberg elliptic systems*, Math. Nachr. **162** (1993), 209–237.

[113] E. Creuse, M. Farhloul, and L. Paquet, *A posteriori error estimation for the dual mixed finite element method for the p-Laplacian in a polygonal domain*, Comput. Methods Appl. Mech. Engrg. **196** (2007), no. 25–28, 2570–2582.

[114] M. Crouzeix and J. Rappaz, *On Numerical Approximation in Bifurcation Theory*, RMA, vol. 13, Masson-Springer, Paris, 1990.

[115] M. Crouzeix and V. Thomée, *The stability in L^p and $W^{1,p}$ of the L^2-projection onto finite element function spaces*, Math. Comp. **48** (1987), no. 178, 521–532.

[116] E. Dari, R. Duran, C. Padra, and V. Vampa, *A posteriori error estimators for nonconforming finite element methods*, RAIRO Modél. Math. Anal. Numér. **30** (1996), no. 4, 385–400.

[117] M. Dauge, *Stationary Stokes and Navier–Stokes systems on two- or three-dimensional domains with corners. I. Linearized equations*, SIAM J. Math. Anal. **20** (1989), no. 1, 74–97.

[118] R. Dautray and J.-L. Lions, *Mathematical Analysis and Numerical Methods for Science and Technology II: Functional and Variational Methods*, Springer, Berlin, 1988.

[119] R. Dautray and J.-L. Lions, *Mathematical Analysis and Numerical Methods for Science and Technology V: Evolution Problems I*, Springer, Berlin, 1992.

[120] G. de Rham, *Variétés Différentiables*, Hermann, Paris, 1960.

[121] A. Demlow and R. Stevenson, *Convergence and quasi-optimality of an adaptive finite element method for controlling L_2 errors*, Numer. Math. **117** (2011), no. 2, 185–218.

[122] P. Destuynder and B. Métivet, *Explicit error bounds for a conforming finite element method*, Math. Comput. **68** (1999), no. 228, 1379–1396.

[123] P. Deuflhard, P. Leinen, and H. Yserentant, *Concepts of an adaptive hierarchical finite element code*, IMPACT Comput. Sci. Engrg. **1** (1989), 3–35.

[124] R. DeVore, *Nonlinear approximation*, Acta Numerica 7 (1998), 51–150.

[125] C. R. Dohrmann and P. B. Bochev, *A stabilized finite element method for the Stokes problem based on polynomial pressure projections*, Internat. J. Numer. Methods Fluids **46** (2004), no. 2, 183–201.

[126] W. Dörfler, *A convergent adaptive algorithm for Poisson's equation*, SIAM J. Num. Anal. **33** (1996), no. 3, 1106–1124.

[127] R. G. Durán, L. Gastaldi, and C. Padra, *A posteriori error estimators for mixed approximations of eigenvalue problems*, Math. Models Methods Appl. Sci. **9** (1999), no. 8, 1165–1178.

[128] R. G. Durán, M. A. Muschietti, and R. Rodriguez, *On the asymptotic exactness of error estimators for linear triangular elements*, Numer. Math. **59** (1991), 107–127.

[129] R. G. Durán, C. Padra, and R. Rodríguez, *A posteriori error estimates for the finite element approximation of eigenvalue problems*, Math. Models Methods Appl. Sci. **13** (2003), no. 8, 1219–1229.

[130] R. G. Durán and R. Rodriguez, *On the asymptotic exactness of Bank–Weiser's estimator*, Numer. Math. **62** (1992), 297–303.

[131] V. Eijkhout and P. Vassilevski, *The role of the strengthened Cauchy–Bujanowski–Schwarz inequality in multilevel methods*, SIAM Rev. **33** (1991), 405–419.

[132] L. El Alaoui and A. Ern, *Residual and hierarchical a posteriori error estimates for nonconforming mixed finite element methods,* M2AN Math. Model. Numer. Anal. **38** (2004), no. 6, 903–929.

[133] G. Engel, K. Garikipati, T. J. R. Hughes, M. G. Larson, L. Mazzei, and R. L. Taylor, *Continuous/discontinuous finite element approximations of fourth-order elliptic problems in structural and continuum mechanics with applications to thin beams and plates, and strain gradient elasticity,* Comput. Methods Appl. Mech. Engrg. **191** (2002), no. 34, 3669–3750.

[134] A. Ern, A. F. Stephansen, and M. Vohralík, *Guaranteed and robust discontinuous Galerkin a posteriori error estimates for convection–diffusion–reaction problems,* J. Comput. Appl. Math. **234** (2010), no. 1, 114–130.

[135] R. Eymard, T. Gallouët, and R. Herbin, *Finite Volume Methods,* Handbook of Numerical Analysis (P. G. Ciarlet and J. L. Lions, eds.), vol. 7, North-Holland, Amsterdam, 2000, pp. 713–1020.

[136] F. Fierro and A. Veeser, *A posteriori error estimators for regularized total variation of characteristic functions,* SIAM J. Numer. Anal. **41** (2003), no. 6, 2032–2055.

[137] F. Fierro and A. Veeser, *A posteriori error estimators, gradient recovery by averaging, and superconvergence,* Numer. Math. **103** (2006), no. 2, 267–298.

[138] F. Fierro and A. Veeser, *A safeguarded Zienkiewicz–Zhu estimator,* Numerical Mathematics and Advanced Applications, Springer, Berlin, 2006, pp. 269–276.

[139] L. P. Franca, S. L. Frey, and T. J. R. Hughes, *Stabilized finite element methods I: Application to the advective–diffusive model,* Comput. Methods Appl. Mech. Engrg. **95** (1992), 253–271.

[140] L. P. Franca, V. John, G. Matthies, and L. Tobiska, *An inf–sup stable and residual-free bubble element for the Oseen equations,* SIAM J. Numer. Anal. **45** (2007), no. 6, 2392–2407.

[141] F. Gardini, *A posteriori error estimates for an eigenvalue problem arising from fluid-structure interaction,* Istit. Lombardo Accad. Sci. Lett. Rend. A **138** (2004), 17–34 (2005).

[142] P. L. George, *Automatic Mesh Generation and Finite Element Computation,* Handbook of Numerical Analysis (P. G. Ciarlet and J. L. Lions, eds.), vol. 5, North-Holland, Amsterdam, 1996, pp. 69–190.

[143] P. L. George and H. Borouchaki, *Delaunay Triangulation and Meshing,* Hermès, Paris, 1998.

[144] V. Girault and P.-A. Raviart, *Finite Element Approximation of the Navier–Stokes Equations,* Lecture Notes in Mathematics, vol. 749, Springer, Berlin, 1979.

[145] V. Girault and P.-A. Raviart, *Finite Element Methods for Navier-Stokes Equations: Theory and Algorithms,* Springer Series in Computational Mathematics, vol. 5, Springer, Berlin, 1986.

[146] M. Grajewski, J. Hron, and S. Turek, *Dual weighted a posteriori error estimation for a new nonconforming linear finite element on quadrilaterals,* Appl. Numer. Math. **54** (2005), no. 3–4, 504–518.

[147] P. Grisvard, *Elliptic Problems in Nonsmooth Domains,* Pitman, Boston, 1985.

[148] S. Grosman, *An equilibrated residual method with a computable error approximation for a singularly perturbed reaction-diffusion problem on anisotropic finite element meshes,* M2AN Math. Model. Numer. Anal. **40** (2006), no. 2, 239–267.

[149] F. Hecht, A. Le Hyaric, K. Ohtsuka, and O. Pironneau, *Freefem++, second edition, v 3.0-1,* Université Pierre et Marie Curie, Paris, 2007, `www.freefem.org/ff++/ftp/freefem++doc.pdf`.

[150] B.-O. Heimsund, X.-C. Tai, and J. Wang, *Superconvergence for the gradient of finite element approximations by L^2-projections,* SIAM J. Numer. Anal. **40** (2002), 1263–1280.

[151] V. Heuveline and R. Rannacher, *A posteriori error control for finite approximations of elliptic eigenvalue problems,* Adv. Comput. Math. **15** (2001), no. 1–4, 107–138.

[152] P. Hood and G. Taylor, *Navier–Stokes equations using mixed interpolation,* Finite Element Method in Flow Problems (J. T. Oden, ed.), UAH Press, 1974.

[153] T. J. R. Hughes and A. Brooks, *Streamline upwind/Petrov–Galerkin formulations for the convection dominated flows with particular emphasis on the incompressible Navier–Stokes equations*, Comput. Methods Appl. Mech. Engrg. **54** (1982), 199–259.

[154] T. J. R. Hughes, L. P. Franca, and M. Balestra, *A new finite element formulation for computational fluid dynamics: V. Circumventing the Babuška–Brezzi condition: A stable Petrov–Galerkin formulation of the Stokes problem accommodating equal order interpolation*, Comput. Methods Appl. Mech. Engrg. **59** (1986), 85–99.

[155] H. Jin and S. Prudhomme, *A posteriori error estimation of steady-state finite element solutions of the Navier–Stokes equations by a subdomain residual method*, Comput. Methods Appl. Mech. Engrg. **159** (1998), no. 1–2, 19–48.

[156] F. Jochmann, *An H^s-regularity result for the gradient of solutions to elliptic equations with mixed boundary conditions*, J. Math. Anal. Appl. **238** (1999), 429–450.

[157] V. John, *Residual a posteriori error estimates for two-level finite element methods for the Navier–Stokes equations*, Appl. Numer. Math. **37** (2001), no. 4, 503–518.

[158] C. Johnson and R. Rannacher, *On error control in CFD*, Nmerical Methods for the Navier–Stokes Equations (Braunschweig) (F.-K. Hebeker *et al.*, eds.), Vieweg, Braunschweig, 1994, pp. 133–144.

[159] G. Kanschat and D. Schötzau, *Energy norm a posteriori error estimation for divergence-free discontinuous Galerkin approximations of the Navier–Stokes equations*, Internat. J. Numer. Methods Fluids **57** (2008), no. 9, 1093–1113.

[160] D. W. Kelly, *The self-equilibration of residuals and complementary a posteriori error estimates in the finite element method*, Internat. J. Numer. Meth. Engrg. **20** (1984), 1491–1509.

[161] K.-Y. Kim and H.-C. Lee, *A posteriori error estimators for nonconforming finite element methods of the linear elasticity problem*, J. Comput. Appl. Math. **235** (2010), no. 1, 186–202.

[162] Y. Kondratyuk and R. Stevenson, *An optimal adaptive finite element method for the Stokes problem*, SIAM J. Numer. Anal. **46** (2008), no. 2, 747–775.

[163] C. Kreuzer, C. A. Möller, A. Schmidt, and K. G. Siebert, *Design and convergence analysis for an adaptive discretization of the heat equation*, IMA J. Numer. Anal. **32** (2012), no. 4, 1375–1403.

[164] M. Křížek and P. Neittaanmäki, *Superconvergence phenomenon in the finite element method arising from averaging of gradients*, Numer. Math. **45** (1984), 105–116.

[165] M. Křížek, P. Neittaanmäki, and R. Stenberg (eds.), *Finite Element Methods. Superconvergence, Post-Processing and A Posteriori Error Estimates*, Lecture Notes in Pure and Applied Mathematics, no. 196, Marcel Dekker, New York, 1998.

[166] D. Kröner, M. Küther, M. Ohlberger, and C. Rohde, *A posteriori error estimates and adaptive methods for hyperbolic and convection dominated parabolic conservation laws*, Trends in Nonlinear Analysis, Springer, Berlin, 2003, pp. 289–306.

[167] D. Kröner and M. Ohlberger, *A posteriori error estimates for upwind finite volume schemes for nonlinear conservation laws in multidimensions*, Math. Comp. **69** (2000), no. 229, 25–39.

[168] G. Kunert, *An a posteriori residual error estimator for the finite element method on anisotropic tetrahedral meshes*, Numer. Math. **86** (2000), 471–490.

[169] G. Kunert and S. Nicaise, *Zienkiewicz–Zhu error estimators on anisotropic tetrahedral and triangular finite element meshes*, M2AN Math. Model. Numer. Anal. **37** (2003), no. 6, 1013–1043.

[170] G. Kunert and R. Verfürth, *Edge residuals dominate a posteriori error estimates for linear finite element methods on anisotropic triangular and tetrahedral meshes*, Numer. Math. **86** (2000), no. 2, 283–303.

[171] P. Labbé, J. Dompierre, and R. Guibault, F. Camarero, *Critères de qualité*, Maillage et Adaptation (Paris) (P. L. George, ed.), Hermès, Paris, 2002, pp. 311–348.

[172] P. Ladevèze and D. Leguillon, *Error estimate procedure in the finite element method and applications*, SIAM J. Numer. Anal. **20** (1983), 485–509.

[173] O. Lakkis and C. Makridakis, *Elliptic reconstruction and a posteriori error estimates for fully discrete linear parabolic problems*, Math. Comp. **75** (2006), no. 256, 1627–1658.

[174] H.-C. Lee and K.-Y. Kim, *A posteriori error estimators for stabilized P1 nonconforming approximation of the Stokes problem*, Comput. Methods Appl. Mech. Engrg. **199** (2010), no. 45–48, 2903–2912.

[175] R. J. LeVeque, *Finite Volume Methods for Hyperbolic Problems*, Cambridge Texts in Applied Mathematics, Cambridge University Press, 2002.

[176] Y. Li, *A posteriori error analysis of nonconforming methods for the eigenvalue problem*, J. Syst. Sci. Complex. **22** (2009), no. 3, 495–502.

[177] H. Liu and N. Yan, *Superconvergence and a posteriori error estimates of nonconforming FEM for boundary control governed by Stokes equations*, Recent Advances in Computational Sciences, World Scientific, Hackensack, NJ, 2008, pp. 133–155.

[178] K. Liu and X. Qin, *A gradient recovery-based a posteriori error estimators for the Ciarlet–Raviart formulation of the second biharmonic equations*, Appl. Math. Sci. (Ruse) **1** (2007), no. 21–24, 997–1007.

[179] W. Liu and N. Yan, *Quasi-norm a priori and a posteriori error estimates for the nonconforming approximation of p-Laplacian*, Numer. Math. **89** (2001), no. 2, 341–378.

[180] W. Liu and N. Yan, *Some a posteriori error estimators for p-Laplacian based on residual estimation or gradient recovery*, J. Sci. Comput. **16** (2001), no. 4, 435–477.

[181] W. Liu and N. Yan, *On quasi-norm interpolation error estimation and a posteriori error estimates for p-Laplacian*, SIAM J. Numer. Anal. **40** (2002), no. 5, 1870–1895.

[182] M. Lonsing, *A posteriori Fehlerschätzer für gemischte Finite Elemente in der linearen Elastizität*, Ph.D. thesis, Ruhr-Universität Bochum, Fakultät für Mathematik, 2002, `www.rub.de/num1/files/theses/diss_lonsing.pdf`.

[183] M. Lonsing and R. Verfürth, *A posteriori error estimators for mixed finite element methods in linear elasticity*, Numer. Math. **97** (2004), no. 4, 757–778.

[184] G. Lube and G. Rapin, *Residual-based stabilized higher-order FEM for a generalized Oseen problem*, Math. Models Methods Appl. Sci. **16** (2006), no. 7, 949–966.

[185] G. Lube and L. Tobiska, *A non-conforming finite element method of streamline diffusion type for solving the incompressible Navier–Stokes equations*, J. Comp. Math. **8** (1989), 147–158.

[186] R. Luce and B. Wohlmuth, *A local a posteriori error estimator based on equilibrated fluxes*, SIAM J. Numer. Anal. **42** (2004), 1394–1414.

[187] J. F. Maitre and F. Musy, *The contraction number of a class of two level methods: an exact evaluation for some finite element subspaces and model problems*, Multigrid Methods, Proceedings, Cologne 1981 (Heidelberg), Lecture Notes in Mathematics, vol. 960, Springer, Heidelberg, 1982, pp. 535–544.

[188] C. Makridakis and R. H. Nochetto, *Elliptic reconstruction and a posteriori error estimates for parabolic problems*, SIAM J. Numer. Anal. **41** (2003), no. 4, 1585–1594.

[189] D. Mao, L. Shen, and A. Zhou, *Adaptive finite element algorithms for eigenvalue problems based on local averaging type a posteriori error estimates*, Adv. Comput. Math. **25** (2006), no. 1–3, 135–160.

[190] G. Matthies, G. Lube, and L. Röhe, *Some remarks on residual-based stabilisation of inf–sup stable discretisations of the generalised Oseen problem*, Comput. Methods Appl. Math. **9** (2009), no. 4, 368–390.

[191] J. M. L. Maubach, *Local bisection refinement for n-simplicial grids generated by reflection*, SIAM J. Sci. Stat. Comput **16** (1995), 210–227.

[192] K. Mekchay, P. Morin, and R. H. Nochetto, *AFEM for the Laplace–Beltrami operator on graphs: design and conditional contraction property*, Math. Comp. **80** (2011), no. 274, 625–648.

[193] K. Mekchay and R. H. Nochetto, *Convergence of adaptive finite element methods for general second order linear elliptic PDEs*, SIAM J. Numer. Anal. **43** (2005), no. 5, 1803–1827.

[194] W. F. Mitchell, *A comparison of adaptive refinement techniques for elliptic problems*, ACM Trans. Math. Software **15** (1989), 326–347.

[195] M. S. Mommer and R. Stevenson, *A goal-oriented adaptive finite element method with convergence rates*, SIAM J. Numer. Anal. **47** (2009), no. 2, 861–886.

[196] P. Morin, R. H. Nochetto, and K. G. Siebert, *Data oscillation and convergence of adaptive FEM*, SIAM J. Numer. Anal. **38** (2000), no. 2, 466–488.

[197] P. Morin, R. H. Nochetto, and K. G. Siebert, *Convergence of adaptive finite element methods*, SIAM Rev. **44** (2002), no. 4, 631–658, Revised reprint of 'Data oscillation and convergence of adaptive FEM' [SIAM J. Numer. Anal. **38** (2000), no. 2, 466–488].

[198] P. Morin, R. H. Nochetto, and K. G. Siebert, *Local problems on stars: a posteriori error estimators, convergence, and performance*, Math. Comp. **72** (2003), no. 243, 1067–1097.

[199] P. Morin, K. G. Siebert, and A. Veeser, *Convergence of finite elements adapted for weaker norms*, Applied and Industrial Mathematics in Italy II, Ser. Adv. Math. Appl. Sci., vol. 75, World Scientific, Hackensack, NJ, 2007, pp. 468–479.

[200] P. Morin, K. G. Siebert, and A. Veeser, *A basic convergence result for conforming adaptive finite elements*, Math. Models Methods Appl. Sci. **18** (2008), no. 5, 707–737.

[201] J. Nečas, *Les Méthodes directes en Théorie des Equations elliptiques*, Masson et Cie, Éditeurs, Paris, 1967.

[202] P. Neittaanmäki and S. Repin, *Reliable Methods for Computer Simulation. Error Control and A Posteriori Error Estimates*, Elsevier, Amsterdam, 2004.

[203] P. Neittaanmäki and S. I. Repin, *A posteriori error estimates for boundary-value problems related to the biharmonic operator*, East–West J. Numer. Math. **9** (2001), no. 2, 157–178.

[204] S. Nicaise and N. Soualem, *A posteriori error estimates for a nonconforming finite element discretization of the heat equation*, M2AN Math. Model. Numer. Anal. **39** (2005), no. 2, 319–348.

[205] S. Nicaise and N. Soualem, *A posteriori error estimates for a nonconforming finite element discretization of the time-dependent Stokes problem*, J. Numer. Math. **15** (2007), no. 2, 137–162.

[206] S. Nicaise and N. Soualem, *A posteriori error estimates for a nonconforming finite element discretization of the time-dependent Stokes problem. II. Analysis of the spatial estimator*, J. Numer. Math. **15** (2007), no. 3, 209–231.

[207] R. H. Nochetto, *Removing the saturation assumption in a posteriori error analysis*, Istit. Lombardo Accad. Sci. Lett. Rend. A **127** (1994), no. 1, 67–82.

[208] R. H. Nochetto, A. Veeser, and M. Verani, *A safeguarded dual weighted residual method*, IMA J. Numer. Anal. **29** (2009), no. 1, 126–140.

[209] M. Ohlberger, *A posteriori error estimate for finite volume approximations to singularly perturbed nonlinear convection–diffusion equations*, Numer. Math. **87** (2001), no. 4, 737–761.

[210] M. Ohlberger, *A posteriori error estimates for vertex centered finite volume approximations of convection–diffusion–reaction equations*, M2AN Math. Model. Numer. Anal. **35** (2001), no. 2, 355–387.

[211] M. Ohlberger, *Error control for approximations of non-linear conservation laws*, Finite Volumes for Complex Applications IV, ISTE, London, 2005, pp. 85–100.

[212] M. Ohlberger, *A review of a posteriori error control and adaptivity for approximations of non-linear conservation laws*, Internat. J. Numer. Methods Fluids **59** (2009), no. 3, 333–354.

[213] A. Papastavrou and R. Verfürth, *A computational comparison of a posteriori error estimators for convection–diffusion problems,* Computational Mechanics (Buenos Aires, 1998), Centro Internac. Métodos Numér. Ing., Barcelona, 1998, pp. CD–ROM file.

[214] A. Papastavrou and R. Verfürth, *A posteriori error estimators for stationary convection–diffusion problems: a computational comparison,* Comput. Methods Appl. Mech. Engrg. **189** (2000), no. 2, 449–462.

[215] B. Patzák, *OOFEM home page,* www.oofem.org, 2000.

[216] B. Patzák and Z. Bittnar, *Design of object oriented finite element code,* Advances in Engineering Software **32** (2001), no. 10–11, 759–767.

[217] L. E. Payne and H. F. Weinberger, *An optimal Poincaré-inequality for convex domains,* Archive Rat. Mech. Anal. **5** (1960), 286–292.

[218] M. Picasso, *Adaptive finite elements for a linear parabolic problem,* Comput. Methods Appl. Mech. Engrg. **167** (1998), 223–237.

[219] O. Pironneau, *On the transport–diffusion algorithm and its applications to the Navier–Stokes equations,* Numer. Math. **38** (1981/82), no. 3, 309–332.

[220] J. Pousin and J. Rappaz, *Consistency, stability, a priori and a posteriori errors for Petrov–Galerkin methods applied to nonlinear problems,* Numer. Math. **69** (1994), 213–231.

[221] W. Prager and J. L. Synge, *Approximations in elasticity based on the concept of function space,* Quart. Appl. Math. **5** (1947), 241–269.

[222] A. Quarteroni, R. Sacco, and F. Salieri, *Numerical Mathematics,* 2nd ed., Texts in Applied Mathematics, vol. 37, Springer, Berlin, 2007.

[223] E. Rank, *A-posteriori Fehlerabschätzungen und automatische Netzverfeinerung für Potential- und Bipotentialprobleme,* Z. Angew. Math. Mech. **65** (1985), no. 5, 281–283.

[224] R. Rannacher, *Error control in finite element computations,* Summer School Error Control and Adaptivity in Scientific Computing (H. Bulgak and C. Zenger, eds.), Kluwer, Dordrecht, 1998, pp. 247–278.

[225] R. Rannacher and L. R. Scott, *Some optimal error estimates for piecewise linear finite element approximations,* Math. Comput. **38** (1982), 437–445.

[226] P.-A. Raviart and J.-M. Thomas, *Introduction à l'analyse numérique des équations aux dérivées partielles,* Collection Mathématiques Appliquées pour la Maîtrise., Masson, Paris, 1983.

[227] S. Repin and S. Sauter, *Functional a posteriori error estimates for the reaction–diffusion problem,* C. R. Math. Acad. Sci. Paris **343** (2006), 349–354.

[228] M. C. Rivara, *Algorithms for refining triangular grids suitable for adaptive and multigrid techniques,* Internat. J. Numer. Meth. Engrg. **20** (1984), 745–756.

[229] M. C. Rivara, *Design and data structure of fully adaptive, multigrid, finite element software,* ACM Trans. Math. Software **10** (1984), 242–264.

[230] J. C. Robinson, *Infinite-dimensional dynamical systems,* Cambridge Texts in Applied Mathematics, Cambridge University Press, 2001, An Introduction to Dissipative Parabolic PDEs and the Theory of Global Attractors.

[231] R. Rodriguez, *Some remarks on Zienkiewicz–Zhu estimator,* Int. J. Numer. Meth. PDE **10** (1994), 625–635.

[232] H. Rui and M. Tabata, *A second order characteristic finite element scheme for convection–diffusion problems,* Numer. Math. **92** (2002), 161–177.

[233] G. Sangalli, *A robust a posteriori error estimator for the residual-free bubbles method applied to advection–diffusion problems,* Numer. Math. **89** (2001), 379–399.

[234] F. Schieweck, *A posteriori error estimates with post-processing for nonconforming finite elements,* M2AN Math. Model. Numer. Anal. **36** (2002), no. 3, 489–503.

[235] A. Schmidt and K. G. Siebert, *Design of Adaptive Finite Element Software. The Finite Element Toolbox ALBERTA*, Springer, Berlin, 2005.

[236] L. R. Scott and M. Vogelius, *Norm estimates for a maximal right inverse of the divergence operator in spaces of piecewise polynomials*, Math. Modell. Numer. Anal. **9** (1985), 11–43.

[237] L. R. Scott and S. Zhang, *Finite element interpolation of nonsmooth functions satisfying boundary conditions*, Math. Comput. **54** (1990), 483–493.

[238] K. G. Siebert, *An a posteriori error estimator for anisotropic refinement*, Numer. Math. **73** (1996), no. 3, 373–398.

[239] K. G. Siebert and A. Veeser, *A unilaterally constrained quadratic minimization with adaptive finite elements*, SIAM J. Optim. **18** (2007), no. 1, 260–289.

[240] C. G. Simader and H. Sohr, *The Dirichlet Problem for the Laplacian in Bounded and Unbounded Domains*, Longman, London, 1996.

[241] R. Stevenson, *Optimality of a standard adaptive finite element method*, Found. Comput. Math. 7 (2007), no. 2, 245–269.

[242] G. Szegö, *Orthogonal Polynomials*, AMS Colloquium Publications, AMS, Providence, Rhode Island, 1959.

[243] R. Temam, *Navier–Stokes Equations*, 3rd ed., North Holland, Amsterdam, 1984.

[244] L. Tobiska and R. Verfürth, *Analysis of a streamline diffusion finite element method for the Stokes and Navier–Stokes equations*, SIAM J. Numer. Anal. **33** (1996), no. 1, 107–127.

[245] Y. Tourigny and E. Süli, *The finite element method with nodes moving along the characteristics for convection–diffusion equations*, Numer. Math. **59** (1991), no. 4, 399–412.

[246] A. Veeser, *Convergent adaptive finite elements for the nonlinear Laplacian*, Numer. Math. **92** (2002), no. 4, 743–770.

[247] A. Veeser and R. Verfürth, *Explicit upper bounds for dual norms of residuals*, SIAM J. Numer. Anal. **47** (2009), no. 3, 2387–2405.

[248] A. Veeser and R. Verfürth, *Poincaré constants for finite element stars*, IMA J. Numer. Anal. **32** (2012), no. 1, 30–47.

[249] R. Verfürth, *Error estimates for a mixed finite element approximation of the Stokes equations*, RAIRO Anal. Numér. **18** (1984), no. 2, 175–182.

[250] R. Verfürth, *Finite element approximation of incompressible Navier–Stokes equations with slip boundary condition*, Numer. Math. **50** (1987), no. 6, 697–721.

[251] R. Verfürth, *A posteriori error estimators for the Stokes equations*, Numer. Math. **55** (1989), no. 3, 309–325.

[252] R. Verfürth, *Finite element approximation of incompressible Navier–Stokes equations with slip boundary condition. II*, Numer. Math. **59** (1991), no. 6, 615–636.

[253] R. Verfürth, *A posteriori error estimators for the Stokes equations. II. Nonconforming discretizations*, Numer. Math. **60** (1991), no. 2, 235–249.

[254] R. Verfürth, *A posteriori error estimates for nonlinear problems. Finite element discretizations of elliptic equations*, Math. Comp. **62** (1994), no. 206, 445–475.

[255] R. Verfürth, *A posteriori error estimation and adaptive mesh-refinement techniques*, Proceedings of the Fifth International Congress on Computational and Applied Mathematics (Leuven, 1992), vol. 50, 1994, pp. 67–83.

[256] R. Verfürth, *The equivalence of a posteriori error estimators*, Fast Solvers for Flow Problems (Kiel, 1994), Notes Numer. Fluid Mech., vol. 49, Vieweg, Braunschweig, 1995, pp. 273–283.

[257] R. Verfürth, *The stability of finite element methods*, Numer. Methods Partial Differential Equations **11** (1995), no. 1, 93–109.

[258] R. Verfürth, *A Review of A Posteriori Error Estimation and Adaptive Mesh-Refinement Techniques*, Teubner-Wiley, Stuttgart, 1996.

[259] R. Verfürth, *A posteriori error estimates for nonlinear problems. L^r-estimates for finite element discretizations of elliptic equations,* RAIRO Modél. Math. Anal. Numér. **32** (1998), no. 7, 817–842.

[260] R. Verfürth, *A posteriori error estimates for nonlinear problems: $L^r(0, T; W^{1,\rho}(\Omega))$-error estimates for finite element discretizations of parabolic equations,* Numer. Methods Partial Differential Equations **14** (1998), no. 4, 487–518.

[261] R. Verfürth, *A posteriori error estimators for convection–diffusion equations,* Numer. Math. **80** (1998), no. 4, 641–663.

[262] R. Verfürth, *A review of a posteriori error estimation techniques for elasticity problems,* Advances in Adaptive Computational Methods in Mechanics (Cachan, 1997), Stud. Appl. Mech., vol. 47, Elsevier, Amsterdam, 1998, pp. 257–274.

[263] R. Verfürth, *Robust a posteriori error estimators for a singularly perturbed reaction–diffusion equation,* Numer. Math. **78** (1998), no. 3, 479–493.

[264] R. Verfürth, *Error estimates for some quasi-interpolation operators,* M2AN Math. Model. Numer. Anal. **33** (1999), no. 4, 695–713.

[265] R. Verfürth, *A note on polynomial approximation in Sobolev spaces,* M2AN Math. Model. Numer. Anal. **33** (1999), no. 4, 715–719.

[266] R. Verfürth, *On the constants in some inverse inequalities for finite element functions,* report 257, Ruhr-Universität, Bochum, May 1999,
`www.ruhr-uni-bochum.de/num1/files/reports/INVEST.pdf`.

[267] R. Verfürth, *A posteriori error estimates for finite element discretizations of the heat equation,* Calcolo **40** (2003), no. 3, 195–212.

[268] R. Verfürth, *A posteriori error estimates for linear parabolic equations,* report, Ruhr-Universität, Bochum, July 2004,
`www.ruhr-uni-bochum.de/num1/files/reports/APEELPE.pdf`.

[269] R. Verfürth, *A posteriori error estimates for non-linear parabolic equations,* report 361, Ruhr-Universität, Bochum, December 2004,
`www.ruhr-uni-bochum.de/num1/files/reports/APNLPE.pdf`.

[270] R. Verfürth, *Robust a posteriori error estimates for nonstationary convection–diffusion equations,* SIAM J. Numer. Anal. **43** (2005), no. 4, 1783–1802.

[271] R. Verfürth, *Robust a posteriori error estimates for stationary convection–diffusion equations,* SIAM J. Numer. Anal. **43** (2005), no. 4, 1766–1782.

[272] R. Verfürth, *A note on constant-free a posteriori error estimates,* SIAM J. Numer. Anal. **47** (2009), no. 4, 3180–3194.

[273] R. Verfürth, *A posteriori error analysis of space–time finite element discretizations of the time-dependent Stokes equations,* Calcolo **47** (2010), no. 3, 149–167.

[274] M. Vohralík, *Residual flux-based a posteriori error estimates for finite volume and related locally conservative methods,* Numer. Math. **111** (2008), no. 1, 121–158.

[275] M. Vohralík, *Guaranteed and fully robust a posteriori error estimates for conforming discretizations of diffusion problems with discontinuous coefficients,* J. Sci. Comput. **46** (2011), no. 3, 397–438.

[276] L. Wahlbin, *Superconvergence in Galerkin Finite Element Methods,* Lecture Notes in Mathematics, no. 1605, Springer, Berlin, 1995.

[277] T. F. Walsh, G. M. Reese, and U. L. Hetmaniuk, *Explicit a posteriori error estimates for eigenvalue analysis of heterogeneous elastic structures,* Comput. Methods Appl. Mech. Engrg. **196** (2007), no. 37–40, 3614–3623.

[278] M. Wang and W. Zhang, *Local a priori and a posteriori error estimate of TQC9 element for the biharmonic equation,* J. Comput. Math. **26** (2008), no. 2, 196–208.

[279] B. D. Welfert, *A posteriori error estimates for the Stokes problem,* Ph.D. thesis, University of California, San Diego, 1990.

[280] M. F. Wheeler and J. R. Whiteman, *Superconvergent recovery of gradients on subdomains from piecewise linear finite element approximations,* Numer. Methods for PDE **3** (1987), 357–374.

[281] J. Xu and Z. Zhang, *Analysis of recovery type a posteriori error estimators for mildly structured grids,* Math. Comput. **73** (2004), no. 247, 1139–1152.

[282] K. Yosida, *Functional Analysis,* Springer, Berlin, 1980.

[283] D. Yu, *Asymptotically exact a posteriori error estimator for elements of bi-even degree,* Chinese J. Numer. Math. Appl. **13** (1991), no. 2, 64–78.

[284] D. Yu, *Asymptotically exact a posteriori error estimators for elements of bi-odd degree,* Chinese J. Numer. Math. Appl. **13** (1991), no. 4, 82–90.

[285] E. Zeidler, *Nonlinear Functional Analysis and its Applications. II/B,* Springer-Verlag, New York, 1990.

[286] J. Z. Zhu and O. C. Zienkiewicz, *A simple error estimator and adaptive procedure for practical engineering analysis,* Internat. J. Numer. Meth. Engrg. **24** (1987), 337–357.

LIST OF SYMBOLS

$\|\cdot\|$ norm of $L^2(\Omega)$, 4
$\|\cdot\|_\omega$ norm of $L^2(\omega)$, 4
$\|\cdot\|_p$ norm of $L^p(\Omega)$, 79
$\|\cdot\|_{p;\omega}$ norm of $L^p(\omega)$, 79
$\|\cdot\|_\gamma$ norm of $L^2(\gamma)$, 4
$\|\cdot\|_{p;\gamma}$ norm of $L^p(\gamma)$, 79
$\|\cdot\|_{H(\mathrm{curl};\Omega)}$ norm of $H(\mathrm{curl}, \Omega)$, 215
$\|\cdot\|_{H(\mathrm{div};\Omega)}$ norm of $H(\mathrm{div}; \Omega)$, 81, 208
$\|\cdot\|_{\frac{1}{2};\Gamma}$ norm of $H^{\frac{1}{2}}(\Gamma)$, 318
$\|\cdot\|_k$ norm of $H^k(\Omega)$, 80
$\|\cdot\|_{k;\omega}$ norm of $H^k(\omega)$, 80
$\|\cdot\|_{\mathcal{L}^2(X,Y,\mathbb{R})}$ norm of $\mathcal{L}^2(X, Y, \mathbb{R})$, 152
$\|\cdot\|_{\mathcal{L}(X,Y)}$ norm of $\mathcal{L}(X, Y)$, 151
$\|\cdot\|_{L^p(a,b;V)}$ norm of $L^p(a, b; V)$, 81
$|||\cdot|||_{\mathcal{T}}$ mesh-dependent norm, 22
$|||\cdot|||$ energy norm, 80, 132, 159, 164, 205, 317
$|||\cdot|||_\omega$ energy norm on ω, 80
$\|\cdot\|_{W^p(a,b;V,W)}$ norm of $W^p(a, b; V, W)$, 81
$\|\cdot\|_{k,p}$ norm of $W^{k,p}(\Omega)$, 80
$\|\cdot\|_{k,p;\omega}$ norm of $W^{k,p}(\omega)$, 80
$\|\cdot\|_X$ norm of X, 151
$|||\cdot|||_*$ dual energy norm, 81, 132, 164, 206, 318
: inner product of matrices, 257
$< \cdot, \cdot >$ duality pairing, 81, 152
$|\cdot|_1$ ℓ^1-norm on $\mathbb{R}^d$, 83
$|\cdot|_\infty$ ℓ^∞-norm on $\mathbb{R}^d$, 83
∂_t partial derivative with respect to time, 81
$(\cdot\,,\,\cdot)_{\mathcal{T}}$ mesh-dependent scalar product, 37
x^α power $x_1^{\alpha_1} \cdot \ldots \cdot x_d^{\alpha_d}$, 83
$\mathcal{A}^s$ approximation class, 280
$\mathbb{A}_E(\cdot)$ average on face E, 257
BDM_{k+1} Brezzi–Douglas–Marini space, 139
B_K anisotropic transformation matrix, 179
$B_X(\varphi_0, R)$ ball with radius R centred at $\varphi_0 \in X$, 282
$C_{A,\cdot,p}(\cdot)$ constant in quasi-interpolation error estimate, 109
$C_{F,p,\zeta}(\cdot)$ Friedrichs' constant, 105
$C_{I,\cdot,k}(\cdot)$ constant in inverse estimate, 112, 113
$C_{I,\cdot,k,p}(\cdot)$ constant in inverse estimate, 112, 125
$\overline{C}_{I,k,p}(\cdot)$ constant in inverse estimate, 126
$\underline{C}_{I,k,p}(\cdot)$ constant in inverse estimate, 126
$C_{P,p,\zeta}(\cdot)$ Poincaré constant, 91
$C_{\mathcal{T}}$ shape parameter of $\mathcal{T}$, 82
$C_{\mathcal{T},a}$ anisotropic shape parameter of $\mathcal{T}$, 178
D^2u Hessian matrix of u, 257
$\mathcal{E}$ faces corresponding to $\mathcal{T}$, 5, 82
$\mathcal{E}_K$ faces of K, 5, 82
$\mathcal{E}_K^+$ outflow faces of K, 174
$\mathcal{E}_\Gamma$ faces on Γ, 5, 82
$\mathcal{E}_{\Gamma_D}$ faces on Γ_D, 5, 82
$\mathcal{E}_{\Gamma_N}$ faces on Γ_N, 5, 82
$\mathcal{E}_\Omega$ interior faces, 5, 82
$\hat{E}_d$ reference face, 116
$\mathbb{F}$ forest of element subdivisions, 266
Γ boundary, 4, 79
Γ_D Dirichlet boundary, 4, 79
Γ_N Neumann boundary, 4, 79
$H^1(\Omega)$ Sobolev space, 4
$H_D^1(\Omega)$ Sobolev space, 4
H_K anisotropic transformation matrix, 179
$H(\mathrm{curl}, \Omega)$ space of L^2-functions having their curl in L^2, 215
$H(\mathrm{div}; \Omega)$ space of L^2-functions having their div in L^2, 48, 81, 208
$H_{\mathcal{T}}(\mathrm{div})$ space of L^2-functions having their div element-wise in L^2, 139
$H^{\frac{1}{2}}(\Gamma)$ trace space, 318
$H^{-\frac{1}{2}}(\Gamma)$ dual trace space, 318
$H^k(\Omega)$ Sobolev space, 54, 80
$\mathcal{I}$ partition of time interval, 86, 310, 318, 330, 336, 348
$I_{\mathcal{T}}$ quasi-interpolation operator, 7, 108, 146
I_n time interval, 86
$\mathcal{J}_{\mathcal{T}}$ quasi-interpolation operator of Raviart–Thomas, 213
J_K Raviart–Thomas or Brezzi–Douglas–Marini interpolation operator, 140
$\mathbb{J}_E(\cdot)$ jump across face E, 9
$\hat{K}$ reference triangle or reference square, 5
$\hat{K}_d$ reference element, 82, 116
K_ϑ squeezed element, 84
$L_0^2(\Omega)$ space of L^2-functions with vanishing mean-value, 237
$L^2(\gamma)$ Lebesgue space, 4
$L^2(\omega)$ Lebesgue space, 4
$\mathcal{L}^2(X, Y, \mathbb{R})$ space of continuous bi-linear mappings of $X \times Y$ into $\mathbb{R}$, 152
$\mathcal{L}(X, Y)$ space of continuous linear mappings of X into Y, 151
L_k Legendre polynomial of degree k, 117
$L^p(a, b; V)$ Lebesgue space of V-valued functions, 81
$L^p(\gamma)$ Lebesgue space, 79
$L^p(\omega)$ Lebesgue space, 79
M_K anisotropic transformation matrix, 179
$M_{\mathcal{T}}$ anisotropic transformation, 179
$\mathcal{N}$ vertices corresponding to $\mathcal{T}$, 5, 82
$\mathcal{N}_E$ vertices of E, 5, 82
$\mathcal{N}_K$ vertices of K, 5, 82
$\mathcal{N}_\Gamma$ vertices on Γ, 5, 82
$\mathcal{N}_{\Gamma_D}$ vertices on Γ_D, 5, 82

$\mathcal{N}_{\Gamma_N}$ vertices on Γ_N, 5, 82
$\mathcal{N}_\Omega$ interior vertices, 5, 82
$N_\mathcal{I}$ number of intermediate times, 86, 310, 318, 330, 336, 348
N_z number of elements sharing vertex z, 178
Ω computational domain, 4, 79
$\mathcal{P}$ collection of all possible refinements of a given coarse partition $\mathcal{T}_0$, 265
$P_\mathcal{T}$ L^2-projection onto $S_D^{1,0}(\mathcal{T})$, 20, 187
Π Stokes projection, 339
$\Psi_{E,1}$ C^1 cut-off function of face E having a prescribed normal derivative, 86
$\mathbb{P}_1$ first-order polynomials, 25
$Q_\mathcal{T}$ restriction or projection operator, 146, 155
R residual, 10, 132
$R_1(K)$ first-order polynomials on element K, 5
$\mathcal{R}_K$ rigid body motions, 233
$\mathrm{RT}_0(K)$ lowest order Raviart–Thomas space on K, 212, 227
$\mathrm{RT}_0(\mathcal{T})$ lowest order Raviart–Thomas space, 174, 212
$R_\mathcal{T}$ Ritz projection, 356
RT_k Raviart–Thomas space, 139
$R_k(K)$ k-th order polynomials on element K, 83
R_z radius of smallest ball with centre z contained in ω_z, 99
$S^{1,0}(\mathcal{T})$ continuous (multi-) linear finite element functions corresponding to $\mathcal{T}$, 5
$S_D^{1,0}(\mathcal{T})$ continuous (multi-) linear finite element functions corresponding to $\mathcal{T}$ vanishing on Γ_D, 5
$S^{k,-1}(\mathcal{T})$ discontinuous k-th order finite element functions corresponding to $\mathcal{T}$, 83
$S^{k,0}(\mathcal{T})$ continuous k-th order finite element functions corresponding to $\mathcal{T}$, 83
$S_D^{k,0}(\mathcal{T})$ continuous k-th order finite element functions corresponding to $\mathcal{T}$ vanishing on Γ_D, 83
$S_0^{k,0}(\mathcal{T})$ continuous k-th order finite element functions corresponding to $\mathcal{T}$ vanishing on Γ, 83
Σ skeleton of $\mathcal{T}$, 5, 82
T final time, 86, 309, 317, 326, 335, 347
$\mathcal{T}$ partition or mesh, 4, 81
$\mathcal{T}_n$ partition corresponding to time t_n, 86, 310, 318, 330, 336, 348
$\mathcal{U}_j(K)$ neighbourhood of element K, 21
V space of solenoidal vector fields, 237, 335
$\mathcal{V}_n$ nodes of a finite element space, 330
V_K auxiliary local finite element space, 29
$\widetilde{V}_K$ auxiliary local finite element space, 28
V_z auxiliary local finite element space, 26
$W_D^{1,p}(\Omega)$ Sobolev space, 80
$W^p(a,b;V,W)$ Sobolev space for time-dependent problems, 81
$W^{k,p}(\omega)$ Sobolev space, 80
$W^{-1,p'}(\Omega)$ dual Sobolev space, 80
X Banach space, 151
X_n finite element space corresponding to time t_n, 86, 310, 318, 330, 348
Y^* dual space of Y, 151
$a_{K,E}$ vertex of K opposite to face E, 50
$c_{A,\cdot}$ constant in quasi-interpolation error estimate, 7
$c_{I,\cdot}$ constant in inverse estimate, 9
χ_A characteristic function of A, 13, 23
$c_p(\cdot)$ Poincaré constant, 135
curl curl-operator, 214, 227
diam diameter, 8, 82, 91
div divergence operator, 81
$\eta_\mathcal{T}$ error indicator, 156, 288
$\widehat{\eta}_\mathcal{T}$ error indicator based on the solution of auxiliary problems, 289
$\eta_{D,K}$ error indicator based on the solution of local Dirichlet problems on ω_K, 28, 172, 236
$\eta_{D,z}$ error indicator based on the solution of local Dirichlet problems on ω_z, 26, 297
η_H hierarchical error indicator, 35
$\eta_{N,K}$ error indicator based on the solution of local Neumann problems, 29, 176, 234, 247, 306
$\eta_{R,E}$ error indicator based on edge residuals, 24, 38, 188
$\eta_{R,K}$ residual error indicator, 15, 160, 168, 184, 197, 204, 207, 222, 224, 232, 244, 250, 256, 260, 263, 294, 304
$\eta_{R,z}$ vertex oriented residual error indicator, 19
η_Z error indicator based on local averages, 37
$\eta_{Z,K}$ error indicator based on local averages, 37
η^n residual error indicator in space and time, 315, 363
$\eta^n_{\mathcal{T}_n}$ spatial error indicator, 314
$f_\mathcal{T}$ L^2-projection of f onto $S^{0,-1}(\mathcal{T})$, 12
$\overline{f}_K$ average of f on K, 12
$\mathbf{f}$ density of forces, 237
$g_\mathcal{E}$ L^2-projection of g onto $S^{0,-1}(\mathcal{E})$, 12
$\overline{g}_E$ average of g on E, 12
γ_E tangential trace on E, 230
$\gamma_{K,E}$ vector field in trace equality, 50, 87
γ_z faces of the boundary of ω_z which do not contain the vertex z, 99
h_E diameter of E, 5, 82
$h_E^\perp$ 'height' above face E, 18, 90, 105, 107, 108, 133, 134, 179, 201
h_K diameter of K, 5, 82
$\hbar_S$ modified diameter of element or face S, 82

$h_{\min,E}$ anisotropic diameter of face E, 179

$h_{\min,K}$ anisotropic diameter of element K, 177, 178

h_z diameter of ω_z, 8, 18, 82

$h_z^\perp$ 'height' above vertex z, 105

$i_\mathcal{T}$ nodal interpolation operator, 54, 112

j face residuals, 11, 132

κ_z eccentricity of ω_z, 99

λ Lamé parameter, 223

λ_z nodal shape function corresponding to vertex z, 7, 83

λ_{z,K_ϑ} nodal shape function of squeezed element, 84

$\ell(K, K')$ length of shortest path connecting elements K and K', 21

$m_\mathcal{T}$ matching function, 182

μ Lamé parameter, 223

μ_d Lebesgue measure in $\mathbb{R}^d$, 7, 79

μ_{d-1} $(d-1)$-dimensional Hausdorff measure in $\mathbb{R}^d$, 79

$\mu_{d,z}$ Lebesgue measure in $\mathbb{R}^d$ with weight function λ_z, 88

μ_ζ measure corresponding to the weight function ζ, 91

$\mathbf{n}_E$ unit normal vector to face E, 9

∇ nabla or gradient operator, 4

$n_j(K)$ number of elements in neighbourhood $\mathcal{U}_j(K)$, 21

ν viscosity, 301

ν_K parameter of element K: d for simplices, 1 for parallelepipeds, 82

n_z number of elements in ω_z, 99

ω_E union of all elements sharing a face with E, 6, 82

$\widetilde{\omega}_E$ union of all elements sharing a point with E, 6, 82

ω_K union of all elements sharing a face with K, 6, 82

$\widetilde{\omega}_K$ union of all elements sharing a point with K, 6, 82

ω_z support of λ_z, union of all elements having z as a vertex, 6, 82

p pressure, 237

π_n L^2-projection onto X_n, 310, 318, 331, 337, 348

p' dual Lebesgue exponent to p, 79

ψ_E cut-off function of face E, 8, 83

$\psi_{E,m}$ C^m cut-off function of face E, 85

$\psi_{E,\vartheta}$ modified cut-off function of face E, 84

ψ_K cut-off function of element K, 8, 83

$\psi_{K,m}$ C^m cut-off function of element K, 85

r element residuals, 11, 132

ρ_K diameter of the largest inscribed ball into K, 82, 108

ρ_z radius of largest ball with centre z contained in ω_z, 99

$\sigma_n(\varphi)$ error of best n-term approximation to φ, 280

σ_z union of all faces having z as a vertex, 6, 82

τ_n length of time interval, 86

$\theta_\mathcal{T}$ data error, 156, 288

θ_K data error indicator, 168, 207, 294

tr trace of a tensor, 223

$\mathbf{u}$ velocity, 237

$\overline{v}_\omega$ average of v on ω, 7

$\overline{v}_z$ weighted average of v on ω_z, 108

ζ weight function with corresponding measure μ_ζ, 91

INDEX